Women in ichthyology: an anthology in honour of ET, Ro and Genie

Developments in environmental biology of fishes 15

Series Editor
EUGENE K. BALON

Women in ichthyology: an anthology in honour of ET, Ro and Genie

Editors:

EUGENE K. BALON, MICHAEL N. BRUTON & DAVID L.G. NOAKES

Reprinted from *Environmental biology of fishes* 41 (1–4), 1994
with addition of species and subject index

SPRINGER SCIENCE+BUSINESS MEDIA, B.V.

Library of Congress Cataloging-in-Publication Data

Women in ichthyology an anthology in honour of ET, Ro, and Genie / editors, Eugene K. Balon, Michael N. Bruton & David L.G. Noakes.
p. cm. -- (Developments in environmental biology of fishes ; 15)
"Reprinted from Environmental biology of fishes 41 (1-4), 1994, with addition of species and subject index."
Includes bibliographical references.
ISBN 978-94-010-4090-7 ISBN 978-94-011-0199-8 (eBook)
DOI 10.1007/978-94-011-0199-8
1. Fishes. 2. Fishes--Reproduction. 3. Fish populations. 4. Ichthyologists--Biography. 5. Women ichthyologists--Biography. 6. Trewavas, Ethelwynn. 7. Lowe-McConnell, R. H. 8. Clark, Eugenie. I. Trewavas, Ethelwynn. II. Lowe-McConnell, R. H. III. Clark, Eugenie. IV. Balon, Eugene K. V. Bruton, M. N. (Michael N.) VI. Noakes, David L. G. VII. Series.
QL615.W62 1994
597--dc20 94-31170

ISBN 978-94-010-4090-7

Cover and logo design by Christine Flegler-Balon and Eugene Balon

Tails of the sexually dimorphic swordtail, *Xiphophorus* — with emphasis on the female tail — have been selected to symbolize the topic of this volume. In the logo the Taoist yin and yang represent the inseparable and harmonious dualism in which one cannot exist without the other. For more details see *Balon, E.K. 1989. The Tao of life: from the dynamic unity of polar opposites to self-organization. pp. 7–40. In: M.N. Bruton (ed.) Alternative Life-History Styles of Animals, Kluwer Academic Publishers, Dordrecht.*

Originally published by Kluwer Academic Publishers in 1994
Softcover reprint of the hardcover 1st edition 1994

(on acid-free paper)

Contents

Environmental Biology of Fishes **41** 7–8, 1994

Prelude to the anthology in honour of women ichthyologists

This woman of mixed cultures has not felt that either her sex or racial background has been a problem in her career

Diane Emberlin[1]
about *Eugenie Clark*

This volume is a collection of papers solicited to honour women in ichthyology. We chose three to represent all the other women ichthyologists and to receive the honours on their behalf, not unlike the explanation sometimes given about Nobel prize recipients. We ignored any ranking and contemplations about who deserves it most – our selection is entirely personal. We chose Ethelwynn Trewavas, Rosemary Lowe-McConnell and Eugenie Clark because we know them best, personally as well as professionally. Each of us, if only by virtue of age differences, knew each of them differently and this reflects itself in the biographical accounts which could not be standardized. After all, these three ichthyologists have had widely varying careers and different temperaments. It took us so long to put this volume together that Ethelwynn did not live to see it published (Noakes[2]), but she knew about it and contributed to its preparation.

The idea for this volume was hatched in 1989 while the three of us were travelling in Eugene's car from Guelph to Peter Stevens' cottage near Dorset in the lake district of northern Ontario, Canada. It was a natural history outing as the two Canadians were planning to show the visiting South African the Precambrian Shield. We also intended to resolve a clash of interests with Peter regarding coelacanth conservation between the Explorer's Club in New York and the Coelacanth Conservation Council (Balon et al.,[3] Musick et al.,[4] Hamlin[5]).

Inevitably, we chatted about the journal *Environmental Biology of Fishes*, and the subject of a dedicated volume on the contribution of women ichthyologists was raised. Could we do this without gender bias or condescension? Would the women whom we sought to honour be embarrassed by our efforts? We decided that it was worth the risk. It was also apparent that, in some ways, women often make better ichthyologists than men – they have the patience to devote decades of their lives to detailed taxonomic studies, the friendly personalities to secure the cooperation of rural fishers in developing countries during their field work, and the resilient physiology to make expert divers, among many other attributes. We were also aware that some may interpret certain issues from a single angle only (see Kass-Simon & Farnes[6]). For example, in reference to the citation of scientific papers, female and male scientists and their accomplishments are most often given credit or ignored for reasons of simple irritation with a novel concept or for conscious politics of clique association, and sex discrimination may be not the primary reason.

We were more than aware of inequities which still exist, like the one Eugenie Clark so nicely pointed out in her interview (p. 122): 'We had to work extra hard, especially on field trips, to prove we could keep up with males; except with Carl Hubbs, who was married to Laura, and took it for granted that females could carry the

[1] Emberlin, D 1977 Contribution of women science Dillon Press, Mineapolis 160 pp
[2] Noakes, D L G 1993 Ethelwynn Trewavas – a charmed life Env Biol Fish 38 295–298
[3] Balon, E K , M N Bruton & H Fricke 1988 A fiftieth anniversary reflection on the living coelacanth, *Latimeria chalumnae* some new interpretations of its natural history and conservation status Env Biol Fish 23 241–280
[4] Musick, J A , M N Bruton & E K Balon (ed) 1991 The biology of *Latimeria chalumnae* and evolution of coelacanths Developments in Environmental Biology of Fishes 12, Kluwer Academic Publishers, Dordrecht 446 pp
[5] Hamlin, J F 1992 Can coelacanths be caught on demand? (with a brief history of the Explorers Club coelacanth project) Env Biol Fish 33 419–425
[6] Kass-Simon, G & P Farnes (ed) 1993 Women of science Righting the record Indiana University Press, Bloomington 398 pp

same loads as males and do the cooking and dishwashing as well.' By virtue of differences in physiology and reproductive biology we approve of deserved compensations for women. We were relieved that our women colleagues agreed with us when asked about this and even added that they occasionally felt embarrassed by the advantages and recognition they received because of being females.

In this volume we are frequently using the nicknames which have become so familiar to us and many others: **ET** for Ethelwynn Trewavas, **Ro** for Rosemary Lowe-McConnell, and **Genie** for Eugenie Clark. The sequence in which their 'life and work' will be presented is according to age. ET, as the oldest (93) will be first, Ro second and Genie, by one year the youngest (72), the third. We feel that by researching and writing the biographical accounts of our female ichthyology colleagues we got to know them better and we consider ourselves even more privileged than before to be among their friends.

It is difficult to write about friends. It is almost impossible to write extensively about their personal and professional accomplishments without a feeling of embarrassment and self consciousness. We are simply too close to them to judge the extent of their accomplishments in many cases. We tend to take our friends and colleagues for granted – they are always there and we assume that everyone knows them and their work as well as we do. Most of us are also reluctant to praise our colleagues and elevate them for special consideration. All our training is to place our science first, and to treat each others as equals.

We knew from the outset that when we decided to proceed with this dedicated issue we would have to deal with the questions of historical inequities and current controversies. Now that the volume is completed we have discovered that these three ichthyologists were and are individuals with differences in experiences and aspirations which they have confronted in their own ways. ET, Ro and Genie are remarkable and worthy of this volume no matter whether they are women or men. There is nothing obvious that ties them in their experiences or their accomplishments together because they were women.

Nonetheless, the wonderful survey article by Pat Brown in this volume shows that there is a large part of our history that is not well known to most, and certainly not to most of our current students. The effort and contributions of women ichthyologists have often differed from those of men, and they have often been acknowledged differently. It is clear that the time is ripe for a comprehensive review and synthesis of the history of ichthyology and ichthyologists, women and men. We already have models in books on women in science (e.g. Kass-Simon & Farnes[6]). Our efforts here have concentrated largely on North America and western Europe, yet we know there is a history of accomplishments and contributions by colleagues in Asia, eastern Europe, South America, Africa and elsewhere. We started this dedicated issue to acknowledge the contributions of a few respected ichthyologists known to us personally. We think we have accomplished that somewhat limited objective. We hope the broader issues we have only touched upon will motivate others to accept the challenge and bring forth the more comprehensive historical and conceptual treatment of our subject. There must be a wealth of material in archives, personal letters and unpublished manuscripts and the stories and insights we can gain from this will be immeasurable.

The three women we have chosen as the focus of this dedicated issue have influenced each of us in a number of ways, directly and indirectly. It happens that they were all senior colleagues to us and so their published papers influenced us in the usual scientific ways. But each of them also influenced us by direct personal contacts, encouragements, positive suggestions and a myriad of other ways. We hope that shows in the brief biographies and personal interviews we have for each of them in this volume.

Guelph, 7 March 1994

Eugene K. Balon
Michael N. Bruton
David L.G. Noakes

Environmental Biology of Fishes **41**: 9–30, 1994.

Early women ichthyologists

Patricia Stocking Brown
Department of Biology, Siena College, Loudonville, NY 12211, U.S.A.

Key words: Marion Griswold Grey, Francesca Raymond LaMonte, Erna Mohr, Canna Maria Louise Popta, Margaret Hamilton Storey, Grace Evelyn Pickford, Cornelia Maria Clapp, Edith Grace White, Helen Irene Battle, Emmeline Moore, Frances Naomi Clark, Rosa Smith Eigenmann, Lucy Wright Smith Clemens, Laura Clark Hubbs, Frances Vorhees Hubbs Miller, Marie Poland Fish

This paper is a brief summary of some of the women who have worked in the field of ichthyology or studied the biology of fishes in a somewhat broader area. All of the women included were born prior to 1920. It does not include Ethelwynn Trewavas since her life and work are covered in a detailed article by David Noakes. The list of women covered here is not meant to be complete, especially with respect to eastern Europe or Asia, and I welcome suggestions of additional women along with biographical information about them for future work.

For purposes of a broader discussion, I have placed these women into the following categories: those who worked in (1) museums of natural history, (2) higher education, (3) government, and (4) women who worked primarily with their husbands who were also zoologists or ichthyologists.

1. Museum work

Museums of natural history have offered special opportunities for a number of early women biologists. Since much of the initial occupation of biologists was with collecting and classifying organisms, museums often functioned as research centers for the discipline. As museums expanded their collections in the late 1800's and early 1900's, many employed women assistants to cope with the growth of the collections. As museums began to be open to the public, women would often be employed to develop the necessary exhibits. Once in such positions, these women might then have additional opportunities to do basic research using the museum collections.

Although research universities were seldom willing to offer women academic positions (and thus provide women with 'a lab of their own'), women could work on a volunteer basis in museums, do professional quality research, and eventually be given a title and be allowed partial or full participation in their discipline. Perhaps the most famous of these women zoologists who had a title but no paying position was the invertebrate zoologist, Libbie Hyman, who spent many years as a Research Associate at the American Museum of Natural History. Hyman supported herself during this time from what eventually became the very minimal royalties earned from her comparative anatomy text. This kind of volunteer but self-directed work would seem to require a field with minimal requirements for expensive research equipment. Thus, work in anatomy and taxonomy on existing museum collections could provide special opportunities for those who commanded few resources. A striking feature of the women described here was their willingness to put their own work aside and help their male colleagues. Both Grey and LaMonte had skill with languages which both gave them access to the scientific literature of other languages and made them particularly valued colleagues.

Additional women who worked in museums include Lillian Dempster (1905–1992) who worked at the California Academy of Sciences, Myvanwy M. Dick (1910–1993) who worked at Harvard University's Museum of Comparative Zoology, and Margaret Mary Smith (1916–1987), the first director of the J.L.B. Smith Institute of Ichthyology (later incorporated as a national museum) and professor at Rhodes University in Grahamstown, South Africa.

Marion Griswold Grey (1911–1964)[1]

1911	**Born in Los Angeles, youngest of 3 children of James and Lucy Griswold**
1929–1931	**Attended Wellesley as member of class of 1933**
1933	**Married Arthur L. Grey, 14 September, and moved to Chicago**
1935–1939	**Three children born – Peter (b. 1935), Lucy (b. 1937) and Sarah (b. 1939)**
1941	**Began to work at Field Museum of Natural History, Chicago, as volunteer**
1943–1946	**In charge of Division of Fishes (unpaid) to Loren Woods' absence during World War II**
1943–1964	**Associate, Division of Fishes, Department of Zoology, Field Museum**

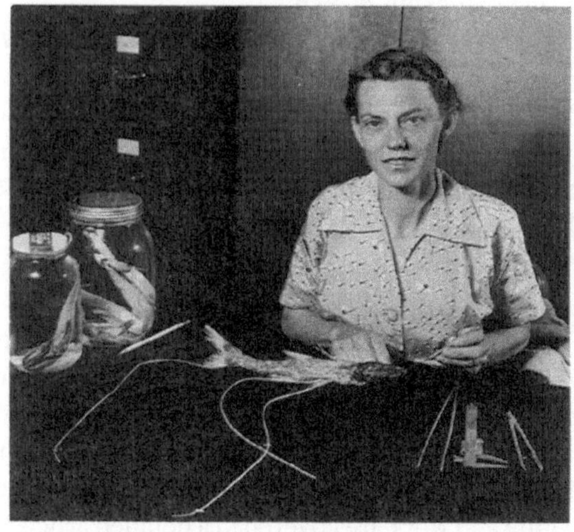

Fig. 1. Marion Griswold Grey. Photograph courtesy of the Field Museum of Natural History, Chicago.

Marion Griswold was born in Los Angeles and moved with her family to Kenosha, Wisconsin when she was nine. She was her father's favorite child and he was a powerful moral influence in her life. From the age of twelve, she wanted to be a writer. Marion's formal education consisted of two years of study in Zoology at Wellesley College. She withdrew from Wellesley because, as she later told her children, many of her friends had had to leave school when their fathers went bankrupt during the Depression, and Marion felt guilty about remaining. Her connection with the Field Museum began when she collected a pipefish on the Maryland shore and brought it to Alfred Weed at the museum to identify. He encouraged her to work at the museum as a volunteer. According to her daughter, 'for the rest of her life, with her husband's grudging permission, she spent every Wednesday at the Museum and much of her free time at home, hunched over a desk inspecting specimens or reading or writing about fish'.

Although she was informally educated by a number of colleagues at the museum (Alfred Week, Loren Woods), her main inspiration came from Karl Schmidt, Chief Curator of Zoology. Only two years after she began her work there, Loren Woods left to serve in the Navy and Karl Schmidt asked Marion to take charge, with the title 'Head of the Fish Division'. It is said that she was paid $ 2.50 per month to cover her transportation to work.

Her research started with the preparation of a *Catalogue of Type Specimens of Fishes in the Chicago Natural History Museum* which was published in 1947. In the summer of 1948 she was a member of the museum's expedition to Bermuda. According to her daughter, this was probably the high point in

[1] Information supplied by letters to P.S. Brown from Margaret Bradbury (25 January 1987); from Marion's daughter, Sarah Grey Thomason, Professor of Linguistics, University of Pittsburgh (1 March 1987; 13 April 1987); and from Myvanwy Dick (1987); a letter to Margaret M. Stewart from John Clay Bruner, 24 June 1986 and an obituary by Rand, A.L. 1964 in The Bulletin, Chicago Natural History Museum. p. 1, 8.

her career for sheer enjoyment. Marion taught herself Russian so she could read the Russian ichthyological literature and she generously translated papers for her colleagues. She entered into a lively correspondence with deep-sea fish workers in Europe, Russia, Japan and elsewhere. In 1953 she presented a paper at the International Zoological Congress in Copenhagen. Two of her most important contributions are *The distribution of fishes found below a depth of 2.000 meters* published in 1956, and *Family Gonostomatidae*, published in Part IV of *Fishes of the Western North Atlantic* which appeared in 1964. She published 21 papers between 1945 and 1964.

Her accomplishments are especially significant considering her lack of formal training, her responsibility in the raising of her family, the limited time she was able to devote to this work (she went to the museum only one day per week for many years) and her relatively short life. Marion had a modest opinion of her own ability and work. According to her daughter, Marion believed that men tended to be more intelligent than women and that women should be careful not to offend men by appearing to be more intelligent. It was not until just before she died that she was finally convinced that the quality of her work was as meritorious as that of her male colleagues. Marion was regarded as a warm-hearted, helpful and generous woman scientist with tremendous enthusiasm and an ever-questing mind by those that worked with her at the museum and were her colleagues elsewhere. She died at the age of 52 from a series of strokes.

Francesca Raymond LaMonte 1895–1982[2]

1895	**Born in Bensberg, Germany**
1918	**B.A. and Certificate of Music, Wellesley College**
1920	**Assistant on Bibliography of Fishes, American Museum**
1925	**Secretary, Department of Ichthyology, American Museum**
1928	**Assistant in Ichthyology, American Museum**
1929	**Assistant Curator of Ichthyology, American Museum**
1930	**One of five official U.S. representatives to International Zoological Congress in Padua, Italy**
1935	**Associate Curator, Department of Living and Extinct Fishes**
1943	**Associate Curator of Fishes and Museum Secretary of International Game Fish Association**
1962	**Associate Curator Emeritus**

Fig. 2. Francesca Raymond LaMonte weighing swordfish ovaries on the Lerner-Cape Breton Expedition, summer 1936. Photograph courtesy of the American Museum of Natural History, New York.

[2] Information supplied to P.S. Brown by Carla Stewart, Wellesley College Archive, January 1987 and by James Atz, American Museum of Natural History, 8 December 1986 and 4 April 1987.

Little is known regarding how family, school or college influenced the development of Francesca La-Monte's interest in natural history. While she was growing up her family lived in Russia and England and vacationed in France and Italy. She went to school in Germany. Originally hired by the American Museum of Natural History to work on Dean's *Bibliography of Fishes* and as the department secretary, she put to good use her ability to translate French, German, Italian, Spanish and Russian. There is no complete bibliography available, but she is said to have published 86 articles and several books on fishes in addition to numerous translations. She participated in expeditions to Cape Breton (1936, 1938), Bimini (1937) and Peru-Chile (1940). She also did field work in South Carolina, Florida, Hawai'i, Brazil, and the Isle of Shoals. Her most extensive work was on the marlin and swordfish. She was responsible for many exhibits in the Fish Hall, including the 'Life History of Swordfish', 'Life History of Eels', 'Life History and Migration of the Salmon' and 'West Indies Fishes'. Her hobby was also big-game fishing. After LaMonte retired in 1962, she spent 15 years reading and recording for the blind before retiring to Biloxi, Mississippi.

Erna Mohr 1894–1968[3]

1894	**Born in Hamburg, Germany**
1914–1934	**High school teacher**
1914	**Published first scientific paper**
1934	**Head of Fish Biology Department, Zoologisches Staatsinstitut und Zoologisches Museum Hamburg**
1936	**Head, Department of Higher Vertebrates, Zoologisches Museum**
1944	**Member of Kaiserliche-Leopoldinische-Karolinische Akademie der Wissenschaften, Halle**
1946	**Curator of the Vertebrate Department**
1950	**Honorary doctorate, Universität München**

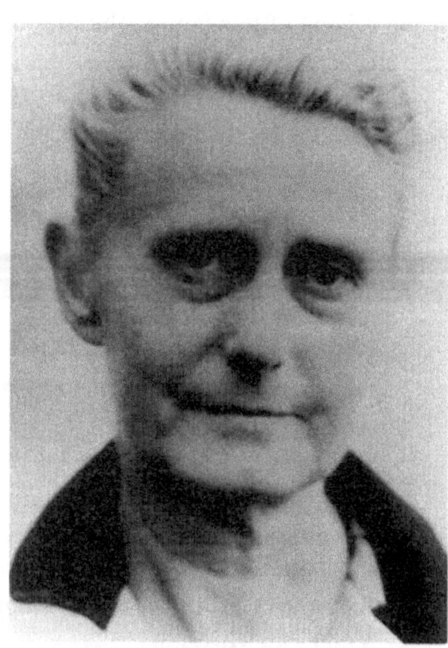

Fig. 3. Erna Mohr. Photograph courtesy of Hamburg Zoologisches Museum.

Erna Mohr, the daughter of a high school teacher, became interested in natural history during childhood. Little is known regarding her formal education. For 20 years she taught at a series of high schools, including one for children with special needs.

---

[3] Information from Herre, W. 1968. Erna Mohr. Rede zur Trauerfeier. Zeitschrift für Säugetierkunde 33: 256–261 (translated by Timo Jarrell); Hubbs, C.L. 1969. Erna Mohr 11 July, 1894 – September 10, 1968. Copeia 1969: 64; Kosswig, D. 1969. Dr. h.c. Erna Mohr, 11.7.1894 – 10.9.1968. Mitt. Hamburg Zool. Mus. Inst. 66: 7–23 (includes full citation of 408 publications).

During the time she was teaching school, she published many articles in popular science magazines. She also pursued research, presumably on a volunteer basis, at the Zoological Museum in Hamburg. She first worked with Professor Ehrenbaum in the Fish Department on problems of fish growth, reproduction and feeding. According to Wolf Herre, she was the first person to determine the age of fish by their ctenoid scales. She later began to work with Professor Dunker on fish systematics. They described the fishes taken by the Südsee-Expedition der Hamburgischen Wissenschaftlichen Stiftung 1908–1909. According to Carl Hubbs, her revisionary studies of synantognathous fishes, particularly the viviparous halfbeaks, and of the Ammodytidae and Centriscidae were especially noteworthy. This work led to international recognition.

When Professor Dunker retired in 1934, Mohr became head of the Fish Biology Department. In 1936 she was put in charge of the Department of Higher Vertebrates. She was responsible for exhibits as well as collections in these departments. In addition to her work in the Hamburg Museum, she became a spokesperson for museums throughout Germany and helped in their development.

Erna Mohr had always been interested in living animals and took many photographs and studied the behavior of numerous mammals in zoos. As this interest developed she began to study the systematics, breeding and preservation of reindeer, the European bison, and the Przewalski wild horse, as well as other species. In addition, she began to study the behavior of mammals both in captivity and in the natural habitat.

Her encouragement of a number of young scientists and her broad interests in natural history, museums, collections, zoos, mammalian behavior, fish systematics, and species preservation – combined with her warm and considerate personality – won her a following from numerous scientists and lay people. She published over 400 articles.

Canna Maria Louise Popta 1860–1929[4]

1860	**Born, probably in Leiden**
1895	**Member of Dutch Zoological Society**
1898	**Doctorate from University of Bern**
1889	**Lab Assistant to curator of reptiles, amphibians and fishes, Rijksmuseum van Natuurlijke Historie, Leiden**
1891	**Curator of Fishes, Rijksmuseum van Natuurlijke Historie**
1928	**Retired, age 68**

Fig. 4. Canna Popta. Photograph supplied by Rijksmuseum van Natuurlijke Historie, Leiden, The Netherlands.

Canna Popta was one of the first women to study at Leiden University, where she obtained a degree in geology, zoology and botany. This degree allowed her to teach in high school. She earned her doctorate in Switzerland at the University of Bern. Her thesis was done in the field of botany (a study of Hemiasci) with E. Fischer as her major professor. After obtaining her position at the museum, she concentrated on the study of fishes. She was known to be extremely helpful to any visiting scientists at

[4] Biographical information and list of Popta's publications supplied to P.S. Brown by Marinus S. Hoogmoed, with the assistance of M.K.P. van Oijen, 23 February 1987.

the university – putting her own work aside completely until the visitor left. She published over 40 papers and several articles for encyclopedias.

Margaret Hamilton Storey 1900–1960[5]

1900	**Born in San Francisco, 31 July 1900**
1922	**A.B. at Cornell**
1932–1935	**Assistant Bibliographer of Fishes, American Museum of Natural History**
1936	**A.M. at Stanford**
1936–1940	**Stanford Natural History Museum, Volunteer Researcher**
1940–1960	**Assistant Curator of Zoological Collections, Stanford Natural History Museum**
1937–1942	**Secretary, Western Division American Society of Ichthyologists and Herpetologists (ASIH)**
1941	**President, Western Division ASIH**

Fig. 5. Margaret Storey. Photograph by R.S. Ferris.

Margaret Storey's parents, Thomas Andrew Storey and Parnie Hamilton, both graduated from Stanford University. Margaret had two younger sisters who became M.D.'s. After obtaining her A.B. degree from Cornell she worked in the theatre as a stage manager. Apparently she was able to use this training in organization and management to plan scientific events later in her career. Margaret also had a long-life interest in athletics. She became the only woman officially appointed a timer of intercollegiate track events. Margaret returned to the west coast and began working at the Stanford Museum as a volunteer. At the insistence of George Myers, Margaret was given a regular staff appointment in 1940.

According to Myers, 'Margaret soon became the busiest and most useful person in the Zoological Division. She acted not only as curator, but also as librarian of the growing zoological library, and editor of *Stanford Ichthyological Bulletin* and *Occasional Papers*, but also as counsellor and helpful assistant to everybody – faculty, graduate students and visiting investigators alike . . . Margaret's research was done mostly before she became a staff member. After that she was too busy . . . although she published comparatively little, few people have ever done so much to further ichthyology, herpetology and zoology; she constantly sacrificed her interests to those of others. She was probably one of the ablest curators of a large zoological research collection in the world. Almost every active American ichthyologist and herpetologist of the past 25 years owes much to her'.

---

[5] Information from Myers, G.S. 1961. Margaret Hamilton Storey. Copeia 1961: 261–263 (includes her publications).

2. Work in higher education

Since the early part of the twentieth century, academic positions in U.S. universities and colleges have required earning a doctorate. Many women earned doctorates in zoology from U.S. graduate schools between 1896 and 1930. For instance, 97 women earned zoology doctorates during this time-period from just six institutions: 26 from Cornell, 9 from Yale, 14 from the University of California at Berkeley, and 16 each from Columbia, Pennsylvania and Chicago. Of these 97 women, only two secured positions at institutions similar to ones where they had done their own graduate work – i.e. Ph.D. granting universities. This contrasted sharply with the careers of the 192 men who earned Ph.D.'s from these same schools during the same time period, 68% (130) of whom found positions in Ph.D. granting institutions. Most of these highly credentialed women scientists who chose to remain in higher education could find positions only in women's colleges, normal schools or high schools – institutions unable to supply the necessary resources for research. In the examples of Grace Pickford, Cornelia Clapp and Grace White there are a number of patterns typical of early women zoologists with doctorates. In spite of the impeccable credentials of both Clapp and White, both taught at women's colleges and were apparently never offered positions at graduate universities. What this meant was that they never had doctoral students of their own and so were unable to influence the field through their training of subsequent generations of graduate students. In spite of this, many of Clapp's students went on for doctoral degrees but could not add significantly to her academic genealogy because they themselves could not find positions where they could mentor doctoral students. Grace Pickford is highly unusual in that she eventually became a professor at Yale. In spite of Pickford's formidable accomplishments in zoology, however, one should note that she first taught many years at a women's college (Albertus Magnus), and that her first faculty appointment at Yale came 28 years after she had earned a doctorate.

Grace Evelyn Pickford 1902–1986[6]

1902	**Born in Bournemouth, England**
1923	**Natural Sciences Tripos, Pt. I, Cambridge University (equivalent to a B.A. degree, which was not then granted to women)**
1925	**Married G. Evelyn Hutchinson**
1925–1927	**Travelling scholar of Newnham College, Cambridge in South Africa**
1931	**Ph.D. at Yale University**
1934–1946	**Research Assistant, Bingham Oceanographic Laboratory, Yale University**
1934–1948	**Assistant Professor, Albertus Magnus College**
1937	**Naturalized U.S. citizen**
1946–1966	**Research Associate, Bingham Oceanographic Laboratory**
1951	**Member, Danish *Galathea* Expedition**
1959–1969	**Associate Professor of Biology, Yale University**
1969–1970	**Professor of Biology, Yale University**
1970–1986	**Distinguished Scientist in Residence, Hiram College, Ohio**
1980	**Pickford Medal established**
1981	**Awarded Wilbur Lucious Cross Medal by Yale**

Fig. 6. Grace E. Pickford on *Galathea* Expedition, 1951.

Grace Pickford was encouraged by her father to study natural history in the area around Bournemouth, England, where she grew up. Her mother believed in career opportunities for women and encouraged Grace to attend Cambridge University. Grace's passion for biology was shown while at Newnham College where she had published one paper and written three others before leaving College. At Cambridge she was founder of the Biological Tea Club, a small undergraduate group, which at that time included Evelyn Hutchinson, Gregory Bateson, Omer-Cooper and Michael Perkins. They met weekly to discuss their own papers. After her marriage to G.E. Hutchinson, she had an opportunity to go with him to South Africa. There she began to work on earthworms and freshwater oligochaetes, and collaborated on limnological studies with Hutchinson. She then went with him to Yale, where she continued her taxonomic and zoogeographic work on oligochaetes as a doctoral dissertation. This work was published in 1937 as a 612 page monograph. Although her marriage to Hutchinson ended in 1934, they remained life-long friends.

After earning her Ph.D., she held two half-time

[6] Information provided by Atz, J.W. 1970. Grace Pickford retires. Discovery 6: 41–43; Ball, J.N. 1987. In Memoriam, Grace E. Pickford (1902–1986). Gen. Comp. Endocrinol. 65: 162–165; Obituary in Newnham College Roll Letter by Anna Bidder and Penelope Jenkin; Brown, P.S. Interview with G.E. Pickford, 12 August 1981; Pickford, G.E. 1973. Introductory remarks. Amer. Zool. 13: 711–717.

positions – a teaching position at Albertus Magnus (a Catholic women's college), and a research assistantship at the Bingham Oceanographic Laboratory. In 1949 she began full-time work at the Oceanographic Lab. She studied cephalopods, publishing more than 20 papers between 1936–1974, and became recognized as the world's authority on the taxonomy and anatomy of octopods. It was this work that led to her discovery that the deep-sea cephalopod, *Vampyroteuthis*, is a living fossil, occupying a unique position between squids and octopuses. This, in turn, led to an invitation to join the Danish *Galathea* Expedition to study living *Vampyroteuthis*. The octopus genus *Pickfordiateuthis* Voss, 1953, is a tribute to her contributions in this field.

Her work with fish did not start until the 1940's and grew from her participation in a World War II project to study the possibility of using so-called 'trash fish' as a food source. She became interested in fish growth and collaborated in a series of studies with A.E. Wilhelmi on the role of hormones in this process. These studies were some of the first on fish hormones and helped to establish the field of comparative endocrinology. A 613 page monograph published in 1957, *The Physiology of the Pituitary Gland of Fishes*, by G.E. Pickford and J.W. Atz is considered an early and still enormously useful classic in this field. Possibly her most important contribution to the evolutionary studies of hormone function was her observation that the killifish, *Fundulus heteroclitus*, requires the pituitary hormone, prolactin, in order to maintain osmotic balance in fresh water. This observation led to 30 years of further work by numerous laboratories, documenting the varied uses to which prolactin has been put during vertebrate evolution.

Grace Pickford was the author or coauthor of 135 publications, of which 78 were concerned with fish. In spite of her life-long accomplishments in research (3 distinct animal groups and including the fields of anatomy, taxonomy, zoogeography and endocrinology), recognition came relatively late in her career. The comparative endocrinologists recognized her contributions by establishing the Pickford medal in 1980. She was awarded the Wilbur Cross Medal by Yale in 1981. Her appointment to Full Professor at Yale occurred in 1969, at age 67. She retired from Yale in 1970 but continued to maintain an active research program at Hiram College in Ohio. Grace had a number of undergraduate, masters, doctoral, and post-doctoral research students and collaborated with many other colleagues world-wide. Those who knew Grace as a teacher or colleague felt privileged to share in her tremendous enthusiasm for the study of zoology and her dedicated friendship.

Cornelia Maria Clapp 1849–1934[7]

1849	**Born in Mantague, Massachusetts**
1868–1872	**Student, Mount Holyoke Seminary**
1888	**Ph.B. at Syracuse University**
1889	**Ph.D. at Syracuse University**
1896	**Ph.D. at University of Chicago**
1872–1896	**Teacher, Mount Holyoke Seminary**
1896–1916	**Professor, Mount Holyoke College**
1906	**Elected Trustee, Marine Biological Laboratory (first woman and only woman trustee until 1950)**
1910	**Designated one of the 150 most important zoologists by 'American Men of Science'**

Fig. 7. Cornelia Clapp. Photograph courtesy of the Marine Biological Laboratory, Woods Hole, Massachusetts.

Cornelia Clapp is one of the most interesting and energetic of the early women zoologists who spent her career teaching in a women's college. She earned both the first and the second zoology doctorate granted to women in the United States, attended the Anderson School of Natural History at Penikese Island in 1874 where she initiated life-long friendships with zoologists, such as David Starr Jordan and C.O. Whitman. She, along with Rosa Smith Eigenmann, went with David Starr Jordan on walking tours through the U.S. and Europe as one of 'Jordan's tramps'. She was the first Ph.D. student to do her work at the Marine Biological Laboratory (MBL) in Woods Hole where she worked with Whitman. Her doctoral research was on the lateral line system of the toadfish and was published in the *Journal of Morphology*. From 1888 to 1922 Clapp spent her summers at the MBL, conducting research primarily on fish development. During all of these years she brought one or more investigators or students with her from Mount Holyoke. Clapp was the MBL Librarian from 1893 to 1907, a Member of the Corporation from 1896 until the end of her life, a member of the staff of the Embryology Class from 1896 to 1906, and Trustee from 1902–1904 and again from 1910 to 1922 and Trustee Emeritus, 1922–1934.

Her entire teaching career was spent at Mount Holyoke where she taught experimental courses in zoology before such an approach was used in most institutions. Many of her students went on to earn doctorates in zoology. Some returned to Mount Holyoke where they continued to maintain one of the strongest undergraduate zoology programs in the country for many decades. Of the women who earned doctorates in zoology from U.S. institutions prior to 1930, the largest percentage had undergraduate degrees from Mount Holyoke. Part of the success of this program was due to the tradition of taking several undergraduates to the MBL every summer where they were initiated into the scientific community.

That Clapp was an engaging and energetic individual can be seen from the many tributes to her by colleagues and former students when a laboratory building at Mount Holyoke was named for her. An

[7] Information from Haywood, C. 1971. Cornelia Maria Clapp in *Notable American Women*, E.T. James, J.W. James & P.W. Boyer (ed.) Cambridge; unpublished address by E.G. Conklin at Centennial celebration of her birth 17 March 1949 at Mt. Holyoke; and Mt. Holyoke Alumnae Quarterly 1934, Vol. 19, p. 1–9.

example from Louise Baird Wallace: 'Her bounding vitality and thirst for knowledge were contagious. I felt then and have felt ever since that I was never fully alive until I knew her'.

Edith Grace White 1890–1975[8]

1890	**Born in Boston, youngest of six children**
1912	**A.B. at Mount Holyoke, majors in English and Zoology**
1913	**Research Assistant with Prof. Dahlgren at Princeton**
1913	**A.M. at Columbia**
1918	**Ph.D. at Columbia on electric organs in *Astroscopus guttatus***
1915–1916	**Research Assistant, Princeton**
1917	**Laboratory Assistant, Milwaukee Downer College and Western Reserve University**
1918–1920	**Faculty Member, Heidelberg College, Ohio**
1920–1923	**Faculty Member, Shorter College, Georgia**
1923–1958	**Professor and Head of Department of Biology, Wilson College, Chambersburg, Pennsylvania**
1929–1930	**Research Fellow, Imperial University, Tokyo; Misaki Marine Lab, and the marine lab at Batavia, Java**
1933–1947	**Research Associate, Department of Fishes, American Museum of Natural History**

Fig 8 E Grace White Photograph courtesy of Wilson College

E. Grace White was an English major at Mount Holyoke when she elected a zoology course from Cornelia Clapp. Clapp, one of the most outstanding college teachers of her time, influenced Grace White first to switch majors from English to Zoology, and then to pursue graduate work on fishes.

Once Grace White had completed her doctorate, she held a series of positions at small colleges. This pattern was typical for women who earned Ph.D.'s at this time. For instance, of the 16 women who earned doctorates from Columbia between 1909 and 1930, nine found positions in women's colleges and none obtained positions at Ph.D. granting institutions. The career path of White's male colleagues differed significantly. Of 18 men who earned doctorates from Columbia between 1909 and 1930, 11 obtained faculty positions at Ph.D. granting institutions while 3 remained at undergraduate institutions for their entire career. White remained at Wilson College (a women's college whose students were primarily the daughters of ministers) for 35 years.

---

[8] Material supplied by Elaine Trehub, Mount Holyoke College Archives, 5 May 1987, Elizabeth Boyd, Wilson College Library, 29 April 1987, and James Atz, American Museum of Natural History, 3 October 1986

White published two textbooks on genetics and general biology in the 1930's and 1940's which were widely adopted and went through several editions. In addition, she continued to do research at the American Museum, where she was eventually given the title of Research Associate. She began studies on elasmobranch anatomy and taxonomy and was invited to study sharks in Japan, China and Java in 1929–1930 by W.K. Gregory of the American Museum. This research resulted in several papers that evaluated the anatomy of dermal denticles, dentition, vertebrae, spiral valves and other aspects of shark morphology. She integrated this information with an analysis of previous classifications and suggested a partly original concept of the origin, history and adaptive radiation of cartilaginous fishes. This was published as a monograph in 1937.

Helen Irene Battle 1903–1994[9]

1903	**Born in London, Ontario**
1923	**B.A. University of Western Ontario**
1924	**M.A. University of Western Ontario**
1928	**Ph.D. University of Toronto**
1929–1934	**Assistant Professor, Zoology, University of Western Ontario**
1934–1949	**Associate Professor, University of Western Ontario**
1949–1972	**Professor, University of Western Ontario**
1956–1958	**Acting Head of the Department**
1961	**Founding member and Vice President, Canadian Society of Zoologists**
1962–1963	**President, Canadian Society of Zoologists**
1975	**Selected by National Museum of Natural Science as one of 19 outstanding women scientists**
1977	**Fry Medal of Canadian Society of Zoologists**
1977	**JCB Grant Award of the Canadian Association of Anatomists**

Fig. 9. Helen Irene Battle. Photograph courtesy of Donald B. McMillan, University of Western Ontario.

Except for her years at the University of Toronto, Helen Battle has lived her entire life in London, Ontario. Her first love was teaching and her teaching career spanned more than 50 years. She was considered one of Canada's most gifted teachers, teaching an introductory biology course to arts students as well as embryology to medical students and advanced courses to honours zoology students. Her enthusiasm and interest in her students generated many life-long relationships. The respect and warmth her students and colleagues felt for her was demonstrated by a banquet organized in her honour 31 March 1967. At that time, a scholarship was funded in her name for zoology students.

In addition to teaching, Helen Battle maintained an active research program over many years. She

---

[9] Information supplied to P.S. Brown by Donald B. McMillan, University of Western Ontario, 8 September 1986.

spent most summers at the Atlantic Biological Station in St. Andrews, New Brunswick where her work included the physiology and embryology of a number of fish species. She published 37 articles between 1926 and 1973. These included work on the embryology of goldfish, Atlantic salmon, zebrafish, goldeye, brown trout and lampreys.

She was committed to the development of a graduate program at the University of Western Ontario. She also was a founding member of the Canadian Society of Zoologists. Her life-long work for her students, the university and the profession was recognized by a series of honors including the Fry Medal, the JCB Grant Award and her selection as one of the 19 outstanding women scientists in Canada.

3. Government work

Both Emmeline Moore and Frances Clark started working in state conservation departments at a time when few biologists in these organizations had earned doctorates. As a consequence, their credentials were outstanding. During this time period, considering the growth of universities, men with these credentials would undoubtedly have been offered attractive academic positions. Moore was almost immediately given a position with major responsibilities when she was put in charge of the New York State Biological Survey.

Emmeline Moore 1872–1963[10]

1872	**Born in Batavia, New York**
1895	**Graduated from Genesco Normal School**
1895–1902	**Taught school in Scottsville, Cattaraugus and Cooperstown, NY**
1905	**A.B. Cornell**
1906	**A.M. Wellesley**
1906–1910	**Instructor, Trenton New Jersey Normal School**
1911	**Exchange Professor, Huguenot College for Women, Wellington, South Africa**
1914	**Ph.D. Cornell**
1914–1917	**Instructor to Assistant Professor, Vassar College**
1917–1919	**Worked on Federal Government World War I project – primary food relations of fish**
1920–1925	**Investigator in Fish Culture for NY State Conservation Department**
1926–1932	**Director of Biological Survey, NY State Conservation Department**
1928	**President, American Fisheries Society (first woman)**
1932–1944	**Chief Aquatic Biologist, NY State Conservation Department**
1939	**Honorary Doctorate, Hobart College, Geneva, NY**
1944–1945	**Research Associate, Bingham Oceanographic Lab, Yale University**

Fig. 10. Emmeline Moore. Photograph courtesy of Janet Moore Thornton.

Emmeline Moore was brought up on a farm in Batavia, New York. She showed an early interest in natural history and recalled, 'I used to go the swamps every Sunday. I remember seeing my first cardinal flowers and ferns there'. After high school graduation she taught in various normal schools and women's colleges for 20 years – taking time off to complete university degrees at Cornell and Wellesley. Her strong interest in natural science and her intense curiosity led to a Ph.D. at Cornell (1914) under George Embody, who is considered the father of fish culture.

A war-time project led directly to her appointment as New York State's first professional investigator in fish culture. This early work resulted in a series of papers on fish diseases. In addition, one of

[10] Information from the following: Carlander, K.D. 1984. Trends in the role of women in the American Fisheries Society, Fisheries 9: 8–9; Heacox, C. Dr. Emmeline Moore. The Conservationist, p. 47; American Fisheries Society Newsletter, October 1963; The Conservationist, Oct–Nov 1963, p. 38; Interview by Johanna Daily with Florence Moore 12 October 1986; Scrapbook of newspaper clippings on E. Moore supplied by family member.

her first projects was a collaborative study (in the summer of 1920) with James Needham, Chauncy Juday, Charles Sibley and John Titcomb on all aspects of the biology, chemistry and physics of Lake George. This study, authorized by the New York State legislature to determine how to increase fish production in this lake, was a convincing success and approval was granted to continue and expand the biological survey to include the entire 60 000 miles of the N.Y.S. watershed. In 1926 Emmeline Moore was appointed Director of this survey.

The original study and subsequent 14 reports (all edited by Moore) published between 1926 and 1939 were the first and remain the most comprehensive scientific examinations of any state's water resources ever conducted. During these years, Emmeline created a bond with numerous universities and colleges around the country by hiring and encouraging great numbers of students and faculty during the summer survey field expeditions. This training ground was important for many biologists who would later become well known in their fields.

In addition to her work on the biological surveys, Moore published technical papers on fish culture and fish diseases as well as many popular articles. After retirement from the N.Y.S. Conservation Department, Emmeline continued publishing, and was given many honors, including having the state research vessel named after her (*The Emmeline M*) in 1958. In retrospect, the careers of the scientists whom she encouraged as students, and the N.Y.S. Biological Survey, which has been relied upon by numerous scientists for the past 50 years, are the greatest tributes to this woman scientist.

Frances Naomi Clark 1894–1987[11]

1894	**Born near St. Edward, Nebraska**
1918	**A.B. Zoology, Stanford University**
1918–1921	**U.S. Bureau Commercial Fisheries, Lab Assistant for C.H. Gilbert**
1921–1923	**Assistant, California Division of Fish and Game**
1924	**M.S. University of Michigan**
1925	**Ph.D. University of Michigan**
1925–1926	**Teacher, San Jose Junior High School**
1926–1941	**California Department of Fish and Game, Assistant Fisheries Biologist, Senior Fisheries Researcher, Supervising Fisheries Researcher**
1941–1956	**Director, California State Fisheries Laboratory at Terminal Island**

Fig 11 Frances N. Clark. Photograph courtesy of Richard S. Croker.

Frances Clark was the younger of two girls. Her older sister was Laura Clark Hubbs. Their father was a farmer in Nebraska who apparently did well enough financially to be able to retire while still in

[11] Information provided by Richard S. Croker to P.S. Brown, 1 June 1987 and 18 June 1987 including obituary written by Croker for the California Cooperative Fisheries Investigation, and retirement notice for Frances N Clark published in California Fish and Game, July 1956 (42: 3); by Robert R Miller to PS Brown, 17 February 1987 including list of Clark's publications and c v, and by Clark Hubbs in phone call to P.S. Brown, 23 February 1994.

his fifties. The family moved to California where both Laura and Frances attended Stanford University. Frances immediately began to work for the U.S. Bureau of Commercial Fisheries. Why Frances decided to go on for a Ph.D. is unclear, but it is thought that either N.B. Scofield, then Chief of Marine Fisheries, or William F. Thompson, a young biologist working at the U.S. Bureau of Commercial Fisheries, encouraged Frances to pursue a doctorate. She went to Michigan where she earned a doctorate with her brother-in-law Carl Hubbs. It is interesting that Frances was awarded a Ph.D. two years before Carl got his doctorate from Michigan.

After one year spent teaching, Frances Clark joined the California Department of Fish and Game. Although there apparently was a period of time where she was the only person in the department with a Ph.D., they were reluctant to appoint her to an administrative post because it was felt that the men would not tolerate a woman supervisor. This attitude apparently did not persist, however, because she served as the Director of the lab at Terminal Island for the 17 years prior to her retirement. Clark lists 62 publications from 1925 through 1962. Her doctoral thesis was on the life history of the grunion, but most of her other research dealt with the life history, dynamics and conservation of the California sardine. Under her leadership the Terminal Island Laboratory initiated research projects on anchovy, yellowtail, surf perch and kelp bass, and expanded the high seas tuna program. In addition to her work in California, she did survey studies in both Peru and New Zealand. In her positions as Senior and supervising Fisheries Researcher as well as Director of the Laboratory, she supervised and trained numerous other researchers who remained indebted to her.

4. Couples: women who worked with their husbands

Many productive men and women scientists have been married to scientists. In each case, what a person brings to the partnership varies, but often includes skills as an editor, artist, photographer, statistician, co-author, organizer, or field collector, as well as being a colleague with whom to discuss science. Men in these partnerships have often held highly visible faculty and administrative positions, whereas the women have often held no title, received little or no pay and, as a consequence, have been almost invisible. Clearly, the husband's reputation and visibility was often enhanced by the increased productivity of two people. In the words of Joseph Needham, 'A man's shadow is lengthened by his own work, but also by the work of his wife and his students'. Frequently, what the men provided to their wives was access to the scientific community, space in a university or college laboratory, microscopes and other research materials, connections to colleagues, and opportunities to participate in scientific discussions in academic settings or at professional meetings. This acceptance by, and access to, the scientific community often appeared to require a connection through a spouse, father or brother since many single women with doctorates failed to gain such access.

Rosa Smith Eigenmann, Lucy Clemens, Laura Hubbs, Frances Miller and Marie Poland Fish represent several different career patterns of women who worked with their husbands. Although Rosa had no formal college degrees, her willingness to join David Starr Jordan on various collecting trips, to hold various jobs, and to identify new species of fish and publish papers on her own indicates that she was an independent spirit. Although she was the single- or joint-author on at least 40 papers, and continued her work after she was married and had several children, she completely stopped work on her own after family responsibilities became overwhelming. Of the couples discussed here, only Lucy and Wilbert Clemens had equivalent educational backgrounds, having both earned doctorates from the same department at Cornell. In spite of having two children, Lucy participated as a full partner in the research at the Pacific Biological Station where clearly they were working as collaborators. Laura Hubbs had both a B.S. and an M.S. degree, but they were in mathematics rather than biology. Her training in biology was through her work with her husband and his colleagues. Laura Hubbs coauthored a number of papers with her husband but did not become identified with a specific area of research of her own. Similarly, Frances Hubbs Miller devoted herself primarily to facilitating her husband's research. However, Marie Poland Fish, who had no more formal training than Laura Hubbs or Frances Miller, carved out her own area of expertise. Although she followed her husband to Buffalo, Rhode Island, Washington, D.C. and back to Rhode Island, she worked on completely separate and independent projects. Of women who worked with their husbands omitted here, the best known is Margaret Mary Smith, wife of the J.L.B. Smith of the coelacanth fame. After his death in 1968 she became an independent ichthyologist and founder of the Institute, as mentioned in the first section of museum workers.

Rosa Smith Eigenmann 1858–1947[12]

1858	**Born in Monmouth, Illinois, last of 9 children**
1880	**Published first scientific paper (age 22)**
1880–1892	**Studied with David Starr Jordan at Indiana University**
1887	**Married Carl H. Eigenmann**
1887–1888	**Studied the Agassiz South American Fish collection at Harvard**
1888	**Helped establish biological station in San Diego**
1890	**Curator of Fishes at California Academy of Sciences**
1890–1902	**5 children – Margaret, Charlotte, Theodore, Adele and Thora**
1895	**Delivered public lecture on 'Women in Science' at National Museum in Washington, D.C.**

Fig 12 Rosa Smith Eigenmann Photograph courtesy of Hubbs Library

Rosa Smith is considered the first woman ichthyologist in the United States. She moved from Illinois to San Diego with her family in 1876, attended the Point Loma Seminary there and a business college in San Francisco. Her family was in the newspaper business and Rosa became the first woman reporter on the *San Diego Union*. Always intensely interested in natural history, she happened to give a paper on a new species of fish she had discovered to the San Diego Society of Natural History while David Starr Jordan was there. He encouraged her to study with him at Indiana University, which she did for two years, living with his family during this time. She spent the summer of 1881 collecting in Europe

[12] Information primarily from Hubbs, C L 1971 Rosa Smith Eigenmann in *Notable American Women*, E T James, J W James & P W Boyer (ed), Cambridge, from two unpublished biographical sketches written by two of their daughters, Thora Eigenmann (from the California Academy of Sciences) and Charlotte Eigenmann (from Indiana University Archives), Myers, G 1928 Carl H Eigenmann, Ichthyologist, Natural History 28 98–101, Stejneger, L 1938 Carl H Eigenmann, 1863–1927 National Academy of Science Biographical Memoirs 18 305–336, and Eigenmann, R S Women in Science, Proceeding of the National Science Club, 13–17 January, 1895 Ogilvie, M B 1986 Women in science Antiquity through the nineteenth century MIT Press, Cambridge

with 'Jordan's tramps' – 34 of Jordan's students and colleagues. Due to family illness, she returned to San Diego before completing an undergraduate degree at Indiana. She continued to collect and correspond with Jordan and other ichthyologists. Through Jordan she met Carl Eigenmann and they were married in 1887. She was 28, four years Carl's senior. By the time of her marriage she had published a list of fish in the San Diego area, been requested by the Smithsonian Museum to make a collection of surf perch from the San Diego area, and published 10 single-authored primary papers in the *Proceedings of the U.S. National Museum.* Immediately after their marriage they left for Harvard where they had arranged to work on the mostly unstudied fish collections that Louis Agassiz had made in Brazil. They both spent part of the summer of 1888 at Woods Hole where the U.S. Fish Commission had a station. During this time they completed several short papers, in addition to a 500 page review of the South American catfishes. Carl and Rosa returned to California in 1889 and together established a biological station in San Diego where they continued their work on fish of the region. They both had appointments as curators at the California Academy of Sciences. In 1891, Jordan had Carl Eigenmann appointed as his replacement as Professor of Zoology at Indiana University when Jordan accepted the position of President of Stanford University. Between 1880 and 1893 Rosa published 37 papers – 12 single-authored and 25 co-authored with her husband. Her last publication was in 1891 and was co-authored. The Eigenmanns had five children. Because one son became mentally ill and one daughter was retarded, raising the family was especially difficult. Rosa retired from active research after 1893 to care for their children. Although she continued to edit Carl's papers, she did not accompany him on his later expeditions. Carl continued to be actively involved in research and his work included studies on the development and reproduction of viviparous fishes on the Pacific Coast, a series of papers on the blind cave vertebrates from southern Indiana and Kentucky, and monographs on the freshwater fishes of South America. In 1895 Rosa gave a lecture on 'Women in Science' at the Smithsonian Museum which was published. This lecture reveals her to be knowledgable about other women scientists, concerned with the opening of education to women, aware of the conflict women had in doing science and raising a family and aware of the indirect and often unacknowledged contributions many women had made to science.

Lucy Wright Smith Clemens 1886–1937[13]

Fig. 13. Lucy Smith Clemens, 1909. Photograph courtesy Mount Holyoke College.

1886	**Born in Sheffield, Massachusetts**
1909	**B.A. Mount Holyoke**
1909–1910	**Research Assistant, Carnegie Institute, Cold Spring Harbor**
1911	**M.A. Cornell**
1914	**Ph.D. Cornell**
1912–1918	**Instructor, Mount Holyoke**
1918	**Married Wilbert A. Clemens**
1919, 1923	**Son, Alvin, and daughter, Ann Morgan, born**
1918–1937	**Joint Director of Pacific Biological Station, Vancouver Island, British Columbia**

Lucy Smith was the third generation of the Smith family to attend Mount Holyoke. She was encouraged by the Mount Holyoke faculty to pursue graduate work in zoology. At Cornell she studied limnology under J.G. Needham in the Entomology Department. Her fellow students included Emmeline Moore and Wilbert Clemens. After completing her doctorate, she returned to teach at Mount Holyoke. She married Wilbert Clemens of Toronto in June 1918 and they spent the summer on a collaborative study of the mutton fish, *Zoarces anguillaris* at the Atlantic Biological Station at St. Andrews. For several years Clemens was on the faculty at the University of Toronto and during the summers, set up a freshwater biological station at Lake Nipigon. Lucy and their children accompanied him during the field work in the summer months even though the living conditions were relatively primitive. In 1924 Wilbert received an offer to be Director of the Biological Station at Departure Bay, Vancouver Island, British Columbia and after much agonizing, decided to accept it. Lucy fully participated in the work of the station including much of the research on salmon. She co-authored the annual reports on salmon with her husband that were published in the 'Report of the Commissioner of Fisheries for British Columbia' from 1925 until her death in 1937. She died when she was only 50.

[13] Material furnished by Mount Holyoke College Archives and by G. Miller, Library, Pacific Biological Station, Nanaimo, British Columbia, including an unpublished autobiography of W.A. Clemens.

Laura Clark Hubbs 1893–1988[14]

Fig. 14. Laura Hubbs next to Carl Hubbs measuring fishes at the University of Costa Rica in 1973. Photograph by E.K. Balon.

1893	**Born near St. Edward, Nebraska**
1915	**B.S. Mathematics Stanford**
1916	**M.S. Mathematics Stanford**
1918	**Married Carl Hubbs, 3 children Clark, Earl and Frances**
1929–1944	**University of Michigan Museum of Zoology, Cataloger, part-time**

Laura Cornelia Clark was born in Nebraska in 1893, one year before her sister Frances Clark was born. The family moved to California and both Laura and Frances attended Stanford University. Although Laura studied mathematics at Stanford, after marrying Carl Hubbs, she devoted her life to collaborating with Carl in his work on fishes. While at the University of Michigan, Laura was on the Museum staff as a part-time employee. However, once Carl became a Professor at Scripps Institute of Oceanography in 1944, nepotism rules prevented Laura from holding a paid position. She continued to work but on a volunteer basis. She did the statistical work for Carl's publications, kept his scientific files, records and library in order, and collaborated on field work. She coauthored 19 of his publications. Two of their children, Clark Hubbs and Frances Hubbs Miller, also devoted their lives to the study of fishes.

[14] Information provided to P.S. Brown by Maria China at Hubbs-Sea World Research Institute (HSWRI), 23 February 1994; in August 1988 H-SWRI Newsletter, *Currents*; and by phone conversation with Clark Hubbs, 23 February 1994.

Frances Vorhees Hubbs Miller 1919–1987[15]

Fig. 15. Frances Hubbs Miller.

1919	**Born in Chicago**
1940	**B.S. Zoology, University of Michigan**
1940	**Married Robert Rush Miller**
1944–1954	**Five children – Frances, Gifford, Roger, Laurence and Benjamin**
1980–1987	**Research Associate, University of Michigan Museum of Zoology**

Frances Hubbs was the oldest child of Laura Clark Hubbs and Carl L. Hubbs. Fran not only grew up in a family where ichthyology was undoubtedly the main topic, she married ichthyologist Robert Rush Miller, and spent much of her life as a collaborator with her husband. Fran's contributions to their research included typing, cataloguing, recording data, data calculation, statistics, organizing field notes and data, and editing. One of their major collaborative research efforts was on the freshwater fishes of Mexico. This included field work in Mexico during 1972, 1974, 1976, 1978 and 1982, work at the U.S. National Museum during 1973–1974, and the organization of these data – all of which Fran was a participant. An additional project of her own was the publication of an annotated bibliography and subject index of her father's work.

[15] Information from Chernoff, B. 1988. Frances Vorhees Hubbs Miller 1919–1987. Copeia 1988: 520–523.

Marie Poland Fish 1902–1989[16]

1902	**Born in Paterson, New Jersey**
1921	**B.A. Smith College**
1922–1927	**Hydrobiologist, U.S. Bureau of Fisheries**
1928–1931	**Curator of Ichthyology, Buffalo Museum of Science**
1928–1931	**Senior Ichthyologist NYS Conservation Department, Investigation of Lake Erie**
1931	**Daughter Marilyn born**
1931–1933	**Research associate, Narragansett Marine Lab**
1939–1942	**State Ichthyologist for Rhode Island**
1942–1943	**Instructor of Zoology, Rhode Island State College**
1944–1946	**Research Associate in Ichthyology, US National Museum**
1946–1948	**Ichthyologist, Pacific Oceanic Biology Project**
1948–1970	**In charge of Office of Naval Research Project on 'Underwater Sound of Biological Origin', Narragansett Marine Laboratory, University of Rhode Island**
1965	**Navy Distinguished Public Service Award for basic research in marine bioacoustics**

Fig 16 Marie Poland Fish Photograph courtesy of Marilyn Fish Munro

Marie (Bobbie) Poland had intended to become a physician. After earning her B.A. degree at Smith College, she worked as a research assistant on cancer problems at the Department of Medical Research of the Carnegie Institute. In 1923 she married Charles J. Fish. Charlie Fish had earned a Ph.D. in 1923 at Brown University studying zooplankton. His major professor, Henry Bigelow – considered the father of oceanography – was at Harvard. Charlie first worked for the Bureau of Fisheries and then directed the Buffalo Museum of Science until 1935. In 1934 he joined the University of Rhode Island faculty. By 1936 he was a full professor in zoology and Director of the Narragansett Marine Laboratory, which he began. During World War II Charlie Fish served on the staff of the Chief of Naval Operations and Bobbie worked at the U.S. National Museum, classifying fishes. After the war, the couple returned to Rhode Island where Bobbie began to do research for the Navy on recording and identifying the sounds of over 100 marine coastal fishes. For 20 years after the war, Bobbie Fish directed the Navy's principal project on biological marine sounds. She published approximately 200 articles including primary research articles and popular scientific work. For four years 1936–1939 she and her husband wrote semi-weekly newspaper columns on popular science but they did not collaborate on their scientific work. Bobbie Fish received the Navy Distinguished Public Service Award in 1966 for her extensive contributions to underwater sound production by marine animals. She passed away on 2 February 1989.

[16] Information provided by Graduate School of Oceanography, University of Rhode Island

Asian women in ichthyology have not been considered in the preceding review. In order to include at least some of them into this volume, we are showing on this and the following page, photographs of female ichthyologists from Japan, Thailand and the Philippines to represent them all.

Gathered under a sakura three in full bloom on 31 March 1994 while attending the 27th Annual Meeting of the Ichthyological Society of Japan, are Dr. Midori Kobayakawa (second from left) of Kyushu University (who studies catfishes) with (left to right) graduate students Tomoko Seki and Yukiko Tanaka from the University of Tokyo (sharks), as well as Fujimi Fukuhara (barnacles), Miwako Kitade (zooplankton feeding by sea bream), Akiko Wakiyama (anabantoids), and doctoral student Akıko Morota (remoras), all from Tokyo University of Fisheries. Photograph by E.K. Balon.

Reiko Baba of Osaka City University studies brood parasitism and egg robbing in several freshwater fishes. Photograph by E.K. Balon, May 1994.

Left – Teodora Bagarinao (with son Carl Emilio) of the Southeast Asian Fisheries Development Center, Aquaculture Department at Iloilo, Philippines, who recently published (in EBF vol. 39, pp. 23–41, 1994) a wonderful review of the biology of the milkfish, and right – Annadel S. Cabanban of the International Center for Living Aquatic Resources Management in Manila, who just returned from a fellowship work on pomacentrid fishes at the Smithsonian Institution, Washington, D.C. and is working on a 'fishbase'-generated check-list of Philippine fishes.

Supap Monkolprasit, dean of the Faculty of Fisheries, Kasetsart University, Bangkok, an authority on Thai mangrove fishes, was recently the chairwoman of the highly successful Fourth Indo-Pacific Fish Conference.

Environmental Biology of Fishes **41**: 33–49, 1994.

The life and work of Ethelwynn Trewavas: beyond the focus on tilapiine cichlids

David L.G. Noakes
Institute of Ichthyology and Department of Zoology, University of Guelph, Guelph, Ontario N1G 2W1, Canada

Key words: Biography, British Museum (Natural History), C. Tate Regan, Cichlids, Sciaenids, Cyprinids, Taxonomy, African lakes, Coelacanth, Freshwater biology, Marine biology

Synopsis

Ethelwynn Trewavas was respected as a person and an ichthyologist by all who knew her. She was recognized as the senior scientist in the Fish Section of the British Museum of Natural History for almost half a century, and known internationally as an authority on several widely different groups of fishes, principally cichlids and sciaenids. Her early career was influenced by C. Tate Regan, her mentor at the Museum. She continued to develop her expertise from those foundations, and established her own reputation and independent collaborations with colleagues from a number of major institutions. She is most widely known for her major revisions of the African cichlids, but she published extensive taxonomic revisions and descriptions on other groups as well. She combined meticulous laboratory studies with extended field trips where she relied on her insights and interviews with local people to understand the biology of the fishes she was studying.

Ethelwynn Trewavas was born in Penzance, Cornwall, England on 5 November 1900. She received her primary elementary education at St Paul's School ('infants') Penzance, Cornwall, 1905–1909 and her secondary education at West Cornwall College for Girls from 1909–1917. Her university education was taken at Reading University College from 1917 to 1921, and lead to her external London Honours B.Sc. in Zoology and a Board of Education Certificate in Teaching. She was awarded the London University Scholarship for 1920–1921, and the Gilchrist Research Scholarship at the University of London from 1925 to 1928. She studied at King's College for Women, University of London, and received her D.Sc. from London in Zoology in 1934, for her published work on the morphology of frogs and the anatomy and classification of fishes. Her life was devoted to ichthyology at the British Museum (Natural History).

She died at her home on 16 August 1993, and so did not see this dedicated issue of Environmental Biology of Fishes (Noakes 1993, Greenwood 1994). Of course she was very much aware of our preparations for this issue, and she contributed freely of her personal recollections, her personal photographs, and a written response to a standardized interview we conducted with all three of our colleagues at the focus of this issue. It would not surprise those who knew Ethylwynn to learn that she was at first surprised and somewhat reluctant when we told her of our intentions to dedicate part of this issue to her. She was always unassuming, gracious and generous to the extreme. She never sought personal honours and always gave credit to others before acknowledging her own achievements.

ET spent her entire scientific career in the British Museum (Natural History), mostly as a senior member of staff of the Fish Section. Her duties involved responsibility for incorporating fishes collected from all over the world into the national col-

Fig 1 Photograph of an informal pastel portrait of ET in 1936 (courtesy of Rosemary Lowe-McConnell)

lection; their correct naming and classification; as well as the maintenance of indexes and catalogues initiated by J.R. Norman facilitating reference to the fishes and the relevant literature. To perform this work it was necessary to constantly review the basis of fish classification, and to carry out research work to this end. For some time ET specialized in the freshwater fishes of Africa. Almost every collection examined caused her to revise our conceptions of their classification. That fundamental work was necessary before she could give answers to the many questions that are addressed to the Museum by those engaged in the fisheries. ET always set the highest standards of scholarship. She was always generous of the time to help others. Much of her research was completed for the love and fun of it, after her official 'retirement', when she was no longer pressed by other Museum duties.

ET published more than 120 scientific and popular articles and books during an active scientific career of over 60 years. Almost half of her publications were after her nominal retirement in 1961. It would be impossible to detect any indication in her rate of publications coincident with that retirement – if anything it may actually have accelerated. Most of us know her for her monumental works on cichlid taxonomy and systematics – but that is largely because most of us are (relatively) young. It is true that her first paper on cichlids was published in 1928 and her first revision of a cichlid genus was published in 1931, but the major work on cichlids for which she is best known was not published until 1983. She was also an authority on sciaenids, various deep sea fishes, and published on cyprinids, and a host of other fishes, including lungfishes and coelacanths.

Ethelwynn became Research Assistant to C. Tate

Fig. 2. C. Tate Regan, J.R. Norman and Ethelwynn Trewavas (left to right), at the British Museum (Natural History) in July 1939. Photograph from personal collection of ET.

Regan in 1928 when he was made Director of the British Museum (Natural History). She did not officially join the BM until 1935 as an Assistant Keeper (Fig. 1), although the BM had its first women members of staff in 1928–1929. Her first work under Regan's guidance was on African cichlids (Trewavas 1, 2, Regan & Trewavas 3*), a harbinger of the great work of her later life. She also worked with him on deep sea fishes from the 'Dana' collection (Regan & Trewavas 4, 5, 8).

Regan was succeeded by J.R. Norman as Head of the Fish Section (Fig. 2), and on Norman's death ET became the senior staff member of the Fish Section (Trewavas 29, 30, 31). She was Deputy Keeper of the Department of Zoology in the Museum from 1958 until her official retirement in 1961. She worked in the Fish Section on a wide range of species over the years (Fig. 3), but developed her own specialization on the freshwater fishes of Africa. In 1935 she was a member of a party with K. Ricardo-Bertram and H.J. Borley conducting a fishery survey of Lake Nyasa (Ricardo-Bertram et al. 26). She was appointed a member of the Fisheries Advisory Committee to the Secretary of State for the Colonies in 1945, and for several years conducted short 3 week courses on fishes of the colonies as part of the training of fishery officers and fishery research officers for territories administered by the Colonial Office.

In 1947 she spent 2 months in Nigeria attached to the Fisheries Development unit, working with Mr. Dowsan, the Fisheries Officer, observing problems and giving advice. While she did not publish much as a result of this trip, she developed an interest in Mugilidae which led to later publications. She discovered that M. Cadenat (ORSTOM, Paris) was specializing in Mugilidae, and so she handed her material on mullets to him. Her later publications on mullets included papers on American species, and with Susan Ingham a key to north east Atlantic and Mediterranean mugilids (Trewavas & Ingham 86, Trewavas 92). She also developed considerable expertise in sciaenid fishes, and eventually published a major monograph on those fishes, based on her use of the swim bladder as a diagnostic feature (Fig. 4).

In 1965 ET convinced (she used the term 'bullied' in a taped interview she provided to Ro) the Colonial Fisheries Advisory Committee to finance an ex-

* See the 'Lifetime list of publications'

Fig. 3. Denys Tucker and ET, photographed in London, England on 21 August 1968, the day that forces of the Warsaw Pact invaded Czechoslovakia. Photograph by J. Holčík.

tensive African visit. She went first to the Rufigi River and its Great Ruaha tributary in Tanzania with Roland Bailey. Then she went to Zambia, to the Kafue with M.A.E. Mortimer, across the Kariba dam (Balon 1974, Trewavas & Stewart 95) to a Joint Fisheries meeting with Zimbabwe (then Rhodesia) fisheries personnel. Finally she went to Malawi, where she visited at Monkey Bay with David Eccles (Eccles & Trewavas 118), who took her to Lower Shire.

ET returned to Africa in 1970 with Jim Green and her niece Sally Corbet. The latter two were both of Westfield College, University of London. They travelled to Cameroon lakes, since ET had put Barombi Mbo in the United Nations Red Data Book for its endemic fishes. A collection of those fishes had come to the BM and ET and Jim Green wanted to look at the ecology of those species in the lake. Upon their arrival they met a local chief who was very helpful in getting fishermen to bring all the fish from their traps in the reeds. This produced nine endemic cichlid species, plus another from midwater and yet another caught 2 weeks later by the fishermen while ET and her party were away visiting Lake Kotto (Trewavas et al. 87). Of course the local fishermen knew all these fishes and the best times for catching them! Despite being in the IUCN Red Data List, Barombi Mbo has subsequently been much changed by cutting of surrounding forests, as Gordon Reid found in a visit there in the 1970's (G. Reid personal communication to Ro).

In 1981 ET enjoyed yet another trip to Africa, this time with Ro, to Kenya to visit Rene Haller's fish culture unit on the coast at Bamburi, near Mombassa. Old coral reef cement works were the site of efforts to attempt the culture of a number of species of plants, fruit trees, and game animals including eland (Fig. 5a). Tilapia culture was being attempted in round tanks with circulating water to discourage territory formation and spawning (Fig. 5b). ET and

Ro stayed in the Director's guest house, on a beautiful palm-fringed white sand beach with a coral reef at their front door. At the age of 81 ET tried snorkelling to observe fish for the first time, a brave effort since it required wading through spiny echinoderms at low tide! She and Ro also watched fish from the precarious confines of a small canoe along the edge of the reef in the lagoon. They visited Lake Jalore on the lower Athi (there called the Sabaki) for further observations on tilapia. They found time to assist with the breach delivery of a baby boy at the road side while the local women from the village held the expectant mother aloft! They visited the head waters of the Athi River trying to find pure *Oreochromis nigra,* but were unsuccessful. They concluded that escapees of introduced tilapias and hybridization appeared to have contaminated all the local species. They went with Ian Parker, an elephant researcher, to Lake Magadi, one of the famous soda lakes, and to the southern Wasanero River, which drains to Lake Natron, another of the soda lakes. They were looking for tilapias reported from there by Copley, but were not successful. ET returned to Africa in 1983, this time with Denis Tweddle (Fig. 6), and yet again in 1985 (Fig. 7) in connection with the book she was preparing on Malawian cichlid fishes (Eccles & Trewavas 118).

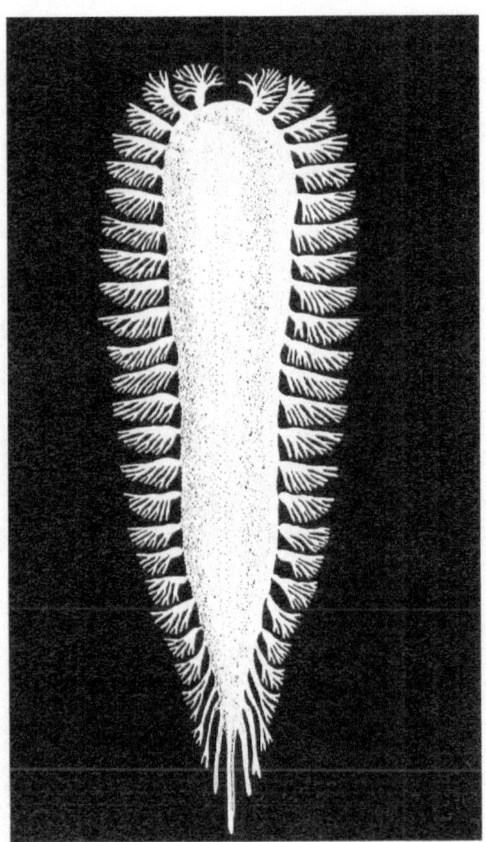

Fig 4 Frontispiece from ET's monograph on sciaenids of the Indo-West-Pacific Reproduced with permission of the Zoological Society of London

In addition to her African field experiences ET also travelled and visited colleagues in other countries. She visited Israel in 1961 to observe tilapias in Lake Kinneret. While in Tel Aviv she went with Adam Ben Tuvia to Megido of Old Testament and King Salomon fame. Lev Fishelson showed her *Tilapia aurea* in his laboratory aquarium (Fig. 8) and Steinitz persuaded ET to describe this species as distinct from *Oreochromis niloticus*. This species had previously been described as *Tilapia exul,* and as *Tilapia dageti* from west Africa by Monod (Trewavas 75). She went to Israel again 1983 to the ICLARM Conference in Nazareth.

ET presented papers at a number of major international conferences during her career. She was an invited contributor at the Bellagio conference on tilapia (Fig. 9), and the published proceedings from that conference were dedicated to her on the occasion of her 80th birthday (Trewavas 109). She continued to travel extensively to the laboratories and field sites of colleagues (Fig. 10) throughout her life. The publication of her monograph on tilapiine cichlids (Trewavas 110) (Fig. 11), and her compilation of Malawian cichlids (Eccles & Trewavas 118), quite appropriately brought the universal recognition she so richly deserved for her lifetime of work on these fishes. Despite increasingly frail health ET continued to maintain contact with her friends and colleagues (Fig. 12, 13, 14, 15, 16) and made her knowledge available to fish hobbyists (Trewavas 99, 102, 103, 105, 107, 117, Ferguson & Trewavas 119, 120).

Her scientific accomplishments were recognized by a number of awards and honours from several different societies and institutions. She received the Linnean Gold Medal of the Linnean Society of London in 1968, and Honourary Doctor of Science degree from the University of Stirling (Scotland) in 1986, and was made a Fellow of the Linnean Society

a

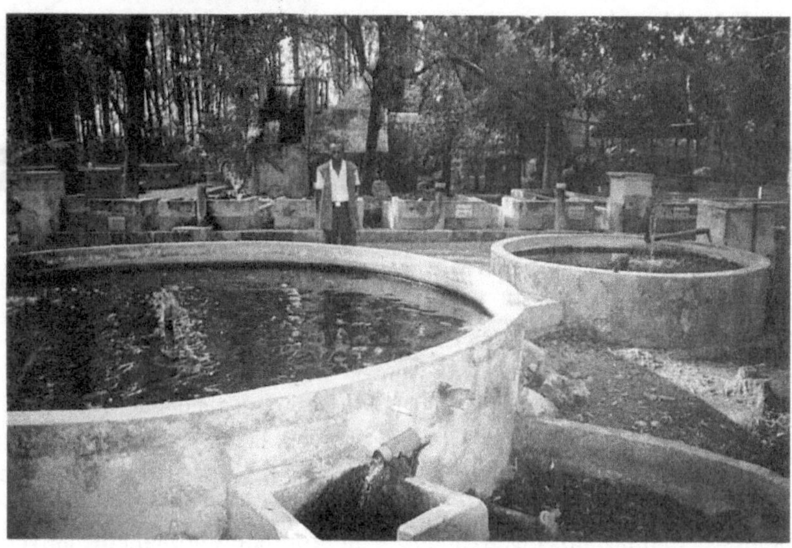

b

Fig 5 a – Domesticated eland at the Bamburi game park b – Ultimately the fish culture section of the game park contained many species and hybrids of tilapiine cichlids from all over Africa Photographs by E K Balon

of London *Honoris causa* in 1991. She was a Fellow of the Zoological Society, a Member of the Systematics Association [on its council for 3 years], a Member of the Institute of the British Council, a Member of the Council of the Freshwater Biological Association [representing the British Museum (Natural History)], a Member of the Institute of Biology, an 'Honorary' Foreign Member of the American Society of Ichthyologists and Herpetologists, an Honorary Member of the American Museum of Natural History, and Honourary President of the British Cichlid Association.

I tried without success to keep this biographical account of ET as conventional as possible. There is simply too much original material in her letters not to let them speak for themselves. The account that

Fig. 6. ET aboard the RV Ethelwynn Trewavas at Monkey Bay, Malawi in 1985 surrounded by members of the Malawi Fisheries Department staff. ET is flanked on her immediate right by Denis Tweddle and on her immediate left by Digby Lewis. Photograph by Rosemary Lowe-McConnell.

Fig. 7. ET swimming with her cichlids in Lake Malawi in 1985. Photograph by Rosemary Lowe-McConnell.

Fig. 8. ET, Lev Fishelson and Ro (left to right) at the British Museum (Natural History) Fish Section in August 1990.

follows is in ET's own words. I simply could not improve upon what she wrote, and it would sound stilted if I tried. It expands considerably upon her own more formal interview.

ET's personal recollections

'After graduating [B.Sc. and Certificate in Teaching] I taught in schools for 2 or 3 years, learning among other things that I was not a brilliant teacher, and longing to go more deeply into zoological studies.

In 1925 the opportunity came to apply for a London University research grant. This I held in the Department of Biology at one of the colleges of the University (Kings College for Women: Household and Social Sciences). The department was staffed by Dr. Phillipa Esdale and Dr. Doris Crofts, both like myself pupils of Professor F.J. Cole at Reading University. I owe much to them, especially to Doris Crofts, a master of microscopic techniques. My research subject however made no impact on the household or social life of the country. It was chosen after consultation with Dr. Tate Regan and Dr. H.W. Parker of the British Museum (Natural History).

At that time the classification of frogs was being debated under the leadership of G.K. Noble of the U.S.A., and it was thought that the anatomy of the hypoid and larynx might contribute to a decision. As I enjoyed fine dissection I embarked on this with enthusiasm. The Museum provided a fine selection of about 60 species of Anura, including some from southeast Asia that had been collected by Dr. Malcolm Smith when he held the post of physician to the King of Siam (now Thailand). The pattern of the larynx on the whole supported Noble's classification, though not the merging of the Bufonidae and Leptodactylidae. The larynx is larger, often much larger, in males than in females. The calls of the frogs have since been recorded by many field workers, but as far as I know the differences have not been correlated with the differences in structure.

Before my thesis [Trewavas 11] was finished, Tate Regan, who had been appointed Director of the Museum, needed a research assistant and offered me the post. Of course I accepted eagerly and in March of 1928 started work at the Museum (NH).

Our first major projects were on some of the deep sea fishes collected on the 'Dana' expeditions during which Johannes Schmidt established the life history of the European eel. We published jointly two reports on the stomiatoides [Regan & Trewavas 4, 5] and one on the deep sea angler fishes [Regan & Trewavas 8]. In elongate Stomiatoides (*Eustomias*)

Fig. 9. ET (left background), and Ro (right background) talking tilapia with Ron Roberts at the Bellagio Conference on Biology and Management of Tilapia 4 September 1980. Photograph by D.L.G. Noakes.

the vertebral column makes a vertical U–shaped bend; we supposed this is an adaptation to swallowing larger prey which may struggle. We also discovered in the angler fish that the lateral line organs were not in a canal but were exposed, and in most species were also at the end of tags or filaments. They took the same course as well known lateral line organs on the head and body, and in one species we did find one or two very delicate sense organs at the tips of the tags, and in the centre of the tags

Fig. 10. ET photographed in front of a tank containing *Labeotropheus trewavasae* at the University of Guelph, Ontario, Canada in May 1973. Photograph by E.K. Balon.

TILAPIINE FISHES
of the genera SAROTHERODON, OREOCHROMIS and DANAKILIA
ETHELWYNN TREWAVAS

Fig. 11. Dust jacket of the monumental monograph on tilapiine cichlids for which ET is best known to most ichthyologists, published in 1983. Reproduced with permission of the Trustees of the British Museum (Natural History).

which ended blindly was a nerve and a tiny blood vessel. Two mistakes were made in the angler fish paper: (1) we misinterpreted the incredible angling mechanism of *Ceratias*, this was corrected later by Dr. Bertelson of Denmark; (2) in saying that the blood systems of the parasitic males had been shown to be continuous with the blood system of the female. Not quite true, but with a sinus system in the gelatinous skin of male and female. I was also given to describe independently the anatomy of some deepsea eels and *Opisthoproctus*, a most curious deep sea creature with telescopic eyes [Trewavas 9].

Another large collection that Tate Regan had begun to sort out but not written about was the Christy collection – two collections in fact: (1) from Lake Nyasa, and (2) from Lake Tanganyika. Dr. Cuthbert Christy was, I believe, a military man who was very keen on big game hunting and collecting fishes was his way of financing his hunting expeditions, the last of which was his fatal encounter with a buffalo. But he was a very good collector, perhaps better than one who knew the fishes well as he did not select; the only trouble was that the fishes had no internal organs – they were gutted for better preservation. But one could almost predict diets from the structure of the teeth, which are so interesting and varied in cichlids.

Before I joined him Tate Regan had discovered a remarkable likeness between the tilapiine 'chambo'

Fig 12 ET (left) and Ro at ET's home in Reading, England in 1990 Photograph by E K Balon

and a fish with distinctive file-like teeth and placed then in the monoptypic genus *Corematodus*. We were later able to find the advantage to *Corematodus* of this mimicry, which was well known to the Lake Nyasa fishermen. They were familiar with the special behaviour of this fish, one or two of which would be found in a shoal of chambo. It would snap at the tails of the chambo, thus hurrying them on like the leader of a gang of workers (or slaves). They named it 'capitao', the Portuguese term for such a gang captain. But the fishermen had mistaken the purpose of this action, and we discovered its true function when we found that the long intestine of the capitao was full of minute scales like those covering the tail fin of the chambo. A curious specialized diet.

In 1931 I published a revision of *Lethrinops* from Lake Nyasa (now Malawi) [Trewavas 7], and continued to work on cichlids, together with other fishes, to complete a revision of the Christy collection of haplochromines from Lake Nyasa. The synopsis was published in 1935 [Trewavas 14], but I did not publish the full paper, rather I called this a synopsis, giving diagnostic characters of the species and a key using these characters. We had had drawings made of all the new species and these and full descriptions were left with the intention of doing similar work on the Lake Tanganyika fishes which Christy had also collected and publishing them together plus full illustrations. But the Tanganyika study never materialised and full descriptions and pictures of Nyasa ones were only published in collaboration with David Eccles in 1989 [Eccles & Trewavas 118].

In 1935 I applied for a post on the established staff of the BM (Natural History), and fortunately a vacancy occurred in the Fish Section at that time. J.R. Norman was in charge of the Fish Section. He confined himself to marine fishes and I specialized on freshwater fishes. Norman was a very good curator, a good organiser of systematic reports. His successors owe a very great deal to him for the establishment of the first generic index, the card index, and his comprehensive work on families and genera of fishes; this was never published but he finished it shortly before his death.

At the end of 1938 the Colonial Office decided to have a Fishery Survey of Lake Nyasa. There was a Nutrition Survey already on the spot working on the diets of the villagers and they wanted to know what resources there were in the lake. They needed a taxonomist to identify the many species being caught and I had the good fortune to be seconded from the BM. Norman would have gone but his health was not suitable [Trewavas 30]. I was delighted at the prospect of seeing alive and fresh the fishes I had known so well as corpses from the Christy collection. The survey was made by three people, John Borley who was an administrative officer in Nyasaland and had had experience on sponge fishery work in the West Indies before being appointed to Nyasaland, and Kate Ricardo (later Bertram) who with her friend Janet Owen had made a remarkable expedition to Lake Rukwa and the Bangweulu swamps in 1936, so her experience of Africa and its fisheries was very valuable. I was fortunate to be able to concentrate on the fishes because John and Kate took care of most of the logistics of the expedition. With the steamboat 'Malombe' [in another

Fig. 13. Ro, Sylvia Marsh (ET's sister) and ET, left to right, at ET and Sylvia's home Pincent's Hill House, in Calcot near Reading, England in 1990. Photograph by E.K. Balon.

note ET gives the name of the 78 foot vessel as "SS Malonda'] and its owner Mr. Hughes we spent 4 months [in the other note ET said 5 months] on the lake and at fishing villages along the Nyasaland and Tanzanian shore, studying the fishermen and letting them teach us what they knew about the fishes. We collected the native names with the scientific ones and learnt about the meaning of many of the local

Fig. 14. Even though her eyesight failed with age, ET always kept in touch with all the literature, including the ornamental and pet fish publications, such as the German language periodical on her reading stand. Photographed at her home in Reading, England in 1990 by E.K. Balon.

Fig. 15. Ro's favourite photograph of ET, taken in 1987.

Fig. 16. ET cutting her 'wonderful birthday cake' on the occasion of her 90th birthday, 10 November 1990. The cake was in the shape of a map of Africa, with lakes and rivers as part of the decorations, and tilapia swimming across it. A long time associate, Barton Worthington, is in the background. Photograph by Rosemary Lowe-McConnell.

names; these were in several local (Bantu) languages as we went round the lake.

The most important food fish was 'chambo'. As long as the local fishermen thought we were complete tyros they would tell us this was 'chambo' – it was a tilapiine fish that grew to a good size, swam in shoals and was a very good table fish. But when chambo were caught in the neighbourhood of Kota Kota we found there were two or three names for them, for male and female of one species, and for another species only represented by elderly females, very often with a mouthful of eggs or young. They called the later one 'lolo' and in the south of the lake they called it 'lidole'. This was very much like chambo, but it was a very lean fish with a big bony head, large mouth and kept the young in its mouth to a greater size. The fishermen insisted it was not just an old female chambo, but another fish 'lolo'. So where was lolo's husband? He is not inshore now, but he will come inshore in October. I looked at these fish more closely (pharyngeal teeth differed from chambo, which had a pale blue breeding male) and when I got back to the BM I looked at the collection again to see if we had both kinds. We had of course, and a third kind (with different dentition) which had to be named – collected by Christy from Karonga. We had not recorded this at all, but it had a black male (like lolo) and while on the lake we had found one or two black males which did not fit into either chambo or lolo. We had asked the fishermen about these, who said it was another one (called chinkulu) which will come inshore in October, we shall catch it then on sandbanks on other beaches. We left the lake in July, but John Borley stayed on and visited the beaches in October and found chinkulu males and females being caught there. Different but very much like the first chambo

Fig 17 Ethelwynn Trewavas at her residence, The Pincent's Hill House at Calcot near Reading, where she lived with her sister Sylvie Marsh Photographed in 1990 by E K Balon

we had seen. Later Ro Lowe-McConnell (1946) went to the lake and found this fourth kind of chambo in many areas. It was called sako, so she named it *Tilapia saka* (Lowe-McConnell 1953). This has some resemblance to the karonga that had been called karongae (but pharyngeal teeth etc. a little different). This is still an unresolved question (George Turner's more recent work). Our report was written mainly by Kate Ricardo. She was responsible for putting it together. John Borley also helped. I had to return to BM duties. In 1939 war broke out and we had to look to safety of BM collections during the war. People were leaving the BM to join the forces (temporarily, in some cases finally). The survey report was published in 1942 by the then Crown Agent for the Colonies [Ricardo–Bertram et al. 26].

During the war I took over compiling the Fishes Section of the Zoological Record – hard work but possible to do it in spare time. At start of war we selected all the types specimens and took them for safety (we hoped) to some underground galleries in the chalk hills at Godstone, Surrey south of London. The labels on the outside of the jars suffered from fungus during their cave stay, but fortunately I had insisted on putting pencilled labels inside the jars with details. After the war Avrion Mitchison, one of the sons of Naomi Mitchison and a nephew of J.B.S. Haldane, called at the Museum to see if he could help. I was very glad to give him the job of writing new outside labels and this he faithfully did. It was a pity that the original labels in the handwriting of Günther, Tate Regan and Norman were lost. Avrion Mitchison later became a professor of Zoology, and his three brothers are also professors.

Fred Irvine who had been teaching at Achimoto College in the Gold Coast came to work at the BM trying to write a book on fishes of the Gold Coast for his pupils at Achimoto College. This was quite difficult for us because specialized recent work was not always available on the groups, and it was not regular museum work to write a geographical account relying on old names and data. Nevertheless J.R. Norman had done the marine fishes and I helped with the freshwater fishes. This took a great deal of time. Although not a good work when finished, it was the only work on the subject and worth doing for the time [Trewavas & Irvine 37]. It brought to light some new species (e.g. schilbeids) and the new cichlid genus adapted to living in rapids I named *Irvinea* after its collector.

Another thing which took up some time during the war was that I became Honorary Secretary of the Council of Women Civil Servants. This was a group of senior women civil servants campaigning for: (1) removal of the marriage-bar for women in government and civil service, (2) equal opportunity, and (3) equal pay for equal work. This was unfamiliar work, but very interesting to get to know some of the senior female civil servants who were rather fine people. In the end all three objectives were gained; the Council of Women Civil Servants had a farewell dinner celebrating success, and left it to individuals to prove the wisdom of the changes.

In 1965 when I visited Tanzania and Zambia I was escorted everywhere by competent fishery research officers, and as member of the Colonial Fisheries Advisory Committee was given VIP treatment and stayed at colleagues homes in great comfort.

In 1970 I had the good fortune to join the Westfield College expedition (Jim Green, Sally Corbet and Professor J. Griffiths) to the crater lakes of Cameroon. We were greatly helped by the resident research team at the Helminthiasis Research Institute at Kumba, especially by the loan of their house usually occupied by a member of their staff who was then on leave. They introduced us to the chief and other members of the only village on the lake shore and lent us their Land Rover for transport to the second lake we visited. Pampered again! This was a most enjoyable expedition ichthyologically. We believed when we started that there were five cichlid species in the lake Barombi Mbo, none of which was known to occur elsewhere. When we haa spent a couple of weeks at the lake and the surrounding rivers and streams we found that there were 10 endem-

Table 1 Fish taxa named in honour of Ethelwynn Trewavas

Family	Genus and species	Reference
Cichlidae	*Aulonocara ethelwynnae*	Meyer et al [1]
	Gobiocichla ethelwynnae	Roberts[2]
	Labeotropheus trewavasae	Fryer[3]
	Petrochromis trewavasae	Poll[4]
	Tilapia trewavasae	Poll & Damas[5]
	Trewavasia (subgenus)	Thys van den Audenaerde[6]
	Tylochromis trewavasae	Stiassny[7]
Cyprinidae	*Garra trewavasae*	Monod[8]
	Garra ethelwynnae	Menon[9]
Citharinidae	*Neolebias trewavasae*	Poll & Gosse[10]
Nemacheilidae	*Triplophysa trewavasae*	Mirza & Ahmad[11]
Sciaenidae	*Atrobucca trewavasae*	Talwar & Satharajan[12]
Pycnodontidae	*Trawavasia* (genus)	White & Moy-Thomas[13]

[1] Meyer, M K , R Riehl & H Zetzsche 1987 A revision of the cichlid fishes of the genus *Aulonocara* Regan 1922 from Lake Malawi, with descriptions of six new species (Pisces, Perciformes, Cichlidae) Cour Forschung Senckenberg 94 7–53

[2] Roberts, T 1982 *Gobiocichla ethelwynnae,* a new species of goby-like cichlid fish from rapids in the Cross River, Cameroon Proc K Ned Akad Wet (Biol Med Sci) 85 575–587

[3] Fryer, G 1956 A new species of *Labeotropheus* from Lake Nyasa, with a redescription of *Labeotropheus fuelleborni* Ahl, and some notes on the genus *Labeotropheus* Rev Zool Bot Afr 54 280–289 (Now synonymized)

[4] Poll, M 1948 Descriptions de cichlidae nouveaux recueilles par la mission hydrobiologique Belge au lac Tanganyika (1946–1947) Bull Mus Hist Nat Belg 24 1–31

[5] Poll, M & H Damas 1939 Exploration du Parc National Albert Mission H Damas (1935–36), Bruxelles 73 pp (Now *Oreochromis leucostictus*)

[6] Thys van den Audenaerde, D F E 1970 The paternal mouthbrooding habit of *Tilapia* (*Coptodon*) *discolor* and its special significance Rev Zool Bot Afr 82 285–300 (proposed subgenus of Tilapia, but but not made available, and preoccupied for fossil genus)

[7] Stiassny, M 1989 A taxonomic revision of the African genus *Tylochromis* (Labroidei, Cichlidae), with notes on the anatomy and relationships of the group K Mus Midden-Afr Tervuren Belg Ann Zool Wet 258 1–161

[8] Monod, T 1950 Sur deux *Garra* d'Afrique occidentale Bull Inst Franc Afr Noire 12 976–983

[9] Menon, A G K 1958 *Garra ethelwynnae,* a new cyprinid fish fom Eritrea (Africa) Curr Sci 27 450–451

[10] Poll, M & J P Gosse 1963 Revision des genres *Nannaethiops* Gunther 1871 et *Neolebias* Steindachner 1894, et description de trois especes nouvelles (Pisces, Citharinidae) Ann Mus Roy Afr Centr Zool 16 1–41

[11] Mirza, M R & S Ahmad 1990 *Triplophysa trewavasae,* new species (Pisces Nemacheilidae) Pak J Zool 22 317–321

[12] Talwar, PK & R Satharajan 1975 A new bathyl fish, *Atrobucca trewavasae* (Pisces Sciaenidae) from the Bay of Bengal J Nat Hist 9 575–580

[13] White, E I & J A Moy-Thomas 1941 Notes on the nomenclature of fossil fishes Part III Homonyms M–Z Ann Mag Nat Hist 11(7) 395–400

ic cichlid species in the lake. Before leaving for a week at Lake Kotto I asked a fisherman if we had now seen all the cichlid species in Barombi Mbo. He said 'No. There is one more, but it will not come into our traps until next month (April).' On our return a fortnight later he brought me a few specimens of number 11, a very distinctive species. Each afternoon and evening I was working on the preservation and description of the fishes and Jim and Sally were examining the stomach contents and gonads. So we were able to publish a joint paper [Trewavas et al. 87] on the systematics and ecology of those fishes, which is what I like.

After that my biggest jobs were on sciaenids [Trewavas 100] and tilapiines [Trewavas 110]. When I got back to the Nyasa haplochromines it was almost too late and they would never have been completed without David Eccles' collaboration [Eccles & Trewavas 118].'

The following text of a letter to me from ET on 16 June 1992 was intended to give some of her personal feelings about her career.

'My official retirement was in 1961. I have had good parents, good teachers, good friends and colleagues, good health and a measure of good luck all the way along. If you have read Dickens' novels, you may remember a likeable character called Mark Tapley, who finding people always friendly insisted that for him there was – 'no credit in being jolly' –. I likewise can claim no credit in being fairly productive and this is what this letter is about.

Congratulations on your Ichthyological Institute where I hope you will have a succession of enthusiastic pupils to tackle the intriguing problems of northern and other fishes.

Yours sincerely,
Ethylwynn T.

A kind friend typed this for us.'

The apparently small symbolism of the word *us*, is exactly typical of ET. It was an act of sharing experience and knowledge, and a kind friend had assisted in that exchange (Fig. 17).

So far as I am aware, there is no list of the fish species named by Ethylwynn Trewavas during her very long, active career except the one presented now by Greenwood (1994). This compilation was complicated by the fact that taxonomy of cichlids, the group with which she was most active, is still very much in a state of flux (Stiassny 1991, Axelrod 1993). However, it is clear that Ethelwynn was reponsible for naming a substantial number of fish taxa, in several families, in addition to her major taxonomic works on cichlids. About a dozen living and fossil fishes were named in her honour (Table 1).

Acknowledgements

Most of this tribute to ET would have been impossible without a tremendous amount of work by Rosemary Lowe-McConnell. She carried out interviews with ET, transcribed tape recordings and written notes, provided original photographs from her own and ET's collections, corrected preliminary drafts of manuscripts, and provided enormous amounts of background information and details in personal conversations and letters. Eugene Balon provided many photographs and a number of factual accounts and details, as did Mike Bruton and Humphry Greenwood. Eugene, Mike and Humphry corrected an earlier draft of this manuscript. My administrative assistant Dorothy Hills worked long hours to transcribe my handwritten notes, edit computer files and generally keep things on schedule. My wife Pat and son Jeff tolerated my long absences from family contacts, commented on earlier drafts and ideas, and contributed a number of helpful suggestions on sources of information and interpretations.

References cited

Axelrod, H.R. 1993. The colored lexicon of cichlids, a colored illustration of every known species. T.F.H. Publications, Neptune City. 853 pp.

Balon, E.K. 1974. Fishes of Lake Kariba, Africa: length-weight relationship, a pictorial guide. T.F.H. Publications, Neptune City. 144 pp.

Greenwood, P.H. 1994. Ethelwynn Trewavas 5 Nov. 1900 – 16 Aug. 1993. Copeia 1994: 565–569.

Greenwood, P H 1994 The generic and infrageneric taxa described by Ethelwynn Trewavas Env Biol Fish 41 55–61

Lowe-McConnell, R H 1953 Notes on the ecology and evolution of Nyasa fishes of the genus *Tilapia,* with a description of *T saka,* Lowe Proc zool Soc London 122 1035–1041

Noakes, D L G 1993 Ethelwynn Trewavas – a charmed life Env Biol Fish 38 295–298

Stiassny, M L J 1991 Phylogenetic intrarelationships of the family Cichlidae an overview pp 1–35 *In* M H A Keenleyside (ed) Cichlid Fishes Behaviour, Ecology and Evolution, Chapman & Hall, London

Ethelwynn Trewavas at the International Zoological Congress in Paris, France in 1948. Photograph by T.C.S. Morrison-Scott (courtesy of Rosemary Lowe-McConnell).

Environmental Biology of Fishes **41** 51–54, 1994

Lifetime list of publications by Ethelwynn Trewavas

David L.G. Noakes
Institute of Ichthyology and Department of Zoology, University of Guelph, Guelph, Ontario N1G 2W1, Canada

Ethelwynn produced more than 120 books, monographs, scientific papers and popular articles from 1922 until the publication of this bibliography. She was the sole or first author of an incredible 95% of those publications. She co-authored publications with some of the foremost investigators in her field, e.g. J. Daget, G. Fryer, P.H. Greenwood, L. Fishelson, C.T. Regan and D. Thys van den Audenaerde. Most of her publications were in English, but some of her more recent ones were in French and her writings on cichlid fish taxonomy were translated into other languages, especially German. Ethelwynn certainly authored and co-authored more publications than are listed here. I included only those publications I could verify personally, but I know a number of less readily available monographs, museum publications and accounts on fish families for encyclopaedias were certainly produced. For example, she was Zoological Recorder from 1939 to 1945 for Recent Fishes. Her publications are listed in chronological sequence over years, but not necessarily within a given year.

1 Trewavas, E 1922 Note on the occurrence of *Echinus esculentus* above low-tide rock on the Cornish coast J Mar Biol Assoc Plymouth N S 12 833–834

2 Trewavas, E 1928 Descriptions of five new cichlid fishes of the genus *Haplochromis* from Lake Victoria Ann Mag Nat Hist 10/2 93–95

3 Regan, C T & E Trewavas 1928 Four new cichlid fishes from Lake Victoria Ann Mag Nat Hist 10/2 224–226

4 Regan, C T & E Trewavas 1929 The fishes of the families Astronesthidae and Chauliodontidae Danish 'Dana' Expedition 1920–1922 Oceanographical Reports 5 1–39

5 Regan, C T & E Trewavas 1930 The fishes of the families Stomiatidae and Melacosteidae Danish 'Dana' Expedition 1920–1922 Oceanographical Reports 6 1–143

6 Trewavas, E 1931 Enteropneusta Great Barrier Reef Expedition 1928–1929 Sci Rep 4(2) 1–67

7 Trewavas, E 1931 A revision of the cichlid fishes of the genus *Lethrinops* Ann Mag Nat Hist 10/7 133–152

8 Regan, C T & E Trewavas 1932 Deep sea angler fishes (Ceratioidea) Danish 'Dana' Expedition 1928–1930 Oceanographic Reports 2 1–113

9 Trewavas, E 1932 A contribution to the classification of the fishes of the order Apodes, based on the osteology of some rare eels Proc Zool Soc London 1932(3) 639–659

10 Trewavas, E 1933 The cichlid fishes of Africa Proc Linn Soc London 1932–1933(2) 75–76

11 Trewavas, E 1933 The hyoid and larynx of the Anura Phil Trans Roy Soc (B) 222 401–527

12 Trewavas, E 1933 Scientific results of the Cambridge expedition to the East African lakes, 1930–1931 II The cichlid fishes J Linn Soc London 38 309–341

13 Trewavas, E 1933 On the structure of two oceanic fishes, *Cyema atrum* Gunther and *Opisthoproctus soleatus* Vaillant Proc Zool Soc London 1933(3) 601–614

14 Trewavas, E 1935 A synopsis of the cichlid fishes of Lake Nyasa Ann Mag Nat Hist 10/16 65–118

15 Trewavas, E 1936 Dr Karl Jordan's expedition to south west Africa and Angola The freshwater fishes Novit Zool Tring 40 63–74

16 Trewavas, E 1937 Fossil fishes of Dr L S B Leakey's expedition to Kenya in 1934–5 Ann Mag Nat Hist 10/19 381–386

17 Trewavas, E 1938 On *Barbus wohlerti* sp n , the 'Zwergsichelbarb' of German aquarists Ann Mag Nat Hist 11/2 63–66

18 Trewavas, E 1938 The Killarny shad or goureen (*Alosa fallax killarnyensis* Regan, 1916) Proc Linn Soc London 150(2) 110–112

19 Trewavas, E 1938 Lake Albert fishes of the genus *Haplochromis* Ann Mag Nat Hist 11/1 435–439

20 Trewavas, E & J R Norman 1939 Notes on the eels of the family Synaphobranchidae Ann Mag Nat Hist 11/3 352–359

21 Trewavas, E 1939 The fishes of Lake Nyasa Waterlife 7 213–2

22 Trewavas, E 1940 New Papuan fishes Ann Mag Nat Hist 11/6 284–287

23 Trewavas, E 1941 British Museum (Natural History) expedition to south-west Arabia 1937–8 3 Freshwater fishes Bull Brit Mus Nat Hist 1/1 7–15

24 Trewavas, E 1941 Nyasa fishes of the genus *Tilapia* and a

new species from Portuguese East Africa Ann Mag Nat Hist 11/7 294–306

25 Trewavas, E 1942 The cichlid fishes of Syria and Palestine Ann Mag Nat Hist 11/9 526–536

26 Ricardo-Bertram, C K , H J H Borley & E Trewavas 1942 Report on the fish and fisheries of Lake Nyasa Crown Agents for the Colonies, London 181 pp

27 Trewavas, E 1943 A new cichlid fish from the Gold Coast Ann Mag Nat Hist 11/10 186–191

28 Trewavas, E 1943 New schilbeid fishes from the Gold Coast, with a synopsis of the African genera Proc Zool Soc London 113B 164–171

29 Trewavas, E 1943 Obituary notice on Dr C Tate Regan Nature 151 188–189

30 Trewavas, E 1944 Obituary notice on J R Norman Copeia 1944 265–266

31 Trewavas, E 1944 Obituary notice on J R Norman Proc Linn Soc London 156 2

32 Trewavas, E 1946 Fishes of the genus *Decapterus* of St Helena Ann Mag Nat Hist 11/12 623–625

33 Trewavas, E 1946 The types of African fishes described by Borodin in 1931 and 1936, and of two species described by Boulenger in 1901 Proc Zool Soc London 116 240–246

34 Trewavas, E 1946 Obituary notice on Dr L Cernosvitov Nature 157 219–220

35 Trewavas, E 1947 Speciation in cichlid fishes of the great African lakes Nature 160 96–97

36 Trewavas, E 1947 An example of 'mimicry' in fishes Nature 160 120

37 Trewavas, E & F R Irvine 1947 Freshwater fishes pp 221–282 *In* F R Irvine (ed) The Fishes and Fisheries of the Gold Coast, Crown Agents for the Colonies, London

38 Trewavas, E 1948 Cyprinodont fishes of San Domingo, Island of Haiti Proc Zool Soc London 118 408–415

39 Trewavas, E & A Amaral-Campos 1949 *Oligosarcus* Gunther, a genus of South American characid fishes and *Paroligosarcus* subgen nov Ann Mag Nat Hist 12/2 157–160

40 Trewavas, E 1949 The origin and evolution of the cichlid fishes of the Great African lakes, with special reference to Lake Nyasa XIII[e] Congr Internat Zool Paris 1948 365–368

41 Trewavas, E 1949 Obituary notice on Dr Vladimir Tchernavin Nature 163 755–756

42 Trewavas, E 1950 The status of the American mullets, *Mugil brasiliensis* and *M curema* Copeia 1950 149

43 Trewavas, E & M Poll 1952 Three new species and two new subspecies of the genus *Lamprologus*, cichlid fishes of L Tanganyika Bull Inst Roy Sci Belg 28 1–16

44 Trewavas, E 1952 Contribution to a discussion of breeding habits as a factor isolating populations of fishes Proc Linn Soc London 163 194

45 Trewavas, E 1953 A new species of the cichlid genus *Limnochromis* of Lake Tanganyika Bull Inst Roy Sci Nat Belg 29 1–3

46 Trewavas, E 1953 Sea-trout and brown-trout Salmon & Trout Mag 139 199–215

47 Trewavas, E 1954 On the presence in Africa east of the Rift Valleys of two species of *Protopterus, P annectens* and *P amphibius* Ann Mus Congo Tervuren in 4°, Zool 1 83–100

48 Fryer, G , P H Greenwood & E Trewavas 1955 Scale-eating habits of African cichlid fishes Nature 157 1089–1090

49 Trewavas, E , E I White, N B Marshall & D W Tucker 1955 Herpetichthyes, Amphibioidei, Choanichthyes or Sarcopterygii (answer to letter from Prof Romer) Nature 176 126–127

50 Trewavas, E 1955 A blind fish from Iraq, related to *Garra* Ann Mag Nat Hist 12/8 551–555

51 Trewavas, E 1956 The Cole Museum of Zoology, University of Reading Nature 177 555–556

52 Trewavas, E 1956 Obituary on Professor Leon Bertin Nature 172 686–687

53 Trewavas, E 1957 Review of 'Exploration Hydrobiologique du lac Tanganyika (1946–7), Poissons Cichlidae' by M Poll Copeia 1957 161–162

54 Trewavas, E 1957 Nominomania Ann Mag Nat Hist 12/10 349–350

55 Trewavas, E 1958 The coelacanth yields its secrets Discovery 19 196–205

56 Trewavas, E 1959 Review of 'Anatomie de *Latimeria chalumnae* Tome 1 squelette, muscles et formation de soutien' by J Millot & J Anthony Nature 183 566

57 Trewavas, E 1959 Obituary on Dr Paul Chabanaud Nature 183 1496–1497

58 Trewavas, E 1960 Obituary on Dr Otto Schindler Nature 186 434

59 Trewavas, E 1960 The characiform fish *Characidium laterale* (Boulenger) Ann Mag Nat Hist 13/2 361–364

60 Trewavas, E & H Matthes 1960 *Petrochromis famula* n sp , a cichlid fish of Lake Tanganyika Rev Zool Bot Afr 61 349–357

61 Trewavas, E & P H Greenwood 1960 Interspecific hybrids of *Tilapia* Nature 188 868–869

62 Shearer, W M & E Trewavas 1960 A Pacific salmon (*Oncorhynchus gorbuscha*) in Scottish waters Nature 188 868

63 Trewavas, E 1961 A new cichlid fish in the Limpopo basin Ann S Afr Mus 46 53–56

64 Trewavas, E 1962 A basis for classifying the sciaenid fishes of tropical West Africa Ann Mag Nat Hist 13/5 167–176

65 Trewavas, E 1962 Fishes of the crater lakes of the northwestern Cameroons Bonn Zool Beitr 13 146–192

66 Trewavas, E (Foreword to R A Whitehead) 1962 The life-history and breeding habits of the west African cichlid fish *Tilapia mariae* and the status of *T meeki* Proc Zool Soc London 139 535–543

67 Trewavas, E 1963 *Sciaena* Linnaeus, 1758 (Pisces) proposed variation under the plenary powers of the ruling given in Opinion 93 concerning the type species Bull Zool Nom 20 349–360

68 Trewavas, E 1964 The sciaenid fishes with a single mental barbel Copeia 1964 107–117

69 Trewavas, E 1964 A revision of the genus *Serranochromis* Regan (Pisces, Cichlidae) Ann Mus Roy Afr Centr Zool 124 1–58

70 Trewavas, E 1964 A new species of *Irvineia*, and African genus of schilbeid fishes Ann Mus Genoa 74 388–396

71 Trewavas, E 1965 Review of 'A study in the classification of the sciaenid fishes of China, with description of new genera and species' by Y-T Chu, Y-L Lo & H-L Wu Copeia 1965 253–254

72 Trewavas, E & G Fryer 1965 Species of *Tilapia* (Pisces, Cichlidae) in Lake Kitangiri, Tanzania, East Africa J Zool 147 108–118

73 Trewavas, E & G M Yazgani 1966 *Chrysochir*, a new genus for the sciaenid fish *Otolithus aureus* Richardson, with consideration of its specific synonyms Ann Mag Nat Hist 13/3 249–255

74 Trewavas, E 1966 Comments on the type species of *Sciaena* Linnaeus 1758 Bull Zool Nom 23 4–5

75 Trewavas, E 1966 *Tilapia aurea* (Steindachner) and the status of *Tilapia nilotica exul, T monodi* and *T lemassoni* (Pisces, Cichlidae) Israel J Zool 14 258–276

76 Trewavas, E 1966 *Chromis aureus* Steindachner, 1864 (Pisces, Cichlidae) proposed addition to the Official List of Specific Names Bull Zool Nom 23 157

77 Trewavas, E 1966 *Otolithus aureus* Richardson, 1846 (Pisces, Sciaenidae) proposed addition to the Official List of Specific Names Bull Zool Nom 23 158–159

78 Trewavas, E 1966 Fishes of the genus Tilapia with four anal spines in Malawi, Rhodesia, Mozambique and southern Tanzania Rev Zool Bot Afr 74 50–62

79 Trewavas, E 1966 A preliminary review of fishes of the genus *Tilapia* in the eastward-flowing rivers of Africa, with proposals of two new specific names Rev Zool Bot Afr 74 394–424

80 Trewavas, E 1968 The name and natural distribution of the Tilapia from Zanzibar (Pisces, Cichlidae) F A O Fisheries Rep 44 246–254

81 Trewavas, E & D F E Thys van den Audenaerde 1969 A new Angolan species of *Haplochromis* (Pisces, Cichlidae) Mitt Hamburg Zool Mus Inst 66 237–239

82 Trewavas, E 1969 Poissons du bassin de l'Ivindo VI Le genre *Tilapia* (Perciformes, Cichlidae) Biol Gabon 4 271–273

83 Trewavas, E 1971 The syntypes of the sciaenid *Corvina albida* Cuvier and the status of *Dendrophysa hooghliensis* Sinha & Rao and *Nibea coibor* (nec Hamilton) of Chu, Lo & Wu J Fish Biol 3 453–461

84 Trewavas, E 1971 The type species of the genera *Phoxinellus, Pseudophoxinus* and *Paraphoxinus* (Pisces, Cyprinidae) Bull Brit Mus Nat Hist Zool 21 359–361

85 Trewavas, E & P K Talwar 1972 On the generic relationship of the sciaenid fish *Bola chaptis* Hamilton, with a description of specimens from Burma J Fish Biol 4 11–16

86 Trewavas, E & S E Ingham 1972 A key to the species of Mugilidae (Pisces) in the northeastern Atlantic and Mediterranean, with explanatory notes J Zool London 167 15–29

87 Trewavas, E, J Green & S A Corbet 1972 Ecological studies on crater lakes in West Cameroon Fishes of Barombi Mbo J Zool London 167 41–95

88 Goren, M, L Fishelson & E Trewavas 1973 The cyprinid fishes of *Acanthobrama* Heckel and related genera Bull Brit Mus Nat Hist Zool 24 293–315

89 Trewavas, E 1973 On the cichlid fishes of the genus *Pelmatochromis* with proposal of a new genus for *P congicus*, on the relationship between *Pelmatochromis* and *Tilapia* and the recognition of *Sarotherodon* as a distinct genus Bull Brit Mus Nat Hist Zool 25 1–26

90 Trewavas, E 1973 A new species of cichlid fishes of rivers Quanza and Bengo, Angola, with a list of the known Cichlidae of these rivers and a note on *Pseudocrenilabrus natalensis* Fowler Bull Brit Mus Nat Hist Zool 25 27–37

91 Trewavas, E 1973 What Tate Regan said in 1925 Systematic Zool 22 92–93

92 Trewavas, E 1973 Mugilidae pp 567–574 *In* J C Hureau & T Monod (ed) CLOFNAM I, Check-list of the fishes of the north-eastern Atlantic and of the Mediterranean, Volume I UNESCO, Paris

93 Trewavas, E 1973 Sciaenidae pp 396–401 *In* J C Hureau & T Monod (ed) CLOFNAM I, Check-list of the fishes of the north-eastern Atlantic and of the Mediterranean, Volume I, UNESCO, Paris

94 Trewavas, E 1974 The freshwater fishes of rivers Mungo and Meme and lakes Kotto, Mboandong and Soden, West Cameroon Bull Brit Mus Nat Hist Zool 26 329–419

95 Trewavas, E & D J Stewart 1975 A new species of *Tilapia* (Pisces, Cichlidae) in the Zambian Zaire Bull Brit Mus Hist Zool 28 191–197

96 Trewavas, E 1975 A new species of *Nanochromis* (Pisces, Cichlidae) from the Ogowe system, Gabon Bull Brit Mus Nat Hist Zool 28 233–235

97 Trewavas, E 1976 Tilapiine fishes from crater-lakes north of Lake Malawi Bull Brit Mus Nat Hist Zool 30 149–156

98 Payne, A I & E Trewavas 1976 A new species of *Hemichromis* (Pisces, Cichlidae) of Sierra Leone and Liberia Bull Brit Mus Nat Hist Zool 30 159–168

99 Trewavas, E 1976 Correction to the foregoing translation in German Aquar Terr Z 30 251

100 Trewavas, E 1977 The sciaenid fishes (croakers or drums) of the Indo-West Pacific Trans Zool Soc London 33 529–541

101 Trewavas, E 1977 *Pennakia* Fowler, 1926 (Pisces, Sciaenidae) request for designation of a type species Bull Zool Nom 34 185–186

102 Trewavas, E 1978 A discussion on *Tilapia* and *Sarotherodon* Cichlidae 1978 127–131

103 Trewavas, E 1978 A discussion on *Tilapia* and *Sarotherodon* Deutsche Cichl Ges 9 181–189

104 Trewavas, E 1979 *Sciaena nibe* Jordan & Thompson, 1911

(Pisces), proposed conservation of the specific name *nibe* by use of the plenary powers Bull Zool Nom 36 155–157

105 Trewavas, E 1980 *Tilapia* and *Sarotherodon*? Buntbarsche Bull 81 1–6

106 Trewavas, E 1981 Nomenclature of the tilapias of southern Africa J Limnol Soc South Africa 7 42

107 Trewavas, E 1981 Addendum to '*Tilapia* and *Sarotherodon*?' Buntbarsche Bull 87 12

108 Trewavas, E 1982 Generic grouping of *Tilapia* used in aquaculture Aquacult 27 79–81

109 Trewavas, E 1982 Tilapias taxonomy and speciation pp 3–13 *In* R S V Pullin & R H Lowe-McConnell (ed) The Biology and Culture of Tilapias, ICLARM, Manila

110 Trewavas, E 1983 Tilapiine fishes of the genera *Sarotherodon, Oreochromis* and *Danakilia* Brit Mus Nat Hist , London 583 pp

111 Trewavas, E 1984 Nouvel examen des genres et sous-genres du complex *Pseudotropheus – Melanochromis* du lac Malawi (Pisces, Perciformes, Cichlidae) Rev fr Aquariol 10 97–106

112 Trewavas, E 1984 Un nom et une description pour l'*Aulonocara* 'sulphur-head', poisson cichlide du lac Malawi Rev fr Aquariol 11 7–10

113 Trewavas, E 1985 The pharyngeal apophysis on the base of the skull of cichlid fishes Neth J Zool 35 716–719

114 Barel, C D N , R Dorit, PH Greenwood, G Fryer, N Hughes, PB N Jackson, H Kawanabe, R H Lowe-McConnell, M Nagoshi, A J Ribbink, E Trewavas, F Witte & K Yamaoka 1985 Destruction of fisheries in Africa's lakes Nature 15 19–20

115 Daget, J & E Trewavas 1986 Sciaenidae pp 333–337 *In* J Daget, J -P Gosse & D F E Thys van den Audenaerde (ed) CLOFFA 2, Check-list of the freshwater fishes of Africa ORSTOM, Paris

116 Ben-Tuvia, A & E Trewavas 1987 *Aerobucca geniae*, a new species of sciaenid fish from the Gulf of Elat (Gulf of Aqaba), Red Sea Israel J Zool 34 15–21

117 Trewavas, E 1988 The genus *Aulonocara* Cichlidae 9 75–76

118 Eccles, D H & E Trewavas 1989 Malawian cichlid fishes, the classification of some haplochromine genera Lake Fish Movies H W Dieckhoff, Herten 334 pp

119 Ferguson, J & E Trewavas 1989 *Aulonocara* species Cichlidae 10 2–15

120 Ferguson, J & E Trewavas 1989 Peacock cichlids The *Aulonocara* review, Part 2 Aquar & Pondkeep 54 15–17

Environmental Biology of Fishes **41**: 55–61, 1994.

The generic and infrageneric taxa described by Ethelwynn Trewavas

P. Humphry Greenwood
J.L.B. Smith Institute of Ichthyology, Private Bag 1015, Grahamstown, 6140 South Africa

Either as the sole author, or in collaboration (especially with C. Tate Regan), Ethelwynn Trewavas was responsible for describing and naming at least 462 genera, subgenera, species and subspecies. Doubtless, in compiling this index (which, intentionally does not include nomina nova she established) I have overlooked Trewavas taxa; for those oversights I can only apologise.

In many respects, as she herself often remarked, Ethelwynn entered the field of taxonomic ichthyology at a fortunate time. It was a time when large collections of fishes were arriving in the British Museum (Natural History), many from the still poorly sampled Great Lakes of Africa, others from the then little explored deeper oceanic waters. Above all, she had the great advantage of having Tate Regan as her guide and mentor. He too was fortunate in having Ethelwynn as his student and assistant, and both were lucky to be working at a period when financially and philosophically the Museums of the world were able to concentrate their activities mainly on research. The times of penury and expensive theme park-like exhibitions were yet to come.

Ethelwynn was not only a tireless worker, but a gifted taxonomist, one not merely of the 'stamp-collecting' variety, but a true biologist who recognised the corpses she handled as once living entities and essential elements of the biotopes from whence they were collected. I have not been able to carry out a detailed analysis to determine the current status of all the taxa in whose recognition and description she was involved; however, for the African Cichlidae (and non-cichlids) my estimate would be that more than 80% are still regarded as valid – and that is despite the activities of the many taxonomists now studying the freshwater fishes of the continent.

Interestingly, and somewhat surprisingly, little of Ethelwynn's work was concerned with higher-level systematics, especially the phyletic inter- and intrarelationships of the groups she studied. This is surprising because she certainly was interested in those aspects of systematics, and read extensively in the literature of evolutionary biology. Those wide-ranging interests were very obvious throughout my long association with her, particularly so during the time when, in the early 1960's, Donn Rosen and I were roughing-out our ideas of a phylogenetic classification for teleosts (Greenwood et al. 1966) and often had long and beneficial discussions with her.

Her few publications in the field of higher-level taxonomy are surprising, too, when one recalls that her major doctoral research (on the hyoid and larynx of Anura) was in large part stimulated by Noble's work on the phylogeny of the group, a study which had paid scant attention to that important aspect of the animals' anatomy. Then, too, her early association with Tate Regan was at a time when he was preparing his major synthetic work on the classification of teleosts (later published in the 14th edition of the *Encyclopaedia Britannica,* 1929), one, together with his earlier papers, long accepted as the definitive work in that field.

Perhaps her reticence in this sphere of systematic ichthyology stemmed from her high regard for Tate Regan's work coupled with the fascination that African cichlids, especially the tilapiines, held for her, and her strong views on the pragmatic importance to fishery biologists of sound alpha-level taxonomy.

Be that as it may, all those who were to be involved later either in cichlid taxonomy or working with those fishes in applied research, are deeply indebted to Ethelwynn, as are the many people who benefited from her tutelage and knowledge.

The list of taxa which follows has been arranged, firstly in higher taxonomic categories, and then within those groups, by families. Within a family the members of each taxonomic rank (i.e. genus, subgenus, species or subspecies) are listed in chronological (i.e. annual) order of their publication. The year of publication is followed, in parentheses, by the number allocated to that particular publication in Noakes (1994). The type species (abbreviated as t.s.) and its author is given for each genus and subgenus.

The names of the taxa are as originally published; no attempt has been made to indicate the present status of a taxon or changes in its nomenclature (i.e. synonymy, allocation to other genera, etc.). Eschmeyer (1990) should be consulted for details about the nomenclatural status of all generic names in the list and for the current familial placement of the genera.

OSTARIOPHYSI

Series Anotophysi

Kneriidae

Sp. nov. 1936 (15) *Kneria polli*

Series Otophysi

Clariidae

Sp. nov. 1936 (15) *Clarias cavernicola*

Mochokidae

Sp. nov. 1974 (94) *Chiloglanis disneyi*

Schilbeidae

Gen. nov. 1943 (28) *Irvineia* (t.s. *I. voltae* Trewavas, 1943)

Spp. nov. 1943 (28) *Irvineia voltae*
1964 (70) *Irvineia orientalis*

Amphiliidae

Sp. nov. 1936 (15) *Amphilius lentiginosus*

Characidae

Sub-gen.nov. 1949 (39) *Oligosarcus (Paroligosarcus)*, t.s. *O. pintoi* Amaral-Campos, 1945

Sp. nov. 1936 (15) *Micralestes argyrotaenia*

Cyprinidae

Gen. nov. 1955 (50) *Typhlogarra* (t.s. *T. widdowstoni* Trewavas, 1955)

Spp. nov. 1973 (88) *Acanthobrama telavivensis*
1936 (15) *Barbus breviceps; dossolineatus; mocoensis*
1938 (17) *Barbus wohlerti*
1941 (23) *Barbus arabicus*
1941 (23) *Garra brittoni; tibanica*
1955 (50) *Typhlogarra widdowstoni*
1974 (94) *Labeo camerunensis*

STOMIIFORMES

Chauliodontidae

Sp. nov. 1939 (4) *Chauliodus danae*

Astronesthidae

Gen. nov. 1929 (4) *Diplolynchus* (t.s. *D. lucifer* Regan & Trewavas, 1929)
Heterophotus (t.s. *H. ophistoma* Regan & Trewavas, 1929)
Rhadinesthes (t.s. *Astronesthes decimus* Zugmayer, 1911)
Neonesthes (t.s. *N. macrolynchnus* Regan & Trewavas, 1929)

Spp. nov. *Astronesthes caulophorus; cyclophotus; filifer; leucopogon; longiceps; luetkeni; neopogon; oculatus*
Borostomias macristius; macrophthalmus; panamaensis; schmidti
Diplolynchus bifilis; lucifer; mononema
Heterophotus ophistoma
Neonesthes macrolynchnus

Melanostomiidae

Gen. nov. 1930 (5) *Chirostomias* (t.s. *C. pliopterus* Regan & Trewavas, 1930)
Haplostomias (t.s. *H. tentaculatus* Regan & Trewavas, 1930)
Thysanactis (t.s. *T. dentex* Regan & Trewavas, 1930)
Trigonolampa (t.s. *Trig. miriceps* Regan & Trewavas, 1930)

Sub-gen. nov. *Bathophilus* (*Notopodichthys*), t.s. *B. brevis* Regan & Trewavas, 1930
Bathophilus (*Trichochirus*), (t.s. *B. paweeni* Regan & Trewavas, 1930)
Eustomias (*Achirostomias*), t.s. *E. lipochirus* Regan & Trewavas, 1930

Eustomias (*Dinematochirus*), t.s. *E. fissibarbis* Pappenheim, 1914
Eustomias (*Haploclonus*), t.s. *E. embarbatus* Regan & Trewavas, 1930
Eustomias (*Nominostomias*), t.s. *E. bibulbosus* Parr, 1927
Eustomias (*Spilostomias*), t.s. *E. braueri* Zugmayer, 1911
Eustomias (*Triclonostomias*), t.s. *E. drechseli* Regan & Trewavas, 1930
Eustomias (*Urostomias*), t.s. *E. macrurus* Regan & Trewavas, 1930
Photonectes (*Melanonectes*), t.s. *P. dinema* Regan & Trewavas, 1930
Photonectes (*Microchirichthys*), t.s. *P. parvimanus* Regan & Trewavas, 1930

Spp. nov. *Bathophilus brevis; chironema; longipes; melas; paweeni; proximus; schizochirus*
Echiostoma guentheri
Eustomias acinosus; barbura; bimarginatus; bituberatus, botrypogon, brevifilis; dactylobolus; dendriticus; drechseli; embarbatus; frondosus; furcifer; globulifer; leptobolus; lipochirus; macronema; macrurus; melanobolus; melanonema; melanostigma; monoclonus; monodactylus; parri; patulus; pyrifer; ramulosus; schmidti; silvescens; simplex; tenisoni; triramus; trituberatus; xenobolus; variabilis
Haplostomias bituberatus; tentaculatus
Lamprotoxus paucifilis, phanobranchus
Leptostomias analis, gracilis; haplocaulus; longibarba; leptobolus; ramosus
Melanostomias albibarba; biseriatus; heteropogon; macrophotus; margaritifer; melanocaulus; melanopogon; spilorhynchus
Pachystomias atlanticus
Photonectes achirus; caerulescens; dinema; fimbria; leucospilus; monodactylus; ovibarba; parvimanus; phyllopogon

Idiacanthidae

Idiacanthus panamensis

Malacosteidae

Spp. nov. *Aristostomias lumifer; polydactylus; uncodentatus; xenostoma*
Malacosteus danae

LOPHIIFORMES

For current family placements, see Eschmeyer (1990).

Gen. nov. 1932 (8) *Amacrodon* (t.s. *Thaumatichthys binghami* Parr, 1927)
Anomalophryne (t.s. *Haplophryne hudsonius* Beebe 1929)
Caranactis (t.s. *Caranactis pumilus* Regan & Trewavas, 1932)
Centrocetus (t.s. *Centrocetus spinulosus* Regan & Trewavas, 1932)
Centrophryne (t.s. *Centrophryne spinulosa* Regan & Trewavas, 1932)
Chirophryne (t.s. *Chirophryne xenolophus* Regan & Trewavas, 1932)
Cryptolychnus (t.s. *Cryptolychnus micractis* Regan & Trewavas, 1932)
Ctenochirichthys (t.s. *Ctenochirichthys longimanus* Regan & Trewavas, 1932)
Nannoceratias (t.s. *Nannoceratias denticulatus* Regan & Trewavas, 1932)
Teleotrema (t.s. *Teleotrema microphthalmus* Regan & Trewavas, 1932)
Trematorhynchus (t.s. *Rhynchoceratias leucorhinus* Regan, 1925)
Tyrranophryne (t.s. *Tyrranophryne pugnax* Regan & Trewavas, 1932)
Xenoceratias (t.s. *Xenoceratias longirostris* Regan & Trewavas)

Sub. gen. nov. *Dolopichthys* (*Leptacanthich-*

thys), t.s. *D. gracilispinis* Regan, 1925
Dolopichthys (*Microlophichthys*), t.s. *D. microlophus* Regan, 1925
Dolopichthys (*Pentherichthys*), t.s. *D. atratus* Regan & Trewavas, 1932

Spp. nov. *Aceratias edentulata*
Caranactis pumilus
Caulophryne ramulosa
Caulophryne acinosa
Centrocetus spinulosus
Centrophryne spinulosa
Chaenophryne atriconus; columnifer; crenata; fimbriata; haplactis; macractis; melanodactylus; melanorhabdus; parviconus; pterolophus; ramifera
Chirophryne xenolophus
Cryptolychnus micractis; paucidens
Cryptosaras normani; pennifer; valdivae
Ctenochirichthys longimanus
Dolopichthys atratus; brevifilis; carlsbergi; cirrifer; claviger; cristatus; diadematus; digitatus; exiguus; flagellifer; frondosus; heteronema; implumis; jubatus; macronema; mirus; multifilis; nigrifilis; pennatus; plumatus; pollicifer; ptilotus; pullatus; simplex; schmidti; thysanophorus; venustus
Edriolychnus macracanthus; radians; roulei
Gigantactis exodon; ovifer; sexfilis
Himantolophus danae
Lasiognathus beebi
Linophryne argyresca; corymbifera; eupogon; racemifera
Lophodolus dinema
Mancalias bifilis; xenistius
Melanocetus cirrifer
Nannoceratias denticulatus
Rhynoceratias altirostris
Teleotrema microphthalmus
Trematorhynchus exiguus; obliquidens
Terranophryne pugnax
Xenoceratias brevirostris; heterorhynchus; laevis; longirostris; micracanthus

ATHERINIFORMES

Atherinidae

Spp. nov. 1940 (22) *Craterocephalus lacustris*

CYPRINODONTIFORMES

Poeciliidae

Sp. nov. 1948 (38) *Mollienisia elegans*

Aplocheilidae

Sp. nov. 1949 (38) *Rivulus roloffi*

Cyprinodontidae

Sp. nov. 1949 (38) *Procatopus lacustris*

PERCIFORMES

Teraponidae

Sp. nov. 1946 (32) *Therapon adamsoni*

Sciaenidae

Gen. nov. 1962 (64)
Miracorvina (t.s. *Sciaena angolensis* Norman, 1935)
Pentheroscion (t.s. *Sciaena mbizi* Poll, 1950)
1964 (68) *Dendrophysa* (t.s. *Umbrina russellii* Cuv., 1830)
1966 (73) *Chrysochir* (t.s. *Otolithus aureus* Richardson, 1846)
1977 (100) *Afroscion* (t.s. *Argyrosomus thorpei* M.M. Smith, 1977)
Austronibea (t.s. *Austronibea oedogenys* Trewavas, 1977)
Boesemania (t.s. *Corvina microlepis* Bleeker, 1858)
Paranibea (t.s. *Nibea semiluctuosa* Cuv., 1830)
Sonorlux (t.s. *Sonorlux fluminis* Trewavas, 1977)

Sub-gen. nov. 1962 (64)
Pseudotolithus (*Fonticulus*), t.s. *Corvina nigrita* Cuv., 1830

Pseudotolithus (*Hostia*), t.s. *Corvina moorii* Günther, 1865

Spp. nov. 1977 (100)

Austronibea oedogenys

Panna heterolepis

Sonorlux fluminus

1987 (116) *Atrobucca geniae*

Cichlidae

Gen. nov. 1935 (14)

Aristochromis (t.s. *Aristochromis christyi* Trewavas, 1935)

Christyella (t.s. *Christyella nyasana* Trewavas, 1935)

Cyathochromis (t.s. *Cyathochromis obliquidens* Trewavas, 1935)

Diplotaxodon (t.s. *Diplotaxodon argenteus* Trewavas, 1935)

Genyochromis (t.s. *Genyochromis mento* Trewavas, 1935

Labidochromis (t.s. *Labidochromis vellicans* Trewavas, 1935)

Lichnochromis (t.s. *Lichnochromis acuticeps* Trewavas, 1935)

Melanochromis (t.s. *Melanochromis melanopterus* Trewavas, 1935)

Petrotilapia (t.s. *Petrotilapia tridentiger* Trewavas, 1935)

Trematocranus (t.s. *Trematocranus microstoma* Trewavas, 1935)

1961 (63) *Chetia* (t.s. *Chetia flaviventris* Trewavas, 1961)

1962 (65) *Barombia* (t.s. *Barombia maclareni* Trewavas, 1962)

Stomatepia (t.s. *Paratilapia mariae* Holly, 1930)

1972 (87) *Konia* (t.s. *Tilapia eisentrauti* Trewavas, 1962)

Myaka (t.s. *Myaka myaka* Trewavas, 1972)

N.B. Although Green and Corbet were co-authors of the paper in which these taxa were described, Trewavas was the sole author of the taxonomic section.

1973 (89) *Pterochromis* (t.s. *Pelmatochromis congicus* Bouleng., 1897)

1989 (118) *Buccochromis* (t.s. *Paratilapia nototaenia* Bouleng., 1902)

Cheilochromis (t.s. *Haplochromis euchilus* Trewavas, 1935)

Copadichromis (t.s. *Haplochromis quadrimaculatus* Regan, 1922)

Ctenopharynx (t.s. *Hemichromis intermedius* Günther, 1893)

Dimidichromis (t.s. *Haplochromis strigatus* Regan, 1922)

Eclectochromis (t.s. *Haplochromis ornatus* Regan, 1922)

Exochromis (t.s. *Cyrtocara anagenys* Oliver, 1984.

Fossorochromis (t.s. *Tilapia rostrata* Bouleng., 1899)

Hemitaeniochromis (t.s. *Haplochromis urotaenia* Regan, 1922)

Maravicochromis (t.s. *Haplochromis ericotaenia* Regan, 1922)

Naevochromis (t.s. *Haplochromis chrysogaster* Trewavas, 1935)

Nimbochromis (t.s. *Hemichromis livingstonii* Günther, 1893)

Nyassachromis (t.s. *Haplochromis breviceps* Regan, 1922)

Placidochromis (t.s. *Haplochromis longimanus* Trewavas, 1935)

Platygnathochromis (t.s. *Haplochromis melanotus* Regan, 1922)

Protomelas (t.s. *Chromis kirkii* Günther, 1893)

Sciaenochromis (t.s. *Haplochromis ahli* Trewavas, 1935)

Stigmatochromis (t.s. *Haplochromis woodi* Regan, 1922)

Taeniochromis (t.s. *Haplochromis holotaenia* Regan, 1922)

Taeniolethrinops (t.s. *Haplochromis praeorbitalis* Regan, 1922)

Tramitichromis (t.s. *Tilapia brevis* Bouleng., 1908)

Tyrannochromis (t.s. *Haplochromis macrostoma* Regan, 1922)

Sub-gen. nov. 1983 (110) *Oreochromis* (*Vallicola*), t.s. *Tilapia amphimelas* Hilgen., 1905.

1984 (111) *Pseudotropheus* (*Tropheops*), t.s. *Pseudotropheus tropheops* Regan, 1921

Spp. nov. 1928 (2) *Haplochromis gowersi; maxillaris; melanopterus; michaeli; obtusidens*

1928 (3) *Haplochromis diplotaenia; eutaenia; plagiodon; tridens*

1931 (7) *Lethrinops alta; christyi; cyrtonotus; furcicauda; furcifer; laticeps; lituris; longimanus; lunaris; macracanthus; microstoma; oculta; parvidens; trilineata; variabilis*

1933 (12) *Haplochromis annectidens; beadlei; dolorosus; elegans; engystoma; labiatus; limax; rudolfianus; taurinus; vellifer; vicarius*

Pelmatochromis exul

Tilapia inducta; leucosticta; vulcani

1935 (14) *Aristochromis christyi*

Aulonocara macrochir; rostrata

Christyella nyasana

Cyathochromis obliquidens

Diplotaxodon argenteus

Genyochromis mento

Haplochromis balteatus; chrysogaster; cynaneus; decorus; epichorialis; euchilus; fenestratus; festivus; formosus; gracilis; heterodon; heterotaenia; incola; insignis; labifer; labiodon; labridens; lobochilus; longimanus; marginatus; mola; mollis; microcephalus; nigritaeniatus; nitidus; obtusus; oculatus; orthognathus; ovatus; pardalis; phenochilus; pholidophorus; pictus; pleurospilus; pleurostigma; polyodon; protostoma; purpurans; semipalatus; serenus; speciosus; spilotichus; taeniolatus; tetraspilus; triaenodon; virgatus

Labidochromis vellicans

Lethrinops intermedia

Lichnochromis acuticeps

Melanochromis brevis; labrosus; melanopterus; perspicax; vermivorous

Petrotilapia tridentiger

Pseudotropheus elegans; fuscus; lucerna; microstoma

Rhamphochromis brevis

Trematocranus auditor; brevirostris; microstoma

1938 (19) *Haplochromis bullatus*

1941 (24) *Tilapia karongae; lidole; placida*

1943 (27) *Gobiochromis irvinei*

1952 (43) *Lamprologus christyi; pleuromaculatus; sexfasciatus*

1953 (45) *Limnochromis christyi*

1960 (60) *Petrochromis famula*

1962 (65) *Barombia maclareni*

Pelmatochromis loennbergi

Tilapia eisentrauti; steinbachi

1964 (69) *Serranochromis janus; spei; stappersi*

1966 (79) *Tilapia mortimeri; ruvumae*

1969 (81) *Haplochromis albolabris*

1972 (87) *Konia dikume*

Myaka myaka

Stomatepia pindu

See note under 1972 (87), page 59 regarding correct authorship of these names.

1973 (89) *Sarotherodon angolensis*

1974 (94) *Chromidotilapia finleyi*

1975 (95) *Tilapia baloni*

1975 (96) *Nanochromis gabonicus*

1976 (98) *Hemichromis fugax*

1984 (112) *Aulonocara maylandi*

Sub-spp. nov. 1935 (14) *Haplochromis marginatus vuae*

Pseudotropheus tropheops gracilior

1936 (15) *Haplochromis philander dispersus*

1952 (43) *Lamprologus savoryı elongatus*
Lamprologus savoryı pulcher
1966 (78) *Tılapıa shırana chılwae*
1983 (10) *Oreochromıs macrochır mweruensıs*
Oreochromıs nılotıcus barıngoensıs
Oreochromıs nılotıcus filoa
Oreochromıs nılotıcus sugutae

Acknowledgement

I am ındebted to Huıbré Tomlınson for her efforts ın preparıng thıs manuscrıpt at very short notıce.

References cited

Eschmeyer, W N 1990 Catalog of the genera of recent fıshes Calıfornıa Academy of Scıences, San Francısco, 697 pp

Greenwood, P H , D E Rosen, S H Weıtzman & G S Myers 1966 Phyletıc studıes of teleostean fıshes, wıth a provısıonal classıfıcatıon of lıvıng forms Bull Amer Mus Nat Hıst 131 339–456

Noakes, D L G 1994 Lıfetıme lıst of publıcatıons by Ethelwynn Trewavas Env Bıol Fısh 41 51–54 (thıs volume)

Environmental Biology of Fishes **41**: 63–65, 1994.

An interview with Ethelwynn Trewavas

David L.G. Noakes
Institute of Ichthyology and Department of Zoology, University of Guelph, Guelph, Ontario N1G 2W1, Canada

Ethelwynn Trewavas died at her home in Reading, England on 16 August 1993. On 19 February 1993 Ethelwynn answered a series of standardized questions prepared by Eugene Balon as the basis for his 'interview with Eugenie Clark' (pp. 121–125). The questionnaire had been delivered to Rosemary Lowe-McConnell by Mike Bruton, and Ro conducted the interview with ET at Ethelwynn's home at Calcot in Reading, England. Ro recorded the conversation, and later transcribed that tape and provided the notes to me. Subsequently ET added some further comments on some of the questions.

EB = questions from Eugene Balon **ET** = Ethelwynn Trewavas

EB: What is the most memorable event of your childhood?

ET: In primary school one of the mistresses drew a lot of coloured pictures for each season on a large sheet of brown paper, illustrating things appropriate to the season, e.g. Jack Frost in winter, for other seasons Mother Nature in brown coat, and then there were pictures of flowers, trees, animals that were particularly active in each season; and the sun at its appropriate place for each season.

EB: Who shaped your early career and especially your attitude to science, or who most affected your life and work?

ET: As it went along I was always encouraged by the botany mistresses at school because I was interested. When I chose to do Natural History and Zoology at University, Dr. Nellie Eales and Prof. Cole were very encouraging and stimulating.

EB: How did you come to specialise in research on fishes?

ET: I first met Tate Regan when as a student I was advised to go to the Natural History Museum in South Kensington (London) to study the classification of fishes. While I was in the fish gallery studying this, Tate Regan, then Keeper of Zoology, later the Director came along. He said, 'Are you still illustrating the water babies? Oh no, I see you are not.' So he became interested and took me into the back rooms and showed me the skeletons of some important fishes and talked about the classification of fishes. After graduating I had 2 or 3 years of school teaching before getting the opportunity to specialise in zoology again. This was given to me by Doris Croft and Phillipa Esdale at Kings College for Women in London. They got in touch with people at the Natural History Museum and Tate Regan suggested I should do a study on the frog to supplement what was known about the classification of the Anura which was then under discussion, especially by G.K. Noble and Melbrook in the US who had some new ideas about it. They suggested I should look at the hyoid. I found that the hyoid and larynx together formed a very interesting unit in the Anura and took that as my thesis using material from the Natural History Museum and working at Kings College for Women. So it was Tate Regan finding me in the fish gallery which led to this.

EB: You joined a male dominated profession, can you recall some relevant details?

ET: I found no difficulty, or very little difficulty at any rate, in keeping my head above water among the males because we were all specialists. We all had something to give, I think that was it.

EB: Were there any unpleasant events which occurred as a female in competition with males?

ET: No.

EB: Which were the most touching moments?

ET: I do not think we have 'touching' moments . . . I find every moment spent on zoological work if not touching at least exciting.

EB: Did some papers you co-authored reflect more than a working relationship?

ET: Every paper reflected, I hope, a very good working relationship.

EB: Of your publications, which are your favourite papers/books?

ET: The deep sea angler fishes, stomatoids, sciaenid fishes and the Cichlidae. What can one say?

EB: They contributed to human knowledge . . . would you mind saying why?

ET: Yes I would! Every bit of research adds a little bit to human knowledge, otherwise it is not research. I always think we should not use the word 'research' in English – in French recherce is OK – but we should use 'search'. Tate Regan always used to call it 'original work', which is better.

EB: Did you ever dream about publishing a widely read, popular book, and if so what did you do about it?

ET: I kept on dreaming.

EB: Did you have contemplative ideas beyond ichthyology? Did you join any philosophical school?

ET: No, but I did think quite a lot about it.

EB: If you could rewind and replay tape of life . . . which part?

ET: . . . all?

EB: Would you like to edit or add some parts?

ET: I would be content to stop and let somebody else do a better job. I think I ought to have related the special jobs I did to wider issues. I thought about it a lot but very rarely put it into papers because I thought that was not the right place.

*

In a personal letter to me on 13 April 1993, Ethelwynn commented on the questionnaire (above) and offered some additional comments. Her letter is in part transcribed first, followed by her comments.

Pincent's Hill House
Calcot
Reading RG2 5TU
Tel (0734) 302440

13 April 1993

Dear David,

I have seen a copy of a questionnaire that Mike Bruton [sic] used for Genie Clark The only part of it that I feel inclined to deal with is on male competition

I enclose a note on this because I think I came into the arena at a favourable time from this point of view and it is only fair to make a note of it

Please excuse a handwritten note If you use it could you let me have a copy of your paraphrases

Best wishes to you and Dr Balon

Yours,
Ethelwynn Trewavas

Male competition

I was never conscious of this as such. My first post at the Museum was not on the Established staff. I replaced a man who had assisted the previous Director, mainly on the Exhibition side. When he retired the Director (C. Tate Regan) wanted to continue his researches. He had been involved in suggesting the topics on which Daphne Aubartin and I were working at a college near the Museum. He was kept in touch with our progress and obtained the Trustees' permission to offer the post to me as Research Assistant to the Director.

Soon after my appointment there were three vacancies on the Established staff of the Museum. I did not apply but all these vacancies were filled by women in a competitive situation. They were Anna Hastings (Bryozoa), Daphne Aubartin (Entomology), and Susan Finnegan (Arachnida). Dr. Regan told me that the members of the interviewing panel had agreed that they all exceeded in calibre the men who applied at the same time.

This was the first time that the higher Civil Service, including the Scientific Civil Service, was open to women. During the 7 years that I held the Research Assistantship I did seriously consider applying for at least one permanent job elsewhere, but nowhere else could offer the same research facilities – great historic collections, a great library and no distractions such as teaching or exhibitions (both of which in themselves would be absorbing) and as I had no family commitments, I waited. Eventually a vacancy arose in the Fish Section. I applied and although over-age for entry, I was successful, thanks I believe to the selection by Dr. Calamn, then Keeper of the Department of Zoology, who knew my work. I do not know who was competing for the post at that time. I had the impression that, in that situation, Tate Regan kept a low profile so that I should not seem to be benefiting from favouritism.

Meanwhile, other women had been appointed to the Museum establishment as scientists in various departments. Some of them were lost upon marriage – perhaps voluntarily but the marriage bar was not removed until after World War Two.

When I was seconded to the Fishery Survey of Lake Nyasa in 1939, I thought it was because I had done a lot of work on African freshwater fishes. But someone told me afterwards that if J.R. Norman's health had not forbidden it he would have been chosen – whether for his sex or seniority I don't know.

Do I dream of writing a popular book?

I believe I have no talent for this. But I like to feel that I am supplying reliable data for those who have.

Environmental Biology of Fishes **41**: 67–80, 1994.

The life and work of Rosemary Lowe-McConnell: pioneer in tropical fish ecology

Michael N. Bruton
J.L.B. Smith Institute of Ichthyology, Private Bag 1015, Grahamstown, 6140 South Africa

Key words: Biography, Behaviour, Seasonality, Predation, Breeding, Feeding, Taxonomy, Evolution, Environmental fluctuations, Tropics, Africa, South America

Synopsis

Rosemary Lowe-McConnell is one of the pioneers of tropical fish ecology. During a colourful and eventful career spanning over 45 years, she has worked in the tropical waters of Africa and South America and contributed significantly to our understanding of the ecology, zoogeography, phenology, evolution and taxonomy of tropical fishes. She has also assisted countless young ichthyologists and fisheries scientists and stimulated ichthyology through her lucid books on fish ecology. She continues to play an active role in the promotion of ichthyology and ecology from her home in Sussex in the English countryside. A brief biography and tribute is given so that her contributions to tropical fish ecology can be more widely appreciated.

Early training and the Africa experience

Rosemary Helen Lowe was born in Liverpool, England, on 24 June 1921 and educated at Howell's School, Denbigh, in Wales. She received her B.Sc., M.Sc. and D.Sc. degrees from the University of Liverpool and from 1942–1945 worked as a biologist on the staff of the Freshwater Biological Association on the migrations of anguillid eels in freshwaters.

Her career in African ichthyology started after the Second World War when she was employed to conduct a survey of the tilapias and their fisheries in the southern part of Lake Nyasa (now Lake Malawi). This study was financed by the United Kingdom Colonial Development & Welfare Fund and was a continuation of a previous survey by Ethelwynn Trewavas, Kate Ricardo Bertram and John Borley in 1939. In 1945 Ro was the only person conducting research on the fishes of Nyasaland (now Malawi) and had to work under arduous conditions with no fisheries research organisation to support her. On the lake she used an old diesel, inboard engine boat, the 'Pelican', with an auxiliary sail, and remembers having 'to plunge flaming bits of newspaper into the engine to get it started'. With the assistance of local fishermen she traversed large tracts of the lake and soon became familiar with the fishery.

In a remarkably short time (three years) she produced, virtually single-handedly, a valuable account of the tilapias and their fishery that laid the foundation for subsequent studies of Malawian cichlids. Five tilapiine species were distinguished and found to have distinct breeding seasons and places. Other economically important fish were also studied, including the endemic potamodromous cyprinid *Labeo mesops*.

In 1947 Ro returned to England via Jinja in Uganda, travelling down the Nile, through Lakes Kyoga and Albert, across the Nile Sudd, and down the Nile River to Alexandria. After writing up the Nyasa work at the Freshwater Biological Association laboratory at Windermere and the British Museum (Natural History) in London, she returned to Afri-

Fig 1 Portrait of Rosemary Lowe McConnell by Hubert Wil liams in 1948

ca by flying boat, touching down in Sicily, Alexandria, Luxor and Port Bell on Lake Victoria She arrived at the newly opened East African Fisheries Research Organisation laboratory in Jinja in 1948 as one of the founder members of E A F R O Her arrival was preceded by a telegram from Barton Worthington which read 'Miss Lowe in hands of Death Up to you to make suitable arrangements' (Steven D'Eathe had been detailed to meet her on arrival) The E A F R O was very ably directed by R S A Beauchamp at the time, later members of staff included PH Greenwood (doing work on haplochromine cichlids) and P Corbet (studying the food of non-cichlids)

From 1948–1953 Ro was employed as a Research Officer by the British Overseas Research Service at the Jinja laboratory (Fig 1, 2), where her main task was to investigate the biology of the various tilapias in East African lakes In Lake Victoria the two en demic tilapias were studied in collaboration with the then Lake Victoria Fisheries Service from their research vessels operating out of Mwanza, Kisumu

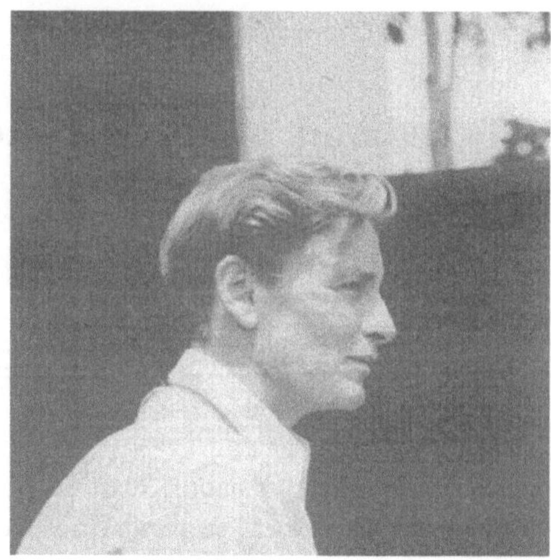

Fig 2 Ro in Jinja in about 1950

and Entebbe She studied the biology of *Oreochromis niloticus* in waters in which it occurred naturally – Lakes Rudolf (now Turkana), Albert (Mobutu) and Edward/George, and in several smaller lakes, ponds and dams into which it had been stocked E B Worthington (Fig 3, 4) had previously found that *O niloticus* dwarfs in lagoons connected to Lake Albert and in crater lakes in Lake Rudolf This switch from growth to reproduction, so significant in fish culture, was investigated further by Ro and other E A F R O staff in this and other tilapia species

In 1955 she described four new species and subspecies of tilapias from Lake Jipe and the upper and lower reaches of the Pangani River in Kenya and Tanzania Here too she produced probably the first account of the physiological switch from growth to reproduction of tilapias in ponds (following Worthington's pioneering field studies on *O niloticus* in Lake Albert) In 1959 she published a major paper on the differences between the substrate-brooding and mouth brooding species of tilapias that was subsequently used by Ethelwynn Trewavas as a basis for the division of the tilapiine fishes into different genera

According to Ethelwynn Trewavas, Ro showed tremendous courage and determination during her field work She found that many fish species were easier to capture at night, and readily braved the

Fig 3 In retirement now, Barton Worthington settled not far from where Ro lives, in a 1200 year-old house near Furnace Green, in which he had the Roman hypocaust rebuilt Photograph by E K Balon on 3 6 1990

marauding hordes of insects and the risks of waterborne diseases in order to achieve her goals (Fig. 5).

Ro also found time to study the breeding behaviour of *Oreochromis karomo* and *O. variabilis* in the clear shallow waters of Lake Tanganyika (visited with Humphry Greenwood in August 1952) and in Lake Kyoga and the Victoria Nile (1950, 1952), respectively. This was in pre-scuba days, and even snorkels were rare in East Africa at the time. After snorkelling in Lake Tanganyika at Kigoma in 1952, Ro's first experimental SCUBA dive was made in Lake Kivu with André Capart's Belgian Expedition in 1953, using rocks tucked into her clothing as weights. She once encountered a large water snake *Boulengerina* sp. while snorkelling in the Malagarasi Swamps, and on several occasions saw crocodiles underwater. These diving experiences led to a passion for underwater fish watching, but later diving opportunities were mainly in the sea: on

Fig 4 Reminiscing on past African days, we spent a charming afternoon at Barton's house

Fig. 5. Ro and Major Gould, Fisheries Officer, Tanganyika, on one of the many field trips in 1950.

the East African coast, Great Barrier Reef, San Blas Islands in Panama, Seychelles and Maldives.

An important feature of life at the E.A.F.R.O. was the visits by experts from whom members of the resident staff learned a great deal. For example, the colourful Cambridge scholar, Hugh Cott, came to Uganda to work on crocodiles, and enlisted Ro's help in dragging the dead reptiles out of the lake and examining their stomach contents. This work later led to a 'most romantic incident' on Lake Turkana when Ro and District Commissioner Dennis McKay were 'pinned together by one crocodile', their fingers caught in its (luckily small) mouth and only released when the DC's wife fetched a large carving knife. Early experiments to assist another distinguished visitor with the detection of electric signals from mormyrid fishes, by trailing wires from a small rowing boat, led to an exciting encounter with an inquisitive and highly vocal hippopotamus.

During this productive period Ro worked in all the East African territories of the United Kingdom, mainly on tilapias, and served as Acting Director of E.A.F.R.O. for a short while. She produced a steady stream of scientific papers and popular reports on the biology, ecology and taxonomy of cichlids and other fishes. Her ecological studies were remarkable for the way in which they combined scientific value with relevance to the fisheries. They provided the essential baseline for later studies that assessed the impact of fishing and other human pressures on the populations of valuable food fishes (Fig. 6).

When she married Richard McConnell, a geol-

Fig. 6. Ro (centre) in Jinja in 1953 with Humphry and Marjorie Greenwood.

ogist, on 31 December 1953 Ro was required to resign from the Colonial Service (because of the 'marriage-bar') but continued working as an Honorary Fisheries Research Officer. (Ethelwynn Trewavas and other women scientists had earlier campaigned against the 'marriage-bar' in the British Civil Service, which was eventually scrapped). Thereafter most of her research and editing was carried out on an expenses-only, voluntary basis, supplemented by occasional contracts, consultancies, teaching assignments and royalties. From 1954–1956 Ro lived in Botswana, where Richard extended his geological work and she collected fishes in the Okavango Delta.

South American experiences

From 1957–1962 the McConnells lived in British Guiana in South America, where Richard directed the Geological Survey. Through the timely visit of C.F. Hickling, Chief Fisheries Advisor to the U.K. Overseas Development Administration, she became associated with the Guiana Department of Agriculture and Fisheries in a survey of the freshwater (and later marine) fish and fisheries. Although she was appointed by the Guiana Department of Agriculture and Fisheries (Fig. 7) for the princely sum of only Guiana $ 1.00 p.a., the appointment did provide her with working facilities, a laboratory, transport, flights to the interior (into which there were no roads) and accommodation on the research ship.

The most exciting freshwater work during this period was a survey on the Rupununi District of the huge floodplain connecting the Essequibo River system seasonally across to the Rio Branco (draining into the Rio Negro in the Amazon drainage), an area previously unsurveyed ichthyologically. Here she had the good fortune to be helped by the McTurk family and other ranchers and their Amerindian cousins, all very good naturalists, well versed in the ways of the fishes on which so many of them depended for their food. The biology of other freshwater fishes was also studied in rivers and estuaries on the coastal plain and in other parts of this fascinating rainforest country.

Fig. 7. Ro with Bertie Allsopp, Fisheries Officer, in British Guiana, 1960.

Later Ro was appointed as the ichthyologist on the R.V. 'Cape St Mary', a vessel that had been brought from West Africa by Hickling to carry out the first survey of the unexplored Guiana shelf between the West Indies and Brazil. This marine field work was supplemented by examining round-the-year catches of fishes on sale in the Georgetown fish market in order to determine seasonal trends in movements as well as breeding and feeding preferences of a variety of species.

During this phase of her career, which she calls her 'marine transgression', she became familiar with over 200 marine fish species in over 70 fish families and greatly expanded her knowledge of tropical fish biology. She studied food partitioning in sciaenid and other fishes, and published a series of papers and reports on the marine fishes of Guiana [one of which is regularly cited in ecology texts,

Lowe (McConnell 19)*]. She also extended her studies on breeding seasonality to egrets (Lowe-McConnell 25). On leaving Guiana in 1962 the then Prime Minister (now President) Chedie Jagan presented Ro with a gift at a formal dinner party in recognition of her contributions to fisheries research in Guiana.

After returning to England in 1968 Ro was appointed as the ichthyologist on the Royal Society of London/Royal Geographical Society Xavantina-Cachimbo Expedition to north-eastern Mato Grosso, Brazil. The aim of this expedition was to assess the environmental impact of a new road extending northwards along the Sierra do Roncador as part of the expanding Amazon highway system. This area was previously poorly explored ichthyologically and provided many challenges to the expedition members. She drew extensively on her field experience in Africa and Guiana during this expedition and made some of the first detailed studies of the synecology of Amazonian fishes. The Mato Grosso is an area of high endemism among the fishes, with very different ichthyofaunae in adjacent rivers. Ro was the first to attempt to explain the differences on ecological grounds, and subsequently became a prolific writer on South American freshwater fishes.

A graphic description of the working conditions in South America was contained in a letter written by Ro and quoted by Anthony Smith in his book on the Mato Grosso expedition (Smith 1971, pp. 203–204): '. . . Of course one was wet through from chest downwards every day and almost every night. I always worked in clothes, a habit developed in Guyana where piranha were bad. Therefore one kept cool while the soil scientists were finding it very hot. (. . .) Sweat bees used to swarm to the damp fish as I was measuring them, mixed with a few stinging feral "Europa" bees. Biting flies were also very bad near the streams.(. . .) One snag about the fish work was the amount of gear one had to carry to catch the fish and preserve them before they went bad, as they do so fast in the tropics.(. . .) One traipsed through the bush looking like the White Knight.(. . .) It was hard work physically as we had to improvize and even set gill nets by swimming (in the absence of any boat) and with circumspection, as I was only too well aware what electric eels and piranha can do. However, the water was clear in many places and one could watch the fish (with polaroid glasses) and assess what there was before disturbing them by trying to catch them'. There were, however, some compensations for a keen angler. Three Brazilian assistants once caught 41 *Aequidens*, 14 *Hoplerythrinus*, 1 *Hoplias*, 6 *Leporinus*, 3 *Crenicichla*, 2 *Acestrorhynchus* and 1 *Moenkhausia* species using hook and line within three hours, as in these virgin waters each baited hook is taken almost immediately after it is thrown in!

* See the 'Lifetime list of publications'.

Back to England

In 1962, on returning to England, Ro was appointed as an Associate of the British Museum (Natural History) (1962–1967) and granted the Appleyard Bequest by the Linnean Society of London (£ 200 p.a.). At the BMNH she was provided with facilities to study the considerable collections of fishes that she had sent back from Africa and South America. She worked in a small office/laboratory with Ethelwynn Trewavas in the new spirit building (Fig. 8). I have fond memories of sharing crustless sandwiches and hot Marmite drinks with them over lunch during my postdoctoral year at the BMNH in 1977/1978. Their office was always a hive of activity, with a constant stream of visitors and non-stop discussions on the ecology and taxonomy of tropical fishes. Melanie Stiassny, Richard Vari and Augustine Baddokwaya were also doing postdoctorals at the British Museum at the time, and Humphry Greenwood, Jim Chambers, Gordon Howes and Keith Banister were just down the corridor. The magnificent BMNH fish collection and library, and this exciting group of people, added up to a most stimulating environment in which to work.

At the Natural History Museum, Ro wrote her first book synthesizing studies on the ecology of freshwater fishes from the tropics of Africa, South America and Asia, *Fish Communities in Tropical Freshwaters, Their Distribution, Ecology and Evo-*

Fig. 8. Ethelwynn Trewavas and Rosemary Lowe-McConnell in the British Museum (Natural History) in the summer of 1985.

lution (Lowe-McConnell 34). This was followed by another synthesis volume, *Ecology of Fishes in Tropical Waters* (Lowe-McConnell 38), which also treated marine fishes; by then she had become a leading authority on this subject. Ro later expanded both books into *Ecological Studies in Tropical Fish Communities* (Lowe-McConnell 50), the definitive title on this subject (Fig. 9). Her penetrating reviews, concise style of writing and generous use of illustrations have made her books accessible to readers of many cultures and educational backgrounds and have greatly stimulated ichthyology in the tropics. Her life-long interest in speciation culminated in the publication of the proceedings of a British Ecological Society/Linnean Society conference on *Speciation in Tropical Environments* (Lowe-McConnell 30), which she initiated and edited.

At this time Ro was able to become involved with many activities as she was not in fulltime employment. She visited Rome at the request of the F.A.O. where she helped to edit papers and participated in the World Symposium on Warm-Water Fish Culture (1966). She was actively involved in the 10-year long International Biological Programme (I.B.P.) from its inception in 1964, contributing a chapter to I.B.P. Handbook 3 on 'Methods for the assessment of fish production in freshwaters' (Lowe-McConnell 28). She also attended I.B.P. regional meetings in Uganda, Malaysia and Latin America. The summary volume from the I.B.P./P.F. Productivity in Freshwaters programme, *The Functioning of Freshwater Ecosystems*, was co-edited with E.D. Le Cren (Le Cren & Lowe-McConnell 43).

During the post-colonial era man-made lakes massively altered tropical rivers, and provided the opportunity to study how riverine fish communities changed into lacustrine ones. These challenges did not escape Ro's attention. In 1964 she edited the proceedings of a conference of the Institute of Biology on man-made lakes (published in 1966, Lowe-McConnell 24), and in 1971 chaired a session in a conference on man-made lakes in Knoxville, U.S.A. (Lowe-McConnell 33). In 1964 she was part of a U.N.D.P. Mission to Ghana to plan research for a new man-made lake on the Volta River, and in 1965 took part in a pre-impoundment survey of Kainji Lake on the Niger River in West Africa. These field excursions led to the publication of a booklet by Ghana Universities Press on the fishes of Volta and Kainji lakes (Lowe-McConnell & Wuddah 32).

In 1974 she attended the First International Congress of Ecology in The Hague and subsequently edited (Fig. 10) with W.H. van Dobben the proceedings of the plenary sessions as a book entitled *Unifying Concepts in Ecology* (Van Dobben & Lowe-

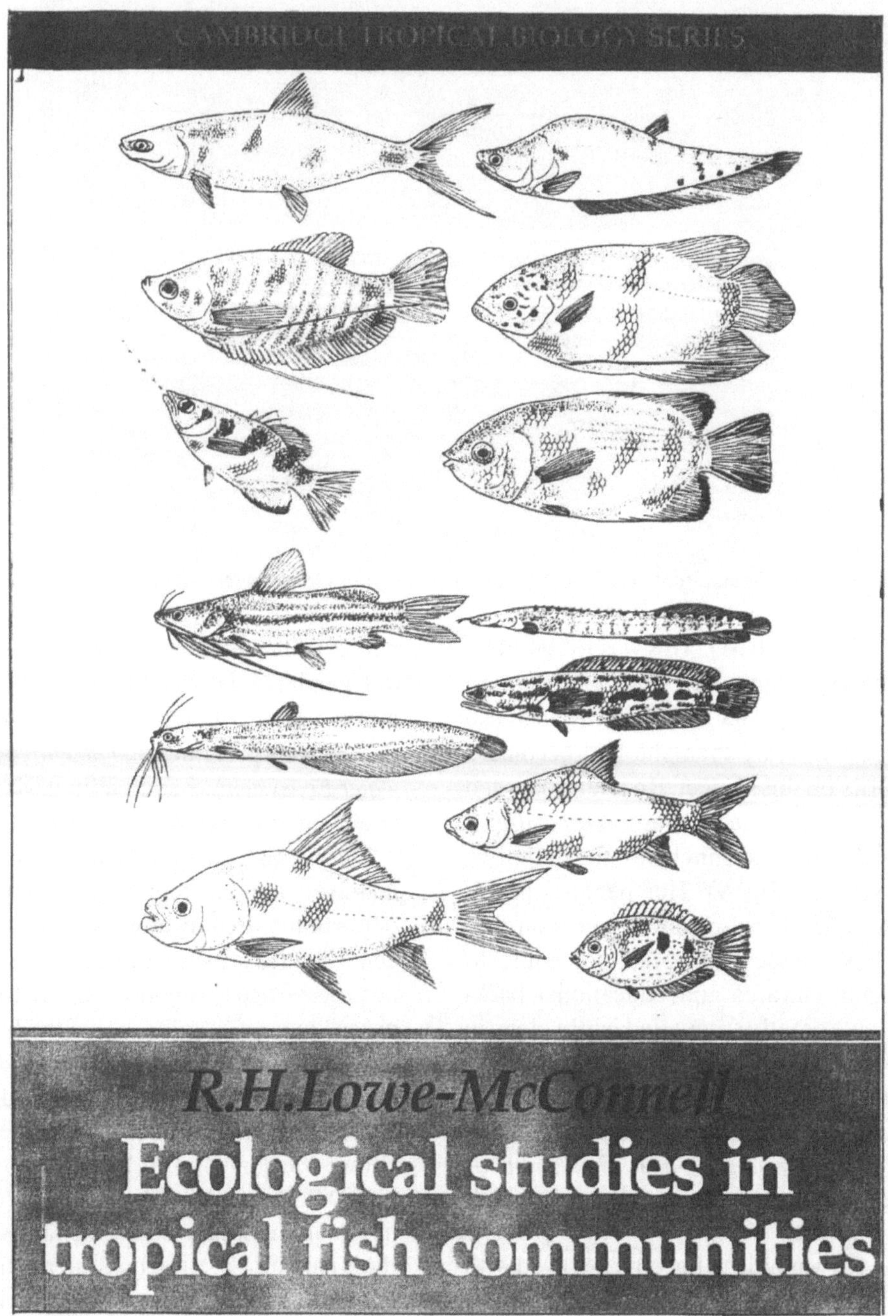

Fig 9 The title page of Ro's latest book

McConnell 36). This book, which was translated into Spanish, was a masterful compilation of the latest in ecological thinking, especially with respect to the comparative productivity, diversity and stability of natural ecosystems. The paragraph written by the editors at the end of the Preface summarises their world view: 'We sincerely hope that the results of the First International Congress of Ecology will

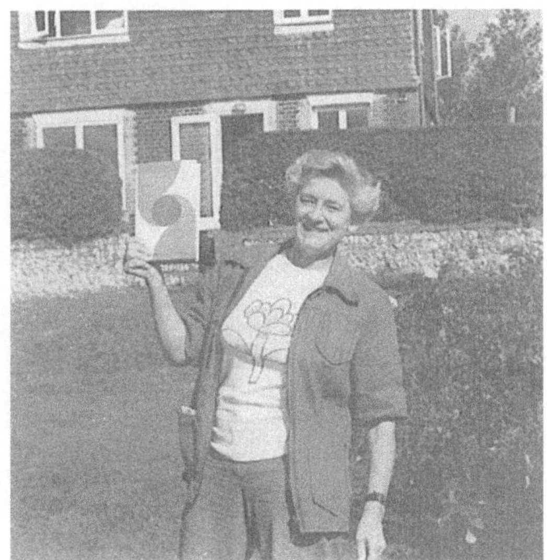

Fig. 10. Ro in 1975 against the background of the house 'Streatwick' during editing of the proceedings of the First International Congress of Ecology (36), holding the logo and wearing a Dr W. Junk Publishers T-shirt.

Fig. 11. Ro and Robin Welcomme study the original E.A.F.R.O. visitors book, signed by many old friends, in Jinja, 1992.

emphasize that the maintenance of vulnerable natural resources demands long-term policies with a sound scientific basis, and that neglect of ecological rules for the sake of immediate profit spells disaster.'

Ro has also been involved with the International Center of Living Aquatic Resources Management (ICLARM) and actively participated in their workshops on tilapias in Bellagio, Italy, 1980 (editing the proceedings with R.S.V. Pullin) and Bangkok in 1987. In 1987 she attended a workshop on 'Community structure and function in temperate and tropical streams' in Flathead, U.S.A. She also contributed a typically incisive overview chapter entitled 'Broad characteristics of the ichthyofauna' to the proceedings of the U.N.E.P. conference on 'African limnology' held in Nairobi in 1979.

Ro has retained her interest in African fishes and in 1987 co-convened the Societas Internationalis Limnologicae (S.I.L.) African Great Lakes Group meeting at a symposium in Burundi on 'Resource use and conservation of the African Great Lakes'. She contributed to the 'First International Conference on the Conservation and Biodiversity of Lake Tanganyika' in Burundi in 1991, and attended a workshop on Lake Tanganyika research in Kuopio in Finland in 1991. Also in 1991, she contributed to a meeting of the S.I.L. Tropical Group in Hong Kong and chaired a session on the 'Conservation and management of tropical inland waters'.

In March 1992 she participated in a conference on 'Biodiversity, fisheries and the future of Lake Victoria' in Jinja, Uganda (Fig. 11), and in August 1992 helped to organise the 'African Great Lakes' session at the S.I.L. congress in Barcelona. She also presented a paper on 'Fish communities in Lakes Malawi, Victoria and Tanganyika' at a seminar on 'Biodiversity – fish populations and communities in Lake Tanganyika' held at the University of Kyoto in Japan in November 1992. Her year was concluded with the presentation of a paper at a conference on 'Biodiversity, production and conservation of African aquatic ecosystems' at the University of Zimbabwe in December 1992. In March 1993 she presented a paper at a workshop on 'Speciation in ancient lakes' in Belgium. In 1993 she also participated in a Great Lakes meeting at the University of Guelph in Canada (Fig. 12) and in the Symposium on the Ecology of Latin American Fish at the A.S.I.H. meeting in Texas.

Throughout her career Ro has been driven by the need to understand the ecology of fishes in order to ensure their sustainable utilisation. She is a respected writer on fish conservation and was invited by the Fisheries Society of the British Isles to summar-

Fig 12 In May 1993 Ro was one of the invited speakers at the Symposium on the Great Lakes of the World a she listens with great concentration to the other speakers and b c later enjoys herself at the Niagara Falls Photographs a b by E K Balon and c by D L G Noakes

ise the proceedings of the international conference on 'The biology and conservation of rare fish' held at the University of Lancaster in England in 1990 (Lowe-McConnell 57)

Rosemary Lowe-McConnell has held important positions in several scientific societies, including serving for five years as Honorary Secretary of the Tropical Group of the British Ecological Society, and as Vice-President of the Linnean Society of London (1967), a member of the editorial board of the Biological Journal of the Linnean Society, and as Convenor of S I L 's African Great Lakes Group (1987–1989) She was also an original member of the Association for Tropical Biology, and was elected a Fellow of the Linnean Society in 1957

Her teaching duties have included courses on fish ecology at Makerere University in Uganda, the tropical fisheries component of M Sc courses at Salford University in the U K , and ecology and environmental concern 'further education' courses at Sussex University She has also acted as a Ph D supervisor at the Open University, U K , and has been external examiner for numerous higher degree theses from universities in various parts of the world

Concluding comments

Rosemary Lowe-McConnell has named six new

Table 1 Fish species and subspecies described by Dr Rosemary Lowe McConnell

Order	Family	Species	Reference
Perciformes	Cichlidae	*Tilapia saka*	Lowe (1953)
		Tilapia girigan	Lowe (1955)
		Tilapia jipe	
		Tilapia pangani	
		Tilapia mossambica karogwe	
Carcharhiniformes	Mustelidae	*Mustelus higmani*	Springer & Lowe McConnell (1963)

species or subspecies of fishes (Table 1), including a new *Tilapia* (now *Oreochromis*) from southern Lake Malawi, three tilapia species and one subspecies from the Pangani River system in East Africa, and the smallest member of the genus *Mustelus*, the smooth dogshark, from the Atlantic coast of South America Two species of aquatic organisms have been named after her, a pelagic catfish from Lake Malawi (*Bathyclarias loweae* Jackson, 1959) and an ephemeropteran (*Afroptilum loweae* Kimmins, 1949) During her fieldwork she collected many other new species of African and South American fishes that were lodged in the British Museum (Natural History) and subsequently sent to specialists throughout the world to be described and named In the Mato Grosso alone she probably collected at least 50 species of fishes that were new to science

Her career has been characterised by an indomitable spirit, a rare ability to identify trenchant biological traits in a fish, and a disregard for personal comfort and well-being in the pursuit of her goals She is an energetic fieldworker, efficient laboratory scientist and enthusiastic conference-goer Probably no other freshwater fish biologist has made such good use of opportunities to study the ichthyofaunas of different parts of the tropics (Fig 13) Ichthyologists and fisheries scientists in developed and developing countries will always be grateful to her for the way in which she has combined her extensive field experience and thorough knowledge of theory to produce a series of outstanding books on tropical fishes, and contributed numerous voucher specimens to museum collections for later study

Ro's wide travels and outgoing personality have resulted in her establishing a wide network of collaborating ecologists and taxonomists These colleagues have aided her greatly in her work and made her multidisciplinary approach possible Her

Fig 13 Ro snorkel diving in the Maldives on a fishwatching holiday in 1989 Photograph by Peggy (M E) Varley (nee Brown)

Fig 14 Ro was a frequent visitor to ET even after the latter moved to live with her sister in Reading Photograph by E K Balon in 1990

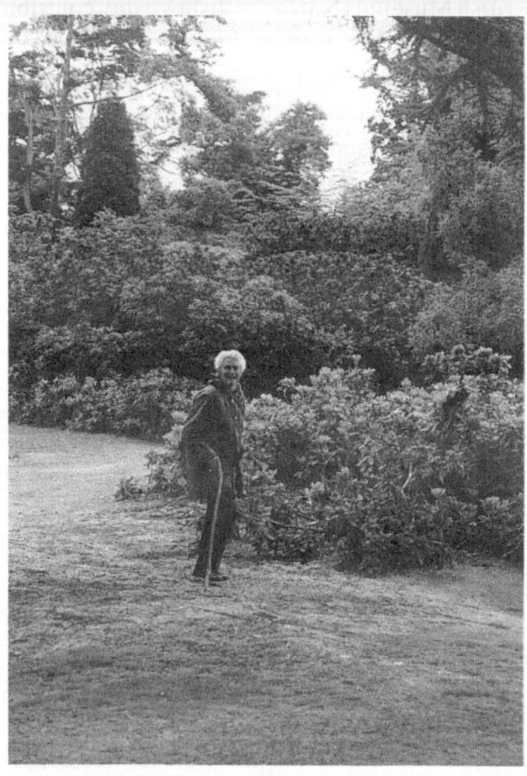

Fig 15 Ro among the magnolias at the Sheffield Park Gardens on 3 June 1990 Photograph by E K Balon

a

b

Fig 16 Rosemary Lowe-McConnell resides in the south of England near Hassocks in a lovely house (a) overlooking a wide valley (b) Photograph by E K Balon, June 1990

partnership with Ethelwynn Trewavas has, in particular, been most fruitful and enduring (Fig. 14, 15, 16). During her career she has worked on many of the great waterbodies of the world – lakes Malawi, Victoria and Turkana, the Okavango Swamps, Amazon and Niger rivers, and has tackled some of the most problematic fish groups, such as the tilapias, loricariids, sciaenids and characids. But her greatest contribution has been her ecological syntheses, the pulling together of threads on a diverse array of themes ranging from seasonality, evolution, predation pressure, reproductive cyclicity, population dynamics, the effect of environmental fluctuations on fish biology and the impact of man-made lakes. She has always regarded ecology and behaviour as opposite sides of the same coin, and is fascinated by the way that they have contributed to evolution. She has gone to considerable lengths to bring fish ecological studies to the attention of general ecologists so that the vast array of data on fishes could be incorporated into the ecological mainstream. Although she worked on such a broad canvas, her contribution has been striking – rarely is a paper on tropical freshwater fishes published without citing her work. Her life's work has truly contributed to unifying concepts in ecology.

Acknowledgements

I am grateful to Ro for the hours of enjoyable discussions and for the letters that we have exchanged during the preparation of this brief biography, which by no means does her accomplishments justice, but is perhaps a starting point for a more comprehensive biographer. I am also grateful to Ethelwynn Trewavas, Peter Jackson, Eugene Balon, Humphry Greenwood, David Noakes, Jean Pote, Carolynn Bruton, Margaret Crampton, Sheila Coutouvidis and Kathy Holden for their assistance.

References cited

Géry, J. 1992. Description de deux nouvelles espèces proches de *Moenkhausia lepidura* (Kner) (Poissons, Characiformes, Tetragonopterinae), avec une revue du groupe. Revue fr. Aquariol. 19: 69–78.

Jackson, P.B.N. 1959. Revision of the clariid catfishes of Nyasaland, with descriptions of a new genus and seven new species. Proc. zool. Soc. Lond. 132 109–128.

Kimmins, D.E. 1949. Ephemeroptera from Nyasaland, with descriptions of new species. Ann. Mag. Nat. Hist. 12· 825–836.

Lowe, R.H. 1953. Notes on the ecology and evolution of Nyasa fishes of the genus *Tilapia*, with a description of *T saka* Lowe. Proc. zool. Soc. Lond 122 1035–1041.

Lowe, R.H. 1955. New species of *Tilapia* (Pisces, Cichlidae) from Lake Jipe and the Pangani River, East Africa. Bull. Brit. Mus. (Nat. Hist.) (Zool.) 2: 349–368

Smith, A. 1971. Mato Grosso. Last virgin land. Michael Joseph, London 969 pp.

Springer, S. & R.H Lowe-McConnell. 1963. A new smooth shark, *Mustelus higmani*, from the equatorial Atlantic coast of South America. Copeia 1963: 241–251.

Rosemary Lowe-McConnell and Mike Bruton at the Freshwater Biological Association's laboratory at The Ferry House, Ambleside, Lake Windermere, England, in July 1990. Photograph by Paul Skelton.

Environmental Biology of Fishes **41** 81–83, 1994

Lifetime list of publications by Rosemary Lowe-McConnell

Michael N. Bruton
J.L.B. Smith Institute of Ichthyology, Private Bag 1015, Grahamstown, 6140 South Africa

Rosemary Lowe-McConnell authored or co-authored at least 65 scientific papers, books, chapters in books and popular articles, and edited or co-edited three books between 1951 and 1993. She was the sole or first author of 88% of her publications and shared authorship in 13 publications with 31 collaborators, but never more than once with a particular author. Surprisingly, Ro co-authored only one publication with her lifelong collaborator Ethelwynn Trewavas, and that was a thirteen-author note to Nature on the destruction of fisheries in African lakes. She published in five different decades, with her most productive decade being the 1970s with 17 publications. Her publication list is notable for the number of books that she authored or co-authored (7) or to which she contributed chapters (7); three of these books were part of International Biological Programme (IBP) initiatives. She published primarily on freshwater fishes (47 publications) but also on marine fishes (6), on fishes in general (4) and on aquatic ecology (7); one paper dealt with birds. Over half of her publications were concerned with aquatic ecology (55%), the rest dealing with taxonomy and speciation (15%), fisheries (14%), conservation (8%) and other subjects. Ro published variously under the names R.H. Lowe, R.H. Lowe (McConnell), R.H. McConnell (Lowe), R.H. McConnell and R.H. Lowe-McConnell. Her publications are arranged chronologically by year but not within years. All publications in this list were verified against an original copy.

1 Lowe, R H 1951 Factors influencing the runs of elvers in the River Bann, Northern Ireland J Cons Int Explor Mer 17 299–315

2 Lowe, R H 1952 The influence of light and other factors on the seaward migration of the silver eel (*Anguilla anguilla* L) J Anim Ecol 21 275–309

3 Lowe, R H 1952 Report on the *Tilapia* and other fish and fisheries of Lake Nyasa, 1945–47 Fishery Publ Colon Off 1(2) 1–126

4 Lowe, R H 1953 Notes on the ecology and evolution of Nyasa fishes of the genus *Tilapia* with a description of *T saka,* Lowe Proc zool Soc Lond 122 1035–1041

5 Lowe, R H 1955 New species of *Tilapia* (Pisces, Cichlidae) from Lake Jipe and the Pangani River, East Africa Bull Br Mus (Nat Hist), Zool 2 349–368

6 Lowe (McConnell), R H 1955 Species of *Tilapia* in East African dams, with a key for their identification E Afr Agric J 20 256–262

7 Lowe (McConnell), R H 1955 The fecundity of *Tilapia* species E Afr Agric J 11 45–52

8 Lowe (McConnell), R H 1956 Observations on the biology of *Tilapia* (Pisces-Cichlidae) in Lake Victoria, East Africa E Afr Fish Res Org Suppl Publ 1 1–72

9 Lowe (McConnell), R H 1956 The breeding behaviour of *Tilapia* species (Pisces Cichlidae) in natural waters observations on *T karomo* Poll and *T variabilis* Boulenger Behaviour (Leiden) 9 141–163

10 Lowe (McConnell), R H 1957 Observations on the diagnosis and biology of *Tilapia leucosticta* Trewavas in East Africa Rev Zool Bot Afr 55 353–373

11 Lowe (McConnell), R H 1958 Observations on the biology of *Tilapia nilotica* Linne in East African waters Rev Zool Bot Afr 57 129–170

12 McConnell (Lowe), R H 1958 Introduction to the fish fauna of British Guiana Timehri, Journal of the Guyana Museum and Zoo of the Royal Agricultural and Commercial Society of British Guiana 34 1–11

13 Lowe (McConnell), R H 1959 Breeding behaviour patterns and ecological differences between *Tilapia* species and their significance for evolution within the genus *Tilapia* (Pisces Cichlidae) Proc zool Soc Lond 132 1–30

14 Mitchell, W G & R H McConnell 1960 The trawl survey carried out by the R/V 'Cape St Mary' off British Guiana 1957–59 Part II The interpretation of the catch records Bulletin no 2, Fisheries Division, Department of Agriculture, Georgetown 53 pp

15 Lowe (McConnell), R H & A R Longhurst 1961 Trawl fishing in the tropical Atlantic Nature 192 620–623

16 Lowe (McConnell), R H 1962 Notes on the fishes in Ge-

orgetown fish markets and their seasonal fluctuations Fish Bull 4, Dept Agriculture, Georgetown 31 pp
17 Lowe-McConnell, R H 1962 The fishes of the British Guiana continental shelf, Atlantic coast of South Africa, with notes on their natural history J Linn Soc (Zool) 44 669–700
18 Springer, S & R H Lowe 1963 A new smooth dogshark, *Mustelus higmani,* from the Equatorial Atlantic coast of South America Copeia 1963 241–251
19 Lowe (McConnell), R H 1964 The fishes of the Rupununi savanna district of British Guiana, South America Part I Ecological groupings of fish species and effects of the seasonal cycle on the fish J Linn Soc (Zool) 45 103–144
20 McConnell (Lowe), R H 1964 Tropical fishery problems and research Ann appl Biol 53 502–503
21 McConnell, R H & E B Worthington 1965 Man-made lakes Nature 208 1039–1042
22 Banks, J W, M J Holden & R H McConnell 1966 Fishery report pp 21–31 *In* E White (ed) The First Scientific Report of the Kainji Biological Research Team, Liverpool University, Liverpool
23 Lowe-McConnell, R H 1966 The sciaenid fishes of British Guiana Bull mar Sci 16 20–57
24 Lowe-McConnell, R H (ed) 1966 Man-made lakes Symposia of the Institute of Biology no 15, Academic Press, London 218 pp
25 Lowe-McConnell, R H 1967 Biology of the immigrant cattle egret *Ardeola ibis* in Guyana, South America Ibis 109 168–179
26 Lowe (McConnell), R H 1967 Some factors affecting fish populations in Amazonian waters Atas do Simposio sobre a Biota Amazônica (Conservação da Natureza e Recursos Naturais) 7 177–186
27 McConnell, R H 1967 The fish fauna of the Rupununi District, Guyana Timehri, Journal of the Guyana Museum and Zoo of the Royal Agricultural and Commercial Society of British Guiana 43 57–72
28 Lowe-McConnell, R H 1968 Identification of freshwater fishes pp 46–77 *In* W E Ricker (ed) Methods for Assessment of Fish Production in Fresh Waters, IBP Handbook no 3, Blackwell Scientific Publications, Oxford
29 Lowe-McConnell, R H 1969 Speciation in tropical freshwater fishes pp 51–75 *In* R H Lowe-McConnell (ed) Speciation in Tropical Environments, Academic Press, London (also in Biol J Linn Soc Lond 1 51–75)
30 Lowe-McConnell, R H (ed) 1969 Speciation in tropical environments Academic Press, London 246 pp
31 Lowe-McConnell, R H 1969 The cichlid fishes of Guyana, S America, with notes on their ecology and breeding behaviour J Linn Soc (Zool) 48 255–302
32 Lowe-McConnell, R H & A A Wuddah 1972 Freshwater fishes of the Volta and Kainji Lakes Keys for the field identification of freshwater fishes likely to occur in or above the new man-made lakes, Lake Volta in Ghana and the Kainji Lake on the River Niger in Nigeria Ghana Universities Press, Accra 22 pp
33 Lowe-McConnell, R H 1973 Summary reservoirs in relation to man – fisheries pp 641–654 *In* W C Ackermann, G F White & E B Worthington (ed) Man-made Lakes Their Problems and Environmental Effects, Geophysical Monograph 17, American Geophysical Union, Washington, D C
34 Lowe-McConnell, R H 1975 Fish communities in tropical freshwaters, their distribution, ecology and evolution Longman, London 337 pp
35 Lowe-McConnell, R H 1975 Freshwater life on the move Man and the changing wildscape IX Geographical Magazine 57 768–775
36 Van Dobben, W H & R H Lowe-McConnell (ed) 1975 Unifying concepts in ecology Report of the plenary sessions of the First International Congress of Ecology, The Hague, 1974 Dr W Junk Publishers, The Hague 302 pp
37 Lowe-McConnell, R H 1976 Review of 'Lake Kariba a man-made tropical ecosystem in Central Africa' J Fish Res Board Can 33 2142–2144
38 Lowe-McConnell, R H 1977 Ecology of fishes in tropical waters Studies in Biology no 76 Edward Arnold, London 64 pp
39 Lowe-McConnell, R H 1977 On environmental stability and its effects on fish populations in tropical freshwaters Actas del IV Simposium Internacional de Ecologia tropical, Marzo 7–11 1977, Panama, Organizacione Patrocianadoras, Panama 2 695–710
40 McConnell, R H 1978 The Amazon Rivers of the World Wayland Publishers Ltd, Howe 65 pp (Translated into German in 1979 as Der Amazonas, Schwager & Steinlein, Nurnberg)
41 Lowe-McConnell, R H 1978 Identification of freshwater fishes pp 48–83 *In* T Bagenal (ed) Methods for Assessment of Fish Production in Fresh Waters, 3rd edition, IBP Handbook no 3, Blackwell Scientific Publications, Oxford
42 Lowe-McConnell, R H 1979 Ecological aspects of seasonality in fishes of tropical waters pp 219–241 *In* PJ Miller (ed) Fish Phenology Anabolic Adaptiveness in Teleosts, Symposia of the Zoological Society of London 44, Academic Press, London
43 Le Cren, E D & R H Lowe-McConnell (ed) 1980 The functioning of freshwater ecosystems International Biological Programme no 22 Cambridge University Press, Cambridge 588 pp
44 Lowe-McConnell, R H & G J Howes 1981 Pisces pp 218–229 *In* S H Hurlbert, G Rodriguez & N D Santos (ed) Aquatic Biota of Tropical South America, Part 2 Anarthropoda, San Diego State University, San Diego
45 Lowe-McConnell, R H 1982 Tilapias in fish communities pp 83–113 *In* R S V Pullin & R H Lowe-McConnell (ed) The Biology and Culture of Tilapias, ICLARM Conference Proceedings 7, Manila
46 Pullin, R S V & R H Lowe-McConnell (ed) 1982 The biology and culture of tilapias International Center for Living Aquatic Resources Management (ICLARM) Conference Proceedings 7, Manila 432 pp

47 Lowe-McConnell, R H 1984 The status of studies on South American freshwater food fishes pp 139–156 *In* T M Zaret (ed) Evolutionary Ecology of Neotropical Freshwater Fishes, Developments in Environmental Biology of Fishes 3, Dr W Junk Publishers, The Hague

48 Barel, C D N , R Dorit, PH Greenwood, G Fryer, N Hughes, PB N Jackson, H Kawanabe, R H Lowe-McConnell, M Nagoshi, A J Ribbink, E Trewavas, F Witte & K Yamaoka 1985 Destruction of fisheries in Africa's lakes Nature 315 19–20

49 Lowe-McConnell, R H 1985 The biology of the river systems with particular reference to the fishes pp 101–140 *In* A T Grove (ed) The Niger and its Neighbours Environmental History and Hydrobiology, Human Use and Health Hazards of the Major West African Rivers, Balkema, Rotterdam

50 Lowe-McConnell, R H 1987 Ecological studies in tropical fish communities Cambridge University Press, Cambridge 382 pp (Translated into Portuguese in 1992, with additional chapters on Brazilian fishes)

51 Benke, A C , C A S Hall, C P Hawkins, R H Lowe-McConnell, J A Stanford, K Suberkropp & J V Ward 1988 Bioenergetic considerations in the analysis of stream ecosystems Journal of the North American Benthological Society 7 480–502

52 Lowe-McConnell, R H 1988 Broad characteristics of the ichthyofauna pp 93–110 *In* C Levêque, M N Bruton & G W Ssentongo (ed) Biology and Ecology of African Freshwater Fishes, Travaux et Documents ORSTOM 216, Paris

53 Lowe-McConnell, R H 1988 Ecology and distribution of tilapias in Africa that are important for aquaculture pp 12–18 *In* R S V Pullin (ed) Tilapia Genetic Resources for Aquaculture, ICLARM Conference Proceedings 16, Manila

54 Lowe-McConnell, R H 1988 Fish of the Amazon system pp 339–351 *In* B R Davies & K F Walker (ed) The Ecology of River Systems, Dr W Junk Publishers, The Hague

55 Lowe McConnell, R H 1988 Concluding remarks II tropical perspectives for future research in river ecology Journal of the North American Benthological Society 7 527–529

56 Lowe-McConnell, R H 1989 Review of 'Rio Negro, rich life in poor water Amazonian diversity and foodchain ecology as seen through fish communities' Trends in Ecology and Evolution 4 120–121

57 Lowe-McConnell, R H 1990 Summary address rare fish, problems, progress and prospects for conservation J Fish Biol 37 (Suppl A) 263–269

58 Lowe-McConnell R H 1991 Ecology of cichlids in South American waters and African rivers pp 60–85 *In* M H A Keenleyside (ed) Cichlid Fishes Behaviour, Ecology and Evolution Chapman & Hall, London

59 Lowe-McConnell, R H 1991 Natural history of fishes in Araguaia and Xingu Amazonian tributaries, Serra do Roncador, Mato Grosso, Brazil Ichthyol Explor Freshwaters 2 63–82

60 Lowe-McConnell, R H 1991 Evolution in tropical lakes review of 'Lake Tanganyika and its life' by G W Coulter (ed) Trends in Ecology and Evolution 6 272–273

61 Lowe-McConnell R H 1991 Ecological roles of littoral fishes in Tanganyika and other African lakes pp 39–41 *In* H Molsa (ed) Proceedings of the International Symposium on Limnology and Fisheries of Lake Tanganyika, Kuopio, Finland (abstract)

62 Lowe-McConnell, R H , R C M Crul & F C Roest 1992 Symposium on resource use and conservation of the African Great Lakes, Bujumbura 1989 Mitt int Verein Limnol 23 1–128

63 Lowe-McConnell, R H 1993 Fish faunas of the African Great Lakes origins, diversity and vulnerability Conservation Biology 7(3) 1–10

64 Lowe-McConnell R H 1993 Threats to, and conservation of, tropical freshwater fishes Mitt internat Verein Limnol 24 1–6

65 Lowe-McConnell R H 1994 The roles of ecological and behaviour studies in understanding fish diversity and speciation in the African Great Lakes a review Arch Hydrobiol (in press)

Ro in the solarium of Balon's Schoolhouse during the Great Lakes meeting, May 1993 Photograph by E K Balon

Environmental Biology of Fishes **41**. 85–87, 1994

An interview with Rosemary Lowe-McConnell

Michael N. Bruton
J.L.B. Smith Institute of Ichthyology, Private Bag 1015, Grahamstown, 6140 South Africa

Rosemary Lowe-McConnell was interviewed in her home Streatwick at Streat near Hassocks, Sussex, in the southern English countryside on 15 February 1993 after we had had a lively lunch with Barton Worthington and his wife, who live nearby. The interview took place amidst piles of papers, books and African and South American memorabilia, and was interrupted by teas and cakes and the telephone. Ro talks like she lives – bubbling over with enthusiasm and energy and full of exclamation marks!! I trust that this discussion provides some insights into her colourful and eventful life.

Ro = Rosemary Lowe-McConnell **MB** = Michael Bruton

MB: Many lifetime contributions to science are encouraged and inspired by personal relationships that are rarely identified in short biographies. Having just finished compiling an account of your 'life and work' for this volume, I am left with some unanswered questions to complete the historical record. Would you allow me, therefore, to ask some more personal questions concerning your work, choice of taxa, location of study, etc.? Let us start with a conventional one: What is the most memorable event from your childhood?

Ro: My godmother was a biologist and gave me very good children's books on natural history. Someone gave my brother Jeffrey a map of Africa and I had one of Australia. I was very envious as I always wanted to go to Africa, attracted by the warm climate and rich natural history. At my first school, when asked what I wanted to be, I replied 'An explorer/naturalist', to which the response was 'Never mind, dear, perhaps you can teach'. But I was very determined so I eventually found a way to do what I had always wanted to do.

MB: You are retired but still active and publishing. Can you identify a person or persons who shaped your early career and especially your attitude to science, or who most affected your life and work?

Ro: Barton Worthington and Ethelwynn Trewavas (ET). Barton was my boss at the F.B.A. and created the opportunity for me to go to Africa. He also tried to teach me not to waver once I had made a decision! A group of us, including Kate Bertram (née Ricardo), Peggy Varley (Brown), Winifred Frost and myself, all worked on fish in Africa and were irreverently dubbed 'Barton's fishwives'. It was terrific fun and I owe a lot to them all.

ET taught me an enormous amount about ichthyology and taxonomy, and provided inspiration and companionship over many years. I also have a great respect for her meticulous way of working. At the F.B.A., where we lived in Wray Castle on Windermere, early influences included members of Council who were distinguished freshwater biologists who visited us to guide the progress of the work. Among them I should mention in particular W.H. Pearsall, who taught me about successions in lakes, J.T. Saunders from Cambridge, and P. Buxton, who had important African connections. Many limnologists from overseas also stayed there. I realise now how very fortunate we were to meet so many of the leading freshwater biologists from many countries in such an informal way. In Africa, Bobby Beauchamp, my Director of the E.A.F.R.O., and his wife Kitty, were splendid mentors and friends.

MB: Why did you specialise in research on fishes?

Ro: Because at that time they would not have a women as an entomologist in the Colonial Service! Tropical fisheries being a new department, they were not so fussy. My father had fishponds and kept tropical fish, and I never wanted to have anything to do with fish! But as a university student in 1941 I went to the F.B.A. at Windermere on an Easter course and was then invited back, with David Le Cren and some others, to work on silver eel migrations. We had an hilarious time cycling around the Lake District during the wartime blackout, manning eeltraps on various rivers. We found that the phases of the moon and water levels controlled the timing and sizes of the migrations. We also constructed an artificial river in the cellar of Wray Castle to calculate the speed at which eels could swim. I was also sent to the River Severn to look at the largest elver runs in Europe. Eventually in 1945 I was offered the opportunity to go to Africa to complete the Lake Nyasa (Malawi) fisheries survey that had been interrupted by the war; Kate Ricardo was to have done this, but she had started a new life after marrying during the war. So ichthyology became my way of reaching Africa!

MB: As a young woman scientist, you joined a male-dominated profession. Were you discriminated against?

Ro: I always saw myself as an individual of *Homo sapiens,* a member of the human race, not as male or female. In 1945 the Colonial Service would not employ a female entomologist, hence the change to fish. (Later I had the pleasure of seeing a female friend appointed as an entomologist to work on termites in Africa – it was rumoured that Fisheries, when consulted, had said nothing 'untoward' had happened as a result of appointing a female, i.e. Ro). In 1954, when a marriage-bar was still in force, I had to resign from the Overseas Research Service. I did not contest this as I was due to leave East Africa anyway. I was much more interested in being free to study biology and publish my findings than being concerned with the status of a job. I had in fact been offered the Directorship of J.R.F.O. (The Joint Fisheries Research Organisation in Central Africa) but turned this down as I was happy with the work at E.A.F.R.O., and Peter Jackson was doing a good job at J.F.R.O. anyway. This showed that I was accepted even though I was a female.

In Malawi I think I benefited as a woman as I was able to communicate informally with people in fishing villages who were perhaps apprehensive about giving information to more official-looking males. The Malawian people were very friendly and helpful. Although they sometimes wondered whether I was male or female, I was accepted as a 'fundi' (= one who knows) on fishes and this helped to overcome any biases the local people had about women in fisheries.

MB: You published numerous scientific papers and books. Which of these are your favourites?

Ro: Probably the work on the Rupununi floodplain in South America, as no-one had worked there before. It was great fun working with the Amerindians, sleeping in hammocks at night, watching them shoot fish with bows and arrows, and being part of a traditional life style. We also witnessed splendid fish migrations and I learned a great deal about natural history.

It was also fun to apply ideas on speciation, evolution and behaviour gained in Africa to South American fishes. The 'Speciation in tropical environment' symposium (1969) was rather an important landmark as we succeeded in putting across ideas that (a) tropical ecology demands some different approaches from those applied in the temperate zone, and (b) that fish studies could make a significant contribution to ecological theory. Most early dogma on tropical biology was based on bird and insect studies and many ecologists were then unaware of the huge amount of data available on fishes. We were able to bring these data to their attention.

MB: I presume that the publications you mentioned contributed significantly to human knowledge. Would you mind explaining in what way?

Ro: It was pioneering work. In Africa we had always followed in someone else's footsteps. It was also fun finding parallels and convergences with our previous work in Africa.

MB: Scientists often dream about publishing a popular bestselling book that would reveal to non-scientists the excitement and joy of our work. What did you do about such dreams?

Ro: The manuscript of a popular book called *Land of Waters,* which describes my natural history travels in British Guiana, is in a suitcase under my spareroom bed awaiting further attention! I also have plans for a book on *African Lakes Revisited* or a more general, fun one... which should I tackle first?

MB: I would like to see you publish all of them, as they all promise to be fascinating accounts. Changing the subject, did you ever have contemplative ideas which go beyond ichthyology?

Ro: No, not really. I have grappled with various concepts to do with evolution, speciation and ecology, but I cannot claim to really belong to any particular philosophical school.

MB: If it were possible to rewind the tape of life, which part would you choose te replay once, or over and over again?

Ro: Most of it! Probably the fieldwork at the F.B.A., in Africa and in South America.

MB: In the light of what you have just said, are you content with most of your life or would you like, given the opportunity, to edit some part out and add some new parts?

Ro: Probably not, as all the bits are necessary to make up the whole (my ecological training is coming out!). I can see now that I should have had a much more critical and experimental approach to many scientific problems.

MB: Would you like to add any comments?

Ro: A very important facet of my scientific life is that I have made good friends with so many scientific colleagues and ex-students over the years. I have also benefited greatly from the stimulation of numerous scientific meetings and conferences as I did not belong to any particular university department or discipline. My fields of interest cross taxa and disciplines, and it was therefore very valuable to interact with a wide range of colleagues and friends.

MB: Ro, thank you very much for your cooperation – this has been great fun.

Environmental Biology of Fishes **41** 89–114, 1994

The life and work of Eugenie Clark: devoted to diving and science

Eugene K. Balon
Institute of Ichthyology and Department of Zoology, University of Guelph, Guelph, Ontario N1G 2W1, Canada

Key words: Biography, Marine biology, Ichthyology, Diving, Coral reefs, Cape Haze Marine Laboratory, Mote Marine Laboratory, Florida, Poeciliids, Plectognaths, Red Sea, University of Maryland, Shark repellent, Sleeping sharks, Submersible dives

Synopsis

Eugenie Clark is an ichthyologist with a talent for communicating about marine life. Her life had three principal periods, (1) studies under Charles Breder, Carl Hubbs, Lester Aronson and Myron Gordon, (2) directorship of the Cape Haze Marine Laboratory sponsored by the Vanderbilts, and (3) professorship and inspired teaching at the University of Maryland. Genie proved that sharks have surprising learning abilities and that, contrary to popular opinion, none are vicious killers. During her studies on reproductive behavior, territoriality, and ecology of tropical marine sand-dwelling fishes of the Caribbean and Red seas, among many other phenomena, she discovered the cross-fertilizing hermaphrodite *Serranus subligarius*, the Moses and peacock soles producing toxins that repel sharks and other predators, and sharks 'sleeping' in underwater caves in Mexico and Japan. She combined a love for swimming and diving with the study of marine fishes – from hard-hat diving and snorkeling to using SCUBA and submersibles. Professor emerita since 1992, she has ridden whale sharks and participated in dives using submersibles to 3 600 m depths. She is a recipient of over 25 honors and awards, participated in 24 television specials, and the current IMAX film on sharks. She is the author of the *Lady with a Spear* and *The Lady and the Sharks* which are of considerable popular fame.

From childhood heroes to the blowfish incident

'When I was a child William Beebe was my hero', exclaimed Eugenie Clark after one of her recent deep-sea dives in a submersible, '. . . I used to read about him going down in the bathysphere and I wanted to do that too; I told my family I would like to go down and be like William Beebe. They said maybe you can take up typing and get to be a secretary to William Beebe or somebody like him. I said, I don't want to be anybody's secretary! I want to be like William Beebe going down . . . and I don't believe it, here I am doing just that . . . in the same place, the same place Beebe went down more than 50 years ago. It's so fantastic, it is a dream come true, it really is' (recorded on video *A Half Mile Down*).

Eugenie Clark has probably SCUBA dived longer than any woman and seen more whale sharks underwater than anyone. After 70 she rode whale sharks off Australia and Mexico and ridden one to 60 m depth. Like her hero William Beebe (Berra 1977) she became a good naturalist, writer and explorer.

Eugenie Clark, called Genie by most, was born in New York on 4 May 1922. After a distinguished scientific career of nearly 50 years, four husbands and four children, she still looks at least 25 years younger. (Sea water, the primordial brine of life and a cheerful disposition, preserve.) On 6 December 1993 she was presented with The Franklin L. Burr Award by the National Geographic Society for her work which has opened the world of life in the sea to

Fig. 1. Genie (right) and her mother Yumiko in 1926.

the general public; the most recent of many awards I will run out of space to mention.

It is not a simple task to write a short 'life and work' of someone who did it very eloquently herself previously in two famous autobiographical books (Clark 25, 58)*, whose profiles appeared in book chapters from 1967 to 1993 at least 30 times (e.g. Ellis 1975, Emberlin 1977, Stein 1982, Stacey 1990, Burns 1992, Samarrai 1992), and whose biography was produced for children by Ann McGovern (1978).

'My own interest in sea life began', wrote Genie in *The Lady and the Sharks*, 'when I was in elementary school in New York. My American father [Charles Clark] died when I was a baby, and my Japanese-born mother (Fig. 1) was working at the cigar and newspaper stand in the lobby of the Downtown Athletic Club. On Saturdays, while she worked, she left me nearby in the old New York Aquarium at Battery Park, where I spent many hours watching the fishes. Afterward we usually went to eat at a charming little Japanese restaurant, Fuji, and gradually become good friends with the owner and cook, Masatomo Nobu, who later became my stepfather (Fig. 2). I was brought up on the Japanese side of my family, but no one, except anthropologists who are quick to spot my Mongolian eyefold, ever thinks I'm part Japanese.

I knew more about produce from the sea than any of my schoolmates, and my reports in school, from kindergarten on, amused and shocked my classmates and teachers. I told them how we ate with chopsticks, had rice and seaweed for breakfast, raw fish, octopus, and sea urchin eggs for supper, and cakes made from sharks. I was the only student of Japanese ancestry in the school where I grew up, in Woodside, Long Island.

Nobusan often visited my family (Grandma Yuriko, Uncle Boya, and Mama) and always brought us some special Japanese delicacies from his restaurant. He already seemed part of the family when he became my stepfather, at the time I graduated from Hunter College. I had majored in zoology. Since my first visits to the Aquarium at Battery Park, I had wanted to be an ichthyologist . . .' (Clark 68, pp. 3–4). Her family, especially her mother, supported a

* See the 'Lifetime list of publications'.

Fig 2 Genie and her stepfather Masatomo Nobu in his trailer in Sarasota, Florida where he lived during part of his retirement Photograph by E K Balon, 1979

Fig 3 Teenage Genie during the field course at Douglas Lake in 1940

child's dream they could hardly understand then (Clark 25).

She went to Hunter College in New York and during summers 1940 (Fig. 3) and 1941 she took field courses at the Biological Station of the University of Michigan on Douglas Lake. Her enthusiasm there and insatiable reading led to a letter to the editor of Natural History, her first publication (Clark 1). She graduated from Hunter College as Bachelor of Arts in Zoology in 1942, and the same year married 'a handsome pilot' Hideo (Roy) Umaki (Fig. 4). Her first marriage lasted seven years but Roy was drafted overseas most of the time.

While the Second World War was on, the only job she could find was as a chemist at the Celanese Corporation of America in Newark, New Jersey (1942–1946). After applying to Columbia University's graduate program and receiving a cold reception as a woman intruding into men's territory, she enrolled at New York University, and in the evenings and weekends was a swimming instructor at the Shelton Hotel's Athletic Club in New York City. While 'flunking' her course in endocrinology that induced sleep rather than lecture notes, she never failed to sit wide awake in ichthyology. It was given at the American Museum of Natural History (AMNH) by Charles M. Breder, Jr. (Fig. 5), curator of the Department of Fishes and former director of the public aquarium at Battery Park. The exhibits and collections at AMNH added to the stimulating atmosphere in which Genie received her Master of Arts degree in 1946. Her thesis, supervised by Charles Breder, dealt with the 'puffing mechanism' of blowfishes. Breder considered Genie's 'results and anatomical drawings' of the plectognath digestive system worth publishing. 'Later when he combined my final Master's thesis with the publication of his own studies', wrote Genie (Clark 25), 'and my name appeared as co-author, my pride was as inflated as the blowfish' (Breder & Clark 2).

From the Hubbs phenomenon to her first book

Genie was introduced to Carl L. Hubbs at the an-

Fig. 4. Genie with her first husband Hideo on Oahu, Hawai'i in 1947. Photograph by Bertha Kon.

nual meeting of the American Society of Ichthyologists and Herpetologists in Pittsburg, and was invited the same year to join him at the Scripps Institute of Oceanography, La Jolla, as his part-time research assistant (Fig. 6). The jovial Hubbs, a good swimmer as well as entertainer, was dead serious about his work on fishes; Genie had to count meristic characters of hundreds of cottid specimens as part of a larger design to understand intraspecific variation and speciation. At the same time she took his course in Marine Vertebrates and other courses which required ship time and handling of oceanographic equipment.

Fig. 5. Genie with Charles Breder and his wife Priscilla at their retirement home in Florida in 1979. Photograph by E.K. Balon.

Carl Hubbs caught a swell shark at sea and gave it to Genie to investigate, one of the rare other fishes outside plectognaths equipped with the ability to blow itself up. She found that the puffing mechanism of swell sharks is based on the highly distensible walls of the stomach, a part of the digestive

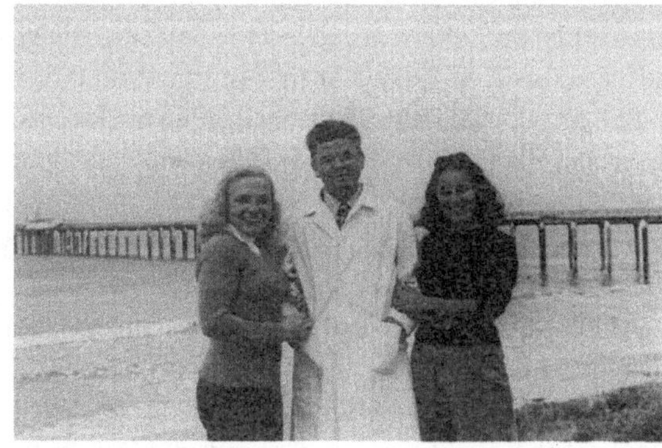

Fig. 6. Carl L. Hubbs flanked by graduate student Betty Kamp and research assistant Eugenie Clark at the Scripps Institution pier in La Jolla, 1946.

system, as in blowfishes. While it was easy to understand why a small puffer would pump itself full of water for defense, it remained a mystery for the swell shark to do the same thing (Clark 3).

Hubbs gave Genie her first opportunity to use a face mask for observing fishes underwater and to walk on the bottom of the sea in a diving helmet. This experience and the mishap with a faulty hose, as well as the advice never to dive in a hard hat after eating hamburger with onions, became a classic story picked-up by most of Genie's profile writers, none of whom, however, managed to convey the true tale the way she did (Clark 25, pp. 27–32). Next, via Hawai'i on an aborted expedition to the Philippines, Eugenie returned to New York. Myron Gordon at the AMNH offered her a research assistantship and the chance to work toward a doctoral degree. She stated later most succinctly: 'My research problem at the museum centered about the reproductive behavior of platies and swordtails, the same kind of fishes that were in my first home aquarium. In the next few years my project was supported by the Department of Animal Behavior (interested in the sexual behavior of these fishes), by the New York Zoological Society (interested in the genetics of these fishes), and finally by the Atomic Energy Commission (interested in sperm physiology of fishes in general). This took care of my needs for living material, laboratory space, and all my expenses. I was very lucky.'

During three years at the museum in collaboration with Myron Gordon and Lester R. Aronson, Genie developed a micropipette method to collect and transfer sperm into living virgin female platies, swordtails and guppies. Her meticulous studies on sperm competition in poeciliids led to her Ph.D. thesis in 1950 and a series of publications on this topic (Clark et al. 4, 5, 10, 27, Clark & Aronson 12, 13, Clark & Kamrin 14, Aronson & Clark 17). She kept in touch with marine fishes at the Biological Station in Woods Hole two summers, and later for several months at the Lerner Marine Laboratory on Bimini, West Indies. Charles Breder was now the director of Lerner Marine Lab, and he wanted her to study local plectognaths (Clark 9). In her spare time she (Fig. 7) assisted Roger Sperry, a visiting neuro-anatomist in his experiments on visual discrimination of gobies (Sperry & Clark 8).

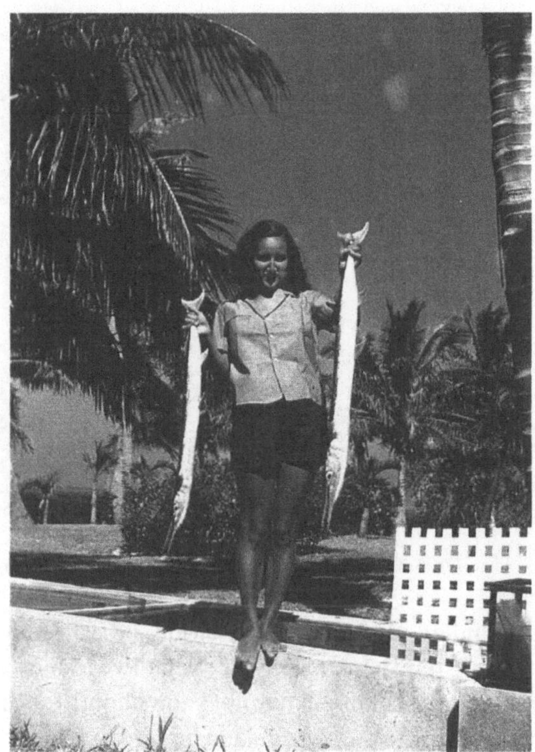

Fig 7 Genie at the Lerner Marine Laboratory on Bimini in 1948 Photograph by R Sperry

As she finished the last experiments for her thesis, she was accepted for Scientific Investigation in Micronesia by the Pacific Science Board and the Office of Naval Research. They were especially interested in poisonous fishes and her studies of plectognaths and her chemistry background proved important assets. Genie took off on 17 June 1949. In Micronesia she perfected her skin-diving skills and learned spearfishing from the Palauan fisherman Siakong (Clark 26). The account of this four months trip forms the major part of her charming first book *Lady with a Spear*. Her next trip to the Red Sea is also included in this first popular book. To illustrate let me use just one incident I like most: In the Red Sea, rotenone collections occasionally made in shallow waters or tidal pools attracted her houseboy, Shehat, otherwise a sea-fearing non-swimmer, who usually ventured only to the shallowest edge. Amazed by the multitude of fishes chased out of

Fig. 8. Genie with her second husband Ilias on the front steps of the AMNH (May 1959) before her Ph.D. defence.

their hiding places by the piscicide, he called the others: 'Come here quickly. See what I've found – a creature Allah himself may not know about! Take it immediately to the *doctora* and be sure to tell her I discovered it' (Clark 25, p. 204).

In 1950 she married in New York's civil court her second husband, then an intern in orthopedics, of Greek origin, Ilias Themistokles Papakonstantinou (Fig. 8), later changed into simplified Americanized version Konstantinu. An article by H.A.F. Gohar, then the director of a lonely Red Sea Marine Biological Station of Fouad University at Ghardaqa, on clown fish and sea anemones caught her attention. She wrote Gohar and was offered all the station's lab facilities. She applied for and got a Fulbright Scholarship to study plectognaths and other poisonous fishes in the Red Sea (Fig. 9). She arrived in Egypt for Christmas 1950. Ilias Konstantinu came for an extended visit towards the end of this first Red Sea experience, and they married officially for the second time in 1951 in a Greek Orthodox church near Khan Khalili bazaar in Cairo. Ilias's mother, related to the well known Onassis family, arrived from Athens for this wedding. During their honeymoon at Ghardaqa, Ilias helped Genie with her work, became quite interested in sharks, and became an enthusiastic spearfisherman.

Ilias returned to the U.S. to take an internship in the Orthopedics Division of the Buffalo General Hospital in Buffalo, New York. Genie soon joined him and wrote her *Lady with a Spear*, during 1952, the same year their first child, Hera was born. The popular, autobiographical book published in 1953 about her resarch and adventures in the Pacific and the Red Sea went through several editions in English and Japanese, and Italian, Danish, German, Swedish, Norwegian, Arabic and Braille. A multitude of excerpts was published in school textbooks. 'I began to realise', said Genie later to one of her many interviewers (Samarrai 1992), 'I had a talent for communicating about the natural world. I came to see that it would be my life's work.'

The Vanderbilts and Cape Haze Marine Laboratory

After publication of her book, Genie became an instructor in the Biology Department of Hunter College, New York, where Ilias had moved to complete his medical training (Fig. 10). She officially became a research associate of the AMNH departments of Animal Behavior and Ichthyology (1954–1981), worked as pharmacologist at the Nepera Corporation in New York and travelled to give lectures at high schools, colleges and universities.

Let her tell you what happened next: 'My introduction to the west coast of Florida was in 1954 after

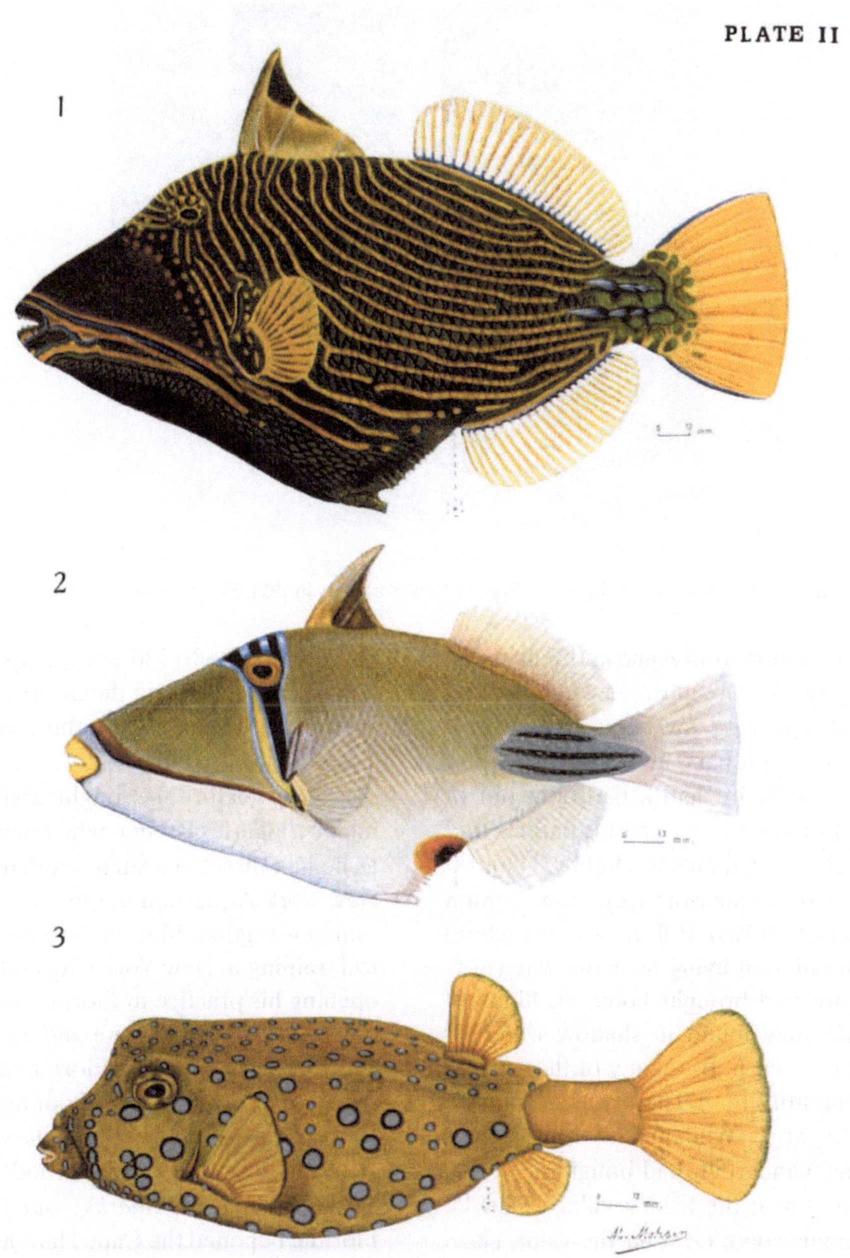

Fig 9 Color plate II from Clark & Gohar (23) with the beautiful illustrations by Moawad Mohsin The amazing high quality color printing was arranged by Gohar Bey in Cairo in 1953 The plate depicts (1) *Balistapus undulatus* (2) *Rhinecanthus assasi* and (3) a juvenile of *Ostracion cubicus*

Fig. 10. Eugenie Clark at Hunter College in New York City in 1954. Photograph by E. Hartmann.

I accepted an invitation from Anne and William H. Vanderbilt to give a lecture in Englewood, Florida. Mrs. Vanderbilt had read my book *Lady with a Spear* and talked her husband into reading it. Their ten-year-old son Bill, Jr., had a bedroom full of aquariums, as I did at his age, and his parents had become fascinated with their son's hobby. Their estate stretched across Manasota Key, from Lemon Bay to the Gulf of Mexico. Bill, Jr., and his school chums, like all children living near the water, explored the shore and brought home all kinds of strange sea life they found in shallow water or washed up on the beach. But many of these items they couldn't identify (. . .) There was no marine biologist in the area. William and his brother, Alfred Gwynne Vanderbilt, had bought a 36,000-acre tract of land near the fishing village Placida, southeast of Englewood, (. . .) on the Cape Haze peninsula'. As we know from previous sections, Genie had experience in marine biology as a visitor and researcher at many marine laboratories. As she wrote in her second book, 'I never dreamed I would have the opportunity to start a laboratory from scratch myself, but later on the evening of my lecture in Englewood I learned that the Vanderbilts had invited me to Florida for just this reason . . .' (Clark 68, pp. 2–4 and the insert of 4 maps).

Eugenie Clark became the founder and executive director of Cape Haze Marine Laboratory, of course, not before she sought advice from her early mentor, Charles Breder, who knew these parts well as he had directed a small biological station of the New York Aquarium in the area in the 1930s. The timing was good. Ilias had just completed his medical training in New York City and liked the idea of opening his practice in Florida 'and of bringing up our young daughter Hera and an expected second child in a beach house in Florida rather than continuing to live in our twelfth-floor apartment on West End Avenue (. . .) In early January 1955, six months after my first visit to Englewood', writes Genie in *The Lady and the Sharks*, 'our family arrived in Florida. I opened the Cape Haze Marine Laboratory as soon as I arranged for a babysitter for two-year-old Hera and her month-old sister, Aya' (Clark 68, p. 10).

A local fisherman, Beryl Chadwick, whom the Vanderbilts recommended to her on her first visit was her co-worker. Beryl build a small wooden building 4 by 6.5 m to serve as a laboratory, a dock, and a 7 m craft was left at her disposal. The details of Genie's 12-year long sojourn in Florida are given in her second popular book (Clark 68). The central activity of this laboratory, acquisition and maintenance of sharks for visiting scientists, and research on sharks, started nearly immediately.

The day after the Konstantinu family had arrived, Genie received a telephone call from John H. Heller, the director of the New England Institute for Medical Research, who needed fresh shark livers for his research and could not find any. Beryl made a shark long line. The Heller's arrived on 24 January, and the next morning they had two large sharks. A week later Genie, John Heller and his wife Terry dissected twelve dusky and sandbar sharks and the reputation of the laboratory was established in the local newspapers.

However, the laboratory's initial mandate was to study the local marine fauna which, as Genie found later, had been well utilized as educational material and shipped all over the States in the 1930s. She soon met Johnny Bass and his wife Barbara who inherited the Bass Biological Station in Englewood from his father. This biological supply house, by then closed, was Florida's equivalent to Ed Ricketts' laboratory on Cannery Row in Monterey, California (Steinbeck 1960). One day the Basses opened the old building for Genie who discovered valuable books and journals, laboratory glassware, chemicals and instruments under cobwebs and dust. Johnny's mother even offered Genie old records her husband had kept. Genie started to understand how useful the supply house had been to scientists, teachers and students while it operated. When she realized that others had been there before, she started to have doubts about the contribution she could make. Even sharks had already received attention here when Stewart Springer was the last manager of the Bass Station before its closure in 1940 (Burgess 1991). During her first mask and snorkel dives offshore she discovered, among other things, a number of small, large-bellied groupers full of eggs with no males around.

During Genie's first year in Florida her parents moved their Japanese restaurant *Chidori* from New York to Grove City, nearly halfway between the Konstantinu residence and the Cape Haze Lab. Idyllic life set in at their beach house in Englewood. By the end of the first year of its existence the laboratory at nearby Placida entertained 28 visiting scientists, hordes of school children and associates. The building expanded, a shark pen was built. Genie started to share a secretary with the Vanderbilts as the Laboratory became a nonprofit organization and the paper work mounted.

On 25 May 1956 Themistokles Alexander Konstantinu was born. When Tak, as he would later be called by his family, was a few weeks old Genie tried out the Cape Haze Marine Lab's new aqualung. She went SCUBA diving to solve the puzzle of the small grouper, *Serranus subligarius*. The fish produced activated eggs without males. The large ovaries full of eggs had white pieces of tissue Genie thought to be fat. Charles Breder on one of his visits suggested she put a piece of that 'fat' in sea water under the microscope. Wriggling multitudes of spermatozoa were revealed. The fish was a functional, self-fertilizing hermaphrodite!

Genie persisted in observing the fish in aquaria and in the sea and discovered a most fascinating style of reproduction: The hermaphroditic grouper spawns in pairs each day from April to September at receding tide and released eggs at each S-curved mating snap. The fish which leads before the mating snap plays the role of a female and is unbanded, dark colored, whereas the fish which follows is strongly banded and performs as a male. At the apex of the snap the unbanded individual becomes banded but in reverse, negative-like banding and soon the banded one becomes unbanded and leading, thus reversing roles. The fish is a simultaneous hermaphrodite all right, but normally cross-fertilizes.

Genie published these findings in 1959 (Clark 35) and exactly 20 years later we arranged to meet at the Mote Marine Laboratory in Sarasota, the successor of her Cape Haze Lab, to try to resolve the last puzzle of this unusual reproductive behavior. She arrived from College Park, Maryland, where she has been professor of zoology for some time, and I from Guelph with two carloads of equipment, technician Marilyn White and graduate student Joan Cunningham (Fig. 11). After setting the incubators and special microscopes (as described in Balon & Flegler-Balon 1985) for detailed studies of

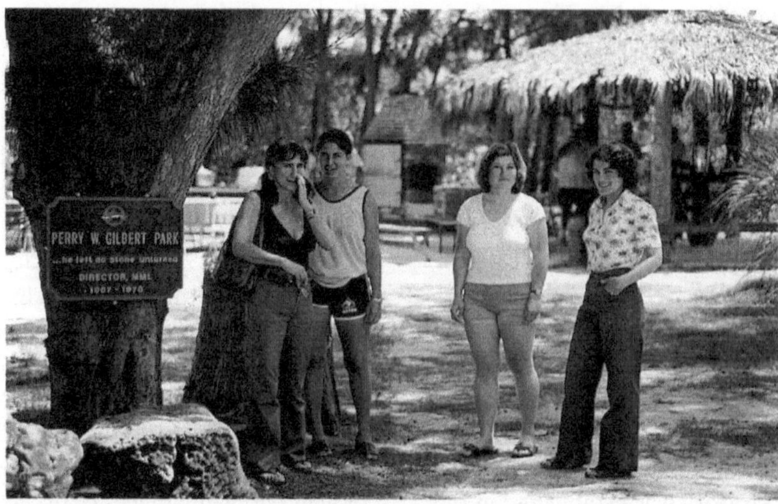

Fig 11 Genie, her daughter Aya, my technician Marilyn White and graduate student Joan Cunningham at the Mote Marine Laboratory on 27 April 1979 Photograph by E K Balon

development of the self-fertilizing *S. subligarius* compared to the cross-fertilizing one, Genie failed to find any of the previously so abundant fish. We SCUBA dived in all places known to her but no *S. subligarius* were found. Thus the puzzle how the progeny of cross-fertilized individuals differ from those of self-fertilizing ones remained unsolved. But a new mystery was added by the absence of this most abundant fish from the waters in that particular area and year; our only chance for the two Eugenes to co-author a paper was lost.

From Placida to Sarasota

Because of the continuous success in keeping various sharks in captivity and the publicity that followed, Genie was soon known as the 'shark lady', even before her experiments with them started. The Cape Haze Marine Laboratory expanded with the additional financial help from the National Science Foundation, the Office of Naval Research, the American Philosophical Society, and the Selby Foundation. Many foreign visitors arrived in addition to Americans interested in studying sharks.

When Lester Aronson came for a visit, Genie designed with his help the first experiments on learning behavior in sharks. In the midst of these shark conditioning experiments, on 20 October 1958, after her SCUBA diving until the last moments, Nikolas Masatomo Konstantinu, the last child of Genie was born. She later wrote that except for feeling like a 'Lady with a Sphere' she was comfortable and believes that because of her swimming and diving all her labor and births were easy.

Along with raising the four children, Hera, Aya, Tak and Niki, in which she was much helped by her mother and stepfather who ran a Japanese restaurant *Chidori* nearby, Genie continued to work with sharks and took part in frequent distractions. For example, she attended meetings (Fig. 12), went on lecture tours and participated in deep scuba dives into Florida's sinkholes where in Little Salt Springs bones of humans 7 000 to 10 000 years old were discovered (Royal & Clark 41).

The first experiments with sharks had proven that these animals can learn to press an underwater target to obtain food (Clark 36). Now they were trained to visually discriminate between targets of different shapes and colors (Clark 44, 47). Clearly sharks can learn like other vertebrates and are not 'stupid' because they are so-called 'lower vertebrates' (Fig. 13).

Ilias' growing orthopedic practice required him to be close to the Sarasota Memorial Hospital. The family moved to a two-story, Spanish-style house on

Fig 12 Genie meets for the first time Jacques-Yves Cousteau (on her right) and Ed Link at a shark seminar of the Zoological Society of Florida in Miami (1959)

Siesta Key in Sarasota (Fig. 14) and Genie had to drive one hour in each direction to the Cape Haze Marine Laboratory. 'When the path of the Intercoastal Waterway was plotted next to our shark pens, we knew we would have to move the site of the Lab', writes Genie later. 'My mother died from a brain hemorrhage in the summer of 1959. I lost much interest in my work, felt I could no longer handle a full-time job, and thought I should stay home with the children and help my stepfather . . . He had never mastered the English language, and with no other Japanese people around, as there had been in New York, he depended on my mother not just as a wife but as his only close verbal companion . . . I thought it would be better for him to move his restaurant to Siesta Key, within walking distance to our house. He agreed and we busied ourselves redecorating a rented store in Oakes' Plaza on Siesta Key into a Japanese restaurant. Even Ilias, who was busy with his medical practice, helped with the painting' (Clark 68, pp. 191–192).

Genie went through a difficult period, tried to resign as director but ultimately found a fine housekeeper and returned to the Laboratory. It had to be moved With luck and the help of enthusiasts, '$8^1/_2$ acres of the most choice real estate on the south end of Siesta Key' was leased, only a five-minute drive from the Konstantinu residence. With grants from the National Science Foundation and funds from the Vanderbilts and the Selby Foundation modern buildings, docks, and shark pens were built for the Cape Haze Marine Laboratory at its new site on Siesta Key, Sarasato. The Lab moved in the winter of 1960.

The laboratory in its new position attracted even more visiting scientists than before. In spite of the Vanderbilts doubling their contribution, the institution was now constantly short of money. 'As Ilias's medical practice grew, I no longer needed a full salary, and as I found I couldn't balance the Lab's budget any other way, I started cutting my own salary', Genie explained later. Luckily an early-retired executive in Sarasota volunteered as adminstrative assistant.

New discoveries and contributions to science started to come to fruition from visiting scientists working at the Lab. For example, John Heller from the New England Institute for Medical Research and his ever larger teams 'had discovered a remarkable substance in shark livers he named 'restim'

Fig 13 The executive director of the Cape Haze Marine Laboratory in her office (1958) once again defending sharks

(short for RES stimulator), which stimulates the body's reaction to fight and resist diseases many times above its normal ability (. . .) to cause an alleviation and even regression in some types of cancer . . .' (Clark 68, p. 197).

During that time Genie went with her diving children on trips to many Caribbean islands and had some remarkable adventures. In 1964 she introduced them to the Dead Sea and Red Sea. She captured garden eels by squirting formalin into their tubes, to the amazement of other divers who thought it impossible. Next day she dived past the garden eel colony and spotted a strange little fish above the sand holding its position by undulating movements. 'When I approached closer, it dived into solid sand and was gone with no hole to mark where it had entered. It was a rare type of sand diver belonging to a family of fish never reported from the Red Sea (. . .) I put my hand net over what I hoped were the peeping eyes of the fish, plunged my free hand into the sand below, and the fish, *Trichonotus*, jumped into my net. Niki was at the shore and we filled his face mask with water and put the fish in it until we located a bucket. It turned out to be a new species of fish (. . .) I named it *Trichonotus nikii* (Table 1), after the youngest member of our expedition to the Red Sea' (Clark 68, pp. 206–208).

In the fall of 1965 Genie went to Japan, and being invited by the ichthyologist crown Prince Akihito, she brought him a small, trained nurse shark in a hat box, including a portable testing apparatus, as a gift. I advise you to read the amusing story towards the end of her *The Lady and the Sharks*. She was even assigned a geisha girl who later in the evening asked: '«Is it really true that sharks have twice as much fun as we do?» I had to think a moment before I could explain that as far as we ichthyologists knew, it is physically impossible for the male shark to insert more than one clasper at a time during copulation' (Clark 68, p. 223).

While the Laboratory grew and things became more complex, the media attention grew as well, and the fact that a woman was studying the 'fearsome' sharks added spice to the publicity. Ilias in the meantime developed a lucrative orthopedic practice in Sarasota but started to be more and more interested in business, buying oil tankers which were given names of their children. Ilias' obsession with money disturbed Genie greatly. When Chandler Brossard, the first American existentialist and author of 17 books came to interview her for his magazine *Look*, the stage was set and Genie left Ilias. She married Chandler in 1967 and moved north with her children to live again in New York City.

Genie wondered what would become of the Cape Haze Marine Laboratory and felt the best successor in the directorship would be Perry Gilbert, then head of the Shark Research Panel of the American Institute of Biological Sciences and professor at Cornell University, Ithaca. Additional support for a director's salary and for the expansion of the laboratory under such distinguished directorship was needed. William R Mote, who had become wealthy in the transportation business, wanted to establish a marine laboratory and had consulted Genie earlier about co-operation with the Cape Haze Marine

Fig 14 Genie and Ilias Konstantinu with their four children, from left to right, Aya, Tak, Hera and Niki, on the beach in front of their new home at the Point of Rocks in Sarasota (1959)

Laboratory. Eventually, Perry Gilbert became the new director in 1967. Bill Mote moved to Sarasota from New York, acquired new land for expansion (Fig. 11) and gradually 'the Cape Haze Marine Laboratory evolved into the Mote Marine Laboratory'.

The University of Maryland years

Somewhere between the brief marriage to Chandler Brossard, and her appointment as associate professor at the Department of Zoology, City University of New York (1966–1967), visiting professorship at the New England Institute for Medical Research (1966–1968) and joining the Department of Zoology, University of Maryland in 1968, Genie wrote her second popular book *The Lady and the Sharks* (Fig. 15) which was published in 1969. After her fourth (and last) marriage to Igor Klatzo of the National Institute of Health in 1970, dissolved in a

Table 1 Fish species described by Eugenie Clark

Order	Family	Species and subspecies	Reference
Tetraodontiformes	Aluteridae	*Paraluteres arqat*	Clark & Gohar 23
		Brachaluteres baueri fahaqa	
	Tetraodontidae	*Lagocephalus suezensis*	
Perciformes	Trichonotidae	*Trichonotus nikii*	Clark & von Schmidt 63
	Malacanthidae	*Asymetricus oreni*	Clark & Ben-Tuvia 79
	Tripterygiidae	*Helcogramma steinitzi*	Clark 100
		Norfolkia springeri	
		Enneapterygius altipinnis	
		Enneapterygius destai	
		Enneapterygius obscurus	
		Enneapterygius pallidus	
		Helcogramma vulcana	Randall & Clark 143

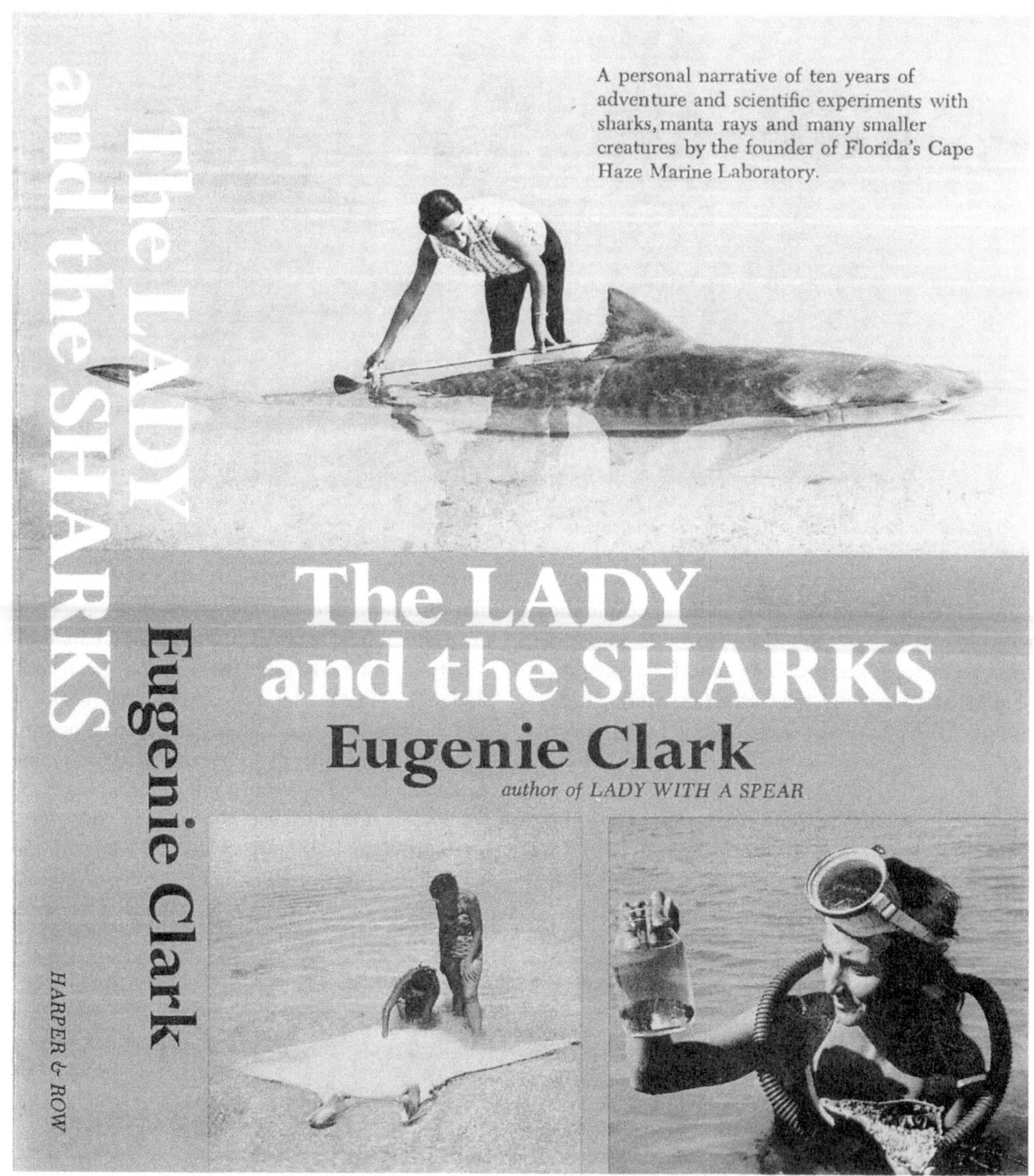

Fig 15 Part of the orıgınal ȷacket of Genıe s second popular bestseller

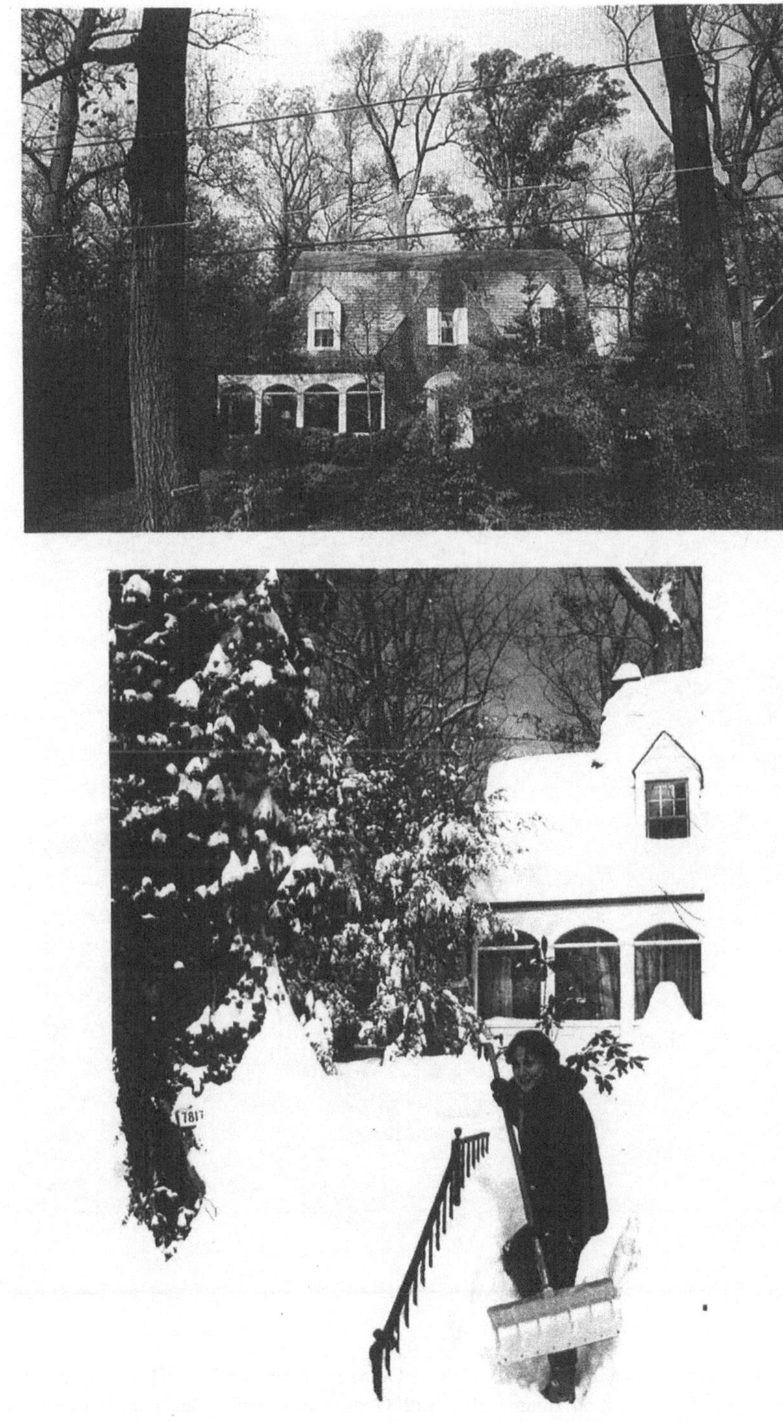

Fig 16 Genie s house off Hampden Lane in Bethesda a – Fall 1993 b – on 19 2 1979 after the largest snow storm in the Washington D C area Genie tries to clear the main entrance Photographs by E K Balon

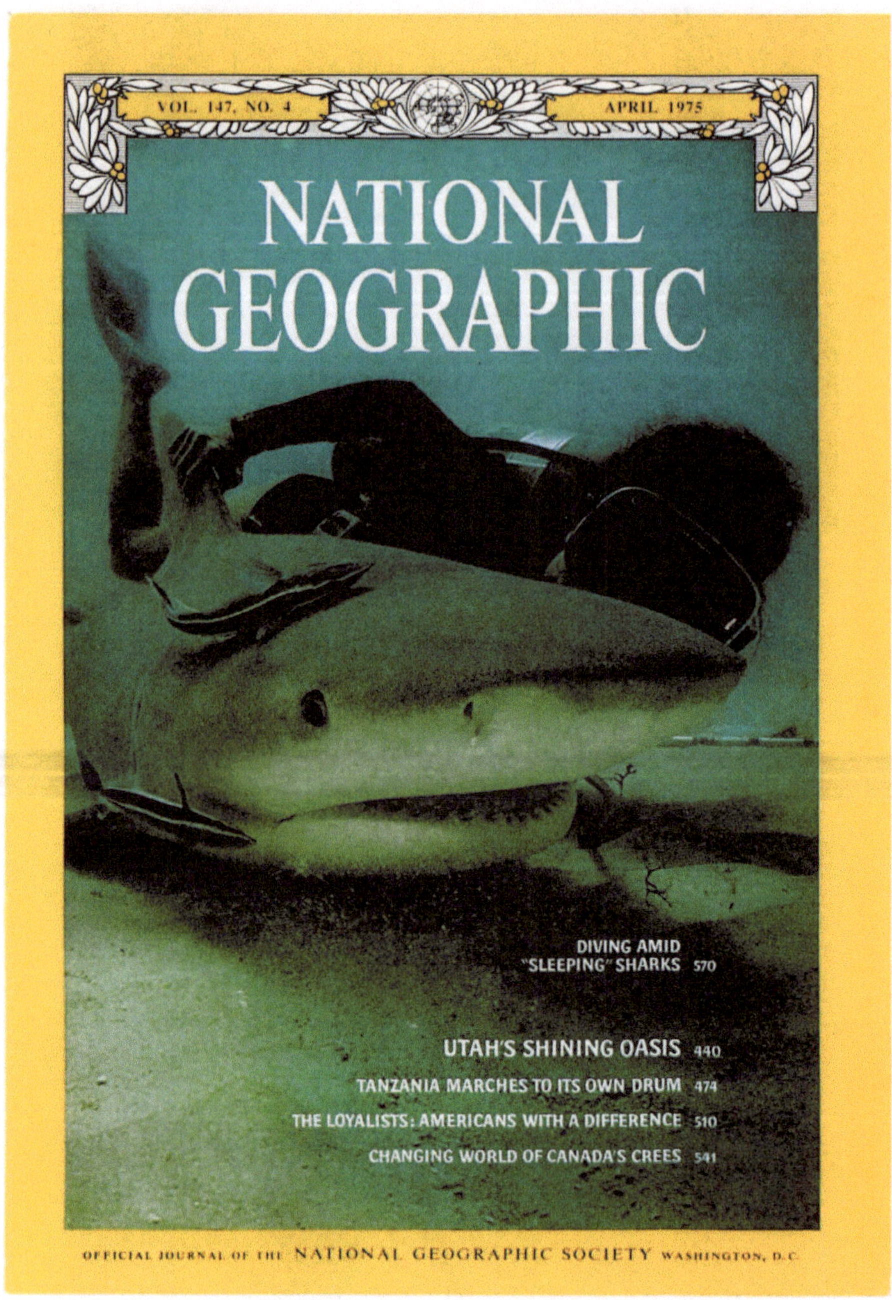

a

Fig 17 a – Cover of National Geographic magazine of April 1975 with Genie examining a shark off Mexico's Yucatan Peninsula b – Cover of National Geographic Research & Exploration, summer 1992, with Genie seen diving at the 'Temple' site in the Red Sea Both photographs by D Doubilet

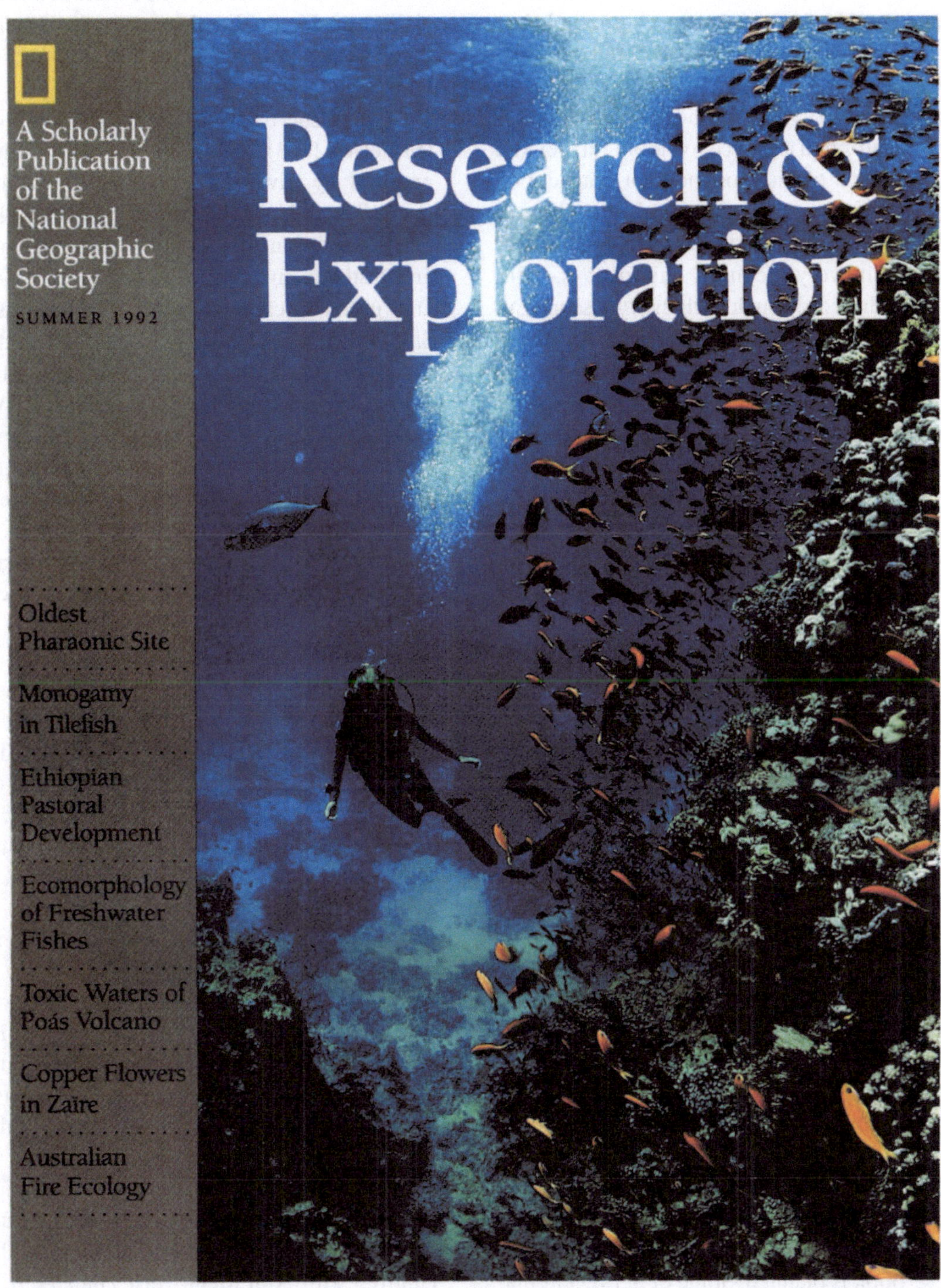
A Scholarly Publication of the National Geographic Society
SUMMER 1992
Research & Exploration
Oldest Pharaonic Site
Monogamy in Tilefish
Ethiopian Pastoral Development
Ecomorphology of Freshwater Fishes
Toxic Waters of Poás Volcano
Copper Flowers in Zaïre
Australian Fire Ecology

b

few years, she settled in Bethesda where she has lived happily alone ever since in a charming house within a picturesque neighborhood of tall oak trees (Fig. 16).

At the University of Maryland, Genie's talent as a communicator had a chance to bloom further and she became 'one of the most popular teachers on the campus', Craig Phillips agrees. 'In 1969 he was Clark's first graduate student. "I remember her classes as being very exciting", Phillips says. "Her enthusiasm shined through with every lesson. Eugenie radiates energy when she talks about fish and her fervor carries over to the student. She's also one of the most gracious and natural people I've ever known." Phillips eventually earned his master's degree in zoology' (Samarrai 1992, p. 18), was director of the National Aquarium in Washington, D.C. for a while and is the illustrator of Genie's latest book, *Desert Beneath the Sea* (McGovern & Clark 138). Genie, after teaching *Ichthyology* (incidentally taught there many years ago by another scholar we honored, Vadim Vladykov, see McAllister 1988) and *Vertebrate Zoology*, introduced her most popular course, *Life in the Oceans* which she presented from 1974 until her retirement in 1992. She also taught *Ecology of Oceans, Marine Vertebrate Zoology*, and just before retirement she started to offer an honors course, *Sea Monsters and Deep Sea Sharks* which she still teaches as professor emerita.

At the University of Maryland she became a full professor in 1973, and along with her teaching career developed an exciting research program carried out in more than 20 countries. Each of her major research discoveries was popularized in an article of National Geographic magazine, 12 in total. She developed close rapport with the executive staff and photographers of the National Geographic Society of which she is justly proud. I guess that the National Geographic articles compensated for the lack of a third popular autobiographical book. She once told me with some aversion that no third 'Lady . . .' book is planned. Some of her recent major scientific papers appeared in the National Geographic Research & Exploraton, and when I tried to solicit some of these papers for this journal she claimed allegiance, free artwork and color reproductions as unbeatable advantages. No wonder that she had been featured on the covers of both National Geographic publications (Fig. 17). She is nevertheless a founding Advisory Editor of *Environmental Biology of Fishes.*

By now she has been to the Red Sea 46 times (Fishelson 1993), with her work beautifully photographed by diver-photographer David Doubilet (Fig. 18) and National Geographic staff photographers. Her teams of highly competent volunteers and students, most of them expert SCUBA divers and underwater photographers (such as David Shen and Ruth Petzold), videographers and observers, even some of her children, study the behavior and ecology of sand fishes with Genie, add valuable documentation to her field studies and are often her co-authors. She has made some wonderful discoveries thanks to her great observational talent and a 'naturalist's nose' I used to tease her about. Her observation talent is complemented by a mastery of SCUBA diving techniques (Fig. 19) and her ability to organize teams of SCUBA divers who assist in her underwater work.

During three diving trips to the Bahamas in February 1986, July 1986 and January 1987 she studied the sand tilefish, *Malacanthus plumieri*, and observed some unusual reproductive behavior (Clark et al. 126, 130). Similar studies on sand fishes were consistently pursued in the Red Sea from 1964 to 1991. Genie observed that the rare tilefish *Malacanthus latovittatus*, in contrast to the polygynous *M. plumieri*, lived in closely bonded pairs, in a monogamous social system (Clark & Pohle 141). In another study on sandfishes, Clark et al. (137) described another social system of a spotted sandperch. *Parapercis hexophtalma* live in social units of one male and two or three females in a stable, guarded territory. Each unit is also a mating group. In one of the many notes to me, Genie complained that «Jane Goodall was allowed to use nicknames for her chimps in scientific papers but I had to call the wonderful 'Boom Boom' as 'fish B'». She rectified this in her book for children with Ann McGovern *The Desert Beneath the Sea*, by naming fish B, what else, *Boom Boom* and his friends *Uncle Albert, Charlie, Fast Freddie, Mr. T.*, and *Elusive Eddie. Fast Freddie* is so fast, he mates with his 3 females in less than 4 minutes (Clark et al. 137, Fig. 11).

Fig 18 Genie with David and Ann Doubilet at the Aburatsubo Marine Park Aquarium in 1979 during the National Geographic assignment to dive for 'sleeping' sharks in Japan's southern islands

Herold & Clark (144) describe close pair-bonding with a monogamous mating system in a tiny Red Sea seamoth, *Eurypegasus draconis*. These rare sand-dwelling fish solved the problem of meeting the opposite sex by staying with a sexual partner for ever. These unusual fish are encased in a skin armor (carapace) which is shed whole every 1 to 5 days. The one page story 13 in the children's book *The Desert Beneath the Sea* describes the creatures admirably.

During these studies on sandfishes, 19 colonies of garden eels *Gorgasia* were observed. One colony on a steep, sandy slope was estimated to contain about 10 000 eels. Mature individuals were in the center and juveniles on the periphery of the ca. 3840 m^2 colony at Râs Mohammed in a density of up to 7 adults eels per m^2 and 14 juveniles eels per m^2, respectively (Clark et al. 134).

Genie's first paper published in this journal was on the toxic Moses sole, *Pardachirus marmoratus* from the Red Sea and the peacock sole, *P. pavoninus* from Japan and adjacent seas (Clark & George 98). During the study of the garden eels in the Red Sea in 1972 Genie discovered the shark-repelling property of the small sole and returned in 1973 to study this in detail. The Moses sole, as it is called locally, releases from glands in the base of fin rays a milky fluid that repels sharks and other predatory fishes (Clark & Chao 78, Clark 80, 88). She later dis-

Fig 19 Genie watching a sand-dwelling pufferfish in the Red Sea (1982) Photograph by D Doubilet

Fig 20 Genie came to Toronto in February 1993 to open the shark exhibit at the Royal Ontario Museum After her lecture she came to my favorite restaurant for a tuna sashimi Photograph by E K Balon

covered the same ability in the peacock sole (Clark 91). Moses sole on a line were observed to have an unusual effect: 'First the sharks swam toward the Moses sole with their mouths open, ready to gobble the little fish. Then, with their jaws still wide open, the sharks jerked away. They thrashed and leaped about the tank, shaking their heads wildly from side to side. All the while, the Moses sole kept swimming, as if nothing unusual was happening' (McGovern 1978). Genie concluded that if the poison can be synthesized a shark repellent may eventually become available that will protect swimmers and divers in 'shark-infested waters'. (Later she realized that such a protection is not practical or needed.) When I first visited Genie in her office at College Park, a telephone interrupted our conversation as someone was inquiring when the shark repellent will be available. I heard her answering that since it took about 40 years to synthetize insulin after its discovery, how can anyone expect the Moses sole toxin to be available in a synthetic form so soon, bearing in mind the much greater significance of insulin. I was very impressed by her answer and even more by her patiently answering such queries day after day.

The 'sleeping' shark story is equally captivating. Reacting to some local divers tales transmitted by an old acquaintance Ramon Bravo, Genie went in 1974 to Mexico but saw none. In 1975 she returned to Isla Mujeres off the Yucatan Peninsula with her 19 year old daughter Aya, student Anita George and David Doubilet to investigate again. The 'sleeping' sharks had been discovered in a cave 20 m below the sea surface. Ultimately, after 99 dives she has seen many open water sharks, mainly requiem, lying motionless on the bottom of caves. They permitted Anita, Aya and Genie close inspection and even handling. Until this discovery, streamlined sharks were thought to need non-stop swimming in order to stay alive. It was energetically

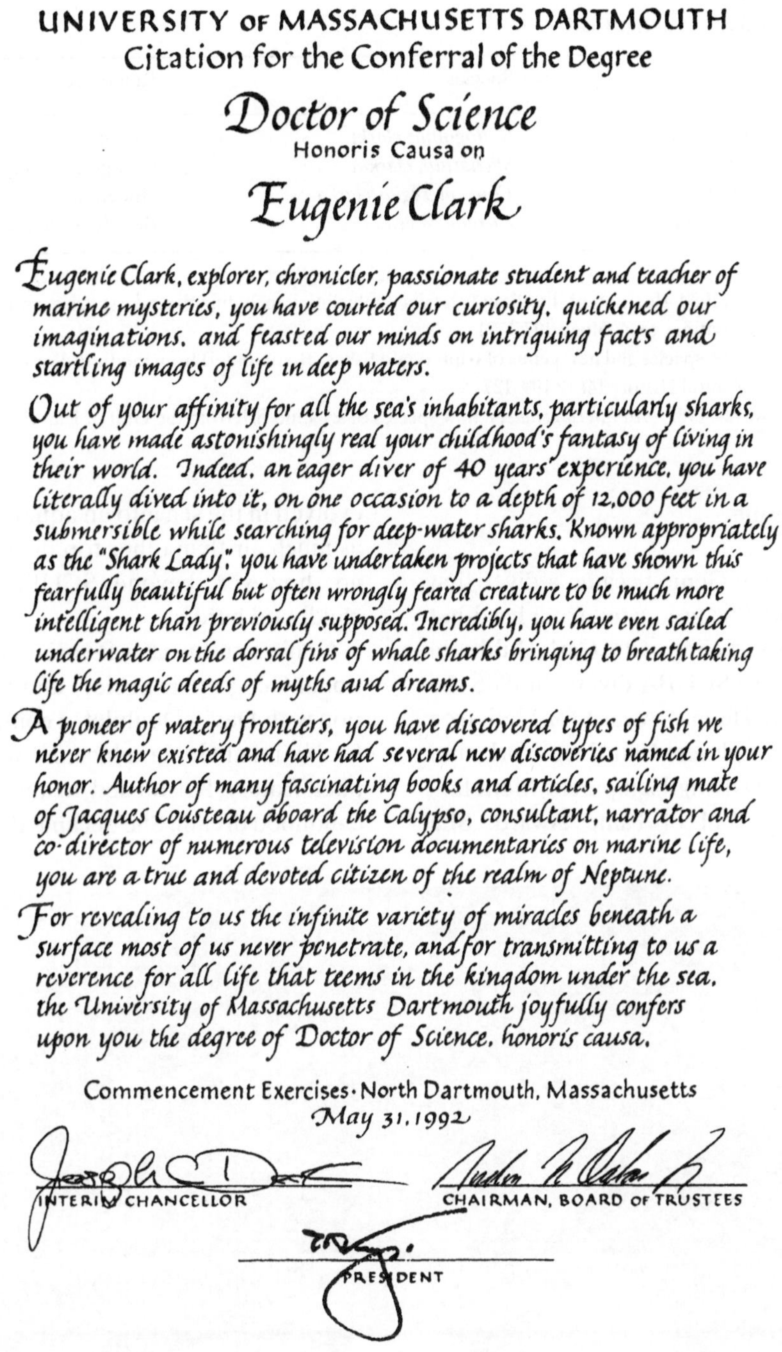

UNIVERSITY OF MASSACHUSETTS DARTMOUTH
Citation for the Conferral of the Degree

Doctor of Science
Honoris Causa on

Eugenie Clark

Eugenie Clark, explorer, chronicler, passionate student and teacher of marine mysteries, you have courted our curiosity, quickened our imaginations, and feasted our minds on intriguing facts and startling images of life in deep waters.

Out of your affinity for all the sea's inhabitants, particularly sharks, you have made astonishingly real your childhood's fantasy of living in their world. Indeed, an eager diver of 40 years' experience, you have literally dived into it, on one occasion to a depth of 12,000 feet in a submersible while searching for deep-water sharks. Known appropriately as the "Shark Lady", you have undertaken projects that have shown this fearfully beautiful but often wrongly feared creature to be much more intelligent than previously supposed. Incredibly, you have even sailed underwater on the dorsal fins of whale sharks bringing to breathtaking life the magic deeds of myths and dreams.

A pioneer of watery frontiers, you have discovered types of fish we never knew existed and have had several new discoveries named in your honor. Author of many fascinating books and articles, sailing mate of Jacques Cousteau aboard the Calypso, consultant, narrator and co-director of numerous television documentaries on marine life, you are a true and devoted citizen of the realm of Neptune.

For revealing to us the infinite variety of miracles beneath a surface most of us never penetrate, and for transmitting to us a reverence for all life that teems in the kingdom under the sea, the University of Massachusetts Dartmouth joyfully confers upon you the degree of Doctor of Science, honoris causa.

Commencement Exercises · North Dartmouth, Massachusetts
May 31, 1992

INTERIM CHANCELLOR

CHAIRMAN, BOARD OF TRUSTEES

PRESIDENT

Fig. 21. The citation for the conferal of the Doctor Honoris Causa degree from the University of Massachusetts.

too costly to pump water over gills. She concluded that freshwater coming out of the bottom of certain caves attracted the sharks. The grip of their ectoparasites loosen and enable the accompanying remoras to do a better cleaning job (Clark 77, 84, 108). This must be worth the higher energy price incurred by pumping water over the gills. She later went to study the same phenomenon in the caves of southern Japan with Anita (Clark 92).

Table 2. Fish species named after Eugenie Clark.

Order	Family	Species	Reference
Perciformes	Gobiidae	*Callogobius clarki*	Goren[1]
	Clinidae	*Sticharium clarkae*	George & Springer[2]
	Tripterygidae	*Enneapterygius clarkae*	Holleman[3]
	Sciaenidae	*Atrobucca geniae*	Ben-Tuvia & Trewavas[4]

[1] Goren, M. 1978. A new gobiid genus and seven new species from Sinai coast (Pisces: Gobiidae). Senckenbergiana biol. 59(3/4): 191–203.
[2] George, A. & V.G. Springer. 1980. Revision of the clinid fish tribe Ophiclinini, including five new species, and definition of the family Clinidae. Smithsonian Contributions to Zoology 307: 1–31.
[3] Holleman, W. 1982. Three new species and new genus of tripterygiid fishes (Blennioidei) from the Indo-West Pacific Ocean. Annals of Cape Provincial Museums Natural History 14(4): 109–127.
[4] Ben-Tuvia, A. & E. Trewavas. 1987. *Atrobucca geniae*, a new species of sciaenid fish from the Gulf of Elat (Gulf of Aqaba), Red Sea. Israel J. Zool. 34: 15–21.

Instead of a conclusion

By her own admission Genie became a diver first and scientist second. She graduated from hard-hat diving and mask-snorkeling with Carl Hubbs in 1946 to one of the best SCUBA divers. In 1986 she received the Lowell Thomas Award for Undersea Exploration by the Explorers Club, New York, and in 1993 the DEMA (Diving Equipment Manufacturers Association) Hall of Fame Award. She evolved in the sea into an aquanaut who needs half as much air than other top divers. Her aspirations go, however, beyond SCUBA. Remember her childhood hero?

Genie's status as a favored scientist of the National Geographic Society and her long collaboration with David Doubilet, a contract photographer for National Geographic, and with staff photographer Emory Kristof led to the realization of her childhood dream. She became the chief scientist on

Fig. 22. During revision of Genie's lifetime list of publications in her laboratory at College Park in November 1993 I had the opportunity to see some of the unfinished manuscripts. Photograph by C. Flegler-Balon.

Fig. 23. Genie and her youngest son Niki at lunch in her kitchen at Bethesda, November 1993. Photograph by E.K. Balon.

many expeditions and 71 dives in seven submersibles of the Beebe Project (Beebe 1934) of Emory Kristof's program in deep-diving manned submersibles. Many other agencies participated to support this ambitious program which started in 1987. 'In these dives', write Clark & Kristof (139, p. 79), 'we lured deep sea sharks with bait to observe and photograph them in their natural habitat (. . .). With the submersible settled on the bottom during dives of 1,000 to 12,000 feet [303 to 3636 m], we set out bait and waited. The long (up to $17^1/_2$ hours), quiet dives and baiting technique have been successful in bringing in as many as 21 individual sharks in one dive. Our dive sites have been located off the coast of Bermuda, Bahamas, Grand Cayman, California's Monterey Canyon, Japan's Suruga Bay, and near the Chagos Archipelago in the Indian Ocean. Through an international effort, we have made dives down as far as 12,000 feet in the U.S. *Alvin*, the Soviet *Mir 1*, the French *Nautile*, and the Canadian-made *Pisces II* and *VI*. For shallower dives of 980 to 1,200 feet we used the PC 1802 and the Johnson Sea Link'.

One of the remarkable findings of these deep dives was the prevalence of the pineal eye in the three types of sharks encountered up to 2000 m, the sixgill shark *Hexanchus griseus*, gulper sharks of the genus *Centrophorus*, and lantern sharks of the genus *Etmopterus* (Clark & Kristoff 133). Probably the pineal eye, so large that they call it rather a pineal window, can sense the amount of light penetration. No sharks were seen below 2000 m depth, but numerous bony fishes, skates and chimeras came to the bait of the deepest dives. 'The largest creature ever seen in the deep sea lumbered in front of the view ports of the submersible *Nautile* on September 13, 1989. Four thousand feet deep [1212 m], the Pacific sleeper shark, *Somniosus pacificus*, crashed into the two-by-two-foot bait cage and pushed it into the mud . . .'. Genie estimated its size as 7 m. Cam-

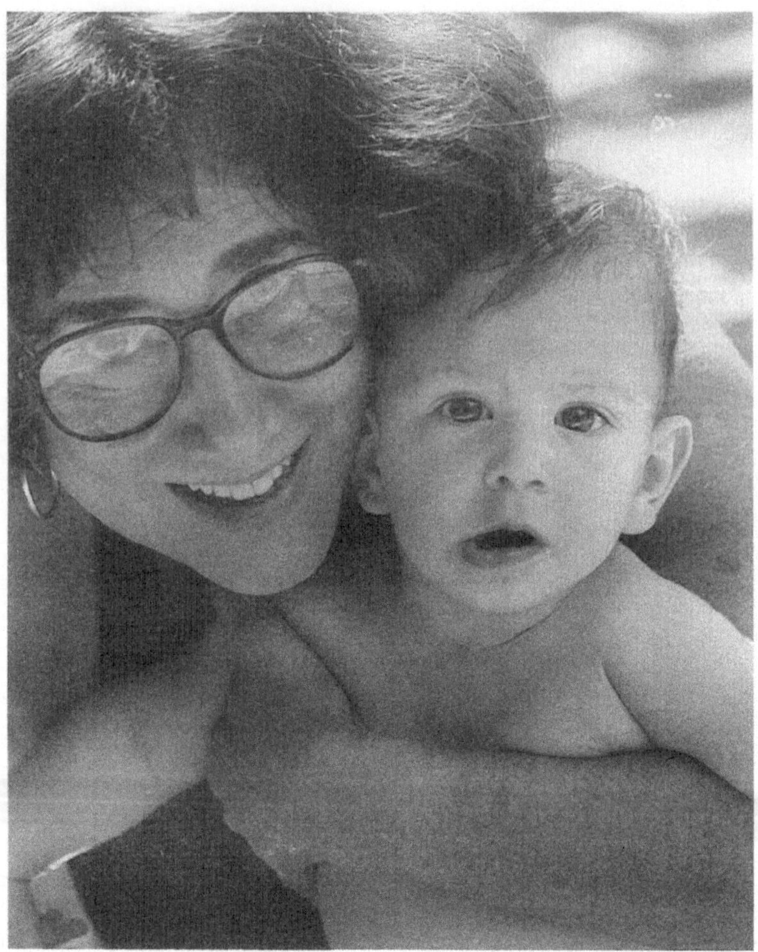

Fig 24 Eugenie Clark and her first grandchild Eli Weiss in the summer of 1991 Photograph by Aya Weiss

eraman Ralph White recalls: '«We saw a fish bump into a wall, and then the wall moved. The sub shook. On the shark's second pass we saw the head. Parasites hung from fluttering gills. As it left, we saw enough of the underside to determine it was female. All we could think was holy mackerel!»' (Doubilet 1990, p. 11).

Genie officially retired in 1992 but did not slow down. As professor emerita she retains her office and laboratory in the Department of Zoology, University of Maryland, continues to give her honor's course, organizes and leads expeditions to the Red Sea, Mexico, Caribbean, Australia and Papua New Guinea, and is invited to give more lectures than she can accept (Fig. 20). Genie has been a consultant, commentator, co-director or principal in 24 television specials about marine life. *The Sharks*, a National Geographic film (1982) holds the highest Nielson rating on PBS; *Reef Watch* was the first live underwater TV documentary to the U.S. with narration also from underwater. It is no surprise that lately she received the Governor's Citation of the State of Maryland, the medal of the President of the University of Maryland and Doctor of Science honoris causa from the University of Massachusetts (Fig. 21). Four species of fishes were named after her (Table 2). In February 1995, she will receive the Doctor of Science honoris causa from the University of Guelph in Canada. Having been the recipient of over 25 medals and awards and the subject of over 100 profiles in magazines and newspapers, she feels that it all far exceeds the significance of her work and that being a female actually gave her unfair advantage over more deserving male colleagues.

Genie witnessed destruction in her favorite reef areas of the oceans and became a conscientious conservationist. Largely through her lobbying efforts, the Râs Mohammed area of the Red Sea, with coral reefs of magnificent beauty, was declared the first Egyptian national park in 1983. When I last visited her she was working on six new manuscripts (Fig. 22). Nobusan died in 1992. She keeps in close touch and dives with all her children (Fig. 23) and is enormously proud of her first grandchild Eli, son of Aya (Fig. 24).

Genie Clark contributed significantly to the exploration and public awareness of underwater life. Her popular books and articles in National Geographic reached children and adults alike, many of whom chose marine biology or related professions because of her. Her scientific papers contributed significantly to the knowledge of fishes and the sea far beyond her native America. Most of all, her radiant personality brought joy of life to all of us who are privileged to know her. Her decision for going on diving, exploring and publishing as long as she lives (Samarrai 1992) is a fascinating challenge. Looking at her willpower, form and appearance at 72, I believe she will achieve it. When the interviewer for Omni (Stein 1982) asked her what was the wildest thing she had ever done, the answer was: 'My ride on a huge whale shark. I was crazy. We wanted to study and photograph her. She was well over forty feet long. Once I got on her, I just couldn't let go. And I went far away from the photographers and the boat. The shark was cruising along steadily at three knots, and, after a while, I thought to myself, Why am I still holding on to the shark, getting farther away from the boat? And I finally let go. I did not ever want to let go'.

I am going to conclude with the words Giles Mead wrote in Tim Berra's (1977, p. 9) bibliography of Genie's hero William Beebe, for these words apply to her as well: Genie 'is best known as a popular writer, but as this work amply demonstrates, [her] popular accounts were well anchored fore and aft in scholarly work of importance'.

Acknowledgements

When I was a young ichthyology graduate, sealed in a communist cage and only dreaming about diving in tropical seas, Genie, then already the executive director of Cape Haze Marine Laboratory, supplied me with pamphlets, reprints and kind words she knows so well to scribble over any blank spaces on mailed material. She became my heroine I never hoped to meet. When we finally met in 1977, it was impossible for Eugenie and Eugene not to become friends. Many thanks to Genie for being here for me, for allowing me to raid her family albums and for correcting the clumsy attempt to present once more her life and work. My wife Christine was the first controller who constrained my temptations to write even more, and David Noakes and Mike Bruton read and corrected the first draft. Many thanks to all.

References cited

Balon, E K & C Flegler-Balon 1985 Microscopic techniques for studies of early ontogeny in fishes problems and methods of composite descriptions pp 33–56 *In* E K Balon (ed) Early Life Histories of Fishes, Developments in Environmental Biology of Fishes, Dr W Junk Publishers, Dordrecht

Beebe, W 1934 A half mile down National Geographic 66(6) 661–704

Berra, T M 1977 William Beebe An annotated bibliography Archon Books, Hamden 157 pp

Burgess, G H 1991 Life history notes on Stewart Springer pp 4–6 *In* S H Gruber (ed) Discovering Sharks, American Littoral Society, Highlands

Burns, R F 1992 Dr Eugenie Clark – the 'Shark Lady' Sea Technology February 1992 72

Doubilet, D 1990 Suruga Bay In the shadow of Mount Fuji National Geographic 178(4) 2–11

Ellis, R 1975 The book of sharks Grosset & Dunlap Publishers, New York 320 pp

Emberlin, D 1977 Contribution of women science Dillon Press, Minneapolis 160 pp

Fishelson, L 1993 Israeli ichthyology in the Red Sea (1951–1992) – a personal perspective Israel J Zool 39 287–291

McGovern, A 1978 Shark lady The adventures of Eugenie Clark Four Winds Press, New York 83 pp

McAllister, D E 1988 Vadim Dimitrievitch Vladykov life of an ichthyologist Env Biol Fish 23 9–20

Samarrai, F. 1992. A life beneath the sea. College Park (Univ. Maryland Alumni Magazine) 3(2): 14–18.

Stacey, P. 1990. Eugenie Clark: without a spear. Calypso Log June 1990: 8–10.

Stein, J. 1982. Eugenie Clark: sweet sharks. Omni 4(9): 94–98, 115–117.

Steinbeck, J. 1960. The log from the Sea of Cortez. Pan Books, London. 320 pp.

Genie Clark (center) collects in Miami *Bathygobius soporator* with Japan's then Crown Prince, now Emperor Akihito (at her left side). With the help of Dick Robins and two graduate students they cornered finally the first specimen in a discarded beer can (29 May 1967).

Environmental Biology of Fishes **41** 115–119, 1994

Lifetime list of publications by Eugenie Clark

Eugene K. Balon
Institute of Ichthyology and Department of Zoology, University of Guelph, Guelph, Ontario N1G 2W1, Canada

Eugenie Clark produced at least 150 scientific papers and popular articles and books from 1947 until the publication of this bibliography. She was a sole or first author in 87% of these and shared authorship in 19 papers with 42 collaborators, 8 of which with L.R. Aronson, 6 with J.S. Rabin and 4 with E. Kristof. I was unable to trace numerous excerpts from her popular books and their translations in spite of her kind help and have omitted manuscripts and reports never published as well as numerous reprints of excerpts from her books, be it in English or other languages, especially in Japanese. Her publications are arranged by year but no effort was made to arrange them chronologically within years. All items in this list were verified against an original copy.

1 Clark, E 1941 Letter to the editor Natural History 50 311–312

2 Breder, C M & E Clark 1947 A contribution to the visceral anatomy, development, and relationships of the plectognathi Bull Amer Mus Nat Hist 88 287–319 + 4 pl

3 Clark, E 1947 Notes on the inflating power of the swell shark, *Cephaloscyllium uter* Copeia 1947 278–280

4 Clark, E , L R Aronson & M Gordon 1948 An analysis of the sexual behavior of 2 sympatric species of poeciliid fishes and their laboratory induced hybrids Anat Rec 101(4) 42 (abstract)

5 Clark, E , L R Aronson & M Gordon 1949 The role of the distal tip of the gonopodium during the copulatory act of the viviparous teleost, *Platypoecilus maculatus* Anat Rec 105(3) 26–27 (abstract)

6 Clark, E 1949 Notes on some Hawaiian plectognath fishes, including a key to the species Amer Mus Novitates 1397 1–22

7 Clark, E & J M Moulton 1949 Embryological notes on *Menidia* Copeia 1949 152–154

8 Sperry, R W & E Clark 1949 Interocular transfer of visual discrimination habits in a teleost fish Physiol Zool 22 372–378

9 Clark, E 1950 Notes on the behavior and morphology of some West Indian plectognath fishes Zoologica 35 159–168

10 Clark, E 1950 A method for artificial insemination in viviparous fishes Science 112(2920) 722–723

11 Clark, E 1950 Fisherman beware! Fishing for poisonous plectognaths in the western Carolines ONR Research Reviews, June 1950 1–6

12 Clark, E & L R Aronson 1950 Sexual behavior in the guppy Anat Rec 108 534 (abstract)

13 Clark, E & L R Aronson 1951 Sexual behavior in the guppy, *Lebistes reticulatus* (Peters) Zoologica 36 49–66 + 7 pl

14 Clark, E & R P Kamrin 1951 The role of the pelvic fins in the copulatory act of certain poeciliid fishes Amer Mus Nov 1509 1–14

15 Clark, E 1951 Field trip to the south seas Natural History 60 8–15

16 Clark, E 1951 Wonders of the deep Cairo Calling 847(2) 12–13

17 Aronson, L R & E Clark 1952 Evidences of ambidexterity and laterality in the sexual behavior of certain poeciliid fishes Amer Nat 86 161–171

18 Clark, E 1952 Spearfishing in the Red Sea The Skin Diver 1(6) 2–3

19 Clark, E 1952 Book review of 'Diving to adventure' by H Haas Natural History 61 5

20 Clark, E 1952 The lost quarry Natural History 61 258–263

21 Clark, E 1952 A scientific journey to the Red Sea, Part 1 Natural History 61 344–349

22 Clark, E 1952 A scientific journey to the Red Sea, Part 2 Natural History 61 414–419

23 Clark, E & H A F Gohar 1953 The fishes of the Red Sea order Plectognathi Cairo University Press, Publ Mar Biol Sta Al Ghardaqa (Red Sea) 8 80 pp + 4 pl

24 Clark, E 1953 Book review of 'The silent world' by J -Y Cousteau & F Dumas Natural History 62 149

25 Clark, E 1953 Lady with a spear Harper & Brothers Publishers, New York 243 pp (a Book-of-the-Month Club selection, translated into seven languages, over 23 foreign editions, Braille and records for the blind – see separate list)

26 Clark, E 1953 Siakong, spear-fisherman pre-eminent Natural History 62 227–234

27. Clark, E., L.R. Aronson & M. Gordon. 1954. Mating behavior patterns in two sympatric species of xiphophorin fishes: their inheritance and significance in sexual isolation. Bull. Amer. Mus. Nat. Hist. 103: 135–225.
28. Clark, E. 1954. The Palaus and the best spearfisherman. pp. 302–303. *In:* A.C. Spectorsky (ed.) The Book of the Sea, Appleton-Century-Crofts, New York (from 'Lady with a spear').
29. Clark, E. 1955. From 'Lady with a spear'. Compact, The Young People's Digest 4(5): 120–130.
30. Clark, E. 1956. Beneath the brine. Book review of 'Man under the sea' by J. Dugan. The Saturday Review (New York) 39: 17–19.
31. Woodburn, K.D., B. Eldred, E. Clark, R.F. Hutton & R.M. Ingle. 1957. The live bait shrimp industry on the west coast of Florida (Cedar Key to Naples). Fl. State Board Cons., St. Petersburg Tech. Ser. 21. 33 pp.
32. Heller, J.H., M.S. Heller, S. Springer & E. Clark. 1957. Squalene content of various shark livers. Nature 179: 919–920.
33. Clark, E. 1958. Jellied condition in *Paralichthys squamilentus* from the Gulf of Mexico. Quart. J. Fl. Acad. Sci. 21: 187–189.
34. Clark, E. 1958. Book review of 'Fishes of the Red Sea and southern Arabia; Vol. 1, Brachiostomida to Polynemida' by H.W. Fowler. Copeia 1958: 156–157 (author's name omitted at the end of the review).
35. Clark, E. 1959. Functional hermaphroditism and self-fertilization in a serranid fish. Science 129(3343): 215–216.
36. Clark, E. 1959. Instrumental conditioning in lemon sharks. Science 130(3369): 217–218.
37. Clark, E. 1959. Instrumental conditioning of sharks. Anat. Rec. 134: 545 (abstract).
38. Clark, E. 1959. Reproductive behavior in hermaphroditic fishes. Anat. Rec. 134: 545–546 (abstract).
39. Clark, E. 1959. The best spearfisherman in the world. pp. 350–357. *In:* J.-Y. Cousteau & J. Dugan (ed.) Captain Cousteau's Underwater Treasury, Harper & Brothers Publishers, New York (from 'Lady with a spear').
40. Clark, E. 1960. Four shark attacks on the west coast of Florida, Summer 1958. Copeia 1960: 63–67.
41. Royal, W. & E. Clark. 1960. Natural preservation of human brain, Warm Mineral Springs, Florida. Amer. Antiquity 26: 285–287.
42. Clark, E. 1960. An ichthyologist at Eylath. Fishermen's Bull., Haifa 3(6): 34–35 (In Hebrew).
43. Clark, E. 1960. Lady with a spear. Edited with notes by Kiyoshi Ikejima. The Hosei University Press, Tokyo. 72 pp (an abbreviated version).
44. Clark, E. 1962. Visual discrimination in lemon sharks. Tenth Pacific Science Congress, Honolulu: 175–176 (abstract).
45. Clark, E. 1962. An ichthyologist at Eilat. Skin Diver Magazine 11: 22–23, 51.
46. Saunders, G.B. & E. Clark. 1962. Yellow-billed cuckoo in stomach of tiger shark. The Auk 79: 118.
47. Clark, E. 1962. Maintenance of sharks in captivity. Part 1. General. Bull. Inst. Océanogr. Monaco, Special No. 1A: 7–13.
48. Clark, E. 1962. Expedition returns from Red Sea. Copeia 1962: 678.
49. Clark, E. 1963. Massive aggregations of large rays and sharks in and near Sarasota, Florida. Zoologica 48: 61–62 + 2 pl.
50. Clark, E. 1963. Cape Haze Marine Laboratory. Amer. Zool. 3: 1–3.
51. Moore, J.C. & E. Clark. 1963. Discovery of right whales in the Gulf of Mexico. Science 141(3577): 269.
52. Clark, E. 1963. The maintenance of sharks in captivity, with a report on their instrumental conditioning. pp. 115–149. *In:* P.W. Gilbert (ed.) Sharks and Survival, Heath and Co., Boston.
53. Davies, D.H., E. Clark, A.L. Tester & P.W. Gilbert. 1963. Facilities for the experimental investigation of sharks. pp. 151–162. *In:* P.W. Gilbert (ed.) Sharks and Survival, Heath and Co., Boston.
54. Clark, E. 1963. The maintenance of sharks in captivity. Part 2. Experimental work on shark behavior. Bull. Inst. Océanogr. Monaco, Special No. 1D: 1–10.
55. Clark, E. 1964. Spinal deformity noted in bull shark. Underwater Naturalist 2: 26–28.
56. Clark, E. 1965. Recent studies of sharks. pp. 243–245. *In:* N.P. Wright (ed.) Fifth Annual Convention, Underwater Society of America, Mexico City.
57. Clark, E. & K. von Schmidt. 1965. Sharks of the central Gulf coast of Florida. Bull. Mar. Sci. 15: 13–83.
58. Clark, E. 1965. Parasitic stone crab? Sea Frontiers 74: 52–53.
59. Clark, E. 1965. Mating of groupers. New studies detect reversal of stripes in hermaphroditic fish. Natural History 74 (6): 22–25.
60. Clark, E. 1965. Massed sharks and rays off Florida. Animals 6(20): 572.
61. Aronson, L.R. & E. Clark. 1966. Instrumental conditioning in young nurse sharks. pp. 34–35. *In:* Eleventh Pacific Science Congress, Tokyo (abstract).
62. Clark, E. 1966. Pipefishes of the genus *Siokunichthys* Herald in the Red Sea with description of a new species. Sea Fish. Res. Sta. Haifa Bull. 41: 3–6.
63. Clark, E. & K. von Schmidt. 1966. A new species of *Trichonotus* (Pisces, Trichonotidae) from the Red Sea. Sea Fish Res. Sta. Haifa Bull. 52: 29–36.
64. Aronson, L.R., F.R. Aronson & E. Clark. 1967. Instrumental conditioning and light-dark discrimination in young nurse sharks. Bull. Mar. Sci. 17: 249–256.
65. Clark, E. 1967. The need for conservation in the sea. Oryx 9: 151–153.
66. Clark, E. 1968. Eleotrid gobies collected during the Israel South Red Sea Expedition (1962), with a key to Red Sea species. Sea Fish. Res. Sta. Haifa Bull. 49: 3–7.
67. Clark, E., A. Ben-Tuvia & H. Steinitz. 1968. Observations

on a coastal fish community, Dahlak Archipelago, Red Sea Sea Fish Res Sta Haifa Bull 49 15–31

68 Clark, E 1969 The lady and the sharks Harper & Row Publishers, New York 269 pp (1972 Japanese edition entitled 'Sea, sun and sharks', and 1990 paperbach edition by Mote Marine Laboratory, Sarasota)

69 Clark, E 1971 Observations on a garden eel colony at Elat Hebrew University of Jerusalem Marine Biological Laboratory Scientific Newsletter 1 5

70 Clark, E 1971 The Red Sea garden eel Underwater Naturalist 7 4–10

71 Clark, E 1971 Useful animals Book review of 'The life of sharks' by P Budker (English edition) Science 174 136

72 Clark, E 1971 The ecology and schooling behavior of the euryhaline cownosed ray, *Rhinoptera bonasus* Second National Coastal and Shallow Water Research Conference (ONR) 41 (abstract)

73 Clark, E 1971 In memoriam Heinz Steinitz Sea Fish Res Sta Haifa Bull 58 2–4

74 Clark, E & S Chao 1972 A toxic secretion from the Red Sea flatfish, *Pardachirus marmoratus* (Lacépede) Hebrew University of Jerusalem, The Heinz Steinitz Marine Biological Laboratory Scientific Newsletter 2 14

75 Clark, E 1972 The Red Sea's garden of eels Nat Geographic 142 724–735

76 Clark, E & W Aron 1972 Heinz Steinitz to realize a dream Israel J Zool 21 131–134

77 Clark, E 1973 'Sleeping' sharks in Mexico Underwater Naturalist, Bull Amer Littoral Soc 8 4–7

78 Clark, E & S Chao 1973 A toxic secretion from the Red Sea flatfish *Pardachirus marmoratus* (Lacepède) Sea Fish Res Sta Haifa Bull 60 53–56

79 Clark, E & A Ben-Tuvia 1973 Red Sea fishes of the family Branchiostegidae with a description of a new genus and species *Asymmetrurus oreni* Sea Fish Res Sta Haifa Bull 60 63–74

80 Clark, E 1974 The Red Sea's sharkproof fish Nat Geographic 146 718–727

81 Clark, E 1974 Houdinis of the Red Sea Internat Wildlife 4(6) 13–17

82 Clark, E 1975 The strangest sea Nat Geographic 148 338–343

83 Clark, E 1975 Acceptance of gold medal Bull Soc Woman Geogr, 50th Anniversary Issue, Part 2 76–78

84 Clark, E 1975 Into the lairs of 'sleeping' sharks Nat Geographic 147 570–584

85 Clark, E & H Kabasawa 1976 Factors affecting the respiration rates of two Japanese sharks, dochizame (*Triakis scyllia*) and nekozame (*Heterodontus japonicus*) Ann Rep Keikyu Aburatsubo Marine Park Aquarium 7–8 14–26 (also published in Sci Bull Dept Navy Off Naval Research 1(1) 1–9)

86 Kabasawa, H & E Clark 1976 Respiration rates and activity patterns of the shark, dochizame (*Triakis scyllia*) in a large circular tank Ann Rep Keikyu Aburatsubo Marine Park Aquarium 7–8 27–38

87 Clark, E 1976 Horrors of the deep pp 13–14 *In* H W Menard (acad coord) Oceans, Our Continuing Frontier (Courses by Newspaper), Publisher's Inc, San Diego

88 Clark, E 1976 Sharkproof! pp 63–66 *In* H W Menard & J L Scheiber (ed) Oceans Our Continuing Frontier, Publisher's Inc, Del Mar

89 Clark, E & A George 1976 A comparison of the toxic soles *Pardachirus marmoratus* of the Red Sea and *P pavoninus* from southern Japan pp 545–546 *In* J-C Hureau & K-E Banister (ed) Actes du 2[e] Congres Europeen Des Ichthyologistes, Paris, Rev Trav Inst Pêches marit 40 (abstract)

90 Clark, E 1976 Foreword (pp 1–3) to book 'Shark's liver oil for health use' by S Abe Yomiuri Shimbun, Tokyo (In Japanese)

91 Clark, E 1976 Morphology and toxic action of the flatfish *Pardachirus pavoninus* Progress Rep Oceanic Biology, U S Off Naval Res ACR-217 49–50 (abstract)

92 Clark, E 1976 Expedition to the Bonin islands Diving World 13 84–85 (In Japanese)

93 Clark, E 1977 Synagogues and sea fans Israel's national parks and nature reserves Nat Parks Conserv Mag 51(4) 13–20

94 Clark, E 1978 Foreword (pp ix–xii) to book 'The man who rode sharks' by W R Royal with R F Burgess Dodd, Mead & Company, New York

95 Clark, E 1978 Sharks that ring bells pp 109–124 *In* V & R Taylor & P Goadby (ed) Great Shark Stories, Harper & Row Publishers, New York

96 Clark, E 1978 Flashlight fish of the Red Sea Nat Geographic 154 718–728

97 Clark, E 1978 A letter from Eugenie Clark pp 92–93 *In* A McGovern (ed) Shark Lady, True Adventures of Eugenie Clark, Scholastic Inc, New York (French edition in 1981)

98 Clark, E & A George 1979 Toxic soles, *Pardachirus marmoratus* from the Red Sea and *P pavoninus* from Japan, with notes on other species Env Biol Fish 4 103–123

99 Clark, E 1979 At the bottom of the sea pp 48–55 *In* Z Sutherland (ed) The Spirit of the Wind, Open Court, La Salle

100 Clark, E 1979 Red Sea fishes of the family Tripterygiidae with descriptions of eight new species Israel J Zool 28 65–113

101 Clark, E 1979 Adventure of the last frontier pp 274–290 *In* T Clymer, K R Green, D Gates & C M McCullough (ed) Measure Me, Sky, Ginn and Company, Lexington (excerpt from 'Lady with a spear')

102 Clark, E 1979 Adventures on the last frontier pp 26–42 *In* The Web of Life, Ginn and Company, Aylebury (excerpt from 'Lady with a spear')

103 Clark, E 1979 Marine life observed pp 57–64 *In* D W Bennett (ed) Fish Stories, Spec Publ 8, Amer Littoral Soc, Sandy Hook

104 Clark, E 1980 Distribution, mobility and behavior of the

Red Sea garden eel Nat Geographic Soc Res Rep 12 91–102

105 Clark, E 1980 Synagogues and sea fans – Israel's national parks and nature reserves pp 143–149 *In* Reading on the Protection and Management of Marine and Submerged Resources of the National Parks, Committee on Energy and Natural Resources, U S Senate, Publ No 96–113

106 Clark, E 1981 Shark repellent effect of the Red Sea Moses sole, *Pardachirus marmoratus* Abstracts of Papers 147th National Meeting AAAS, Toronto 49

107 Clark, E 1981 Shark repellent effect of the Red Sea Moses sole, *Pardachirus marmoratus* Nat Geographic Soc Res Rep 13 177–186

108 Clark, E 1981 Why sharks sleep Anima (Magazine of Natural History) 6 6–12 (In Japanese)

109 Clark, E 1981 A Japanese shark repellent for the shark lady pp 82–84 *In* B Moore (ed) Starting Points in Reading, Level D, Ginn and Company, Aylesbury

110 Clark, E 1981 Sharks magnificent and misunderstood Nat Geographic 160 138–187 (reprinted in J A Quatrini 1985 Speed reading Arco Publishing, New York, pp 172–180)

111 Clark, E 1982 Secrets of the Red Sea's Sci Digest 90(4) 46–53

112 Clark, E 1983 Sand-diving behavior and territoriality of the Red Sea razorfish, *Xyrichtys pentadactylus* Bull Inst Oceanogr Fish (Cairo) 9 225–242

113 Vanderbilt, H C & E Clark 1983 A scientific and economic conceptual endeavor for the development of coastal marine parks and mariculture Bull Inst Oceanogr Fish (Cairo) 9 477–478

114 Clark, E 1983 Shark repellent effect of the Red Sea sole pp 135–150 *In* B J Zahuranec (ed) Shark Repellents from the Sea, AAAS selected symposium 83, Westview Press, Boulder

115 Clark, E 1983 Hidden life of an undersea desert Nat Geographic 164 128–144

116 Clark, E 1984 Japan's Izu Oceanic Park Nat Geographic 165 462–491

117 Clark, E 1985 Preface to book 'The Shark Watchers' Guide' by G Dingerkus Julian Messner, New York

118 Clark, E , P Pemberton & R Leen 1985 Population density of a colony of bluespotted jawfish off Baja California Underwater Naturalist 15(2) 3–7

119 Clark, E & D Shen 1986 Territoriality of Red Sea and sand-diving fishes of the genera *Xyrichtys* and *Trichonotus* p 937 *In* T Uyeno, R Arai, T Taniuchi & K Matsuura (ed) Indo-Pacific Fish Biology, The Ichthyological Society of Japan, Tokoy (abstract)

120 Rabin, J S & E Clark 1986 Films in marine biology and related sciences J College Sci Teaching 15(4) 264–266

121 Clark, E , E Kristof & D Lee 1986 New eyes for the dark reveal the world of sharks at 2,000 feet Nat Geographic 170 680–691

122 Clark, E 1987 Fisheries and use of deep-sea sharks pp 221–225 *In* S Cook (ed) Sharks, an Inquiry into Biology, Behavior, Fisheries, and Use, Proc Confer NOAA, Oregon State University Extension Service, Portland

123 Clark, E & E Kristof 1987 Observations on sixgill sharks, *Hexanchus griseus* and *H vitulus*, from submersibles Papers of Amer Elasmobranch Soc 3rd Annual Meeting, Albany 39 (abstract)

124 Clark, E 1987 Foreword (p 11) to book 'Red Sea Fish Guide' by R Deuvletian Nubar Printing House, Cairo

125 Clark, E 1987 Women scientific divers pp 134–143 *In* W Fife (ed) Women in Diving, Proc 35th Undersea and Hyperbaric Medical Soc Workshop, Bethesda

126 Clark, E , J S Rabin & S Holderman 1988 Reproductive behavior and social organization in the sand tilefish, *Malacanthus plumieri* Env Biol Fish 22 273–286

127 Rabin, J S , L M Benveniste & E Clark 1988 Flexibility of burrow-building by the sand tilefish, *Malacanthus plumieri* an adaptation for survival in changing habitats Proc Abstr 59th Ann Meet Eastern Psychol Assoc 9

128 Clark, E 1988 Down the Cayman wall Nat Geographic 174 712–731

129 Clark, E 1989 Rad Mohammed, Egypt's National Park, still in peril CMAS (World Underwater Federation) Congress, Nagoya 57 (abstract in Japanese)

130 Clark, E , J S Rabin, E Bunyan, Jr , I Murdock, D Shen & R Petzold 1989 Social behavior in the Caribbean tilefish Underwater Naturalist 18(2) 20–23

131 Rabin, J S & E Clark 1989 Burrow-building strategies in the sand tilefish 21st Internat Ethol Conf , Utrecht 139 (abstract)

132 Clark, E 1989 At the bottom of the sea pp 48–55 *In* Z Sutherland & M F Cunningham (ed) Sound of the Sea, Open Court, La Salle (excerpt from 'Lady with a spear')

133 Clark, E & E Kristof 1990 Deep-sea elasmobranchs observed from submersibles off Bermuda, Grand Cayman, and Freeport, Bahamas pp 269–284 *In* H L Pratt, Jr , S H Gruber & T Taniuchi (ed) Elasmobranchs as Living Resources, Advances in the Biology Ecology, Systematics, and the Status of the Fisheries, NOAA Technical Report NMFS 90

134 Clark, E , J F Pohle & D C Shen 1990 Ecology and population dynamics of garden eels at Râs Mohammed, Red Sea Nat Geographic Res 6 306–318

135 Clark, E 1990 Preface (pp ix–xiii) to 2nd edition of 'The Lady and the Sharks' Mote Marine Laboratory, Sarasota

136 Clark, E 1990 Dispatches from a distant world Nat Geographic 178(4) 12–19

137 Clark, E , M Pohle & J Rabin 1991 Spotted sandperch dynamics Nat Geographic Res Exploration 7 138–155

138 McGovern, A & E Clark 1991 The desert beneath the sea Scholastic Inc , New York 48 pp (paperback edition in 1993)

139 Clark, E & E Kristof 1991 How deep do sharks go? Reflections on deep sea sharks pp 79–84 *In* S H Gruber (ed) Discovering Sharks, Underwater Naturalist 19(4), 20 (1), Spec Publ 14 Amer Littoral Soc , Highlands

140 Clark, E 1991 Preface to 'Watching Fishes' by R Wilson & J Q Wilson Harper & Row, New York

141 Clark, E & J F Pohle 1992 Monogamy in tilefish Nat Geographic Res Exploration 8 276–295

142 Clark, E 1992 Gifted guidance to Egypt's wondrous reefs Sea Frontiers 38(5) 20–27

143 Clark, E 1992 Gentle monsters of the deep whale sharks Nat Geographic 182(6) 120–139

144 Herold, D & E Clark 1993 Monogamy, spawning and skin-shedding of the sea moth, *Eurypegasus draconis* (Pisces Pegasidae) Env Biol Fish 37 219–236

145 Clark, E 1993 The biggest fish in the world Reader's Digest 142(854) 100–105 (condensed from Nat Geographic)

146 Randall, J E & E Clark 1993 *Helcogramma vulcana*, a new triplefin fish (Blennioidei Tripterygiidae) from the Banda Sea, Indonesia Revue fr Aquariol 20 27–32

147 Clark, E 1994 Foreword (pp iii–iv) to 'Fossil Sharks of the Chesapeake Bay Region' by B W Kent Egan Rees & Boyer, Columbia

148 Clark, E & S C Nemtzov 1994 Intraspecific egg predation by male razorfishes (Labridae) during broadcast spawning filial cannibalism or intrapair parasitism? Bull Mar Sci 55 133–141

149 Clark, E 1994 Foreword (p 5) to 'Whale Sharks The Giants of Ningaloo Reef' by G Taylor Angus & Robertson, Sydney

150 Clark, E & M Pohle 1994 *Trichonotus halstead*, a new sand-diving fish from Papua New Guinea Nat Geographic Res Exploration (in press)

151 Carey, F G & E Clark 1995 Depth telemetry from the sixgill shark, *Hexanchus griseus*, at Bermuda Env Biol Fish 42 (in press)

Foreign and special language editions of 'Lady with a spear':

Clark, E 1953 Lady with a spear Harper Bros , New York (Braille edition, 3 volumes, BRJ 00780)

Clark, E 1954 Una donna sotto i mari Edizioni A P E , Milano 240 pp

Clark, E 1955 Spydfiskeri i tre oceaner Grafish Forlag, Copenhagn 197 pp

Clark, E 1955 Lebensratsel des Meeres Auf Forschungsreisen in tropischen Gewassern Verlag Ullstein, Wien 224 pp

Clark, E 1955 Amason pa havsbottnen Natur och Kultur, Stockholm 193 pp

Clark, E 1957 Dame med spyd Gyldendal Norsk Forlag, Oslo 173 pp

Clark, E 1957 Fi 'amaq al mohitat Dar Al Hilal, Cairo 216 pp and many official (i e arranged with the author) or unofficial editions in Japanese, of which I have seen at least 5 but could not decifer even the year of publication

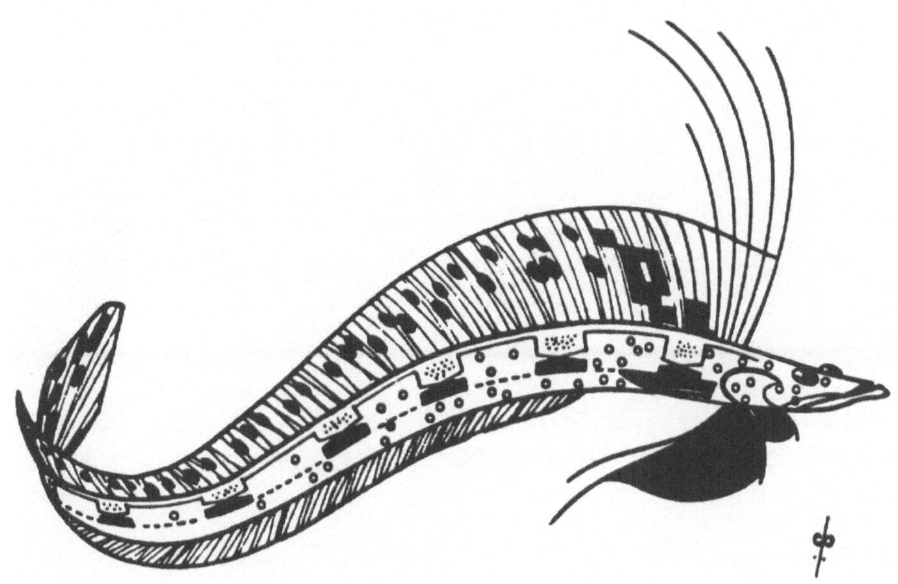

A caricature of the newly described *Trichonotus halstead* drawn by Craig Phillips

Genie examines a frill shark at the ISU Oceanic Park Japan 1984 Photograph by David Shen

Environmental Biology of Fishes **41** 121–125, 1994

An interview with Eugenie Clark

Eugene K. Balon
Institute of Ichthyology and Department of Zoology, University of Guelph, Guelph, Ontario N1G 2W1, Canada

Eugenie Clark who now lives in Bethesda, U.S.A. was interviewed at my residence near Guelph on 15 February 1993. The interview provides some revealing insights into the development of Genie's personality.

EB = Eugene Balon **Genie** = Eugenie Clark

EB: Many lifetime contributions to science are encouraged and inspired by personal relationships and friendships rarely identified or explained in most short biographies. Having just finished compiling an account of your 'life and work' for this volume, am left with some unanswered questions to complete the historical record. Would you allow me, therefore, some more personal questions as concerns their effects on your work, choice of taxa, location of study, etc.? The questions are not meant to be condescending but, in view of this volume's purpose, of quite some interest to communicate to others. Let us start with a conventional one: What is the most memorable event from your childhood?

Genie: At the age of nine I visited the old New York Aquarium at Battery Park on the southwestern end of Manhattan. My mother worked half a day on Saturdays at the nearby Downtown Athletic Club taking care of the cigar, cigarette, magazine and newspaper stand in the lobby of this tall building. I used to sit behind the counter all morning until she was free. Then we went to lunch at a Japanese restaurant 'Fuji' where the owner (who fell in love with my mother and later became my stepfather) showered me with presents and exotic gourmet dishes. After lunch was treat time when I had the undivided attention of my mother and she suggested various things we could do, the movies, the zoo, shopping etc. But when she took me to the aquarium I was 'hooked' on fishes. Every Saturday afternoon I wanted to watch the fishes. Her attention span was not as long as mine and I couldn't get enough of peering into the fish tanks, especially the big tank that held the sharks (sand tiger sharks) in water slightly murky green, where I could press my face against the glass by hanging over the railing. I couldn't see the back or sides of the tank and I pretended I was on the sea bottom with sharks swimming around me. I thought the sharks were beautiful, graceful, and magnificent. It was my dream to learn more about them and all the other beautiful and wondrous smaller fish.

I convinced by mother to drop me off in the aquarium in the morning and pick me up at noon instead of sitting behind her work counter at the Club. It became our regular Saturday routine. On rainy days the bums from Battery Park came into the free Aquarium and I made friends with them. Soon they came in to meet me on sunny Saturdays and I would tell them what I had learned from reading about the different fishes we watched together alive. On Saturday afternoons my mother took me to nearby aquarium supply stores and soon I had aquariums in my bedroom at home with my own living pet jewel fishes – but no sharks, of course. My dream to be underwater with sharks and swim with fishes came many years later. [**Note**: I wrote about the above in more detail in the first chapter of *Lady With a Spear* and Ann McGovern rewrote it for children in the first chapter of *Shark Lady*.]

I have mixed feelings about keeping fishes in glass 'cages' today but I realize my exposure to this as a child set the course for my life.

EB: You are now retired but still active, diving and

publishing. Can you identify a person who shaped your early career and especially your attitude to science, or who most affected your life and work?

Genie: When I was a child, William Beebe was my hero. I loved reading his books: *Jungle Days, Edge of the Jungle, Galapagos: Worlds End*, but above all his books and National Geographic articles about diving under the sea – walking on the sea bottom wearing a diving helmet set on his shoulders connected to a hose that pumped air from a boat, and diving deep into the sea in a bathysphere. I loved the way he wrote about animals and his great understanding about them.

My biology teachers and my family encouraged my enthusiasm for fishes even though my family suggested I also study typing so that I might get a job as secretary to someone like William Beebe. They were convinced I could never make a living studying fishes. But they loved me and wanted me to be happy so they bought me the books I wanted and fishes for my aquarium and enjoyed my fascinating 'hobby' themselves, thinking this was enriching my childhood, keeping me off the streets, and would evolve into something more practical as a way of life and earning a living.

After majoring in Zoology at Hunter College I started graduate school with Dr. Charles M. Breder who became my mentor, my greatest inspiration as a teacher and confidant in ichthyology. He saw nothing wrong about a young woman wanting to spend her life studying fishes. I dedicated my second book, *The Lady and the Sharks*, to him (the first was dedicated to my mother and stepfather). In the preface to the second edition of this book (published by the Mote Marine Laboratory) I express in detail what it meant to me and my children to know this brilliant ichthyologist, whose life was devoted to studying the behavior, morphology and ecology of fishes. He introduced me to plectognath fishes.

EB: How did you come to specialize in research on fishes?

Genie: When I graduated, World War II had started and there was a shortage of chemists. I worked four and a half years as a chemist, while going to night school to start graduate studies, before I was qualified for grants to study fishes. In 1948–1950 I made three grant applications as I was finishing my Ph.D. on the genetics and behavior of freshwater poeciliid fishes: to the Atomic Energy Commission to study sperm physiology and competition in fishes; to the Pacific Science Board to study poisonous plectognath fishes in Micronesia; and for a Fulbright Scholarship to study plectognath fishes in the Red Sea where I first used SCUBA. All three grants were approved and these field and laboratory studies combining my love for studying fishes and swimming, together with my basic experiences studying fishes under great teachers (Breder, Myron Gordon, Lester Aronson) at the American Museum of Natural History in New York, launched my career in ichthyology.

EB: As a young woman scientist you joined a male-dominated profession. Can you recall some relevant details?

Genie: Betty Kamp and I were the first women students in graduate school at Scripps Institute of Oceanography (1946). The great oceanographer, Harold Sverdrup was the Director. He was charming and gallant but did not allow Betty and me to go on overnight trips in our oceanography class. All day trips were OK. So we missed the trips on the high seas and to the Galapagos but in all other major aspects we were treated as equals. We had to work extra hard, expecially on field trips, to prove we could keep up with males; except with Carl Hubbs, who was married to Laura, and took it for granted that females could carry the same loads as males and do the cooking and dishwashing as well. It amused me that when I did do some of the things (e.g. diving in caves with 'sleeping' sharks) considered 'macho male accomplishments' that I was given *more* credit than males for doing the same thing they did. It helped to balance some of the prejudices against females.

When I applied to graduate school at Columbia University, the Chairman of the Zoology Department (a famous geneticist) told me, 'Well, I guess we could take you but to be honest, I can tell by looking at you, if you do finish you will probably get married, have a bunch of kids, and never do anything in science after we have invested our time and money in you'. I went instead to NYU where the Chairman, Harry Charipper, welcomed enthusiastic zoologists, regardless of sex and where I had the good fortune to come under the magical wing of Charles Breder and the wonderful professors and students there, so many of whom became part of my lifelong friends in ichthyology. I took William King Gregory's great course in the evolution of vertebrates, as a special student, and that was enough of Columbia University for me.

EB: Can you remember any unpleasant events which occurred because you were a female in competition with males?

Genie: I already answered this but might add that early in life, perhaps because of my intense desire to be an ichthyologist, I somehow sensed that it was not wise to ever play up my role as a female or to encourage flirtations, no matter how attractive, with people I studied fishes with. It made it easier and less complicated to concentrate on my desire to learn more about fishes while keeping the respect of my colleagues. To take advantage of my femaleness, would be disrespectful to the great men I have held in awe and loved as ichthyologists. Perhaps I was just lucky that in my formative years my consuming love of ichthyology did not coincide with an ichthyologist as the great love of my life.

EB: On the other hand, which are the most touching moments you remember from your early days as a scientist?

Genie: When Dr. Breder came to visit me in Placida, Florida (1955) to see how I was coming along as 'Executive Director' of the tiny, new Cape Haze Marine Lab and I asked him about this strange little grouper colony I found where all the individuals were females, their bellies swollen with ovulated eggs and no males were around to fertilize them. I showed him a female, I had just dissected, with a big bilobed ovary full of eggs, some oozing out of the oviduct. The ovary had a white wavy band around it which on my drawing I labelled 'fat?'.

'You make good drawings', Breder commented, 'Why don't you look at a pinch of that 'fat?' under a microscope in a drop of sea water?' It was swarming with spermatozoa. I discovered where the 'males' were, as Dr. Breder continued to be my teacher in his gentle way that pointed out my errors, as always, in a most complimentary way that made me feel good while still learning. His great insight into the ways of fishes and his humble informal way of teaching and encouraging me always touched me. I worshipped him.

I was once called to examine a sea monster, a strange creature that washed up on an out of the way sand spit. It proved to be a rare beaked whale, the only one of its kind ever reported from the Gulf of Mexico. Dr. John Moore at the Smithsonian identified it from photos I sent him of the skeleton we were cleaning and asked if I'd give the rare specimen to the Smithsonian. The only thing missing was the pair of enlarged ivory teeth. We searched in the sand in vain, then advertised in the local newspaper for them and learned a fisherman had found them and given them to his 12 year old son who considered them the greatest treasure in his collection of animal oddities in what his parents (they lived in a crowded trailer) called their son's 'Junkorama'. I asked the boy to bring in the teeth so I could photograph them. Dr. Moore had authorized me to try to buy them for any reasonable (even unreasonable) sum of money so the Smithsonian could have the complete skeleton.

I showed the boy and his parents around our Cape Haze Marine Lab and explained the work we did studying marine life and showed him the skeleton of the beaked whale we had carefully cleaned – complete except for the pair of ivory teeth. He told me how much he loved and was fascinated by all the strange animals and how delighted he was to own the rare teeth I had photographed. With some misgivings I told him about the offer from Dr. Moore and asked if he would consider selling them and what he would charge.

He conferred with his parents, then came back to me. 'I don't want to sell them' he stated firmly 'but I'd like to donate them to science'. He put the 2 precious treasures in my hands and smiled proudly as he left with his parents.

Dr. Moore was as touched as I was and told me to find out if there was anything they might have in surplus at the Smithsonian that the boy might like to have as an exchange, 'thank-you' gift. There was no hesitation when I phoned the boy. 'I'd like to have a grizzly bear skin' and Dr. Moore saw to it that a skin was soon hanging in the Junkorama.

EB: Did some papers you co-authored reflect more than a working relationship?

Genie: Yes, deep friendship and on my part love and admiration for many of my co-authors. Only once did I co-author a paper with a scientist who was not an ichthyologist and get involved to a point where we became engaged. I was recently divorced from a handsome pilot, my first great love affair and I thought, the only one of my life. So I decided I should make an intelligent second marriage with someone I admired. Then I met the second great love of my life who ultimately became the father of my four children. I broke off the calculated engagement, probably best for both parties. My co-author went on to be very successful in his own field of science, winning a Noble Prize, and took a wife who wasn't absorbed in looking at fish.

EB: You published numerous scientific papers and books. Which of these are your absolute favorites?

Genie: My first book because it was easy (based in large part on letters to my family which they fruitfully saved) and was a joy to reminisce the delightful experience of my formative years as an ichthyologist, and because it surprisingly made a lot of money, was a Book-of-the-Month selection, and I could dedicate it to my mother and stepfather. My first major scientific article (on plectognath fishes) together with Dr. Breder in 1947. My paper on instrumental conditioning of sharks in Science (1959) because it showed that sharks were not stupid, mindless, man-eaters.

EB: I presume that those you mentioned contributed significantly to human knowledge. Would you mind explaining in what way?

Genie: Not really. I don't consider any of my published books or articles to be significant contributions to human knowledge. In total my popular and scientific articles have helped dispel some of the myths about sharks that are so unfair to sharks, added a piece to the puzzle here and there towards our ultimate understanding about fishes, and inspired young people, especially girls, to study sciences.

EB: Scientists often dream, at least I do, about publishing a widely read popular book, a bestseller, which would reveal also to the non-scientists the excitement and joy of our work. What did you do about such dreams?

Genie: I feel my first book *Lady with a Spear* did this in a way that has been very satisfying. I did not dream it would be as successful as it was but was especially pleased that, after I was criticized for accepting a contract that would 'prostitute science', it got good reviews and is generally respected.

EB: Did you ever have contemplative ideas which go far beyond ichthyology? Did you join any philosophical school?

Genie: No, not like you, Eugene. I'm a simple ichthyologist who tries to get the facts straight, analyze my data as carefully as I can but I don't get philosophical. Love fish. Love sharks. Keep the water and their habitats as clean and protected as possible.

EB: If it were possible to rewind the tape of life, which part would you choose to replay once, or over and over again?

Genie: None. What's done is done. I don't like repetitions. A repetition is never as fresh and delightful as the first time. It's fun to write about it but not to replay any part. There is still so much to learn and experience. Even at 71. I look forward to my next experience. And time is getting short. I don't want to replay any one part of it. So much of it has been wonderful to live through. Thinking about it is enough and there is not enough time for all of this. It is almost more fun to discuss and analyze the past with friends and family. To rehash it exactly as it was, no thanks. I'd rather think ahead even to death and the remote possibility of 'life' after death. Wouldn't that be a surprise? It's tantalizing but how could heaven be happy with 4 ex-husbands? And would there be whale sharks to ride? And how can you ever recapture the thrill of the first time.

EB: In light of what you have just said, are you content with most of your life or would you like, given the opportunity, to edit some parts out and add some new parts?

Genie: I wouldn't want to change any of it.

EB: Would you advise young women to become ichthyologists?

Genie: Sure. I can't imagine a better life. But it's not as easy as it seems to many. There's a lot of hard work, many years of schooling, but it is worth it. Don't give up the full life – marriage, kids, for science. You can have it all if you choose a mate that is equally busy and not jealous of your work.

EB: Maybe you would like to add and answer some important questions I failed to ask?

Genie: No, I'm bushed.

EB: Thank you most cordially for your delightful cooperation.

Genie: Any egotist would cooperate for such a complimentary cause.

Environmental Biology of Fishes **41**: 127–145, 1994.

A review of hybridization in marine angelfishes (Perciformes: Pomacanthidae)

Richard L. Pyle & John E. Randall
Bernice P. Bishop Museum, Box 19000-A, Honolulu, HI 96817, U.S.A.

Received 15.3.1994 Accepted 25.3.1994

Key words: Coral reef fish, Behavioral ecology, Mate-choice, Protogyny, Polygyny, Zoogeography

Synopsis

Although hybridization of terrestrial and freshwater organisms has been well-studied, very little work has focused on hybridization among coral reef fish species. In the present paper, eleven examples of probable hybrids between marine angelfishes (Pomacanthidae) are reviewed. Evidence is presented which strongly suggests that the nominal species *Apolemichthys armitagei* is invalid and that specimens previously identified as this species represent hybrids between *A. trimaculatus* and *A. xanthurus.* Of the remaining ten probable pomacanthid hybrids, five are in *Centropyge* (*C. eibli* × *C. flavissimus, C. eibli* × *C. vrolikii, C. flavissimus* × *C. vrolikii, C. loriculus* × *C. potteri,* and *C. multifasciatus* × *C. venustus*); one in *Holacanthus* (*H. bermudensis* × *H. ciliaris*), and four in *Pomacanthus* (*P. arcuatus* × *P. paru, P. chrysurus* × *P. maculosus, P. maculosus* × *P. semicirculatus,* and *P. sexstriatus* × *P. xanthometapon*). An additional five examples of possible pomacanthid hybrids are described, two in *Centropyge*, two in *Chaetodontoplus* and one in *Pomacanthus*. Examination of hybrids may provide clues on reproductive behavior, dispersal capabilities, and phylogenetic relationships of species. More studies on hybridization in coral reef fish species, particularly those involving molecular techniques, are needed.

Introduction

Although there is a rich literature regarding hybridization of terrestrial plants and animals (e.g., Harrison 1993) and freshwater fishes, comparatively little work has been done on marine fish hybrids. Randall (1956) cited an unpublished list of New World hybrid fishes prepared by Robert R. Miller. Of 134 hybrids listed by Miller, only five are from the marine environment. Slastenenko (1957) made a compilation of natural hybrid fishes of the world; of his total of 212 hybrids, only 30 are marine. Schwartz (1972) reviewed the world literature on hybridization in fishes, citing 1810 references. These include papers on hybrids observed under natural conditions and those created artificially by cross fertilization or by exposure in aquaria of related species that would never co-occur in the wild. He made no analysis of natural vs. artificial hybrids or of the relative number of hybrids that are known from marine, estuarine, or freshwater environments. However it is clear that examples of freshwater species far outnumber those of marine species. He reported hybrids from 56 families of fishes, adding that the bulk of them occur in seven freshwater families. An increasing number of hybrid fishes have been documented in recent years, the great majority of which continue to be freshwater crosses.

Chaetodontidae (butterflyfishes) is the family of marine fishes with the greatest number of reported hybrids. Fifteen natural chaetodontid crosses have been recorded (Burgess 1974, 1978, Randall et al.

1977, Steene 1978, Allen 1979, Randall & Fridman 1981, Sano et al. 1984, Clavijo 1985), and we are aware of at least twelve others which have not yet been documented. The family Pomacanthidae (marine angelfishes), once classified as a subfamily of the Chaetodontidae, contains the next greatest number of reported hybrids for marine species. Allen (1979) listed five probable hybrids between pomacanthid species. Three others were described in subsequent publications (Krupp & Debelius 1990, Pyle 1992a, b). The purpose of the present paper is to review literature on pomacanthid hybrids, and add additional examples ranging from highly probable to speculative. The hybrids are discussed below in alphabetical order. Methods of counts and measurements follow Pyle & Randall (1993), except standard length was measured from the anteriormost tip of the premaxilla to the point where a crease forms when the caudal peduncle is bent (rather than from radiographs), and measurements of snout length, snout to dorsal-fin origin, snout to anal-fin origin, and snout to pelvic-fin insertion were made to the medial point of the ventral edge of the nasal bone (rather than to the anteriormost tip of the premaxilla). Specimens of hybrids and presumed parent species examined by us were deposited in the Bernice P. Bishop Museum, Honolulu (BPBM), and Senckenberg Museum, Frankfurt (SMF).

Following the hybrid accounts, we discuss some implications of observed patterns in pomacanthid hybridization and how they relate to questions of behavioral ecology, biogeography, dispersal, and phylogenetic relationships, and suggest future avenues for research.

Presumed and probable pomacanthid hybrids

Apolemichthys trimaculatus × *A. xanthurus*

Smith (1955) described the pomacanthid *Apolemichthys armitagei* on the basis of a single specimen collected in the Seychelles. This specimen closely resembles *Apolemichthys trimaculatus* Lacépède (Fig. 1a) but differs in certain aspects of color, most notably in having a prominent black region over much of the dorsoposterior portion of the soft dorsal fin and a dusky coloration of the head (Fig. 1b). Because of the similarity in color between this specimen and *A. trimaculatus,* Randall & Maugé (1978, p. 302) stated: 'the possibility that *armitagei* is an aberrant color form of *trimaculatus* cannot yet be discounted'. Allen (1979) subsequently placed *armitagei* in the synonymy of *trimaculatus;* however, Heemstra (1984, p. 6) wrote: 'Although *A. armitagei* is still only known from one specimen, I believe that it is a valid species.' Allen (1986), Pyle (1989), Randall (1992), and Randall & Anderson (1993) all illustrated *Apolemichthys armitagei* (the latter three recording it from the Maldives) and followed Heemstra in considering it as a valid species.

Several photographs of subadult *Apolemichthys* from the Maldives closely matching *A. armitagei* were published in the Japanese aquarium magazine *Tropical Marine Aquarium* (Takeuchi 1984, Anon. 1989). One of the photographs also shows an adult *A. trimaculatus* close by. Initially, these reports noted the similarity between the unidentified *Apolemichthys* and *A. xanthurus* Bennett (Fig. 1c), and suggested that the fish may either be an aberrant *A. xanthurus,* or a hybrid between *A. xanthurus* and *A. trimaculatus.* In subsequent issues of the magazine, however, the fish are identified as *A. armitagei.*

The senior author began to suspect that specimens identified as *A. armitagei* are examples of hybrids between *A. trimaculatus* and *A. xanthurus,* and discussion with the junior author reinforced this belief. A photograph of a fish matching *A. armitagei* was sent to the senior author by J. Charles Delbeek, who reported that the owner of the fish, Joe Randazzo, also concluded that it was a hybrid between these two species (Delbeek personal communication). As is apparent in Figures 1 a–c, the coloration of *A. armitagei* is intermediate to these two species. With *A. trimaculatus, A. armitagei* shares a predominantly yellow body and fins, soft anal-fin color with black distally and pale yellow proximally, bright yellow spots at the center of each scale, black spot on the nape, and bluish color on the lips. With *A. xanthurus,* it shares a dark region on the dorsoposterior portion of the soft dorsal fin, white or bluish white margin on the dorsal and anal fins, dusky head and thorax, and generally dusky body

Fig 1 a – *Apolemichthys trimaculatus*, Seychelles (J Randall), b – *A trimaculatus* × *A xanthurus*, Maldives (J Randall), c – *A xanthurus*, Sri Lanka (J Randall), d – *Centropyge eibli*, Thailand (J Randall), e – *C eibli* × *C flavissimus*, Cocos-Keeling (R Pyle), f – *C flavissimus*, Cocos-Keeling (R Pyle), g – *C vrolikii*, Indonesia (J Randall), h – *C flavissimus* × *C vrolikii*, Majuro (S Michael), i – *C flavissimus*, Ogasawara Islands (R Pyle), j – *C loriculus*, Hawaiian Islands (R Pyle), k – *C loriculus* × *C potteri*, at time of collection, Hawaiian Islands (R Pyle), l – *C potteri*, Hawaiian Islands (J Randall)

ground color. Some of these color characteristics are intermediate to the two presumed parent species. For example, the head and throat of the presumed hybrid are dusky, but not as dark as on *A. xanthurus.* The color of the lips on the presumed hybrid (dusky blue) is intermediate to the bright blue lip color of *A. trimaculatus* and the dusky brown of *A. xanthurus.* The ground color of the body and fins of the presumed hybrids is also intermediate to that of the presumed parent species.

Examination of two specimens formerly identified as *A. armitagei* (BPBM 32805 and BPBM 36307), revealed several morphological characters which are intermediate to the two presumed parent species. Heemstra (1984) chose to recognize *A. armitagei* as distinct from *A trimaculatus* primarily because his specimen of the former has the first four haemal spines expanded in the median plane, whereas only the first two haemal spines are expanded in all six specimens of *A trimaculatus* which

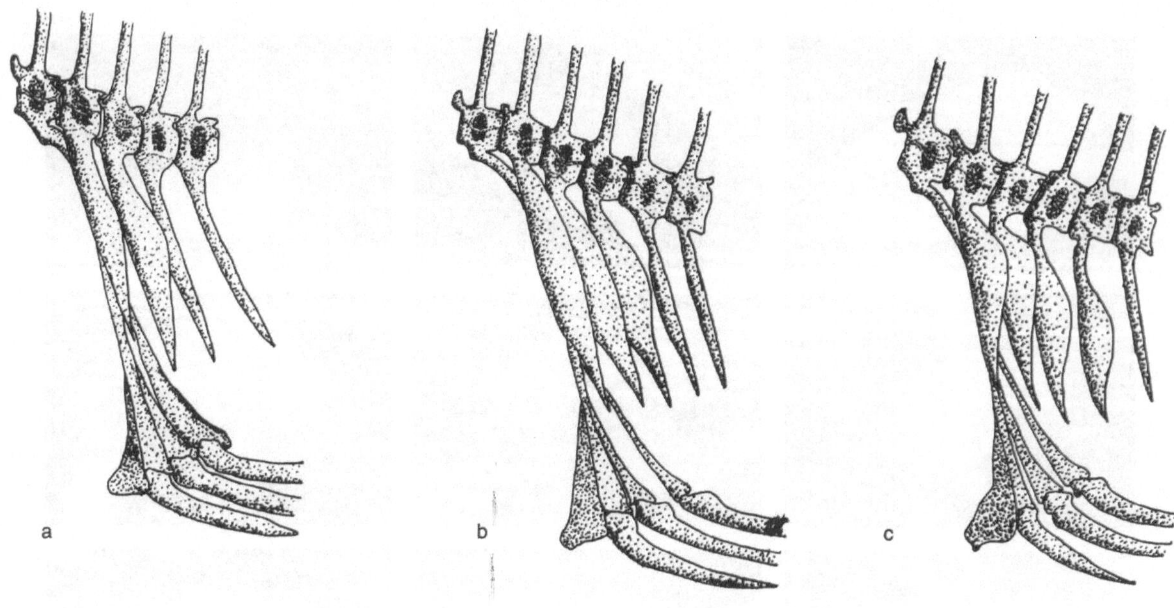

Fig. 2. a – Posteriormost precaudal vertebra plus anterior four caudal vertebrae to show expansion of first two haemal spines of *Apolemichthys trimaculatus,* 100 mm SL (re-drawn from Heemstra 1984); b – Posteriormost precaudal vertebra plus anterior five caudal vertebrae to show expansion of first three haemal spines of *A. trimaculatus* × *A. xanthurus,* BPBM 36307, 96.3 mm SL; c – Posteriormost precaudal vertebra plus anterior five caudal vertebrae to show expansion of first four haemal spines of *A. xanthurus,* BPBM 27206 (1 of 3 specimens), 103.7 mm SL.

he examined (Fig. 2a). Heemstra also noted that the bulge over the predorsal bone on the nape of *A. armitagei* is less pronounced than it is on *A. trimaculatus,* and the supracleithrum is more oblong in the former species than in the latter. We examined radiographs of six adult *A. trimaculatus* and found, as did Heemstra, that only the first two haemal spines are expanded (the third spine is very slightly expanded in two specimens). In both of our specimens of *A. 'armitagei'*, the first three haemal spines are expanded in the median plane (Fig. 2b; the 4th haemal spine slightly expanded in one of the specimens). We also examined radiographs of three specimens of *A. xanthurus* and found that the first four haemal spines are expanded in all of them (Fig. 2c). Thus, this character of *A. 'armitagei'* is intermediate to *A. trimaculatus* and *A. xanthurus.*

The nape profile and supracleithrum shape of our specimens of *A. 'armitagei'* are also intermediate to the two presumed parent species. The predorsal bulge on our specimens of *A. 'armitagei'* is less pronounced than on our adult specimens of *A. trimaculatus,* and the bulge is absent on *A. xanthurus.* The supracleithrum of *A. trimaculatus* is relatively large and circular, whereas in *A. xanthurus* it is small and oblong; it is intermediate in shape and size in our specimens of *A. 'armitagei'.*

The ranges of meristic data are broadly overlapping in our specimens of *A. trimaculatus, A. xanthurus,* and *A. 'armitagei'.* Although the ranges of most proportional measurements are similarly overlapping among these specimens, there are a few slight differences (Table 1). When ranges differ substantially between adult *A. trimaculatus* and *A. xanthurus,* the range for *A. 'armitagei'* is either more similar to that of *A. trimaculatus* (head length, depth of body, anal fin base), more similar to *A. xanthurus*

Material examined

A. trimaculatus – Maldives: BPBM 18944, 104.6 mm SL; Great Barrier Reef: BPBM 13668, 132.5 mm SL; Solomon Islands: BPBM 16146, 112.0 mm SL; Guam: BPBM 5843, 128.8 mm SL; BPBM 6357, 127.2 mm SL; BPBM 8553, 116.5 mm SL. *A. xanthurus* – Sri lanka: BPBM 27206, 3: 98–103.7 mm SL. *A. trimaculatus* × *A. xanthurus* – Maldives: BPBM 32805, 65.5 mm SL; BPBM 36307, 96.3 mm SL.

(predorsal length, preanal length, second and third anal spine length, and preopercle spine length), or intermediate (length of dorsal fin base).

Additional evidence supporting the conclusion that *A. armitagei* is a hybrid is its rare occurrence in nature. Such a level of scarcity might be expected of deep-dwelling species or species restricted to remote localities, such as *Apolemichthys guezei* Randall & Maugé (known only from the holotype taken at a depth of 70 m at Réunion). However, all observed or collected specimens of *A. 'armitagei'* have been at moderate depths (10–40 m) in the Seychelles and Maldive islands, localities regularly visited by divers. Furthermore, to our knowledge, only solitary individuals have ever been observed.

Centropyge eibli × *C. flavissimus*

Allen & Steene (1979) first reported this hybrid in their list of fishes from Christmas Island, Indian Ocean. Allen (1979, p. 315) reported collecting a 70 mm specimen of this hybrid, and presented an underwater photograph of it taken at a depth of 18 meters at Christmas Island. The fish in his photograph is clearly intermediate in color to *C. eibli* Klausewitz (Fig. 1d) and *C. flavissimus* (Cuvier) (Fig. 1f). A second individual with similar coloration was observed but not collected (Allen personal communication). In 1991, the senior author purchased an identically-colored *Centropyge* from an aquarium-fish importer (BPBM 36308; Fig. 1e; Pyle 1992a). This fish was collected in the Cocos-Keeling Islands, west of Christmas Island in the Indian Ocean. We can find no substantial meristic or morphological differences between *C. eibli*, *C. fla-*

Material examined

C eibli – Indonesia BPBM 17567, 58 mm SL, BPBM 19236, 51 mm SL *C flavissimus* – Coral Sea BPBM 33663, 2 32&64 mm SL, Vanuatu BPBM 1074, 83 mm SL, Guam BPBM 190, 93 mm SL, BPBM 9198, 68 mm SL, Marcus BPBM 7095, 34 mm SL, BPBM 7096, 3 26–63 mm SL, Wake BPBM 4221, 75 mm SL, BPBM 15344, 73 mm SL, Enewetak BPBM 6392, 2 53&87, BPBM 8027, 48 mm SL, BPBM 8218, 2 22&23 mm SL, BPBM 8855, 5 55–76 mm SL, BPBM 10784, 2 60&75 mm SL, BPBM 17749, 2 26&44 mm SL, BPBM 27804, 67 mm SL, BPBM 29118, 68 mm SL BPBM 29225, 2 60&74 mm SL, Samoa BPBM 5182, 76 mm SL, Line Islands BPBM 7535, 4 66–92 mm SL, BPBM 7562, 40 mm SK, BPBM 14069, 18 mm SL, BPBM 14119, 2 34&65 mm SL, Cook Islands BPBM 5634, 3 77–96 mm SL, BPBM 10831, 33 mm SL, Society Islands BPBM 6044, 3 23–75 mm SL, BPBM 6048, 4 18–57 mm SL, BPBM 8356, 2 60&71 mm SL, BPBM 10278, 2 67&83 mm SL, BPBM 10947, 4 57–79 mm SL, BPBM 11279, 2 17&33 mm SL, BPBM 13568, 61 mm SL, BPBM 15154, 10 56–94 mm SL, Tuamotu Archipelago BPBM 13082, 3 76–80 mm SL, BPBM 15158, 2 57&75 mm SL, Austral Islands BPBM 13710, 88 mm SL, Marquesas Islands BPBM 10318, 73 mm SL, BPBM 11690, 2 21&74 mm SL, BPBM 12150, 7 28–63 mm SL, BPBM 12316, 86 mm SL, Rapa BPBM 12982, 88 mm SL, Pitcairn BPBM 13241, 5 43–60 mm SL, BPBM 16940, 2 44&71 mm SL, BPBM 30201, 2 39&62 mm SL, Ducie BPBM 12240, 83 mm SL, Unknown locality BPBM 4220, 85 mm SL, BPBM 11090, 70 mm SL *C eibli* × *C flavissimus* – Cocos-Keeling BPBM 36308, 78 5 mm SL

Table 1 Ranges for selected proportional measurements of six specimens of *Apolemichthys trimaculatus*, two specimens of *A 'armitagei'*, and three specimens of *A xanthurus* All values except standard length are expressed as a percentage of standard length

Character	*A trimaculatus*	*A 'armitagei*	*A xanthurus*
Standard length (mm)	104 6–132 5	65 5–96 3	98 0–103 7
Head length	25 5–26 4	26 2–27 3	24 4–26 0
Body depth	53 1–57 9	55 0–56 3	58 3–61 5
Snout to dorsal fin origin	27 1–30 2	30 4–32 5	30 6–31 7
Snout to anal fin origin	66 4–70 4	61 5–64 1	64 0–66 0
Dorsal-fin base	70 0–73 7	72 5–75 0	74 8–75 8
Anal-fin base	32 9–36 3	24 7–36 6	37 6–38 2
Length of 1st anal-fin spine	11 9–13 7	14 5–15 6	12 5–14 0
Length of 2nd anal-fin spine	17 5–19 0	20 6–21 8	19 0–20 4
Length of 3rd anal-fin spine	21 6–23 5	23 5–25 4	23 5–24 5
Length of preopercular spine	14 1–16 7	9 9–11 8	10 7–14 7

vissimus, or the probable hybrid. Meristic data and proportional measurements for the hybrid are included in Table 2. To our knowledge, no other specimens of this hybrid have been reported.

Centropyge eibli × *C. vrolikii*

In his book on Indonesian fishes, Kuiter (1992) noted that hybrids between *Centropyge eibli* (Fig. 1d) and *C. vrolikii* Bleeker (Fig. 1g) are common in Indonesia, and several aquarium fish importers have reported individuals of *Centropyge* intermediate in color to these two species, all apparently originating from Indonesia. However, this hybrid has not been formally documented. The senior author has seen several of these presumed hybrids in aquarium fish shops, and he obtained a specimen of one (BPBM 36309) which was intermediate in color to the two presumed parent species (Fig. 1d, g). Meristic and morphological data of this specimen are included in Table 2. These data fall within the ranges for our specimens of *C. eibli* and *C. vrolikii.* Although we do not have photographs of this presumed hybrid, individuals apparently resulting from this cross have been illustrated elsewhere. For example, figure 162 of Steene (1978, p. 109) is labelled as a subadult *C. vrolikii,* but it seems to be a hybrid of this species and *C. eibli* (the same photograph, also labelled as *C. vrolikii,* appears in Takeshita 1976). The illustrated fish resembles *C. vrolikii,* but the thorax, pelvic fins, and interspinous membranes are orange, and faint dark bars are visible on the body. These color characters are typical of *C. eibli* but are not observed in *C. vrolikii.* Another fish, illustrated in Shirai (1986, p. 234), also appears to be a hybrid of these two species. Like the fish illustrated by Steene, it is most similar to *C. vrolikii.* Although it lacks the orange coloration on the thorax and dorsal fin, dark bars are clearly evident on the body, and the black coloration of the tail does not extend anteriorly on the body as far as on *C. vrolikii* (black on *C. eibli* is limited to the caudal fin, caudal peduncle, and posterior portion of the soft dorsal fin). Also, faint orange markings on the anal fin (reminiscent of those on *C. eibli*) are evident.

Kuiter (personal communication) informs us that these intermediate forms are more common at Flores in the Lesser Sunda Islands in Indonesia, where 'pure' *C. eibli* is scarce, than at Bali to the west, where individuals of typical *C. eibli* are more often observed. He also suggests that there is a gradation in the frequency of intermediates across the Lesser Sunda region, indicating that it may represent a true 'hybrid zone' (sensu Harrison 1993; see Discussion below).

Centropyge flavissimus × *C. vrolikii*

Takeshita (1976) was the first to report on the hybrid of these two species from fish imported to Modern Pet Shop in Honolulu from Majuro, Marshall Islands. He wrote: 'Dr. John E. Randall's records show that he obtained a *C. flavissimus* × *C. vrolikii* hybrid from Modern Pet Shop in Honolulu in 1969. In 1970 he went to Majuro and observed that these hybrids were very common there in all intermediate stages. Dr. Randall will be publishing a scientific paper on this hybrid in the near future (personal communication, Nov 1975).' Because Takeshita documented this hybrid well, there was no subsequent paper by the junior author.

Takeshita quoted field observations of Colin Young in Majuro on the depth at which hybrids were collected (between 1.5 and 7 m) and their relative abundance. In areas where these two angelfishes predominate, approximately 10% are hybrids. Of these, 20% are what Young called 'true' hybrids (i.e., directly intermediate in color between the two parent species) and the rest he called 'intermediate' hybrids (individuals closer in color to one or the

Material examined

C. eibli – see listing above. *C. vrolikii* – Indonesia: BPBM 32169, 61 mm SL; BPBM 34167, 67 mm SL; Philippines: BPBM 10927, 5: 58–75 mm SL; Ryukyu Islands: BPBM 6827, 77 mm SL; BPBM 19155, 83 mm SL; Great Barrier Reef: BPBM 14454, 3: 64–65 mm SL; Solomon Islands: BPBM 16161, 2: 50&54 mm SL; Lord Howe: BPBM 14817, 95 mm SL; Palau: BPBM 6890, 2: 62&70 mm SL; BPBM 8089, 49 mm SL; BPBM 9231, 2: 50&63 mm SL; BPBM 9362, 57 mm SL; BPBM 9806, 3: 50–52 mm SL; BPBM 10191, 53 mm SL; BPBM 13493, 28 mm SL; Pohnpei: BPBM 9670, 55 mm SL. *C. eibli* × *C. vrolikii* – Indonesia: BPBM 36309, 54.8 mm SL.

other parent species). These hybrids are regularly imported into the aquarium trade from Majuro and Pohnpei (Ponape). A wide variety of color forms have been observed, ranging from nearly 'pure' *C.*

Table 2 Proportional measurements and meristic data of four suspected hybrid *Centropyge* specimens All values except standard length and meristic data are expressed as a percentage of standard length

Character	*C eibli* × *C flavissimus* BPBM 36308	*C eibli* × *C vrolikii* BPBM 36309	*C loriculus* × *C potteri* BPBM 36310	*C heraldi* × *C bispinosus*? SMF 23661
Standard length (mm)	78 5	54 8	55 2	66 5
Head length	26 1	28 5	27 9	28 9
Diameter of orbit	8 7	10 2	10 0	9 6
Snout length	6 6	5 8	7 2	6 2
Bony interorbital width	8 3	8 4	9 1	7 7
Body width	21 5	17 9	21 6	18 2
Body depth	49 7	53 3	52 7	51 9
Caudal peduncle depth	14 1	14 8	13 8	14 1
Caudal peduncle length	14 3	14 1	13 8	12 6
Snout to dorsal-fin origin	32 6	33 9	36 2	33 7
Snout to pelvic-fin insertion	36 3	36 1	36 1	36 4
Snout to anal-fin origin	60 4	60 9	59 4	58 9
Longest pectoral-fin ray	27 3	27 4	30 8	23 5
Caudal fin length	23 6	24 3	23 7	24 1
Base of dorsal fin	67 1	65 3	66 1	68 9
Base of spinous dorsal fin	43 9	42 9	40 9	46 8
Base of soft dorsal fin	23 2	24 6	24 1	21 4
Base of anal fin	31 8	33 0	34 2	35 2
Pelvic spine length	19 0	20 4	19 0	18 8
Pelvic-fin length	31 2	33 4	33 3	23 6
First dorsal-fin spine length	9 6	8 8	8 2	9 0
Second dorsal-fin spine length	12 7	12 4	12 5	13 7
Third dorsal-fin spine length	14 8	14 8	14 9	15 3
Fourth dorsal-fin spine length	16 3	16 2	16 3	16 5
Fifth dorsal-fin spine length	17 3	17 2	16 8	16 7
Last dorsal-fin spine length	21 8	20 4	18 7	21 2
First anal-fin spine length	12 9	13 3	12 3	13 4
Second anal-fin spine length	17 8	18 1	16 7	19 1
Third anal-fin spine length	24 1	22 4	21 6	24 2
Preopercle-spine length	14 4	13 0	11 8	11 7
Dorsal fin	XIV, 16	XIV, 16	XIV 18	XV, 16
Anal fin	III, 16	III, 16	III, 18	III, 18
Caudal fin	10 + 11	10 + 11	10 + 11	10 + 11
Pectoral fin	16	16	17	17
Pelvic fin	I, 5	I, 5	I, 5	I, 5
Lateral-line scales	28	34	31	34
Scale rows above lateral line	7	7	6	8
Scale rows below lateral line	25	22	20	24
Gill rakers	5 + 13	5 + 12	7 + 17	5 + 12
Vertebrae	10 + 13	10 + 13	10 + 13	10 + 13

flavissimus to nearly 'pure' *C. vrolikii,* with every combination in-between. There is an additional color anomaly of bright blue spots on the posterior portion of the body and median fins which is often, but not always, associated with this hybrid (Fig. 1h). We have nine specimens of the blue-spotted color form, eight of which were collected at Majuro (the other at Enewetak, Marshall Islands). We can find no consistent differences in the meristic data and proportional measurements between the hybrids and either of the parent species (ranges for the parent species broadly overlap). Additional color illustrations showing several varieties of this hybrid have been published in Takeshita (1976: fig. on pp. 30 and 31), Steene (1978: fig. 163, 164), and Wedge (1984). This hybrid has also been observed or collected at Guam (Myers 1989), Kosrae (Donald Baker personal communication) and in the Ryukyus Islands (Hiroyuki Tanaka personal communication).

Centropyge loriculus × *C. potteri*

The first known specimen of this hybrid, a juvenile of about 30 mm standard length (SL), was collected at Oahu in 1990 by aquarium fish collector Dennis Yamaguchi. It was found living among a group of *C. loriculus.* A photograph of it appeared in a Japanese aquarium magazine. A second, larger individual (about 40 mm SL) was collected less than a year later by Ivan Bonilla, also at Oahu. This one, nearly identical in color to the first, was found with a group of *C. potteri.* It was donated to the senior author (Pyle 1992b), who maintained it in an aquarium for approximately one year. When first obtained, this fish more closely resembled *C. loriculus* (Günther) (Fig. 1j) in color (Fig. 1k), but also showed elements of color pattern characteristic of *C. potteri* Jordan & Mets (Fig. 1l). As it grew, it became increasingly similar to *C. potteri* until the time of its death (Fig. 3a). A third individual was subsequently collected at the island of Hawaii by aquarium fish collector Pete Basabe, who sent the senior author photographs of the fish. The color of this fish was nearly identical to the first two. Randy Fernley of Coral Fish Hawaii recently informed the senior author of a fourth individual, also taken at Oahu. Meristic data and proportional measurements of one of these specimens (BPBM 36310) are included in Table 2. The ranges for these data are nearly identical for our Hawaiian *C. loriculus, C. potteri,* and the hybrid specimen.

Centropyge multifasciatus × *C. venustus*

Krupp & Debelius (1990) described and illustrated two pomacanthids from the Philippines which appear to represent natural hybrids between *Centropyge multifasciatus* (Smith & Radcliffe) and *C. venustus* (Yasuda & Tominaga). In their description of this hybrid, Krupp & Debelius briefly commented on the generic status of the presumed parent species; at the time, *C. venustus* was assigned to the genus *Holacanthus* and *multifasciatus* to *Centropyge* (thus these hybrids would have been inter-generic). Burgess (1991) described the new genera *Paracentropyge* and *Sumireyakko* for *C. multifasciatus* and *C. venustus,* respectively. Pyle & Randall (1993), however, cast doubt on the validity of the latter genus, presenting evidence that *multifasciatus* and *venustus* (along with their recently described *Centropyge boylei*) are congeners. Until a systematic study of Pomacanthidae is completed, we prefer to place these species in *Centropyge;* however, they may belong to *Paracentropyge* (the latter genus having page-priority over *Sumireyakko*).

Pyle & Randall (1993) noted that at least three

Material examined

C. flavissimus × *C. vrolikii* – Guam: BPBM 17335, 68 mm SL; Enewetak: BPBM 8880, 62 mm SL; Majuro: BPBM 8780, 71 mm SL; BPBM 19209, 56 mm SL; BPBM 19557, 61 mm SL; BPBM 19679, 6: 55–72 mm SL; BPBM 26831, 2: 68&70 mm SL; BPBM 26832, 2: 58&77 mm SL; BPBM 28178, 69 mm SL. C. flavissimus – see listing above. *C. vrolikii* – see listing above.

Material examined

C. loriculus – Hawaiian Islands: BPBM 6001, 2: 54.2–55.4 mm SL; BPBM 6998, 47 mm SL; BPBM 8898, 46.6 mm SL; BPBM 10944, 55.8 mm SL. *C. potteri* – Hawaiian Islands: BPBM 6007, 55.6 mm SL; BPBM 6989, 42.5 mm SL; BPBM 9337, 57.3 mm SL; BPBM 22389, 44.1 mm SL. *C. loriculus* × *C. potteri* – Hawaiian Islands: BPBM 36310, 55.2 mm SL.

Fig. 3. a – *Centropyge loriculus* × *C. potteri,* at time of death, Hawaiian Islands (J. Randall); b – *C. heraldı* × *C. bıspınosus?*, Philippines (Courtesy *Tropical Marine Aquarium* Magazine); c – *C. heraldi,* Ryukyu Islands (R. Pyle); d – *C. bıspınosus* (J. Randall); e – *C. bispinosus* × *C. shepardi?,* Guam (R. Pyle); f – *C. shepardi,* Guam (J. Randall); g – *Chaetodontoplus caeruleopunctatus,* Philippines (R. Pyle); h – *C. caeruleopunctatus* × *C. septentrionalis*?, southern Japan (Y. Ohkata); i – *C. septentrionalis,* southern Japan (J. Randall); j – *C. melanosoma,* Indonesia (J. Randall); k – *C. 'chrysocephalus',* southern Japan (J. Randall); l – *C. 'chrysocephalus'* with two *C. septentrionalis,* southern Japan (Y. Ohkata).

other individuals of this apparent cross have been imported into the United States, all apparently originating from the Philippines. Most reports suggest that the hybrids were collected in the northern Philippines, which is likely to be the only area of sympatry of these two species. *Centropyge venustus* ranges from the Ryukyu islands southward to Taiwan and northern Philippines, and *C. multifasciatus* is broadly distributed from the Society Islands to the western Pacific and the Cocos-Keeling islands, but is not known from the Ryukyus.

Holacanthus bermudensis × *H. ciliaris*

This is the first angelfish hybrid that was reported in the literature. The queen angelfish, *Holacanthus ciliaris* (Linnaeus), and the blue angelfish, *H. ber-*

mudensis Goode, occur in the tropical western Atlantic. *H. isabelita* (Jordan & Rutter) is the name used for the blue angelfish in papers prior to 1970. Bailey et al. (1970), however, showed that it is a junior synonym of *H. bermudensis.* As noted by Feddern (1968), *H. bermudensis* is rare in the West Indies where *H. ciliaris* is common, but in the Florida Keys it is more common than *H. ciliaris.* Nichols & Mowbray (1914) described *H. townsendi* as a new species of angelfish from Key West. Longley in Longley & Hildebrand (1941) concluded that it is a probable hybrid of *H. bermudensis* and *H. ciliaris,* and Feddern (1968) confirmed this. Color illustrations of this hybrid have appeared in Allen (1979: Fig. 425, 426) and Robins et al. (1986: pl. 37).

Pomacanthus arcuatus × *P. paru*

The gray angelfish, *Pomacanthus arcuatus* (Linnaeus) and the French angelfish, *Pomacanthus paru* (Bloch) are also tropical Atlantic species. A hybrid of the two was reported by Moe (1976); however this was produced under laboratory conditions. In correspondence to the authors, Moe (in lit. 26.2.1994) wrote: 'We made artificial hybrids of some Atlantic species of angelfish while I was running the fish farm in the Florida Keys. We crossed them by extracting and mixing the gametes; no natural courtship behavior was involved. Males and females of *P. arcuatus* and *P. paru* were crossed both ways. There was no apparent difference in survival or development either way. In coloration, some juveniles seem to favor *P. paru* and others, *P. arcuatus.*' Commenting on the possibility that this hybrid may occur naturally, he continued: 'I don't know of any [Atlantic] *Pomacanthus* hybrids occurring naturally, either juveniles or adults, although one occasionally finds a specimen that is suspicious. The hybrids [produced under laboratory conditions] were variable in coloration and, as juveniles, not clearly discernible from juvenile *P. arcuatus.* Even after close observation in an aquarium, one might be suspicious of a natural juvenile but there was no feature that I saw in the known hybrids that would give a positive hybrid ID to a wild juvenile. Tank reared marine fish also exhibit a lot of malformation and unnatural coloration due, I suspect, to problems with water quality and nutrition in the mid to late larval stage. These artifacts of the rearing process further confuse the hybrid issue.'

Pomacanthus chrysurus × *P. maculosus*

Allen (1979) illustrated an angelfish photographed in Kenya which appears to represent a hybrid of *Pomacanthus chrysurus* (Cuvier) and *P. maculosus* (Forsskål). About this fish Allen (p. 305) wrote: 'The color pattern is a compromise between that of the two presumed parents. The anal fin has the coloring of *P. maculosus,* but the caudal is bright yellow as in *P. chysurus.* The curved yellow bars on the sides are also inherited from the *chrysurus* parent, but close examination reveals the presence of dusky scale margins on the anterior half of the body, a typical feature of *maculosus.* The head coloration is also similar to *maculosus* and the fish lacks the bright yellow ring on the upper portion of the gill cover which is found in *chrysurus.* In addition, the anal fin shape represents a compromise between the presumed parents. It is intermediate in shape between the rounded fin of *chrysurus* and the very pointed and elongate fin of *maculosus.*'

Condé (1990) also illustrated this apparent hybrid, and included a photograph of a second individual from Kenya. The latter photograph also appeared in Debelius (1993). Condé used the genus *Pomacanthodes* Gill for *chrysurus* and *Pomacanthops* Smith for *maculosus;* the hybrid would thus be intergeneric. However, at present, we prefer to regard these two genera as subgenera of *Pomacanthus.*

Pomacanthus maculosus × *P. semicirculatus*

Condé (1990) described a specimen collected at Mombasa, Kenya, which appears to represent a hybrid of *Pomacanthus maculosus* and *P. semicirculatus* (Cuvier). His specimen, which was reared in captivity from a juvenile to an adult and lived nearly ten years, more closely resembled the former species as a juvenile, and the latter as an adult. Another adult individual of this presumed hybrid was observed and photographed at southern Oman by Gary Rhodes. Although this fish exhibits a prominent yellow bar on the side of the body, it more closely resembles *P. semicirculatus* than it does *P. maculosus.*

Pomacanthus sextriatus × *P. xanthometapon*

Steene (1978, p. 123) illustrated an apparent hybrid of *Pomacanthus sexstriatus* (Cuvier) and *P. xanthometapon* (Bleeker) (he classified both species in the genus *Euxiphipops* Fraser-Brunner, which we regard as a subgenus of *Pomacanthus*). The specimen was collected off Queensland, Australia, and was found paired with a similarly-sized *P. sextriatus.* Pairs of both presumed parent species were also observed in the area. To our knowledge, no other examples of this hybrid have been reported.

Possible pomacanthid hybrids

In addition to the presumed and probable pomacanthid hybrids listed above, five other examples which may represent hybrids are described below:

Centropyge bispinosus × *C. heraldi?*

Figure 3b is an individual of the genus *Centropyge* which was imported to Japan from the Philippines through the marine aquarium trade in 1985. The same photograph was published in the Japanese aquarium magazine *Tropical Marine Aquarium* (No. 11, Winter 1985, p. 58) along with the following text (translated from Japanese): 'This angelfish must be a *Centropyge* and it has a great possibility of being a hybrid. The color of the belly is brownish yellow, the dorsal region appears somewhat darker purple, and the head color and the color of the fins are also purple. The anal fin is unusual with orange lines, similar to [*C. heraldi* Woods & Schultz]. It must be a hybrid between [*heraldi*] and [*C. tibicen* (Cuvier)].'

The same photograph appeared in a subsequent issue of the magazine, but was identified as a hybrid between *C. heraldi* (Fig. 3c) and *C. bispinosus* Günther (Fig. 3d). In 1989, Friedhelm Krupp of the Senckenberg Museum in Frankfurt, sent on loan an unidentified specimen of *Centropyge* collected in the Philippines (SMF 23661) to the senior author for examination. Although the specimen had lost it's life color while in preservative, enough of the original pattern remained to confirm that it was nearly identical to the fish discussed above. In early 1992, Julian Sprung sent photographs of a similar *Centropyge* which he purchased from an aquarium fish shop in Florida. It was larger in size than either of the other fish, and was proportionally deeper-bodied (it is not unusual for species of *Centropyge* to become proportionally deeper-bodied as they grow large). This fish had a more extensive dark region on the dorsoposterior body and on the soft anal fin. Also, Sprung described the dark color in life as greenish blue, rather than purple. He postulated that the fish may be a cross between *C. heraldi* and *C. bicolor* (Bloch), or perhaps *C. heraldi* and *C. vrolikii.* Because of the deep body, he also proposed that it may be a cross between *C. heraldi* and *C. venustus,* or *C. heraldi* and a species of *Apolemichthys.* We believe it is most likely a hybrid of *C. heraldi* and *C. bispinosus.* However, the possibility remains that these fish represent aberrant individuals of *C. heraldi* or some other *Centropyge.* Meristic data and proportional measurements for SMF 23661 are presented in Table 2.

Centropyge bispinosus × *C. shepardi?*

Moyer (1981) reported on observations of natural spawning between a male *Centropyge shepardi* Randall & Yasuda (Fig. 3f) and a female *C. bispinosus* at Guam, and included underwater photographs of these two individuals interacting (one of these photographs is reproduced in Thresher 1984). While diving in Guam in 1990, the senior author photographed an unusual *Centropyge* at a depth of

---

Material examined
C. bispinosus × *C. heraldi* – Philippines: SMF 23661, 66.5 mm SL.

about 45 meters (Fig. 3e). This may represent a hybrid between *C. bispinosus* and *C. shepardi.* The fish is intermediate in color to the postulated parent species. At first glance it seemed most similar to *C. bispinosus;* however, it shares with *C. shepardi* a translucent yellowish tail and a dark blue patch on the side of the body posterior to the operculum and dorsal to the base of the pectoral fin. In their original description of *C. shepardi,* Randall & Yasuda (1979) stated that the species was most similar to *C. bispinosus* [however, Moyer (1989) suggested that *shepardi* was closer to *C. ferrugatus* Randall & Burgess]. With regard to *C. shepardi,* Myers (1989, p. 161) noted: 'It is quite variable in color: the ground color may range from an almost red to a light apricot and the barring can be reduced to a small patch behind the operculum, or in rare cases, be entirely absent.' Similarly, *C. bispinosus* is also extremely variable in color (e.g., Kosaki & Toyama 1987). Thus, it is not certain whether the fish in Figure 3e is a hybrid, or is an aberrant individual of either *C. bispinosus* or *C. shepardi.*

Chaetodontoplus caeruleopunctatus × *C. septentrionalis?*

The photograph reproduced here as Figure 3h was taken by Youji Ohkata in Kashiwajima, Japan. It shows a fish with a body coloration like that of *C. caeruleopunctatus* Yasuda & Tominaga (Fig. 3g), but with head coloration similar to *Chaetodontoplus septentrionalis* (Temminck & Schlegel) (Fig. 3i). Thus, it may represent a hybrid between these two species. Because we have no specimens upon which to base conclusions, we can only speculate about the possible origin of this fish's coloration. It should be noted that the distribution of the two postulated parent species overlap, and they are apparently indistinguishable except for color (Yasuda & Tominaga 1976).

Chaetodontoplus melanosoma × *C. septentrionalis?*

The pomacanthid *Chaetodontoplus chrysocephalus* Bleeker (Fig. 3k) was described from a single specimen collected in Indonesia, and has remained poorly known ever since. Most specimens have been observed or collected in the region between Taiwan and southern Japan. Because no juveniles have ever been reported, and because numerous individuals are known which exhibit intermediate color to this species and *C. septentrionalis,* it has been suspected by several authors (e.g., Allen 1979, Masuda et al. 1984) that specimens of *C. chrysocephalus* represent the adult male form of *C. septentrionalis.* Photos like the one presented here as Figure 3l, as well as observations by the senior author and others (Hajime Masuda personal communication) of the two forms socially interacting, lend support to this hypothesis. However, we propose an alternate hypothesis; that specimens of *C. chrysocephalus* may represent hybrids between *C. septentrionalis* and *C. melanosoma* (Bleeker) (Fig. 3j). We base this hypothesis primarily on the observation that the '*chrysocephalus*' form is somewhat intermediate to these two species in color. Allen (1979) noted that *C. melanosoma* occasionally exhibits an unusual color form with an entirely yellow caudal fin (unlike the more usual brownish black caudal fin with yellow margin). The caudal fin of *C. septentrionalis* is all yellow, so the 'unusual' form of *melanosoma* may represent a back-cross between hybrid '*chrysocephalus*' and *C. melanosoma.* Aberrant *C. septentrionalis* that are occasionally observed may also represent back-crosses of the hybrid with *C. septentrionalis.* The absence of juvenile *C. chrysocephalus* could be explained by the fact that juveniles of the two proposed parent species (*melanosoma* and *septentrionalis*) are very similar in color to each other (underscoring their possible close relationship) until they begin to mature. Hybrid juveniles, intermediate between these similar forms, would probably not be noticed or would be assumed to be a slightly aberrant form of one of these two species. In any case, more research is needed on these enigmatic species of *Chaetodontoplus.*

Pomacanthus navarchus × *P. xanthometapon?*

Myers (1989: pl. 72E) includes a color illustration of a fish which appears to represent a hybrid of *Poma-*

canthus navarchus and *P. xanthometapon*. The fish shares color characteristics of both possible parent species, and was maintained on display at the Steinhart Aquarium.

★

In addition to the probable and possible examples of pomacanthid hybrids listed above, several other observed anomalous pomacanthids may represent hybrids as well. The senior author observed a fish in the Ogasawara Islands (formerly Bonin Islands) which may represent a hybrid of *Centropyge ferrugatus* and *C. shepardi*. Aquarium fish collector Fenton Walsh reported that he collected two specimens of what appeared to him to be a hybrid of *Centropyge loriculus* and *C. shepardi* at Pohnpei, Caroline Islands. Rohan Petiyagoda sent the authors photographs of an unusual *Centropyge* collected in Sri Lanka that generally resembles *C. flavipectoralis* Randall & Klausewitz, but differs in having yellow areas on the median fins and lips. It may represent a hybrid between *C. flavipectoralis* and the closely-related *C. multispinis* Playfair, both of which occur in Sri Lanka. Other possible pomacanthid hybrids reported by aquarium fish collectors include *Holacanthus clarionensis* × *H. passer*, *Pomacanthus asfur* × *P. maculosus*, and *P. imperator* × *P. semicirculatus;* however these await confirmation.

In addition to the example of *Pomacanthus arcuatus* × *P. paru* described above, several other attempts or observations of captive cross-breeding between different pomacanthid species have been made. Moe (1976) reported on an attempt to cross-breed a male *Holacanthus bermudensis* with a female *P. arcuatus*. In correspondence to the authors, Moe (in lit. 27.2.1994) wrote: 'We obtained about 50% fertilization, and after 15 hours, only about 100 eggs still displayed embryonic development. Most were abnormal in development, the embryos were shrunken and abnormal cell masses occurred at the tail of the embryo. Hatching evidently did not take place.' Moe (1990) also attempted to produce captive hybrids between a female *Pomacanthus arcuatus* and a male *Holacanthus tricolor* (Bloch). Although fertilization and embryo development through hatching occurred, the free embryos did not survive more than three days. Other observations of heterospecific pomacanthid spawnings in aquaria include *Centropyge argi* Woods & Kanazawa × *C. resplendens* Lubbock & Sankey (J. Charles Delbeek personal communication) and *C. argi* × *C. acanthops* (Norman) (based on photos published in several issues of the Japanese aquarium magazine *Tropical Marine Aquarium*).

Discussion

Identification of hybrids

Most marine fish hybrids are initially recognized by their unusual or intermediate color pattern. In order to obviate the possibility that the different color is not just an aberrant pattern of one of the presumed parent species, corroborative evidence of intermediate morphology between the parent species should be sought. Because proportional morphology often broadly overlaps between closely-related pomacanthid species, color patterns are often used as diagnostic characters in this family of fishes. Many of the probable pomacanthid hybrids described above are between two closely-related species, so there may be no obvious morphological differences of a magnitude whereby a clear intermediate state can be documented. One of the main difficulties in establishing the status of an unusually-colored pomacanthid as a hybrid, therefore, is distinguishing probable hybrids from aberrant color forms of a single species. Aberrant and anomalous pomacanthids have been reported for numerous species, including *Genicanthus watanabei* Yasuda & Tominaga and *Holacanthus ciliaris* (Allen 1979), *Pygoplites diacanthus* (Maugé & Randall 1979), *C. bispinosus* and *C. potteri* (Kosaki & Toyama 1987), *Centropyge multispinis* (Krupp & Debelius 1990), and several others. As discussed above, it is not clear whether the unusual *Centropyge* from Sri Lanka represents a hybrid between *C. flavipectoralis* and *C. multispinis*, or simply an aberrant *C. flavipectoralis*. An all-yellow variety of the latter species has been illustrated in a Japanese aquarium magazine. In some cases, evidence supporting hybridization in anomalous pomacanthids comes from observations over time in aquaria. For example, the *Pomacanthus maculosus* × *P. semicir-*

culatus described by Condé (1990) was initially collected as a juvenile which mostly resembled juvenile *P. maculosus,* but over the course of several years in captivity, the same fish grew to resemble an adult *P. semicirculatus.* Similarly, the *Centropyge loriculus* × *C. potteri* hybrid maintained by the senior author initially most closely resembled *C. loriculus,* but grew to resemble *C. potteri.* The conclusion of hybridization in the latter case is also supported by the fact that the first suspected hybrid of this cross was collected among *C. loriculus,* and the second with *C. potteri.*

Commenting on the preponderance of freshwater fish hybrids over marine, Randall (1956) suggested that conditions are more favorable for hybridization of fishes in the freshwater environment, and there is greater opportunity to sample populations from freshwater habitats. Two questions are brought forth in this suggestion: (1) what biological and/or environmental factors are conducive for hybridization; and (2) how variable is the ability of researchers to observe, recognize, and document naturally occurring hybridization events? The latter may have affected the number of marine fish hybrids which have been described. As mentioned, the marine fish genus for which the greatest number of hybrids have been reported is the butterflyfish genus *Chaetodon* Linnaeus (Chaetodontidae). Fifteen probable natural crosses in this genus have been documented and at least twelve others have been observed. Randall & Fridman (1981, p. 116) wrote: 'It is of interest to speculate whether this genus has a predilection for hybridization, perhaps by virtue of so many closely related species existing together in the coral reef habitat, or whether the hybrids are simply more easily recognized than most other marine fishes because of the distinctive color patterns. An additional factor is the greater exposure these showy fishes receive because they are sought by aquarists, underwater photographers, and fish watchers.' Likewise, species of pomacanthids are also distinctively colored and sought after by aquarists; therefore, hybrids of this family are more apt to be easily recognized. Indeed, many of the hybrids described above were brought to the attention of the authors by aquarists or aquarium fish collectors. It is conceivable that many other hybrids between marine species occur in nature, but are overlooked because differences between similar species are more subtle and less visually obvious (thus, hybrids with intermediate appearances would not be easily noticed). Grant & Grant (1992) point out similar patterns and problems in reported hybrids of birds.

Factors favoring hybridization

The question of which ecological and social situations are most conducive for hybridization to occur is also worthy of consideration. In discussing hybrids among chaetodontid species, several authors cite Reese (1975), who demonstrated that some butterflyfishes form essentially permanent (life-long) pairs, while others occur alone or in aggregations. Allen (1979, p. 314) wrote: 'Most of the butterflyfishes and angelfishes that are known to hybridize belong to solitary or pair-forming species.' While this may be true of many of the butterflyfish species and some of the larger pomacanthids, most species of *Centropyge* form harems comprised of a single male and from two to seven or more females (Moyer & Nakazono 1978, Moyer 1990, Lutnesky 1992a).

Many authors (e.g., Hubbs 1961, Reese 1975, Meyer 1977, Randall et al. 1977, Allen 1979, Moyer 1981) note that marine fish hybrids may occur when one species of a closely related pair is rare and the other abundant. If the rare species cannot locate a suitable conspecific mate, it may spawn with a closely-related, more abundant species. This hypothesis was used by Allen (1979) to account for the *Centropyge eibli* × *C. flavissimus* hybrid at Christmas Island, Indian Ocean (and also applies to the same hybrid collected at Cocos-Keeling). Because *C. eibli* is very rare at these localities, solitary individuals may be unable to locate other conspecifics at spawning time, and therefore might spawn instead with *C. flavissimus.* Moyer (1981) invoked the same hypothesis in discussing his observation of spawning between *C. bispinosus* and *C. shepardi* at Guam, where the former species is very rare and the latter species is ubiquitous. It may also apply to the *C. loriculus* × *C. potteri* cross in Hawaii, where the former species is rare and the latter species is abundant. Of the eight examples of probable or possible natural pomacanthid hybrids described above for which the relative abundance of the presumed parent species is known, all involve cases where one presumed parent is rare and the other common.

A problem with this hypothesis is the question of why a member of a common species (such as *C. flavissimus* at Christmas Island, *C. shepardi* at Guam, or *C. potteri* in Hawaii) would chose to spawn with the 'rare' heterospecific individual, if conspecific mates are readily available. In the case of hybrids between species of *Centropyge*, the social system of the species may play an important role in the opportunity for hybrid formation. Species of this genus which have been studied form polygynous harems and undergo protogynous sex-reversal (see review by Moyer 1990). Moyer (1981), citing differences in the energetic costs of producing eggs versus sperm (see Trivers 1972), proposed that when interspecific spawning occurs among species of this and other haremic genera, it is most likely to be between a male of a common species and a female of a rare species. Moyer's assertion is supported by his field observations of spawning between a female *C. bispinosus* and male *C. shepardi* in Guam. It is further supported by his observation of a female *C. bispinosus* among a harem of *C. tibicen* at Miyake-jima (where the former is rare and the latter abundant); by Patrick Colin's observation of a single female *C. bicolor* among a harem of *C. heraldi* at Enewetak (Moyer 1981); and by the senior author's collection of a single female *C. multicolor* Randall & Wass among a harem of the closely-related *C. nahackyi* Kosaki at Johnston Atoll (Kosaki et al. 1991).

On the other hand, there is reason to believe the converse may be true, i.e., that a rare *Centropyge* at a given locality is more likely to spawn as a male. Lutnesky (1992a) investigated the mating behavior of *Centropyge potteri* in Hawaii. He found this species to be haremic and proposed a 'Temporal-Threshold Model of Polygynous Mating' (TMPM) to account for temporal patterns of spawning among different females within a harem (Lutnesky 1992a, b). He found that females within a harem may be ranked according to a dominance hierarchy whereby the most dominant females monopolize the male during optimal spawning times, and subordinant females spawn at sub-optimal times. It is conceivable, therefore, that a subordinant female within a harem of *Centropyge* would choose to spawn with a heterospecific male at the optimal spawning time, rather than with a conspecific male at a suboptimal time. Thus, a rare *Centropyge* might be more likely to spawn with heterospecifics as a male, than as a female. One problematic issue with the aformentioned scenario is how a solitary rare *Centropyge* would ever become a male. Lutnesky (1992, 1994) found that isolated females of the protogynously hermaphroditic *C. potteri* did not change sex even after 100 days of isolation, and concluded that the presence of additional females was a crucial factor in the initiation of sex-change in this species. It seems unlikely, therefore, that a solitary individual *Centropyge* would ever be induced to change sex in the absence of conspecific females. Nevertheless, a solitary *C. potteri* collected at Johnston Atoll (BPBM 29606; Randall et al. 1985) was found to be male (although the testes were greatly reduced). Lutnesky (personal communication) suggests that stimulation for sex-change from heterospecific females requires further investigation.

The question of whether a rare *Centropyge* is more likely to spawn with a heterospecific as a female or as a male might best be answered using molecular techniques. As mentioned above, the hybrid *Centropyge flavissimus* × *C. vrolikii* is regularly observed at both Majuro and Kosrae. Whereas in Majuro, *C. flavissimus* is the more abundant species, the reverse is true at Kosrae. Because mtDNA is inherited exclusively from the female parent, it might be possible to determine which species was the 'mother' species of each individual hybrid by comparing mtDNA sequences of the hybrids with each of the parent species (assuming consistent differences in the mitochondrial genome between the two parent species were evident). If rare *Centropyge* are more likely to obtain spawnings with heterospecifics as females (i.e., female-choice outweighted male-choice), then we would expect to find a tendency for the hybrids to share mtDNA sequences with the rare species (i.e., we would expect a tendency for the female parent of the hybrids in Majuro to be *C. vrolikii,* and in Kosrae to be *C. fla-*

vissimus). Conversely, if rare *Centropyge* are more apt to spawn with heterospecifics as males, then there would be a tendency for hybrids to share mtDNA sequences with the common species. A similar comparison could be made with *C. loriculus* × *C. potteri* hybrids between Hawaii and Johnston Atoll (*C. loriculus* is rare in Hawaii and *C. potteri* is common, and the converse is true for Johnston). Unfortunately, the latter hybrid is very rare.

Dispersal and biogeography

Centropyge loriculus × *C. potteri* hybrids are so rare, in fact, that they may be useful for examining other biological questions, particularly with regard to patterns of dispersal. Despite several decades of active aquarium fish collecting activity, this hybrid had never been reported until the four hybrid specimens were collected in the Hawaiian Islands, all within three years of each other. Furthermore, the first collected was the smallest; the second was larger; and the third was larger still (the size of the fourth is not known). It is possible that the three Oahu hybrids resulted from the same spawning event or from the same pair of fish cannot be discounted. If it could be demonstrated that these specimens were, in fact, from the same parents [perhaps through comparison of otolith growth rings in conjunction with DNA fingerprinting techniques such as RAPD analysis (see Dinesh et al. 1993)], it would shed light on the dispersal potential of larvae within the Hawaiian Islands. Similar 'burst' patterns of reported hybrids have occurred in Hawaii before: two specimens of the hybrid butterflyfish *Chaetodon miliaris* × *C. tinkeri* were collected within a month of each other, but have never been reported since (Randall et al. 1977); and several *C. miliaris* × *C. multicinctus* hybrids, all of approximately equal size, were collected or observed within a several-year span, but have not been reported since (Burgess 1978 and observations by the senior author). One compounding factor is that there appears to be extensive back-crossing between hybrids of *Cenropyge flavissimus* × *C. vrolikii* and either parent species. In the Marshall Islands, hybrids are extremely variable in color and range in appearance from nearly 'pure' *C. flavissimus* to nearly 'pure' *C. vrolikii* (Pyle 1992a referred to this as an example of 'spectral' hybridization). In other examples of pomacanthid hybridization (e.g., *C. eibli* × *C. vrolikii* and *C. loriculus* × *C. potteri*), hybrids are nearly identical to each other and are centrally intermediate between the two presumed parent species, suggesting that back-crossing does not occur (Pyle 1992a referred to these as examples of 'discrete' hybridization). It is unclear, however, whether the lack of observed back-crossing in these cases is due to hybrid infertility or the scarcity of the hybrids.

Most of the published work on hybridization has focused on the 'hybrid zone' concept; i.e., zones or regions characterized by populations of organisms that exhibit a relatively steep cline in character states between two otherwise distinct (often parapatric) populations. Such zones are usually characterized by hybrid 'swarms' (Harrison 1993). As noted earlier, the vast majority of this work has involved terrestrial or freshwater organisms. The physiography of insular marine localities limits opportunities for such clines between populations of coral-reef species to exist; hybrid 'zones', when they occur in the coral reef environment, are usually limited to single islands or island groups. Indeed, the majority of marine fish hybrids are extremely rare in nature, and are most likely the result of very unusual circumstances.

Clearly, however, the relatively high percentage of *C. flavissimus* × *C. vrolikii* hybrids at Majuro is exceptional. Similarly, the abundance of *C. eibli* × *C. vrolikii* hybrids in Indonesia reported by Kuiter may also represent a 'hybrid swarm'. Pyle (1992b) discussed these two hybrids and noted that they, along with *C. eibli* × *C. flavissimus,* represent a triangle of species between which all possible hybrid combinations had been reported. Figure 4 shows the distribution of these three *Centropyge* species, with symbols to indicate where hybrids have been recorded. Hybrids have been reported from almost every location where two of these species are sympatric. Two apparent exceptions to this rule are the Ogasawara Islands and the Great Barrier Reef/ Coral Sea region, where *C. flavissimus* and *C. vrolikii* are sympatric but no hybrids have been reported. At the former locality, *C. vrolikii* is known only

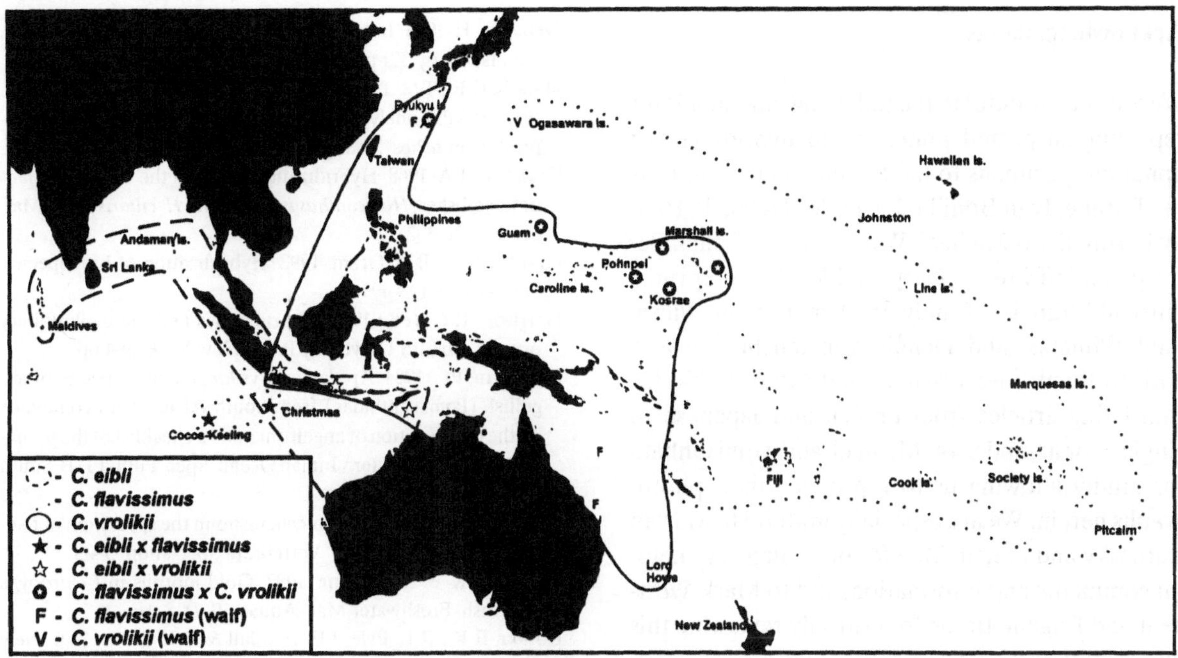

Fig. 4. Distribution of *Centropyge eibli, C. flavissimus,* and *C. vrolikii,* and locations where hybrids have been reported.

on the basis of a single observed individual; and at the latter locality, *C. flavissimus* is exceedingly rare. We suspect that no hybrids have been reported from either of these localities either because of insufficient collecting effort, or because inadequate numbers of the 'rarer' of the parent species are present. These essentially parapatric distributions, along with the exceptional preponderance of hybridization in areas of sympatry, led Pyle (1992b) to propose that these three species represent a monophyletic clade of closely-related, recently diverged species. An additional complexity in this example is the unusually disjunct distribution of *C. flavissimus.* Except for two populations in the eastern Indian Ocean (at Christmas Island and the Cocos-Keeling Islands), this species is essentially restricted to the central Pacific Ocean. It is noteworthy that individuals from Indian Ocean populations differ from Pacific *C. flavissimus* in lacking the blue ring around the orbit, and exhibiting a distinct black bar posterior to the operculum (Fig. 1f, i). We predict that a systematic evaluation of this complex of *Centropyge* species will provide supporting evidence of their close affinities.

Conclusions

Although modern molecular techniques have not been applied in any of the above cases of pomacanthid hybrids, the use of this technology could provide much greater insight into hybrid formation in these fishes. Barton & Gale (1993) review genetic techniques which have been used to further elucidate the evolutionary significance of hybrids and 'hybrid zones'. McMillan & Palumbi (unpublished) have used similar techniques in studying gene flow between closely-related species of chaetodontids, and their work allows a glimpse of what direction future research on marine fish hybridization might take. Genetic analysis of hybrids in conjunction with an understanding of the social dynamics and distribution of their parent species could provide much additional insight into the evolution and speciation of coral-reef fishes. As Harrison (1993, p. 10) pointed out: 'Ultimately, genetic analysis of hybrid zones leads to a better appreciation of the processes involved in the origin of adaptations and the origin of species'.

Acknowledgements

We are very grateful to the following individuals for reporting suspected pomacanthid hybrids and/or donating specimens to the authors: Peter and Bertha Basabe, Ivan Bonilla, J. Charles Delbeek, Randy Fernly, Scott Michael, Wayne Miura, Rohan Petiyagoda, Jeff Preble, Gary Rhodes, Julian Sprung, Hiroyuki Tanaka, Randy Walker, Fenton Walsh, Paul Williams, and Dennis Yamaguchi. We also wish to thank Lisa Privitera and Miho Preble for translating articles from French and Japanese to English for us, and Scott Michael, and Youji Ohkata for kindly allowing us to reproduce their photographs herein. We are especially grateful to Marvin Lutnesky and Martin Moe for providing very helpful comments and information, and to Mark Westneat and Eugene Balon for critically reviewing this manuscript.

References cited

Allen, G R 1979 Butterfly and angelfishes of the world Volume 2 John Wiley and Sons, New York 149–352 pp

Allen, G R 1986 Appendix butterfly and angelfishes of the world – vol 2 Mergus Publishers Hans A Baensch, Melle 7 pp

Allen, G R & R C Steene 1979 The fishes of Christmas Island, Indian Ocean Australian Government Publishing service, Canberra 81 pp

Anon 1989 Message of explanation from newly arrived fish Trop Mar Aquar 25 (Summer 1989) 48–49 (In Japanese)

Bailey, R M , J E Fitch, E S Herald, E A Lachner, C C Lindsey, C R Robins & W B Scott 1970 A list of common and scientific names of fishes from the United States and Canada (third edition) Spec Publ Amer Fish Soc 6 1–150

Barton, N H & K S Gale 1993 Genetic analysis of hybrid zones pp 13–45 *In* R G Harrison (ed) Hybrid zones and the evolutionary process, Oxford University Press, New York

Burgess, W E 1974 Une forme atypique de *Chaetodon* de Ceylan Rev fr Aquariol 2 37–40

Burgess, W E 1978 Butterflyfishes of the world T F H Publications, Neptune City 832 pp

Burgess, W E 1991 Two new genera of angelfishes, family Pomacanthidae Trop Fish Hobbyist, 39 (7) 68–70

Clavijo, I E 1985 A probable hybrid butterflyfish from the western Atlantic Copeia 1985 235–238

Conde, B 1990 Elevage et evolution d'un hybride présume entre *Pomacanthops maculosus* et *P semicirculatus* (Perciformes, Pomacanthidae) Rev fr Aquariol 16(4) 117–122

Debelius, H 1993 Indian Ocean tropical fish guide Aquaprint, Neu Isenberg 321 pp

Dinesh, K R , T M Lim, K L Chua, W K Chan & V P E Phang 1993 RAPD analysis an efficient method of DNA fingerprinting in fishes Zool Sci 10 849–854

Feddern, H A 1968 Hybridization between the western Atlantic angelfishes, *Holacanthus isabelita* and *H ciliaris* Bull Mar Sci 18 351–382

Grant, P R & B R Grant 1992 Hybridization of bird species Science 256 193–197

Harrison, R G (ed) 1993 Hybrid zones and the evolutionary process Oxford University Press, New York 364 pp

Heemstra, P C 1984 *Apolemichthys kingi,* a new species of angelfish (Pomacanthidae) from South Africa, with comments on the classification of angelfishes and a checklist of the pomacanthids of the western Indian Ocean Spec Publ J L B Smith Inst Ichth 35 1–17

Hubbs, C L 1961 Isolating mechanisms in the speciation of fishes Univ of Texas Symp Vertebrate Speciation 5–23

Kosaki, R K & D Toyama 1987 Gold morphs in *Centropyge* angelfish Freshwater Mar Aquar 10(7) 8–11

Kosaki, R K , R L Pyle, J E Randall & D K Irons 1991 New records of fishes from Johnston Atoll, with notes on biogeography Pac Sci 45 186–203

Krupp, F & H Debelius 1990 The hybrid of *Centropyge multifasciatus* × *Holacanthus venustus* from the Philippines and notes on aberrant colour forms of *Centropyge multispinis* from the Maldives and the Red Sea Rev fr Aquariol 17(2) 53–56

Kuiter, R H 1992 Tropical reef-fishes of the western Pacific, Indonesia, and adjacent waters Garmedia Pustaka Utama, Jakarta 314 pp

Longley, W H & S F Hildebrand 1941 Systematic catalogue of the fishes of Tortugas, Florida Carnegie Inst Wash Publ 535 331 pp

Lutnesky, M M F 1992a Behavioral ecology of reproduction in the pomacanthid angelfish *Centropyge potteri* Ph D Dissertation, University of Hawaii, Honolulu 155 pp

Lutnesky, M M F 1992b A temporal-threshold model of polygynous mating in cyclical environments Amer Nat 139 1102–1115

Lutnesky, M M F 1994 Density-dependant protogynous sex change in territorial-haremic fishes models and evidence Behavioral Ecology (in press)

Masuda, H , K Amaoka, C Araga, T Uyeno & T Yoshino (ed) 1984 The fishes of the Japanese Archipelago Tokai University Press, Tokyo 437 pp

Maugé, L A & J E Randall 1979 A propos d'une forme aberrante de *Pygoplites diacanthus* (Boddaert, 1772) Pisces Pomacanthidae Rev fr Aquariol 6(4) 97–100

Meyer, K A 1977 Reproductive behavior and patterns of sexuality in the Japanese labrid fish *Thalassoma cupido* Japan J Ichthyol 24 101–112

Moe, M A , Jr 1976 Rearing Atlantic angelfish Mar Aquarist 7(7) 17–26

Moe, M A 1990 The invisible nursery Sea Frontiers 36(3) 28

Moyer, J T 1981 Interspecific spawning of the pygmy angelfishes

Centropyge shepardi and *C bispinosus* at Guam Micronesica 12(1–2) 119–124

Moyer, J T 1989 On the blinding nature of experience how many species of pygmy angelfishes are there? Trop Fish Hobbyist 37(7) 86–96

Moyer, J T 1990 Social and reproductive behavior of *Chaetodontoplus mesoleucus* (Pomacanthidae) at Bantayan Island, Philippines, with notes on pomacanthid relationships Japan J Ichthyol 36 459–467

Moyer, J T & A Nakazono 1978 Population structure, reproductive behavior, and protogynous hermaphroditism in the angelfish *Centropyge interruptus* at Miyake-jima Japan Japan J Ichthyol 25 25–39

Myers, R F 1989 Micronesian reef fishes Coral Graphics, GMF, Guam 298 pp

Nichols, J T & L L Mowbray 1914 A new angel-fish (*Angelichthys townsendi*) from Key West Bull Amer Mus Nat Hist 33(37) 581–583

Pyle, R L 1989 Rare and unusual marines the armitage angelfish *Apolemichthys armitagei* Smith Freshwater Mar Aquar 12(4) 26–27

Pyle, R L 1992a Rare and unusual marines a hybrid angelfish *Centropyge flavissimus* × *eibli* Freshwater Mar Aquar 15(3) 98–110, 212

Pyle, R L 1992b Rare and unusual marines another hybrid angelfish *Centropyge loriculus* × *potteri* Freshwater Mar Aquar 15(8) 40–45

Pyle, R L & J E Randall 1993 A new species of *Centropyge* from the Cook Islands, with a redescription of *Centropyge boylei* Rev fr Aquariol 19(4) 115–124

Randall, J E 1956 *Acanthurus rackliffei*, a possible hybrid surgeonfish (*A achilles* × *A glaucopareius*) from the Phoenix Islands Copeia 1956 21–25

Randall, J E 1992 Divers guide to fishes of Maldives IMMEL Publishing, London 193 pp

Randall, J E , G R Allen & R C Steene 1977 Five probable hybrid butterflyfishes of the genus *Chaetodon* from the central and western Pacific Rec West Austral Mus 6 3–26

Randall, J E & R C Anderson 1993 Annotated checklist of the epipelagic and shore fishes of the Maldive Islands J L B Smith Inst Ichthyol Bull 59 1–47

Randall, J E & D Fridman 1981 *Chaetodon auriga* × *Chaetodon fasciatus*, a hybrid butterflyfish from the Red Sea Rev fr Aquariol 7(4) 113–116

Randall, J E , PS Lobel & E H Chave 1985 Annotated checklist of the fishes of Johnston Atoll Pac Sci 39 24–80

Randall, J E & L A Mauge 1978 *Holacanthus guezei*, a new angelfish from Reunion Bull Mus natn Hist nat Paris 353 (514) 297–303

Randall, J E & F Yasuda 1979 *Centropyge shepardi*, a new angelfish from the Mariana and Ogasawara Islands Japan J Ichthyol 26 55–61

Reese, E S 1975 A comparative field study of the social behaviour and related ecology of reef fishes of the family Chaetodontidae Z Tierpsychol 37 37–61

Robins, C R G C Ray & J Douglass 1986 A field guide to Atlantic coast fishes of North America Houghton Mifflin Co , Boston 354 pp

Sano, M , K Okuzawa, T Yamakawa & K Mochizuki 1984 A probable hybrid butterflyfish of the genus *Chaetodon* from the Ogasawara Islands Japan J Ichthyol 31 79–82

Schwartz, F J 1972 World literature to hybrids, with analysis by family, species, and hybrid Publ Gulf Coast Res Lab Mus 3 1–328

Shirai, S 1986 Ecological encyclopedia of the marina animals of the Indo-Pacific Vol 1 Vertebrata Shin Nippon Kyoiku, Tosho Co , Okinawa 352 pp

Slastenenko, E P 1957 A list of natural fish hybrids of the world Hidrobiologi, ser B, 4(2–3) 76–97

Smith, J L B 1955 The fishes of the family Pomacanthidae in the western Indian Ocean Ann Mag nat Hist 8(12) 377–384

Steene, R C 1978 Butterfly and angelfishes of the world Volume 1 A H & A W Reed Pty Ltd Sydney 144 pp

Takeshita, G Y 1976 An angel hybrid Mar Aquarist 7(1) 27–35

Takeuchi, K 1984 Paradise of the Indian Ocean diving in the Maldives Archipelago Trop Mar Aquar 11 (Winter 1985) 3–11 (In Japanese)

Thresher, R E 1984 Reproduction in reef fishes T F H Publications, Neptune City 399 pp

Trivers, R L 1972 Parental investment and sexual selection pp 136–179 *In* B Campbell (ed) Sexual Selection and the Descent of Man 1871–1971 Adine Atherton Chicago

Wedge, J M 1984 The mating game Sea Frontiers 30(5) 308–311

Yasuda, F & Y Tominaga 1976 A new pomacanthid fish, *Chaetodontoplus caeruleopunctatus*, from the Philippines Japan J Ichthyol 23 130–132

Environmental Biology of Fishes **41**: 147–170, 1994.

Geographical variation of some taxonomically important characters in fishes: the case of the bitterling *Rhodeus sericeus*

Juraj Holčík[1] & Ladislav Jedlička[2]
[1] *Institute of Zoology and Ecosozology, Slovak Academy of Sciences, Dubravska cesta 9, 842 06 Bratislava, Slovakia*
[2] *Department of Zoology, Comenius University, Mlynská dolina B1, 842 15 Bratislava, Slovakia*

Received 28.5.1993   Accepted 7.12.1993

Key words: Meristic characters, Clines, Geographical coordinates, Altitude, Temperature, Fish size, Subspecies, Bergmann's rule, Ostariophysi, Cyprinidae

Synopsis

Geographical variation of three meristic characters, the number of pored scales in lateral line (LL), the number of branchial spines on the first gill arch (SPBR) and the number of transverse scale rows (SQU), were studied in the cyprinid fish *Rhodeus sericeus* and their validity for subspecific discrimination was tested. Counts investigated were taken from population samples at 23 localities covering an area of about 68° geographical longitude (2°40 W – 49°30 E and 128°00 E – 143°00 E), and about 15° geographical latitude (37°30 N – 52°00 N). All three characters manifested a distinct clinal variation. LL, SQU and SPBR closely followed the species' essentially longitudinal distribution. The number of segments was also related to latitude, elevation, mean annual air temperature and fish size. Bergmann's rule was fully demonstrated for fishes and at the species level. Local regional differences were found within particular regions, indicating the existence of *subclines.* Variation of truly isolated populations seems to be predictable and dependent on the clines found in adjacent populations. None of the counts investigated, including a key character (number of pored scales in LL), can be used to distinguish *amarus* from *sericeus.* Owing to sufficient evidence that meristic and morphometric characters used to distinguish fish subspecies succumb to the clinal variation controlled by various factors, the concept of subspecies and trinomial nomenclature is inefficient, superfluous and misleading, and should be excluded from taxonomy, nomenclature and ecology.

Introduction

A cyprinid species, the bitterling *Rhodeus sericeus* (Pallas, 1776), is the only Eurasian member of the subfamily Acheilognathinae. Its distribution covers a vast area in Europe, Transcaucasia and Asia Minor. Exceptions where this species is absent or not native are the British Isles, Denmark, Iceland, Scandinavia, Finland, Italy and in the Iberian, Peloponesian and Chalcidician peninsulas, the River Ural and rivers entering the eastern coast of the Caspian Sea. The bitterling also occurs in the basin of the River Amur, the Sakhalin Island, the rivers entering Peter de Great Bay and the rivers of the northern part of Korea entering the Sea of Japan. As may be seen, this area is not continuous but disjunct over the vast space between the Ural Mountains of the west and the catchment of the River Amur in the east, which is not inhabited by this species (Fig. 1).

The bitterling found in the east were described by Pallas (1776) as *Cyprinus sericeus,* whereas in the

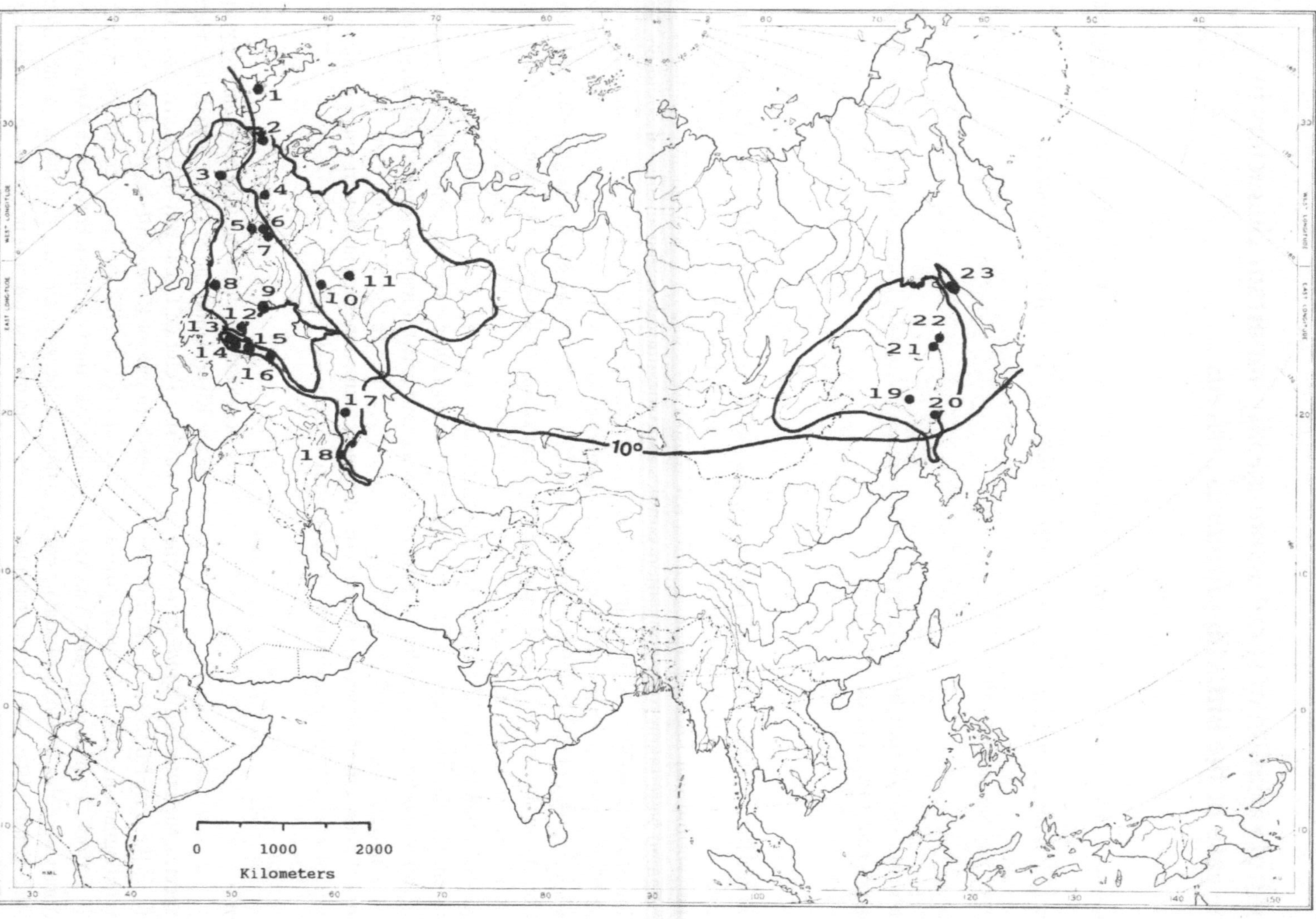

Fig. 1. Geographical distribution of *Rhodeus sericeus* (ssp. *amarus* in west, nominotypic form *sericeus* in east). Dots and figures indicate sampling sites (see Table 1). 10° C isotherm is also shown.

west it was described by Bloch (1782) as *Cyprinus amarus*. Both taxa were considered to be valid species until the beginning of the 1930's, when Svetovidov & Eremeev (1935) made a comparison of two samples of bitterlings, one from the Dnieper River and the other from the River Amur. They found them to be conspecific, both forms were lumped together and relegated to a subspecific status (*Rhodeus sericeus amarus* in the west and *Rhodeus sericeus sericeus* in the east). A third form was subsequently described by Karaman (1924) from the River Vardar (Aegean Sea catchment) under the name *R.s.*var.*meridionalis*, afterwards elevated to a subspecies *R.s.meridionalis* (Karaman 1955). Further intraspecific taxa, all from Europe, were described by Karaman (1955; *R.s.meridionalis* forma *strumicae* from the Strumica River in Macedonia) and

Table 1 Characteristics of the *Rhodeus sericeus* samples used in this study and sites they were taken from AZ = Azerbaijan, GB = Great Britain, NL = Netherlands, CH = Switzerland, CS = Czecho-Slovakia, D = Germany, IR = Iran, KO = Korea, R = Romania, RUS = Russia, TJ = China, TR = Turkey, UKR = Ukraine, YU = Yugoslavia

No	**Region** locality	n	*SL* ranges (mm)	Geographical longitude	Geographical latitude	Elevation (m)	Mean air temperature (° C)
	Rhodeus sericeus amarus						
I	**Western Europe**						
1	Old Black Brook Canal, St Helens, Lancashire, GB	2	47–48	2°40 W	53°30 N	61	9 6
2	Ditches near Groeningen, NL	17	47–78	6°35 E	53°12 N	3	9 2
3	Pond near Genthod, Basel, CH	13	36–51	7°35 E	47°30 N	343	10 1
II	**Elbe River Basin**						
4	Oxbows and ponds near Čelákovice and Prague	125	37–69	14°36 E	50°07 N	191	9 4
III	**Danube River Basin**						
5	Wurm River, Bavaria, D	6	50–62	11°23 E	49°00 N	439	9 2
6	Dyje River (Znojmo), CS	50	45–61	16°02 E	48°51 N	236	9 2
7	Šúr Reserve, Bratislava, CS	100	37–75	17°10 E	48°10 N	151	10 2
IV	**Mediterranean Sea**						
8	Skadar Lake, YU	20	33–46	19°26 E	42°13 N	5	15 1
V	**Black Sea**						
9	Siut Ghiol Lake, R	100	38–59	28°35 E	44°15 N	14	11 3
VI	**Eastern Europe**						
10	Dnieper and Yuzhnyi Bug rivers, UKR	50	36–59	32°30 E	49°10 N	15	17 2
11	Svapa River, Kursk, RUS	3	44–55	35°00 E	52°00 N	235	5 2
VII	**Asia Minor**						
12	Terkos Lake, TR	2	28–32	28°00 E	41°23 N	35	13 7
13	Manyas Lake, TR	15	34–49	28°00 E	40°10 N	99	14 4
14	Abulyont Lake, TR	28	39–53	28°30 E	40°10 N	99	14 4
15	Sapanca Lake, TR	2	37–37	30°20 E	40°40 N	77	14 2
16	Terme Lake, TR	2	37–39	36°59 E	41°12 N	40	14 3
VIII	**Middle East**						
17	Soyukhbulag River, AZ	52	30–60	45°14 E	41°19 N	452	12 2
18	Anzali Lagoon, IR	45	31–53	49°30 E	37°30 N	– 21	15 8
	Rhodeus sericeus sericeus						
IX	**Amur River Basin**						
19	Stream near I-Mien-Po, TJ	3	51–71	128°00 E	45°05 N	20	2 4
20	Amur River, middel course, RUS	50	37–73	135°00 E	48°40 N	50	0 9
X	**Primorye**						
21	Kabar Lake, Yelabuga, RUS	16	51–74	135°55 N	48°50 N	50	0 9
22	Tuymen-Ula, KO	3	34–78	130°41 E	42°48 N	15	5 6
XI	**Sakhalin**						
23	Tym River, Katangli, RUS	12	38–76	143°00 E	51°30 N	105	– 1 8

Table 2. Comparison of count ranges and principal measurements in two subspecies of *Rhodeus sericeus*. Values shown represent our data combined with those given by Svetovidov & Eremeev (1935), Berg (1949), Holčík (1959), Nikol'skii (1956), Abdurakhmanov (1962), Holčík & Duyvené de Wit (1964), Zhukov (1965), Ivanović (1973) and Kim (1982). The most frequent values of counts (more than 75% of the total frequency) are in parentheses.

Character	*Rhodeus sericeus sericeus* (n = ~ 200)	*Rhodeus sericeus amarus* (n = ~ 700)
Maximum standard length (mm)	80	77
Unbranched rays in D	2–3(3)	2–4(3)
Branched rays in D	8–11(9)	7–11(9)
Unbranched rays in A	2–3(3)	2–4(3)
Branched rays in A	8–11(9)	6–11(9)
Pored scales in l.l.	4–10(6–7)	0–9(4–6)
Transverse rows of scales	33–44(36–40)	32–45(37–40)
Branchial spines	9–16(12–14)	9–13(10–12)
In % of standard length:		
Head length	21–27	20–28
Head depth	16–22	16–26
Preorbital length	5–10	5–9
Horizontal diameter of eye	5–11	5–11
Interorbital distance	7–13	6–12
Maximum body depth	29–42	24–45
Minimum body depth	7–14	7–14
Caudal peduncle length	21–29	20–31
Caudal peduncle depth	14–18	10–19
Predorsal distance	45–57	44–58
Preanal distance	57–66	49–70
Preventral distance	42–50	40–65
Length of D	17–25	15–26
Length of A	12–22	10–20
Length of P	15–23	10–22
Length of V	14–19	11–21
Length of C	21–27	15–27
Depth of D	13–22	13–26
Depth of A	13–20	11–23
Distance between P and V bases	19–26	18–28
Distance between V and A bases	12–20	12–22

then by Holčík (Golchik 1959[1]; *R.s.amarus* natio *danubicus* from the Middle Danube in Slovakia, and *R.s.amarus* natio *svetovidovi* from the Dnieper River in Ukraine). Later on, Holčík & Duyvené de Wit (1964) found that some characters of the bitterling, both counts and measurements, display regular geographical variation, so the validity of bitterling subspecies, as well as other intraspecific taxa, raised some doubts (Holčík 1983), especially because the difference between *amarus* and the nominotypic *sericeus* is only in a single character: in the number of pored scales of the lateral line, which is incomplete in this species as well as in the genus *Rhodeus*. According to the original data by Svetovidov & Eremeev (1935), which was then slightly modified by Berg (1949), their number should be 4–6 (7) in *amarus* and 5–10 in the nominotypic form. In ssp. *meridionalis,* the only differences cited by Karaman (1924, 1955) to discriminate it from the other two subspecies were some mensural characters.

Over the past three decades, the senior author has gathered more samples of bitterling from various parts of its distribution area. At the same time, more data were published on the species, thus enabling an evaluation of its intraspecific status. Moreover, the unique distribution of the bitterling, covering almost 95° of geographical longitude and approximately 25° of geographical latitude (Fig. 1), sheds more light on the geographical variation both in this species and fishes in general, and its consequences for taxonomy. We have in mind the concept of subspecies which is still widely used especially in Europe, both in its eastern and western part (e.g. Kazancheev 1981, Hoestland 1991).

Material and methods

This paper is based on data taken from literature (Table 1; samples 4, 6, 7, 11 – Holčík 1959; 8 – Ivanović 1973, 10 and 20 – Svetovidov & Eremeev 1935; 12 – 16 – Holčík & Duyvené de Wit 1964; 17 – Abdurakhmanov 1962) and on examinations of subsequent samples received from other localities. The material examined is housed in the ichthyological collections of the following institutions: British Museum (Natural History), London (sample 1, 2, 3 in part, and 5), U.S. National Museum, Washington, D.C. (19), Museum of Natural History, Geneva (3 in part), Department of Systematic Zoology, Charles University, Prague (4, 6, 7, 9, 11 and 21 in part), Slovak National Museum, Bratislava (12–16, 18 and 21

[1] As Golchik (or Gol'chik) is a Russian transliteration of Holčík, the original, (i.e. not transliterated) spelling is used further in this paper.

in part), Museum of Zoology, Moscow State University, Moscow (10 and 20) and the Institute of Zoology, Academy of Sciences of the Russian Federation, Sankt Petersburg (22 and 23). Furthermore, one additional sample from Romania was also used for some statistical calculations.

Based on previous results, experience (Holčík & Duyvené de Wit 1964, Holčík 1983) and a comparison of data dealing with particular characters, only three counts are evaluated in this paper: the number of pored scales on the lateral line (LL), the number of transverse scales rows (SQU), and the number of branchial spines (SPBR). The values of all other characters (Table 2), both counts and measurements, are virtually the same in both subspecies. Some wider ranges of particular characters in *amarus* in comparison with those in *sericeus* are probably due to differences in the number of samples, which is more than twice as high in the former than in the latter. Moreover, the sampling sites for *amarus* cover a considerably larger geographical area, being spread over an area of approximately 51° of geographical longitude. Whereas in *sericeus*, the sampling area covers only about 15° of longitude (Fig. 1).

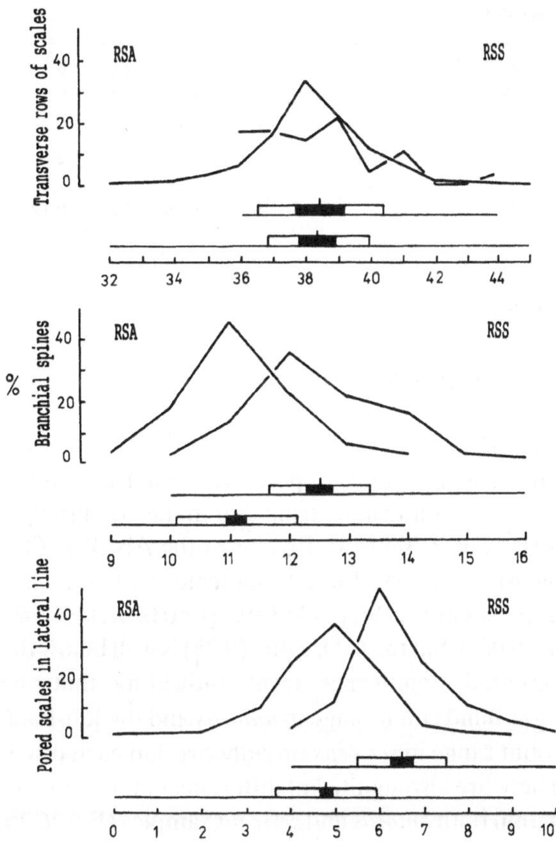

Fig. 2. Frequency (in %) of the pored scales in lateral line, branchial spines and transverse scale rows in two subspecies of bitterling: subspecies *amarus* (RSA) and nominotypic form *sericeus* (RSS). The horizontal scale gives the total observed range; the white rectangle gives an interval of one standard deviation to either side of the mean and the black rectangle represents twice the standard deviation to either side of the mean; the mean is shown as a vertical crossbar. The same construction was used in Figure 3.

Standard length (SL) was measured as the distance between the tip of the snout and the rear edge of the last scale on the caudal peduncle (Holčík et al. 1989). Counts investigated were examined as previously (Holčík 1959). For LL, all scales perforated by a sensory canal were counted regardless of whether they formed a continuous row or if this row was interrupted with complete (i.e. unperforated) scale(s). All transverse rows of scales were considered for SQU, including the first two scales behind the nape (lying one above the other for the first row) and those forming the full row running across the caudal peduncle, beginning from the upper base of the caudal fin for the last row. SPBR represents all gill rakers including rudimentary ones, found on the outer side of the first fill arch. All counts were taken from the left side of the fish.

Statistical analyses were applied to particular samples, which were treated separately, but also as clusters grouped according to particular river basins and/or geographical regions (Table 1). Differences among means were tested using Student's *t* test, one-way analysis of variance (ANOVA) and by the least significant difference multiple range test (LSD). Regression and correlation analyses were used to find the relationships between particular counts and some environmental factors. Geographical longitude (LONG) and latitude (LAT), altitude above sea level (ELEV) and the mean annual air temperature (TEMP) of the sampling site or the area closest to it were examined as direct or abiotic factors, whereas SIZE was considered to be an indirect or biotic factor. Data for ELEV and TEMP were taken from Wernstedt (1972). The air temperatures were included because water temperatures of the sampling sites were unavailable. In this re-

Table 3 Number and frequency of pored scales in the lateral line in *Rhodeus sericeus* from particular regions

Region	n	0	1	2	3	4	5	6	7	8	9	10	x̄	s	$s_{\bar{x}}$
	Rhodeus sericeus amarus														
Western Europe	31	–	–	3	3	8	11	4	2	–	–	–	4 52	1 31	0 24
Elbe River basin	124	–	1	2	30	47	37	5	–	1	1	–	4 15	1 07	0 10
Danube River basin	156	3	1	5	15	36	61	28	6	1	–	–	4 63	1 32	0 11
Mediterranean	20	–	–	–	–	1	7	10	2	–	–	–	5 65	0 74	0 17
Black Sea	100	–	–	1	6	18	46	21	6	2	–	–	5 06	1 07	0 11
Eastern Europe	53	–	–	–	–	8	23	21	1	–	–	–	5 28	0 74	0 10
Asia Minor	45	–	–	–	–	17	21	7	–	–	–	–	4 78	0 70	0 10
Middle East	97	1	–	1	6	16	29	38	4	1	1	–	5 17	1 25	0 13
Sum total and grand average	626	4	2	12	60	151	235	134	22	5	1	–	4 78	1 19	0 05
	Rhodeus sericeus sericeus														
Amur River basin	69	–	–	–	–	1	2	37	17	7	4	1	6 62	1 03	0 13
Primorye	3	–	–	–	–	–	–	1	2	–	–	–	6 67	0 57	0 33
Sakhalin	12	–	–	–	–	1	5	4	1	1	–	–	5 67	1 07	0 31
Sum total and grand average	84	–	–	–	–	2	7	42	20	8	4	1	6 49	1 09	0 12

spect, we follow Hubbs (1922) and Schlesinger & Regier (1982), who used air temperatures of the region as a substitute for water temperatures. In addition, the possible influence of the standard length (i.e., SIZE) was tested on a sample of 125 specimens of various SL caught in Romania (locality not known). Equations of relationships calculated are summarized in Appendix 1.

Table 4 ANOVA multiple range test for pored scales in lateral line (LSD, 95% confidence intervals)

Region	n	Average	Homogeneous groups
Elbe River basin	124	4 1532258	X
Western Europe	31	4 5161290	XX
Danube River basin	156	4 6346154	X
Asia Minor	45	4 7777778	XX
Black Sea	100	5 0600000	XX
Middle East	97	5 1752577	XXX
Eastern Europe	53	5 2830189	XX
Mediterranean	20	5 6500000	XX
Sakhalin	12	5 6666667	XXX
Amur River basin	69	6 6231884	X
Primorye	3	6 6666667	XX

Results

Variation pattern

Lateral line

The mean and mode of pored scales in LL for *amarus* differ significantly from that in the nominotypic form (Fig. 2, Table 3). Results of the ANOVA (Table 4) and t test (Table 5) indicate the two means differ significantly ($F = 153.193$, $p < 0.0001$, $t = 12.49$, $p < 0.001$), but the C.D. value (0.75) is well below the accepted significance level, indicating that the right-hand count range in *amarus* and the left-hand count range in *sericeus* broadly overlap each other. There are also significant differences among means within both *amarus* and *sericeus* samples ($F = 26.25$, $p < 0.0001$), but the C.D. values were under the level of statistical significance in all cases (Table 5). When particular samples and groups of populations are arranged in accordance with LONG of their localities (Table 3, Fig. 3), one can see a continuous, although not always regular, increase in this count from West to East, thus showing a typical meridional clinal variation, as has been found earlier (Holčík

& Duyvené de Wit 1964). This is also confirmed by the LSD test (Table 4). The highest mode and mean number of pored scales was found in samples from the Amur River. Interestingly, further eastward, the number of segments in this count showed some decrease. Generally within the whole variation range, there were no significant gaps among particular populations and all sample subsets overlapped each other at their extreme values at least.

Transverse scale rows

Significantly wider and irregular variation was found for SQU, which tended to follow a mosaic pattern (Fig. 2, Table 6). Differences among samples were highly significantly (Tables 7, 8; F = 10.50, p < 0.0001), but the C.D. test again did not show a significant difference between the two subspecies or among samples. The LSD test indicates the existence of clinal variation. The number of elements in SQU increased eastward, although this trend was not as clearly expressed as in LL. In comparison with LL and SPBR, the overlapping of particular subsets was more conspicuous and the differences within the whole variation range were less expressed.

Branchial spines

The same pattern as in LL was also found in SPBR, as there were differences between both subspecies in ranges, modes and means (Fig. 2, Table 9). Contrary to LL, however, the ranges of segments in this character were wider in *sericeus,* and the shape of the curve suggests two peaks, with the mode and mean significantly shifted to the left (Fig. 2). This is due to the sudden decrease of SPBR elements in samples from eastward of the River Amur, being more pronounced than in LL (Fig. 3). This indicates that the decrease in both counts for samples from Sakhalin and Primorye, which may be considered geographically separated from the Amur River basin, is rather regular and not accidental. Differences among means within both subspecies, as well as among particular subsets, were statistically highly significant in most cases (Table 11; ANOVA, F = 14.96, p < 0.0001). However, the C.D. test again did not reveal differences either within or among particular subspecies and samples respectively (Table 10). This character shows distinct clinal variation, similar to that in LL, with increasing numbers of branchial spines eastward (Fig. 3).

Controlling factors

Geographical longitude

(Fig. 4–6; Appendix 1: equations 1–3, 13–15, 25–27)

The segments of all three characters depend on

Table 5. t test (*t*) and coefficient of difference (*C.D.*) among means of pored scales in lateral line in *Rhodeus sericeus* from particular regions (see Table 1 for region codes I–X). Statistically significant values printed in bold.

		C.D.											
		I	II	III	IV	V	VI	VII	VIII	IX	X	*amarus*	*sericeus*
	I	–	0.16	0.04	0.55	0.23	0.37	0.13	0.25	0.89	0.57		
	II	1.75	–	0.21	0.86	0.44	0.65	0.37	0.46	1.20	0.84		
	III	0.42	**3.40**	–	0.49	0.18	0.32	0.07	0.21	0.84	0.52		
	IV	**3.51**	**6.37**	**3.37**	–	0.33	0.25	0.60	0.24	0.54	0.12		
t	V	**2.32**	**6.58**	**2.72**	**2.35**	–	0.12	0.16	0.05	0.73	0.38		
	VI	**3.39**	**7.30**	**3.39**	1.89	1.89	–	0.35	0.06	0.74	0.33		
	VII	1.12	**3.90**	0.73	**4.52**	**4.52**	**3.40**	–	0.20	1.04	0.62		
	VIII	0.39	1.22	0.71	0.23	0.23	0.08	0.28	–	0.63	0.30		
	IX	**7.65**	**16.02**	**11.01**	**3.82**	**3.82**	**7.84**	**10.43**	1.29	–	0.35		
	X	**2.88**	**6.21**	**3.51**	0.72	0.72	**2.46**	**4.55**	0.29	**2.53**	–		
	amarus											–	*0.75*
	sericeus											**2.49**	–

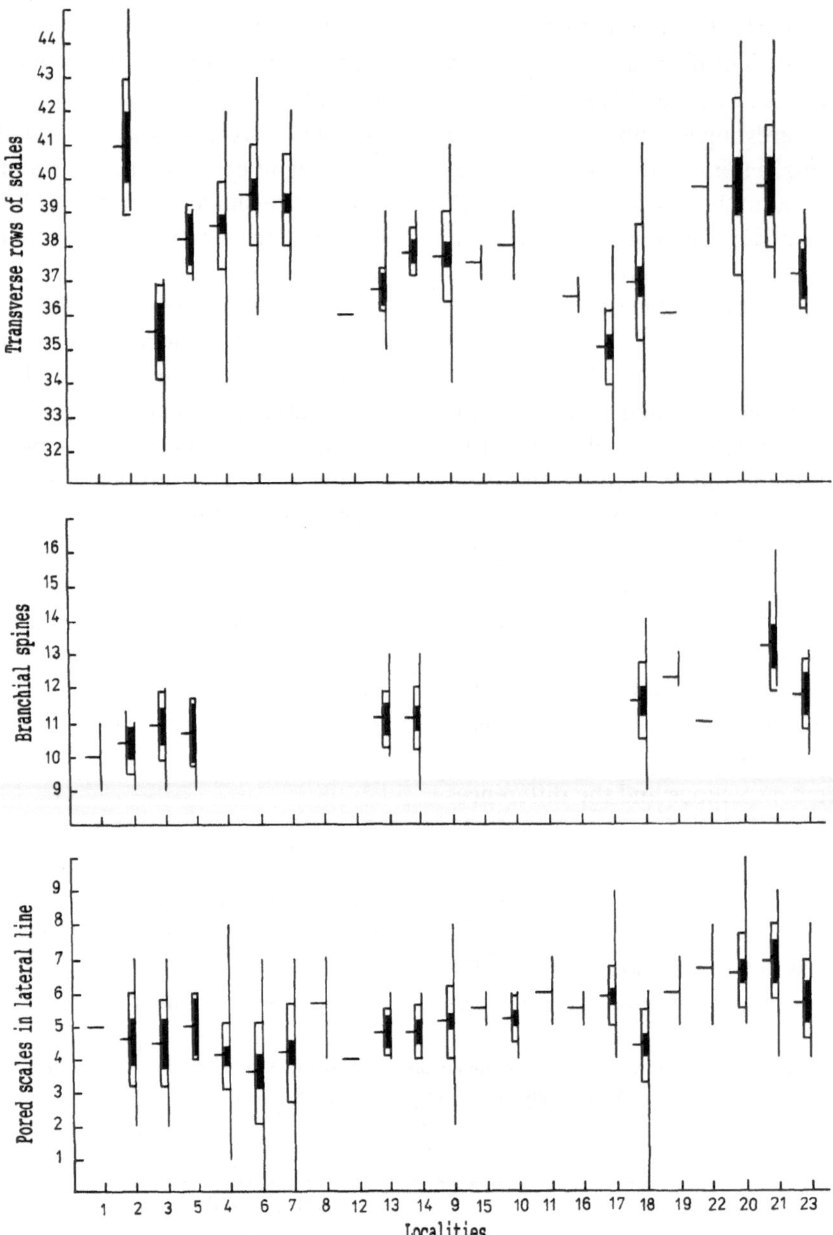

Fig. 3. Variation with geographical longitude in the number of pored scales in lateral line, branchial spines and transverse scale rows, respectively. Numbers on the abscissa indicate localities samples (see Table 1).

LONG, which is most distinct in LL and SPBR (Fig. 4, 5) and weak in SQU (Fig. 6). According to correlation coefficients (Appendix 1) about 54%, 60%, and 0.1% of variation in LL, SPBR and SQU respectively, may be attributed to the variation in LONG. The number of segments in LL and SPBR increases with increasing LONG, whereas in SQU the relationship is reversed. The reverse dependence upon LONG is also found in both subspecies, as LL in *sericeus* decreases and SQU increases with increasing LONG. We hypothesise that the number of SPBR segments in the nominotypic form also decreases, but the small number of samples, the great dispersion of data, and the small area from which *sericeus* samples came, hide the true course of this relationship.

Table 6 Frequency of transverse scale rows in the various *Rhodeus sericeus* populations

Region	n	32	33	34	35	36	37	38	39	40	41	42	43	44	45	x̄	s	s_x
	Rhodeus sericeus amarus																	
Western Europe	30	1	–	1	4	4	3	1	4	2	5	1	2	1	1	38 57	3 24	0 59
Elbe River basin	124	–	–	1	1	4	12	44	38	15	8	1	–	–	–	38 56	1 28	0 11
Danube River basin	401	–	–	1	8	21	54	124	99	60	32	1	1	–	–	38 50	1 46	0 07
Black Sea	100	–	–	1	6	9	21	47	12	3	1	–	–	–	–	37 60	1 21	0 12
Eastern Europe	3	–	–	–	–	–	1	1	1	–	–	–	–	–	–	38 00	–	–
Asia Minor	33	–	–	–	2	2	9	16	4	–	–	–	–	–	–	37 54	1 00	0 17
Middle East	42	–	2	1	4	7	16	6	3	2	1	–	–	–	–	36 90	1 68	0 26
Sum total and grand average	733	1	2	5	25	47	116	239	161	82	47	3	3	1	1	38 25	1 55	0 06
	Rhodeus sericeus sericeus																	
Amur River basin	16	–	–	–	–	2	1	1	5	4	2	–	–	1	–	39 25	1 98	0 50
Primorye	3	–	–	–	–	–	–	1	–	1	1	–	–	–	–	39 67	–	–
Sakhalin	10	–	–	–	–	3	4	2	1	–	–	–	–	–	–	37 10	0 99	0 31
Sum total and grand average	29	–	–	–	–	5	5	4	6	5	3	–	–	1	–	38 55	1 94	0 36

Geographical latitude
(Fig. 4–6; Appendix 1: equations 4–6, 16–18, 28–30)

LAT seems to have no effect on LL and SPBR, as there is no statistically significant correlation between these variables (Figs 4, 5, Appendix 1: equations 6, 18). However, a highly significant ($r = 0.595$, $p < 0.01$) positive dependence of SQU upon LAT appeared (Fig. 6, Appendix 1: equation 30). A reverse pattern in this relationship may be seen in both forms for all three counts, showing an decrease of SPBR, but increase of LL and SQU, with increasing LAT.

Table 7 ANOVA multiple range test for transverse rows of scales (LSD, 95% confidence intervals)

Region	n	Average	Homogeneous groups
Middle East	42	36 904762	X
Sakhalin	10	37 100000	XX
Asia Minor	33	37 545455	XX
Black Sea	100	37 600000	X
Eastern Europe	3	38 000000	XXX
Danube River basin	401	38 498753	XX
Elbe River basin	124	38 556452	XXX
Western Europe	30	38 566667	XXX
Amur River basin	16	39 250000	X X
Primorye	3	39 666667	XXX

Elevation
(Fig. 4–6; Appendix 1: equations 7–9, 19–21, 31–33)

In all three counts the number of segments decreased with increasing ELEV, although this correlation was not significant. Here, too, an inverse correlation between *sericeus* and *amarus* was found, indicating decreasing LL and increasing SPBR in the former, and vice versa in the latter. In SQU, the correlation between variables was negative in both taxa, however, more distinct in *sericeus*.

Temperature
(Fig. 4–6); Appendix 1: equations 10–12, 22–24, 34–36)

There is a negative correlation between TEMP and all three counts (Fig. 4, 6), which is statistically significant in SPBR but insignificant in LL and SQU.

Table 8 t test (*t*) and coefficient of difference (*C D*) among means of transverse scales rows in *Rhodeus sericeus* from particular regions Statistically significant values printed in bold

		CD									
		I	*II*	*III*	*V*	*VII*	*VIII*	*IX*	*XI*	*amarus*	*sericeus*
	I	–	0.00	0.01	0.22	0.24	0.34	0.13	0.35		
	II	0.03	–	0.02	0.38	0.45	0.56	0.21	0.64		
	III	0.24	0.42	–	0.34	0.39	0.51	0.21	0.12		
	V	**2.49**	**5.73**	**5.87**	–	0.03	0.24	0.52	0.23		
t	VII	1.74	**4.25**	**3.83**	0.26	–	0.24	0.57	0.22		
	VIII	**2.85**	**6.70**	**6.87**	**2.80**	1.93	–	0.64	0.07		
	IX	0.76	1.89	**2.05**	**4.59**	**4.03**	**4.53**	–	0.72		
	XI	1.40	**3.52**	**3.12**	1.27	1.22	0.36	**3.17**	–		
	amarus									–	0.09
	sericeus									1.01	–

Again, the reverse dependence of SQU and SPBR upon ELEV may be seen in both subspecies. In LL, this relationship in *sericeus* and *amarus* is rather unusual; although in both forms the number of segments is increasing with increasing TEMP, the general correlation is negative. This is because data for *amarus* follow distinctly lower levels than those of *sericeus*.

Size of fish
(Fig. 7, 8; Appendix 1: equations 37–48)

The effect of fish size was tested in one *amarus* population from Romania consisting of 125 fishes, 31 to 76 mm in SL, where LL and SQU were known. A significantly high ($p < 0.05$) positive correlation was found between bitterling size and their respective LL and SQU (Table 7). It is supposed that the same is also true for SPBR, as reported for other groups of fishes including clupeids, engraulids, cyprinids, salmonids and gadids (e.g. Vladykov 1934, Čihař 1958, Reshetnikov 1961, Holčík & Nagy 1987). Their findings prompted us to correlate the mean SL of particular samples with their respective LL, SQU

Table 9 Number and frequency of branchial spines in *Rhodeus sericeus* from particular regions

Region	n	9	10	11	12	13	14	15	16	x̄	s	s_x
	Rhodeus sericeus amarus											
Western Europe	25	3	8	9	5	–	–	–	–	10.64	0.95	0.19
Danube River basin	6	1	–	5	–	–	–	–	–	10.67	0.82	0.33
Asia Minor	38	1	7	20	8	2	–	–	–	11.08	0.85	0.14
Middle East	41	1	4	15	13	5	3	–	–	11.63	1.13	0.18
Sum total and grand average	110	6	19	49	26	7	3	–	–	11.16	1.05	0.10
	Rhodeus sericeus sericeus											
Amur River basin	19	–	–	1	6	5	5	1	1	13.10	1.24	0.28
Primorye	1	–	–	1	–	–	–	–	–	11.00	–	–
Sakhalin	11	–	2	2	5	2	–	–	–	11.64	1.03	0.31
Sum total and grand average	31	–	2	4	11	7	5	1	1	12.52	1.36	0.24

Table 10 ANOVA multiple range test for branchial spines (LSD, 95% confidence intervals)

Region	n	Average	Homogeneous groups
Western Europe	25	10 640000	X
Danube River basin	6	10 666667	XX
Asia Minor	38	11 078947	XX
Middle East	41	11 634146	X
Sakhalin	11	11 636364	XX
Amur River basin	19	13 105263	X

and SPBR. There was a generally positive correlation between these variables (Fig. 8), although in the *amarus* samples correlations for LL and SPBR were negative.

Relationships among abiotic environmental factors
(Fig. 9; Appendix 1: equations 49–59)

As shown above, the correlation between environmental factors and the meristic characters studied was ambiguous in some cases. This also indicates that one may expect different relationships between particular environmental factors. We tested this hypothesis and found TEMP, amongst abiotic factors, to be negatively correlated with LONG, LAT, and also with ELEV. A weak total correlation between ELEV and TEMP in this case was due to high temperature differences (– 1.8° to 15.8° C) at the relatively narrow elevation amplitude between the lowest and the highest situated sampling sites involving less than 500 meters above sea level (Anzali Lagoon – 21 m, Soyukhbulag 452 m). However, we found that there were significant differences in both slopes and intercepts between variables tested separately for different groups of sampling sites, which we classified as geographical regions, including temperate (localities 1 to 7, and 9 to 11), southern (8, and 12 to 18) and northern (19 to 23) regions, respectively. Except for temperate region (where temperature amplitude is narrow, but the elevation wide), the correlation coefficients were significant, and suggest specific climatic conditions for each region.

Effect of environmental factors upon fish size
(Fig. 10; Appendix 1: equations 60–79)

An increase in LONG, LAT and ELEV was correlated with growth of SIZE, which is statistically very significant in the case of geographical coordinates, but insignificant, as expected, because of a narrow amplitude, in altitude above sea level. However, SIZE is inversely and very significantly correlated with temperature. This indicates that the dependence of size (of fish) on geographical coordinates, and especially on temperature, may be taken for granted in spite of contradictions among trends in some subsets.

Table 11 t test (*t*) and coefficient of difference (*C D*) among means of branchial spines in *Rhodeus sericeus* from particular regions Statistically significant values printed in bold

C D									
		I	III	VII	VIII	IX	XI	*amarus*	*sericeus*
	I	–	0 02	0 24	0 48	1 12	0 50		
	III	0 07	–	0 24	0 49	1 18	0 52		
	VII	1 92	1 10	–	0 28	0 97	0 29		
t	VIII	**3.94**	1 99	**2.42**	–	0 62	0 00		
	IX	**7.44**	**4.46**	**7.22**	**4.53**	–	0 65		
	XI	**2.81**	1 97	1 80	0 00	**3.31**	–		
	amarus							–	**0.58**
	sericeus							**6.08**	–

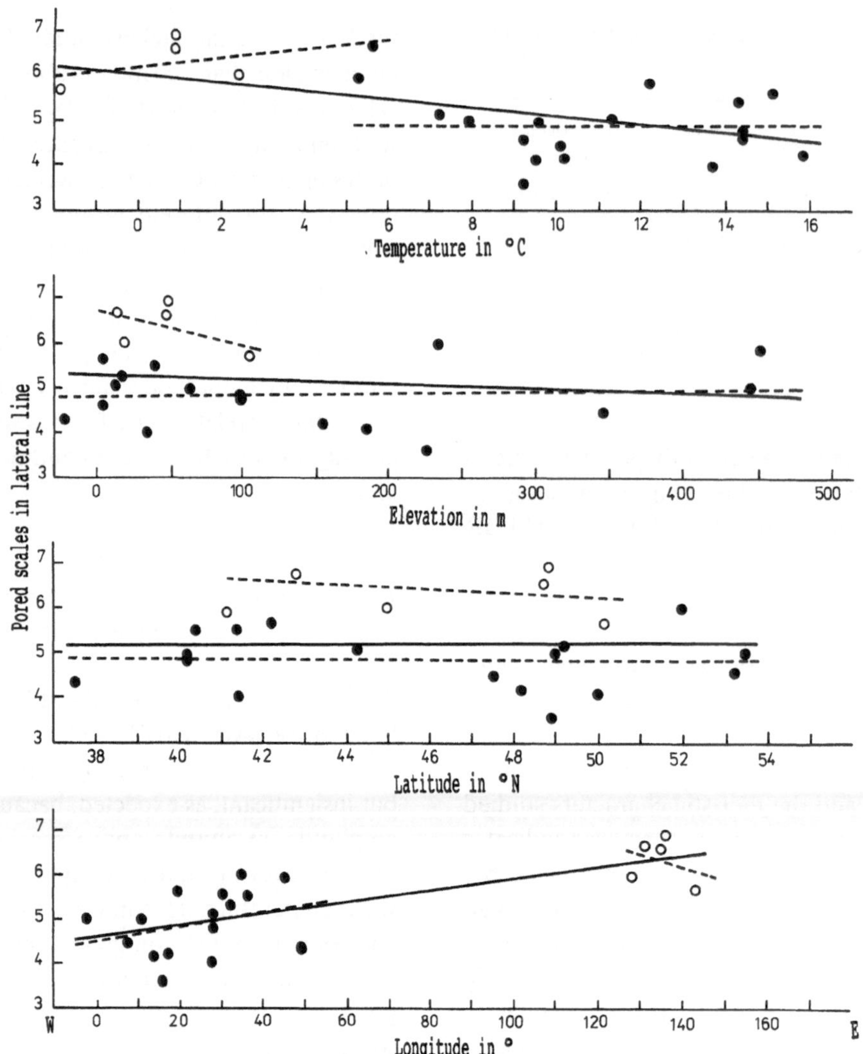

Fig. 4. Relationship of the number of pored scales in lateral line to the geographical longitude, latitude, elevation and the mean annual air temperature in the bitterling samples. In this diagram, as well as in Figs 5 and 6, the full lines give the total calculated regressions for both subspecies, dashed lines give regressions calculated separately for *amarus* (black dots) and *sericeus* (open circles), respectively. Regression equations, calculated correlation coefficients and significance levels for this and also following figures of a similar type are summarized in Appendix 1.

Discussion

All three examined counts in *Rhodeus sericeus* show the typical clinal variation, i.e. there is a gradient or continuum between two extreme values, which typically coincide with extremes of this species range. Clinal variation in both counts and measurements of fishes is well known (e.g. Jordan 1891, Moenkhaus 1895, 1898, Huntsman 1919, Hubbs 1922, Lindsey 1953, Himberg 1970, Holčík & Skořepa 1971, Holčík & Nagy 1987), although not always named under the term cline. In fishes, clinal variation is usually related with geographical latitude (for extensive literature review see Vladykov 1934), examples of longitudinal or meridional clines are few (Holčík & Duyvené de Wit 1964, Pivnička 1970).

The west-east gradient was also revealed in a principal component analysis (PCA) of the correlation matrix we used for ordination of both localities and populations; the variables were standardized by range for their equal weighting and the elimination of size influences. Analysis without standardization revealed the high variance of ELEV (about 86% of pooled variance). For PCA with standardization, eight eigenvalues were extracted. Among them, only the first three were greater than 1 and significant, accounting for 69.9, 15.0, and 12.5% of the variance, respectively. The first component was highly positively correlated with SQU, SPBR, SIZE, LAT, TEMP, LL, and negatively with ELEV; correlation with LONG was not significant. The second axis was highly positively correlated with LONG, and correlations with other variables were not significant. The position of Anzali Lagoon was extreme, because of its low elevation (– 21 m below sea level). All other localities were ordinated along one eastward cline. To elucidate the influence of variables without ELEV, PCA with seven variables was performed. From seven positive eigenvalues the first accounted for 79.7% and the second 13.7% of the variance. The first component was highly positively correlated with LONG only (however, this could be a known mathematic artifact of the correlation of the second component with the first). Correlations of variables with the first axis confirm the presence of the west-east gradient along which the values of characters change.

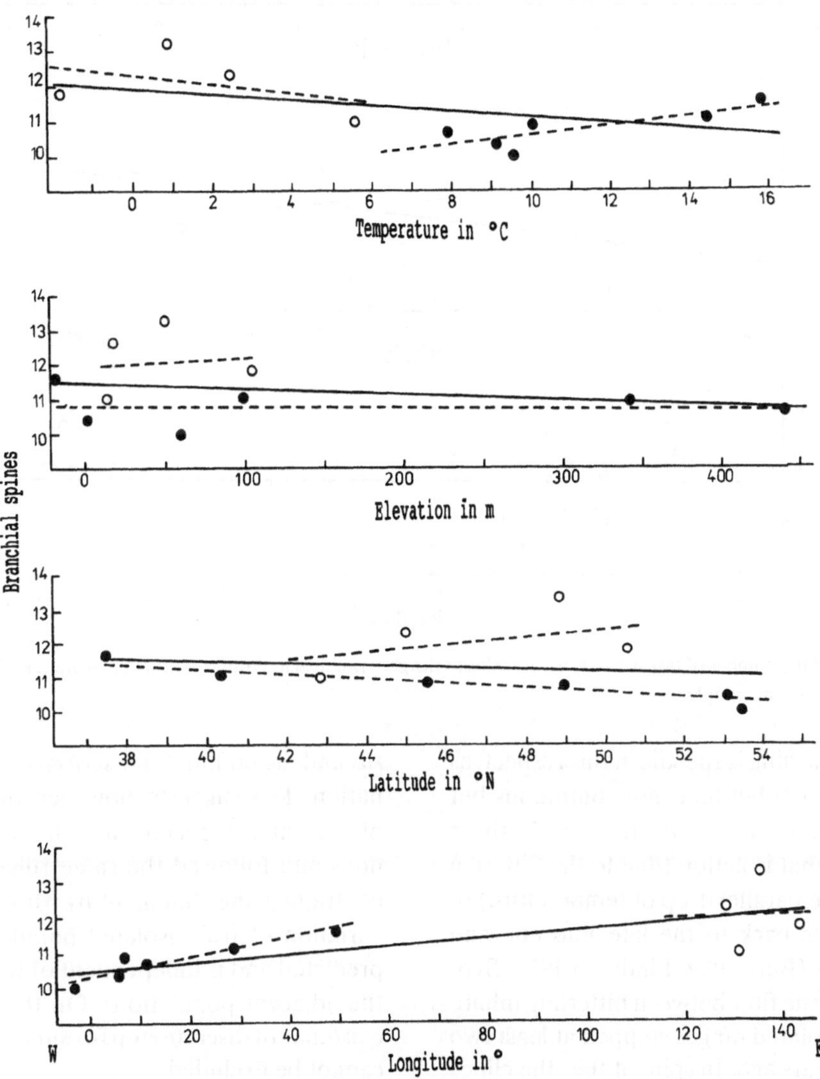

Fig 5 Relationship of the number of branchial spines to the geographical longitude, latitude, elevation and the mean annual air temperature in the bitterling samples

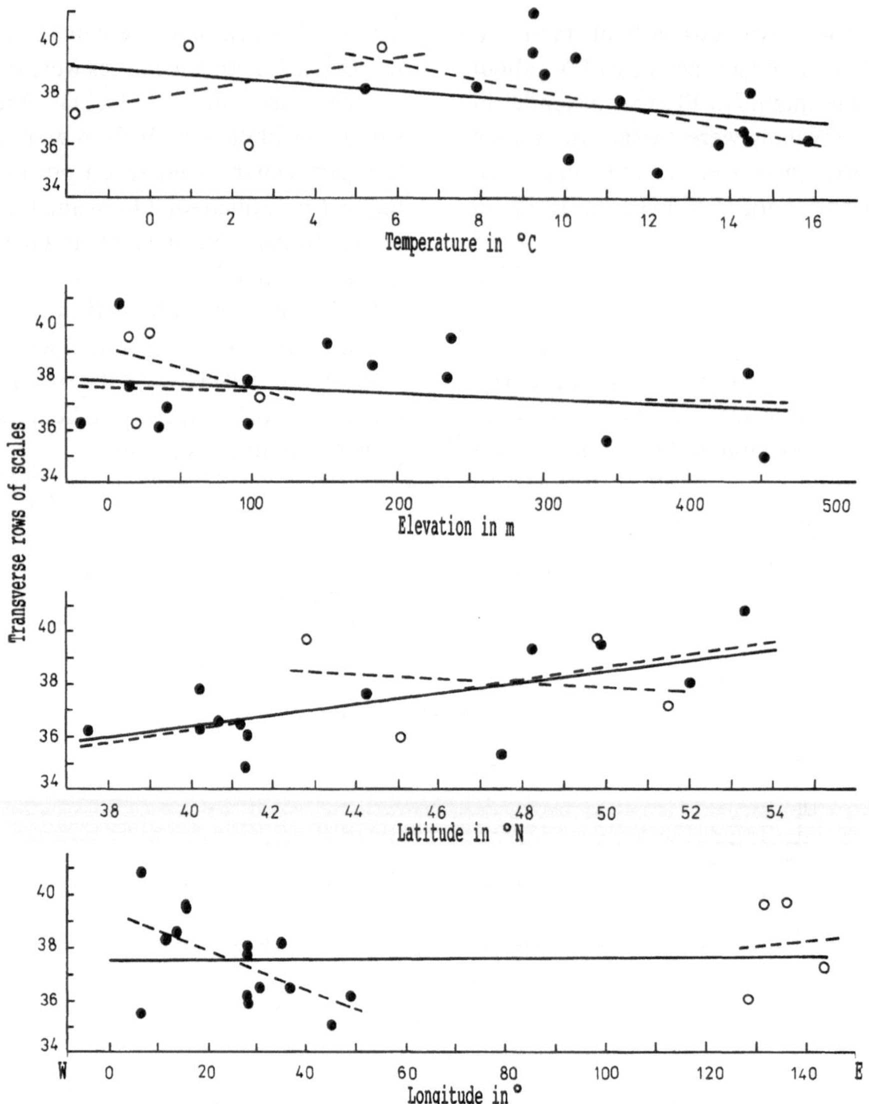

Fig. 6. Relationship of the number of transverse rows of scales to the geographical longitude, latitude, elevation and the mean annual air temperature in the bitterling samples.

The case of bitterling is specific in this respect, as its geographical distribution is not continuous but split into two distant and isolated regions. As there is firm evidence that isolation (due to the Siberian glaciation and the parallel drop of temperature) of both ranges dates back to the late Pliocene and early Pleistocene (Berg 1949, Lindberg 1972, Svoboda 1983), the gene flow between bitterling inhabiting these two isolated ranges stopped at least two to four million years ago. In spite of this, the clines in all three characters do not show any visible steps establishing a boundary between subspecies *amarus* and the nominotypic *sericeus* and their discrimination. This suggests, however, that the evolution of bitterling characters in both ranges was synchronous and followed the same rules. However, this contradicts the statement by Mayr (1963), that the variation of truly isolated populations cannot be predicted and is independent of the clines found in the adjacent populations. On the other hand, the existence of discrete steps or *subclines* within a cline cannot be excluded.

In Fig. 11, regression lines showing the relationships between counts investigated and three main

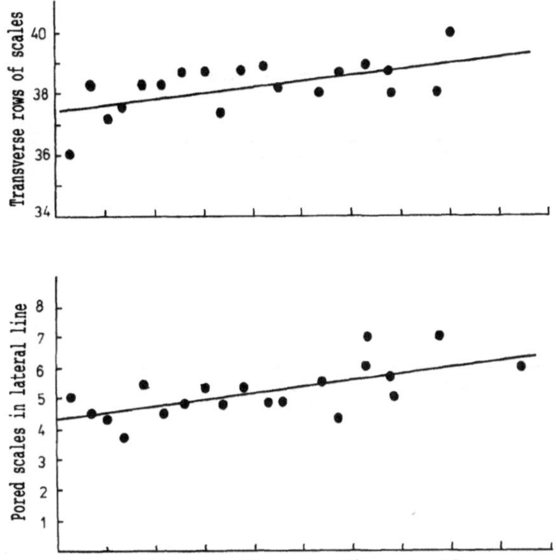

Fig. 7. Relationship of the number of pored scales in lateral line and transverse rows of scales to standard length in the *amarus* sample. Data from 125 specimens were grouped by two mm size classes.

variables, calculated separately for the temperate, southern and northern regions are summarized. Only SPBR displays a continuous, and more or less smooth, gradient with LONG. The other two counts show particular trends for each region, separated from, and in some cases even opposite to, the trend of populations inhabiting the neighbouring region.

Also in this case, the groups of populations in these three geographical regions are mutually isolated, because the gene flow among them ceased during the Piacenza transgression of the World Ocean (about 4 million years ago; Lindberg 1972). In addition, it seems clear from the geology and geography of the regions that no gene flow now occurs among most of the populations inhabiting the same geographic region. This is true even for the closely adjacent populations (e.g. between bitterling populations inhabiting various streams entering the Black, Caspian, Adriatic, Baltic etc. seas). This situation seems to be common in the freshwater ani-

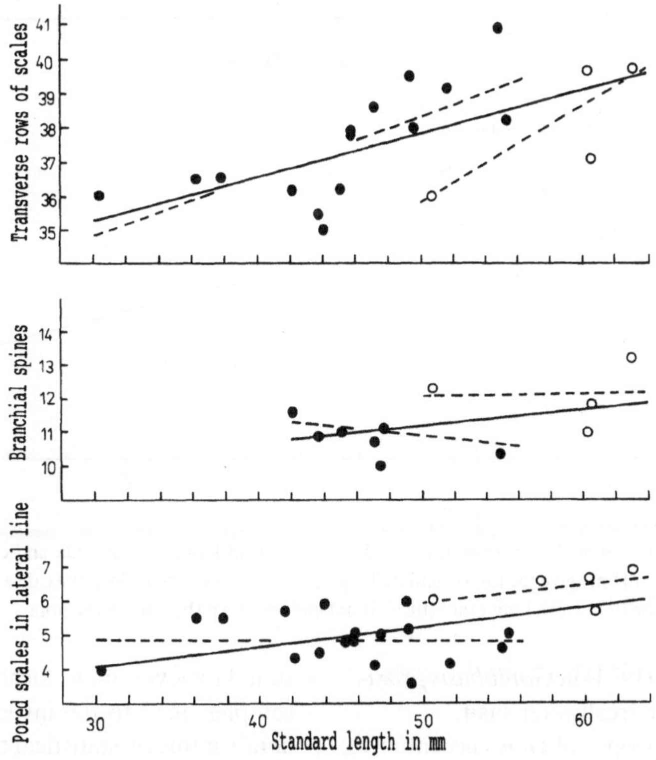

Fig. 8. Relationship of the number of transverse rows of scales and pored scales in lateral line to standard length in the bitterling samples. In each sample the mean standard length was correlated to the means of corresponding three meristic characters.

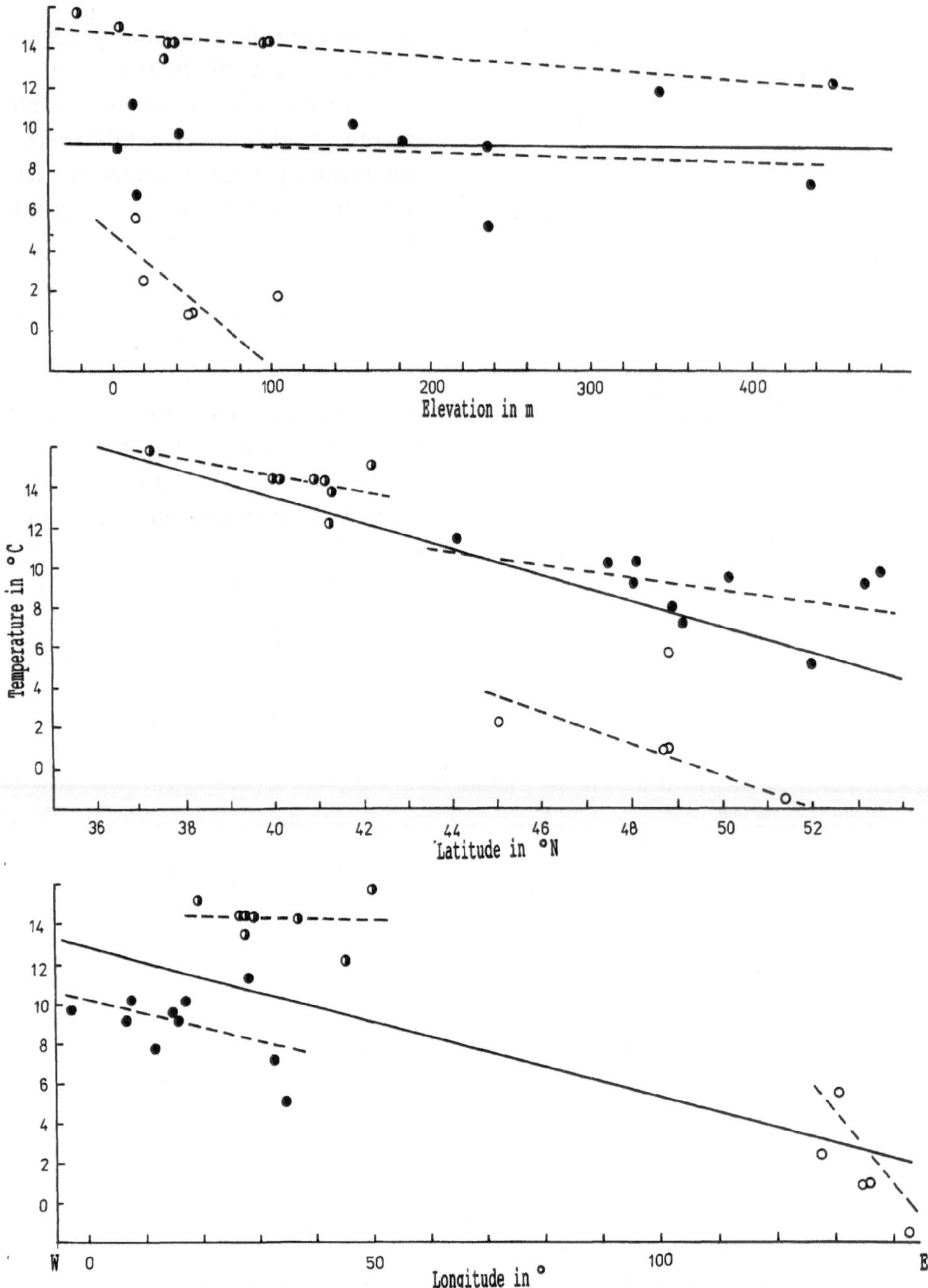

Fig. 9. Relationship of the mean annual air temperature to the geographical longitude, latitude and elevation. Data from sampling localities (Table 1) were used. Full lines give the total calculated regressions for all sampling localities, dashed lines illustrate relationships for localities in the northern (open circles), temperate (black dots) and southern (half-black) regions.

mals as found by Dillon (1984) in *Goniobasis proxima,* the North American freshwater snail.

Different trends and slopes of clines accentuate the possible existence of discrete subclines which express the adaptive pattern of geographic variation. However, more localities and more complete samples need to the investigated and analysed to confirm this by statistical tests.

As shown above in the bitterling sample from Sakhalin the number of segments in counts was less than in more centrally sampled populations, although Sakhalin is the most eastern range of this species. This phenomenon was previously noted by Holčík & Duyvené de Wit (1964) who ascribed it to the territory's past history. There is no doubt that *Rhodeus sericeus* populated the rivers of Sakhalin from the Amur River through which they were united in the past. Together with rivers of Primorye (catchment of the Gulf of Peter the Great) they have formed the common basin of the Paleoamur (Lindberg 1955, 1972). After the rise of the World Ocean during the Sicilian transgression, the formerly continuous gene flow was interrupted and bitterling populations inhabiting the Sakhalin and Primorye were fully isolated from those populating the Amur River basin.

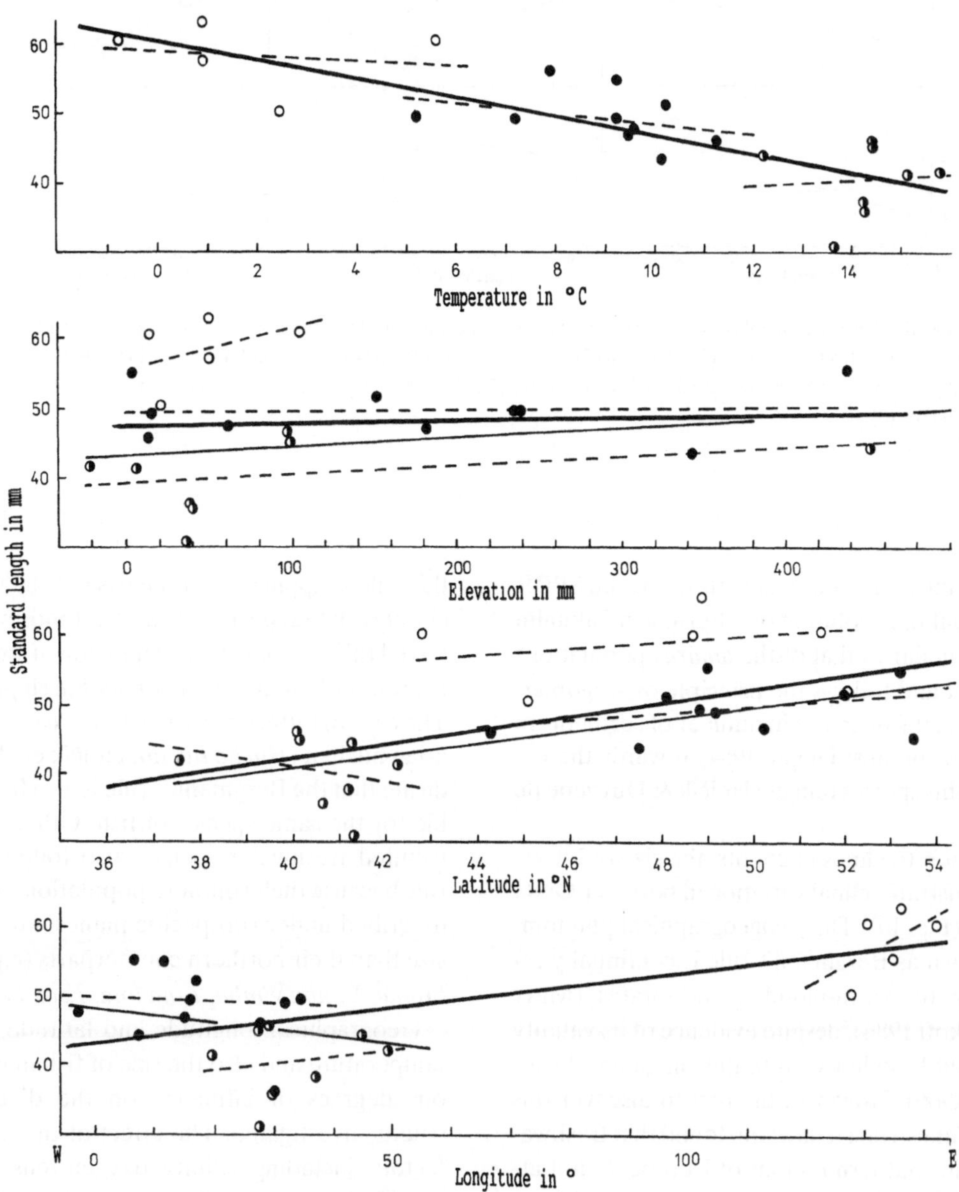

Fig 10 Relationship of the mean standard length of the bitterling samples to the geographical longitude, latitude, elevation and the mean annual air temperature Full thick lines illustrate regressions for all samples, full thin lines for *amarus* (standard length-mean annual air temperature regression for *amarus* is almost identical to the total regression and therefore not depicted), dashed lines (marked as in Fig. 9) indicate regressions for the northern (= nominotypic subspecies), temperate and southern regions

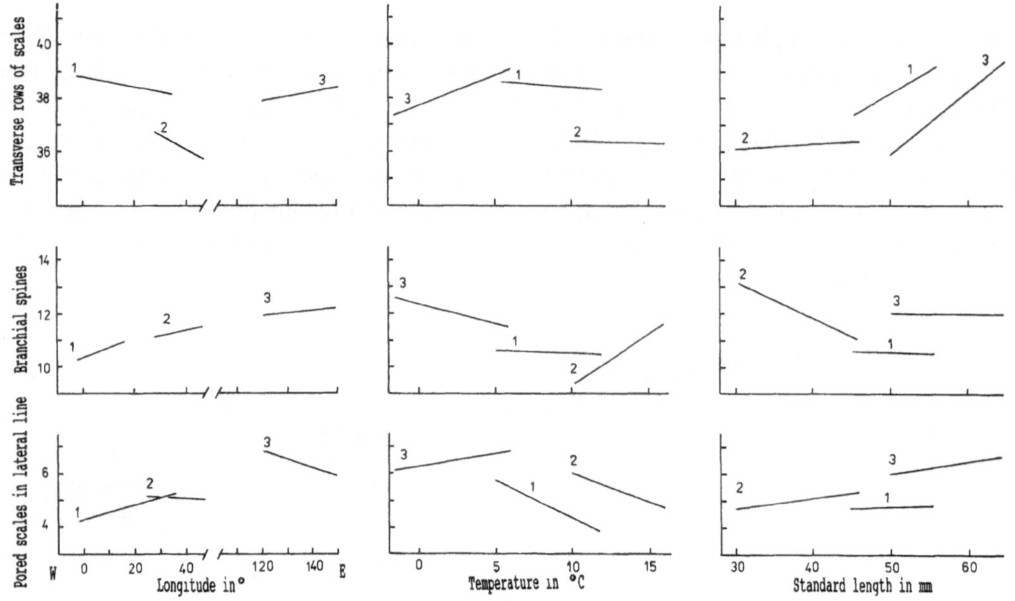

Fig. 11. Relationships between the meristic characters and geographical longitude, mean annual air temperature and standard length, respectively, for samples from temperate (1), southern (2) and northern (3) regions. Corresponding regression equations and correlation coefficients (Appendix 1): 80, 83, 86, 89, 92, 95, 98, 101 and 104 for (1); 81, 84, 87, 90, 93, 96, 99, 102 and 105 for (2); and 82, 85, 88, 91, 94, 97, 100. 103 and 106 for (3).

Decreasing number of segments in LL and SPBR suggests that the evolution of bitterling in Sakhalin has been similar to that of the *amarus* populations in the West and follows the principle of '*oligomerization'* (i.e., the decrease in number of segments in homologous organs; Dogel' 1954) towards the extremes of this species ranges (Holčík & Duyvené de Wit 1964).

In addition to clines in counts, the size of bitterling demonstrates clinal variation in both to LONG and LAT (Fig. 10). This ecogeographical phenomenon, known as Bergmann's rule is continually related only to warm-blooded vertebrates (Mayr 1963, Minkoff 1984), despite evidence of its validity for the cold-blooded vertebrates, including fishes (Lindsey 1966). However, the first to discover this rule was Canestrini (1866), who found that freshwater fishes in southern regions of Europe do not attain so large a size as European specimens from more northern localities. Vladykov (1934) referred to other authors who observed this phenomenon in fishes, both freshwater and marine, and stated that this rule is applicable in the case of the same or of closely related species. However, Lindsey (1966) referred this rule only to a proportion of species within groups of families and also within single families. The case of bitterling as well as the case of the Eurasian huchen (*Hucho hucho*; Holčík et al. 1988), indicate that the Bergmann's rule is also fully applicable for the same species of fish. Other widely distributed freshwater species also follow the same rule because their southern populations, sometimes described under subspecific names, are of smaller size than their northern counterparts (e.g. *Abramis brama, Vimba vimba, Lota lota, Esox lucius*).

Geographical longitude and latitude, elevation, temperature and also the size of fish manifest various degrees of influence on the divergence of counts investigated. The effect of these and other factors, including salinity, oxygen tension, carbon dioxide, amount and type of food, on the meristic characters of fishes is well documented (e.g., Walter 1913, Huntsman 1919, Schmidt 1921, Svetovidov 1932, Vladykov 1934, Čihař 1958, Tåning 1952, Barlow 1961, Reshetnikov 1961, Holčík & Nagy 1987, and references therein).

The effect of temperature is best known and also experimentally proven (e.g Tåning 1952, Tatarko 1968). However, our results indicate that temperature cannot be considered as a simple factor affecting fish metabolism in field conditions. In nature, temperature is a function of the local conditions, including geographical coordinates, elevation, proximity of the sea coast and many others. In other words, temperature is a complex climatic factor, synergistically accumulating the effect of geographical longitude, latitude, elevation, and other local environmental conditions. The effect of the size of fish is even more complex, as this factor is a function of metabolism, affected to a great extent by temperature.

Our results on the bitterling indicate again that the category subspecies is in principle subjective, an artifact and not a unit of evolution (Mayr 1954, 1963). Bitterling populations that inhabit distant, and for a long time, isolated areas in the West and East, known under the subspecific names *amarus* and *sericeus*, respectively, did not differ taxonomically by diagnostic morphological characters. They cannot be separated as the C.D. values for LL, SQU and SPBR are 0.75, 0.09 and 0.58 respectively (Tables 5, 8, 11), which is far below the accepted value 1.28 and corresponding 90% of the total overlap of characters (Mayr et al. 1953). The chromosomal set of various populations appeared to be identical (Bozhko et al. 1976, Hafez et al. 1978, Hong et al. 1983, Vasil'ev 1985, Aref'ev 1988). The same is true for their ecology including ontogeny, though the eggs of *sericeus* are incubated in quite different host mussels than those of *amarus* (Holčík 1993).

In general, the bitterling resembles the Eurasian huchen, *Hucho hucho*, as distribution and history of both are similar. In spite of the long time of isolation, the Eurasian huchen also does not show any significant morphological, caryological and ecological difference. Their separation into the nominotypic form *hucho* and subspecies *taimen*, inhabiting the western and eastern range, respectively, is based on slight colour differences only (Holčík et al. 1988).

The most important characters used to distinguish particular subspecies in fishes are the number of scales in the lateral line, the number of transverse scale rows, the number of rays in the dorsal and anal fin, number of branchial spines and also the number of pyloric caeca. As examples we cite *Rutilus frisii, R.f. meidingeri* and *R.f.kutum, Capoeta capoeta, C.c.gracilis,* and *C.c.heratensis, Alburnoides bipunctatus, A.b.rossicus, A.b.fasciatus*, and *A.b.eichwaldi, Vimba vimba* and *V.v.persa, Lota lota, L.l.lacustris* and *L.l.asiatica*, and many others. All these species show clear clinal variation in their counts. However, many body proportions are also often used as discriminating and even key characters, and show clinal variation (e.g. maximum body depth, head length and depth, caudal peduncle length and depth; Vladykov 1934, Holčík & Duyvené de Wit 1964, Holčík & Skořepa 1971). In our opinion, sufficient data now exist to allow us to state that, in general, most of meristic characters used till now to separate and to discriminate subspecies are controlled by environmental factors and succumb to clinal geographical or other types of variation, regardless of the continuous or separated range, and regardless of the distance or the time of isolation.

In spite of repeated attempts to conserve the category of subspecies in taxonomy (see Hubbs 1943, Starret 1958, Mayr 1963 and references therein), and its still frequent use in the European zoological literature, we agree with Wilson & Brown (1953), Gosline (1954) and Terent'ev (1957) that this concept and the trinomial nomenclature (in the cold-blooded vertebrates at least) has proved to be inefficient, superfluous and misleading and should be excluded from taxonomy, nomenclature and ecology.

Acknowledgements

The senior author is indebted to many persons for invaluable help during the long period of gathering samples, provided access to collections and literature, advise, information, interest and encouragement, especially in the early phases of his work: the late Vít'azoslav Mišík, Georgii V. Nikol'skii, Anatol N. Svetovidov, Alexandra A. Svetovidova, Ethelwynn Trewavas, and Peter L.P. Whitehead as well as Nurdin Hosseinpour, Ivan Löbl, Volker Mahnert,

Ilja Okáli, and my teacher Ota Oliva. My sons Martin and Juraj Holčík Jr. assisted with data processing and measuring of some samples, respectively. Graphs were drawn by Tatiana Besedová and Jana Weisová and tables were typed by Agnesa Malkowitsová. Thanks are extended to all. Last, but not least, the senior author expresses his sincere and deep thanks to Rosemary H. Lowe-McConnell, P.H. Greenwood and the late Ethelwynn Trewavas and Peter J.P. Whitehead, not only for their interest in his work when he visited the British Museum (Natural History) in 1966 and 1968, but especially for their kindness, sincere sympathy, hospitality and all support they manifested to him during the last days of the memorable August 1968, when Soviet invasion to Czechoslovakia prevented him from returning home and he was forced to stay in London longer than originally planned. Part of this work was accomplished in the laboratory of Herbert Fernando, University of Waterloo, Canada, and the senior author is deeply indebted to him for the creation of an enthusiastic atmosphere and working facilities. He and Gordon Copp made linguistic corrections of this paper for which we thank them very much. Last but not least we are very indebted to Dalibor Povolný, our colleague Zbyšek Šustek and two anonymous reviewers for their critical comments, notes and suggestions.

References cited

Abdurakhmanov, Yu A 1962 Freshwater fishes of Azerbaijan Izd Akademii Nauk Azerb SSR, Baku 406 pp (in Russian)

Aref'ev, V A 1988 Karyotypes of two species of cyprinid fish – asp *Aspius aspius* (L) and bitterling *Rhodeus sericeus amarus* (Bloch) Buyll Mosk Obshch Isp Prir , otd biol 93 57–61 (in Russian)

Barlow, G W 1961 Causes and significance of morphological variation in fishes Syst Zool 10 105–117

Berg, L S 1949 Freshwater fishes of the USSR and adjacent countries Vol 2, 3 469–925, 926–1382 Izd Akademii Nauk SSSR, Moskva-Leningrad (in Russian)

Bloch, M 1782 Oekonomische Naturgeschichte der Fische Deutschlands Hesse, Berlin 234 pp

Bozhko, S I , A Horvath & B Meszaros 1976 Karyological examinations of four species of Cyprinidae from Hungary Acta Biol Debrecina 13 237–256

Canestrini, G 1866 Prospetto critico dei pesci d'acqua dolce d'Italia Arch Zool Nat 4, 1895, Fasc 1 141 pp

Čihař, J (Chigarzh, J) 1958 Notes to the systematics of the crucian carp (*Carassius carassius*) Voprosy Ikhtiologii (11) 136–141 (in Russian)

Dillon, Jr , R T 1984 Geographic distance, environmental difference and divergence between isolated populations Syst Zool 33 69–82

Dogel', V D 1954 Oligomerization of the homologous organs Izd Leningradskogo universiteta, Leningrad 368 pp (in Russian)

Gosline, W A 1954 Further thoughts on subspecies and trinomials Syst Zool 3 83–93

Hafez, R , R Labat & R Quillier, 1978 Etude cytogenetique chez quelquez especes de cyprinides de la region Midi-Pyrenees Bull Soc d'Hist Natur Toulouse 114 122–159

Himberg, K J M 1970 A systematic and zoogeographic study of some north European coregonids pp 219–250 *In* C C Lindsey & C S Woods (ed) Biology of Coregonid Fishes, University of Manitoba Press, Winnipeg

Hoestland, H 1991 (ed) Clupeidae, Anguillidae The freshwater fishes of Europe, Vol 2, AULA-Verlag, Wiesbaden 448 pp

Holčík, J (Golchik, Yu) 1959 Systematic status of the European bitterling, *Rhodeus sericeus amarus* (Bloch, 1783) Voprosy Ikhtiologii (13) 39–50 (in Russian)

Holčík, J 1983 *Rhodeus sericeus* (Pallas, 1776) *In* O Oliva & V Baruš (ed) Fauna ČSSR, Fishes Academia, Praha (in press)

Holčík, J 1993 *Rhodeus sericeus* (Pallas, 1776) *In* P Banarescu (ed) The Freshwater Fishes of Europe, Vol 5, Cyprinidae, AULA-Verlag, Wiesbaden (in press)

Holčík, J , P Banarescu & D Evans 1989 General introduction to fishes pp 18–147 *In* J Holčík (ed) The Freshwater Fishes of Europe, Vol 1, part II, AULA-Verlag, Wiesbaden

Holčík, J & J J Duyvene de Wit 1964 Systematic status of the bitterling from Asia Minor and notes on the geographical variability of *Rhodeus sericeus* (Pallas, 1776) in the area of its distribution Zeitschr f Wissenschaft Zoologie 169 396–412

Holčík, J , K Hensel, J Nieslanik & L Skacel 1988 The Eurasian huchen, *Hucho hucho*, largest salmon of the world Dr W Junk Publishers, Dordrecht 239 pp

Holčík, J & S Nagy 1987 Burbot (*Lota lota* [Linnaeus, 1758]) from the Turiec River Folia Zoologica 36 85–96

Holčík, J & V Skořepa 1971 Revision of the roach, *Rutilus rutilus* (Linnaeus, 1758), with regard to its subspecies Annot Zool et Bot 64 60 pp

Hong, V , M Zhou & T Zhou 1983 Studies on the karyotypes of Chinese cyprinid fishes III Comparative analysis of the chromosomes of seven species of acheilognathid fishes J Wuhan Univ (Nat Sci) 1983 (2) 96–102 (in Chinese)

Hubbs, C L 1922 Variations in the number of vertebrae and other meristic characters of fishes correlated with the temperature of water during development Amer Nat 56 560–372

Hubbs, C L 1925 Racial and seasonal variation in the Pacific herring, California sardine and California anchovy Fish Bull (California) 8 25 pp

Hubbs, C L 1943 Criteria for subspecies, species and genera, as determined by researches on fishes Ann N Y Acad Sci 44 109–121

Huntsman, A G 1919 Variation of fishes according to latitude Science 50 592

Ivanovic, B M 1973 Ichthyofauna of Skadar Lake Biological Station Titograd 146 pp

Jordan, D S 1891 Relations of temperature to vertebrae among fishes Proc Nat Mus 14 (815) 107–120

Karaman, St 1924 Pisces Macedoniae Split 90 pp

Karaman, St 1955 Die Fische der Strumica (Struma-System) Acta Mus Maced Scient Nat 3 181–207

Kazancheev, E N 1981 Fishes of the Caspian Sea Legkaya i Pishchevaya Promyshlennost', Moskva 183 pp (in Russian)

Kim, I S 1982 A taxonomic study of the Acheilognathinae fishes (Cyprinidae) in Korea Ann Rep of Biol Research 3 1–18 (in Korean)

Lindberg, G U 1955 The Quaternary in the light of biogeographic data Izd Nauka, Moskva-Leningrad 334 pp (in Russian)

Lindberg, G U 1972 Great fluctuations of the world ocean in the Quaternary Izd Nauka, Leningrad 548 pp (in Russian)

Lindsey, C C 1953 Variation in anal fin ray count of the redside shiner *Richardsonius balteatus* (Richardson) Can J Zool 31 211–222

Lindsey, C C 1966 Body sizes of poikilotherm vertebrates at different latitudes Evolution 20 456–465

Mayr, E 1954 Notes on nomenclature and classification Syst Zool 3 86–89

Mayr, E 1963 Animal species and evolution The Belknap Press of Harvard University Press, Cambridge 797 pp

Mayr, E, E G Linsley & R L Usinger 1953 Methods and principles of systematic zoology McGraw-Hill, New York 385 pp

Minkoff, E C 1984 Evolutionary biology Addison-Wesley Publ Comp, Reading 627 pp

Moenkhaus, W J 1895 Variation of North American fishes II Variation of *Etheostoma caprodes* Rafinesque in Turkey Lake and Tippecanoe Lake Proc Indiana Acad Sci 1895 278–296

Moenkhaus, W J 1898 Material for the study of the variation of *Etheostoma caprodes* Rafinesque and *Etheostoma nigrum* Rafinesque in Turkey Lake and Tippecanoe Lake Proc Indiana Acad Sci 1897 207–228

Nikol'skii, G V 1956 Fishes of the Amur River Izd Akademii nauk SSSR, Moskva 551 pp (in Russian)

Pallas, PS 1776 Reise durch verschiedene Provinzen des Russischen Reichs 3 Academia Scient, St Petersburg 19 + 760 + 25 pp

Pivnička, K 1970 Morphological variation in the burbot (*Lota lota*) and recognition of the subspecies a review J Fish Res Board Can 27 1757–1765

Reshetnikov, Yu S 1961 On connection between the number of gill rakers and the diet characteristics in charrs of the genus *Salvelinus* Zoologicheskii Zhurnal 40 1574–1577 (in Russian)

Schlesinger, D A & H A Regier 1982 Climatic and morphoedaphic indices of fish yield from natural lakes Trans Amer Fish Soc 111 141–150

Schmidt, J 1921 Racial investigations VII Annual fluctuations of racial characters in *Zoarces viviparus* L Comptes-Rendus Trav Lab Carlsberg 14 14 pp

Starret, A 1958 What is the subspecies problem? Syst Zool 7 111–113

Svetovidov, A N 1932 On the dependence between the diet characteristics and the number of pyloric caeca in clupeids Doklady Akademii Nauk SSSR (Ser A) (8) 202–204 (in Russian)

Svetovidov, A N & G K Eremeev 1935 On the European and Amur bitterling (*Rhodeus sericeus*) Doklady Akademii Nauk SSSR (1) 582–587

Svoboda, J (ed) 1983 Encyclopaedic dictionary of the geological sciences 2 vols Academia, Praha 917 + 852 pp (in Czech)

Taning, A V 1952 Experimental study of meristic characters of fishes Biol Reviews 27 169–193

Tatarko, K I 1968 The effect of temperature on the meristic characters of fishes Voprosy Ikhtiologii 8 425–439 (in Russian)

Terent'ev, PV 1957 The 'subspecies' concept's utility in study of the intraspecific variation Vestnik Leningradskogo Universiteta (21) 75–80 (in Russian)

Vasil'ev, VP 1985 Evolutionary karyology of fishes Izd Nauka, Moskva 300 pp (in Russian)

Vladykov, VD 1934 Environmental and taxonomic characters of fishes Trans Royal Canad Inst 20 (Part 1) 99–140

Walter, E 1913 Einfuhrung in die Fishkunde unserer Binnengewassern Quelle und Mezer, Leipzig 364 pp

Wernstedt, FL 1972 World climatic data Climatic Data Press, Lemont 552 pp

Wilson, E O & W L Brown, Jr 1953 The subspecies concept and its taxonomic application Syst Zool 2 97–111

Zhukov, PI 1965 Fishes of Belorussia Izd Nauka i Tekhnika, Minsk 416 pp (in Russian)

Appendix 1 Relationships, sets and regression equations including correlation coefficients pertinent to Figures 4–11 LONG – geographical longitude (°), LAT – geographical latitude (°), ELEV – elevation (m), TEMP – mean annual air temperature (° C), SL – standard length (mm), LL – pored scales in lateral line, SPBR – branchial spines, SQU – transverse scale rows, RSA – *Rhodeus sericeus amarus,* RSS – *Rhodeus sericeus sericeus,* RS – *Rhodeus sericeus,* TR – temperate region, SR – southern region, NR – northern region, EA – Eurasia * = p < 0 05, ** = p < 0 01, *** = p < 0 001

No	Relationship	Set	a	+	bx	r	n	Fig
1	LONG – LL	RSA	4 45339	+	0 01753	0 358	18	4
2		RSS	10 31377	+	0 02939	– 0 333	5	4
3		RS	4 54461	+	0 01359	0 734***	23	4
4	LAT – LL	RSA	5 51678	+	0 01408	– 0 107	18	4
5		RSS	8 77560	–	0 05099	– 0 346	5	4
6		RS	5 03690	+	0 00350	0 019	23	4
7	ELEV – LL	RSA	4 92055	–	0 00403	– 0 018	18	4
8		RSS	6 71556	–	0 00741	– 0 525	5	4
9		RS	5 31271	–	0 00099	– 0 153	23	4
10	TEMP – LL	RSA	4 92055	–	0 004027	– 0 018	18	4
11		RSS	6 20757	+	0 09527	0 509	5	4
12		RS	6 09014	–	0 09700	– 0 550**	23	4
13	LONG – SPBR	RSA	10 31980	+	0 0275	0 929**	7	5
14		RSS	10 43293	+	0 01233	0 086	4	5
15		RS	10 57182	+	0 01178	0 773**	11	5
16	LAT – SPBR	RSA	14 25259	–	0 07472	– 0 931**	7	5
17		RSS	7 01434	+	0 10787	0 441	4	5
18		RS	12 89576	–	0 03478	– 0 210	11	5
19	ELEV – SPBR	RSA	10 86862	–	0 00029	– 0 097	7	5
20		RSS	11 96112	+	0 00272	0 118	4	5
21		RS	11 46935	–	0 00167	– 0 266	11	5
22	TEMP – SPBR	RSA	9 26710	+	0 13403	0 796*	7	5
23		RSS	12 33394	–	0 13743	– 0 448	4	5
24		RS	11 95289	–	0 08283	– 0 522	11	5
25	LONG – SQU	RSA	39 37805	–	0 07599	– 0 602	15	6
26		RSS	35 95094	+	0 01629	0 058	4	6
27		RS	37 53100	+	0 00133	0 037	19	6
28	LAT – SQU	RSA	26 56994	+	0 24157	0 727**	15	6
29		RSS	42 04930	–	0 0831	0 172	4	6
30		RS	28 08470	+	0 20919	0 595***	19	6
31	ELEV – SQU	RSA	37 71178	+	0 00167	– 0 156	15	6
32		RSS	38 53066	–	0 00822	– 0 183	4	6
33		RS	37 89154	–	0 00222	– 0 192	19	6
34	TEMP – SQU	RSA	40 94725	–	0 30552	– 0 551*	15	6
35		RSS	37 73160	+	0 23009	0 383	4	6
36		RS	38 70834	–	0 11828	– 0 352	19	6
37	SIZE – LL	RSA	3 01870	+	0 04264	0 652*	19	7
38	SIZE – SQU	RSA	36 25640	+	0 03954	0 611*	19	7
39	SIZE – LL	RSA	5 18689	–	0 00687	– 0 205	18	8
40		RSS	3 60657	+	0 04721	0 444	4	8
41		RS	2 47205	+	0 05579	0 504	23	8
42	SIZE – SPBR	RSA	13 68102	–	0 0560	0 602	7	8
43		RSS	11 62813	+	0 00788	0 046	4	8
44		RS	8 92383	+	0 04566	0 371	11	8

Appendix I Continued

No	Relationship	Set	a	+	bx	r	n	Fig
45	SIZE – SQU	RSA	29 60062	+	0 17350	0 711**	15	8
46		RSS	22 08900	+	0 27379	0 011	4	8
47		RS	31 45355	+	0 12780	0 649**	19	8
48	LONG – TEMP	TR	10 06648	–	0 06748	– 0 467	10	9
49		SR	14 73580	–	0 01386	– 0 132	8	9
50		NR	58 50602	–	0 42125	– 0 764	5	9
51		EA	12 91837	–	0 0773	– 0 739***	23	9
52	LAT – TEMP	TR	23 58648	–	0 29539	– 0 473	10	9
53		SR	29 95382	–	0 38639	– 0 523	8	9
54		NR	37 96924	–	0 76767	– 0 974*	5	9
55		EA	40 54451	–	0 68183	– 0 647***	23	9
56	ELEV – TEMP	TR	9 39123	–	0 00269	– 0 229	10	9
57		SR	14 84096	–	0 00604	– 0 869**	8	9
58		NR	4 88421	–	0 06842	– 0 907*	5	9
59		EA	9 28494	–	0 00073	– 0 020	23	9
60	LONG – LL	TR	4 30000	+	0 02582	0 464	10	10
61		SR	5 11054	–	0 00167	– 0 025	8	10
62		NR	see regression No	2				4, 10
63	LONG – SPBR	TR	10 18367	+	0 05489	0 831	4	10
64		SR	10 34920	+	0 02587	0 999**	3	10
65		NR	see regression No	14				5, 10
66	LONG – SQU	TR	38 80136	–	0 02072	– 0 497	7	10
67		SR	37 94570	–	0 04641	– 0 497	7	10
68		NR	see regression No	26				6, 10
69	TEMP – LL	TR	6 87700	–	0 24004	– 0 623	10	10
70		SR	8 28385	–	0 22619	– 0 351	8	10
71		NR	see regression No	11				4, 10
72	TEMP – SPBR	TR	10 64622	–	0 01616	– 0 039	4	10
73		SR	5 42286	+	0 39286	1 00***	3	10
74		NR	see regression No	23				5, 10
75	TEMP – SQU	TR	38 92243	–	0 05247	– 0 061	8	10
76		SR	36 39870	–	0 00275	– 0 152	7	10
77		NR	see regression No	35				6, 10
78	SIZE – LL	TR	4 38669	+	0 00697	0 039	10	10
79		SR	3 53784	+	0 03761	0 206	8	10
80		NR	see regression No	42				8, 10
81	SIZE – SPBR	TR	10 99033	–	0 00978	– 0 142	4	10
82		SR	18 35964	–	0 16007	– 0 987	3	10
83		NR	see regression No	45				8, 10
84	SIZE – SQU	TR	24 68991	+	0 27686	0 720*	8	10
85		SR	35 40381	+	0 0226	0 154	7	10
86		NR	see regression No	48				8, 10
87	LONG – SIZE	RSA	50 16789	–	0 19897	– 0 431	18	11
88		RSS = NR	– 0 795437	+	0 49379	0 587	5	11
89		RS	43 46770	+	0 09864	0 587**	23	11

Appendix I. Continued.

No	Relationship	Set	a	+	bx	r	n	Fig.
90		TR	49.87876	–	0.02748	– 0.089	10	11
91		SR	37.46063	+	0.08655	0.162	8	11
92	LAT – SIZE	RSA	6.06184	+	0.8627	0.693	18	11
93		RSS = NR	34.69830	+	0.49966	0.362	5	11
94		RS	2.76290	+	0.98857	0.585**	23	11
95		TR	24.59136	+	0.50074	0.376	10	11
96		SR	83.39043	–	1.06100	– 0.089	8	11
97	ELEV – SIZE	RSA	43.46552	+	0.01422	0.335	18	11
98		RSS = NR	55.68246	+	0.05620	0.421	5	11
99		RS	47.86440	+	0.00310	0.053	23	14
100		TR	49.05718	+	0.00216	0.084	10	11
101		SR	39.15078	+	0.01268	0.356	8	11
102	TEMP – SIZE	RSA	61.59993	–	1.43369	0.694**	18	11
103		RSS = NR	58.85375	–	0.29609	– 0.167	5	11
104		RS	60.27488	–	1.31154	– 0.817**	23	11
105		TR	55.755881	–	0.70870	– 0.333	10	11
106		SR	34.63215	+	0.39967	0.079	8	11

Environmental Biology of Fishes **41**: 171–190, 1994

Structure and resilience of a tidepool fish assemblage at Barbados

Robin Mahon[1] & Susan D. Mahon[2]
[1] *Fisheries and Environmental Consulting, 48 Sunset Crest, St. James, Barbados*
[2] *Environment, People and Information, Durants Ridge No. 3, St. James, Barbados*

Received 19 10 1993 Accepted 1 3 1994

Key words: Caribbean, Community structure, Coral reef fishes, Nurseries, Stability, Recolonisation

Synopsis

Fish collections from 19 tidepools on a rock plateau at Martins Bay, on the east coast of Barbados, taken on three occasions (1981, 1983 and 1987) contained 2078 individuals of 63 species. The number of species, individuals and total biomass increased with pool size. Partial residents, primarily juveniles of reef species, comprised 44% of species, 36% of numbers, and 26% of biomass. True and partial residents were of similar sizes. Most of the latter grow to larger sizes than those observed in the pools, indicating that the use of tidepools by fishes is size-dependent. Species richness, numbers of individuals and biomass in individual pools was positively associated with pool size. These relationships did not vary among sampling occasions. Species composition and relative abundance was also found to be similar among sampling occasions, leading to the conclusion that the tidepool assemblages are resilient and stable.

Introduction

Intertidal habitats have generally been viewed as ecotones where marine organisms encounter harsh and highly variable physical conditions, such as wave action, periodic desiccation, high temperatures, and are also vulnerable to predation (e.g. Nybakken 1982). In the intertidal zone, tidepools provide a refuge for marine organisms which become concentrated in them during periods of tidal exposure. Owing to easy access, the intertidal zone has been intensively studied. Although these studies have included fishes, tropical tidepools have received little attention (Chadwick 1976, Gibson 1982).

In studies of tidepools in temperate areas, local distribution and abundance of fishes appear to depend largely on the area and the depth of the pool, its degree of isolation from the sea and exposure to wave action (Gibson 1969, 1972, Green 1971, Yoshiyama 1981, Bennett & Griffiths 1984). Size and the degree of isolation of pools determine the extent of stressful fluctuations such as in temperature and salinity. Isolation is determined primarily by the height of tidepools above mean low water.

A common feature of tidepool fish assemblages is the presence of two main groups of fishes: true residents, and partial residents (Gibson 1969). True residents are generally small, benthic fishes such as the blennies and gobies. Partial residents are primarily sublittoral, but occur in tidepools, particularly as juveniles. This pattern is supported by several studies which have used slight variations of these categories (Thompson & Lehner 1976, Chang et al. 1977, Lee 1980a, 1980b, Grossman 1982). A third group, transients, usually comprises a small proportion of species in tidepools, and is sometimes included with the partial residents.

We investigated a tidepool fish assemblage from the east coast of Barbados, in the western tropical Atlantic, with the objectives of describing the species composition, distribution and abundance of fishes in tidepools of various sizes. We sampled the assemblage on three occasions at the same time in different years to evaluate the extent to which the observed characteristics are persistent through time.

The study area

The study area is located at Martins Bay on the east coast of Barbados. Barbados lies about 150 km east of the Lesser Antilles island arc in the eastern Caribbean. A general description of rocky-shore faunas and habitats of Barbados is provided by Lewis (1960). Since the prevailing winds and currents are easterly, the coast on which the study area is located is exposed to heavy wave action. At many places there are eroded limestone plateaus on which tidepools occur. These plateaus are exposed at low tide, isolating the pools. In contrast the south and west coasts are primarily sandy beaches with intermittent low limestone cliffs and sublittoral fringing reefs. On these sheltered coasts tidepools are scarce. Studies of rocky-shore zonation in the Caribbean have not resulted in a consistent classification (John & Price 1979, Brattstrom 1980). However, the slate pencil urchin, *Echinometra lucunter* and *Sargassum* spp. consistently characterise the low extreme of regular tidal exposure.

At Barbados, tides are mixed, semidiurnal with no marked inequality in amplitude (Lewis 1960). There is an amphidromic point near Barbados and consequently tidal amplitude is low (King 1975). Over an 18 month period, mean tidal range and diurnal range were 0.7 and 1.1 m, respectively (Lewis 1960).

At Martins Bay the limestone plateau is protected from the full force of the waves by headlands and by an offshore reef. The plateau is about 50 m in width at the widest point, and about 200 m in length along the shore. Shoreward it is bounded by a beach. At the seaward edge, where it drops off, the plateau is honeycombed with burrows of *Echinometra lucunter*; the surf zone of Lewis (1960). The surface of the plateau is covered with a mat of algae and dotted with depressions which become pools at low tide. The sides of the pools are usually vertical with holes and undercuts. In the pools a variety of substrates may occur: bare sand, sand with turtle grass, *Thalassia testudinum*, rock covered with algae, and limestone rubble.

The incoming tide floods the entire plateau in a period of about 30 minutes. Thus there is little variation in isolation among pools. This is in contrast to areas studied by Green (1971), Gibson (1972), and Yoshiyama (1981) where tidal range was high and pool isolation varied considerably with elevation.

Methods

The fishes were collected at low tide from 19 tidepools at Martins Bay at about the same time of year in each of three years: 17–19 June 1981, 9 June 1983, 13 June 1987. Each pool was treated with an ichthyocide, ProNoxfish, in which the active ingredient is rotenone, and was searched until no more fishes could be found. Most fishes succumbed within 5–10 minutes but some eels took up to an hour to emerge from hiding. The fishes were preserved in 4% formaldehyde and identified using keys in Böhlke & Chaplin (1968) and Fischer (1978). Identifications were checked at the Royal Ontario Museum, Toronto where the fishes have been deposited.

The length (L), width (W), and maximum depth (D) of each pool was measured and pool area ($A = LxW$) and volume ($V = A \times 0.5\ D$) were estimated. The percent of each pool covered by each substrate type – sand, sand with turtle grass, *Thalassia testudinum*, rock covered with algae, and limestone rubble – was estimated by eye. Habitat diversity was estimated using Simpson's index, $D = \Sigma p_i^2$, where p_i is the probability of encountering the *i*th habitat type or the proportion of that habitat type (Southwood 1978).

Each species of fish was designated as either a true or partial resident on the basis of information given by Böhlke & Chaplin (1968), Randall (1968), Fischer (1978) and references cited therein. True residents were those which could complete their

Table 1 Numbers and weights of species collected from pools (ALL = full sample, RP = repeated pools)

Species	Numbers of individuals					Weight (g)				
True residents	81 All	81 RP	83 RP	87 RP	Total	81 All	81 RP	83 RP	87 RP	Total
Labrisomus bucciferus (puffcheek blenny)	231	18	188	154	573	365 3	289 6	231 4	291 9	888 6
Malacoctenus erdmani (imitator blenny)	149	11	106	175	430	25 4	20 1	21 4	39 7	86 5
Malacoctenus gilli (dusky blenny)	139	10	31	12	182	57 2	44 1	13 6	5 8	76 6
Ogilbia spp (brotula)	50	46	31	38	119	25 5	23 3	13 1	15 2	53 8
Gymnothorax spp (moray eels)	18	16	50	49	117	56 1	55 9	212 7	308 1	576 9
Ginsburgellus novemlineatus (ninelined goby)	55	38	40	16	111	1 9	1 3	1 0	1 3	4 2
Paraclinus nigripinnis (blackfin blenny)	54	33	7	46	107	11 0	8 2	1 3	12 2	24 5
Arcos rubiginosus (red clingfish)	46	46	17	27	90	11 2	11 2	4 3	64 5	80 0
Moringua edwardsi (spaghetti eels)	21	18	12	12	45	99 6	83 7	35 6	49 1	184 3
Labrisomus nuchipinnis (hairy blenny)	8	6	10	18	36	42 7	39 5	90 2	199 7	332 6
Apogon maculatus (flamefish)	14	10	17	5	36	2 0	1 6	9 4	13 1	24 5
Stathmonotus stahli (eelgrass blenny)	14	13	21		35	1 3	1 2	1 7		3 0
Labrisomus nigricinctus (spotcheek blenny)	11	9	15	7	33	3 1	2 9	5 2	3 2	11 5
Starksia sp (blenny)	0		25		25			4 2		4 2
Enchelychore spp (moray eels)	14	11	4	6	24	442 5	402 4	244 7	234 0	921 2
Bathygobius curacao (notchtongue goby)	9	9	5	9	23	20 0	20 0	25 8	15 1	60 9
Malacoctenus triangulatus (saddled blenny)	3	3	11	7	21	1 7	1 7	4 1	1 3	7 1
Malacoctenus aurolineatus (goldline blenny)	4	4	4	7	15	0 8	0 8	1 7	3 1	5 6
Scorpaena plumieri (spotted scorpionfish)	2	1	1	3	6	16 5	15 6	0 5	179 1	196 1
Echidna catenata (chain moray eel)	4	3		2	6	311 4	299 0		137 6	449 0
Ophioblennius atlanticus (redlip blenny)	1	1	2	2	5	0 6	0 6	2 3	5 3	8 2
Barbulifer antennatus (barbulifer)	1	1		4	5	0 1	0 1		0 1	0 2
Labrisomus gobio (goggle eye blenny)	4	4		1	5	12 5	12 5		0 2	12 7
Paraclinus cingulatus (coral blenny)				5	5				0 3	0 3
Labrisomus guppyi (mimic blenny)	4	4	1		5	3 9	3 9	35 0		38 9
Starksia sluteri (? blenny)	3	3			3	0 2	0 2			0 2
Gobiosoma hildbrandi (? goby)	0		3		3			0 4		0 4
Ahlia egmontis (keyworm eel)	1	1	2		3	2 7	2 7	3 7		6 4
Myrichthys acuminatus (sharptail eel)					3	79 7				79 7
Lythrypnus sp (?)	2	2			2	0 2	0 2			0 2
Myrophis sp (worm eel)	2	2			2	0 7	0 7			0 7
Cerdale floridana (pugjaw wormfish)	1	1			1	1 0	1 0			1 0
Hypsoblennius exstochilus (longhorn blenny)	0			1	1				1 7	1 7
Stegastes partitus (?)	0		1		1			0 3		0 3
Subtotal	868	68	604	606	207	1596 8	1344 0	963 6	1581 6	4142 0
% for spp used in concordance test	97	97	98	97	97	73	75	96	79	81

adult lives in the pools. Partial residents were those which would be more common in sublittoral habitats adjacent in tidepools than in tidepools themselves. This includes species which inhabit tidepools only as juveniles, and therefore includes the category of transient species considered by others (Thompson & Lehner 1976, Chang et al. 1977, Lee 1980a, 1980b)

The extent to which the tidepool fish assemblage varied among sampling years was explored using Kendall's coefficient of concordance, W (Siegel 1956). We tested the significance of W differently for sample sizes > 7 and < 8. We calculated values of W separately for true residents and partial residents for each pool and for the entire set of pools. The relationships of number of species, biomass, and

Table 1 Continued

Species	Numbers of individuals					Weight (g)				
Partial residents	81 All	81 RP	83 RP	87 RP	Total	81 All	81 RP	83 RP	87 RP	Total
Acanthurus bahianus (ocean surgeonfish)	213	20	123	192	528	317.3	308.9	265.6	406.8	989.7
Eupomacentrus dorsopunicans										
(dusky damselfish)	92	76	33	57	182	84.0	69.1	50.5	96.1	230.6
Halichoeres bivittatus (slippery dick)	40	38	105	35	180	24.9	24.3	70.3	42.6	137.8
Thalassoma bifasciatum (bluehead wrasse)	35	30	88	35	158	17.9	16.6	20.0	13.6	51.5
Abudefduf spp (sergeants)	4	2	61	44	109	1.1	0.8	6.2	8.9	16.2
Sparisoma spp (parrotfishes)	43	38	19	24	86	16.2	15.6	14.7	20.7	51.6
Adioryx vexillarius (dusky squirrelfish)	15	11		4	19	20.5	15.5	0.0	11.0	31.5
Chaetodon striatus (banded butterflyfish)	13	10	3	2	18	9.3	7.6	7.3	8.3	24.9
Halichoeres maculipinna (clown wrasse)	1	1	8	2	11	1.3	1.3	3.9	1.2	6.4
Mugil liza (liza)			9		9			4.6		4.6
Syngnathus dunkeri (pugnose pipefish)	1	1		6	7	0.1	0.1		0.8	0.9
Pomacanthus paru (French angelfish)				7	7				1.7	1.7
Haemulon aurolineatum (tomtate)	0			6	6				16.0	16.0
Rypticus saponaceus (soapfish)	2	2	3	1	6	13.1	13.1	0.4	4.6	18.1
Allanetta harringtonensis (reef silverside)			1	3	4			0.4	0.7	1.1
Antennarius multiocellatus (longlure frogfish)	1	1		2	3	0.1	0.1		0.5	0.6
Canthigaster rostrata (sharpnose pufferfish)	1	1		2	3	0.4	0.4		4.1	4.5
Opistognathus maxillosus (mottled jawfish)	1	1	1	1	3	6.2	6.2	0.6	4.9	11.7
Epinephelus adscenionis (rock hind)				3	3				23.9	23.9
Scorpaenodes caribbaeus (reef scorpionfish)	1	1		1	2	1.7	1.7		3.5	5.2
Halichoeres radiatus (puddingwife)			2		2			2.9		2.9
Halichoeres pictus (painted wrasse)	1	1			1	0.1	0.1			0.1
Holocentrus rufus (squirrelfish)				1	1				30.3	30.3
Pseudupeneus maculatus (spotted goatfish)			1		1			1.7		1.7
Sphoeroides spengleri (bandtail pufferfish)	1	1			1	0.4	0.4			0.4
Adioryx bullisi (deepwater squirrelfish)	1	1			1	3.3	3.3			3.3
Caranx latus (yellowjack)	1	1			1	2.7	2.7			2.7
Stromateidae			1		1			0.1		0.1
Holocentrus ascensionis (longjaw squirrelfish)			1		1			6.6		6.6
Subtotal	467	42	459	428	135	520.6	487.8	455.8	700.2	1676.6
% for spp. used in concordance test	98	97	96	92	95	95	94	96	87	92
Total	1335	11	106	103	343	2117.4	1831.8	1419.4	2281.8	5818.6

number of individuals with pool volume were compared among sampling years, and between true and partial residents by regression analysis using dummy variables (Kleinbaum et al. 1988).

Prior to, and during the process of collecting the fishes, there was considerable opportunity to observe the way in which the species were distributed among the habitats within the pools, and their position in the water column.

Results

Species composition and relative abundance

Altogether we collected 2 078 fishes of 63 species from 19 tidepools (Table 1). The numbers of species, individuals and weight of fishes caught in each pool on each sampling occasion is shown in Table 2, together with the pool dimensions, and the percent coverage by each substrate type.

Table 2 Pool dimensions (L = length, W = width, D = depth in cm), habitat composition, numbers of species individuals and weights of fishes collected from pools in 1981, 1983 and 1987

Pool no	Pool characteristics								Number of species					
	Dimensions			Habitat (% cover)					Total			True Res		
	L	W	D	Sand	Bare rock	Algal rock	Turt grass	Rubble	81	83	87	81	83	87
1	1000	280	40	20	0	50	10	20	31	28	28	20	16	18
2	800	280	40	10	0	0	50	40	26	32	29	18	18	13
3	510	170	35	20	0	40	20	20	24	14	10	15	9	7
4	660	120	30	30	0	50	0	20	24	22	19	15	15	11
5	470	190	25	10	0	60	10	20	12	9	9	7	7	5
6	400	90	25	10	0	20	0	70	16	16	14	11	11	8
7	290	100	25	10	0	0	0	90	10	–	–	6	–	–
8	180	200	20	10	0	60	10	20	11	21	15	8	14	7
9	180	130	25	0	30	20	0	50	11	13	8	6	10	4
10	175	110	25	20	0	60	0	20	9	8	9	7	6	6
11	200	110	20	10	70	0	10	10	9	–	–	7	–	–
12	150	90	30	50	0	30	0	20	9	–	–	5	–	–
13	200	60	25	40	0	30	0	30	14	–	–	8	–	–
14	250	50	20	20	0	30	0	50	10	13	6	6	9	5
15	150	60	20	20	0	0	0	80	11	–	–	8	–	–
16	140	80	15	30	0	10	0	60	11	–	–	8	–	–
17	150	55	20	0	0	0	100	5	–	–	2	–	–	
18	70	60	15	40	0	20	0	40	5	–	–	4	–	–
19	90	50	10	30	0	40	0	30	6	–	–	5	–	–

Pool no	Number of individuals						Weight of fish (g)					
	Total			True Res			Total			True Res		
	81	83	87	81	83	87	81	83	87	81	83	87
1	294	256	436	177	96	219	360 2	340 5	963 3	226 7	146 6	559 3
2	242	272	261	127	137	142	550 6	429 7	802 3	399 6	259 6	504 6
3	175	52	31	98	39	28	497 7	22 3	22 8	386 3	20 5	20 8
4	117	152	123	80	101	87	128 0	319 6	206 0	90 1	274 3	78 3
5	87	50	37	58	40	28	66 8	38 2	11 3	35 0	34 1	9 1
6	73	48	33	53	33	21	84 4	82 4	45 8	69 4	72 9	38 4
7	35	–	–	27	–	–	52 8	–	–	45 4	–	–
8	32	143	62	27	87	38	53 1	106 9	175 1	50 4	78 8	141 7
9	30	29	14	19	24	8	25 6	20 3	25 4	18 6	18 0	23 2
10	45	28	25	41	21	21	48 4	10 6	15 5	45 0	10 1	13 1
11	26	–	–	23	–	–	45 5	–	–	42 5	–	–
12	32	–	–	25	–	–	22 9	–	–	18 3	–	–
13	39	–	–	29	–	–	23 2	–	–	16 3	–	–
14	27	33	12	17	25	11	19 9	48 9	14 3	9 8	48 2	14 1
15	26	–	–	21	–	–	20 5	–	–	16 6	–	–
16	21	–	–	15	–	–	101 8	–	–	98 6	–	–
17	7	–	–	4	–	–	8 3	–	–	4 3	–	–
18	11	–	–	10	–	–	0 9	–	–	0 8	–	–
19	16	–	–	15	–	–	6 8	–	–	6 6	–	–

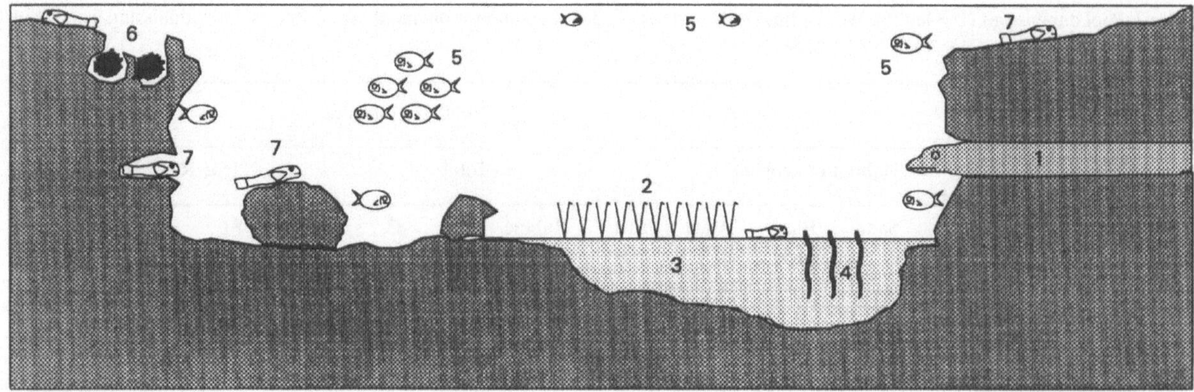

Fig. 1. The distribution of fishes in a typical tidepool in cross-section: 1 – eel, 2 – turtle grass, 3 – sand deposit, 4 – spaghetti eels, 5 – partial resident juvenile reef fishes, 6 – *Echinometra lucunter* in burrows which are inhabited by nineline goby and red clingfish, 7 – cryptic/benthic true resident.

As observed for tidepool fish assemblages in other parts of the world, true and partial residents both contribute substantially to the numbers of species and individuals in our collections. The most common partial residents were juveniles of reef species e.g. dusky damselfish, parrotfishes, ocean surgeonfish and banded butterflyfish. These fishes are usually nektonic and conspicuous. The most common true residents were benthic, cryptic species such as the puffcheek, imitator and dusky blennies (Table 1).

Within the category of true residents, there were at least five species of eels, which were treated separately in subsequent analyses. The uncertainty in the identity of the genera *Gymnothorax* and *Echylychore* pertains to the small specimens. In both cases, the larger individuals belonged to a single species, *G. moringa* and *E. nigricans,* respectively, and it is likely that the juveniles were predominantly these species.

Observations on the distribution of fishes in pools

The common partial resident juveniles swam in small schools in the middle of the pools (ocean surgeonfish, sergeant majors) or singly at the edges under rock overhangs (wrasses, dusky damselfish) (Fig. 1). These species are easily visible to the casual observer. Most true resident species are distributed singly in the shallow pool margins, or on rocks and rubble in the pools and are extremely difficult to see (blennies and gobies). This is partly because of their cryptic coloration, but also because they seek cover in crevices, holes, under ledges, and in algae very rapidly when they detect the slightest movement outside the pool. Other true resident species have very specific microhabitats within the pools. For example, spaghetti eels occur only in pools with sand deposits (Tables 1, 2), red clingfish and nineline gobies occur only in association with *Echinometra lucunter,* as described by Teytaud (1971), and moray eels inhabit holes in the sides of the pools.

Variation in assemblage composition among years

For the entire set of pools, Kendall's W was significant at the p = 0.01 and p = 0.05 levels for the true residents and partial residents respectively (Table 3). For the former, W was significant at the 0.05 level in four pools, whereas for the latter, W was significant in only two pools. In both instances the significant concordance was found in the largest pools.

The effect of pool size on assemblage similarity between pairs of samples was explored by plotting the probabilities of the Spearman rank correlation coefficients (Table 3), versus pool size (Fig. 2). The figure shows that the probability of occurrence of the calculated r_s value decreases markedly with increasing pool size. This indicates that assemblage similarity increases with pool size.

Table 3 Calculation of Spearmans rank correlation (R_s) between pools, Kendalls W and corresponding Chi Square (X^2) and S values for testing

Pool	R_s			N	Probability of R_s			W	X^2	S	Prob value
	Time 1 vs 2	Time 1 vs 3	Time 2 vs 3		Time 1 vs 2	Time 1 vs 3	Time 2 vs 3				
True residents											
1	0 536	0 732	0 577	13	0 06	0 01	0 05	0 743	26 8	1217 6	< 0 01
2	0 515	0 403	0 723	13	0 07	0 16	0 01	0 698	25 1	1143 3	0 02 > 0 01
3	0 145	0 371	0 783	6	0 75	0 41	0 08	0 622	NA	98 0	> 0 05
4	0 776	0 443	0 205	8	0 04	0 24	0 59	0 650	13 6	245 6	0 01 > 0 05
5	0 486	0 441	0 883	6	0 28	0 32	0 05	0 736	NA	115 9	0 05 > 0 01
6	0 718	0 103	0 158	5	0 15	0 84	0 75	0 551	NA	49 6	> 0 50
8	0 164	0 413	0 631	7	0 69	0 31	0 12	0 602	NA	151 7	> 0 50
9				< 3					NA		
10	− 0 500	0 500	− 1 000	3	0 48	0 48			NA		
14	0 500	1 000	0 500	3	0 48	1 00			NA		
All	0 630	0 709	0 460	18	0 01	0 00	0 06	0 733	37 4	3196 7	< 0 01
Partial residents											
1	0 714	0 429	0 429	6	0 11	0 34	0 34	0 683	NA	107 5	0 05 > 0 02
2	− 0 214	0 900	− 0 126	7	0 60	0 03	0 76	0 458	NA	115 4	0 05 > 0 02
3	− 0 688	0 104	0 301	6	0 12	0 82	0 50	0 270	NA	42 6	> 0 05
4	0 177	0 677	0 191	6	0 69	0 13	0 67	0 566	NA	89 1	> 0 05
5				< 3					NA		
6	− 0 564	0 205	0 526	5	0 26	0 68	0 30	0 370	NA	33 3	> 0 05
8	0 429	− 0 237	− 0 939	6	0 34	0 60	0 04	0 167	NA	26 4	> 0 05
9	− 0 866	0 000	− 0 500	3	0 22	1 00	0 48	0 030	NA		
10				< 3					NA		
14				< 3					NA		
All	0 569	0 696	0 790	9	0 11	0 05	0 03	0 790	19 0	426 6	0 02 > 0 01

Relationships of species richness, abundance and biomass with pool size

Fishes were found in even the smallest pools (Table 2). Of various transformations of estimated area and volume, the log of volume was found to explain the highest proportion of variation in the number of species found in the pools (Fig. 3). The relationship between the number of species per pool and pool size differed significantly between true and partial residents ($F = 41.96$, $p < 0.01$). There were consistently more true residents than partial residents in pools of all sizes. However, as the fitted regression lines in Figure 3 show, the proportional contribution of partial resident species to total number of species increases with pool size.

There was no significant association between habitat diversity and the residuals from the regression of number of species on pool size. Rubble would be expected to provide hiding places for fishes. However, there was no significant association between the residuals and the proportion of rubble in each pool.

Numbers of individuals and biomass were also significantly related to pool size, and the relationships were different for true and partial residents (Numbers – $F = 33.91$, $p < 0.01$, Biomass – $F = 8.22$, $p < 0.01$). With regard to numbers of individuals the proportional contribution of partial resident species to total numbers increases rapidly with pool size, being about 50% at the largest observed pool sizes (Fig. 4). The biomass of partial residents was higher than that of true residents at all pool sizes, and increased more rapidly with pool size (Fig. 5).

Figures 4 and 5 show that although eels contributed relatively few individuals, their contribution to biomass was often similar to that of the true and partial resident groups. The biomass contribution

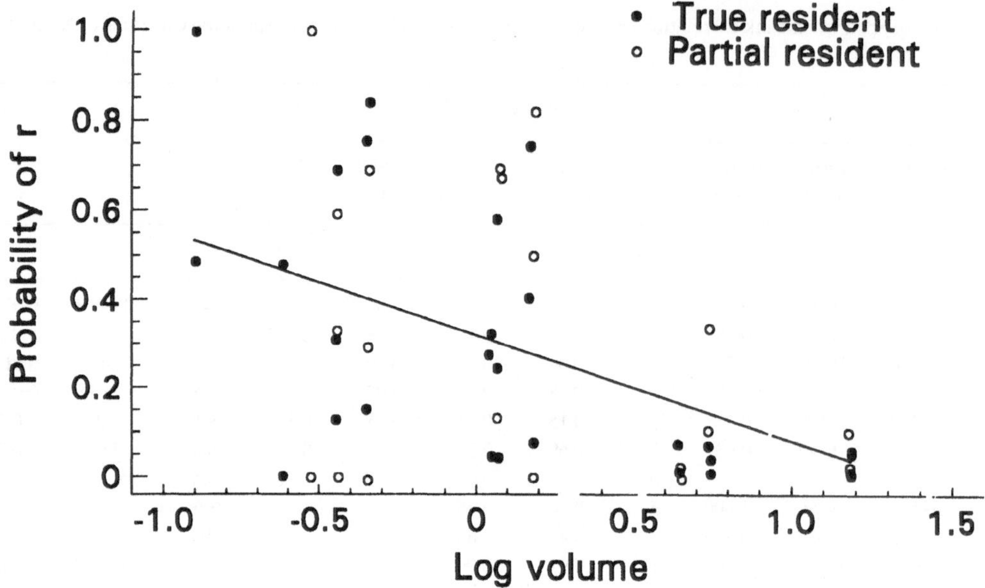

Fig. 2. The probability of Spearman's correlation coefficient between ranked abundances on different sampling occasions in pools of various sizes, and for the entire collection.

of eels was also highly variable, being determined by the presence or absence of one or two large individuals.

The relationships of number of species, number of individuals and biomass with pool size did not differ significantly among the three sampling years.

The size of fishes

The mean weight of specimens (total weight/total numbers), excluding eels, increases with estimated pool volume (Fig. 6). This could have been due to intraspecific changes in mean fish size as found by

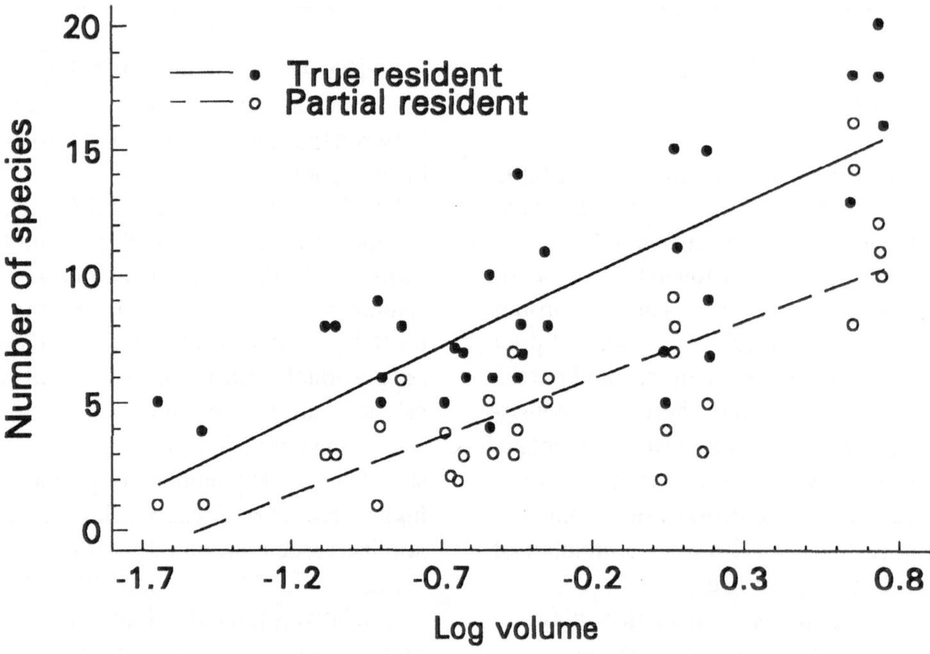

Fig. 3. The relationship between the numbers of species and pool size for true and partial residents.

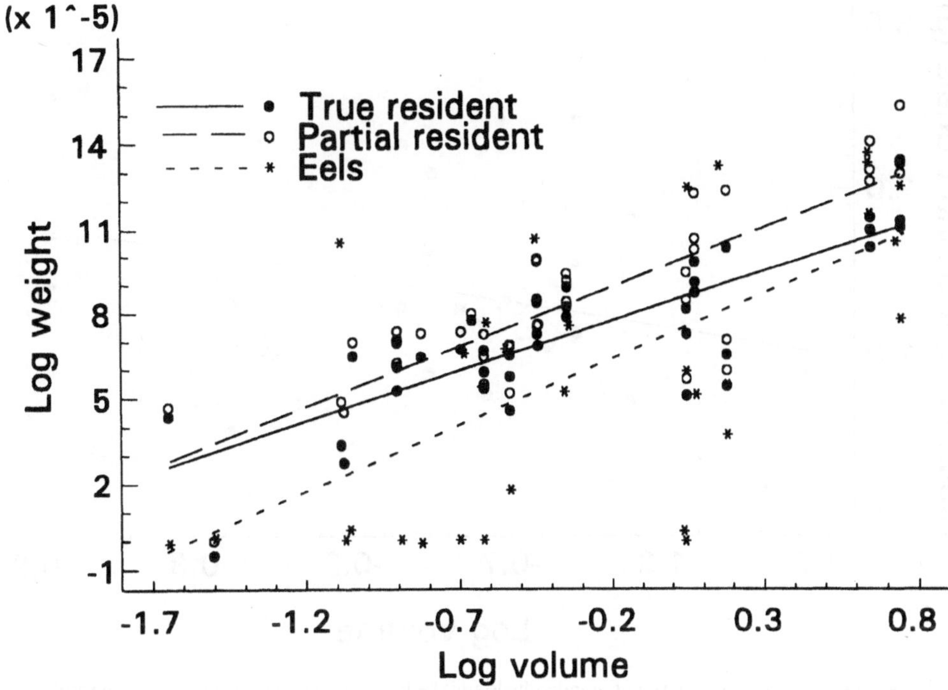

Fig. 4. The relationship between the numbers of individuals and pool size for true residents, partial residents, and eels.

Gibson (1972) in tidepools on the coast of France, or to shifts in species composition with larger species being added as pool size increased. The former explanation is certainly the case here. The correlations between mean weight and log of pool volume were calculated for 14 species for which there was

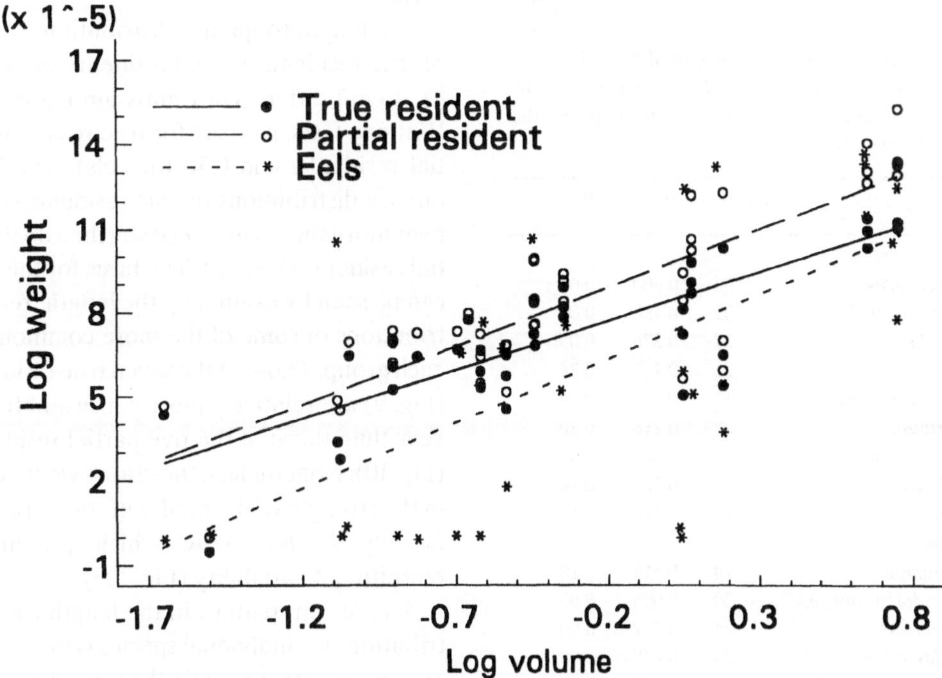

Fig. 5. The relationship between biomass and pool size for true residents, partial residents, and eels.

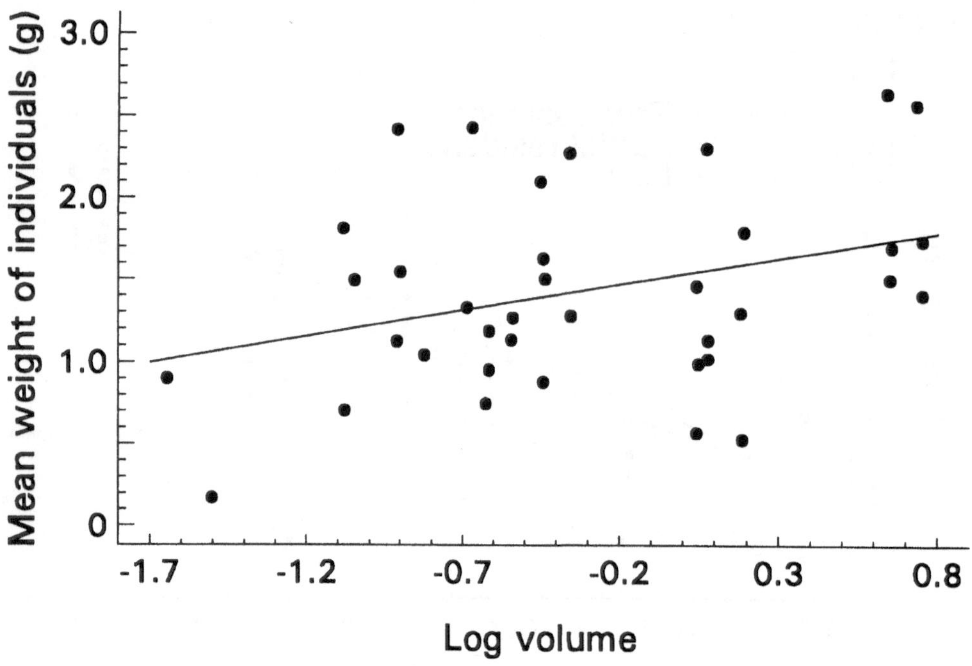

Fig 6 The relationship between the mean weight of individuals and the log of pool size for true and partial residents combined

sufficient data (occurrence in 10 or more collections and a total of 30 or more individuals). All the relationships were positive, 5 were significant at $p < 0.01$ and 6 at $p < 0.2$ (Table 4). The size ranges of *Paraclinus nigripinnis* and *Ginsburgellus novemlineatus* were so small that no relationship would be expected. Eels were not included in the analysis as the size of available holes for hiding is probably more important than pool size in determining the size of individuals.

Table 4 Relationships between mean weight of individual species in pools and the log of pool volume (N = number of occurrences used in calculating the correlation coefficient R, P = the probability value associated with R)

Species	N	R	P
True residents			
Labrisomus bucciferus	38	0 300	0 06
Malacoctenus erdmani	32	0 488	0 00
Malacoctenus gilli	27	0 376	0 05
Ogilbia spp	23	0 247	0 24
Ginsburgellus novemlineatus	13	0 023	0 94
Paraclinus nigripinnis	28	0 218	0 26
Labrisomus nuchipinnis	18	0 547	0 01
Apogon maculatus	17	0 366	0 14
Labrisomus nigricinctus	15	0 380	0 15
Partial residents			
Acanthurus bahianus	19	0 315	0 18
Eupomacentrus dorsopunicans	30	0 098	0 60
Halichoeres bivittatus	24	0 782	0 00
Thalassoma bifasciatum	25	0 648	0 00
Sparisoma spp	23	0 669	0 00

The length frequency distributions of all species of true residents, partial residents and eels (Fig. 7, 8), do not differ significantly among years (Mann-Whitney tests, $p = 0.47$ for true residents, 0.58 partial residents, and 0.70 for eels). The length frequency distributions of true residents (Fig. 7a) appear more similar among years than do those of partial residents (Fig. 7b). The basis for this difference can be seen by examining the length frequency distributions of some of the more common species in each group. Those of the seven true resident species (Fig. 9) are relatively more consistent from year to year than those of the five partial resident species (Fig. 10). Nonetheless, there are evident differences in the strength of the modes of some true residents, notably *L. bucifferus,* which probably reflect recruitment variability (Fig. 10e).

A recurrent pattern in the length frequency distributions for individual species is that the individuals appear larger in 1983 than in 1981 or 1987. This could be due to interannual differences in timing of

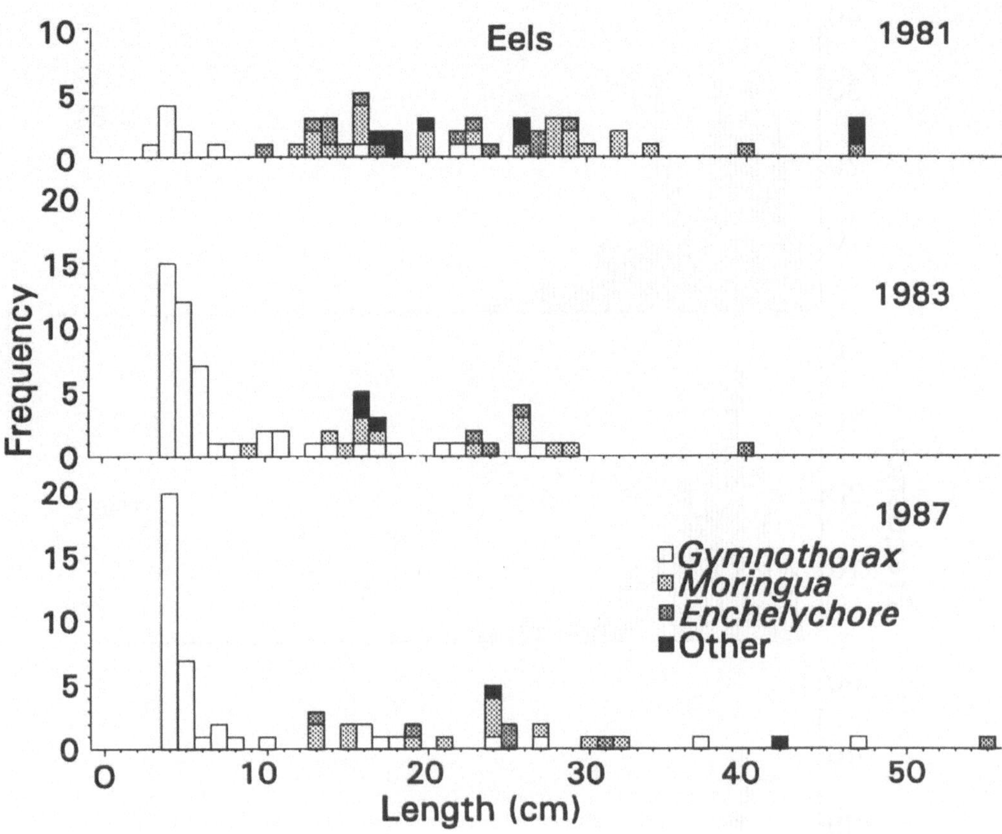

Fig 8 Length frequency distributions on different sampling occasions for eels

reproduction, or growth rate. Monthly mean air temperatures for the 12 months preceding each sampling occasion were examined for evidence of interannual differences which could have affected either of the two above processes. Although there are differences in seasonal temperature pattern in the three periods, there was no obvious relationship to the observed size pattern, and the data were inappropriate for more detailed analysis.

Discussion

In single sample surveys such as ours, the timing of recruitment could affect the composition, particularly for juveniles of reef species which probably only inhabit the pools for a few months. Lee (1980a) observed seasonal changes in the fish assemblage composition in tidepools in Taiwan. Although Caribbean reef fishes breed year-round there are distinct peaks in recruitment (Munro et al. 1973, Powles 1975, Luckhurst & Luckhurst 1977). Around Barbados, catches of larvae of inshore fishes (e.g. labirds, scarids, pomacentrids) showed two annual peaks, from March to May and from August to October, but some larvae were present at all times (Powles 1975). Therefore, we would expect some seasonal succession in species composition and abundance in these tidepools. Whereas we attempted to control for this by sampling at the same time in each year, the timing of biological events is likely to be determined by climate, which varies from year to year. Therefore, some of the variation observed between years is likely due to the timing of sampling.

Another factor to be considered in comparing the assemblages among years is the effect of removal of fishes in previous samples on subsequent samples. Other studies have found that recolonisation of defaunated tidepools is rapid relative to the timescale of our sampling (Bussing 1972, Thomson &

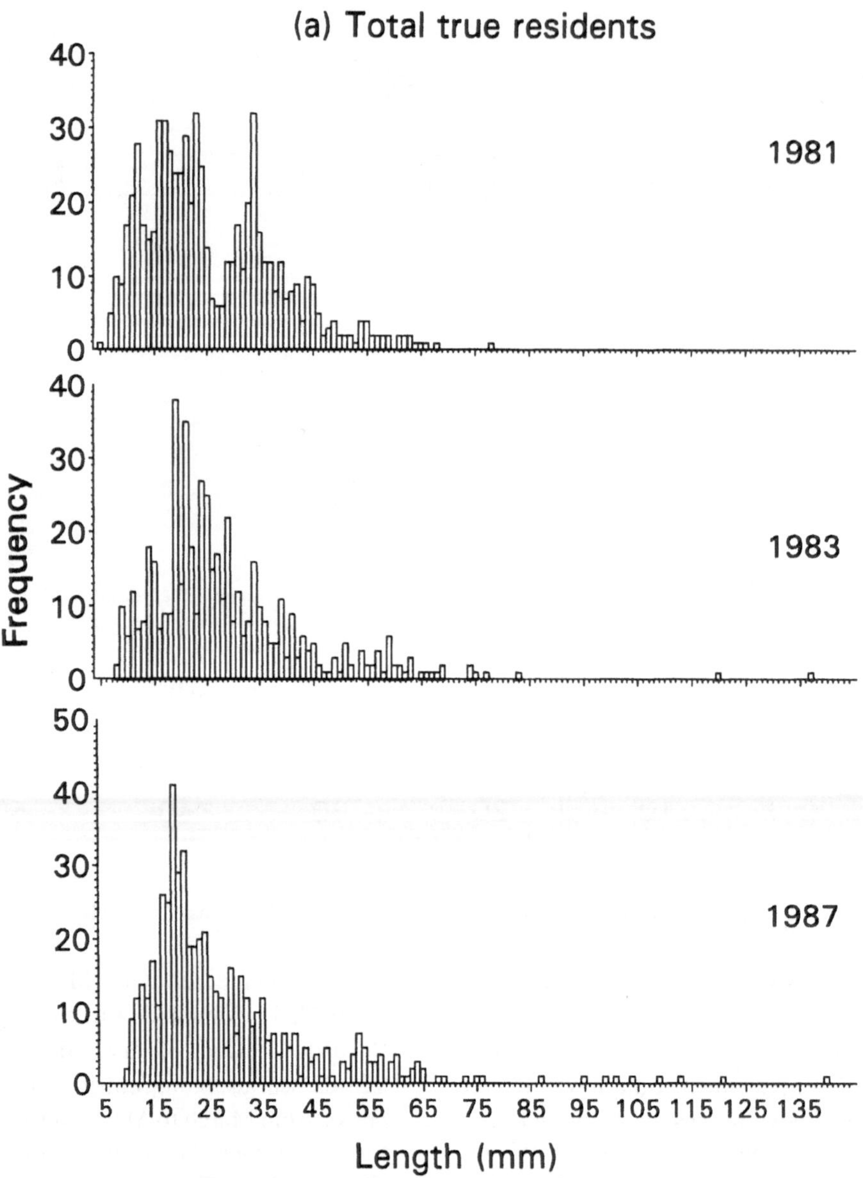

Fig. 7. Overall length frequency distributions on different sampling occasions for (a) true and (b) partial residents.

Lehner 1976, Beckley 1985). We observed that the tidepools in this study had been recolonised by juveniles of reef species within a few days of sampling. Furthermore, the lifespan of all true resident species (except the eels) is sufficiently short that complete recolonisation could have occurred by recruitment alone in the interval between samples. In the case of the eels which are longer lived, we did not find any indication of a decline in abundance over the three samples. We therefore conclude that the sampling did not significantly affect subsequent assemblage structure.

Tidepools in Barbados are similar to those elsewhere in the world in that true and partial residents both contribute significantly to the assemblage. Quantitative comparison of the relative contributions of true and partial residents between studies would be useful, since the former would likely be controlled by local factors, whereas the latter may be controlled by factors external to the tidepool habitat. However, methodological differences

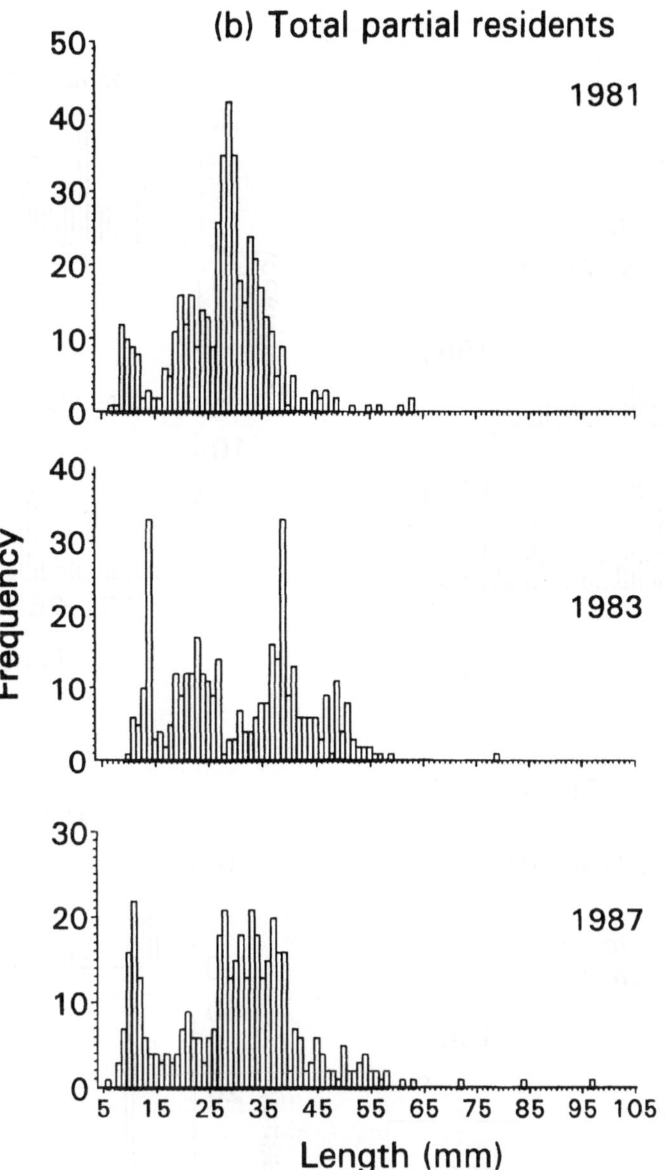

Fig. 7. Continued.

among studies make direct comparison of published results difficult. First, there may be inconsistency among authors in designation of species as true or partial residents. Second, repeated sampling will probably yield a higher proportion of partial residents than will single samples. In 13 collections over 29 months Grossman (1982) collected all resident species in the first two collections but continued to find new 'non-resident' species up to the eighth collection. Third, is the possible bias due to pool size, as demonstrated in this study? Comparison between studies would require that these biases be accounted for.

The fish assemblage structure in the tidepools changed continuously with pool size. The proportion of partial residents, and their contribution in numbers and biomass increased steadily with pool size. The size of individuals also increased with pool size. This suggests that the tidepool fish assemblage may grade into the subtidal fish assemblage in which the partial residents of the tidepools predominate. This contrasts with the situation described by

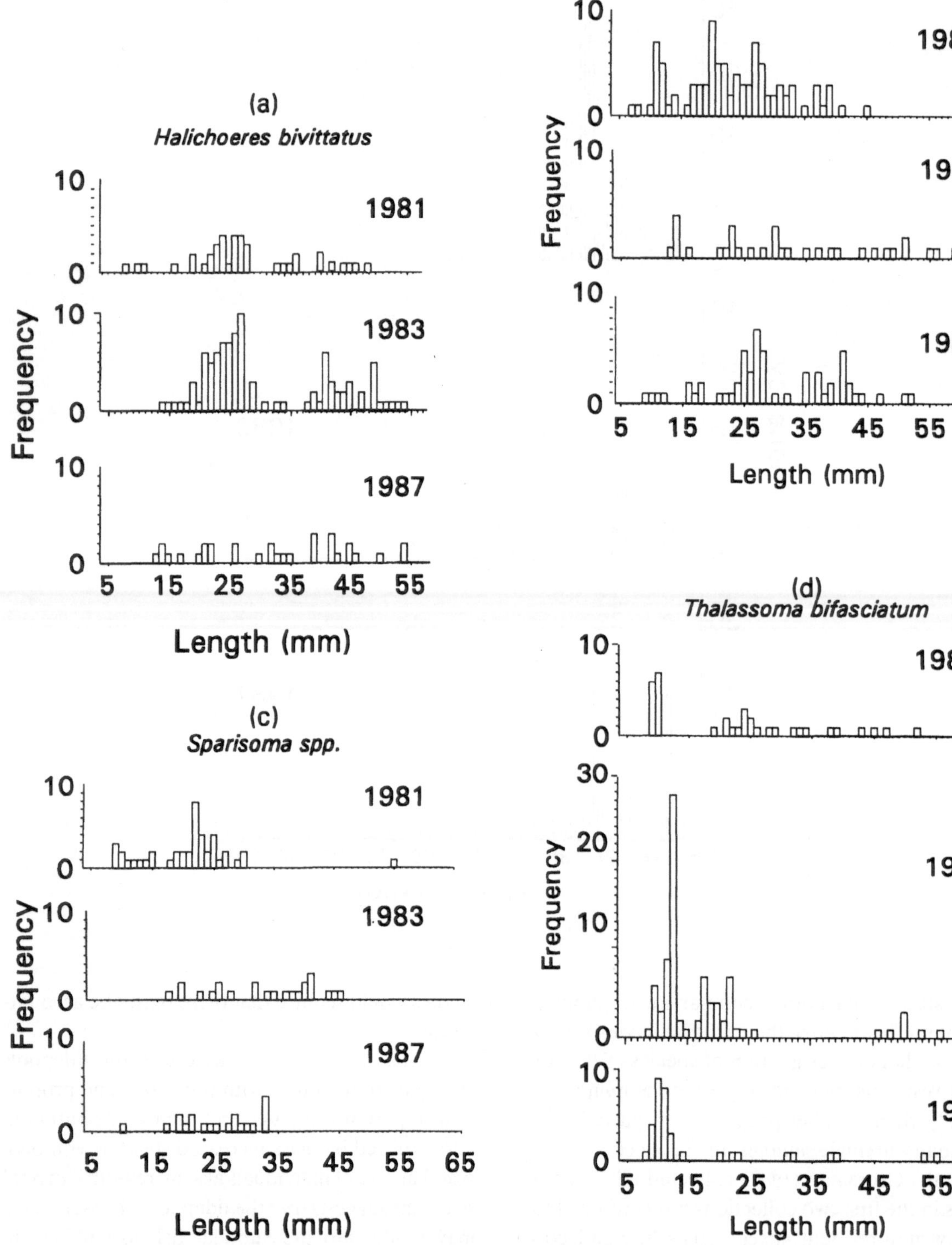
(a)
Halichoeres bivittatus
1981
1983
1987
Frequency
Length (mm)
(b)
Eupomacentrus dorsopunicans
1981
1983
1987
Frequency
Length (mm)
(c)
Sparisoma spp.
1981
1983
1987
Frequency
Length (mm)
(d)
Thalassoma bifasciatum
1981
1983
1987
Frequency
Length (mm)

Fig. 9.

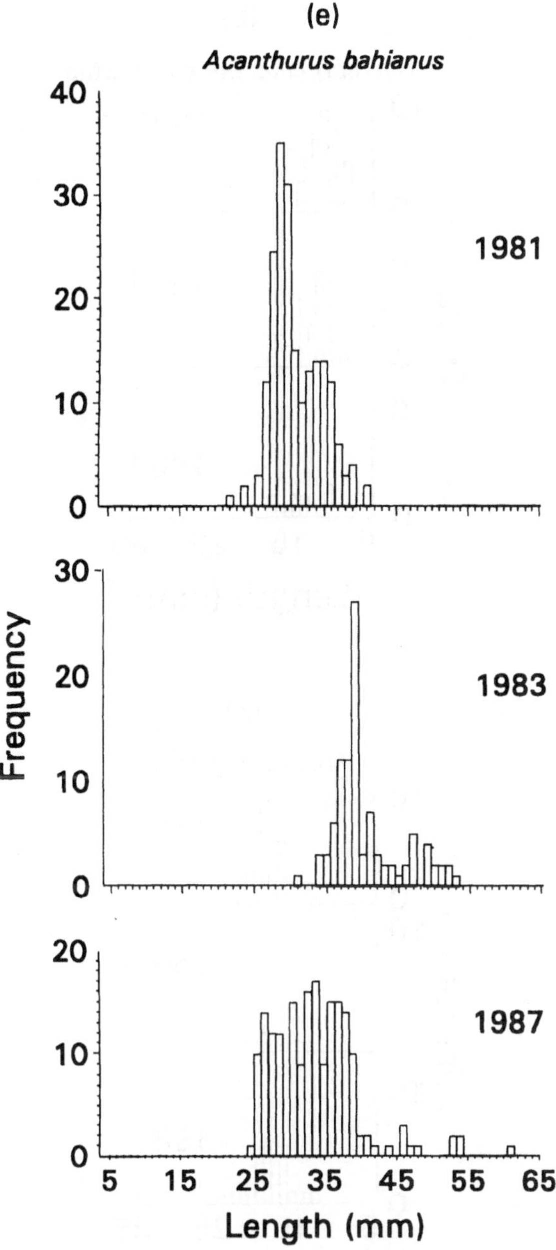

←

Fig 9 Length frequency distributions on different sampling oc casions for the most common true residents

Yoshiyama et al (1986) who found distinct intertidal and subtidal fish assemblages

The observed increase in mean size of individuals with increasing pool size for several species of both true and partial residents could be due to increased vulnerability to predation, or to the physical action of current or waves in small pools The increase in both depth and cover with pool size would probably be especially important in offering protection for the highly visible, nektonic partial resident fishes During low tide, predation by birds is probably intense in shallow pools as green herons, *Butorides virescens,* were frequently observed feeding in them During high tide the area would be accessible to predatory fishes from adjacent habitats, as is known to occur in intertidal areas elsewhere (Gibson 1978) There were also several species of eels, some of which are known to be piscivorous (Randall 1967), however, there is no information on which to evaluate the extent to which pool size would affect their effectiveness as predators

The similarity in size frequency distribution of both true and partial residents suggests that there is an upper limit to size in the tidepool habitat Many of the partial residents, frequently juveniles of reef species, grow considerably larger in subtidal habitats than in the tidal pools as shown in Figure 11, where sizes observed in the pools, and the maximum sizes reported in the literature are compared The sizes of the eels, which live in holes, are not constrained in the same way as that of other fishes

The use of Kendall's *W* as an indication of assemblage similarity over time has been the subject of considerable debate (Ebeling et al 1990) Because it is based on rank ordering of species abundances, the primary concern has been that species abundances may fluctuate considerably without changes in rank Thus the finding of a significant value of *W* can only be taken to indicate that there was no significant reordering of ranks among sampling occasions

Species rank ordering in abundance, as well as relationships of species richness, abundance and biomass with pool size did not differ significantly among the three sampling occasions spaced over six years, a time period which is significantly longer than the average lifespan of most true residents, and the residence period of partial residents Length frequency distributions of the most abundant species were also similar among sampling occasions Viewed together, these findings indicate that the fish assemblages in the tidepools at Martin's Bay are resilient, and therefore would probably be stable (sensu Connell & Sousa 1983), in that

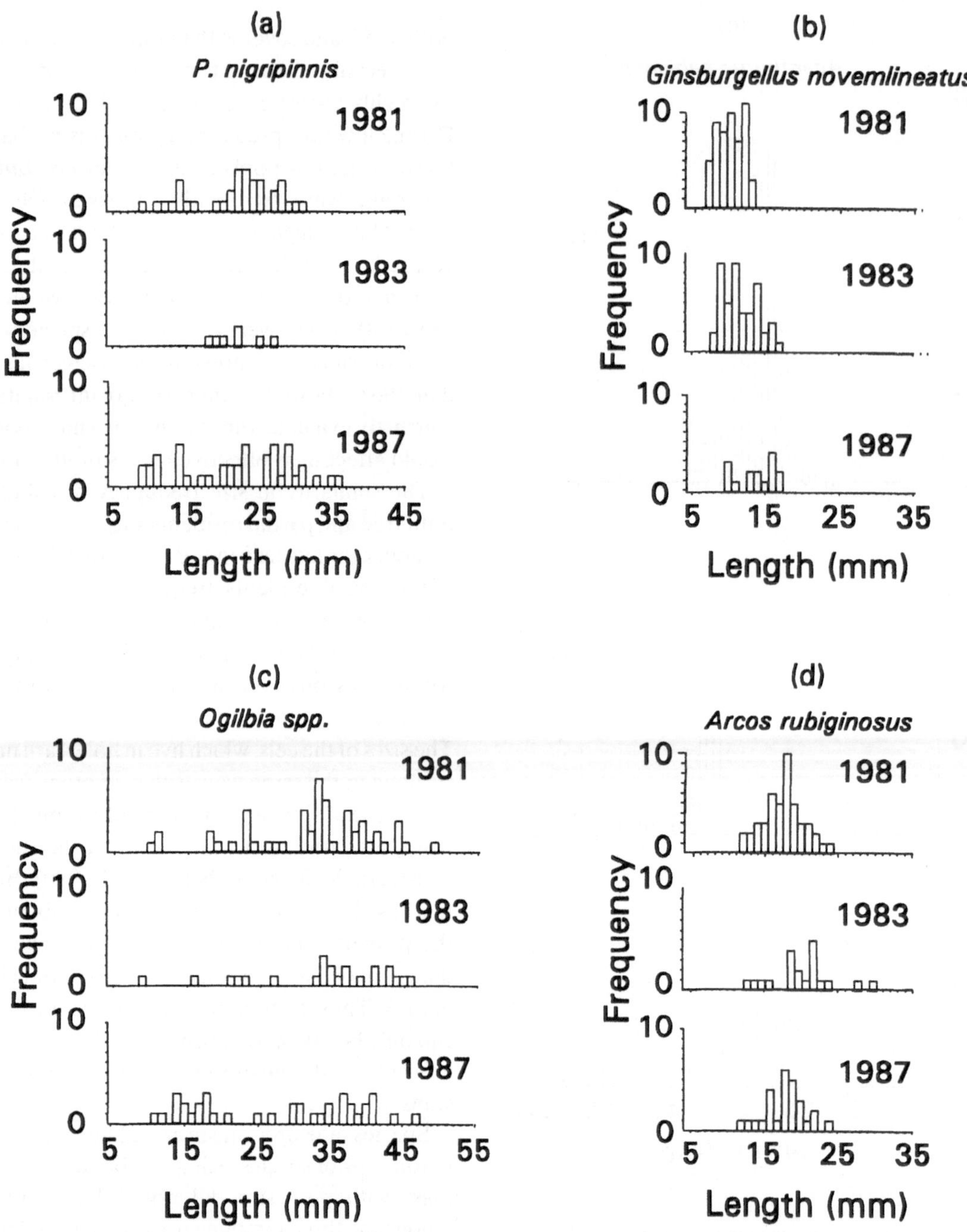

Fig. 10. Length frequency distributions on different sampling occasions for the most common partial residents.

they would remain constant over time. The results also indicate that there was less stability in the partial resident component of the assemblage than in the true resident component.

Several other studies of tidepool fish assemblages have also found them to be stable and resilient (Thompson & Lehner 1976, Grossman 1982, Beckley 1985, Collette 1988). Bennett & Griffiths (1984) observed stable population structure in South African tidepools for seasonal and longer periods. These findings are inconsistent with the classical ecological view of intertidal habitats as physically stressed and variable, such that the assemblages in

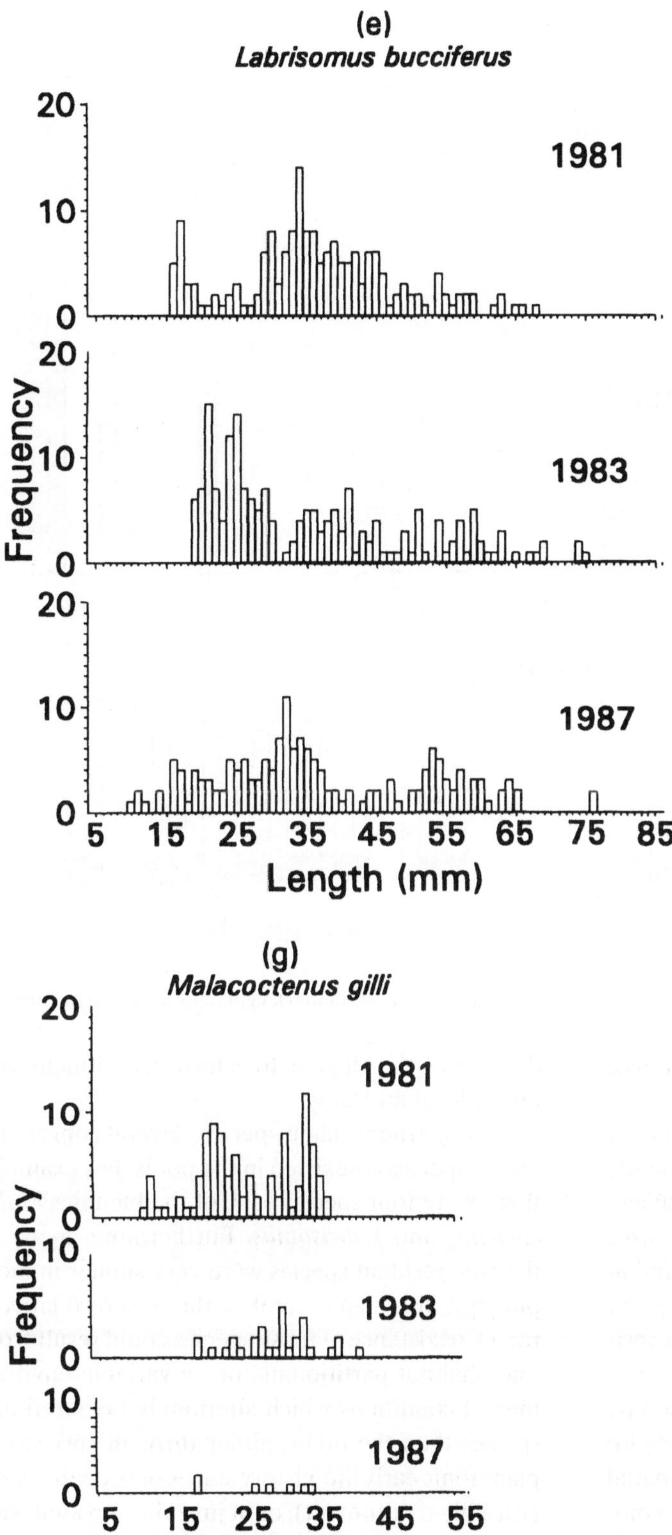

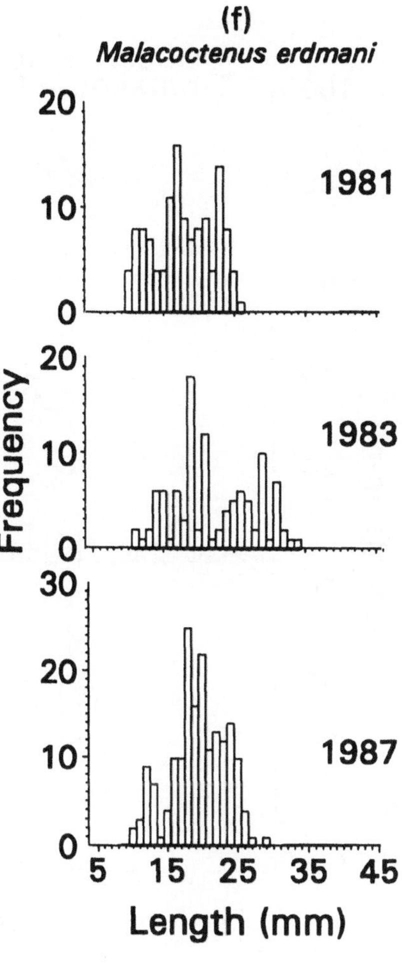

Fig. 10. Continued.

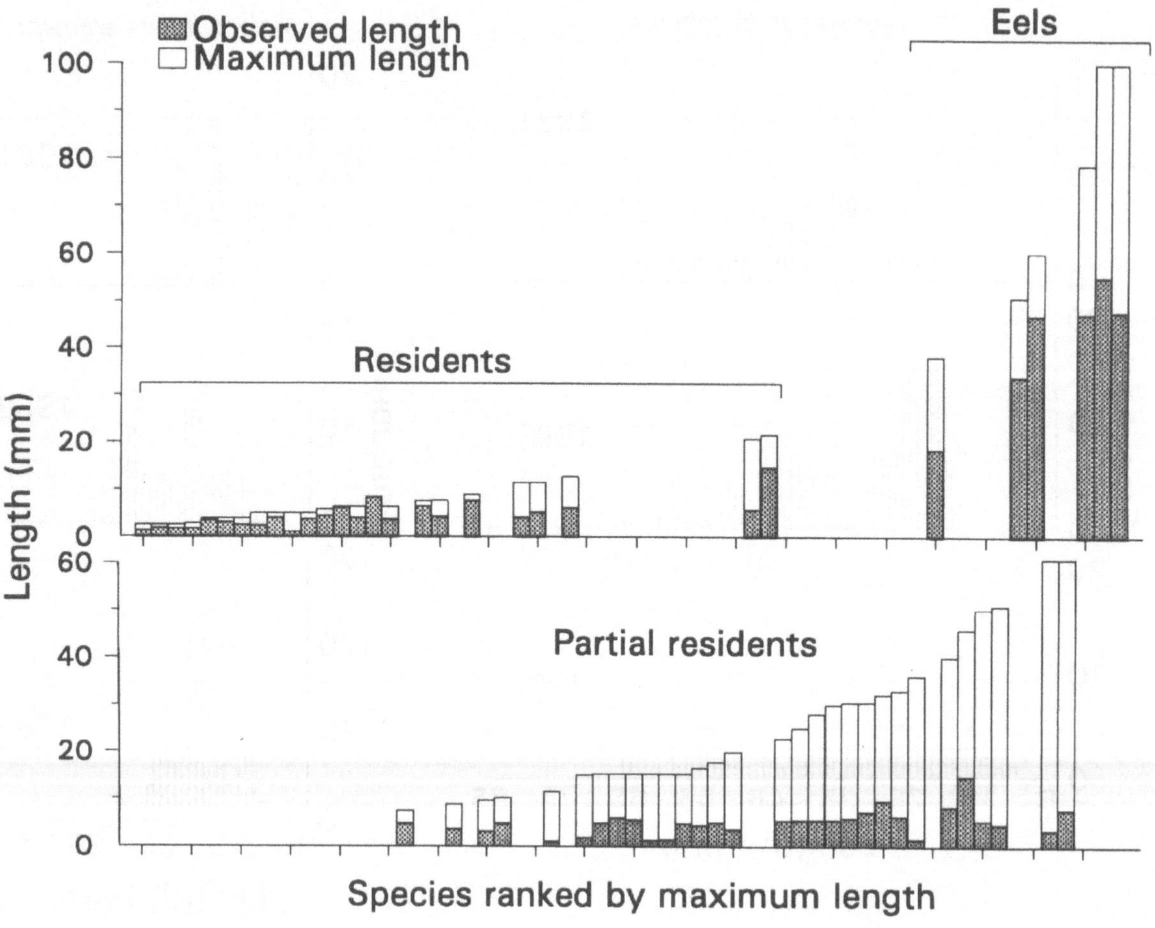

Fig. 11. The maximum length of species of fishes observed in the tidepools relative to the maximum length reported in the literature.

them exhibit unpredictable fluctuations (Mahon & Smith 1989).

The observed stability may be due to the linkages of the tidepool populations with adjacent habitats. The partial resident component of the assemblage is certainly largely determined by population sizes and physical conditions in the subtidal area, and in the plankton where the early life-history stages of most of these fishes occur. Most of the true residents also have planktonic early life history stages, and their abundance may be more determined by rates of recruitment from the plankton, than by local variability in adult survival. The relative spatial scales of the populations contributing to the common recruitment pool, and the factors responsible for post settlement mortality in the tidepools will determine the degree to which recruitment can buffer local mortality.

Among true resident species, several congeneric sets of species co-existed in the pools. For example, there were four species each of the blennies *Malacoctenus* and *Labrisomus*. Furthermore, many of the true resident species were very similar in morphology, although several of them were relatively rare. Coexistence of these species could result from microhabitat partitioning, or by variable environmental conditions which alternately favoured one species then the other, either through survival of planktonic early life-history stages outside the pools (variable recruitment), or of juvenile and adult survival in the pools (Chesson 1986). For the common true residents, the constancy of relative abundance among collection years suggests that coexistence is mediated by the former mechanism. The occurrence of rare species can be accounted for by continuous recolonisation of the tidepools from adjacent habitats.

Other studies of tidepool fishes have focused on resource partitioning. Cross (1981) noted that tidepool fishes from three assemblages appeared to be partitioning the food resource. In tidepool fishes of South Africa, Bennett et al. (1983) found considerable resource subdivision due to vertical and horizontal distribution, microhabitat use and morphology. They observed that the fishes frequently fed outside of the pools when the tide was in but noted that pool habitat may control density. Therefore, further studies aimed at evaluating the mechanisms of coexistence of the assemblages of fishes in our tidepools should include their activities during the flood tide.

The observation that the tidepool fish assemblages contain a high proportion of partial residents which are juveniles of reef species leads to the question of their importance as nurseries and reservoirs of recruitment to the adjacent reef fish assemblages. One possibility is that the juveniles in tidepools are merely an overflow from a glut of juveniles considered typical of reefs (Smith & Tyler 1975) and that they die there. However, it is not clear that saturation of reef habitats by juveniles is normal (Mapstone & Fowler 1988) and the presence of juveniles inshore suggests that they may prefer these habitats as juveniles. If, as suggested by Sale (1978), living space on the reef does become available throughout the year and is obtained by whichever individuals are available, these inshore habitats may provide a continuous supply of recruits, and play a significant role in structuring fish assemblages in adjacent subtidal habitats (Ayal & Safriel 1982).

Acknowledgements

We are especially grateful to Michele and Michael Brown for their assistance with collecting the fishes. Thanks to Alan Emery who checked our identification of fishes and to the Royal Ontario Museum where this was done. We are also grateful to Wayne Hunte, Alan Emery, and John Green for their comments on an early draft of the manuscript.

References cited

Ayal, Y & U N Safriel 1982 Species diversity of the coral reef – a note on the role of predation and of adjacent habitats Bull Mar Sci 32 787–790

Beckley, L E 1985 Tide-pool fishes recolonisation after experimental elimination J Exp Mar Biol Ecol 85 287–295

Bennett, B, C L Griffiths & M-L Penrith 1983 The diets of littoral fish from the Cape Peninsula S Afr J Zool 18 343–352

Bennett, B A & C L Griffiths 1984 Factors affecting the distribution, abundance and diversity of rock-pool fishes on the Cape Peninsula, South Africa S Afr J Zool 19 97–104

Bohlke, J E & C C G Chaplin 1968 Fishes of the Bahamas and adjacent tropical waters Livingston Publ, Wynnewood 771 pp

Brattstrom, H 1980 Rocky shore zonation in the Santa Marta area, Colombia Sarsia 65 163–226

Bussing, W A 1972 Recolonisation of a population of supratidal fishes at Eniwetok Atoll, Marshall Islands Atoll Res Bull 154 1–4

Chadwick, E M P 1976 A comparison of growth and abundance for tidal pool fishes in California and British Columbia J Fish Biol 8 27–34

Chang, K H, S C Lee & W L Wu 1977 Fishes of reef limestone platform at Maopitou, Taiwan diversity and abundance Bull Inst Zool, Acad Sinica 16 9–21

Chesson, P L 1986 Environmental variation and the coexistence of species pp 240–256 *In* J Diamond & T J Case (ed) Community Ecology, Harper & Row, New York

Collette, B B 1988 Resilience of the fish assemblage in New England tidepools US Fish Bull 84 200–204

Connell, J H & W P Sousa 1983 On the evidence needed to judge ecological stability of persistence Amer Nat 121 789–824

Cross, J M 1981 Resource partitioning in three rocky intertidal fish assemblages pp 142–150 *In* G M Cailliet & C A Simenstad (ed) Gutshop '81, Washington Sea Grant, University of Washington, Seattle

Ebeling, A W, S J Holbrook & R J Schmitt 1990 Temporally concordant structure of a fish assemblage bound or determined Amer Nat 135 63–73

Fischer, E (ed) 1978 FAO fishery identification sheets for fishery purposes western central Atlantic Vol I–VI Food & Agriculture Organisation, Rome

Gibson, R N 1969 The biology and behaviour of littoral fish Oceanogr Mar Biol Ann Rev 7 367–410

Gibson, R N 1972 The vertical distribution and feeding relationships of intertidal fish on the Atlantic coast of France J Anim Ecol 4 189–207

Gibson, R N 1978 Lunar and tidal rhythms in fish pp 201–213

In: J. Thorpe (ed.) Rhythmic Activity of Fishes, Academic Press, New York.

Gibson, R.N. 1982. Recent studies on the biology of intertidal fishes. Oceanogr. Mar. Biol. Ann. Rev. 20: 363–414.

Green, J.M. 1971. Local distribution of *Oligocottus maculosus* Girard and other tidepool cottids of the West Coast of Vancouver Island, British Columbia. Can. J. Zool. 49: 1111–1128.

Grossman, G.D. 1982. Dynamics and organisation of a rocky intertidal fish assemblage: the persistence and resilience of taxocene structure. Amer. Nat. 119: 611–637.

John, D.M. & J.H. Price. 1979. The marine benthos of Antigua (Lesser Antilles). I. Environment, distribution and ecology. Bot. Mar. 22: 313–326.

King, C.A.M. 1975. Introduction to physical and biological oceanography. Edward Arnold, London. 372 pp.

Kleinbaum, D.G., L.L. Kupper & K.E. Muller. 1988. Applied regression analysis and other multivariable methods. PWS-Kent, Boston. 718 pp.

Lee, S.C. 1980a. Intertidal fishes of the rocky pools at Lanyu (Botel Tobago), Taiwan. Bull. Inst. Zool., Acad. Sinica 19: 1–13.

Lee, S.C. 1980b. Intertidal fishes of a rocky pool of the Sanhsientai, eastern Taiwan. Bull. Inst. Zool., Acad. Sinica 19: 19–26.

Lewis, J.B. 1960. The fauna of rocky shore of Barbados, West Indies. Can. J. Zool. 38: 391–435.

Luckhurst, B.E. & K. Luckhurst. 1977. Recruitment patterns of coral reef fishes on the fringing reef of Curacao, Netherlands Antilles. Can. J. Zool. 55: 681–689.

Mahon, R. & R.W. Smith. 1989. Demersal fish assemblages on the Scotian Shelf, Northwest Atlantic: spatial distribution and persistence. Can. J. Fish. Aquat. Sci. 46 (Suppl. 1): 134–152.

Mapstone, B.D. & A.J. Fowler. 1988. Recruitment and the structure of assemblages of fish on coral reefs. Trends in Ecology and Evolution 3: 72–77.

Munro, J.L., V.C. Gaut, R. Thompson & P.H. Reeson. 1973. The spawning seasons of Caribbean reef fishes. J. Fish Biol. 5: 69–84.

Nybakken, J.W. 1982. Marine biology an ecological approach. Harper & Row Publishers, New York. 446 pp.

Powles, H. 1975. Abundance, seasonality, distribution, and aspects of the ecology of some larval fishes off Barbados. Ph.D. Thesis, McGill University, Montreal. 227 pp.

Randall, J.E. 1967. Food habits of reef fishes of the West Indies. Stud. Trop. Oceanography 5: 665–847.

Randall, J.E. 1968. Caribbean reef fishes. TFH Publications, Neptune City. 318 pp.

Sale, P.F. 1978. Coexistence of coral reef fishes – a lottery for living space. Env. Biol. Fish. 3: 85–102.

Siegel, S. 1956. Nonparametric statistics for the behavioural sciences. McGraw Hill, New York. 312 pp.

Smith, C.L. & J.C. Tyler. 1975. Succession and stability in fish communities of dome-shaped patch reefs in the West Indies. Amer. Mus. Novit. 2572: 1–18.

Southwood, T.R.E. 1978. Ecological methods, with particular reference to the study of insect populations. English Language Book Society, Cambridge. 524 pp.

Teytaud, A.R. 1971. Food habits of the goby, *Ginsburgellus novemlineatus*, and the clingfish, *Arcos rubiginosus,* associated with echinoids in the Virgin Islands. Carib. J. Sci. 11: 41–45.

Thompson, D.A. & C.E. Lehner. 1976. Resilience of a rocky intertidal fish community in a physically unstable environment. J. Exp. Mar. Biol. Ecol. 22: 1–29.

Yoshiyama, R.M. 1981. Distribution and abundance patterns of rocky intertidal fishes in central California. Env. Biol. Fish. 6: 315–332.

Yoshiyama, R.M., C. Sassaman & R.N. Lea. 1986. Rocky intertidal fish communities of California: temporal and spatial variation. Env. Biol. Fish. 17: 23–40.

Environmental Biology of Fishes **41**: 191–205, 1994.

Age, growth, and sex ratio among populations of least brook lamprey, *Lampetra aepyptera*, larvae: an argument for environmental sex determination

Margaret F. Docker[1] & F. William H. Beamish
Department of Zoology, University of Guelph, Guelph, Ontario N1G 2W1, Canada
[1] *Present address: Fisheries and Oceans Canada, Pacific Biological Station, Nanaimo, British Columbia V9R 5K6, Canada*

Received 5.10.1992 Accepted 23.9.1993

Key words: Density-dependent sex determination, Adaptive significance, Sex-specific recruitment, Agnatha, Cyclostomata

Synopsis

Sex ratios of least brook lamprey, *Lampetra aepyptera,* larvae varied widely among 12 geographically-diverse streams of the eastern United States. The extremes were 29 and 71% male, and the proportion of males increased significantly with relative population density, which was estimated among the streams from the number of larvae collected per m^2 of substrate. The skewed sex ratios were not likely due to differential mortality between the sexes or differential recruitment to the adult stock, since they were established at the time of gonadal differentiation (at ca. 2 years of age) and remained relatively constant over the subsequent 2–3 years of larval life. Furthermore, although females seemed to predominate in the oldest larval age class, thus appearing to metamorphose later than males, their numbers were small and were omitted from the overall sex ratio. Sex ratio did not vary significantly with water hardness, pH, annual thermal units, or latitude. The possible adaptive significance of density-dependent sex determination in lampreys, however, remains elusive. It has been proposed that growth-promoting conditions might yield female-biased sex ratios as a tactic for ensuring that relatively large individuals become females, thereby increasing their fecundity. As predicted, larval size at a given age was generally greater in low-density populations, but there was no relationship between sex ratio and larval size, and female larvae were not consistently larger than the males.

Introduction

In some organisms, sex is determined by the environment that an individual encounters, rather than being fixed by genotype at conception (Charnov & Bull 1977, Conover 1984). Such environmental sex determination (ESD) has been known to occur in invertebrates such as parasitic nematodes (Christie 1929, Ellenby 1954), copepods (Christie 1929), and echiurid worms (Jaccarini et al. 1983), and skewed sex ratios in some isopods (e.g., Legrand & Juchault 1972, Williams & Franks 1988) have led to similar speculations of ESD. Numerous experimental studies have also demonstrated that ESD is common among many poikilothermic vertebrates. Incubation temperature, for example, has conclusively been shown to influence sex determination in many turtles (e.g., Bull & Vogt 1979, Pieau & Dorizzi 1981, Gutzke & Paukstis 1984, Mohanty-Hejmadi & Dimond 1986), an alligator (Ferguson & Joanen 1982), and a teleost fish (Conover & Kynard 1981). Population density, likewise, altered sex ratio in the European eel, *Anguilla anguilla* (D'Ancona 1950), and threespine stickleback, *Gasterosteus aculeatus* (Lindsey 1962), and low water pH produced a predominance of male progeny in 6 species of livebearing teleosts (Rubin 1985).

Sex differentiation in lampreys may also be sensitive to environmental influence, especially during their prolonged period of sexual indeterminacy (Okkelberg 1921, Hardisty 1965a, b). A small but variable excess of males has long been noted among spawning adult lampreys (e.g., Dean & Sumner 1898, Young & Cole 1900, Wigley 1959, Zanandrea 1961), and a positive relationship between the proportion of males and adult abundance has been observed (Hardisty 1954, 1961a). In particular, among the landlocked sea lamprey, *Petromyzon marinus,* of the upper Great Lakes, the sex ratio of both adults and larvae varied widely with abundance (Purvis 1979). As lamprey numbers were drastically reduced following treatment of their natal streams with the lampricide 3-trifluoromethyl-4-nitrophenol (TFM), the proportion of males correspondingly declined and a predominance of female larvae and adults was soon observed (Smith 1971, Purvis 1979). Since the sex composition of transformed lampreys collected during the treatment was nearly identical to that of untreated lampreys migrating from upstream (Manion & Smith 1978), TFM appears not to be differentially toxic to males and females. It has therefore been suggested that sex differentiation in lampreys, at least the sea lamprey of the upper Great Lakes, is influenced by population density (Purvis 1979).

Consequently, the objective of the current study is to determine whether, despite life cycle and geographic differences, a similar relationship between sex ratio and population density exists in another species of lamprey, the least brook lamprey, *Lampetra aepyptera* (Abbott, 1860). This nonparasitic lamprey, which is distributed in small, cool

Table 1. Collection sites for least brook lampreys.

Stream	Date	River system	County and State	Latitude and longitude
Unicorn Branch	13 Apr 1987 15 Dec 1988[1]	Chester R.	Queen Anne's Co., MD	39°13′-39°15′N 75°50′-75°51′E
Jordan Branch	16 Apr 1987	Chester R.	Kent Co., DE	39°13′-39°15′N 75°41′-75°43′E
Cod Creek	17 Apr 1987	Nanticoke R.	Sussex Co., DE	38°31′-38°33′N 75°40′-75°42′E
Butler Mill Branch	16 Apr 1987	Nanticoke R.	Sussex Co., DE	38°31′N 75°39′-75°40′E
Chapel Branch	16 Apr 1987	Nanticoke R.	Sussex Co., DE	38°38′-38°40′N 75°39′E
Garey Mill Pond Branch	15 Apr 1987 16 Dec 1988[1]	Choptank R.	Kent Co., DE	38°56′N 75°43°-75°46′E
Little Whippoorwill Cr.	16 Jun 1987[1] 27 Apr 1988	Red R.	Logan Co., KY	36°44′-36°46′N 86°50′-86°53′E
Sinking Creek	27 Apr 1988	Red R.	Logan Co., KY	36°42′-36°43′N 86°47′-86°51′E
Dry Fork Creek	27 Apr 1988	Red R.	Logan Co., KY	36°46′-36°43′N 87°59′E
Cane Creek	19 Jun 1987	Cumberland R.	Putnam Co., TN	36°03′-36°10′N 85°31′-85°37′E
Harid Creek	1 May 1987 2 Feb 1988[1]	Black Warrior R.	Tuscaloosa Co., AL	33°11′N 87°39′-87°41′E
Jay Creek	30 Apr 1987 2 Feb 1988[1]	Black Warrior R.	Tuscaloosa Co., AL	33°14′N 87°26′E

[1] Samples examined for sex ratio only; others for sex, length, and age.

streams throughout the Atlantic slopes of the U.S.A. and as far west as Missouri and Arkansas (Rohde & Jenkins 1980), has not been subjected to TFM treatment. Furthermore, since the physical and chemical properties of streams inhabited by lampreys can be expected to vary greatly, this study will also investigate the influence of several other environmental factors on the sex ratio of least brook lamprey larvae.

Although a relationship between density and sex ratio may suggest ESD in lampreys, it does not preclude density-dependent differences between the sexes with respect to mortality or recruitment to the adult population (Hardisty 1961a). The sex ratio of each age class will therefore be determined: if there are sex-specific differences in mortality, for example, a progressive decrease in the proportion of one sex with age is expected. Age in lampreys has traditionally been estimated from length-frequency distributions based on a large number of animals. Although this technique is satisfactory for younger larvae, the cumulative effect of individual variations in growth results in an overlapping of the length distributions of the older age classes (Potter 1980). In the present study, length-frequency age classes are consequently compared to ages determined from statoliths, structures analogous to the teleost otolith (Medland & Beamish 1987, Beamish & Medland 1988).

If ESD in least brook lampreys is indicated, its possible adaptive significance will be investigated. Environmental sex determination has been proposed, for example, as a tactic for ensuring that an individual of a relatively large size will become the sex in which the rewards for being large are greatest (Conover 1984). If such is the case in least brook lampreys, the following is expected: sexual dimorphism with respect to size; a relationship between growth and the environmental variable to which sex determination is sensitive; and consequently, a relationship between sex ratio and growth.

Methods

Least brook lamprey larvae were collected using electrical fishing gear from the substrate of 12 streams in Maryland, Delaware, Kentucky, Tennessee, and Alabama between April 1987 and December 1988 (Table 1). Identification to species (Vladykov & Kott 1980) was confirmed with adults collected from 8 of the streams. Chemical and physical characteristics of the streams were measured (Table 2), and annual thermal units, expressed as ° C days, were estimated for these or nearby streams (US De-

Table 2. Density estimates and physicochemical data for streams from which least brook lampreys were collected. The area given is the product of the length of the stream surveyed and its average width. Density ratings are: 1 = fewer than 0.01 larvae m^{-2}; 2 = 0.01–0.05; 3 = 0.05–0.10; 4 = 0.10–0.30; 5 = 0.30–0.50 larvae m^{-2}; an asterisk indicates a qualitative estimate only. Values for pH and total hardness represent 1–3 measurements throughout the year.

Stream	Area (m^2)	Density		Annual thermal units (° C days)	pH	Total hardness (mg l^{-1} as $CaCO_3$)
		larvae m^{-2}	rating			
Unicorn	3645	0.044	2	6275	7.26	60
Jordan			3*	6275	7.22	66
Cod	225	0.251	4	5480	6.57	39
Butler Mill			4*	5480	6.48	39
Chapel			4*	5480	6.52	46
Garey Mill	2700	0.078	3	5088	6.86	35
L. Whippoorwill	22500	0.003	1	5775	7.71	260
Sinking	6000	0.009	1	5775	7.98	238
Dry Fork	24750	0.003	1	5775	8.17	215
Cane	400	0.462	5	5012	7.76	157
Harid	500	0.072	3	5850	7.56	51
Jay	1215	0.148	3	5850	6.92	19

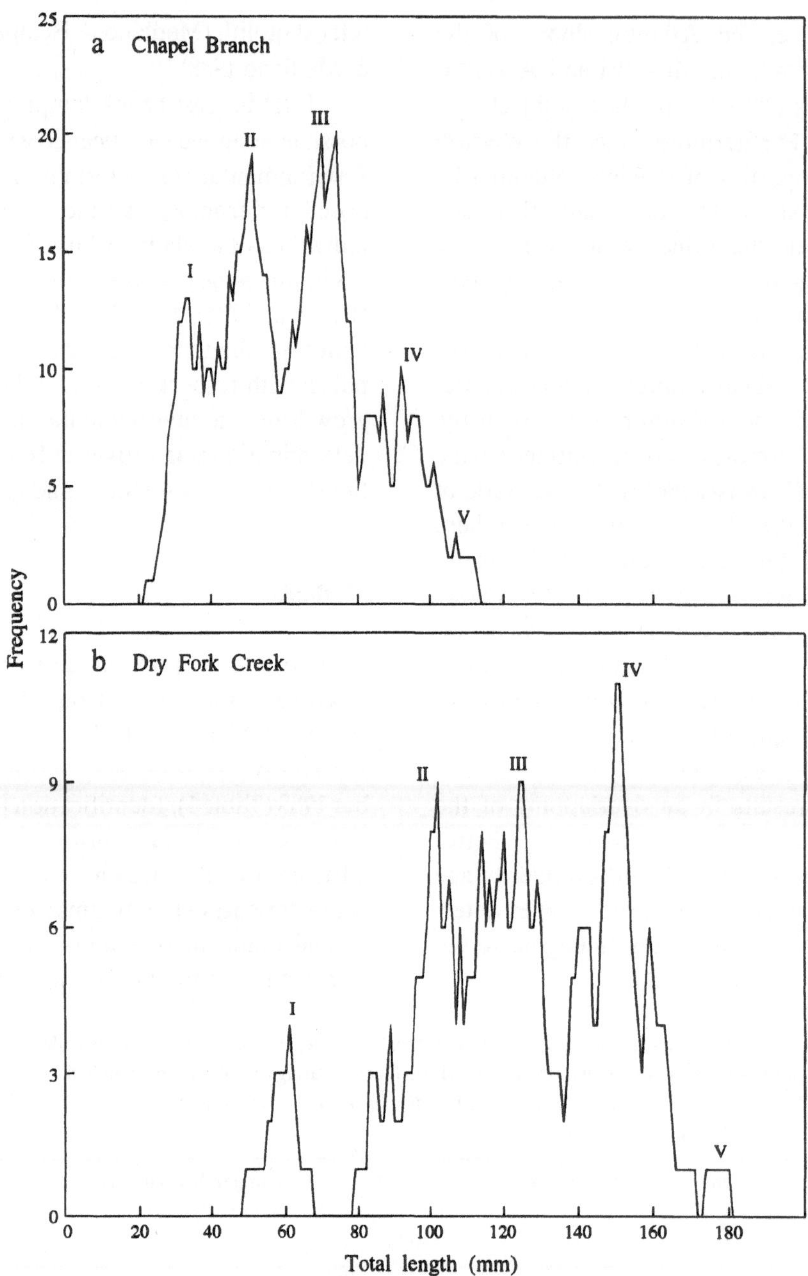

Fig. 1. Age class discrimination from length-frequency distributions for: a – Chapel Branch (n = 123); and b – Dry Fork Creek (n = 71). Age class frequency peaks, and hence modal lengths, are indicated by the position of the respective Roman numerals. The lowest points between peaks demarcate the age classes; the length range of each age class is presented in Figure 2.

partment of Commerce, National Technical Information Service; Water resources data for MD, DE, KY, TN, and AL water year 1987). The average width of the 12 streams ranged from 2.5 to 10.0 m, and the length of stream surveyed varied depending on the number of larvae collected. In most streams, distances of 75–675 m yielded large numbers of lamprey larvae; only in Little Whippoorwill and Dry Fork Creeks were distances in excess of 2 km necessary to collect samples of sufficient size. Relative population density was estimated from the number of larvae collected per m^2 of substrate (Table 2), and the densities were ranked on a scale of 1 (lowest) to 5 (highest) to allow for the inclusion of 3 streams where sampling area was imprecisely known. The relative density rating (D_r) was directly related to the logarithm of actual density (D_a, larvae m^{-2}) in the streams where the latter was available; the regression equation (± SE) is:

$$D_r = 1.7 \pm 0.2 \log D_a + 4.9 \pm 0.3$$
$$(n = 9, p < 0.0001).$$

Immediately after capture, all larvae were killed by an overdose of tricaine methanesulfonate (> 100 mg l^{-1}). Length was measured to the nearest 1 mm, and the larvae were frozen for later removal of the statoliths (Beamish & Medland 1988). Larvae were not preserved in formalin because it promotes dissolution of these calcareous structures. After thawing, length and weight of all larvae were recorded to the nearest 1 mm and 1 mg, respectively, and these measurements were used throughout the study unless otherwise indicated. The relationship between fresh length (L_f) and thawed length (L_t), both in mm, was determined for larvae from 6 populations and is described by the regression (± SE):

$$L_f = 1.03 \pm 0.003\ L_t + 1.13 \pm 0.42$$
$$(n = 343, p < 0.0001).$$

Statoliths were removed from the otic capsules, stored in immersion oil for at least 10 days to intensify the banding pattern, and the number of annuli was subsequently counted under a dissecting microscope. Each statolith was aged independently 2–3 times by the same individual, and was numerically coded so that the estimation of age was not biased by the length of the larva or by an age previously assigned. Annuli did not form in larvae from some populations, presumably due to a relatively constant growth rate throughout the year (Beamish & Medland 1988). For age class discrimination in these populations, length-frequency distributions were constructed by taking a sliding average over 7 mm to enhance the frequency peaks. Modal lengths were assigned to each age class and, from the logarithmic relationships between length and weight for each stream, a modal weight was also assigned. To determine the degree to which length-frequency analysis was reliable, age classes were similarly estimated in populations where statolith annuli were apparent, and the 2 techniques were compared.

The sex of least brook lamprey larvae could be distinguished in all individuals at least 55 mm in length. Two 5-mm long segments were taken from the midregion of each individual, fixed in 5% formalin for a minimum of 48 hours, and embedded in paraffin. The embedded tissues were sectioned to a thickness of 8 µm, stained with Harris' hematoxylin and eosin (Drury & Wallington 1980), and sex was determined following histological examination.

To evaluate postlarval sex ratios and sex-specific differences in body size, metamorphosing and adult least brook lampreys were collected from 8 of the 12 streams, as described previously (Docker & Beamish 1991). Length and weight were measured to the nearest 1 mm and 1 mg, respectively, and sex was determined by visual inspection of the gonad under a dissecting microscope. Only 5 postlarval lampreys were collected from Butler Mill Branch, however, and these were used for identification purposes only.

Data were analyzed using linear regression, analysis of variance (ANOVA), or chi-squared analysis (X^2, contingency tables) (Statgraphics 1986). Significant deviations in sex ratio from parity were identified from a table of 95% confidence limits for proportions (Beyer 1968). The level of significance used was $p < 0.05$ unless otherwise stated.

Results

Adult least brook lampreys were captured in 7 of the streams between mid-December and late February (Docker & Beamish 1991), and ripe and spent lampreys were found in 2 streams in mid-April. It is thus reasonable to assume that the young-of-the-year larvae emerge between late April and mid-May (Piavis 1971). According to convention (Chilton & Beamish 1982), however, lamprey ages were assigned assuming a January 1 birthdate. That is, the young-of-the-year larvae constitute age class 0 only until December 31.

Statolith annuli were observed in the 7 popula-

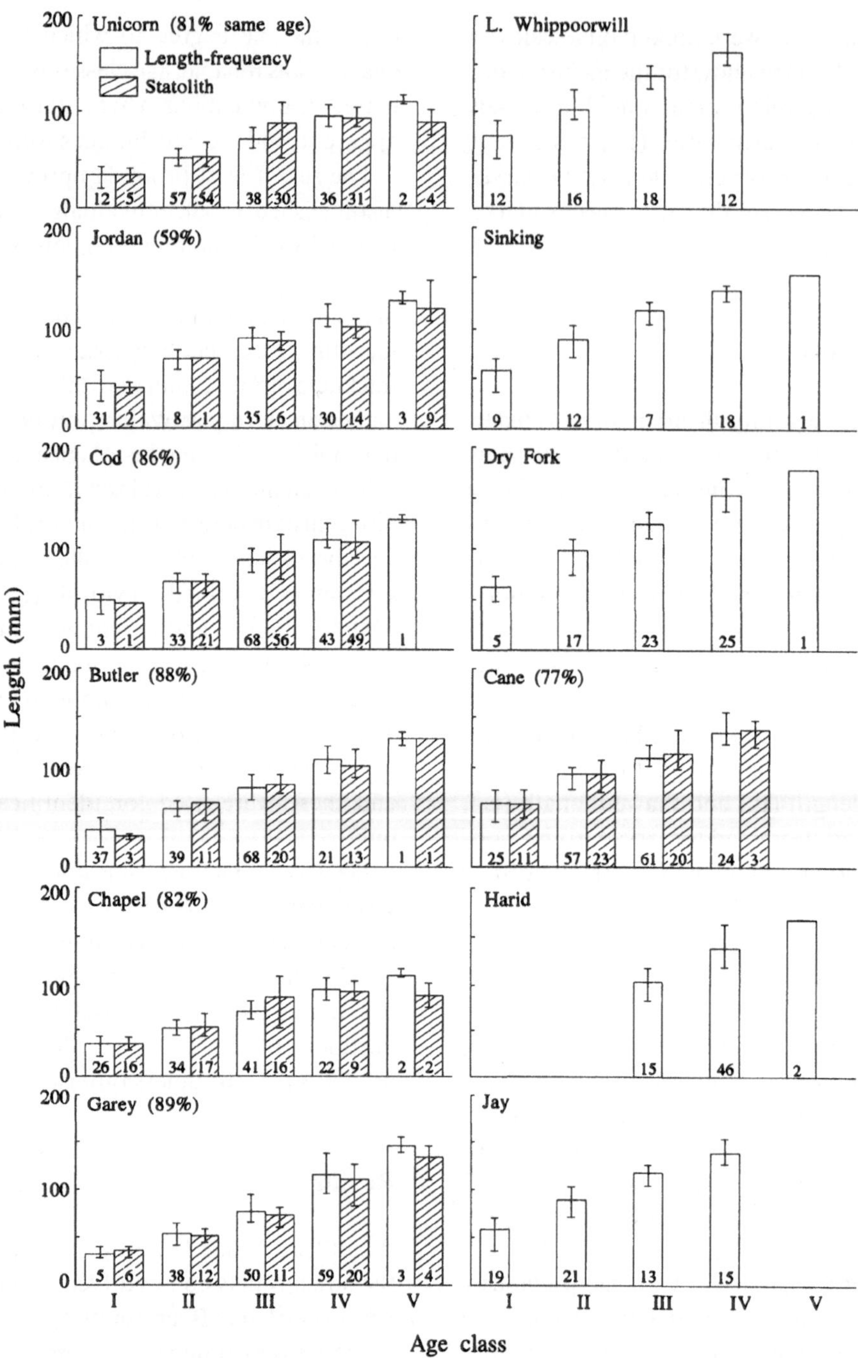

Fig. 2. Modal length of each age class from 12 populations of the least brook lamprey, determined by length-frequency analysis, and compared to mean length of statolith-derived age classes; the percentage of larvae assigned the same age by both techniques is indicated. Sample sizes are given, and vertical bars indicate the length ranges of each age class.

tions from Maryland, Delaware, and Tennessee. Repeated ageing of a given individual resulted in assignment of the same age more than 90% of the time, and the number of annuli increased significantly with lamprey total length (linear regression, $p < 0.001$). There were consistently 5 year classes (Fig. 2), although only 4.0% of the 501 larvae that were aged belonged to the fifth year class.

Length-frequency distributions for these populations showed 4 main peaks (e.g., Fig. 1a), which were assumed to be the modal lengths of each of the first 4 age classes; in all but Cane Creek, a minor fifth peak was also evident. Individuals of a given length were assigned an age based on these curves, with the lowest points between the peaks demarcating the age classes. In 82.0% of the larvae aged by both techniques, the length-frequency age corresponded with the statolith annuli number (Fig. 2). Length-frequency curves were fitted to the 5 remaining populations, in which statolith annuli did not form or were indistinct, and these also showed 4 main frequency peaks (e.g., Fig. 1b).

For larvae of age class I, modal length and weight ranged from 32 mm and 80 mg in Garey Mill Pond Branch to 75 mm and 741 mg in Little Whippoorwill Creek (e.g., Fig. 2). Little Whippoorwill Creek was sampled extensively to ensure that a smaller size class had not been overlooked: the smallest larva found was 53 mm and only 5 larvae < 70 mm in length were observed. The modal length and weight of age class IV ranged from 95 mm and 1475 mg in Chapel Branch to 163 mm and 7759 mg in Little Whippoorwill. Modal length and weight corresponded well with mean length and weight of the statolith-derived age classes (Fig. 2).

Size at a given age varied with both population density and the physicochemical characteristics of the stream (linear regression; Table 3). In each age class, length and weight increased significantly with both water hardness and pH (total hardness and pH were highly correlated; p = 0.001), and length decreased significantly with latitude. A decrease in modal length and weight with increasing density was significant in age classes III and IV. Neither length nor weight were significantly related to annual thermal units. Due to their limited numbers, the effect of the environment on the modal length and weight of age class V larvae was not analyzed.

The sex of least brook lampreys could only be identified following a prolonged period of sexual indeterminacy. Prior to sex differentiation, the minute gonad contained only a small number of undifferentiated germ cells (Fig. 3a). Germ cell proliferation, beginning in larvae measuring 35–45 mm in length, subsequently led to the appearance of numerous undifferentiated germ cells and oocytes in both future males and females. In these smaller larvae, identification of the sexes was still ambiguous. By 55 mm total length, however, differentiated females were recognized by the virtual absence of any undifferentiated germ cells and the persistence of a number of nucleated, basophilic oocytes (Fig. 3b). Although differentiation of germ cells into spermatocytes in lampreys does not occur until the end of the larval period (Hardisty 1965a, b), presumptive males were identified by the degeneration of all but a few small oocytes and the persistence of large numbers of undifferentiated germ cells (Fig. 3c).

Although the sex of age class I larvae could often be distinguished in the faster-growing Kentucky populations, sex could not be consistently determined until age class II Consequently, the sex ratio

Table 3 Levels of significance for regressions between modal length (mm) or weight (mg) at a given age, and stream properties hardness, pH, latitude, density, and annual thermal units (° C days) The sign of each regression coefficient (slope) is given An asterisk indicates significance at p < 0 05

Factor	Slope	Significance levels (p-values)							
		Age I[1]		Age II[1]		Age III		Age IV	
		mm	mg	mm	mg	mm	mg	mm	mg
Hardness	+	0 002*	< 0 001*	0 002*	< 0 001*	0 001*	< 0 001*	0 002*	< 0 001*
pH	+	0 011*	0 005*	0 002*	< 0 001*	0 003*	0 002*	0 001*	< 0 001*
Latitude	–	0 017*	0 097	0 012*	0 055	0 035*	0 141	0 026*	0 082
Density	–	0 202	0 199	0 109	0 112	0 030*	0 022*	0 022*	0 017*
° C days	±	0 997	0 797	0 808	0 963	0 716	0 857	0 844	0 885

[1] n = 11 modal length and weight for age classes I and II not available for Harid Creek

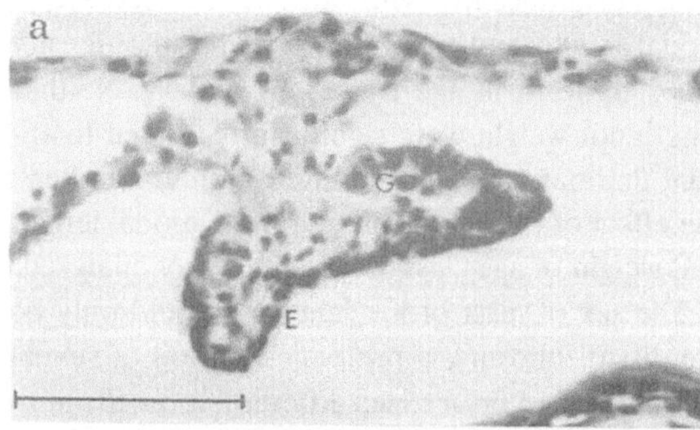

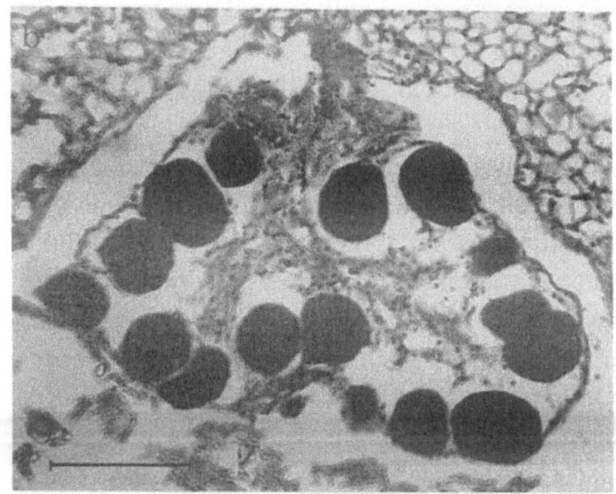

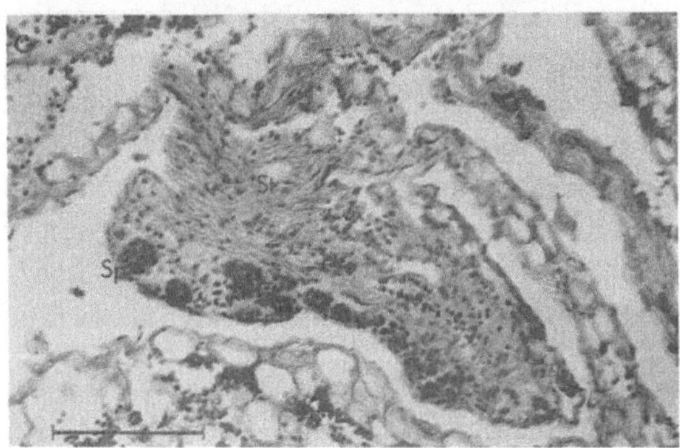

Fig 3 a – Undifferentiated gonad of a 51 mm least brook lamprey larva, containing germ cells (G) enclosed by a peritoneal epithelium (E) Scale bar = 50 µm b – Ovary from a 118 mm larva Scale bar = 200 µm c – Testis from a 104 mm larva, illustrating cysts of presumptive spermatogonia (Sp) and extensive stromal tissue (St) Scale bar = 100 µm

of age class I larvae could not be routinely calculated. Given the small sample size of age class V, these larvae were also omitted from overall sex ratio calculations. The sex ratios of age class II to IV larvae varied widely among the 12 streams, from 28.7% male in Unicorn Branch to 70.9% male in Cod Creek (Table 4). Cod Creek showed a significant preponderance of males (95% CL for proportions), and a significant excess of females was noted in Unicorn, Garey Mill, and Little Whippoorwill Creeks. In each of the 5 streams from which a second collection of larvae was made (Table 1), sex ratios did not differ significantly between collections (X^2 test).

In only Cod Creek and Butler Mill Branch did sex ratios vary significantly among length-frequency age classes (X^2 test): the proportion of males decreased with age in Cod Creek and increased with age in Butler Mill Branch (Table 4). Sex ratios in statolith-derived age classes were similar, but differences were no longer significant (e.g., 86, 65, and 65% male, respectively, in ages II, III, and IV in Cod Creek; 60, 80, and 64% male in ages II–IV in Butler Mill). It is interesting to note that the sex ratio of age class V, although based on a small number of individuals, was overall female-biased. Of a total of 36 age class V larvae, 34.6% were male. In contrast, age classes II to IV averaged 42.3 to 49.3% male.

Sex ratio variations among the populations were related to differences in larval density: the proportion of males increased significantly with relative density (Fig. 4). Sex ratio was not significantly related to water hardness, pH, annual thermal units, or latitude. Sex ratio likewise was not correlated with larval size at a given age: neither overall sex ratio nor the sex ratio of the corresponding age class was related to the modal length or weight of each age class. Sex-specific differences in length and weight were inconsistent, varying among streams and age classes (Table 5). Overall, male and female size differed significantly only among age class V larvae, with females being longer than males (ANOVA).

The sex ratio of postlarval least brook lampreys (Table 6) also varied significantly among streams (X^2 test). Chapel and Unicorn Branches, however, at 79.2 and 75.0% male, respectively, were significantly male-biased (95% CL for proportions). No streams showed a significant excess of females; the smallest proportion of males, 46.2%, was found in Little Whippoorwill Creek. The postlarval sex ratio of each stream was unrelated to the sex ratio of the larval population (linear regression).

The size of the postlarval lampreys likewise varied significantly among streams (ANOVA): the mean length and weight for each stream ranged from 93 mm and 1.5 g in Chapel Branch to 147 mm and 5.9 g in Little Whippoorwill Creek. The smallest and largest postlarval lampreys were 82 and 165 mm, from Chapel and Little Whippoorwill Creeks, respectively. Unlike sex ratio, the length of the postlarval lampreys in a stream was directly related to the modal length of each larval age class (linear regression). In all but Cane Creek, female postlarval lampreys were larger than their male counterparts (Table 6). Accounting for the size variation among streams, the overall sex-specific differences in both length and weight were significant (2-factor ANOVA).

Discussion

Sex ratios of larval least brook lampreys varied widely among the 12 populations examined in this study. The extremes were 29 and 71% male, and the proportion of males increased significantly with population density. A similar relationship between larval sex ratio and adult abundance has been observed in the landlocked sea lamprey (Smith 1971, Purvis 1979). Density-dependent sex determination in lamprey larvae, acting during a prolonged period of sexual indeterminacy, is consequently proposed.

The relationship between sex ratio and abundance has long been observed among spawning adult lampreys, in both the landlocked sea lamprey (Wigley 1959, Hardisty 1961a) and the nonparasitic European brook lamprey, *Lampetra planeri* (Hardisty 1954, 1961a). Hardisty (1961a), however, proposed that there were density-dependent differenc-

Table 4 Overall sex ratio (percent males) of least brook lamprey larvae in 12 streams, and the sex ratio in length-frequency derived age classes II–IV Age classes I and V have been excluded from overall sex ratio calculations due to their small sample sizes Significantly male- or female-biased sex ratios are identified by a superscript 'm' or 'f' (95% CL for proportions), an asterisk indicates significant variation in sex ratio among age classes (X^2 test) The number of sexed individuals represents only those in which sex could be unequivocally determined

Stream		Overall		Age II		Age III		Age IV	
		n Sexed	%M	n	%M	n	%M	n	%M
Unicorn	Apr 1987	128	25 8^{f}	54	25 9^{f}	38	23 7^{f}	36	27 8^{f}
	Dec 1988	81	33 3^{f}	13	7 7^{f}	60	38 3^{f}	8	37 5^{f}
	Total	209	28 7^{f}	67	22 4^{f}	98	32 7^{f}	44	29 5^{f}
Jordan		94	48 9	8	50 0	37	45 9	49	51 0
Cod		134	70 9^{m}*	32	90 6^{m}	61	65 6^{m}	41	63 4
Butler Mill		102	51 9*	28	32 1	56	57 1	18	66 7
Chapel		96	57 3	34	50 0	40	70 0^{m}	22	45 5
Garey Mill	Apr 1987	123	35 8^{f}	24	37 5	42	35 7	57	35 1^{f}
	Dec 1988	174	43 1			124	45 2	50	38 0
	Total	297	40 1^{f}	24	37 5	166	42 8	107	36 4^{f}
L Whippoorwill	Jun 1987	16	12 5^{f}	9	11 1^{f}	3	33 3	4	0 0
	Apr 1988	44	36 4	16	31 3	17	41 2	11	36 4
	Total	60	30 0^{f}	25	24 0^{f}	20	40 0	15	26 7
Sinking		47	36 2	12	41 7	17	47 1	18	22 2^{f}
Dry Fork		63	50 8	17	41 2	23	52 2	23	56 5
Cane		142	50 8	56	44 6	60	51 7	26	57 7
Harid	May 1987	61	60 3			15	80 0^{m}	46	54 4
	Feb 1988	47	53 2	3	66 7			44	52 3
	Total	108	57 4	3	66 7	15	80 0^{m}	90	53 3
Jay	Apr 1987	43	37 2	17	47 1	11	45 5	15	20 0^{f}
	Feb 1988	23	65 2	3	66 7	13	61 5	7	71 4
	Total	66	47 0	20	50 0	24	54 2	22	36 4

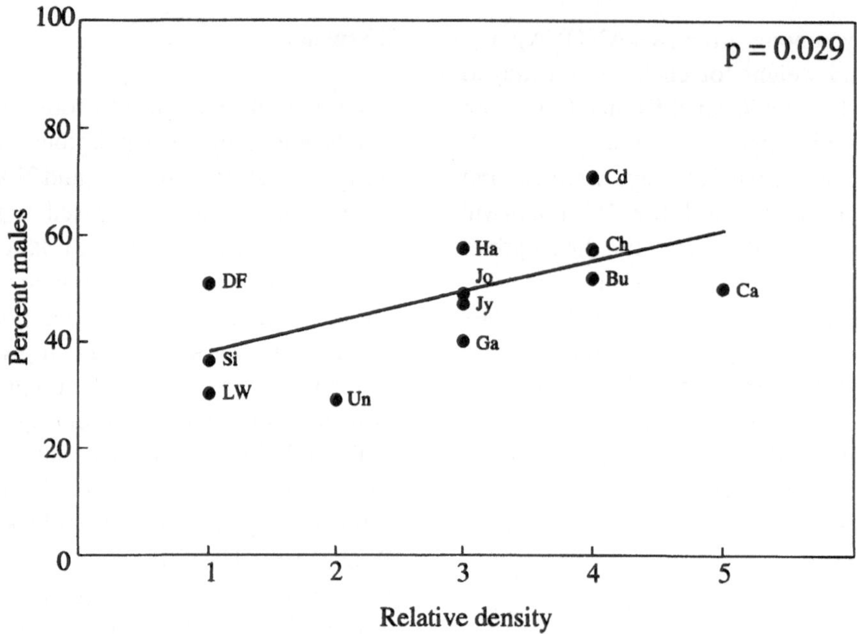

Fig. 4. Relationships between sex ratio (percent males) and relative density of larvae, ranked on an ascending scale of 1 to 5. Bu = Butler, Ca = Cane, Cd = Cod, Ch = Chapel, DF = Dry Fork, Ga = Garey, Ha = Harid, Jo = Jordan, Jy = Jay, LW = Little Whippoorwill, Si = Sinking Creek, Un = Unicorn.

es between the sexes with respect to mortality or recruitment to the adult population. An environmental effect on sex differentiation was dismissed on the finding of approximately equal proportions of male and female larvae in landlocked sea lampreys, European brook lampreys, and the European river lamprey, *Lampetra fluviatilis* (Hardisty 1960). That Hardisty (1960) did not observe skewed sex ratios in larval lampreys may have been due to his examination of only 1 or 2 populations for each species. In the present study, the sex ratios of 8 of the 12 sites examined did not differ significantly from parity, and the sampling of 1 or 2 random sites may similarly have found equal proportions of male and female least brook lampreys. Furthermore, small changes in density may not produce appreciable differences in sex ratio. For example, the relative density of Cod Creek, in which the bias towards males was greatest, was more than 80 times higher than that of Little Whippoorwill Creek. In the landlocked sea lamprey of the upper Great Lakes, the fluctuations in lamprey abundance have also been of great magnitude (Smith 1971). More extensive surveys of the European brook and river lampreys may yet show support for density-dependent sex determination.

Table 5. Significant sex-specific differences in length and weight of least brook lampreys within each stream and age class; LF and St refer to length-frequency and statolith-derived age classes, respectively.

Stream	Males larger		Stream	Females larger	
	Age class	Variable		Age class	Variable
Chapel	II (LF)	weight	Jordan	V (LF)	length, weight
			Garey	II (LF, St)	length
			Garey	IV (LF)	length, weight
			L. Whippoor.	IV (LF)	length, weight
			Dry Fork	IV (LF)	weight

Furthermore, the present study addressed the issue of sex-specific differences in mortality and recruitment. With regards to differential recruitment, it is both possible and likely that female least brook lampreys metamorphosed at a greater age than males: the oldest larval age class appeared to be female-biased, and females were larger than males among the oldest larvae and the metamorphosing and adult lampreys. That females metamorphose at an older age than males has also been indicated in a number of other lamprey species (e.g., Hardisty 1965b, Purvis 1970, Beamish & Austin 1985, Murdoch et al. 1992). Differential recruitment to the adult stock, however, was not responsible for the differences in larval sex ratios. Similarly, the sex ratio variations were not likely the result of differential mortality between the sexes: sex ratios varied among streams from the time of gonadal differentiation, and remained relatively constant thereafter. Sex ratios differed significantly, but inconsistently, among age classes in only 2 streams. Furthermore, among sea lamprey larvae maintained at various densities for over 3 years, there was no evidence of sex-specific mortality (Docker 1992).

The systematic position of lampreys makes their mode of sex differentiation of interest to comparative biologists, since environmental sex determination has already been demonstrated in some invertebrates and other so-called lower vertebrates. For example, although there have only been a few reports of an effect of population density on sex ratio (Christie 1929, Ellenby 1954, D'Ancona 1950, Lindsey 1962), temperature-dependent sex determination is common. Incubation temperature has been found to alter sex ratio in at least 3 teleosts (D'Ancona 1959, Conover & Kynard 1981, Sullivan & Schultz 1986), and is widely known to influence sex differentiation in many reptiles. For example, low incubation temperatures in some turtle species produce males and higher temperatures yield females (e.g., Bull & Vogt 1979, Morreale et al. 1982), whereas in lizards (e.g. Bull 1980) and an alligator (Ferguson & Joanen 1982), the pattern is reversed. In addition, low environmental pH resulted in a predominance of male progeny in 6 species of livebearing teleosts (Rubin 1985). In the present study, lamprey sex ratio was not significantly related to annual thermal units, latitude, or pH.

The adaptive significance of density-dependent sex determination was investigated in the least brook lamprey. It has been suggested that environmental sex determination should be favoured when an environmental factor is more advantageous to one sex or the other (Charnov & Bull 1977). Such control of sex differentiation has been proposed, for example, as a tactic for ensuring that an individual of a relatively large size will become the sex in which the rewards for being large are greatest (Charnov & Bull 1977, Conover 1984). The environmental variables to which sex determination is sensitive may thus act as cues to indicate conditions of favourable growth. For example, an inverse relationship between lamprey density and larval size was significant in the 2 older age classes. Such a decrease in growth at high densities is well-known in

Table 6 Sex ratio of postlarval least brook lampreys, and mean (± SE) length and weight of each sex. Lampreys were collected between October and February, and weights were adjusted to correspond to those captured in February (Docker & Beamish 1991). Length and weight after thawing are given for consistency with larval data

Stream	n	% males	Males		Females	
			Length (mm)	Weight (g)	Length (mm)	Weight (g)
Unicorn Br	20	75.0	104 ± 3	2.1 ± 0.2	106 ± 4	2.2 ± 0.2
Chapel Br	24	79.2	92 ± 1	1.4 ± 0.1	98 ± 3	1.8 ± 0.2
Garey Mill	25	60.0	103 ± 2	2.4 ± 0.1	108 ± 2	3.1 ± 0.2
L. Whippoorwill	26	46.2	144 ± 3	5.7 ± 0.2	149 ± 2	6.1 ± 0.3
Cane Cr	23	52.2	136 ± 2	5.6 ± 0.3	133 ± 3	5.1 ± 0.3
Harid Cr	38	63.2	130 ± 2	3.7 ± 0.2	138 ± 2	4.4 ± 0.2
Jay Cr	30	56.7	129 ± 2	3.6 ± 0.2	131 ± 3	3.9 ± 0.2

teleosts (Allen 1974, Trzebiatowski et al. 1981), and has been observed in other lamprey species (Mallatt 1983, Morman 1987, Murdoch et al. 1992). That the relationship was not consistently significant among the 12 streams examined is likely due to the modifying effects of other environmental factors. Modal length and weight increased with water hardness and pH, and length varied inversely with latitude.

If growth-promoting environmental factors favour differentiation into the sex benefitting most from large size, however, a sexual dimorphism with respect to size would be expected. For example, in parasitic nematodes, where females are universally the larger sex, crowding inhibits growth and results in differentiation of males (Christie 1929, Ellenby 1954). In the Atlantic silverside, *Menidia menidia,* low fluctuating temperatures characteristic of the early breeding season produce a high proportion of females and the higher temperatures representative of late in the season yield an excess of males (Conover & Kynard 1981). Female Atlantic silversides, having a longer growing season, are consequently the larger sex. In nematodes and teleosts, males therefore appear to be penalized less by small size (Charnov & Bull 1977, Conover 1984). For example, whereas gamete production in teleosts may be relatively unconstrained by male body size (Kazamov 1981), fecundity is known to increase with maternal size (e.g., Bagenal 1966, Docker et al. 1986, Hay & Brett 1988). Egg number and female length are likewise correlated in lampreys (Hardisty 1964, Kott 1971, Beamish & Thomas 1983, Docker & Beamish 1991). In the least brook lamprey of the present study, however, female larvae were not consistently larger than the males. In particular, there were no consistent sex-specific differences in size at the approximate time of gonadal differentiation. Females were regularly the larger sex only in the oldest larval age class and in the adult population, presumably due to later metamorphosis as was discussed above.

Furthermore, there was no relationship between larval size at a given age and the sex ratio of the population: streams in which sexual differentiation occurred at relatively large sizes were not consistently female-biased. Although a high proportion of females was indeed found in Little Whippoorwill Creek where larval size at a given age was greatest, both Unicorn and Garey Mill Pond Branches showed a significant preponderance of females despite the small size of the individuals in the population.

A plausible explanation for the evolution of environmental sex determination in lampreys is consequently lacking. In many reptiles, where temperature-dependent sex determination is well-documented (e.g., Bull & Vogt 1979, Bull 1980, Ferguson & Joanen 1982), its adaptive significance likewise remains elusive (Conover & Heins 1987). It may be that, although fertility is known to increase with body size in females, large body size may also be important to male fitness. Size-specific mortality, for example, may favour large size regardless of sex. Among sea lamprey larvae maintained under controlled conditions for over 3 years, mortality was highest among the smallest of 3 size classes (Docker 1992), and Potter (1980) suggested that a similar pattern exists in nature. Furthermore, the right-skewed length-frequency curves constructed by Manion & Smith (1978) indicate that, even within a single age class, smaller larvae suffer greater mortality.

Further speculation on the possible adaptive significance of environmental sex determination in lampreys clearly requires a better understanding of their ecology. In addition to the above considerations of fertility and survival, for example, future studies should examine the demographics of the spawning stock. For example, highly-skewed sex ratios could lead to frequency-dependent selection for the rarer sex (Charnov & Bull 1977, Conover & Heins 1987). Consequently, the environmental cue to which sex differentiation is sensitive would be expected to be spatially or temporally variable to ensure adequate production of both sexes (Bull 1980, Conover 1984). Mating between individuals from different densities would then result in spawning sex ratios closer to parity.

The extent to which least brook lamprey density is spatially variable within its range of dispersal, however, appears limited. For example, although there is evidence that some sea lampreys do not home to their natal streams (Potter et al. 1974), mating between nonmigratory least brook lampreys from different streams is unlikely. Furthermore, an unpublished study by F.W.H. Beamish suggests that density and sex ratio may be relatively constant within a given stream. Alternatively, there may be annual differences in population density and sex ratio, and individual or sex-specific differences in age at metamorphosis (Manion & Smith 1978) would result in spawning among lampreys from different age classes. The populations in the present study, however, appeared to show relatively minor annual variations in sex ratio. The 5 populations sampled up to 20 months apart showed little change in sex ratio between collections, and there was little variation among year classes.

Larval sex ratios, however, are not necessarily representative of sex ratios at spawning. In the present study, there was a clear bias towards males among spawning least brook lampreys and, relative to the larval sex ratios, this excess of males was not consistent. Hardisty (1960) found a similar disparity between larval and adult sex ratios, which he has attributed to higher female mortality at some time between metamorphosis and spawning (Hardisty 1961b). Alternatively, it could be that the disproportionate percentage of males at spawning was artifactual, due either to small samples sizes or to the time or location of collection. Lamprey sex ratios, for example, have been shown to fluctuate during the spawning season (Zanandrea 1951, Hardisty 1961b, Stier & Kynard 1986), and may be attributed either to differential migration rates between the sexes (Applegate & Thomas 1965) or shorter female residence time on the spawning grounds (Farlinger & Beamish 1984). A better understanding of the relationship between larva and adult sex ratios is crucial to further hypotheses regarding the evolutionary significance of environmental sex determination in lampreys.

In summary, all 3 female-biased populations of the least brook lamprey were characterized by low larval density, and among the 12 geographically-diverse streams, the proportion of males increased significantly with density. Although females appear to metamorphose later than males, the number of age class V larvae was minimal and the observed differences in sex ratio were not the result of differential recruitment. Likewise, differential mortality between the sexes was unlikely. Rather, the variable sex ratios were established at the time of gonadal differentiation and remained relatively constant over the subsequent 2–3 years of larval life. Although only experimental studies can verify a causal relationship between density and sex ratio, density-dependent sex determination is indicated in the least brook lamprey. Similarly, although the possible adaptive significance of ESD in lampreys remains elusive, further experimental and field studies may shed some light on density-related differences between the sexes.

Acknowledgements

Financial assistance was provided through an operating grant (F.W.H.B.) and postgraduate scholarship (M.F.D.) from the Natural Sciences and Engineering Research Council of Canada. Scientific collecting permits were kindly supplied by Alabama Department of Conservation, Delaware Department of Natural Resources and Environmental Control, Kentucky Department of Fish and Wildlife Resources, Maryland Department of Natural Resources, and Tennessee Wildlife Resources Agency. T.E. Medland performed the statolith ageing, for which we are especially grateful.

References cited

Allen, K O 1974 Effects of stocking density and water exchange rates on growth and survival of channel catfish *Ictalurus punctatus* (Rafinesque) in circular tanks Aquaculture 4 29–39

Applegate, V C & M L H Thomas 1965 Sex ratios and sexual dimorphism among recently transformed sea lampreys, *Petromyzon marinus* Linnaeus J Fish Res Board Can 22 695–711

Bagenal, T B 1966 The ecological and geographical aspect of fecundity of plaice J Mar Biol Assoc UK 46 161–186

Beamish, F W H & L S Austin 1985 Growth of the mountain brook lamprey *Ichthyomyzon greeleyi* Hubbs and Trautman Copeia 1985 881–890

Beamish, F W H & T E Medland 1988 Age determination for lampreys Trans Amer Fish Soc 117 63–71

Beamish, F W H & E J Thomas 1983 Potential and actual fecundity of the 'paired' lampreys, *Ichthyomyzon gagei* and *I castaneus* Copeia 1983 367–374

Beamish, FWH & E J Thomas 1984 Metamorphosis of the southern brook lamprey, *Ichthyomyzon gagei* Copeia 1984 502–515

Beyer, WH 1968 CRC handbook of tables for probability and statistics Chemical Rubber Company, Cleveland 642 pp

Bull, J J 1980 Sex determination in reptiles Quart Rev Biol 55 3–21

Bull, J J & R C Vogt 1979 Temperature-dependent sex determination in turtles Science 206 1186–1188

Charnov, E L & J J Bull 1977 When is sex environmentally determined? Nature 266 828–830

Chilton, D E & R J Beamish 1982 Age determination methods for fishes studied by the groundfish program at the Pacific Biological Station Can Spec Publ Fish Aquat Sci 60 120 pp

Christie, J R 1929 Some observations on sex reversal in the mermithidae J Exp Zool 53 59–76

Conover, D O 1984 Adaptive significance of temperature-dependent sex determination in a fish Amer Nat 123 297–313

Conover, D O & S W Heins 1987 The environmental and genetic components of sex ratio in *Menidia menidia* (Pisces Atherinidae) Copeia 1987 732–743

Conover, D O & B E Kynard 1981 Environmental sex determination interaction of temperature and genotype in a fish Science 213 577–579

D'Ancona, U 1950 Determination et differentiation du sexe chez les poissons Arch Anat Microsc Morphol Exp 39 274–292

D'Ancona, U 1959 Distribution of the sexes and environmental influence in the European eel Arch Anat Microsc Morphol Exp 48 (suppl) 61–70

Dean, B & FB Sumner 1898 Notes on the spawning habits of the brook lamprey (*Petromyzon wilderi*) Trans NY Acad Sci 16 321–324

Docker, M F 1992 Labile sex determination in lampreys the effect of larval density and sex steroids on gonadal differentiation Ph D Thesis, The University of Guelph, Guelph 269 pp

Docker, M F & FWH Beamish 1991 Growth, fecundity, and egg size of least brook lamprey, *Lampetra aepyptera* Env Biol Fish 31 219–227

Docker, M F, TE Medland & FWH Beamish 1986 Energy requirements and survival in embryo mottled sculpin (*Cottus bairdi*) Can J Zool 64 1104–1109

Drury, R A B & E A Wallington 1980 Carleton's histological technique Oxford University Press, Oxford 520 pp

Ellenby, C 1954 Environmental determination of the sex ratio of a plant parasitic nematode Nature 174 1016–1017

Farlinger, S P & R J Beamish 1984 Recent colonization of a major salmon-producing lake in British Columbia by Pacific lamprey (*Lampetra tridentata*) Can J Fish Aquat Sci 41 278–285

Ferguson, M WJ & T Joanen 1982 Temperature of egg incubation determines sex in *Alligator mississippiensis* Nature 296 850–853

Gutzke, W H N & G L Paukstis 1984 A low threshold temperature for sexual differentiation in the painted turtle, *Chrysemys picta* Copeia 1984 546–547

Hardisty, M W 1954 Sex ratio in spawning populations in *Lampetra planeri* Nature 173 874–875

Hardisty, M W 1960 Sex ratios of ammocoetes Nature 186 988–989

Hardisty, M W 1961a Sex composition of lamprey populations Nature 191 116–117

Hardisty, M W 1961b Studies on an isolated spawning population of brook lampreys (*Lampetra planeri*) J Anim Ecol 30 339–355

Hardisty, M W 1964 The fecundity of lampreys Arch Hydrobiol 60 340–367

Hardisty, M W 1965a Sex differentiation and gonadogenesis in lampreys I The ammocoete gonads of the brook lamprey, *Lampetra planeri* J Zool , Lond 146 305–345

Hardisty, M W 1965b Sex differentiation and gonadogenesis in lampreys II The ammocoete gonads of the landlocked sea lamprey, *Petromyzon marinus* J Zool , Lond 146 346–387

Hay, D E & J R Brett 1988 Maturation and fecundity of Pacific herring (*Clupea harengus pallasi*) an experimental study with comparisons to natural populations Can J Fish Aquat Sci 45 399–406

Jaccarini, V, L Aguis, PJ Schembri & M Rizzo 1983 Sex determination and larval sexual interaction in *Bonellia viridis* Rolando (Echiura Bonelliidae) J Exp Mar Biol 66 25–40

Kazamov, R V 1981 Peculiarities of sperm production by anadromous and parr Atlantic salmon (*Salmo salar*) and fish cultural characteristics of such spawn J Fish Biol 18 1–8

Kott, E 1971 Characteristics of pre-spawning American brook lamprey from Big Creek, Ontario Ont Field-Nat 85 235–240

Legrand, J -J & P Juchault 1972 Mise en evidence dans une population d'*Armadillidium vulgare* Latr (Crustace Isopode Oniscoide) de deux types de lignees arrhenogenes en relation avec des facteurs epigenetiques a effet respectivement masculinisant et feminisant C r hebd Seanc Acad Sci Paris (Ser D) 276 2313–2316

Lindsey, C C 1962 Experimental study of meristic variation in a population of threespine stickleback, *Gasterosteus aculeatus* Can J Zool 40 271–312

Mallatt, J 1983 Laboratory growth of larval lampreys (*Lampetra (Entosphenus) tridentata* Richardson) at different food concentrations and animal densities J Fish Biol 22 293–301

Manion, PJ & B R Smith 1978 Biology of larval and metamorphosing sea lampreys, *Petromyzon marinus,* of the 1960 year class in the Big Garlic River, Michigan, Part II, 1966–72 Great Lakes Fish Comm Tech Rep 30, Ann Arbor 35 pp

Medland, TE & FWH Beamish 1987 Age validation for the mountain brook lamprey, *Ichthyomyzon greeleyi* Can J Fish Aquat Sci 44 901–904

Mohanty Hejmadi P & M T Dimond 1986 Temperature dependent sex determination in the Olive Ridley turtle pp 159–162 *In* H C Salvkin (ed) Progress in Developmental Biology, Part A, Liss, New York

Morman, R H 1987 Relationship to density to growth and metamorphosis of caged larval sea lampreys, *Petromyzon marinus* Linnaeus, in Michigan streams J Fish Biol 30 173–181

Morreale, S J , G J Ruiz, J R Spotila & E A Standora, 1982 Temperature-dependent sex determination current practices threaten conservation of sea turtles Science 216 1245–1247

Murdoch, S P, M F Docker & F W H Beamish 1992 Effect of density and individual variation on growth in sea lamprey (*Petromyzon marinus*) larvae in the laboratory Can J Zool 70 184–188

Okkelberg, P 1921 The early history of the germ cells in the brook lamprey, *Entosphenus wilderi*, up to and including the period of sex differentiation J Morph 35 1–152

Piavis, W G 1971 Embryology pp **61–400.** *In* M W Hardisty & I C Potter (ed) The Biology of Lampreys, Volume 1, Academic Press, London

Pieau, C & M Dorizzi 1981 Determination of temperature sensitive stages for sexual differentiation of the gonads in embryos of the turtle, *Emys orbicularis* J Morphol 170 373–382

Potter, I C 1980 Ecology of larval and metamorphosing lampreys Can J Fish Aquat Sci 37 1641–1657

Potter, I C , F W H Beamish & B G H Johnson 1974 Sex ratios of adult sea lampreys, *Petromyzon marinus,* from a Lake Ontario tributary J Fish Res Board Can 31 122–124

Purvis, H A 1970 Growth, age at metamorphosis and sex ratio of Northern brook lampreys in a tributary of southern Lake Superior Copeia 1970 326–332

Purvis, H A 1979 Variation in growth, age at transformation, and sex ratio of sea lampreys re-established in chemically treated tributaries of the Upper Great Lakes Great Lakes Fish Comm Tech Rep 35, Ann Arbor 49 pp

Rohde, F C & R E Jenkins 1980 *Lampetra aepyptera* (Abbott), least brook lamprey p 21 *In* D S Lee, C R Gilbert, C H Hocutt, R E D E McAllister & J R Stauffer (ed) Atlas of North American Freshwater Fishes, North Carolina State Museum of Natural History, Raleigh

Rubin, D A 1985 Effects of pH on sex ratio in cichlids and a poecillid (Teleostei) Copeia 1985 233–235

Smith, B R 1971 Sea lampreys in the Great Lakes of North America pp 207–247 *In* M W Hardisty & I C Potter (ed) The Biology of Lampreys, Volume 1, Academic Press, London

Statgraphics 1986 Statistical graphics system, user's guide STSC, Inc , Rockville

Stier, K & B Kynard 1986 Abundance, size and sex ratio of adult sea-run sea lampreys *Petromyzon marinus* in the Connecticut River USA US Nat Mar Fish Serv Fish Bull 84 476–548

Sullivan, J A & R J Schultz 1986 Genetic and environmental basis of variable sex ratios in the laboratory strains of *Poeciliopsis lucida* Evolution 40 152–158

Trzebiatowski, R , J Filipiak & R Jakubowski 1981 Effect of stock density on growth and survival of rainbow trout (*Salmo gairdneri* Rich) Aquaculture 22 289–295

Vladykov, V D & E Kott 1980 Description and key to metamorphosed specimens and ammocoetes of Petromzonidae found in the Great Lakes Region Can J Fish Aquat Sci 37 1616–1625

Vogt, R C , J J Bull, C J McCoy & T W Houseal 1982 Incubation temperature influences sex determination in kinosternid turtles Copeia 1982 480–482

Wigley, R L 1959 Life history of the sea lamprey of Cayuga Lake, New York US Fish Wildl Serv , Fish Bull 59 559–617

Williams, T & N R Franks 1988 Population and growth rate, sex ratio and behaviour in the ant isopod, *Platyarthrus hoffmannseggi* J Zool , Lond 215 703–717

Young, R T & L J Cole 1900 On the nesting habits of the brook lamprey (*L wilderi*) Amer Nat 34 617–620

Zanandrea, G 1951 Rilievi e confronti biometrici e biologici sul *Petromyzon (Lampetra) planeri*, Bloch nelle acque della marca Trevigiana Bollettino Pesca Piscicoltura Idrobiologia 6 53–67

Zanandrea, G S J 1961 Studies of European lampreys Evolution 15 523–534

Environmental Biology of Fishes **41**: 206, 1994.

Fish imagery in art 68: Brown's *Eve with Fish and Snake*

Marilyn A. Moyle
612 Eisenhower St., Davis, CA 95616, U.S.A.

Joan Brown (1938–1990) was a professor of art at the University of California at Berkeley. A student of San Francisco Bay area artists Elmer Bischoff and David Park, at age 22 she was the youngest artist in the prestigious Whitney Annual (New York) of 1960. *Eve with Fish and Snake* is from a series created in 1970 called *Paradise Series # 1.* Brown's use of bright color and symbolic figurative imagery documents important moments in her life. In her paintings she celebrated birthdays and her passion for animals. Fish appear to be a personal symbol, reflecting in part her love of swimming and her attempts to swim from Alcatraz Island to San Francisco (she was successful in 1975). Her paintings may appear to be 'naive' to some viewers, but Brown was completely aware of what she was doing. During the 1980s Brown devoted much of her energy to public sculptures. She died tragically in 1990 while installing a colorful tiled obelisk at the Eternal Heritage Museum in Prasanthinalayam, India (Adams 1992).

Eve with Fish and Snake (1970, oil on masonite, 2.4 × 1.2 m) is used courtesy the Seattle Art Museum (photo by Gary T. Sutto).

Adams, B. 1992. Alternate lives. Art in America 80: 86–89.

Environmental Biology of Fishes **41**: 207–245, 1994.

Reproduction in the North Atlantic oceanic ichthyofauna and the relationship between fecundity and species' sizes

Nigel R. Merrett
Department of Zoology, The Natural History Museum, Cromwell Road, London SW7 5BD, U.K.

Received 22.4.1993 Accepted 10.1.1994

Key words: Fish reproductive style, Fish phylogeny, Vertical distribution, Pelagic, Demersal, Size spectra, Biomass spectra

Synopsis

The pelagic (589 spp.) and demersal (505 spp.) oceanic ichthyofaunas of the North Atlantic Basin have very different compositions at ordinal and family level. Yet the pattern of relationships between species' maximum size and maximum fecundity from data available (10% of the pelagic, 19% of the demersal species) was similar. A positive relationship between fecundity and weight was confirmed among most teleosts, but was not followed by the elasmobranchs represented. Species' reproductive styles are reviewed in ordinal groupings within a framework of the overall body size/fecundity distribution. Species size (maximum weight) spectra were synthesized for both pelagic and demersal assemblages to assess the allocation of potential reproductive effort throughout the North Atlantic oceanic ichthyofauna. The only available examples of species size spectra and biomass spectra from the pelagic and demersal ichthyofauna in this ocean basin imply geographic and bathymetric variation in overall reproductive effort among fishes whose fecundity is size dependent. Further implications concerning reproductive effort are discussed in the light of food availability.

Introduction

The first substantial review of reproduction among deep-sea fishes was published thirty years ago (Mead et al. 1964). Since then additions have been made at the species level (e.g. Kawaguchi & Marumo 1967, Badcock & Merrett 1976, Gjosaeter 1981, Fisher 1983, Nazarov 1983, Marshall 1984, Sulak et al. 1985, Wenner 1984, Crabtree et al. 1985, Oven 1985, Badcock 1986, Crabtree & Sulak 1986, Miya & Nemoto 1986a, b, 1987a, b, 1991, Mazhirina & Filin 1987, Pankhurst et al. 1987, Pankhurst 1988, Silverberg et al. 1987. Alekseyev et al. 1992) and family level (e.g. Nielsen et al. 1968, Nielsen 1969, Marshall 1973, Pietsch 1976, Stein 1980, Stein & Pearcy 1982, Kawaguchi & Mauchline 1987) and for broader groupings (e.g. Marshall 1971, 1979, Gordon 1979a, b, Clarke 1983, 1984, Golovan & Pakhorukov 1984).

Mead et al. (1964) reviewed reproduction among oceanic fishes along phylogenetic lines. Almost 100 years ago Woodward (1898) wrote of the antiquity of the deep-sea fish fauna, declaring that 'those out-of-the date forms of life which can no longer compete with the vigorous shore-dwelling races, are compelled to retreat to the freshwaters on the one hand, or to the deep-sea on the other'. Andriyashev (1953) expanded this view by recognizing groupings of ancient and secondary deep-water fishes. These works were general and only a few attempts have been made to approximate the total number of species and phylogenetic composition of the oceanic ichthyofauna overall (e.g. Cohen 1970, Parin 1984).

The aim of this study is to update the review of

Mead et al. (1964) and, through consideration of the ichthyofauna of a specific ocean basin, define patterns of reproductive styles and discuss underlying functionality. The basin considered is the North Atlantic, as part of an on-going study of the region and its ichthyofauna (Haedrich & Merrett 1988, 1992, Merrett et al. 1991a, b). Of necessity the results are preliminary, because detailed information on the reproduction of the majority of species represented still remains unknown. The most common parameters recorded are size [unfortunately, usually in terms of length; 'fish weights are almost never given' (Clarke 1984)] and fecundity. Since the size range of teleost eggs, at least pelagic ones, is not large in comparison with the size range of females (Ware 1975), fecundity can be used as a comparable index of reproductive effort.

The depth and associated gradients of the deep-sea (e.g. see Mann & Lazier 1991), especially the attenuation of sunlight, are major factors influencing oceanic life. A vast midwater realm, where food supply and predation pressure are relatively low, separates much thinner surface and benthic boundary layers where both these parameters are considerably higher. Depth influences the temperature structure strongly, resulting in the development of permanent and seasonal thermoclines. The structure of the seasonal thermocline has a substantial bearing upon the hatching time of embryos from pelagic eggs (Pauly & Pullin 1988), the evolution of which was the key to the efficient ichthyofaunal colonization of the deep-sea (Duarte & Alcaraz 1989). Balon (1984) pointed out that hatching time itself is of little survival value, but at higher temperature more rapid progress is made through vulnerable early larval stages.

Wootton (1992) used a theoretical approach to view the positions occupied by fish species in a multi-variate space whose dimensions were defined by life-history traits. He summarized a substantial array of life-history patterns. One of the several constraints on these that he considered was size. His remarks on body size, fecundity and egg size are all very relevant to the current work. He noted the substantial gap in egg volume between the minimum for oviparous elasmobranchs and the maximum for teleosts (by a factor of almost forty) and that each of these extremes was rare. Yet he observed that elasmobranchs were able to produce ova well within the teleost size range but that these forms were viviparous and so the propagule released was still large.

Expanding on the fact that the egg mass produced by a female teleost at any one time is constrained by body size, Wootton (1992) reviewed the advantages gained by producing a large egg at the cost of a reduction in fecundity. He concluded that larger propagules should have a lower mortality rate, but that there was little or no advantage to pelagic spawners in producing larger eggs. Duarte & Alcarez (1989) assessed the advantages of producing many small or few large eggs from among 51 species of mixed marine (mostly neritic) and freshwater fishes. They found no evidence of a phylogenetic trend towards greater egg sizes. In reworking the data used by Duarte & Alcaraz (1989), Elgar (1990) concluded, among other things, that the way in which reproductive output is partitioned between the size and number of offspring varies independently of body size. The general rule seems to be simply that pelagic spawning marine fish produce many small eggs (< 2 mm diameter, e.g. Duarte & Alcaraz 1989, Ware 1975) and demersal spawners produce fewer large eggs [but, in addition, Elgar (1990) pointed out that this trade-off can also be found across taxa]. Yet, if the efficient colonization of the oceanic environment did involve the evolution of pelagic eggs (Duarte & Alcaraz 1989), it follows that consideration must largely focus on the dependence on small eggs. Indeed, the data given by Crabtree & Sulak (1986) for deep demersal teleosts, at least, is consistent with this view. Hence the variation investigated is tacitly constrained within one of these two basic fecundity patterns.

Adams (1980) demonstrated that marine fish conform with the hypothetical r- and K- correlates in their life-history parameters. He showed that species on the r- end of the continuum inclined towards small body size, low maximum age, low age at first maturity, high natural mortality and high growth rate. Conversely, species towards the K- end of the spectrum were opposite in these respects. Such conformity, with its dependence on food availability, links these correlates through body size to fecundity. (Because of a commensurate mouth size, the smaller the teleost the lower the trophic level it exploits. In a food-poor environment like the deep-sea this is an important consideration.) A measure of potential reproductive investment per species is therefore obtainable from species' maximum adult size and maximum fecundity. The gross reproductive investment can then be assessed along phylogenetic lines in relation to vertical/bathymetric distribution and therefore to broad resource categories.

The r- and K- concept in fish life-history styles has been considerably refined by Balon (1988, 1989, 1990). He postulates homeorhetic states of altricial (small eggs, little yolk, smaller less developed young) and precocial (large amount of dense yolk, larger more developed young) life-history styles, which are the cause of evolutionary patterns and relate to a complex classification of reproductive guilds (e.g. Balon 1990). Such direct and indirect developmental alternatives were discussed in detail by Flegler-Balon (1989), who elaborated on specific reproductive parameters for a wide variety of such guilds. She pointed out that indirect development was typified by altricial species, which was especially common among pelagic marine species. Increasing parental care from such egg scatterers to internal bearers resulted in yolkier and less numerous eggs and culminated in direct development. Flegler-Balon (1989) gave a zoarcid and a dasyatid ray as examples of different precocial styles relevant here. While the current investigation concerns largely egg scattering altricial species (see above), general detailed categorization is precluded here by lack of information for most species. Known noteworthy styles, however, are categorized where appropriate.

My investigation was structured from a data base of species' maximum size (weight) for the known pelagic and demersal fishes of the North Atlantic Basin. This provided the phylogenetic composition of the pelagic and demersal ichthyofauna at ordinal and familial rank, together with the means of synthesising the overall species size spectra of the two ichthyofaunal elements. Where available, the maximum fecundity per species was incorporated into the data base for a comparison with species' size classes among the major orders and as a framework for consideration of the various arrangements for maximizing fecundity (e.g. sexual dimorphism, synchronous hermaphroditism, viviparity, 'batch' spawning, seasonal synchrony, egg rafts, parental care, etc.). Finally, species' size and biomass spectra from known populations of pelagic and demersal fishes were used to assess potential basinwide patterns of reproduction. No simple picture of factors explaining the evolution of reproductive arrangements among oceanic fishes emerged. Instead a web of possible correlations is indicated to form a framework for future research.

Materials and methods

For the purposes of this paper the area considered as the North Atlantic Basin extends from the Arctic Basin southward to the equator. It includes the Caribbean and Gulf of Mexico but excludes the Mediterranean Sea. The northerly limit of this region abutts the rim of the Arctic Basin, with its distinct faunistic composition (see Andriyashev 1953 for review). This biogeographic boundary runs from the Hebridean slope edge, along the Wyville Thomson Ridge and Faroe-Iceland Ridge to the Icelandic slopes and across the Denmark Strait to the Greenland slope. The oceanic realm has been taken as that region seaward of the 200 m isobath which broadly marks the shelf-slope break. Thus the pelagic species considered include only those whose geographic ranges extend far beyond the limits of the continental shelf waters. The demersal species taken are those whose depth (sounding) ranges commence at, or extend substantially beyond, the 200 m isobath.

The term demersal is used here to combine both the benthic and benthopelagic categories of deep bottom-living fishes (sensu Marshall & Merrett 1977). The lack of buoyant means in benthic fishes sets them apart with a very different lifestyle from those that are benthopelagic. Nevertheless this group [listed by Marshall & Merrett (1977) as being most: chlorophthalmids, ipnopids, synodontids (Aulopiforms), ogcocephalids (Lophiiformes), scorpaenids, cottids (Scorpaeniformes), zoarcids (Perciformes)] form a small proportion of the demersal fauna overall and are widely distributed through all depth zones. Thus they have not been treated as a separate category, but attention is drawn to them where appropriate.

The inventory of pelagic and demersal oceanic species was culled from a variety of sources, from the literature [notably FNAM (Whitehead et al. 1984, 1986a, b); CLOFETA (Quero et al. 1990a, b); FWNA (Fishes of the western North Atlantic, various editors); Haedrich & Merrett (1988)], unpublished data from Woods Hole Oceanographic Institution (WHOI) midwater trawl samples, the Institute of Oceanographic Sciences Deacon Laboratory's (IOSDL) and my own demersal trawl samples. The inventory was arranged phylogenetically following Nelson (1984).

Distributional differences within both the pelagic and demersal ichthyofaunas were examined by partitioning the data by depth zone. The pelagic fauna was divided into two strata (0–399 m and > 400 m), based upon presumed natural distributional limits of surface-living (epipelagic) fishes and the meso- and bathypelagic forms. The usual concept of the epipelagic layer as a 200 m thick euphotic zone has been broadened here to achieve a more distinct separation between forms associated with the near-surface waters (e.g. tuna) and the truly deep-sea fauna. Subdivision of the lower pelagic zone, although desirable, was impractical owing to the small proportion of species sampled with mouth opening/closing nets. The demersal fauna, on the other hand, could be considered in four strata (200–399 m, 400–1999 m, 2000–3999 m, 4000 + m), conforming with a generalized concept of outer shelf, continental slope, continental rise and abyss. Since vertical ontogenetic descent from epipelagic larvae is commonplace among deep-sea fishes, species were allocated to zones according to their known limits of adolescent-adult depth distribution [daytime distribution in the case of pelagic species]. Species allocation was to the shallowest zone of known abundance, in view of the constraints to distributional knowledge given below. The data were not subdivided geographically.

The great variation in body shape of the species considered dictated that, for comparative purposes with fecundity, length was an unrealistic measure of size. Fish weight was considered more appropriate for such comparison, but is rarely available in the literature. Some data could be obtained in this way, however, and these were augmented by unpublished Institut für Seefischerei, Hamburg (ISH), International Center for Living Aquatic Resources Management, Manilla (ICLARM), Scottish Association for Marine Science (SAMS), Woods Hole Oceanographic Institution (WHOI) and personal data. Maximum sizes attained in the North Atlantic Ocean basin were taken for comparison with maximum known fecundity, to offset bias in sampling techniques and as a measure of potential for each species. In many cases maximum weights were unobtainable and so had to be estimated from length in comparison with species of analogous size and shape. Sizes were grouped into $\log_2$ classes.

Instantaneous fecundity of a species was taken to be the total number of the most advanced cohort of eggs present in an ovary pair. This overlooks the functional aspects of determinate or indeterminate fecundity, which have not been assessed among deep-sea species. Maximum fecundities were taken, both from the literature and from my unpublished data. Only data originating in the North Atlantic were used, to offset any inter-ocean variation in widely distributed species. Fecundity was reported by number classed according to $\log_2$.

A wide variety of limitations are implicit in this study, to caution care in the interpretation of results. Among the main sources of inaccuracy is the likely incompleteness of the inventories of pelagic and demersal species. This is compounded by the 'boundary effect' for species inclusion, at the shelf-slope break and other zonal boundaries chosen. Allocating a species to a single zone on the basis of its overall limit of depth distribution is a somewhat crude approach, since within this range there usually exists a principal distribution layer of maximum density. This is known for such a small proportion of either assemblage, however, that the upper limit of the adolescent-adult distributional range was used here for consistency. It should be pointed out that fish density in the deeper of the two pelagic ichthyofaunal zones considered here is considerably attenuated below about 3000 m. It remains very sparse until some tens of metres above the bottom (IOSDL unpublished data).

Geographic variations in occurrence and depth

range could not be taken into account for most species and so were disregarded. Similarly, intra-specific differences occur in fecundity in some species on an ocean basin scale (Badcock & Merrett 1976, Clarke 1984, Kawaguchi & Mauchline 1987, McKelvie 1989) but, for the majority of species there are no comparative data and so such variation also had to be ignored. The necessary estimation of weight data for so many species is a source of inaccuracy in itself, but this is ameliorated by grouping the data into $\log_2$ size classes. The maximum size of many species is also open to doubt. In mid-water, especially, sampling has largely been by small nets (with the notable exception of the collections made by the 'Walther Herwig', e.g. see Krefft 1974), where net avoidance by larger specimens must be expected (e.g. Clarke 1973, Clarke & Wagner 1976). Furthermore, according to Childress et al. (1980) mature individuals are so short-lived once they reach maximum size that they comprise only a small fraction of the population at any given time. No attempt has been made to consider lifetime fecundity among those species for which estimates are available. While some of these species are known to be semelparous (spawning once per lifetime), the assumption of iteroparity (repeat spawnings per lifetime) in all the remainder may be unjustified. Fecundity counts give no indication of whether the estimate represents 'all-at-once' (single cohort of eggs spawned together) or 'batch' (several cohorts of eggs spawned sequentially) spawning styles. However, wherever possible these aspects are discussed in the accounts of the orders.

Table 1 Vertical distribution of oceanic fishes by family and species in the North Atlantic Basin, a – subdivided by major zones with the proportion of families restricted to each indicated, b – subdivisions of the demersal zone

		Total number of families	Total number of species	Proportion restricted to zone
(a) Pelagic	0–399 m	28	80	89%
	> 400m	66	509	79%
Demersal	total	72	505	78%
(b)	200–399 m	30	74	
Demersal –	400–1999 m	58	347	
by strata	2000–3999 m	14	64	
	> 4000 m	7	20	

Results and discussion

Using the criteria outlined above, the oceanic ichthyofauna of the North Atlantic basin currently stands at 1094 species, representing 143 families and 25 orders. There are rather more pelagic species (589) than demersal (505), distributed among 93 pelagic and 72 demersal families. A remarkably low level of congruity exists at family level among the upper pelagic, lower pelagic (mesopelagic + bathypelagic) and overall demersal zones (Table 1). In all zones more than three quarters of the families are restricted to a single zone, with only the Gempylidae represented in all. Members of this family are often pseudoceanic (e.g. see Merrett 1986) and therefore concentrated around the oceanic rim and oceanic islands (Nakamura & Parin 1993), where zonal intermixing is most likely. Three families occur in both the upper pelagic and the demersal zones, while 13 are common to the lower pelagic and demersal.

Distribution of species within orders

Overall, the demersal assemblage contains more orders (22) than does the pelagic assemblage (18). Moreover, species representation in the upper pelagic zone (0–399 m) contrasts strongly with that of the lower pelagic zone (> 400 m) (Fig. 1a). Dominant in the upper zone are perciform [order no. 22 (see Appendices 1, 2); 43% zonal frequency], lamniform (no. 4; 19%), cyprinodontiform (no. 16; 16%), and lampriform (no. 17; 10%) species. Rajiform (no. 5), lamniform, cyprinodontiform and tetraodontiform (no. 24) species are unreported from the lower pelagic zone.

Again, there is a striking contrast in species representation by order between the lower pelagic and demersal assemblages which, taken together, constitute the true deep-sea fauna (Fig. 1b). Among the lower pelagic assemblage, the dominant orders are the Stomiiformes (no 10; 29%), Myctophiformes

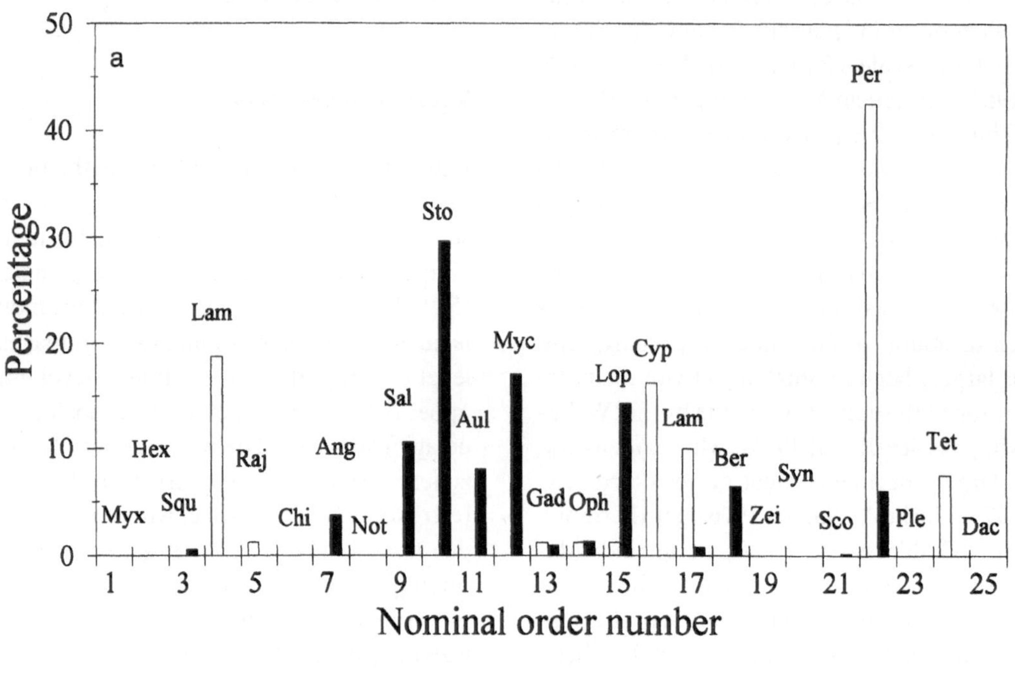
a
Percentage
Nominal order number
Myx
Hex
Squ
Lam
Raj
Chi
Ang
Not
Sal
Sto
Aul
Myc
Gad
Oph
Lop
Cyp
Lam
Ber
Zei
Syn
Sco
Per
Ple
Tet
Dac
0-399 m, n = 80 spp.
400+ m, n = 509 spp.

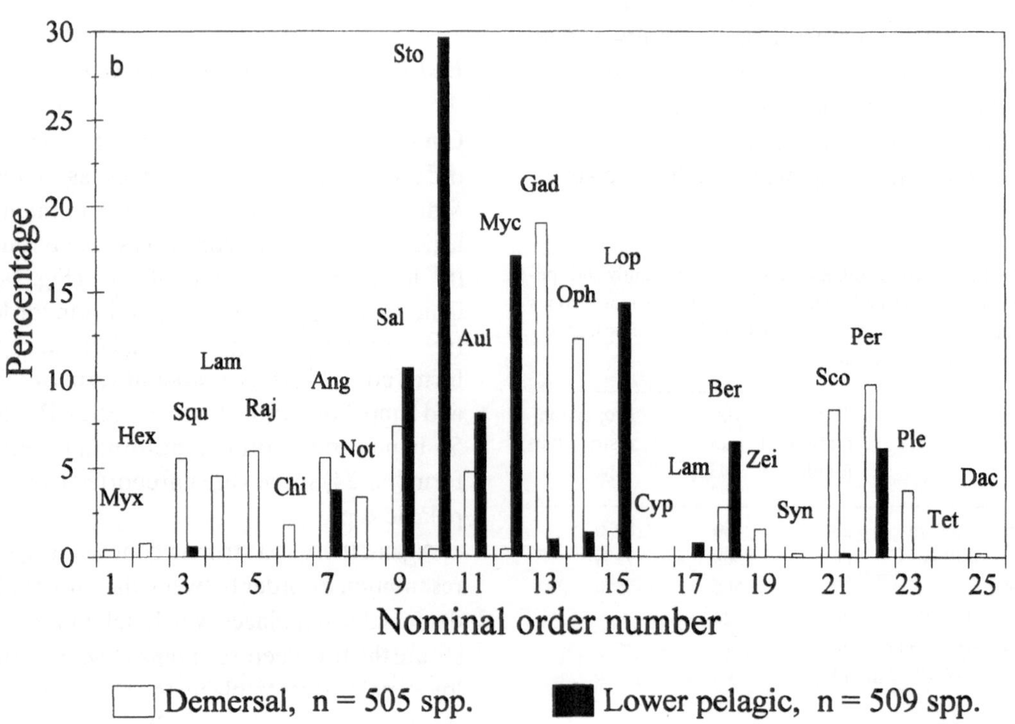
b
Percentage
Nominal order number
Myx
Hex
Squ
Lam
Raj
Chi
Ang
Not
Sal
Sto
Aul
Myc
Gad
Oph
Lop
Cyp
Lam
Ber
Zei
Syn
Sco
Per
Ple
Tet
Dac
Demersal, n = 505 spp.
Lower pelagic, n = 509 spp.

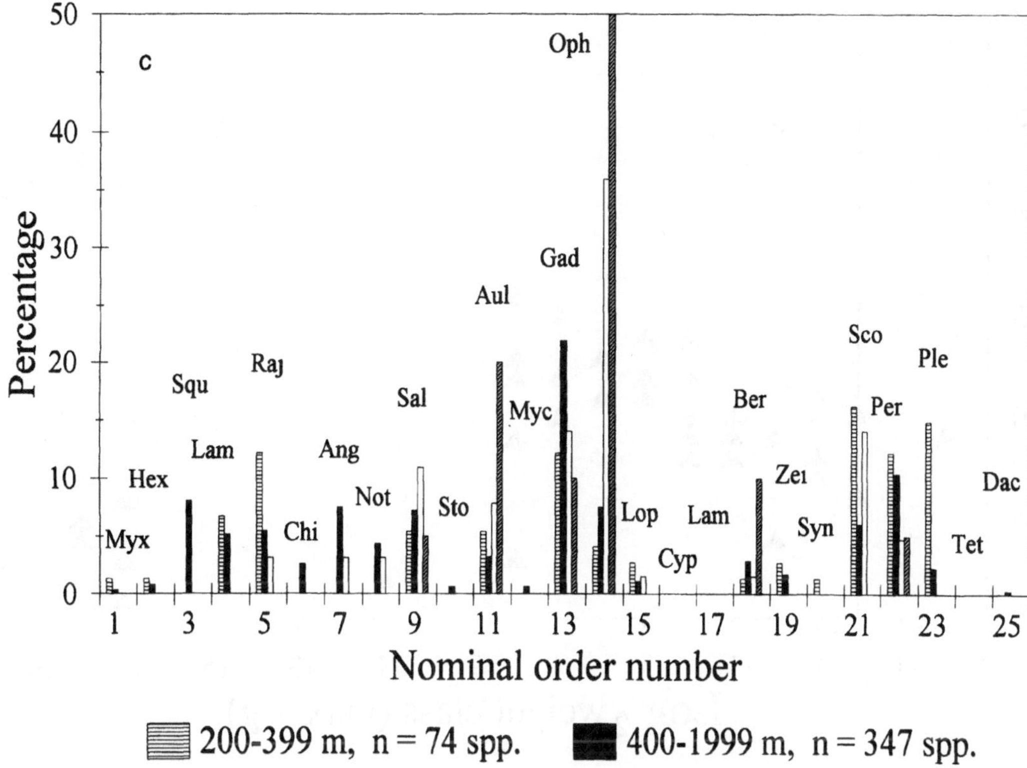

←

Fig. 1. Relative distribution of species per order among the oceanic fishes of the North Atlantic Basin. a – Pelagic ichthyofauna (589 spp. from two depth zones), b – Comparison of the deep pelagic ichthyofauna (509 spp.) with the demersal (505 spp.), c – depth stratified representation of the demersal ichthyofauna. (Nominal order numbers and labels refer to order listing in Appendix 1.)

(no. 12; 17%) and Lophiiformes (no. 15; 14%). Few orders, besides the Anguilliformes (no. 7; 4%), Salmoniformes (no. 9; 11%), Aulopiformes (no. 11; 9%), Beryciformes (no. 18; 6%) and Perciformes (no. 22; 6%) approach similar species frequency in both assemblages. (Note, however, that the perciform species in the lower pelagic comprise only 6%, in comparison to 43% in the upper pelagic assemblage.) Dominant among the demersal assemblage are gadiform (no. 13; 19%), ophidiiform (no. 14; 12%) and scorpaeniform (benthic – no. 21; 8%) species, as well as the perciform (9%) and salmoniform (7%) species (Fig. 1b). A far greater diversity among the orders represented is evident from the demersal assemblage when compared with the lower pelagic assemblage (22 vs. 13). Thus, apart from three squaliform (no. 3) species, the Agnatha (no. 1) and Chondrichthyes (no. 2, 4–6) are unknown from the lower pelagic, as are the Notacanthiformes (no. 8), Zeiformes (no. 19), Syngnathiformes (no. 20), Pleuronectiformes (benthic – no. 23) and the Dactylopteriformes (benthic – no. 25).

Bathymetric sub-division of the demersal assemblage indicates differences in ordinal composition with increased depth (Fig. 1c). The frequency of ordinal representation increases from 15 orders in the upper slope zone (200–399 m – a quasi-shelf zone) to 21 orders in the mid-slope zone (400–1999 m). Representation is then much reduced in the rise (2000–3999 m; 11 orders) and abyssal zones (4000 + m; 6 orders). Some of this reduction arises from the evident absence of the Agnatha (no. 1) and chondrichthyans [no. 2–6, with the exception of a small representation of Rajiformes (benthic – no. 5) on the rise] at the two lower levels. Furthermore no demersal Anguilliformes (no. 7) nor Notacanthiformes (no. 8) are found predominantly at abyssal depths. Species' representation within orders is also variable among zones. The percentages of the benthic orders Rajiformes (no. 5; 12.5%), Scorpaeni-

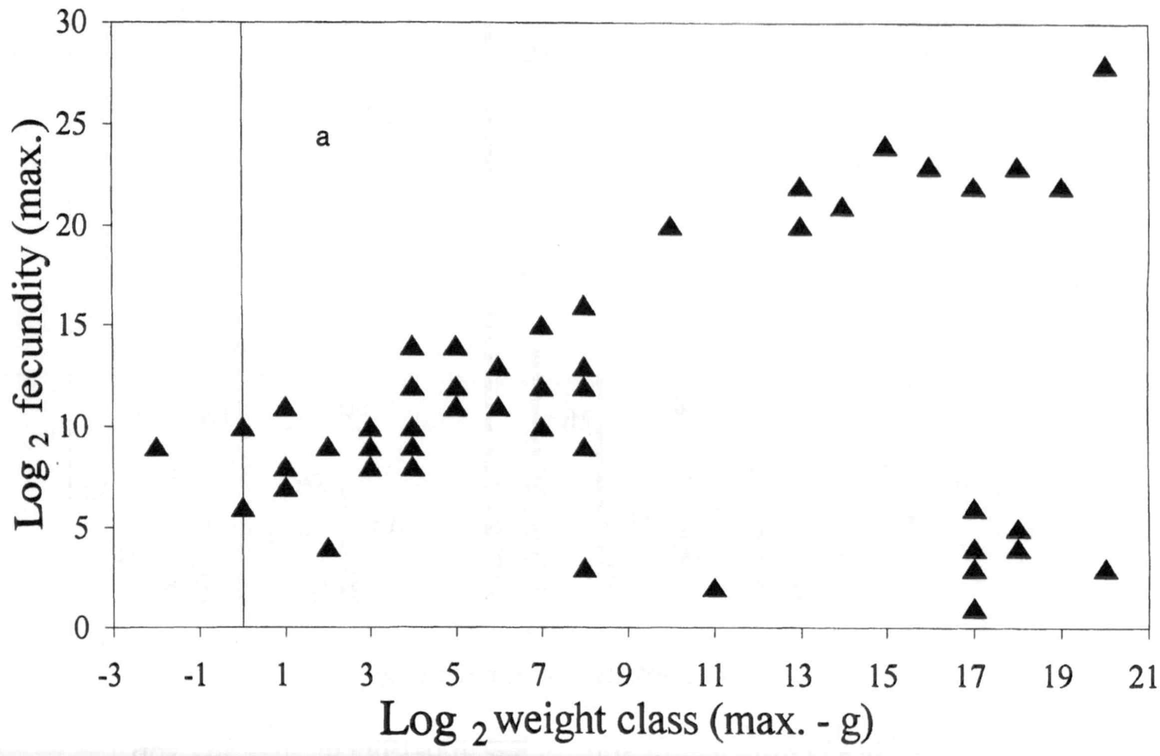

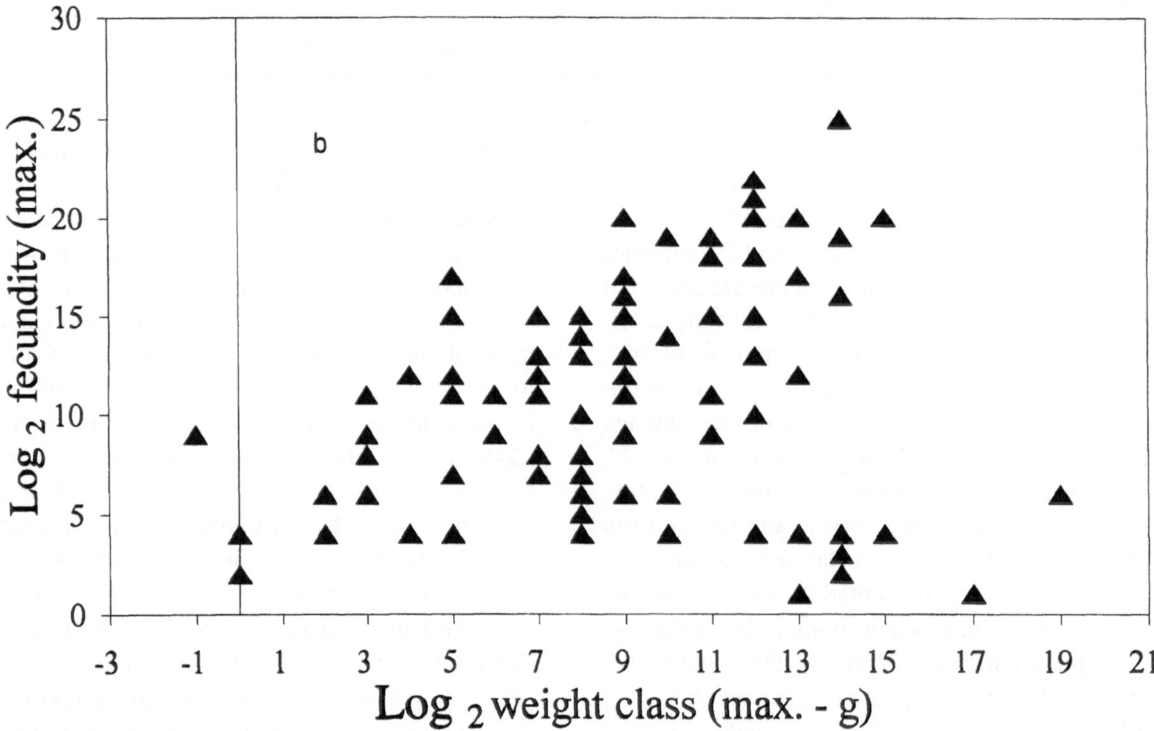

Fig. 2. Relationship between species' maximum fecundity and maximum size among the oceanic fishes of the North Atlantic Basin (log_2 groupings) for a – the pelagic total (n = 60) and b – demersal total (n = 95) ichthyofauna.

formes (no. 21; 16.7%) and Pleuronectiformes (no. 23; 15.2%) are higher on the upper slope than elsewhere. The Squaliformes (no. 3; 8.1%) are concentrated in the mid-slope zone, where the Anguilliformes (no. 7; 7.6%) and the Gadiformes (no. 13; 22.1%) peak also. The Salmoniformes (no. 9; 10.9%) reach their highest proportion in the rise zone, where the Ophidiiformes (no. 14; 35.9%) and the Scorpaeniformes (no. 21; 14.1% – all family Liparididae) are also important. While the numbers of species centred in the abyssal zone are low (20), the Ophidiiformes (no. 14; 50%), Aulopiformes (benthic – no. 11; 20%) and Beryciformes (no. 18; 10%) all assume their highest proportions at this level.

By excluding the predominantly secondary deep-water fauna of the upper pelagic zone and the Arctic Basin from consideration, it appears that Andriyashev's (1953) primary ancient deep-water fauna of the oceanic North Atlantic Basin can therefore be further sub-divided on the basis of the above evidence. Sufficient differences are apparent at the ordinal level between the demersal and lower pelagic assemblages, and also among the bathymetrically sub-divided demersal assemblage, of the overall ichthyofauna to suggest that specialization and divergence probably occurred early on in the colonization of the deep-sea by fishes (Fig. 1b, c).

Body size and its relationship to fecundity among the major orders represented in the North Atlantic Basin

Fecundity data are available for some 10% and 19%, respectively, of the pelagic and demersal ichthyofauna currently listed for the North Atlantic Basin. Despite this low proportion, apparently similar trends are evident from the total representation of both assemblages. As expected, the majority of species (pelagic spawners) display a positive relation between (maximum) fecundity and (maximum) body mass (Fig. 2a, b). This observation agrees with those from other assemblages (e.g. neritic and freshwater, Duarte & Alcaraz 1989). The regressions of $\log_2$ fecundity against $\log_2$ weight derived from various orders comprise the general relationship (Anguilliformes/Notacanthiformes, Salmoniformes/Aulopiformes, Stomiiformes, Myctophiformes, Gadiformes – Fig. 8, 9, 4, 5, 10), however, all differ significantly from one another at least to the 0.05 level. This suggests a phylogenetic element to the overall scatter plots (Fig. 2), but subject to the constraints of the fecundity data given above. In Fig. 2a, b another group can be identified whose fecundity is evidently independent of body mass. These comprise the live-bearing chondrichthyans and the large egg/low fecundity, benthic spawning teleosts.

The aim here is to consider the arrangemert of data from the major orders and/or families within the overall fecundity/body size distribution. Thus any emergent pattern can be related to what is known about species' reproductive styles and relative abundance on a basin-wide scale.

Pelagic representatives

Chondrichthyes

Squaliformes. – Pelagic squaloids are represented by relatively few species in the North Atlantic and these occur largely at mesopelagic levels (weight range 2^8 to 2^{11} g; Appendix 1b). At the bottom of the range is *Squaliolus laticaudus* (ref. no. 14 – Fig. 3), possibly the smallest known shark (Compagno 1984). All pelagic squaloids are smaller than their demersal counterparts. They are all ovoviviparous (sensu Heemstra & Greenwood 1992; specifically the 'yolksac viviparity' reproductive style of Compagno 1990) with low fecundity in the range 2^2 to 2^3 young (Fig. 3). Compagno (1984a, b, 1990) gave their reproductive details, so far as these are known, and discussed this style within the reproductive phylogenetic progression.

Lamniformes. – This group of pelagic sharks comprises large, top predators. Their representation in the oceanic realm of the North Atlantic is within the weight range 2^{11} to 2^{25} g (Appendix 1a). Despite their size, they have low fecundity (2^1 to 2^6 eggs; Fig. 3) and depend on large offspring size attained through ovoviviparity to counter juvenile mortality.

The order Lamniformes of Nelson (1984) encompasses several orders in the classification of Com-

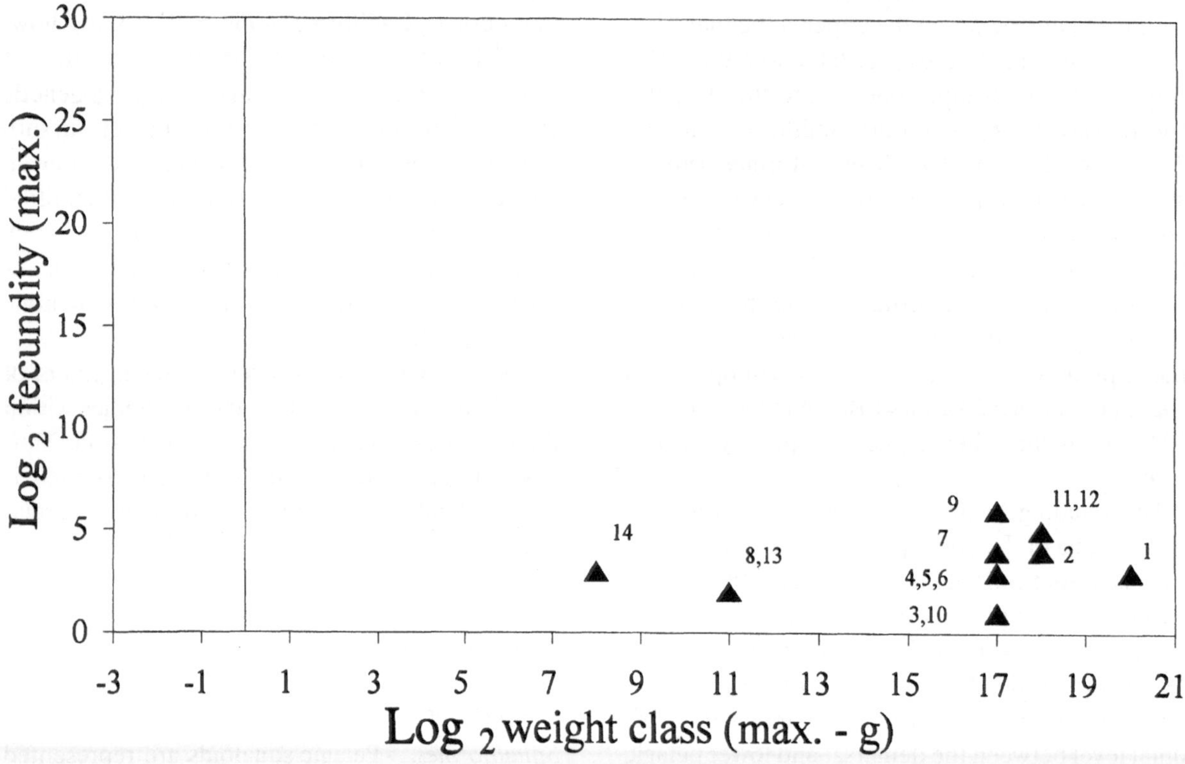

Fig. 3. Relationship between species' maximum fecundity and maximum size among the pelagic oceanic sharks (n = 14; orders Squaliformes and Lamniformes) of the North Atlantic Basin (log_2) groupings). (Species are identified by number from Appendix 3a).

pagno (1984a, 1990), with those species found in the North Atlantic Basin being grouped into the orders Orectolobiformes, Lamniformes and Carcharhiniformes. It is instructive to consider detailed reproductive arrangements according to Compagno's (1990) scheme. Thus the single orectolobiform species represented, the world's largest living fish, the whale shark *Rhincodon typus*, may be oviparous (Compagno 1984). The lamniforms and carcharhiniforms, however, are viviparous. Development in all lamniforms is by what Compagno (1990) termed 'cannibal viviparity' and in the carcharhiniforms represented by 'placental viviparity'. Compagno (1990) considered both styles phylogenetically derived and discussed these specializations in detail.

Osteichthyes

Stomiiformes. – This is the most speciose order among the pelagic fishes of the North Atlantic (Fig. 1a), with peak representation in the weight range 2^4 to 2^8 g (Fig. 4, Appendix 1b). Fecundity estimates are available for only a few species representing the smaller end of the size class range. The known reproductive arrangements in this group are complex. Several species are 'batch' spawners [e.g. *Valenciennellus tripunctulatus* (ref. no. 21 – Fig. 4) – Badcock & Merrett 1976, Howell & Krueger 1987], some are semelparous [*Cyclothone braueri* (ref. no. 16 – Fig. 4) – Badcock & Merrett 1976, *C. alba* (ref. no. 15 – Fig. 4) – Miya & Nemoto 1986b], while others are wholly or partially protandrous hermaphrodites [*C. microdon* (ref. no. 17 – Fig. 4) – Badcock & Merrett 1976, *Gonostoma bathyphilum* – Badcock 1986, *G. elongatum* – Fisher 1983]. Clarke (1983) reported that most central Pacific Stomiiformes were sexually dimorphic in size or abundance. Sexual dimorphism in size is evident among some North Atlantic species [e.g. *Argyropelecus hemigymnus* (ref. no. 24 – Fig. 4), *Argyropelecus aculeatus, Valenciennellus tripunctulatus* – Howell & Krueger 1987, *Cyclothone braueri, C. pseudopallida* (ref. no. 19, Fig. 4), *C. pallida* (ref. no. 18 – Fig. 4) – Badcock & Merrett

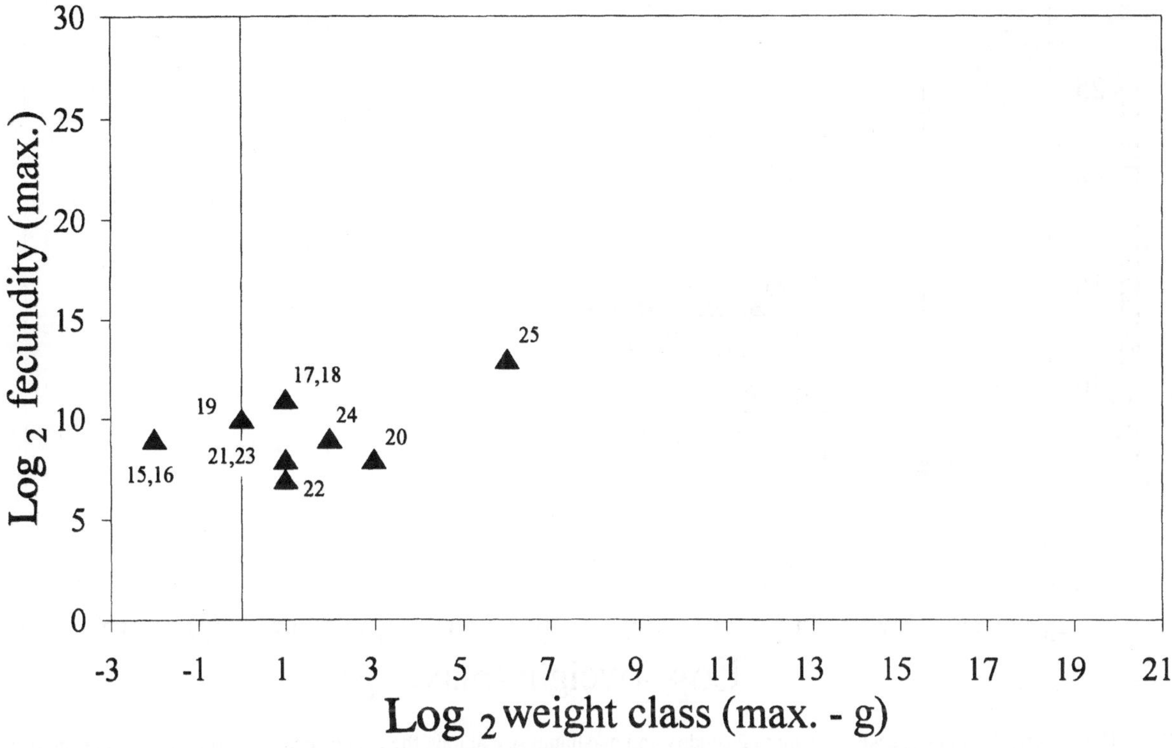

Fig 4 Relationship between species' maximum fecundity and maximum size among the pelagic oceanic representatives of the order Stomiiformes (n = 11) of the North Atlantic Basin (log_2 groupings – fecundity = 8 99 + 0 34 weight, r^2 = 0 19) (Species are identified by number from Appendix 3a)

1976, *Idiacanthus fasciola* (ref. no. 25 – Fig. 4) – Beebe 1934, *Stomias* spp. Gibbs 1969]. Sexual differences in depth distribution and abundance were noted among North Atlantic *Cyclothone* by Badcock & Merrett (1976). Females in the more advanced stages of maturity were caught deeper than males and less advanced females in *C. braueri*, *C. microdon* and *C. pallida*. In *C. braueri*, the overall sex ratio was skewed in favour of females in depths where the highest proportion of running ripe females occurred. Yet a higher proportion of males were ripe (identified by advanced nasal rosette development) in this part of the distributional range.

Marshall (1984) concluded that semelparity in mesopelagic *Cyclothone* was an adaptive feature accompanying paedomorphism. He contrasted their non-migrant life-style with their rapidly growing competitors, the lanternfishes. He concluded that without such a tendency, a two-year life cycle of *Cyclothone* might be too costly in terms of mortality to sustain its life history. Furthermore, Marshall pointed out that the bathypelagic, protandric *C. microdon* is substantially more fecund than the mesopelagic, semelparous *C. braueri*. He related this to both size and relative abundance, a feature that will be returned to later. Miya & Nemoto (1991) expanded on this theme and noted that protandry was a means of maintaining high fecundity under food-poor conditions.

Myctophiformes. – Most lanternfishes are nocturnal vertical migrants and together have by far the greatest biomass of any midwater fish group (Marshall 1984). Yet their individual adult size is generally small. The North Atlantic representatives are largely embraced by the weight range 2^3 to 2^7 g and fecundity range 2^6 to 2^{15} eggs (Fig. 5, Appendix 1b). Most species are short-lived. Karnella (1987) commented on the longevity of 28 of the 63 species of myctophids sampled in Ocean Acre, off Bermuda.

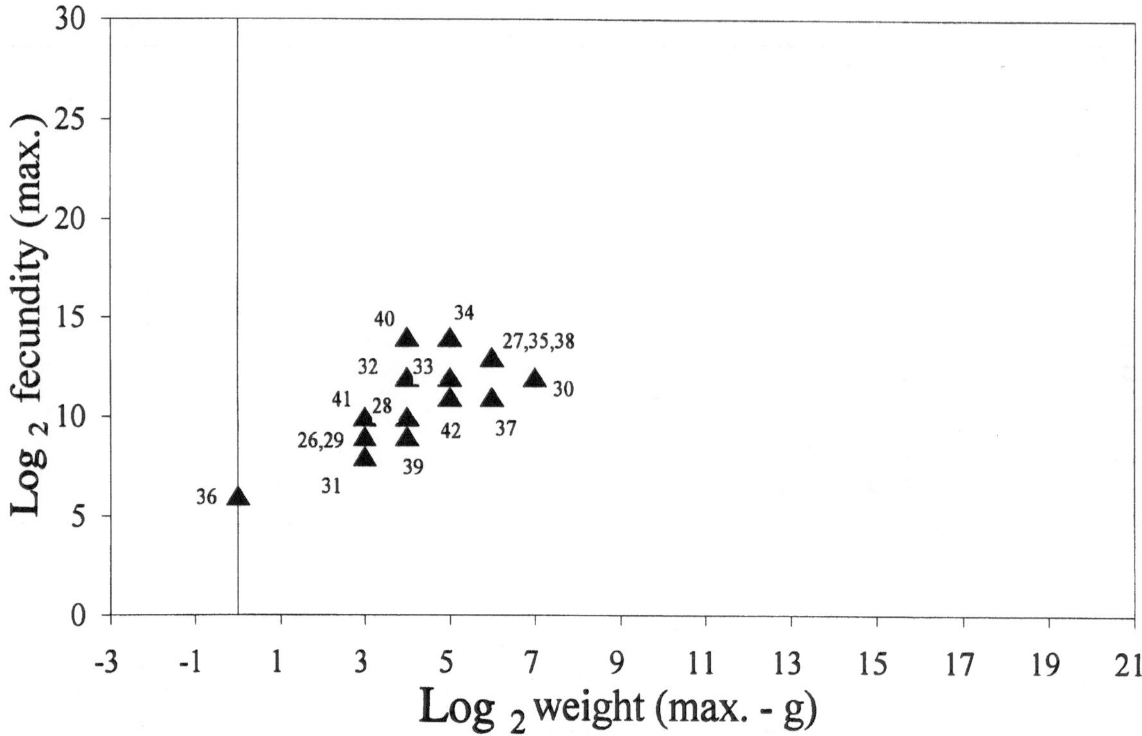

Fig 5 Relationship between species' maximum fecundity and maximum size among the pelagic oceanic representatives of the order Myctophiformes (n = 17) of the North Atlantic Basin (log_2 groupings – fecundity = 6 45 + 1 03 weight, r^2 = 0 60) (Species are identified by number from Appendix 3a)

Of these 54% (15) were annual species, 32% (9) had a 2-year life span and only 14% (4) lived longer than 2 years. As in the case of some stomiiforms, some Pacific myctophid females are either larger or more abundant than males, so that the biomass of mature females in the populations is greater than that of mature males (Clarke 1983). In Atlantic collections from off Bermuda, Karnella (1987) found skewed sex ratios among four of the 13 most abundant species, all in favour of males. Yet, in some 7 of these species, females were found to grow to a larger size than males, an observation in agreement with the Pacific data. In addition, sexual dimorphism is widely expressed in additional luminous tissue and organs borne by male lanternfish (see Nafpaktitis et al. 1977, Kawaguchi & Mauchline 1982). It is noteworthy, however, that while such caudal sexual dimorphism is the norm among the subfamily Myctophinae, it is rarely found in the subfamily Lampanyctinae (Paxton 1972). There is no correlation with either size or fecundity between these two subfamilies in the North Atlantic data.

Miscellaneous species. – Included in the plot of miscellaneous species are the scombroid fishes, an important group of 'higher' teleosts. These are large (weight range 2^{13} to 2^{19} g, Fig. 6), highly fecund (range 2^{21} to 2^{24} eggs), epipelagic predators. The remaining species represent widely differing categories. At one extreme is one of two pelagic teleosts known to be live-bearers [Parabrotulidae: *Parabrotula plagiophthalma* (ref. no. 51 – Fig. 6), 2^2 g, 2^4 eggs resulting in the birth of advanced young and categorizing the species as a 'histotrophic livebearer' in the scheme of Balon (1990) from evidence of trophotaeniae observed by Turner (1936), discussed in Wourms & Cohen (1975); sexually dimorphic in size, with the smaller males producing spermatophores as a possible adaptation to low density in a pelagic environment (Nielsen 1968, Nielsen et al. 1990)]. At the other is the gigantic and highly fe-

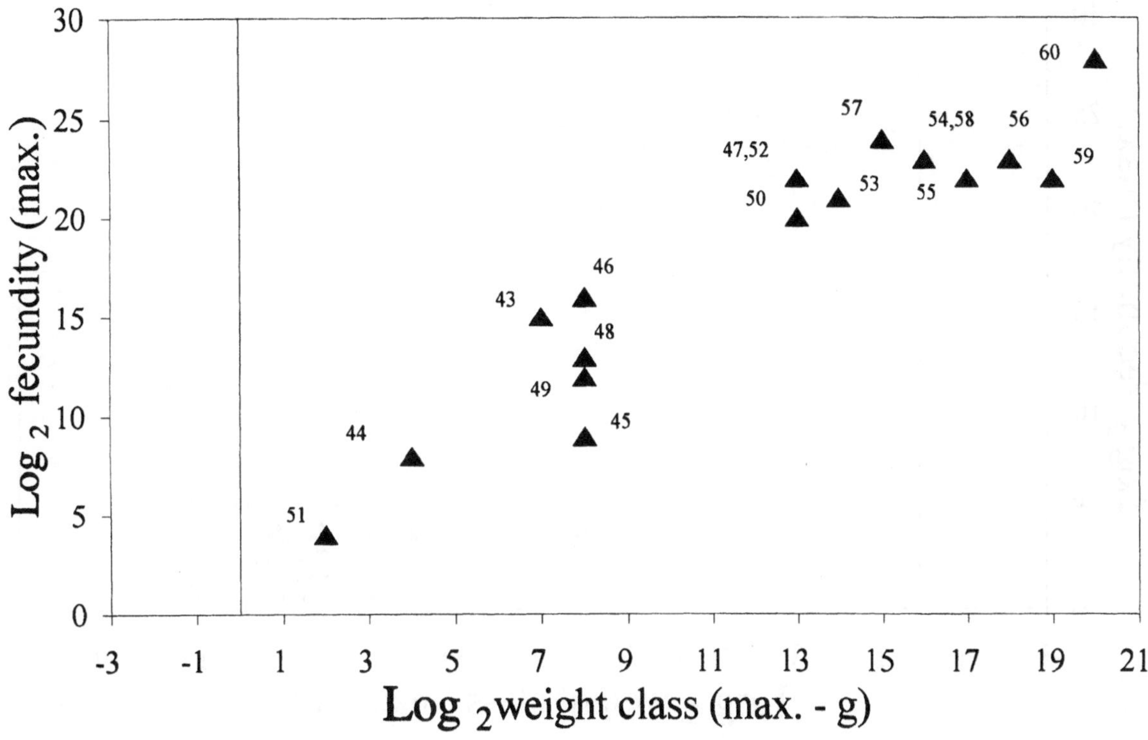

Fig 6 Relationship between species' maximum fecundity and maximum size among miscellaneous pelagic oceanic representatives of the orders Anguilliformes, Salmoniformes, Gadiformes, Lophiiformes, Cyprinodontiformes and Perciformes (n = 18) of the North Atlantic Basin (log_2 groupings) (Species are identified by number from Appendix 3a)

cund ocean sunfish [Molidae: *Mola mola* (ref. no. 60 – Fig. 6), weight 2^{20} g, fecundity 2^{28} eggs]. Intermediate in fecundity are *Monognathus taaningi* (ref. no. 44 – Fig. 6) and the gulper eel, *Eurypharynx pelecanoides* (ref. no. 43 – Fig. 6), which, together with the nemichthyids and saccopharyngids, are semelparous and display considerable sexually dimorphic changes with increasing ripeness (Nielsen & Smith 1978, Gartner 1983, Nielsen & Bertelsen 1985, Bertelsen & Nielsen 1987, Nielsen et al. 1989). While fecundity is known for only two of the 100 or so species of ceratioid anglerfishes [*Ceratias holboelli* (ref. no. 47 – Fig. 6) and *Haplophryne mollis* (ref. no. 48 – Fig. 6)], the reproductive adaptations of this suborder are striking. Pietsch (1976) reviewed ceratioid reproductive styles and stated, 'they (ceratioids) are most strikingly characterized by having an extreme sexual dimorphism in which males are dwarfed and, in some species, become parasitically attached to the body of a relatively gigantic female'. Egg protection may be a further aspect of ceratioid style, based on evidence that the egg clutch is embedded in a mucous sheet in at least one species (*Linophryne arborifera* – Bertelsen 1980). Such an 'egg veil' is also produced in the shelf and upper slope genus *Lophius,* indicating that this is likely to be a phylogenetic, rather than an environmental, adaptation. In the classification of reproductive guilds devised by Balon (1990), such a mode could perhaps be aligned with 'froth nesters' although evidence of guarding is unknown.

Demersal representatives

Chondrichthyes. – Knowledge of this group is restricted to the demersal sharks representing the orders Hexanchiformes, Squaliformes and Lamniformes. They are comprised of species with maximum sizes within the weight range 2^7 to 2^{20} g (Appendix 2a, b). In common with the pelagic sharks, their demersal counterparts included here are all

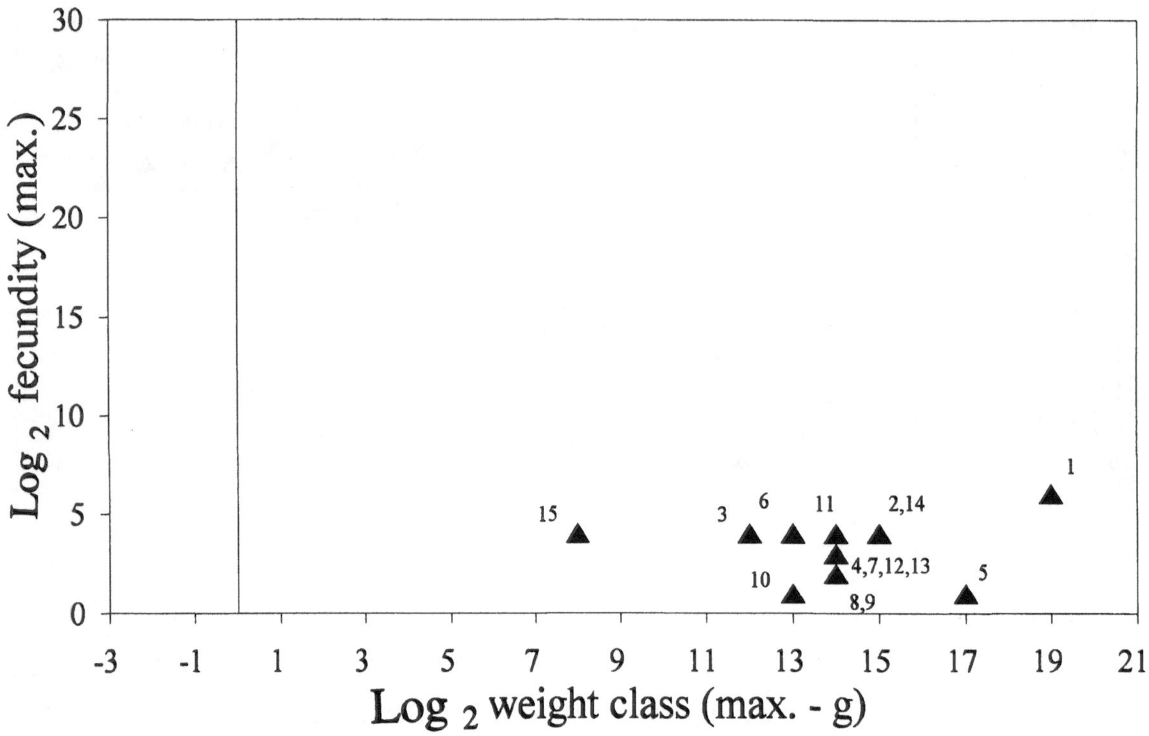

Fig. 7. Relationship between species' maximum fecundity and maximum size among the demersal oceanic sharks (n = 15; orders Hexanchiformes, Lamniformes and Squaliformes) of the North Atlantic Basin (log_2 groupings). (Species are identified by number from Appendix 3b).

ovoviviparous. They all have typically low fecundity (range 2^1 to 2^4 (2^6) young – Fig. 7), demonstrating again the size independent nature of fecundity in this group of fishes. The details of their ovoviviparity are given by Compagno (1990). The Hexanchidae, Chlamydoselachiidae and Squalidae all utilize 'yolksac viviparity', while the lamniforms Odontaspidae, Mitsukurinidae and probably Pseudotriakidae (see Yano 1992) utilize 'viviparous cannibalism'. Unrepresented in Figure 7 are those chondrichthyans (Scyliorhinidae, Rajiformes and Chimaeriformes) which according to Compagno (1990) undergo 'extended oviparity', but for which there are as yet no fecundity data available for North Atlantic species.

Anguilliformes. – Demersal anguilliforms are represented by the family Synaphobranchidae, whose members span a substantial size range (2^4 to 2^{13} g; Appendix 2b, c). Fecundity estimates are few, but counts are available for *Synaphobranchus kaupi* (ref. no. 18 – Fig. 8), the dominant species of the slopes of the North Atlantic Basin (see Haedrich & Merrett 1988, 1990). In relation to its rise/abyssal relative, *Histiobranchus bathybius* (ref. no. 16 – Fig. 8), it is of smaller size and lower fecundity. Bruun (1937) traced the development of *S. kaupi* as a leptocephalus larva from an origin in the Sargasso Sea, implying a semelparous life history. The pattern in *H. bathybius* is unknown, although the presence of ripening females over a wide size range (575–1370 mm standard length (SL), personal observation) suggests possible iteroparity. This is supported by the capture of two recently spent females (1330 and 1370 mm SL) at 4580–4540 m in 21° N, 31° W (September–October, 1993 – personal observation). Neither of these fish exhibited obvious morphological degeneration typical of semelparous eels (e.g. Nielsen & Bertelsen 1985, Nielsen & Smith 1978).

Notacanthiformes. – While the largest notacanths

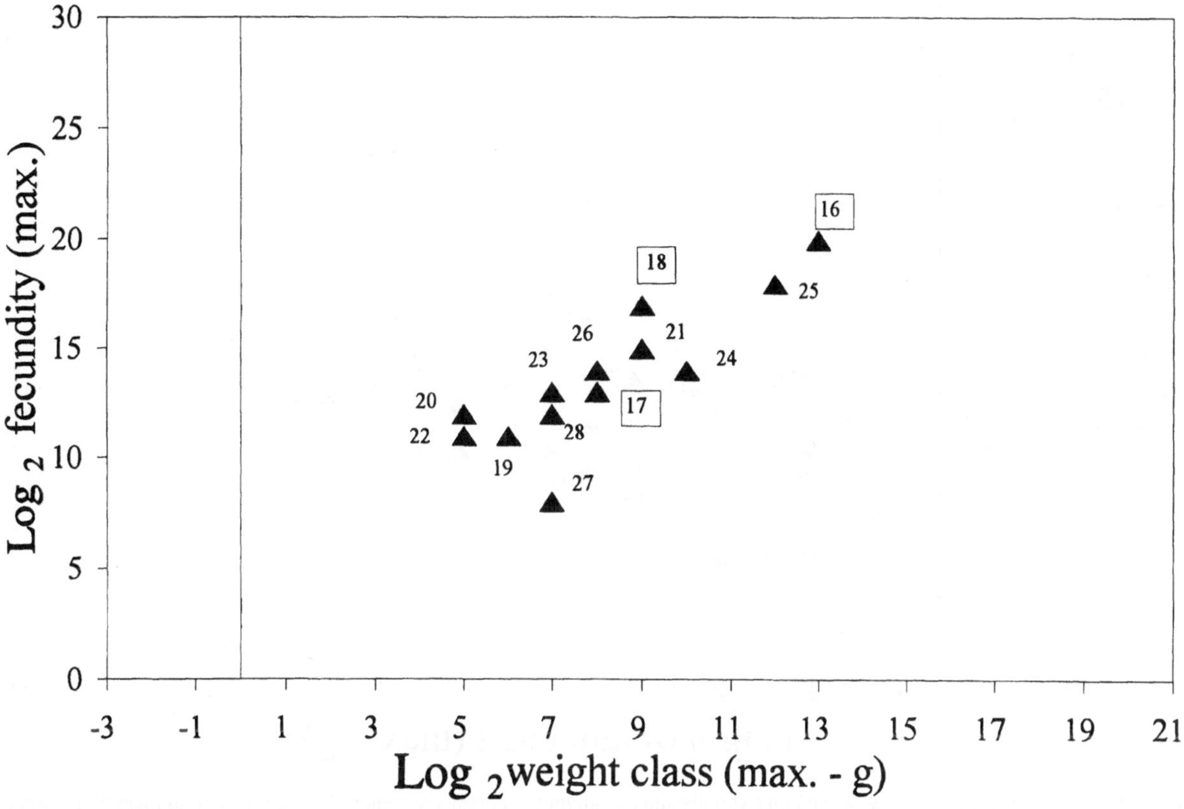

Fig. 8. Relationship between species' maximum fecundity and maximum size among the demersal oceanic eels, halosaurs and spiny eels of the orders Anguilliformes (boxed ref. numbers) and Notacanthiformes (n = 13) of the North Atlantic Basin ($\log_2$ groupings – fecundity (combined) = 4.52 + 1.12 weight; $r^2 = 0.73$). (Species are identified by number from Appendix 3b).

are of similar size to the largest synaphobranchids, the peak in their size frequency distribution is narrower and at smaller size (range 2^4 to 2^{12} g, peak 2^4 to 2^7 g; Appendix 2b, c). Few details are available on their reproductive arrangements; what is known suggests that they are iteroparous 'all-at-once' spawners. Crabtree et al. (1985) have shown that females predominate in populations of *Polyacanthonotus merretti* (ref. no. 27 – Fig. 8), *P. rissoanus* (ref. no. 28 – Fig. 8) and *P. challengeri* (ref. no. 26 – Fig. 8). Sulak demonstrated a similar skewing of the sex ratio in favour of females in four species of the halosaur, *Aldrovandia.* Olfactory sexual dimorphism has been reported in both notacanths and halosaurs and may be typical of breeding males of notacanthiform fishes (McDowell 1973, Sulak 1977, Crabtree et al. 1985). Among those species whose fecundities are plotted here, *P. rissoanus* and *Notacanthus bonapartei* (ref. no. 24 – Fig. 8) are ranking species on the slope, while *Halosauropsis macrochir* (ref. no. 21 – Fig. 8) is a ranking species on the lower slope and rise in the Porcupine Seabight region of the eastern North Atlantic (Merrett et al. 1991a, b).

Salmoniformes. – The dominant salmoniformes in the deep North Atlantic are members of the family Alepocephalidae. Together they cover a wide size range (2^4 to 2^{13} g; Appendix 2a–d). All known ripe ovarian eggs are large, (2)3 to 4(8) mm in diameter (Golovan & Pakhorukov 1984, Crabtree & Sulak 1986, personal observation), deposited sometimes in batches and sometimes synchronously by implication from multi-modal size frequencies in ripe ovaries in some species. The large eggs size suggests direct development and a precocial life-history style (Flegler-Balon 1989). This is reflected in the placement of the family in the general trend of size-related fecundity (Fig. 9, cf. Fig. 2b). It is notewor-

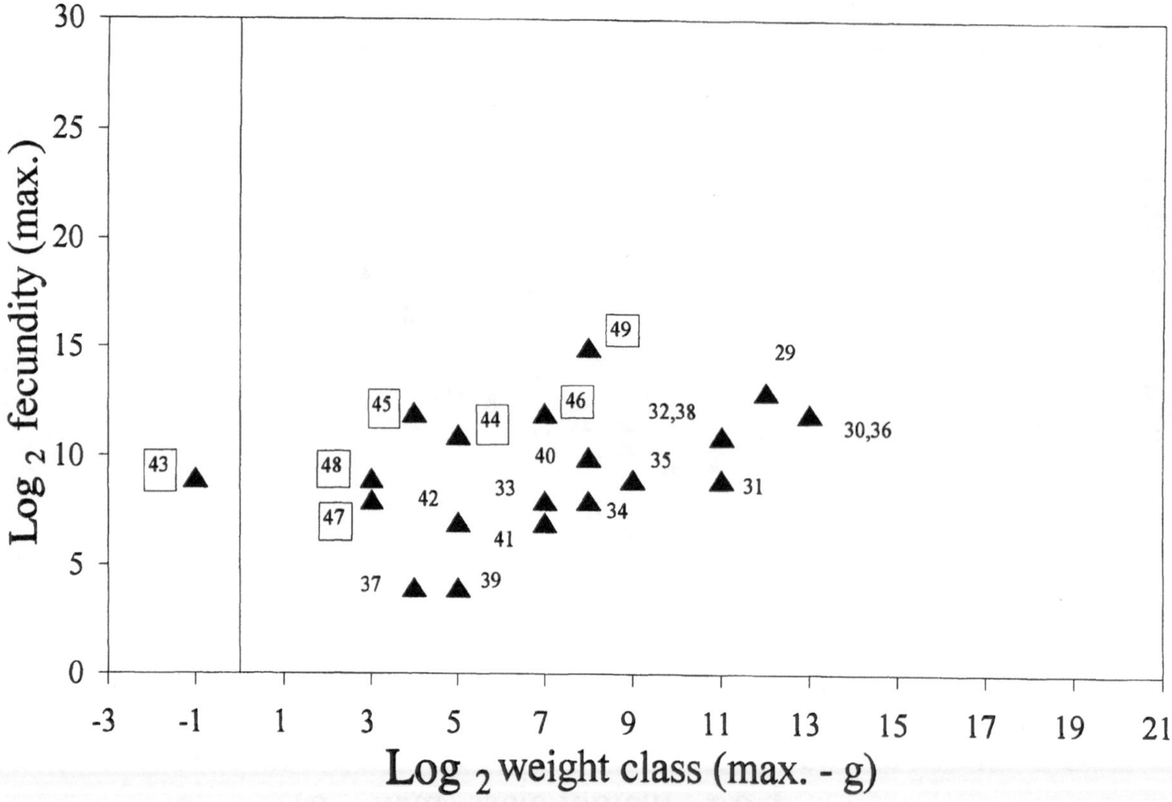

Fig. 9. Relationship between species' maximum fecundity and maximum size among the demersal oceanic representatives of the orders Salmoniformes and Aulopiformes (n = 21; boxed ref. numbers) of the North Atlantic Basin ($\log_2$ groupings – Salmoniformes: fecundity = 1.16 + 0.87 weight; r^2 = 0.86; Aulopiformes: fecundity = 9.08 + 0.20 weight; r^2 = 0.09). (Species are identified by number from Appendix 3b).

thy that success measured by relative numerical abundance is not necessarily correlated with high fecundity in Salmoniformes. In the Porcupine Seabight area, *Alepocephalus bairdi* (ref. no. 30 – Fig. 9) and *A. rostratus* (ref. no. 32 – Fig. 9) dominate on the slope and are relatively highly fecund. *Rinoctes nasutus* (ref. no. 39 – Fig. 9), on the other hand, ranks as a dominant at abyssal depths but with very modest fecundity (Merrett et al. 1991a, b). Crabtree & Sulak (1986) reviewed what is known about alepocephalid reproduction in a consideration of the biology of *Conocara.*

Aulopiformes. – Overall, deep demersal aulopiforms from the North Atlantic are represented by benthic fishes from a wide range of sizes (2^1 to 2^{13} g; Appendix 2a–d). Available fecundity counts from the area, however, are restricted in all but one case [*Bathysaurus ferox* (ref. no. 49 – Fig. 9) – Synodontidae] to the family Ipnopidae (sensu Hartel & Stiassny 1986) (Fig. 9). All are synchronous hermaphrodites, enabling them to maximize fecundity potential in their sparse populations and relatively immobile, benthic life-style. Some, certainly, are 'batch' spawners [*Ipnops meadi, Bathymicrops regis* (ref. no. 43 – Fig. 9); Nielsen 1966, Nielsen & Merrett 1993]. Moreover, ipnopid species of small adult size are more dominant among abyssal demersal populations beneath oligotrophic surface waters (*B. regis*) than they are beneath eutrophic surface conditions [*Bathypterois longipes* (ref. no. 44 – Fig. 9)] (Merrett 1987, 1992).

Gadiformes. – The order Gadiformes is the most speciose among deep demersal fishes in the North Atlantic basin (Fig. 1b) and is represented by a wide range of size classes (2^1 to 2^{14} g; Appendix 2a–d). While the family Macrouridae has the greatest species richness among demersal fishes around the oceanic rim (continental slope and rise) of the

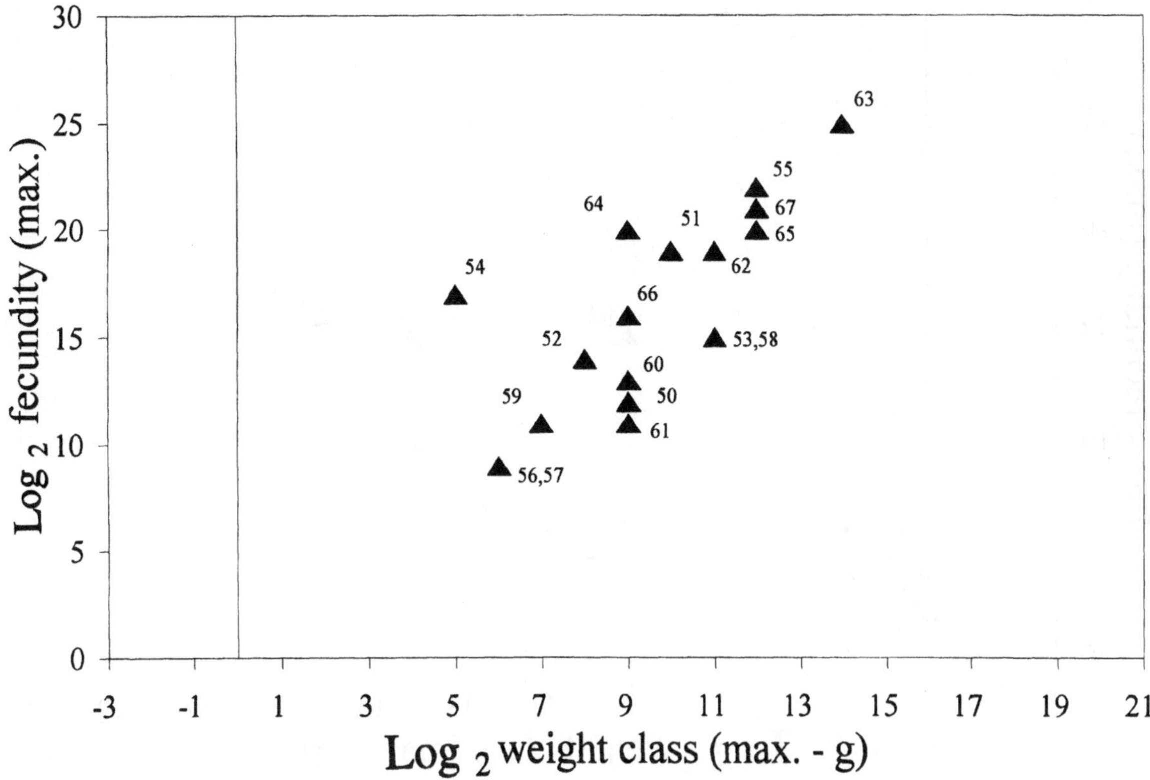

Fig. 10. Relationship between species' maximum fecundity and maximum size among the demersal oceanic representatives of the order Gadiformes (n = 18) of the North Atlantic Basin (log_2 groupings – fecundity = 2.37 + 1.44 weight; r^2 = 0.56). (Species are identified by number from Appendix 3b).

world ocean, the majority are smaller and less fecund than other oceanic gadiforms (Fig. 10). As pointed out above, there seems little relationship between size/fecundity and numerical abundance. *Coryphaenoides (N.) armatus* (ref. no. 55 – Fig. 10)is a highly fecund species (a possible semelparous spawner – Stein 1985) dominating on the rise and abyss beneath eutrophic surface conditions. In the eastern North Atlantic it is sexually dimorphic in size (females larger than males) and abundance, with sex ratios weighted in favour of females (Merrett 1992). *Echinomacrurus mollis* (ref. no. 56 – Fig. 10) is a ranking species ('batch' spawner – Merrett 1987) beneath oligotrophic surface waters (Haedrich & Merrett 1988, 1990, Merrett 1987, 1992, Merrett et al. 1991b). Both *C. (C.) rupestris* (ref. no. 53 – Fig. 10) and *Macrourus berglax* (ref. no. 58 – Fig. 10) are commercially exploited species of medium size and fecundity. *Nezumia aequalis* (ref. no. 59 – Fig. 10 – 'batch' spawner) and *C. (C.) guentheri* (ref. no. 52 – Fig. 10) are small species ranking high in relative abundance on the mid-slope and lower slope-rise, respectively (Merrett et al. 1991a, b), but with relatively modest fecundity (Fig. 10). Another species of small adult size but with contrasting high fecundity is *C. (Lionurus) carapinus* (ref. no. 54 – Fig. 10). This is a ranking species on the continental rise. Haedrich & Polloni (1976) report a small egg diameter (0.5 mm) in this species relative to *N. aequalis* and *C. (C.) guentheri* (1.6 mm and 0.8–0.9 mm, respectively, personal observation). Merrett (1989) showed that macrourids, at least, underwent a transitory ontogeny with a vestige of a larva called an alevin. Adult males in reproductive condition of several species of the Macrouridae are macrosmatic, with enlarged nasal rosettes as in the notacanths (see above).

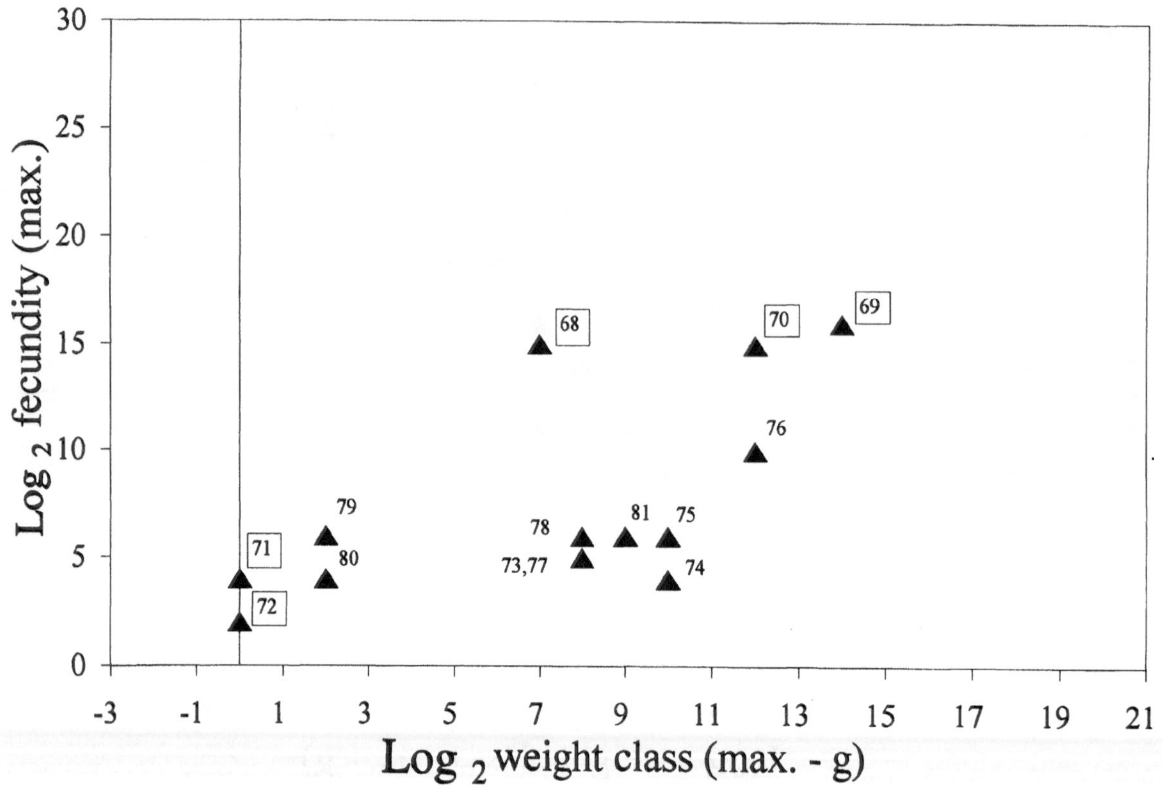

Fig. 11. Relationship between species' maximum fecundity and maximum size among the demersal oceanic representatives of the orders Ophidiiformes (boxed ref. numbers) and Perciformes (Zoarcidae) (n = 14) of the North Atlantic Basin (log_2 groupings). (Species are identified by number from Appendix 3b).

Ophidiiformes. – Knowledge of fecundity within this group, which spans a substantial size range (2^0 to 2^{14} g; Appendix 2 a–d), is poor. The two suborders within the order Ophidiiformes are in part distinguished by their reproductive arrangements. The Ophidioidei are oviparous fishes, while the Bythitoidei are ovoviviparous, a style confirmed by the presence of an intromittent organ in males (Cohen & Nielsen 1978). Spermatophores found in a range of species within the latter group are believed to function in a storage capacity. This is a valuable feature among an apparently sparsely distributed group (Nielsen et al. 1968). Known fecundity among bythitoid females varied widely. In the small, seemingly paedomorphic, family Aphyonidae up to three relatively large clutches of eggs may be distinguished within the ovary, although the number of developing embryos found is in the range 2^2 to 2^4 (Nielsen 1969) (Fig. 11). Conversely, in the bythitoid species *Cataetyx laticeps* (ref. no. 70 – Fig. 11), much larger adult size coupled with evident parturition in the free embryo (yolk-sac) phase, results in a fecundity in one example of 2^{15} embryos. Such a fecundity is comparable with that of known oviparous ophidioids (Fig. 11).

The advanced development of the free embryos found in the Aphyonidae (Nielsen 1969), coupled with the evidence provided by Wourms & Cohen (1975) that embryos of the related bythitid, *Oligopus longhursti,* possessed trophotaeniae, suggests that they too might be classified as 'histotrophic live bearers' in the reproductive guilds of Balon (1990). *Cataetyx laticeps,* on the other hand, is clearly classed as an 'obligate lecithotrophic live bearer'.

Perciformes – Zoarcidae. – While fecundity data are sparsely represented in Perciformes in general and in zoarcids in particular (Appendix 2a–d), a differ-

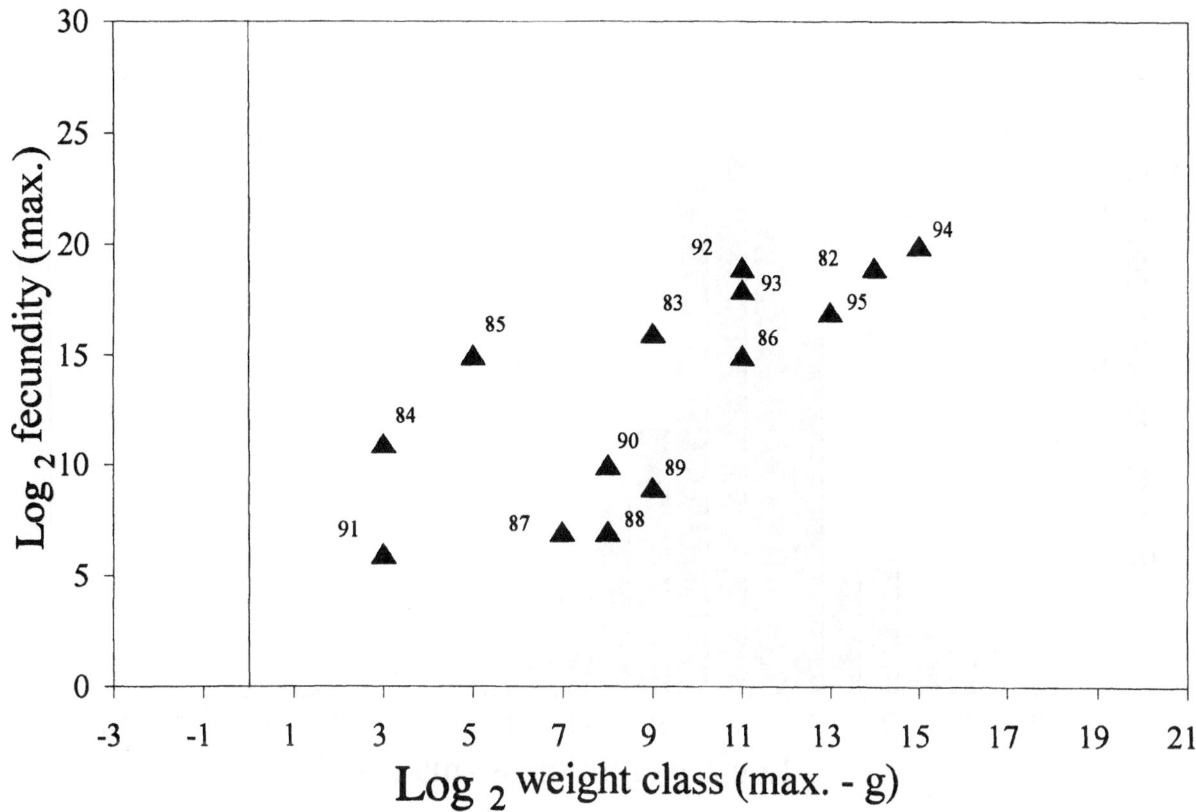

Fig. 12. Relationship between species' maximum fecundity and maximum size among the demersal oceanic representatives of the orders Lophiiformes, Beryciformes, Syngnathiformes, Scorpaeniformes and Pleuronectiformes (n = 14) of the North Atlantic Basin ($\log_2$ groupings). (Species are identified by number from Appendix 3b).

ent relationship of known fecundity with size is seen within seven oceanic representatives of the largely benthic family Zoarcidae (Fig. 11, ref. no. 73–81). Little observed difference in fecundity (range 2^4 to 2^6) occurs over some 8 $\log_2$ orders of magnitude in size, while the eighth species, *Lycodes esmarkii* departs somewhat from this trend (2^{10} – ref. no. 76 – Fig. 11). This relatively low fecundity accompanies large egg size (3 mm + diam., Johnsen 1921, Musick et al. 1975, Markle & Wenner 1979, Silverberg et al. 1987). As Mead et al. (1964) pointed out, some form of parental care is implied by this style. Indeed *Melanostigma atlanticum* (ref. no. 80 – Fig. 11), which has taken to a pelagic existence, was postulated to spawn demersally (Markle & Wenner 1979). Recently it was collected from a box-core sample in a sub-surface burrow complete with eggs (Silverberg et al. 1987). The large investment per egg in this group typifies the fundamental patterns of low relative fecundity correlated with a benthic life-style. This implies direct development associated with a precocial life-history style (Flegler-Balon 1989), exemplifying a 'hole nesting guarder' in Balon's (1990) scheme of reproductive guilds. It is noteworthy that Andriyashev (1953) treated the diverse deep-sea zoarcids as secondarily deep-water forms and pointed out that features of this group are their benthic habit and few eggs spawned on the bottom.

Miscellaneous species. – While fishes from a variety of orders (i.e. 15, 18, 20, 21, 23 – Appendix 2) comprise this group (Fig. 12), the trend among those represented appears to show a roughly linear increase in fecundity with weight, similar to the majority of teleost groups. Noteworthy is *Sebastes norvegicus* (ref. no. 86 – Fig. 12) for its ovoviviparity [an 'obligate lecithotrophic live bearer' in Balon's (1990) classification], with high fecundity on a scale

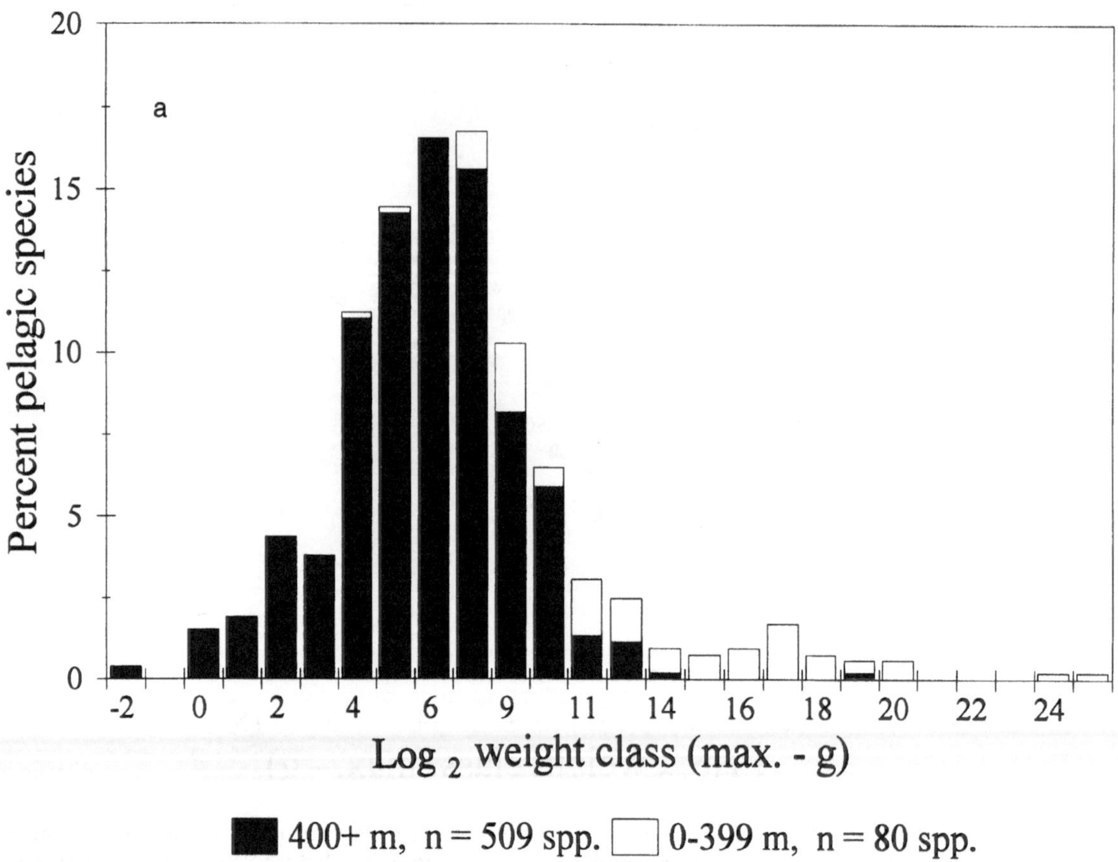

Fig. 13. Overall size spectra based upon maximum known sizes attained by oceanic species of the North Atlantic Basin: a – 589 pelagic species from the upper and lower depth zones, b – 505 demersal species from four depth zones.

equivalent to that of the oviparous bythitoid, *Cataetyx laticeps.*

Bathymetric influence on reproductive effort, implied from the overall size spectra of pelagic and demersal fishes in the North Atlantic

Species size spectra indicate responses to evolutionary pressure, while giving some idea of the range of food sources available for exploitation. Haedrich & Merrett (1992) developed this theme, showing that most demersal deep-sea species are foragers/scavengers. They exploit allochthonous food sources from shallower layers, migrating midwater fauna and, especially among species of small adult size at abyssal depths, small particles within the benthic boundary layer. Haedrich & Merrett (1992) pointed out that the broader, flatter spectra indicated variation in food sources, while 'spikey' spectra suggested concentration on a certain particle size.

From the point of view of reproductive styles on an ocean basin scale, therefore, an analysis of species size spectra provides an insight to the composition of reproductive contributions among the species of an assemblage, assuming the relationship between body size and fecundity confirmed here. When the size spectra for pelagic and demersal species are plotted as frequency distributions they are found to differ, especially when considered by depth (Fig. 13a, b).

Taken overall, the pelagic size spectrum has an asymmetric distribution, peaking in the weight class range 2^4 to 2^9 g (16 to 511 g maximum size) and with an extended 'tail' to size class 2^{25} g (33.5 to 67.1 t –

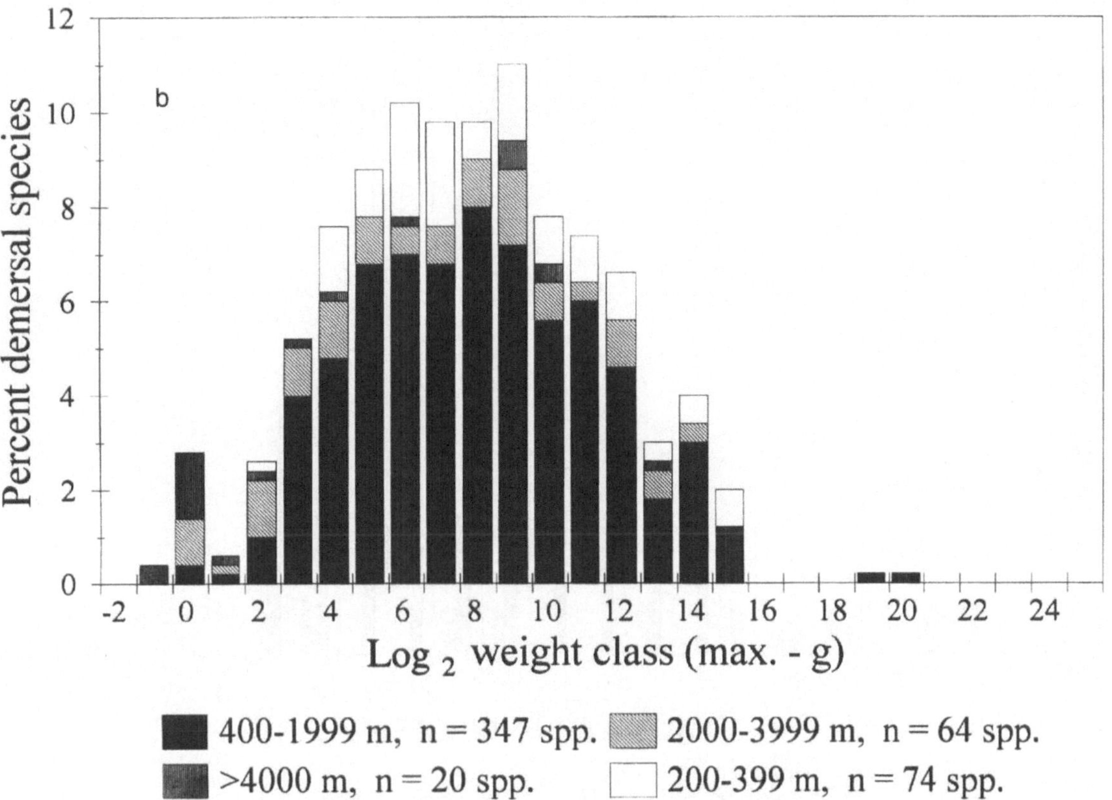

Fig. 13. Continued.

Fig. 13a). The peak of the demersal size spectrum is broader (range 2^4 to 2^{12} g: 16 to 8191 g), without extension into the higher size categories (Fig. 13b). Note, however, that these figures are directly comparable only below the 400 m level (see Materials and methods).

Considered by depth strata, it is evident that those pelagic, secondarily deep water (sensu Andriyashev 1953 – see p. 207), species dwelling in the upper 400 m dominate the larger weight classes (Fig. 13a). (The counterpart to the upper 200 m of this zone in the demersal habitat falls outside the limits of this study. It is unknown whether inclusion of demersal species from the shelf would extend the size spectrum into the larger size classes similarly.) The lower pelagic size spectrum (400 + m) contributes almost entirely to the peak at sizes 2^4 to 2^9 g. [The single representative at 2^{19} g is the swordfish, *Xiphias gladius* (ref. no. 59 – Fig. 6). Recent sonic tagging work has shown that this species, in contrast to other billfish, spends most of its day below 400 m (Block 1991) and is therefore included in the lower zone.] Subdivision of the demersal size spectrum, on the other hand, broadly maintains similar distributions among zones (Fig. 13b). Within this overall distribution, however, there is a perceptible shift in size class dominance between the two shallower strata (larger classes dominating) and the two deeper strata (smaller classes dominating). Most notably, the abyssal species (> 4000 m) dominate the smallest size classes.

While Figure 13 represents the overall situation, do the assemblages found at particular localities follow the same pattern? Spectra constructed from the data given by Merrett et al. (1991a, b) for the Porcupine Seabight/Abyssal Plain (PSB/PAP) assemblage in the eastern North Atlantic, together with that given by Merrett (1987) for the assemblage at

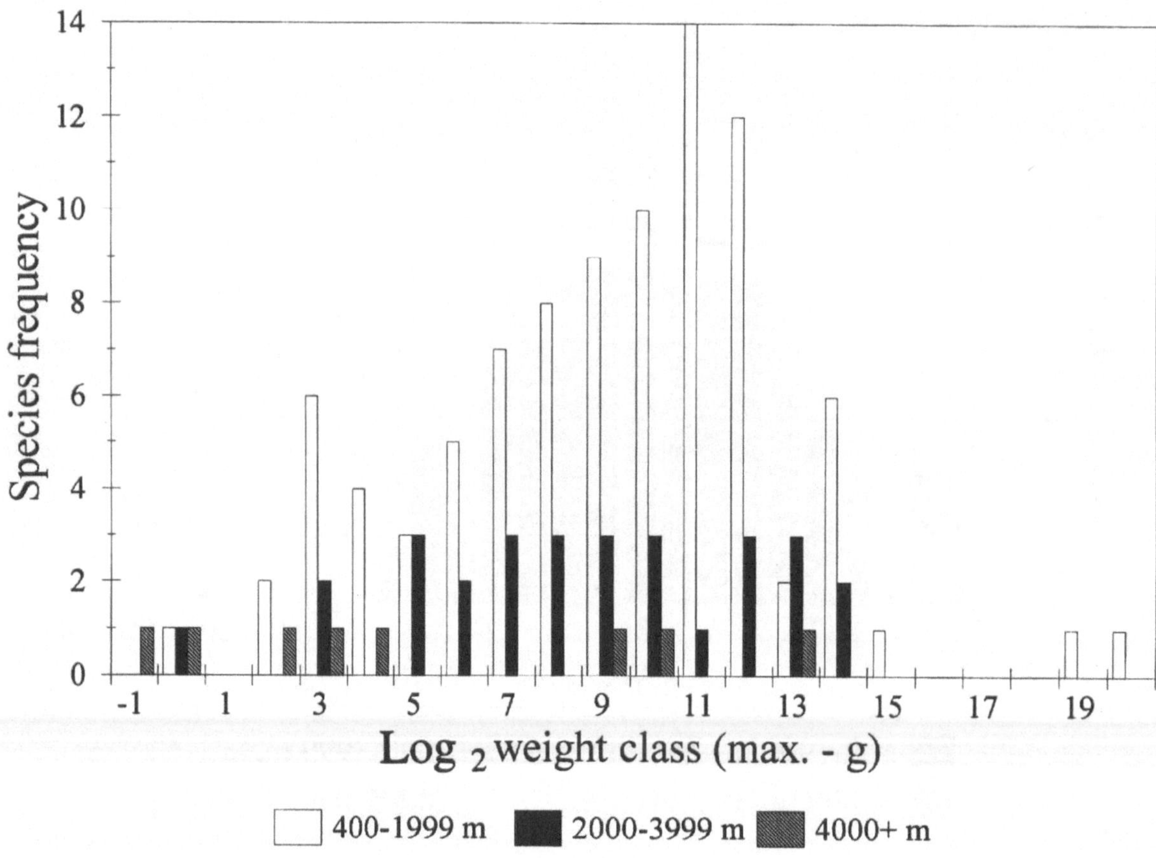

Fig. 14. Size spectra based upon species' adult maximum size of the demersal ichthyofauna sampled from the Porcupine Seabight/Abyssal Plain (PSB/PAP) and the Madeira Abyssal Plain (MAP), eastern North Atlantic, by depth strata (from Merrett 1987, Merrett 1992a, b).

5400 m in the Madeira Abyssal Plain (MAP) and based upon species adult maximum size are given in Figure 14. By implication from the body size-fecundity relationship, the pattern of reproductive effort reflected by species size classes is broadest on the slope. A narrow peak occurs among species of 2^3 to 2^4 g and a much broader one is evident among species of larger adult size which peaks in the 2^{11} to 2^{12} g size classes. At continental rise depths species representation among the largest size classes is curtailed. In the abyss representation is concentrated within the smaller categories (2^{-1} to 2^4 g) and the larger ones (2^9 to 2^{13} g). It is evident that this specific example differs most markedly in species size representation at slope depths in comparison with the basin-wide situation (Fig. 13b). Whatever the cause, the species composition of the size classes, with respect to reproductive style, shows that each class contains a mix of family representation. Such is to be expected with relatively few size classes in relation to the number of species represented. Yet, in general, each family is represented across a wide range of size classes (e.g. the 11 species of macrourid occupying the 400–2000 m depth range occur in the following $\log_2$ size classes: – 2, 3, 6 (2), 7, 8 (2), 9 (2), 10, 11).

To illustrate the complexity of the overall situation, Table 2 indicates the composition of each size class in terms of presumed life-style, reproductive mode and known or presumed fecundity pattern of the major zones, the slope and rise. Relative to the ubiquity of the benthopelagic species among the size classes, benthic species, although sparsely distributed over a wide size range, are most prevalent

Table 2. Assemblage composition and reproductive variables from the Porcupine Seabight demersal ichthyofauna.

a Slope

Size class	*Systematic* family representation	Teleost species	Chondrichthyan species	*Ecological* Benthopel.	Benthic	*Reproductive variables* Gonochor.	Hermaphrod.	Oviparous	Viviparous	Sm. egg. low fec.	Sm. egg high fec.	Lge egg low fec.	Lge egg med. fec.
0	1	1		1		1			1				
1													
2	2	2		2		2		2			2		
3	6	6		5	1	5		5	1	2	3	1	
4	4	4		4		4		4			4		
5	3	3		2	1	2	1	3		1	2		
6	4	5		5		5		5			5		
7	7	5	2	3	4	7		7			5	2	
8	6	8		8		8		8			4	4	
9	8	8	1	7	2	9		9			5	4	
10	10	9	1	5	5	9	1	10			5	4	1
11	10	8	6	12	2	14		11	3		5	3	3
12	10	6	6	11	1	12		9	3		5	4	
13	2	1	1	2		2			1				1
14	3	3	3	5	1	6		3	3		3		
15	1		1		1	1						1	
16													
17													
18													
19	1		1	1		1			1				
20	1		1	1		1			1				

b Rise

Size class	*Systematic* family representation	Teleost species	Chondrichthyan species	*Ecological* Benthopel.	Benthic	*Reproductive variables* Gonochor.	Hermaphrod.	Oviparous	Viviparous	Sm. egg low fec.	Sm. egg high fec.	Lge egg low fec.	Lge egg med. fec.
0	1	1		1		1			1				
1													
2													
3	1	2			2	2		2				2	
4													
5	3	3		2	1	2	1	3		1	1	1	
6	1	2		2		2		2			2		
7	1	3		3		3		3				3	
8	3	3		2	1	3		3			2	1	
9	3	3		2	1	3		3			1	2	
10	3	2	1	2	1	3		3			1	2	
11	1	1		1		1		1					1
12	3	3		3		3		3			2		1
13	3	3		2	1	2	1	3			2		1
14	2	1	1	1	1	2		2			1	1	

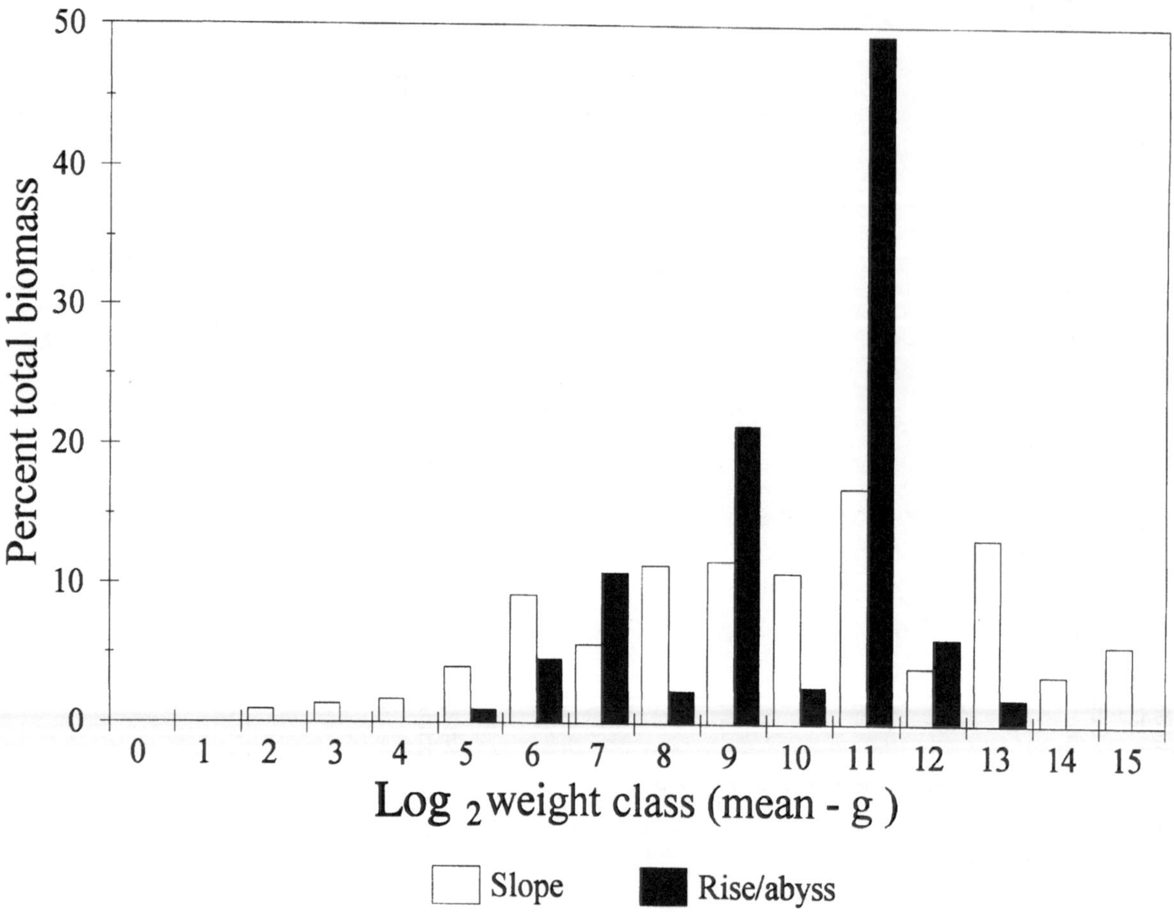

Fig. 15. Biomass spectra for demersal fish sampled from the eastern North Atlantic, from PSB continental slope depths (200–2000 m) and PSP, PAP & MAP continental rise and abyssal depths (2000–5400 m) (after Haedrich & Merrett 1992).

within the 2^7 to 2^{10} g span. Most species are gonochoristic, with only two hermaphroditic species each on the slope and rise. Most species fall into the small egg/high fecundity category and span all but the largest size groups. Species with small eggs and low fecundity are limited to 3 species only (2 benthic) within the size range 2^3 to 2^5 g. Species producing large eggs usually attain large adult size (2^7 to 2^{10} g). Most have relatively low fecundity, but five species with large eggs and medium fecundity occur within the 2^{10} to 2^{13} fecundity range. On the slope, chondrichthyans occupy the largest size classes and the viviparous ones are represented in categories larger than 2^{11} g. Viviparity, small egg/high fecundity and large egg/low or medium fecundity are roughly equal styles among species of adult size of 2^{10} to 2^{24} g. At rise levels evident trends seem similar, apart from the general lack of chondrichthyans (2 rays only).

Using biomass spectra from known populations of pelagic and demersal fishes, what reproductive arrangements appear to dominate on a basin-wide scale?

Species size spectra indicate the allocation of potential reproductive effort across the spectrum of diversity for the overall fish assemblage. They do not take the relative abundance of the constituent species into account, however, and therefore shed no light on the total assemblage effort realized. Biomass spectra provide the frequency distribution of individuals by weight within each size class represented in the assemblage (Gaedke 1992). Such an approach includes all individuals of the assemblage, of course, both pre-reproductive and reproductive. Nevertheless, in this preliminary investigation, it is a useful comparative indicator of any regional or bathymetric variation in the distribution of energy flow (see Gaedke 1992) and reproductive effort on a basin-wide scale. In broad terms, the shape of the biomass spectrum reflects the section of the species size spectrum potentially contributing the most to the total fecundity of the assemblage overall.

Quantifying such an estimate for an entire ocean basin is far beyond the scope of the information currently available. Well-sampled populations do provide information on biogeographic and bathymetric variation. In general, inferences on reproductive style can be drawn from the overall shape of biomass spectra. Those that have been called 'flat' (Sheldon et al. 1977) suggest a wide range of reproductive arrangements across the altricial and precocial interspecific range. Spectra that have been called 'spikey' (Sprules & Knoechel 1984, Haedrich 1986) imply a concentration of potential reproductive effort over a narrow range of size classes. Thus, spectra which are spikey in the smaller size classes suggest a preponderance of species which probably incline towards altricial life-history traits and reproductive arrangements (Adams 1980, and see p. 208). Spikey spectra in the larger size classes probably indicate a more precocial pattern.

Biomass spectra were also constructed for the deep demersal fishes of the PSB/PAP/MAP area (Haedrich & Merrett 1992). Such similarity was found among the patterns displayed among the chosen strata both at slope and at rise to abyssal depths that the information was combined (200–2000 m and 2000–5400 m, here Fig. 15). The biomass spectrum for slope depths can be termed flat, which indicates that potential reproductive effort is relatively evenly spread across the spectrum of individual fish sizes and, by analogy, species diversity. Deeper, on the continental rise and abyss, however, the biomass spectrum is spikey with a very distinct mode at size 2^{11} g (2 kg). Thus at these levels there is a favoured reproductive unit size whose effort is channelled through relatively high fecundity.

Similar data are available from one pelagic study, that of Haedrich (1986), which provides an insight to the changing shape of biomass spectra on a regional scale. Haedrich found that, among the mesopelagic fish assemblage of the North Atlantic Basin, biomass spectra of very different patterns occurred in different regions (Fig. 16). A very spikey distribution occurred in the sub-Arctic, with by far the most biomass concentrated in the 2^{1} g (0.5 to 0.9 g) size class (Fig. 16a). Conversely, both the northern and southern Sargasso data produced a flat biomass spectrum, with biomass distributed much more evenly over the total size range of fishes (Fig. 16b). The Azores-Britain and, to some degree, the Guinean regions were found to be intermediate between these two extremes (Fig. 16a). The implications here are that reproductive effort in the sub-Arctic region is concentrated among (few) species of small adult size. The emphasis of one size class is reduced in the other areas, broadening the size class (and therefore fecundity) allocation to reproductive effort.

Reproductive style and food availability

While potential ichthyofaunal fecundity levels can be associated with various oceanic zones of productivity as above, teleost conformity to the idealized concepts of an altricial and precocial range (the 'alprehost' concept of Balon 1990) means that variations in reproductive style within major taxa are similarly associated with such zones.

Pelagic

The exponential decline of biomass with depth in midwaters (Angel & Baker 1982) limits that majority of pelagic fishes which tend towards the precocial end of the range to the upper zone (e.g. scombroid fishes and pelagic sharks). Both examples are top predators, but display very different reproductive patterns which reflect phylogenetic origins and constraints. Scombroids are highly fecund and subject to high juvenile mortality. They exploit all size classes of food supply during their life history. Pe-

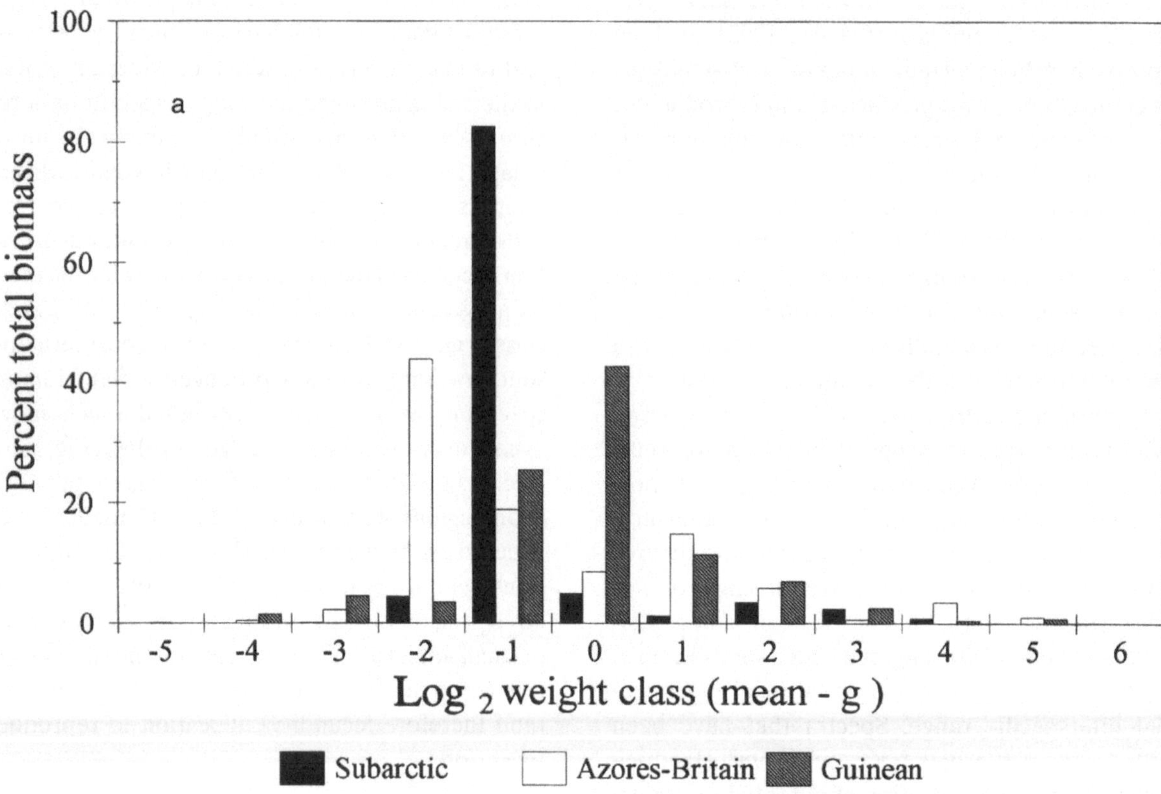

Fig. 16. Biomass spectra for oceanic mesopelagic fishes sampled in the North Atlantic: a – Atlantic sub-Arctic, Azores-Britain and Guinean regions; b – northern and southern Sargasso Sea regions (after Haedrich 1986 and personal communication).

lagic sharks, on the other hand, are viviparous with a relatively low fecundity resulting in a style which by-passes juvenile mortality and a dependence on small size classes of food. The large young are produced directly into the top predator feeding category.

Those fish that occur deeper in midwaters, however, attain considerably smaller adult size relative to their shallower-living counterparts (cf. Fig. 13a). The midwaters, in daytime between depths of ca. 400 m from the surface to 100 m off the sea-bed, are the headquarters of the so-called 'Lilliputian fauna' of the deep-sea (Murray & Hjort 1912).

The vertical effect on food availability results in an overall shift among species towards the altricial end of the range. A variety of reproductive styles are employed among this ichthyofaunal assemblage where resources are limited. Here species exploiting the lower trophic levels are both more numerous in the fauna and individually most abundant. They are typically of very small adult size and short lifespan (e.g. sternoptychids – Howell & Krueger 1987; *Cyclothone* – Badcock & Merrett 1976, Marshall 1984, Miya & Nemoto 1986a, b, 1987a, b, 1991, McKelvie 1989; myctophids – Kawaguchi & Mauchline 1982, Oven 1985, Karnella 1987, Albikowskaya 1988; melamphaids – Keene et al. 1987) and constrained by their body size to low fecundity. Only among predators higher up the food chain does body size allow for higher fecundity. Even so, it is probably trophic limitation on the largely bathypelagic ceratioid angler species, which causes some at least to become demersal at fully adult size (e.g. *Thaumatichthys* – Bertelsen & Struhsaker 1977, *Gigantactis* – personal observation).

Childress et al. (1980) demonstrated contrasting

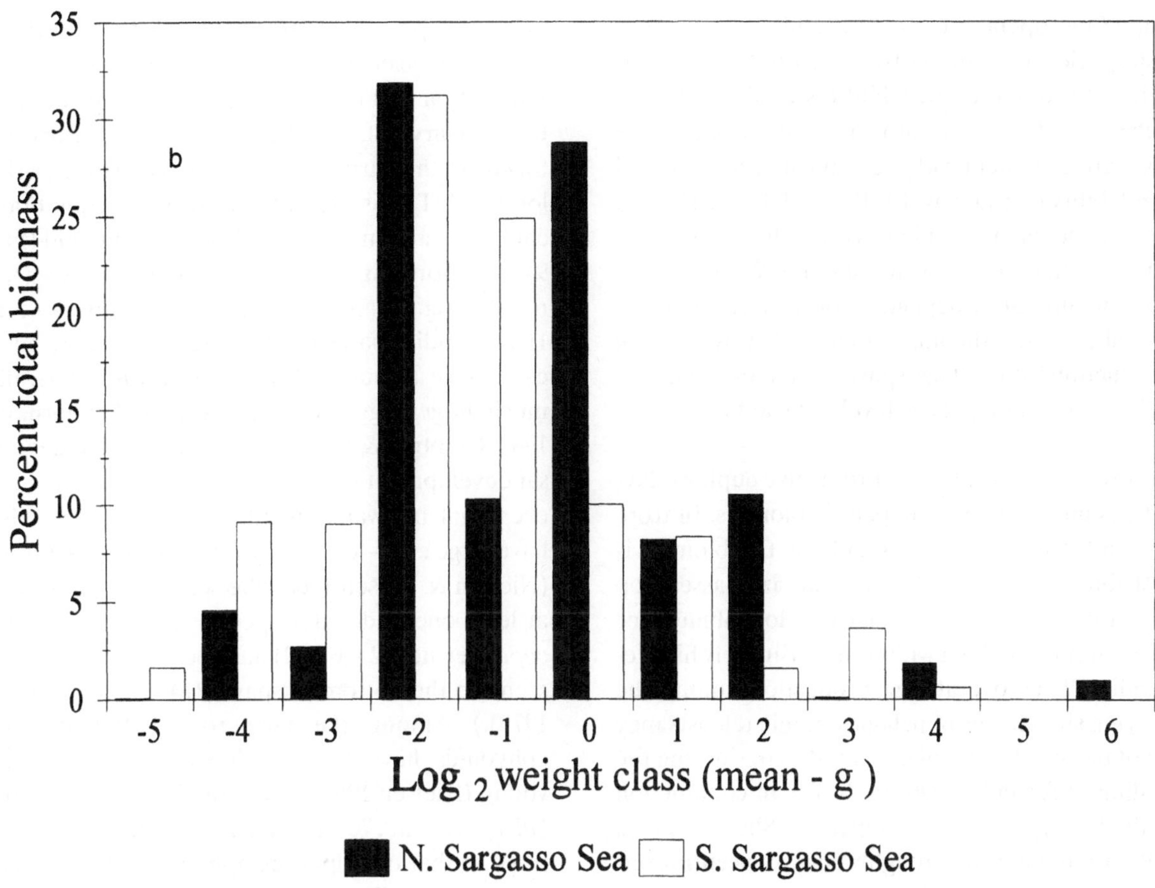

Fig. 16. Continued.

patterns in growth, energy utilization and reproduction among epi-, meso- and bathypelagic fishes in waters off southern California. They found that epipelagic (sensu stricto) fishes were adapted to large size, rapid growth, long life and early, repeated reproduction. Mesopelagic fishes shared the long life and early, repeated reproduction, but were characterized by small size and slow growth. Bathypelagic fishes, on the other hand, generally showed large size and rapid growth which was achieved by high relative growth efficiencies facilitated by low metabolic rates. They also had somewhat shorter lives and late reproduction (perhaps semelparity). (While the area illustrated is in a possibly atypical part of the Pacific Ocean, these results are noteworthy).

In contrast to the mesopelagic species studied by Childress et al. (1980) which were all diurnal migrants, Marshall (1984) discussed the evident paedomorphic trend in the non-migrant, mesopelagic *Cyclothone.* Paedomorphism, he argued, might result from competition with migrating species, which dominate the fauna and are annual fishes in the tropics (Childress et al. 1980). According to Marshall (1984), as non-migrants these *Cyclothone* have access to a lesser food supply for growth and development than do migrant species. Thus neoteny might offer them the only opportunity to mature in one year, competitively with migrant species exploiting the richer surface waters. In agreement with the trend found by Childress et al. (1980) among (non-migrant) bathypelagic fishes, Marshall (1984) pointed out that bathypelagic *Cyclothone* species are both larger in adult size and have greater fecundity than their mesopelagic congeners. In addition, using the North Atlantic data of Badcock & Merrett (1976), he showed that the abundance of bathypelagic *C. microdon* was approximately half that of mesopelagic *C. braueri.* While bathypelagic fish species may demonstrate the life-history characteristics observed by Childress et al. (1980), reduced standing stocks of fish (i.e. biomass) in this zone are consistent with the vertical decline in food availability (sensu Angel & Baker 1982, see above). This reduction in fish biomass, and therefore abundance, with depth is a general trend. With egg production ultimately dependent on species adult size and abundance, the implication is that overall egg production from pelagic spawning teleosts at mesopelagic and bathypelagic levels may approach parity.

Towards the surface, the productive euphotic layers sustain the maximum pelagic biomass. In tropical and temperate regions, at least, this biomass is distributed rather evenly among all size classes. Top predator teleosts are in relatively low abundance (e.g. istiophorids) but with commensurately high fecundity. Thus overall egg production among the varying size classes of pelagic oceanic teleosts may be of the same order of magnitude, considering the scaling of fecundity with body size in conjunction with the hypothesis developed by Sheldon et al. (1972) and others that roughly equal concentrations of material occur at all particle sizes from bacteria to whales.

Demersal

The seabed acts as the ultimate sink for oceanic production. Food particles of all sizes sink and accumulate there. This offers the diverse ichthyofauna far greater food supplies than are encountered in the overlying midwaters. Ichthyofaunal demersal biomass, abundance and mean fish size decrease with increased soundings from a peak at mid-slope depths (Merrett et al. 1991b). Only in deep mid-ocean basins beneath oligotrophic surface waters, however, is the size spectrum dominated by species of very small adult size (2^0 g – Figs 14 and 13b).

Food supplies are sufficient for some top predators to extend to abyssal levels. In contrast with the pelagic realm, large demersal elasmobranchs (e.g. rajids and squalids) are distributed down to 3000 m or so. While the rajids are oviparous and of unknown fecundity, squalids have the potential to bear up to about 30 well-developed young (Fig. 7) into the top predator trophic level down to these depths. Typically teleost fecundity is high, with the implication that the small eggs develop pelagically through larval stages, the indirect or transitory ontogeny of the altricial life-history style (Flegler-Balon 1989). This is exemplified by the synaphobranchid, halosaur and notacanth (Fig. 8), ipnopid (Fig. 9), gadiform (Fig. 10) and ophidiid [*Dicrolene* (ref. no. 68 – Fig. 11) and *Spectrunculus* (ref. no. 69 – Fig. 11)] fecundity patterns. The alepocephalids, however, have a more precocial style with generally much larger eggs (e.g. Golovan & Pakhorukov 1984, Crabtree & Sulak 1986), with implied demersal development (Fig. 9). It should be noted, however, that midwater members of this family also have large eggs – e.g. *Bathylaco* (ref. no. 45 – Fig. 9) (Nielsen & Larsen 1968, Nielsen 1972). Moreover, at least one mid-water species (*Xenodermichthys copei*, ref. no. 42 – Fig. 9) has been suggested to descend to the bottom to spawn (Markle & Wenner 1979).) Among the viviparous bythitoids, the aphyonids have low clutch sizes and advanced young (Nielsen 1969), while the bythitid, *Cataetyx laticeps* (ref. no. 70 – Fig. 11), has a fecundity comparable to that of oviparous ophidioids, but releases free embryos. Zoarcids display a precocial pattern with fecundities which seem to mirror those of the squalid sharks, in that egg number seems relatively uncorrelated with species adult size (Fig. 11). Since one species, at least, is known to reproduce in burrows [*Melanostigma atlanticum* (ref. no. 80 – Fig. 11) – Silverberg et al. 1987 and see Kendall et al. 1983], this may be a trend among other deep living members of this family (M.E. Anderson personal communication). The protection thus afforded would allow for greater energetic investment in a few large eggs for release into the environment as advanced offspring.

Conclusion

The pelagic oceanic fauna clearly separates into two phylogenetic groupings. The surface waters, down to about 400 m, contain one group representing 'more advanced' orders consistent with the neritic ichthyofauna. Below this layer resides the true deep-sea ichthyofauna composed of representatives of generally 'less advanced' orders. The ordinal representation by species is, again, strikingly different between the pelagic and demersal assemblages. Changes occur also within the demersal ichthyofauna alone. They indicate, for instance, that chondrichthyans are important constituents to little deeper than slope waters, while the salmoniforms, aulopiforms and ophidioids characterize the rise and abyssal assemblages.

Close to the surface and to the sea-bed, where the environment is relatively rich in food, a wide range of fish sizes (by species) may co-occur to exploit a wide range of relatively easily accessible and abundant particle sizes. Abundant, diminutive species, which are planktivorous throughout their life history, can compete successfully with large teleosts which soon pass through a planktivorous state as young and grow to become top predators. Adult large teleosts also compete with sharks whose young never compete for particles of small size, but are born directly into the top predator trophic level. Indeed, the advanced modes of viviparity developed by the oceanic lamniform and carcharhiniform sharks of the North Atlantic maximize the size at which the young are born and hence the advantage the young have relative to other organisms in the oceanic food web (Compagno 1990). In the midwaters, where the depth dimension strongly influences food availability, the pelagic ecosystem declines in species richness and biomass to minimal levels around 3000 m. Only particles of a relatively small size and size range may be sufficiently concentrated to sustain a relatively sparse ichthyofauna in which large top predators are (energetically) denied a living. In the shallower parts of this region, therefore, the prevalent pattern is obligatorily modest fecundity constrained by small body size, but substantially compensated for by a high abundance (if not density). Yet reproductive adaptations displayed by a particular taxon (body size/egg size/fecundity balance, in this case) are often a closer reflection of the evolutionary history of that group than of selection pressures in the contemporary context (Crabtree & Sulak 1986). Hence, while such parameters as those above, together with species abundance and generation time, may have long-standing phylogenetic origins, many arrangements for maximizing fecundity and survival (e.g. sexual dimorphism, synchrony in hermaphroditism, teleost viviparity, 'batch' spawning, seasonal synchrony, egg rafts, parental care, etc.) may be associated with more contemporary selection pressures and provide trade-off and alternative life history opportunities.

The current review indicates a variety of reproductive trends at ordinal or lower systematic levels within the North Atlantic Basin. Among the chondrichthyans, the successful deepsea forms utilize 'extended oviparity' (Scyliorhiniformes, Rajiformes and Chimaeriformes) or 'yolk-sac viviparity' (Hexanchiformes, Squaliformes), while the 'viviparous cannibalism' of the Lamniformes dominates in the food-rich near-surface waters.

Altricial non-guarding egg scatterers (sensu Balon 1990) are the norm among the teleosts. The Anguilliformes are characterized by considerable sexually dimorphic changes with increased ripeness, in pelagic forms at least, and semelparity over a probably extended life-span in most species. They and the Notacanthiformes share leptocephalous larvae. The latter group, however, appears to be iteroparous. In contrast to the high fecundity and small egg size displayed by these orders, the Salmoniformes are characterized by low to medium fecundity and large egg size, typical of a more precocial style. 'Batch' spawning is usual in both pelagic and demersal salmoniforms, but the whereabouts of spawning is unknown. The Aulopiformes considered here are synchronous hermaphrodites, a style which may compensate for the reduction in sexual encounters resulting from their benthic life-style. Most are rather small fishes, with low fecundity and small, 'batch'-spawned, pelagic eggs.

The vast majority of the representatives of the orders Stomiiformes and Myctophiformes are pelagic and dominate this assemblage in species richness and abundance. Both orders are composed of altricial, short-lived species of small adult size, with low fecundity and small egg size. Many myctophiforms are nocturnal vertical migrants and exploit the near-surface waters, to facilitate completion of the life-cycle in one year in many species. Sexual dimorphism is expressed by size, sex ratio and photophore distribution. Many stomiiforms are also nocturnal migrants. The reproductive arrangements of the highly abundant non-migrant species are best known and in this respect are found to be more versatile than are the myctophiforms. For instance, the species of the genus *Cyclothone,* together with others, may display paedomorphic tendencies, protandric hermaphroditism and sexual dimorphism in size, sex ratio and olfactory development. Stomiiforms are also short-lived, 'batch' or 'all-at-once' spawners.

The order Gadiformes is represented by dominant families in the deep demersal ichthyofaunal assemblage, the Macrouridae and Moridae. Reproductively these are altricial generalists of medium to large adult size, with high fecundity and usually small eggs. The majority of species are iteroparous, but possibly at least one of the continental rise/abyssal species may be semelparous, despite its large size and likely longevity. The males in several species are macrosmatic. Females may be 'all-at-once' or 'batch'-spawners.

The Ophidiiformes are largely demersal in the deep-sea. They are composed of one sub-order of oviparous species, the Ophidioidei, and another of ovoviviparous species, the Bythitoidei. Ophidioid species occupy a wide range of size classes but are generally altricial, displaying high fecundity and small egg size. The bythitoid species also cover a wide size range with precocial reproductive styles involving direct development, which span small neotenic species bearing few advanced young (family Aphyonidae; benthopelagic and demersal), to much larger species which produce great numbers of free embryos (Bythitidae). (This latter adaptation is one shared with some species of the order Scorpaeniformes). Sexual dimorphism in the Bythitoidei is expressed in the intromittent organ of males. The spermatophores produced by males are believed to function in a storage capacity in impregnated females.

Deep-sea Lophiiformes are mostly bathypelagic lie-and-wait predators of the sub-order Ceratioidei. They have exploited extremes of sexual dimorphism in size, culminating in male parasitism in some families. Species of this order are oviparous and display high fecundity and small egg size. In common with certain shallow lophiiform representatives, at least one species of ceratioid protects its egg mass in a mucus veil.

Deep-sea members of the order Perciformes are dominated in the North Atlantic by the sub-order Zoarcoidei. Most speciose is the benthic family Zoarcidae, which are oviparous and of moderate size. Their precocial lifestyle, with low fecundity and large eggs, is contrary to oceanic teleost trends. Some evidence suggests that burial of eggs occurs for protection, which could be expected to enhance the survival rate. The Parabrotulidae are diminutive pelagic ovoviviparous zoarcoids that produce few advanced young.

Finally, this review of deep-sea fish reproduction since the work of Mead et al. (1964), coupled with an endeavour to describe ichthyofaunal organization on an ocean basin scale via a variety of parameters associated with reproduction, has highlighted the extent of the basic information still outstanding. Further wide-ranging input (data on early ontogenetic structures; species size data; accurate, size-related, distributional data; patterns of oogenesis in the majority of species, etc.) is now necessary before the web of possible correlations advanced here can be clarified.

Acknowledgements

I dedicate this paper, in particular, to the fond memory of ET (Ethelwynn Trewavas) in recognition of her outstanding contributions and lifelong devotion to ichthyology. Certain weight-length data were generously provided by R. Froese (ICLARM), J.D.M. Gordon (SAMS), R.L. Haedrich (Memorial University, Newfoundland – WHOI data), A. Post and Christine Karrer (ISH). I am very grateful to them and to the many people with whom I have had discussions on all aspects of this work. R.H. Beverton was especially helpful in the early stages of the paper and R.L. Haedrich, N.B. Marshall, J. Nielsen (Zoological Museum, Copenhagen), and J.R. Paxton (Australian Museum) put much effort into improving the final draft of the manuscript. I thank them all particularly.

References cited

Adams, PB 1980 Life history patterns in marine fishes and their consequences for fisheries management US Fish Bull 78 1–11

Albikovskaya, L K 1988 Some aspects of the biology and distribution of glacier lanternfish (*Benthosema glaciale*) over the slopes of Flemish Cap and Eastern Grand Bank NAFO Sci Coun Studies 12 37–42

Alekseyev, F Ye , Ye I Alekseyeva & A N Zakharov 1992 Vitellogenesis, nature of spawning, fecundity, and gonad maturity stages of the roundnose grenadier, *Coryphaenoides rupestris,* in the North Atlantic J Ichthyol 32(3) 32–45

Andriyashev, A P 1953 Ancient deep-water and secondary deep-water fishes and their importance in a zoogeographical analysis pp 58–64 *In* G U Lindberg (ed) Notes on Special Problems in Ichthyology, Akad Nauk SSSR, Ikhtiol Kom , Moscow (in Russian)

Angel, M V & A de C Baker 1982 Vertical distribution of the standing crop of plankton and micronekton at three stations in the Northeast Atlantic Biol Oceanogr 2 1–30

Badcock, J 1986 Aspects of the reproductive biology of *Gonostoma bathyphilum* (Gonostomatidae) J Fish Biol 29 589–603

Badcock, J & N R Merrett 1976 Midwater fishes in the eastern North Atlantic – I Vertical distribution and associated biology in 30° N 23° W, with developmental notes on certain myctophids Prog Oceanog 7 3–58

Balon, E K 1984 Reflections on some decisive events in the early life of fishes Trans Amer Fish Soc 113 172–185

Balon, E K 1988 Tao of life universality of dichotomy in biology 1 The mystic awareness Rivista di Biologia/ Biology Forum 81 185–231

Balon, E K 1989 The epigenetic mechanisms of bifurcation and alternative life-history styles pp 467–501 *In* M N Bruton (ed) Alternative Life-History Styles of Animals, Kluwer Academic Publishers, Dordrecht

Balon, E K 1990 Epigenesis of an epigeneticist the development of some alternative concepts on the early ontogeny and evolution of fishes Guelph Ichthyol Rev 1 1–48

Beebe, W 1934 Deep-sea fishes of the Bermuda Oceanographic Expeditions Family Idiacanthidae Zoologica 16 149–241

Bertelsen, E 1980 Notes on Linophrynidae V a revision of the deep-sea anglerfishes of the *Linophryne arborifera*-group (Pisces, Ceratioidei) Steenstrupia 6 29–70

Bertelsen, E & P Struhsaker 1977 The ceratioid fishes of the genus *Thaumatichthys*, osteology, relationships, distribution and biology Galathea Rep 14 7–40

Bertelsen, E & J G Nielsen 1987 The deep sea eel family Monognathidae (Pisces, Anguilliformes) Steenstrupia 13 141–198

Block, B A 1991 Endothermy in fish thermogenesis, ecology and evolution pp 269–311 *In* I M Hochachka & A B Mommsen (ed) Biochemistry and Molecular Biology of Fishes, Elsevier, London

Bruun, A F 1937 Contributions to the life histories of the deep sea eels Synaphobranchidae Dana Rep 9 1–31

Childress, J J , S M Taylor, G M Cailliet & M H Price 1980 Patterns of growth, energy utilization and reproduction in some meso- and bathypelagic fishes off Southern California Mar Biol 61 27–40

Clarke, T A 1973 Some aspects of the ecology of lanternfishes in the Pacific Ocean near Hawaii US Fish Bull 71 401–434

Clarke, T A & PJ Wagner 1976 Vertical distribution and other aspects of the ecology of certain mesopelagic fishes taken near Hawaii US Fish Bull 74 635–645

Clarke, T A 1983 Sex ratios and sexual differences in size among mesopelagic fishes from the central Pacific Ocean Mar Biol 73 203–209

Clarke, T A 1984 Fecundity and other aspects of reproductive effort in mesopelagic fishes from the North Central and Equatorial Pacific Biol Oceanogr 3 147–165

Cohen, D M 1970 How many recent fishes are there? Proc Calif Acad Sci 38 341–346

Cohen, D M & J G Nielsen 1978 Guide to the identification of genera of the fish order Ophidiiformes with a tentative classification of the order NOAA Tech Rep NMFS Circ (417) 1–72

Compagno, L J V 1984a FAO species catalogue, Vol 4 Sharks of the world An annotated and illustrated catalogue of shark species known to date Part 1 Hexanchiformes to Lamniformes FAO Fish Synop 125 1–249

Compagno, L J V 1984b FAO species catalogue, Vol 4 Sharks of the world An annotated and illustrated catalogue of shark species known to date Part 2 Carcharhiniformes FAO Fish Synop 125 251–655

Compagno, L J V 1990 Alternative life-history styles of cartilaginous fishes in time and space Env Biol Fish 28 33–75

Crabtree, R E , K J Sulak & J A Musick 1985 Biology and distribution of species of *Polyacanthonotus* (Pisces Notacanthiformes) in the western North Atlantic Bull Mar Sci 36 235–248

Crabtree, R E & K J Sulak 1986 A contribution to the life history and distribution of Atlantic species of the deep-sea fish genus *Conocara* (Alepocephalidae) Deep-Sea Res 33 1183–1201

Duarte, C M & M Alcaraz 1989 To produce many small or few large eggs a size-dependent reproductive tactic of fish Oecologia 80 401–404

Elgar, M A 1990 Evolutionary compromise between a few large and many small eggs comparative evidence in teleost fish Oikos 59 283–287

Fisher, R A 1983 Protandric sex reversal in *Gonostoma elongatum* (Pisces Gonostomatidae) from the eastern Gulf of Mexico Copeia 1983 554–557

Flegler-Balon, C 1989 Direct and indirect development in fishes – examples of alternative life-history styles pp 71–100 *In* M N Bruton (ed) Alternative Life-History Styles of Animals, Kluwer Academic Publishers, Dordrecht

Gaedeke, U 1992 Identifying ecosystem properties a case study using plankton biomass size distributions Ecol Modelling 63 277–298

Gartner, J V, Jr 1983 Sexual dimorphism in the bathypelagic

gulper eel *Eurypharynx pelecanoides* (Lyomeri Eurypharyngidae), with comment on reproductive strategy Copeia 1983 560–563

Gibbs, R H , Jr 1969 Taxonomy, sexual dimorphism, vertical distribution and evolutionary zoogeography of the bathypelagic fish genus *Stomias* (Stomiatidae) Smithson Contr Zool 31 1–25

Gjosaeter, J 1981 Growth, production and reproduction of the myctophid fish *Benthosema glaciale* from western Norway and adjacent seas FishDir Skr (Havundersok) 17 79–108

Golovan, G A & N P Pakhorukov 1984 Some new information on the reproduction of bathyal fishes J Ichthyol 24 113–120

Gordon, J D M 1979a Lifestyle and phenology in deep sea anacanthine teleosts Symp Zool Soc Lond 44 327–359

Gordon, J D M 1979b Seasonal reproduction in deep-sea fish pp 223–229 *In* E Naylor & R G Hartnoll (ed) Cyclical Phenomena in Marine Plants and Animals, Pergamon Press, Oxford

Haedrich, R L 1986 Size spectra in mesopelagic fish assemblages pp 107–111 *In* A C Pierrot-Bults, S van der Spoel, B J Zahuranec & R K Johnson (ed) Pelagic Biogeography, UNESCO Tech Pap Mar Sci 49

Haedrich, R L & N R Merrett 1988 Summary atlas of deep-living demersal fishes in the North Atlantic Basin J Nat Hist 22 1325–1362

Haedrich, R L & N R Merrett 1990 Little evidence for faunal zonation or communities in deep-sea demersal fish faunas Prog Oceanog 24 239–250

Haedrich, R L & N R Merrett 1992 Production/biomass ratios, size frequencies, and biomass spectra in deep-sea demersal fishes pp 157–182 *In* G T Rowe & V Pariente (ed) Deep-Sea Food Chains and the Global Carbon Cycle, Kluwer Academic Publishers, Dordrecht

Haedrich, R L & PT Polloni 1976 A contribution to the life history of a small rattail fish, *Coryphaenoides carapinus* Bull So Calif Acad Sci 75 203–211

Hartel, K E & M L J Stiassny 1986 The identification of larval *Parasudis* (Teleostei, Chlorophthalmidae) with notes on the anatomy and relationships of aulopiform fishes Breviora 487 1–23

Heemstra, PC & PH Greenwood 1992 New observations on the visceral anatomy of the late-term fetuses of the living coelacanth fish and the oophagy controversy Proc R Soc Lond B, 249 49–55

Howell, WH & WH Krueger 1987 Family Sternoptychidae, marine hatchetfishes and related species pp 32–50 *In* R H Gibbs, Jr & WH Krueger (ed) Biology of Midwater Fishes of the Bermuda Ocean Acre, Smithsonian Contrib Zool 452, Washington, D C

Johnsen, S 1921 Ichthyologiske notiser I Bergens Museums Aarbok, 1918–19 Naturvid rakke 6 1–95

Karnella, C 1987 Family Myctophidae, lanternfishes pp 51–168 *In* R H Gibbs, Jr & WH Krueger (ed) Biology of Midwater Fishes of the Bermuda Ocean Acre, Smithsonian Contrib Zool 452, Washington, D C

Kawaguchi, K & R Marumo 1967 Biology of *Gonostoma gracile* (Gonostomatidae) I Morphology, life history and sex reversal Inf Bull Planktol Jap 1967 53–69

Kawaguchi, K & J Mauchline 1982 Biology of myctophid fishes (family Myctophidae) in the Rockall Trough, northeastern Atlantic Ocean Biol Oceanogr 1 337–373

Kawaguchi, K & J Mauchline 1987 Biology of sternoptychid fishes, Rockall Trough, northeastern Atlantic Ocean Biol Oceanogr 4 99–120

Keene, M J , R H Gibbs, Jr & WH Krueger 1987 Family Melamphaidae, bigscales pp 169–185 *In* R H Gibbs, Jr & WH Krueger (ed) Biology of Midwater Fishes of the Bermuda Ocean Acre, Smithson Contrib Zool 452, Washington, D C

Kendall, A W Jr , C D Jennings, TM Beasley, R Carpenter & B L K Somayajulu 1983 Discovery of a cluster of unhatched fish eggs of a zoarcid buried 10 to 12 cm deep in continental slope sediments off Washington State, USA Mar Biol 75 193–199

Krefft, G 1974 Investigations on midwater fishes in the Atlantic Ocean Ber dt wiss Kommn Meeresforsch 23 226–254

McDowell, S B 1973 Order Heteromi (Notacanthiformes) pp 1–228 *In* D M Cohen (ed) Fishes of the Western North Atlantic, Mem Sears Found Mar Res 1, Part 6

McKelvie, D S 1989 Latitudinal variation in aspects of the biology of *Cyclothone braueri* and *C microdon* (Pisces Gonostomatidae) in the eastern North Atlantic Ocean Mar Biol 102 413–424

Mann, K H & J R N Lazier 1991 Dynamics of marine ecosystems, Blackwell, Boston 466 pp

Markle, D F & C A Wenner 1979 Evidence of demersal spawning in the mesopelagic zoarcid fish *Melanostigma atlanticum* with comments on demersal spawning in the alepocephalid fish *Xenodermichthys copei* Copeia 1979 363–366

Marshall, N B 1971 Explorations in the life of fishes Harvard University Press, Cambridge 204 pp

Marshall, N B 1973 Family Macrouridae pp 496–665 *In* D M Cohen (ed) Fishes of the Western North Atlantic, Mem Sears Found Mar Res 1, Part 6

Marshall, N B 1979 Developments in deep-sea biology Blandford Press, Poole 566 pp

Marshall, N B 1984 Progenetic tendencies in deep-sea fishes pp 91–101 *In* G W Potts & R J Wootton (ed) Fish Reproduction Strategies and Tactics, Academic Press, Oxford

Marshall, N B & N R Merrett 1977 The existence of a benthopelagic fauna in the deep-sea pp 483–497 *In* M V Angel (ed) A Voyage of Discovery George Deacon 70th Anniversary Volume, Suppl to Deep-sea Res 24

Mazhirina, G P & A A Filin 1987 Gonad development and spawning of *Notoscopelus kroeyeri* in the northwest Atlantic, with observations on other biological characteristics J Northw Atl Fish Sci 7 99–106

Mead, G W , E Bertelsen & D M Cohen 1964 Reproduction among deep-sea fishes Deep-sea Res 11 569–596

Merrett, N R 1986 Biogeography and the oceanic rim a poorly known zone of ichthyofaunal interaction pp 201–209 *In* A C Pierrot-Bults, S van der Spoel, B J Zahuranec & R K Johnson (ed) Pelagic Biogeography, UNESCO Tech Pap Mar Sci 49

Merrett, N R 1987 A zone of faunal change in assemblages of abyssal demersal fish in the eastern North Atlantic a response to seasonality in production? Biol Oceanogr 5 137–151

Merrett, N R 1989 The elusive macrourid alevin and its seeming lack of potential in contributing to intrafamilial systematics pp 175–185 *In* D M Cohen (ed) Papers on the Systematics of Gadiform Fishes, Los Angeles County Nat Hist Mus Sci Ser 32

Merrett, N R 1992 Further evidence on abyssal demersal ichthyofaunal distribution in the eastern North Atlantic, with special reference to *Coryphaenoides (Nematonurus) armatus* (Macrouridae) J Mar Biol Ass U K 72 5–24

Merrett, N R, J D M Gordon, M Stehmann & R L Haedrich 1991a Deep demersal fish assemblage structure in the Porcupine Seabight (eastern North Atlantic) slope sampling by three different trawls compared J Mar Biol Ass U K 71 329–358

Merrett, N R, R L Haedrich, J D M Gordon & M Stehmann 1991b Deep demersal fish assemblage structure in the Porcupine Seabight (eastern North Atlantic) results of single warp trawling at lower slope to abyssal soundings J Mar Biol Ass U K 71 359–373

Miya, M & T Nemoto 1986a Reproduction, growth and vertical distribution of the mesopelagic fish *Cyclothone pseudopallida* (family Gonostomatidae) pp 830–837 *In* T Uyeno, R Arai, T Taniuchi & K Matsuura (ed) Indo-Pacific Fish Biology Proceedings of the Second International Conference on Indo-Pacific Fishes, Ichthyol Soc Japan, Tokyo

Miya, M & T Nemoto 1986b Life history and vertical distribution of the mesopelagic fish *Cyclothone alba* (family Gonstomatidae) in Sagami Bay, central Japan Deep-Sea Res 33 1053–1068

Miya, M & T Nemoto 1987a The bathypelagic gonostomatid fish *Cyclothone obscura* from Sagami Bay, central Japan Jap J Ichthyol 33 417–418

Miya, M & T Nemoto 1987b Some aspects of the biology of the micronektonic fish *Cyclothone pallida* and *C acclinidens* (Pisces Gonostomatidae) in Sagami Bay, central Japan J Oceanogr Soc Jap 42 473–480

Miya, M & T Nemoto 1991 Comparative life histories of the meso- and bathypelagic fishes of the genus *Cyclothone* (Pisces Gonostomatidae) in Sagami Bay, central Japan Deep-Sea Res 38 67–89

Murray, J & J Hjort 1912 The depths of the ocean, Macmillan, London 821 pp

Musick, J A, C A Wenner & G R Sedberry 1975 Archibenthic and abyssobenthic fishes of Deepwater Dumpsite 106 and the adjacent area National Oceanographic and Atmospheric Administration Dumpsite Evaluation Report 75-1, Baseline Investigation of Deepwater Dumpsite 106, NOAA S/T 76-1870(a) 229–269

Nafpaktitis, B G, R H Backus, J E Craddock, R L Haedrich, B H Robison & C Karnella 1977 Family Myctophidae pp 13–265 *In* R H Gibbs, Jr (ed) Fishes of the Western North Atlantic Mem Sears Found Mar Res 1, Part 7

Nakamura, I & N V Parin 1993 Snake mackerels and cutlassfishes of the world (families Gemylidae and Trichiuridae) FAO Species Catalogue 15, FAO, Rome 136 pp

Nazarov, N A 1983 Data on the reproduction of *Alepocephalus bairdi* (Alepocephalidae) from the northeastern Atlantic J Ichthyol 23(5) 29–35

Nelson, J S 1984 Fishes of the world 2nd Ed Wiley, New York 523 pp

Nielsen, J G 1966 Synopsis of the Ipnopidae (Pisces, Iniomi) with description of two new abyssal species Galathea Rep 8 49–75

Nielsen, J G 1968 Redescription and reassignment of *Parabrotula* and *Leucobrotula* (Pisces, Zoarcidae) Vidensk Meddr Dansk Naturh Foren 131 225–250

Nielsen, J G 1969 Systematics and biology of the Aphyonidae (Pisces, Ophidioidea) Galathea Rep 10 7–88

Nielsen, J G 1972 Ergebnisse der Forschungsreisen des FFS 'Walther Herwig' nach Suedamerika XX Additional notes on Atlantic Bathylaconidae (Pisces, Isospondyli) with a new genus Archiv FishWiss 23 29–36

Nielsen, J G, A Jespersen & O Munk 1968 Spermatophores in Ophidioidea (Pisces, Percomorphi) Galathea Rep 9 239–254

Nielsen, J G & E Bertelsen 1985 The gulper-eel family Saccopharyngidae (Pisces, Anguilliformes) Steenstrupia 11 157–206

Nielsen, J G, E Bertelsen & A Jespersen 1989 The biology of *Eurypharyncx pelecanoides* (Pisces, Eurypharyngidae) Acta Zool Stockholm 70 187–197

Nielsen, J, J Badcock & N R Merrett 1990 New data elucidating the taxonomy and ecology of the Parabrotulidae (Pisces Zoarcoidei) J Fish Biol 37 437–448

Nielsen, J G & V Larsen 1968 Synopsis of the Bathlaconidae (Pisces, Isospondyli) with a new eastern Pacific species Galathea Rep 9 221–238

Nielsen, J & N R Merrett 1993 Taxonomy and biology of *Bathymicrops* (Pisces, Ipnopidae), with description of two new species Steenstrupia 18 149–167

Nielsen, J G & D G Smith 1978 The eel family Nemichthyidae (Pisces, Anguilliformes) Dana Rep 88 1–71

Oven, L S 1985 Comparative analysis of reproductive biology of some lanternfishes (Myctophidae) from the tropical zone of the Atlantic Ocean J Ichthyol 25(3) 50–60

Pankhurst, N W 1988 Spawning dynamics of orange roughy, *Hoplostethus atlanticus*, in mid-slope waters of New Zealand Env Biol Fish 21 101–116

Pankhurst, N W, P J McMillan & D M Tracey 1987 Seasonal reproductive cycles in three commercially exploited fishes from the slope waters off New Zealand J Fish Biol 30 193–211

Parin, N V 1984 Oceanic ichthyologeography an attempt to review the distribution and origin of pelagic and bottom fishes outside continental and neritic zones Archiv FischWiss 35 5–41

Pauly, D. & R.S.V. Pullin. 1988. Hatching time in spherical, pelagic, marine fish eggs in response to temperature and egg size. Env. Biol. Fish. 22: 261–271.

Paxton, J.R. 1972. Osteology and relationships of the lanternfishes (family Myctophidae). Bull. Nat. Hist. Mus. Los Angeles Co. (13): 1–81.

Pietsch, T.W. 1976. Dimorphism, parasitism and sex: reproductive strategies among deepsea ceratioid anglerfishes. Copeia 1976: 781–793.

Quero, J.C., J.C. Hureau, C. Karrer, A. Post & L. Saldanha (ed.) 1990a. Checklist of fishes of the eastern tropical Atlantic, JNICT, Lisbon 1: 1–519.

Quero, J.C., J.C. Hureau, C. Karrer, A. Post & L. Saldanha (ed.) 1990b. Checklist of fishes of the eastern tropical Atlantic, JNICT, Lisbon 2: 520–1080.

Sheldon, R.W., A Prakash & W.H. Sutcliffe, Jr. 1972. The size distribution of particles in the ocean. Limnol. Oceanogr. 17: 327–340.

Sheldon, R.W., W.H. Sutcliffe & M.A. Paranjape. 1977. Structure of pelagic food chains and relationship between plankton and fish production. J. Fish. Res. Board Can. 34: 2344–2353.

Silverberg, N., H.M. Edenborn, G. Ouellet & P. Beland. 1987. Direct evidence of a mesopelagic fish, *Melanostigma atlanticum* (Zoarcidae) spawning within bottom sediments. Env. Biol. Fish. 20: 195–202.

Sprules, W.G. & R. Knoechel. 1984. Lake ecosystem dynamics based on functional representations of trophic components. *In:* D.G. Myers & R. Stricker (ed.) Trophic Interactions within Aquatic Systems, Amer. Assoc. Adv. Sci., Select Symp. 85: 383–403.

Stein, D.L. 1980. Aspects of reproduction of liparid fishes from the continental slope and abyssal plain off Oregon, with notes on growth. Copeia 1980: 687–699.

Stein, D.L. 1985. Towing large nets by single warp at abyssal depths: methods and biological results. Deep-Sea Research 32: 183–200.

Stein, D.L. & W.G. Pearcy. 1982. Aspects of reproduction, early life history, and biology of macrourid fishes off Oregon, U.S.A. Deep-Sea Res. 29: 1313–1329.

Sulak, K.J., C.A. Wenner, G.R. Sedberry & L. Van Guelpen. 1985. The life history and systematics of deep-sea lizard fishes, genus *Bathysaurus* (Synodontidae). Can. J. Zool. 63: 623–642.

Turner, C.L. 1936. The absorptive processes in the embryos of *Parabrotula dentiens*, a viviparous, deep-sea brotulid fish. J. Morph. 59: 313–321.

Ware, D.M. 1975. Relation between egg size, growth and natural mortality of larval fish. J. Fish. Res. Board Can. 32: 2503–2512.

Wenner, C.A. 1984. Notes on the ophidioid fish *Dicrolene intronigra* from the middle Atlantic continental slope of the United States. Copeia 1984: 538–541.

Whitehead, P.J.P., M.-L. Bauchot, J.-C. Hureau, J. Nielsen & E. Tortonese (ed.) 1984. Fishes of the north-eastern Atlantic and the Mediterranean, UNESCO, Paris 1: 1–510.

Whitehead, P.J.P., M.-L. Bauchot, J.-C. Hureau, J. Nielsen & E. Tortonese (ed.) 1986a. Fishes of the north-eastern Atlantic and the Mediterranean, UNESCO, Paris 2: 517–1007.

Whitehead, P.J.P., M.-L. Bauchot, J.-C. Hureau, J. Nielsen & E. Tortonese (ed.) 1986b. Fishes of the north-eastern Atlantic and the Mediterranean, UNESCO, Paris 3: 1015–1473.

Woodward, A.S. 1898. The antiquity of the deep-sea fish-fauna. Nat. Sci. 12: 257–260.

Wootton, R.J. 1992. Constraints in the evolution of fish life histories. Netherlands J. Zool. 42: 291–303.

Wourms, J.P. & D.M. Cohen. 1975. Trophotaeniae, embryonic adaptations, in the viviparous ophidioid fish, *Oligopus longhursti*: a study of museum specimens. J. Morph. 147: 385–402.

Yano, K. 1992. Comments on the reproductive mode of the false cat shark *Pseudotriakis microdon.* Copeia 1992: 460–468.

Appendix 1 Log_2 (maximum) weight class frequency distribution of pelagic oceanic species among the orders (numbered arbitrarily from less to more advanced) represented in the North Atlantic Basin a – 0–399 m, b – 400 + m

Ord no	Orders	Log₂ weight class −2	−1	0	1	2	3	4	5	6	7	8	9	10	11	12	13	14	15	16	17	18	19	20	21	22	23	24	25
a PELAGIC – 0–399 m																													
1	Myxiniformes																												
2	Hexanchiformes																												
3	Squaliformes																												
4	Lamniformes														1						8	3		1				1	1
5	Rajiformes															1													
6	Chimaeriformes																												
7	Anguilliformes																												
8	Notacanthiformes																												
9	Salmoniformes																												
10	Stomiiformes																												
11	Aulopiformes																												
12	Myctophiformes																												
13	Gadiformes											1																	
14	Ophidiiformes														1														
15	Lophiiformes										1																		
16	Cyprinodontiformes							1				5	6	1															
17	Lampriformes													1	2	1	1	1		2									
18	Beryciformes																												
19	Zeiformes																												
20	Syngnathiformes																												
21	Scorpaeniformes																												
22	Perciformes							1		4	1		4	2	5	1	4	3	3	3	1	1	1						
23	Pleuronectiformes																												
24	Tetraodontiformes									1							1		1				1	2					
25	Dactylopteriformes																												
b PELAGIC – 400+ m																													
1	Myxiniformes																												
2	Hexanchiformes																												
3	Squaliformes												1	1	1														
4	Lamniformes																												
5	Rajiformes																												
6	Chimaeriformes																												
7	Anguilliformes						1	3	1	3	5	1	3	2															
8	Notacanthiformes																												
9	Salmoniformes							15	18	8	7	6																	
10	Stomiiformes	2		3	7	5	4	13	21	23	33	26	5	8	1														
11	Aulopiformes					6	1	4	11	13		3	1		1	1													
12	Myctophiformes			1	1	4	9	14	16	21	15	6																	
13	Gadiformes						1	1		1			1	1															
14	Ophidiiformes			2								2	1	1	1														
15	Lophiiformes					1				6	11	13	24	14	3		1												
16	Cyprinodontiformes																												
17	Lampriformes				1	1	1			1																			
18	Beryciformes					4	2	4	1	5	8	3	4	2															
19	Zeiformes																												
20	Syngnathiformes																												
21	Scorpaeniformes						1																						
22	Perciformes			2					2	8	2	2	4	3		1	5	1					1						
23	Pleuronectiformes																												
24	Tetraondontiformes																												
25	Dactylopteriformes																												

Appendix 2 Log_2 (maximum) weight class frequency distribution of demersal oceanic species among the orders (numbered arbitrarily from less to more advanced) represented in the North Atlantic Basin a – 200–399 m, b – 400 – 1999 m, c – 2000–3999 m, d – > 4000 m

Ord no	Orders	Log-2 weight class																											
		-2	-1	0	1	2	3	4	5	6	7	8	9	10	11	12	13	14	15	16	17	18	19	20	21	22	23	24	25
a DEMERSAL – 200–399 m																													
1	Myxiniformes										1																		
2	Hexanchiformes																		1										
3	Squaliformes																												
4	Lamniformes												3	1					1										
5	Rajiformes										1	1			2	3		2											
6	Chimaeriformes																												
7	Anguilliformes																												
8	Notacanthiformes																												
9	Salmoniformes								1			1		1															
10	Stomiiformes																												
11	Aulopiformes									3			1																
12	Myctophiformes																												
13	Gadiformes							1	2		1			2		1	1		1										
14	Ophidiiformes					2		1			1																		
15	Lophiiformes											1					1												
16	Cyprinodontiformes																												
17	Lampriformes																												
18	Beryciformes														1														
19	Zeiformes							1					1																
20	Syngnathiformes								1																				
21	Scorpaeniformes							2	1	6	1		2																
22	Perciformes							1				1	1	1	2	1		1	1										
23	Pleuronectiformes								1	3	5				1	1													
24	Tetraodontiformes																												
25	Dactylopteriformes																												
b DEMERSAL – 400–1999 m																													
1	Myxiniformes										1																		
2	Hexanchiformes															1		1					1						
3	Squaliformes											2	4		5	4	5	6	1					1					
4	Lamniformes										1	1	8	4	1				3										
5	Rajiformes										1	2		7	4	1		3	1										
6	Chimaeriformes													1	3	4	1												
7	Anguilliformes							4	5	3	5	5	3			1													
8	Notacanthiformes							2	5	4	2			1		1													
9	Salmoniformes							3	2	2	4	4	4	2	3		1												
10	Stomiiformes						1	1																					
11	Aulopiformes						2	1	4	1	1		1																
12	Myctophiformes										1	1																	
13	Gadiformes					2	8	3	7	14	6	17	7	4	2	3	1	2											
14	Ophidiiformes			2	1	2	3		3	7	3		3	1		1													
15	Lophiiformes									1		1			1			1											
16	Cyprinodontiformes																												
17	Lampriformes																												
18	Beryciformes								1				2	1	1	5													
19	Zeiformes							2				1	2		1														
20	Syngnathiformes							3																					
21	Scorpaeniformes					1	2		1	1	4	3	2	2	2														
22	Perciformes						4	4	4		6	3	1	3	7	2		2											
23	Pleuronectiformes								2	2				1	1		1		1										
24	Tetraodontiformes																												
25	Dactylopteriformes							1																					

Ord no	Orders	Log-2 weight class																											
		−2	−1	0	1	2	3	4	5	6	7	8	9	10	11	12	13	14	15	16	17	18	19	20	21	22	23	24	25
c DEMERSAL – 2000–3999 m																													
1	Myxiniformes																												
2	Hexanchiformes																												
3	Squaliformes																												
4	Lamniformes																												
5	Rajiformes													1				1											
6	Chimaeriformes																												
7	Anguilliformes											1					1												
8	Notacanthiformes											1	1																
9	Salmoniformes								1		2		1		1	1	1												
10	Stomiiformes																												
11	Aulopiformes						2		1			1					1												
12	Myctophiformes																												
13	Gadiformes				1				1	2			1	1		3													
14	Ophidiiformes			5				6	2	1		2	1	3	2	1	1		1										
15	Lophiiformes												1																
16	Cyprinodontiformes																												
17	Lampriformes																												
18	Beryciformes						1																						
19	Zeiformes																												
20	Syngnathiformes																												
21	Scorpaeniformes					6	2		1																				
22	Perciformes									1		1	1																
23	Pleuronectiformes																												
24	Tetraodontiformes																												
25	Dactylopteriformes																												
d DEMERSAL – 4000+ m																													
1	Myxiniformes																												
2	Hexanchiformes																												
3	Squaliformes																												
4	Lamniformes																												
5	Rajiformes																												
6	Chimaeriformes																												
7	Anguilliformes																												
8	Notacanthiformes																												
9	Salmoniformes													1															
10	Stomiiformes																												
11	Aulopiformes		2	1			1																						
12	Myctophiformes																												
13	Gadiformes									1							1												
14	Ophidiiformes			5		1		1					2	1															
15	Lophiiformes																												
16	Cyprinodontiformes																												
17	Lampriformes																												
18	Beryciformes			1	1																								
19	Zeiformes																												
20	Syngnathiformes																												
21	Scorpaeniformes																												
22	Perciformes												1																
23	Pleuronectiformes																												
24	Tetraodontiformes																												
25	Dactylopteriformes																												

Appendix 3 Key to species' representation in the fecundity per weight plots of a – the pelagic species in Fig 3–6 and b – the demersal species in Fig 7–12

	Species	Ref No
a		
Figure 3		
Lamniformes	*Carcharodon carcharias*	1
	Isurus oxyrinchus	2
	Isurus paucus	3
	Carcharhinus falciformis	4
	Carcharhinus longimanus	5
	Carcharhinus obscurus	6
	Carcharhinus signatus	7
	Pseudocarcharias kamoharai	8
	Prionace glauca	9
	Alopias superciliosus	10
	Sphyrna leweni	11
	Sphyrna mokharran	12
Squaliformes	*Isistius brasiliensis*	13
	Squaliolus laticaudus	14
Figure 4		
Stomiiformes	*Cyclothone alba*	15
	Cyclothone braueri	16
	Cyclothone microdon	17
	Cyclothone pallida	18
	Cyclothone pseudopallida	19
	Maurolicus muelleri	20
	Valenciennellus tripunctulatus	21
	Vinciguerria attenuata	22
	Vinciguerria poweriae	23
	Argyropelecus hemigymnus	24
	Idiacanthus fasciola	25
Figure 5		
Myctophiformes	*Protomyctophum arcticum*	26
	Symbolophorus rufinus	27
	Hygophum benoiti	28
	Lampanyctus macdonaldi	29
	Lampanyctys alatus	30
	Lobianchia dofleini	31
	Myctophum affine	32
	Myctophum asperum	33
	Myctophum nitidulum	34
	Myctophum punctatum	35
	Notolychnus valdivae	36
	Notoscopelus krøyeri	37
	Notoscopelus resplendens	38
	Benthosema glaciale	39
	Ceratoscopelus maderensis	40
	Diaphus garmani	41
	Diaphus rafinesquei	42
Figure 6		
Anguilliformes	*Eurypharynx pelecanoides*	43
	Monognathus taaningi	44
Salmoniformes	*Bathylaco nigricans*	45
Gadiformes	*Micromesistius poutassou*	46
Lophiiformes	*Ceratias holboelli*	47
	Haplophryne mollis	48
Cyprinodontiformes	*Hirundichthys* sp	49
Perciformes	*Coryphaena hippurus*	50
	Parabrotula plagiophthalma	51
	Katsuwonus pelamis	52
	Thunnus alalunga	53
	Thunnus albacares	54
	Thunnus obesus	55
	Thunnus thynnus	56
	Istiophorus albicans	57
	Tetrapterus albidus	58
	Xiphias gladius	59
Tetraodontiformes	*Mola mola*	60

b

	Species	Ref No
Figure 7		
Hexanchiformes	*Hexanchus griseus*	1
	Hexanchus vitulus	2
	Clamydoselachus anguineus	4
Lamniformes	*Pseudotriakis microdon*	5
Squaliformes	*Oxynotus centrina*	6
	Centrophorus granulosus	7
	Centrophorus lusitanicus	8
	Centrophorus squamosus	9
	Centrophorus uyato	10
	Centroscymnus coelolepis	11
	Centroscymnus owstoni	12
	Dalatias licha	13
	Echinorhinus brucus	14
	Etmopterus spinax	15
Figure 8		
Anguilliformes	*Histiobranchus bathybius*	16
	Ilyophis brunneus	17
	Synaphobranchus kaupi	18
Notacanthiformes	*Aldrovandia affinis*	19
	Aldrovandia phalacra	20
	Halosauropsis macrochir	21
	Halosaurus johnsonianus	22
	Halosaurus ovenii	23
	Notacanthus bonapartei	24
	Notacanthus chemnitzii	25
	Polyacanthonotus challengeri	26
	Polyacanthonotus merretti	27
	Polyacanthonotus rissoanus	28
Figure 9		
Salmoniformes	*Alepocephalus agassizi*	29
	Alepocephalus bairdi	30
	Alepocephalus productus	31
	Alepocephalus rostratus	32
	Bathytroctes microlepis	33
	Conocara macroptera	34
	Conocara murrayi	35
	Conocara salmonea	36
	Leptoderma macrops	37
	Narcetes stomias	38
	Rinoctes nasutus	39
	Rouleina attrita	40
	Rouleina maderensis	41
	Xenodermichthys copei	42
Aulopiformes	*Bathymicrops regis*	43
	Bathypterois longipes	44
	Bathypterois quadrifilis	45
	Bathytyphlops sewelli	46
	Ipnops agassizii	47
	Ipnops murrayi	48
	Bathsaurus ferox	49
Figure 10		
Gadiformes	*Caelorinchus labiatus*	50
	Coryphaenoides (Chalinura) mediterraneus	51
	(C) Coryphaenoides guentheri	52
	(C) Coryphaenoides rupestris	53
	(C) Lionurus carapinus	54
	(C) Nematonurus armatus	55
	Echinomacrurus mollis	56
	Gadomus longifilis	57
	Macrourus berglax	58
	Nezumia aequalis	59
	Trachyrincus murrayi	60
	Trachyrincus scabrus	61
	Merluccius merluccius	62
	Molva dypterygia	63
	Phycis chesteri	64
	Antimora rostrata	65
	Lepidion eques	66
	Mora moro	67
Figure 11		
Ophidiiformes	*Dicrolene intronigra*	68
	Spectrunculus grandis	69
	Cataetyx laticeps	70
	Barathronus bicolor	71
	Nybelinella erikssoni	72
Perciformes (Zoarcidae)	*Lycenchelys alba*	73
	Lycenchelys paxilla	74
	Lycodes atlanticus	75
	Lycodes esmarkii	76
	Lycodes pallidus	77
	Lycodes vahlii	78
	Lycodonus mirabilis	79
	Melanostigma atlanticum	80
	Pachycara bulbiceps	81
Figure 12		
Lophiiformes	*Lophius piscatorius*	82
Beryciformes	*Hoplostethus mediterraneus*	83
	Acanthochaenus luetkeni	84
Syngnathiformes	*Macrorhamphosus scolopax*	85
Scorpaeniformes	*Sebastes norvegicus*	86
	Artediellus atlanticus	87
	Cottunculus microps	88
	Cottunculus thompsoni	89
	Leptagonus decagonus	90
	Paraliparis hystrix	91
Pleuronectiformes	*Glyptocephalus cynoglossus*	92
	Hippoglossoides platessoides	93
	Hippoglossus hippoglossus	94
	Reinhardtius hippoglossoides	95

Environmental Biology of Fishes **41**: 246, 1994.

Fish imagery in art 69: Japanese Netsuke *Coiled fish*

Peter B. Moyle & Marilyn A. Moyle
Department of Wildlife and Fisheries Biology, University of California, Davis, CA 95616, U.S.A.

Netsuke are tiny carvings from Japan that were originally tied on to the end of a cord attached to a pouch, so the pouch could hang from the belt of a kimono. They evolved into a distinctive art form: exquisite carvings made from ivory, boxwood, and a wide variety of other materials, valued for their own sake. The subjects of netsuke carvings are as varied as Japanese art itself, but animals are a favorite subject, often in whimsical poses. Netsuke usually have meaning beyond the subject, representing traditional legends, stories, or symbols. The fish shown here is a carp, a common subject of netsuke because of its pleasing shape and symbolism of success or wealth. Carp are particularly conspicuous as a symbol in Japan on May 5 (Boy's Day), when families having sons fly carp banners over their rooftops, 'in hope the boys will grow up to be as strong and perseverant as the carp. . . (Symmes 1990, p. 178)'.

Coiled fish was carved in the early 19th century from ivory, with black stones for eyes (1.1 × 3.8 × 4.1 cm). It is used courtesy the Seattle Art Museum (Gift of Duncan MacTavish Fuller).

Symmes, E.C., Jr. 1990. Netsuke: Japanese life and legend in miniature. C.E. Tuttle Co., Rutland. 199 pp.

Environmental Biology of Fishes **41**: 247–268, 1994.

The comparative reproductive styles of two closely related African minnows (*Pseudobarbus afer* and *P. asper*) inhabiting two different sections of the Gamtoos River system

Jim A. Cambray
Albany Museum, Somerset Street, Grahamstown, 6140, South Africa

Received 25.8.1993 Accepted 9.4.1994

Key words: Alternative life-history, Altricial, Cyprinidae, Egg size, Fecundity, Generalist, Life-history, Multiple spawn, Precocial, Seasonality, Sister species, Spawning, Specialist, Water regulation

Synopsis

Two species of a monophyletic lineage of flexible-rayed redfin minnows, *Pseudobarbus afer* and *P. asper,* were studied to establish if there were any significant differences in their reproductive styles. They are sister species with few morphological or meristic differences. *P. afer* and *P. asper* are open substrate benthic spawners on coarse bottoms (rocks) and non-guarders of non-adhesive eggs. Their young are photophobic as free embryos. Riverine spawning sites indicated a conservative tendency and represented a phylogenetic constraint as compared to the more variable attributes, such as egg size, which were under environmental control within the limits expressed by the genotype. The combination of life-history attributes, gonadosomatic index, fecundity, egg size, investment per clutch, number of clutches per season and reproductive lifespan was found to be different for *P. afer* and *P. asper. P. asper* is derived and atypical of other *Pseudobarbus* species studied to date. Differences between *P. afer* and *P. asper* are directly related to the two distinct environments inhabited by these species – coastal Cape Fold Belt mountain streams and the inland Karoo streams of the Gamtoos River system. *P. asper* may have reverted, by juvenilization, to a more altricial form to survive the turbid, intermittent Karoo stream.

Introduction

In southern Africa there are 60 *Barbus* and *Pseudobarbus* species of which 48 are less than 150 mm total length. There have been few biological studies on these small barbs (Cambray 1992). The genus *Pseudobarbus,* separated from *Barbus* by Skelton (1988), consists of seven species of flexible-rayed redfin minnows. This is the largest non-cichlid monophyletic lineage in southern Africa. The distribution of *Pseudobarbus* is confined to a geographically distinct region, the Cape Fold mountain drainage, except for *P. quathlambae* which is confined to the Lesotho highlands. The present day distributions of the seven species are, with one exception, entirely complementary and allopatric.

P. afer and *P. asper,* sister species (Fig. 1) in the Gamtoos River system, have few morphological and meristic differences (Skelton 1980, 1988). *P. afer* is restricted to stable, clear, acidic, perennial mountain streams whereas *P. asper* occurs in widely fluctuating, turbid, alkaline, seasonal rivers of Karoo origin (Fig. 2), such as the Kariega and Groot rivers (Fig. 3). The presence of two closely-related species in two extreme environments without a physical barrier lent itself to a comparative life-history study.

Fig. 1. Male *Pseudobarbus afer* (above) and male *P. asper* (below).

Fig 2 Wit River a clear mountain stream (above) and Groot River a turbid, intermittent Karoo stream (below)

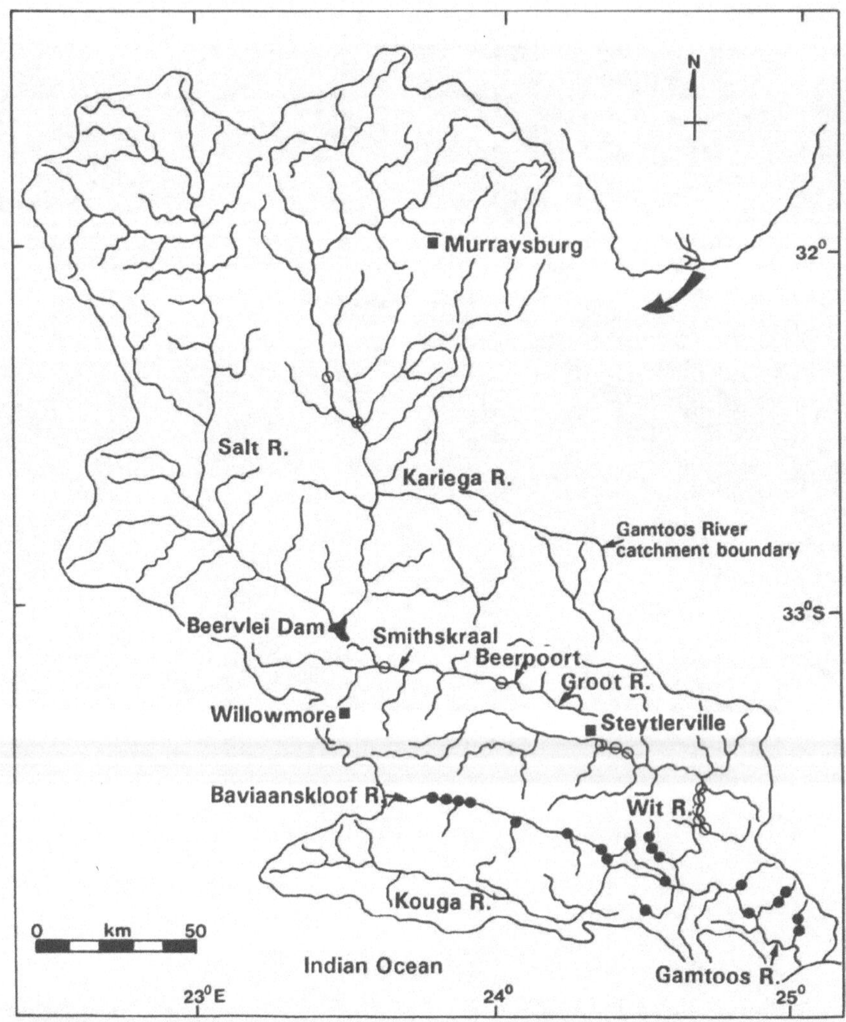

Fig. 3. Distribution records of *Pseudobarbus afer* (closed circles) and *P. asper* (open circles) in the Gamtoos River system (Albany Museum records).

There is also a paucity of life-history information for this monophyletic group of fishes, with five of the seven species in the Red Data Book (Skelton 1987).

Fishes have evolved diverse reproductive styles (for reviews see Breder & Rosen 1966, Balon 1975, 1981a). Fitness of a reproductive style is sensitive to energy allocation parameters and the mortality of juveniles and adults of the life cycle. To establish which reproductive style is most adaptive requires knowledge of life-history characteristics of the population and the effect of environmental factors on the species (Ware 1984).

A combination of morphological, developmental, behavioural and ecological criteria was evaluated on a comparative basis in order to place *P. afer* and *P. asper* in a particular reproductive guild. The framework of the classification of reproductive styles of Balon (1975, 1981a) provided useful information on evolutionary trajectories. Balon (1988a, b) suggested that the main trends of reproductive guilds are shaped by epigenetic bifurcations and heterochronies during development. Since there are very few meristic or morphological differences between *P. afer* and *P. asper* (Skelton 1988) and no physical barrier separating them they might have

Table 1 Developmental stage of the gonads of *Pseudobarbus afer* and *Pseudobarbus asper* (modified from Cambray & Bruton 1984)

Stage	Description
Stage 1 (juveniles)	Gonads small, ovary thin, translucent, without visible oocytes (40 × magnification) Testes very thin, thinner than ovaries and almost transparent
Stage 2 (inactive)	Includes immature virgins and recovering spent fish Ovaries small to moderate, oocytes (< 0 2 mm) and distinguishable at 10 × magnification Testes thin, white and strap-like
Stage 3 (maturing)	Ovaries noticeably enlarged and almost fill the width of the abdominal cavity Oocytes are now visible to the naked eye The largest (> 0 6 mm) are yellowish, some have the nucleus obscured by yolk deposition, and the smallest (< 0 2 mm) are still translucent Testes are swollen and whiter
Stage 4 (late maturing)	Ovaries almost fill the entire body cavity There are noticeably more mature oocytes scattered throughout the ovary than at stage 3 Ovaries are a distinct yellow colour Testes whiter and more enlarged occupying more than half the width of the body cavity Sexual products are not extruded when light pressure is applied
Stage 5 (ripe-running)	Gonads are at their maximum size Ovaries now distend the body cavity Oocytes large, translucent yellow, and when handled usually shed a few eggs with only slight pressure on the abdomen Testes are brilliant white, width similar to stage 4, and milt is easily extruded under slight abdominal pressure
Stage 6a (females – partially spent)	Ovary noticeably smaller than a stage 5 ovary with a smaller number of mature oocytes present Mature ova are still scattered throughout the ovary Recruitment oocytes, that is smaller yolked oocytes are more dominant than in previous stage Difficult to interpret as similar to stage 4
Stage 6b (spent males)	In males the partially spent condition could not be distinguished from stage 5 As in *B anoplus* the freshly collected testes were slightly pink, but in preserved material only the reduced size could be noticed in some specimens Spent males had thin testes, which were flaccid and translucent
Stage 7 (fully spent)	An additional stage was added for the females of both species since ovaries could be classified as partially spent (stage 6) or fully spent Ovaries small and flaccid and some retained only a few large oocytes which were granular in structure and a pale yellow (resorbing) and irregularly arranged in the ovary

been phenotypic options or what Balon (1990) has termed the 'first draft of evolution'.

In their study of the life-history phenotypic options of the sculpin, *Cottus gobio*, Mann et al. (1984, p. 179–181) posed the question, 'Do the different reproductive strategies in this species arise from genetic (ultimate) differences or from environmental (proximate) influences?' A reciprocal transfer experiment gave support to the thesis that there was a strong environmental influence upon life-history phenotypic options. Results suggested that genetically determined life-history phenotypic options are overshadowed by the effects of productivity and temperature. *P. afer* and *P. asper* may be ecophenotypes or a genetic explanation may exist for the few morphological and meristic differences. A comparative study assesses the variability of reproductive life-history attributes such as egg size, fecundity, age at maturity and seasonal and lifetime reproductive effort of *P. afer* and *P. asper*.

Fish species in stable environments with predictable environmental changes have a distinct suite of characters associated with a precocial life style (Bruton 1989, his table 1) Species which occur in an unstable environment with unpredictable environmental changes have a suite of characters associated with an altricial life style. Therefore the hypothesis tested in the present study was whether the *Pseudobarbus* species in the more stable mountain stream would adopt more precocial life-history phenotypic options than the species in the highly variable Karoo stream.

Methods

Both species were collected monthly over a 31 month period (October 1986 to April 1989) from the Gamtoos River system by means of minnow seine and scoop nets (Cambray & Bruton 1984).

The reproductive cycle was determined by measuring changes in gonadal condition. Data recorded were: fork length (FL) (mm), standard length (SL) (mm), total mass (g), gonad mass (g), gonadal maturation stages, monthly development of ova, nuptial tubercle formation. Eggs and newly hatched free embryos were collected. The raw data and samples of *P. afer* and *P. asper* were lodged at the Albany Museum, Grahamstown.

After blotting the fish, mass was measured to an accuracy of 10^{-3} g. Gonads of both sexes were removed, blotted dry and their mass measured to an accuracy of 10^{-4} g to assess seasonal changes in sizes of ovaries which reflected the growth of the developing oocytes.

The relative condition factor (K_n) was defined as:

$$K_n = Mo/Mp,$$

where Mo was the observed mass of the fish; Mp was the predicted mass from the length-mass relationship (LeCren 1951); K_n measured the deviation from the mass predicted for a fish of a given length for that population. K_n would be specific to a population and useful when quantifying changes in condition within a population but not for comparing the condition of fish from two or more populations (Wootton 1990). In the present study K_n was used to follow differences in total and somatic relative condition of the minnows on a monthly basis (Cambray & Bruton 1984) and not for comparing relative condition of *P. afer* to *P. asper.*

The relationship between length and mass for *P. afer* (n = 770) and *P. asper* (n = 460) was analyzed by measuring a sample of fish collected during the winter period when somatic and total mass readings were closest. The relationship was expressed as:

$$M = aL^b,$$

where M = mass in grams, L = SL in mm, a and b are constants estimated by regression analysis. These curvilinear relationships for *P. afer* and *P. asper* were fitted as recommended by Sokal & Rohlf (1973).

Significance of differences between mean monthly total and somatic condition at the 95% confidence level for all length groups for both sexes of both species was tested using the Student's *t*-test. A *t*-test was used because the data set was unbalanced. The level of significance for each *t*-test was reduced by dividing the nominal level of significance by the number of individual *t*-tests performed to ensure that the overall level of significance was not higher than 5% (Miller 1980). The data for October 1986, October 1988 and October 1989 were subjected to a one-way analysis of variance to establish whether there were differences in the prespawning condition of *P. afer* and *P. asper* females. When there were significant differences the source of variation between months was analyzed by the Scheffe multiple range test (Zar 1984).

The gonadosomatic index (GSI) was calculated for each specimen using the equation:

$$GSI = (\text{gonad mass} \times 100)/\text{total mass of fish}.$$

This index was used to describe the gonadal maturation cycle of *P. afer* and *P. asper.* In order to determine whether there were significant differences in the mean GSI values between months the data were subjected to a one-way analysis of variance for both males and females of each species. The source of variation between months was analyzed by the Scheffe multiple range test (Zar 1984).

Maturation stages were assigned (Table 1) after viewing the gonads at a magnification of 10–40 X through a stereoscopic microscope (Nikolsky 1963). An additional stage was added for females called 'partially spent' (stage 6), followed by spent (stage 7) (Cambray & Bruton 1984). Median size at maturity was the length at which 50% of the catch was mature.

Preserved ovaries were examined monthly from each of the ten largest female *P. asper* and *P. afer* to follow changes in mean egg size. The ovaries were put into petri dishes and the eggs near the posterior end of the ovary were separated and scanned with an eyepiece micrometer in a stereoscopic microscope. The ten largest eggs were selected and their diameter measured (accuracy 0.05 mm) for each fish giving a total of 100 eggs per month.

Fecundity is the product of the number of spawnings and the mean number of eggs per spawning within a breeding season (Cambray 1982). The term *multiple-spawning* used here refers to repeated reproduction within a season or year (Hubbs 1985; Burt et al. 1988). Ideally the fish should be collected immediately before spawning commences, providing that all the oocytes destined for later spawnings are mature enough to be counted. The main problem in studies on fecundity of multiple-spawning fish is to distinguish between reserve and developing oocytes (Bagenal 1978). A common criterion has been presence or absence of yolk, with a count of yolked eggs giving the fecundity for the season. *Absolute fecundity* was defined as the number of yolked eggs (mature and immature) in the ovary just prior to spawning.

Fish fecundity has traditionally been estimated by: direct counts of eggs in ovaries, counts or estimates made when females are stripped of their eggs, counting the eggs in a given mass or volume of the ovary, determining the total mass or volume of the ovary and then estimating the total number of eggs present by proportion (Carlander 1950).

In most studies on North American minnows, direct counts of eggs, rather than estimates, have been made which usually excluded minute and yolkless eggs. Phillips (1969) compared the accuracy of volumetric and gravimetric methods to direct counts of eggs for a minnow species and concluded that the best approach for small fish was to decide which eggs are mature and then determine their numbers by actual counts.

There were relatively few eggs in each *Pseudobarbus* female so the direct count approach was used to determine the absolute fecundity. In addition closer examination could be made of different egg sizes. Fish of different lengths were selected from data sheets whilst the mass and gonad mass section of the data sheet was covered. In many studies only the females with fully extended abdomens are chosen for fecundity studies (e.g. Heins et al. 1980). Cambray (1982) suggested that this method for selecting females might be misleading as only the most fecund fishes in the collection are studied.

A range in sizes of adult females for both species (*P. afer* n = 14; *P. asper* n = 14) was selected from the October collections. Ovaries were removed, measured to an accuracy of 10^{-4} grams and placed in vials containing 5% formalin. After hardening, the vials were shaken. The separated eggs were placed in a grooved perspex counting tray (Cambray 1982). Eggs were counted at 20x. Oocytes with yolked nuclei or fully yolked eggs (≥ 0.2 mm), were measured with a calibrated eyepiece micrometer in a stereoscopic microscope (accuracy 0.05 mm) and placed in seven 0.2 mm egg size categories (from ≥ 0.2 mm to ≥ 1.4 mm).

Regressions were calculated for each egg size category (where applicable) related to length and mass for both species. The following formula was used:

$$F_{batch} = aL^b,$$

where F_{batch} is batch fecundity, L is standard length, a and b are constants. Similarly the relationship between egg number (F_{batch}) and fish mass (M) was described by the function:

$$F_{batch} = aM^b.$$

Curvilinear regression lines expressing the relationship between standard length or mass were fitted (Sokal & Rohlf 1973). The equality of the regression lines for both length and mass of both species was tested for each egg size group (Zar 1984).

Early ontogeny methodology

Fish were collected during their breeding seasons so that the mature eggs were nourished in the field and not under artificial conditions and diets. In the laboratory they were injected with carp pituitary at 18:00 h and left overnight in covered aquaria. The parental groups consisted of 5 to 7 males and 10 to 15 females to ensure heterogeneity. Eggs were stripped into petri dishes and the testes were mashed in a separate container and immediately added to the eggs. A saline solution was not used for sperm because it might have influenced hardening and final size of egg envelopes, especially of *P. afer* which comes from an area of low conductivity. The eggs were then put into nylon meshed hatching trays and incubated at 23° C ± 1° C in aquaria with continuous water currents passing over the eggs and airstones under the hatching trays. The photoperiod was controlled by an automatic timer at 13 light : 11 dark hours during incubation. Aquaria were covered with black plastic sheeting to stimulate the natural spawning habitats where the eggs would be in dark areas between and under boulders. Once hatching commenced half of each aquarium was uncovered. When the fish showed positive phototaxis the entire aquaria were exposed to the

13:11 h photoperiod to approximate spawning habitat photoperiods. Fish larvae were fed a laboratory culture of rotifers and finely ground tropical fish food.

Sampling was continuous for the first hour after activation but decreased as the fish grew and developmental events slowed. For the purpose of this study behaviour of free embryos at hatching, age at first swim-up, and age and size at first feeding were seen as important comparative life-history events, and are reported here. Detailed comparative early ontogeny of these two species will be published separately.

Life-history terminology

Use of the terms tactic, trait and strategy have been criticised (Balon 1990). Strategies and tactics imply rational planning which is inappropriate in the context of evolutionary biology (Wootton 1984). The terms style, attribute and phenotypic option are used here instead of strategy, trait and tactic.

Within a life-history style there are a number of life-history attributes, such as age at first maturity. In different environments an individual may have different phenotypic options such as early or delayed maturity. The range of the expression of these phenotypic options is under genetic control. The variability of expression of a life-history attribute is called phenotypic variability. Phenotypic variability is a term used to describe a range of phenotypic options available to an individual such as changes in egg size, growth rate, age at maturity etc. which are used to confront a particular environment.

Variable phenotypic trajectory is synonymous with plastic trajectory as used by Stearns & Crandall (1984). This trajectory reflects the phenotypic options which an individual has in its genome. Stearns & Crandall (1984) considered two life-history attributes (age and size at maturity) and established that organisms matured along a trajectory of age and size which depended on demographic conditions.

Balon (1990) adopted the terms *altricial* and *precocial* from the accepted terminology for birds (Ricklefs 1979). These terms are used in the present study to indicate parental investment per progeny which was seen as an important difference between *P. afer* and *P. asper* which possibly influences their life-history trajectories. The main attributes of the altricial form are relatively smaller and incompletely developed young compared to larger, better developed young in the precocial forms. Balon (1990) believed that the differences between altricial and precocial forms in ontogeny are small, with the generalist a little more inclined towards attributes of altriciality and the specialist more inclined towards the attributes of precociality. These terms will be used in an interspecific sense when referring to *P. afer* and *P. asper*. These terms reflect the effect of the mechanism of epigenetic bifurcation and emphasize the importance of the interplay between the environment and the genome (Balon 1990).

Alprehost stands for 'Altricial = Precocial Homeorhetic States' which Balon (1990, p. 31) has suggested '... is the cause of evolutionary patterns'. The ontogeny of each taxon is created in a sequence of these alternative altricial to precocial states. *Alprehost* is best described in Bruton (1989) and Balon (1990, his fig. 20).

Reproductive style terminology for *P. afer* and *P. asper* follows the comprehensive classification of Balon (1975, 1981a) based on spawning site, adaptations of eggs and embryos to the site and degree of parental care.

Results

Wit and Groot River environments

Photographs of the environment (Fig. 2) clearly show the distinctive differences between the Cape Fold Mountain Belt stream (Wit River) and the Karoo stream (Groot River). The photoperiod, differences in temperature and rainfall between the two collection sites during the study period are summarised in Figure 4. The chemical differences (Table 2) can readily be seen in the fluctuations of the conductivity readings in the Groot River as compared to those in the Wit River (Fig. 5). The Wit River is a perennial system whereas the Groot River flows intermittently (Cambray 1991).

Condition factor

The condition factor for *P. afer* collected in winter

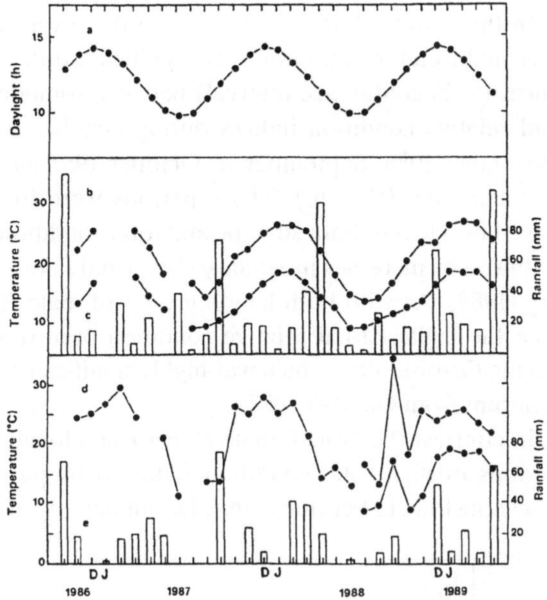

Fig 4 a – Mean number of daylight hours, b – maximum and minimum water temperatures, Wit River, c – monthly rainfall Wit River, d – maximum and minimum water temperatures Groot River, e – monthly rainfall, Groot River at Fullarton

was described by the equation: $M = -11.1866L^{3\ 13683}$ (n = 772, r = 0.9977) and for *P. asper* by: $M = -11.0504L^{3\ 08024}$ (n = 460, r = 0.9919). The total and somatic relative condition factors during the study period for the length group 35–45 mm SL and > 45 mm SL showed seasonal trends (Fig. 6, 7).

In male *P. afer* (35–45 mm SL) there was no significant difference between the somatic and total condition (Fig. 6a). A similar seasonally fluctuating pattern occurred in the small female *P. afer* with no significant differences between somatic and total condition for the entire period (Fig. 6b). The changes in condition factor of the small *P. asper* males followed the same seasonal pattern with winter lows and early summer highs and no significant difference between the total and somatic condition (Fig. 6c). In contrast the small female *P. asper* showed significant differences (95% confidence interval) between somatic and relative condition from September to December 1987 but remained relatively stable in the winter months (Fig. 6d).

Larger male *P. afer* (> 45 mm SL) showed no significant differences between somatic and total condition during the study period but a very similar seasonal trend to that seen in the smaller males was apparent with the low August 1987 and high October 1988 condition factors (compare Fig. 6a, 7a). In both years there was a decrease in condition during December (post breeding period) followed by an increase in January. There was also a seasonality in the condition of the large male *P. asper* with the highest condition during November followed by a decrease during the breeding season of both somatic and total condition (Fig. 7c). There were no significant differences between somatic and relative conditions during the study period.

Table 2 Water analysis of the Wit and Groot rivers giving the extreme ranges of the monthly readings during the period 22 February 1987 to 23 April 1989

	Wit River	Groot River	Difference between highest reading, taking Wit River as one
pH	6 6–7 2	8 0–8 5	–
Electrical conductivity (mSm^{-1})	7 7–10 94	118 0–1006 0	92
Total dissolved solids (mgl^{-1})	44 0–159 0	763 0–5063 0	32
Turbidity (ntu)	0 2–3 0	1 7–36 0	12
Secchi disk (cm)	500 plus*	8–129	–
Chloride (mgl^{-1})	12 408–23 47	255 9–3172 8	135
Alkalinity as $CaCo_3$ (mgl^{-1})	2 5–30 0	10 0–245 0	8
Calcium (mgl^{-1})	0 5–2 0	29 0–235 0	118
Magnesium (mgl^{-1})	1 4–2 9	31 0–313 0	108
Sodium (mgl^{-1})	10 0–16 0	50 0–1500 0	94
Potassium (mgl^{-1})	0 4–0 9	0 5–27 8	31

* Secchi disk was always visible to bottom of deepest pool

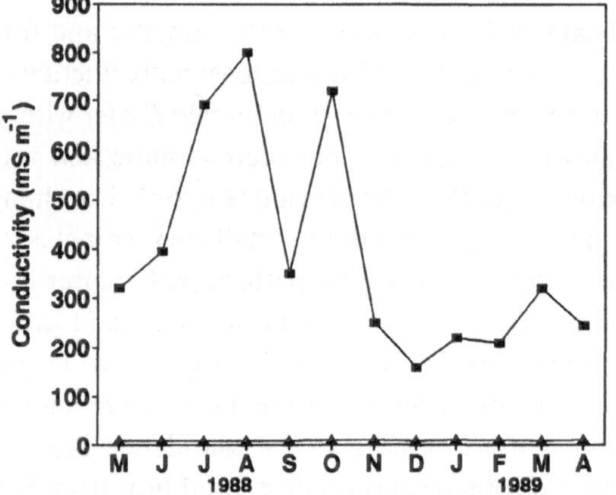

Fig 5 Fluctuations in the conductivity readings for the Groot River (squares) and the Wit River (triangles), Gamtoos River system between May 1988 and April 1989

In the female *P. afer* (> 45 mm SL) there was a seasonal trend in condition with significant differences (95% confidence interval) between somatic and relative condition indices during October to November 1986, September to October 1987 and October 1989 (Fig. 7b). These periods were followed by a loss of both somatic and total condition followed by increases in January (1987) and February (1989). As with both length groups of male *P. afer*, the highest total relative condition occurred during October 1988 which was highly significantly different from October 1987.

In contrast the large female *P. asper* had longer periods of high total condition followed by post spawning lows in February 1987, December – Janu-

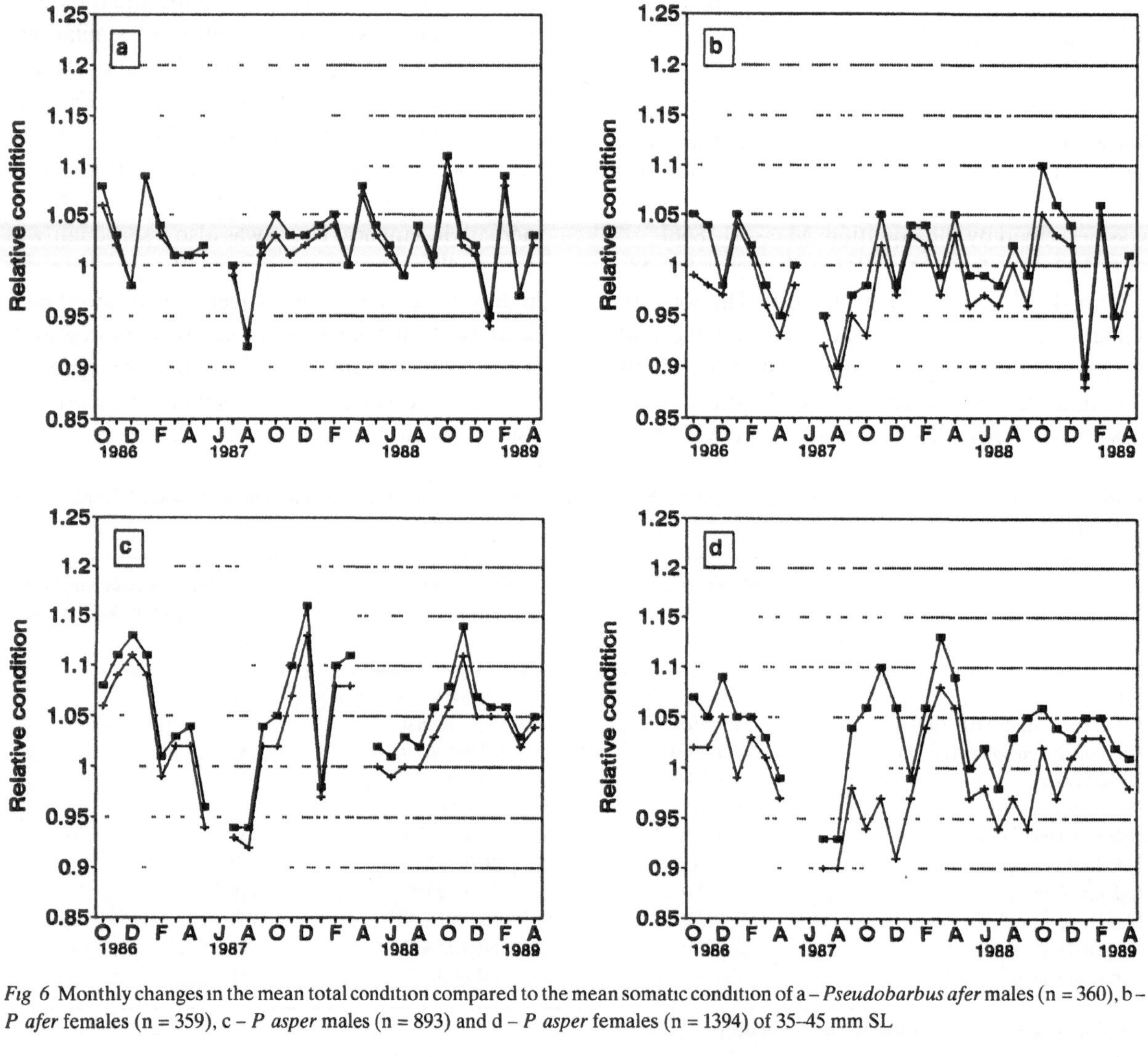

Fig 6 Monthly changes in the mean total condition compared to the mean somatic condition of a – *Pseudobarbus afer* males (n = 360), b – *P afer* females (n = 359), c – *P asper* males (n = 893) and d – *P asper* females (n = 1394) of 35–45 mm SL

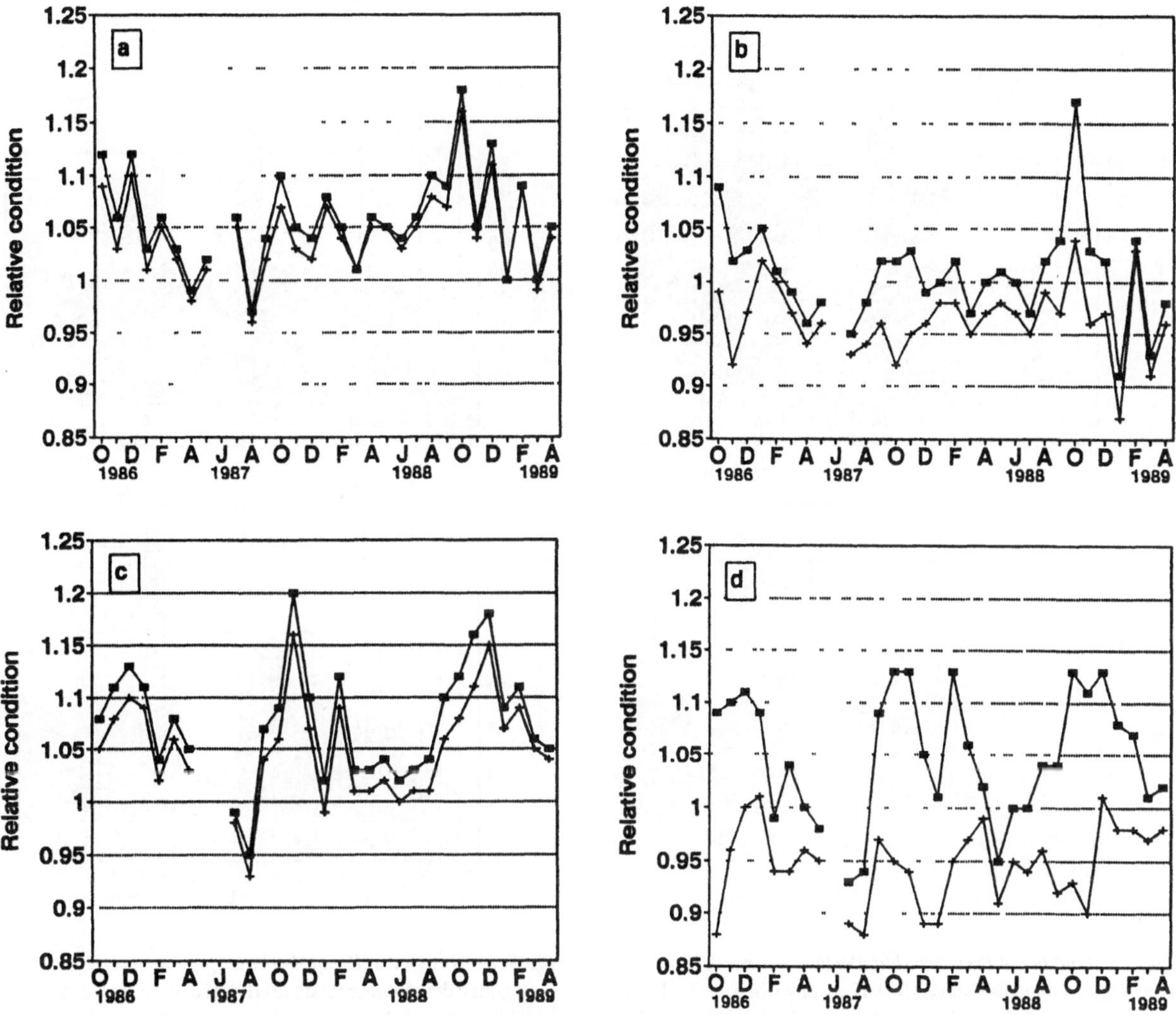

Fig 7 Monthly changes in the mean total condition compared to the mean somatic condition of a – *Pseudobarbus afer* males (n = 297), b – *P afer* females (n = 1166), c – *P asper* males (n = 561) and d – *P asper* females (n = 1524) greater than 45 mm

ary 1988 and November 1988 (Fig. 7d). There were significant differences (95% confidence interval) for 20 months (October 1986 to December 1986, August 1987 to March 1988, May 1988 to December 1988 and February 1989) between the somatic and total conditions. There were no significant differences (95% confidence level) between the pre-spawning high total condition indices for the three years for female *P. asper* (> 45 mm SL) unlike the significant differences between October 1987 and October 1988 for the total condition of the *P. afer* females.

Maturation stages

Seasonality of both species was clearly apparent and *P. asper* had a longer breeding season. *P. afer* could breed from October to February (Fig. 8) whereas *P. asper* could breed from October to as late as April (Fig. 9).

Gonadosomatic index (GSI)

In both sexes of both species there was a seasonal pattern of gonadal activity. The proportion of ripe fish was greatest from October to February 1987 and October to January 1988 for *P. asper* (Fig. 10a) and from October to November 1987 and 1988 for *P. afer* (Fig. 10b). In both species the mean GSI values for females were significantly higher (95% confidence interval) during the breeding season than at other times of the year. In the males and females of

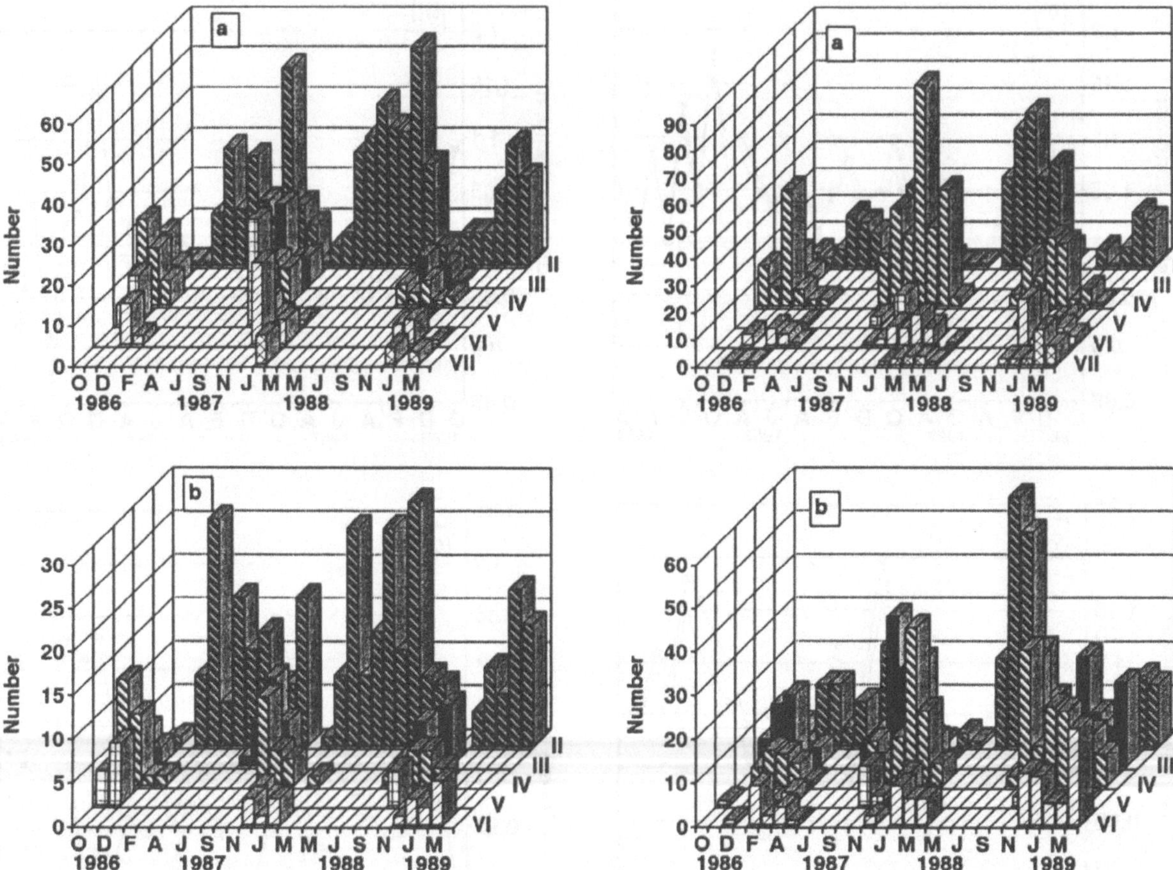

Fig. 8. a – Number of *Pseudobarbus afer* females (n = 914) at each gonad developmental stage (II–VII as described in Table 1), caught per month, b – number of *P. afer* males (n = 432) at each gonad developmental stage (II–VI), caught per month. Maturity stage I (immatures) are not included here as they are uninformative for seasonal maturity trends.

Fig. 9. a – Number of *Pseudobarbus asper* females (n = 1044) at each gonad developmental stage (II–VII as described in Table 1), caught per month, b – number of *P. asper* males (n = 901) at each gonad developmental stage (II–VI), caught per month. Maturity stage I (immatures) are not included here as they are uninformative for seasonal maturity trends.

P. afer and *P. asper* there were no significant differences between the three October periods in GSI values.

An analysis of the differences between the mean monthly GSI values of female *P. afer* and *P. asper* showed that there were significant differences (95% confidence interval) for most months (except for November 1986, January, March and July 1987 and January 1989). The difference was not as pronounced but still evident in males with significant differences (95% confidence interval) for 15 months (February 1987, April to August 1987, November 1987 to February 1988, May 1988 to September 1988, November 1988, February and March 1989). *P. asper* males and females therefore allocated significantly more resources to their gonads than did *P. afer* males and females.

The highest GSI recorded was 28.13 for a *P. asper* female (69.7 mm SL) collected during October 1986. In contrast the highest GSI for a *P. afer* female (57.8 mm SL) was 15.46 for a specimen collected during October 1987. The area under the GSI curve for *P. asper* (Fig. 10a) between September 1987 and April 1988 is 2.87 times that of the area under the *P. afer* GSI curve for the same period (Fig. 10b). In the following season, September 1988 to April 1989, this figure was 2.32, indicating higher reproductive effort in *P. asper*. The pattern in the reproductive sea-

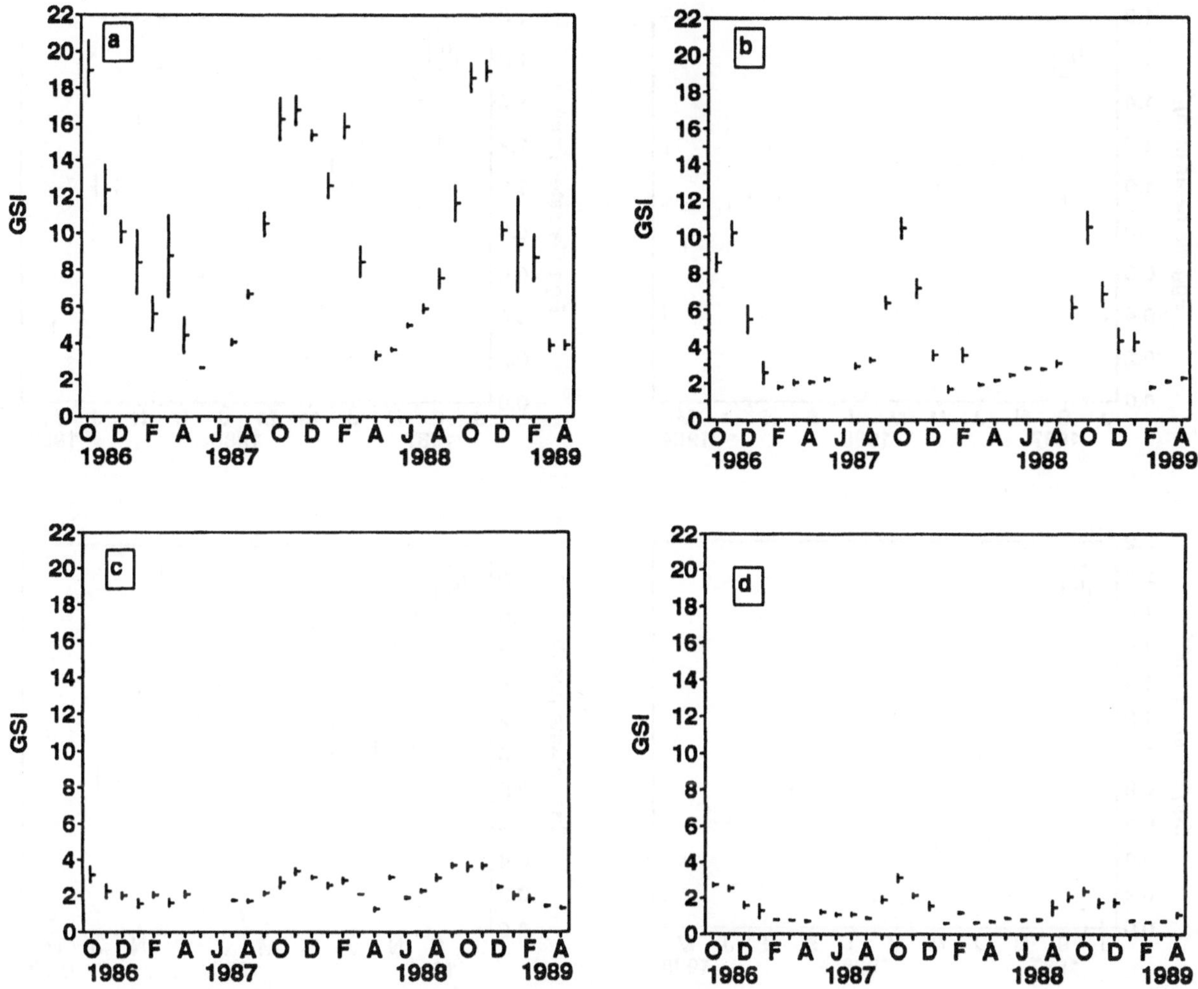

Fig 10 Monthly mean gonadosomatic indices for mature female a – *Pseudobarbus asper* (n = 790) collected in the Groot River, b – mature female *P afer* (n = 615) collected in the Wit River, Gamtoos River system, October 1986 to April 1989, c – mature male *P asper* (n = 563) collected in the Groot River and d – mature male *P afer* (n = 299) collected in the Wit River, Gamtoos River system, October 1986 to April 1989 Mean and ± one standard error

sonality of female *P. asper* showed an increase in gonad mass which commenced in May and reached a peak in the October – November period during which spawning occurred. In the summer of 1987/1988 the protracted spawning period was from October to April. The possibility of a late spawning was confirmed in April 1989 when an artificial release of water from Beervlei Dam was sufficient to trigger spawning (Cambray 1991). *P. asper* was therefore able to spawn over a 6–7 month period. In contrast the *P. afer* population in the Wit River had briefer reproductive seasons (Fig. 10b). The population usually finished breeding by the end of January (1987, 1989) although some individuals were still capable of spawning in February 1988. The reproductive season was therefore limited to 4–5 months with the major spawning during October-November.

There was sexual dimorphism in gonad sizes of both species, and less energy was put into the maturation of male gonads (Fig. 10c, d). The highest GSI recorded for a male *P. asper* (44.4 mm SL) collected during November 1987 was 5.58. The highest GSI for a male *P. afer* (70.8 mm SL) collected in October 1987 was 5.1. The higher GSI for *P. asper* females was not apparent in the *P. asper* compared to *P. afer* males. However, the area under the GSI curves (Fig. 10c) for *P asper* males between September

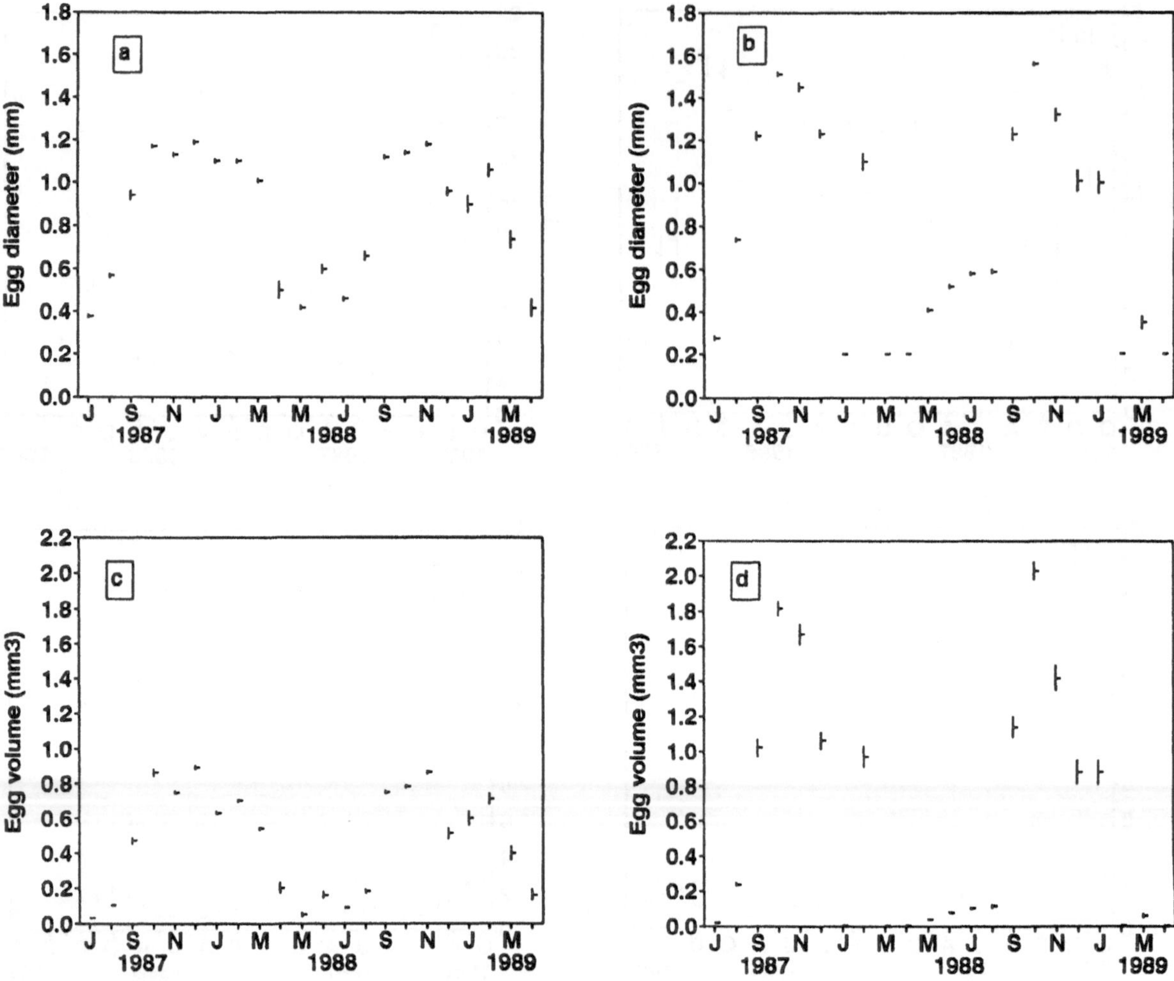

Fig. 11. Seasonal changes in diameter of the largest eggs in a – *Pseudobarbus asper* (n = 219 fish, n = 2190 eggs) and b – *P. afer* (n = 203 fish, n = 2030 eggs) and seasonal changes in volume of the largest eggs in c – *P. asper* (n = 219 fish, n = 2190 eggs) and d – *P. afer* (n = 203 fish, n = 2030 eggs) over the time period 12 July 1987 to 23 April 1989. Mean and ± one standard error.

1987 to April 1988 was 1.85 times that of *P. afer* males (Fig. 10d) for the same time period. In the following season, September 1988 to April 1989 the figure was 1.83.

Seasonality of ova sizes

P. afer had consistently larger ova with diameters up to 1.8 mm whereas the largest ova for *P. asper* was 1.3 mm (Fig. 11). There was a highly significant difference between the egg diameters of *P. asper* and *P. afer* during both October 1987 ($t = 20.684$, $p < 0.001$, DF = 198) and October 1988 ($t = 28.283$, $p < 0.001$, DF = 198).

Mature ova of *P. afer* in October 1987 had a mean egg diameter of 1.51 mm (n = 100, SE = 0.01) and in October 1988 1.56 mm (n = 100, SE = 0.01), whereas *P. asper* were 1.12 mm (n = 100, SE = 0.001) in October 1987 and 1.14 mm (n = 100, SE = 0.01) in October 1988. Assuming that the egg was spherical, a 1.5 mm diameter egg would have a volume of 1.77 mm^3 whereas that of a 1.1 mm egg of *P. asper* would be 0.7 mm^3. A *P. afer* egg of 1.5 mm in diameter would be 2.53 x the volume of a 1.1 mm *P. asper* egg. The largest mean egg volume for the 10 largest eggs from the 10 largest *P. afer* occurred in October 1987 (1.82 mm^3) and in October 1988 (2.03 mm^3) (Fig. 11d). For *P. asper* the highest mean egg volume occurred in December 1987 (0.89 mm^3) and in November 1988 (0.86 mm^3) (Fig. 11c). For October 1987 the difference in mean egg volume

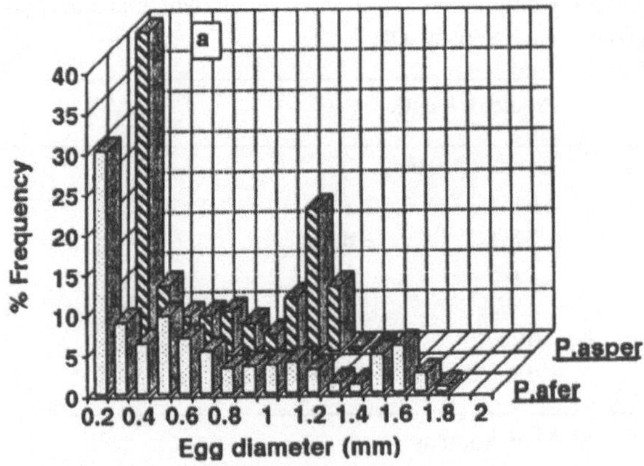

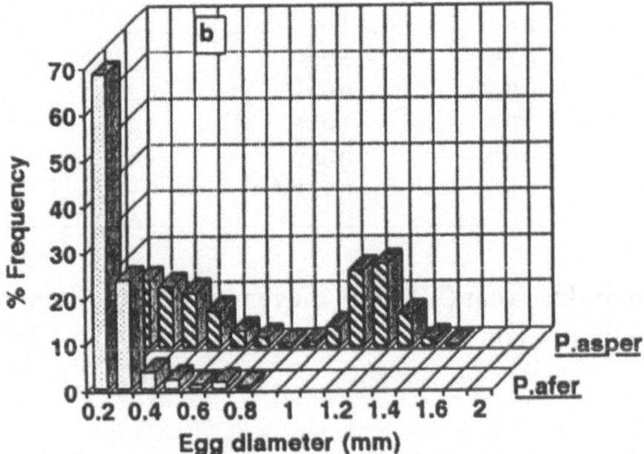

Fig. 12. Egg size frequencies of a – *Pseudobarbus asper* (63.7 mm SL, n = 2272 eggs) collected on 28 October 1988 in the Groot River and *P. afer* (61.1 mm SL, n = 1368 eggs) collected on 26 October 1988 in the Wit River and b – egg size frequencies of *P. asper* (76.5 mm SL, n = 3058 eggs) collected on 18 February 1989 in the Groot River and *P. afer* (64.3 mm SL, n = 239 eggs) collected on 19 February 1989 in the Wit River

was a ratio of 2:1 (*P. afer : P. asper*) and for 1988 this ratio was 2.4:1. There were highly significant differences in egg volume between the two species in October 1987 ($t = 20.777$, $p < 0.001$, DF = 198) and October 1988 ($t = 23.722$, $p < 0.001$, DF = 198). In the extreme range the largest *P. afer* egg had a volume of 3.05 mm^3 whereas the same figure for *P. asper* was 1.15 mm^3 (Fig. 11c, d) which was a ratio of 2.7:1 similar to the mean egg volume ratios mentioned above.

Egg size frequencies of ripe female *P. afer* and *P. asper* collected during October 1988 demonstrated the smaller egg size of *P. asper* as compared to that of *P. afer* (Fig. 12a). In each species there was a high percentage of oocytes with yolked nuclei which would have had the potential to become mature eggs. A similar comparison of egg size frequency differences for fish collected during February 1989 indicated that there was a higher percentage of oocytes with yolked nuclei in *P. afer* than *P. asper* (Fig. 12b).

Length – fecundity relationship

The regression statistics for body length and number of eggs in various size classes indicated a highly significant relationship in all groups especially those greater than 0.4 mm for both *P. afer* and *P. asper* (Table 3, 4). When all the yolked nuclei ≥ 0.2 mm in diameter were counted, the highest absolute fecundity recorded was 4771 in a 63.7 mm SL *P. asper* collected on 28 October 1989. The highest absolute fecundity of a *P. afer* (73.1 mm SL) was 3922 of which only 783 (20%) were ≥ 0.8 mm. For the above-mentioned *P. asper* this figure was 1662 (35%). As indicated in the previous section, many of the recruitment eggs with yolked nuclei of *P. afer* may go unutilized. It may therefore be more meaningful to compare only fully yolked 'mature' egg counts (≥ 0.8 mm) between the two species rather than counts including recruitment eggs (Table 3). The main problem here was that *P. afer* has larger mature eggs and therefore one was including all the eggs from 0.8 to 1.5 mm compared to *P. asper* where only the eggs from 0.8 to 1.1 mm were included. There was therefore a tendency to include eggs in more size classes for *P. afer* than *P. asper* although overall *P. asper* still had more eggs.

Mass – fecundity relationship

At the 95% confidence level there was a highly significant relationship between body mass and fecundity in both *P. afer* and *P. asper* (Table 3, 4) for all egg size groups. The *P. afer* and *P. asper* fecundity regression lines were not equal for the five egg size categories related to body length and mass for both species (Table 5). The least significant difference was seen in the egg size group ≥ 0.2 mm.

Spawning sites and reproductive guild

P. asper spawned after an increased river flow either caused by rainfall or by water released from a dam

Table 3 Regression statistics of the relationship between fecundity (F) including different ova sizes and standard length and mass in *Pseudobarbus afer* (n = 14), F = ax^b (s e = standard error, r = correlation coefficient, *** p < 0 001)

Pseudobarbus afer				Predictor length		
Egg size (mm)						
	≥0 2	≥0 4	≥0 6	≥0 8	≥1 0	≥1 2
a	− 5 755	− 7 05	− 7 716	− 7 875	− 8 166	− 10 928
s e of a	1 98	1 149	1 207	1 126	1 07	2 107
b	3 217	3 347	3 43	3 393	3 384	3 994
s e of b	0 504	0 292	0 307	0 286	0 272	0 536
r	0 88***	0 957***	0 955***	0 968***	0 963***	0 907***
Pseudobarbus afer				Predictor mass		
Egg size (mm)						
	≥0 2	≥0 4	≥0 6	≥0 8	≥1 0	≥1 2
a	5 754	4 907	4 529	4 238	3 914	3 325
s e of a	0 217	0 113	0 109	0 099	0 09	0 206
b	0 98	1 035	1 068	1 056	1 054	1 247
s e of b	0 159	0 083	0 08	0 073	0 066	0 152
r	0 872***	0 963***	0 968***	0 972***	0 977***	0 922**

(Cambray 1991) and *P. afer* spawned after an increase in water flow during rains. Eggs were located under boulders in mid-channel for both species and there was no observed parental guarding. Fish bred just above (< 2 m) riverine pools. Late free embryos and early larvae drifted from riffles into pools where they were observed feeding in the pelagic zone.

At hatching free embryos of both *P afer* and *P asper* exhibited negative phototaxis and during the swim-up period exhibited positive phototaxis after 5 days for *P. afer* and 4.25 days for *P. asper.* There was almost a difference of three days in the time to first feeding between *P. afer* (10.21 days) and *P. asper* (7.25 days). The laboratory reared *P. afer* larvae were also larger at first feeding (8.99 mm TL) than *P. asper* (6.6 mm TL) (Table 6). The reproductive guild of both species is nonguarding rock and gravel spawners with benthic larvae (A.1.3) (Balon 1990).

Size at sexual maturity

P. afer males and females matured at 39–40 mm SL (Fig. 13a). *P. asper* males matured at 41–42 mm SL whereas *P asper* females matured at 43 mm SL (Fig. 13b). Lengths between 39–43 mm SL appeared to be the size at which 50% of the population of both species became sexually mature. *P. afer* were in their third year (2^+) when they matured and *P asper* in their second year (1^+) (Cambray 1992).

Table 4 Regression statistics of the relationship between fecundity (F) including different ova sizes and standard length and mass in *Pseudobarbus asper* (n = 14), F = ax^b (s e = standard error, r = correlation coefficient, *** p < 0 001)

Pseudobarbus asper				Predictor length	
Egg size (mm)					
	≥0 2	≥0 4	≥0 6	≥0 8	≥1 0
a	− 1 286	− 1 879	− 4 479	− 6 437	− 7 372
s e of a	1 364	1 279	1 355	1 424	1 51
b	2 261	2 297	2 878	3 299	3 48
s e of b	0 335	0 314	0 333	0 35	0 371
r	0 89***	0 904***	0 928***	0 939***	0 938***
Pseudobarbus asper				Predictor mass	
Egg size (mm)					
	≥0 2	≥0 4	≥0 6	≥0 8	≥1 0
a	6 746	6 291	5 76	5 299	5 01
s e of a	0 16	0 154	0 162	0 166	0 179
b	0 722	0 727	0 908	1 043	1 098
s e of b	0 095	0 092	0 096	0 098	0 106
r	0 91***	0 917***	0 939***	0 951***	0 948***

Table 5 Test for equalıty of *P afer* and *P asper* regressıon lınes for number of ova ın dıfferent sıze classes to length and mass (95% confıdence levels)

Pseudobarbus afer and *Pseudobarbus asper* Length		
Egg sıze (mm)	F	p
≥0 2	11 42	<0 001
≥0 4	61 572	<0 001
≥0 6	62 248	<0 001
≥0 8	71 892	<0 001
≥1 0	90 669	<0 001
Pseudobarbus afer and *Pseudobarbus asper* Mass		
≥0 2	10 534	0 001 < p < 0 002
≥0 4	68 306	<0 001
≥0 6	70 145	<0 001
≥0 8	86 832	<0 001
≥1 0	98 32	<0 001

Discussion

'Each individual fish has a suite of reproductive attributes which are determined by its genotype and hence by the evolutionary history of the gene pool of which the fish is a member' (Wootton 1990, p. 160). The reproductive style is a combination of the reproductive attributes of individuals belonging to the same gene pool. Thıs has been diagrammatically represented for *P. afer* and *P. asper* to indicate the range of phenotypic expression and phenotypic op-

Table 6 Summary of several ımportant early lıfe-hıstory events of *P afer* and *P asper*

	P afer	*P asper*
Egg sıze (mm)	1 5	1 1
Egg volume (mm³)	1 77	0 7
Fırst cleavage (mın)	66	50
Fırst heart-beat (days)	1 85	1 6
Negatıve phototaxıs (days)	2 42	2 25
Posıtıve phototaxıs (days)	4 96	4 25
Swım-up (days)	4 96	4 25
Fırst gulpıng movements (days)	6 54	5 0
Swımbladder ınflated (days)	9 54	6 67
Fırst feedıng (days)	10 21 (8 99 mm TL)	7 25 (6 6 mm TL)

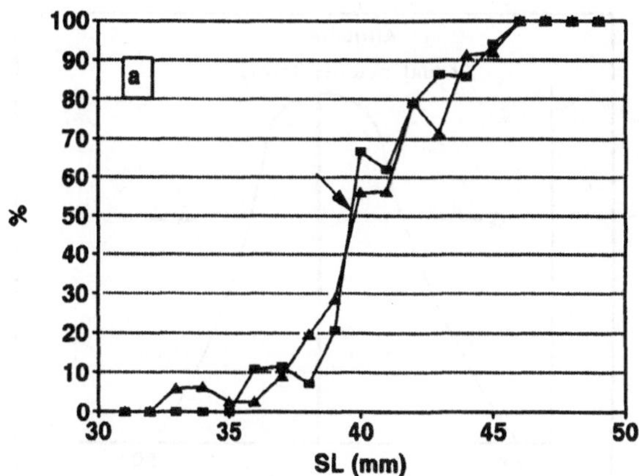

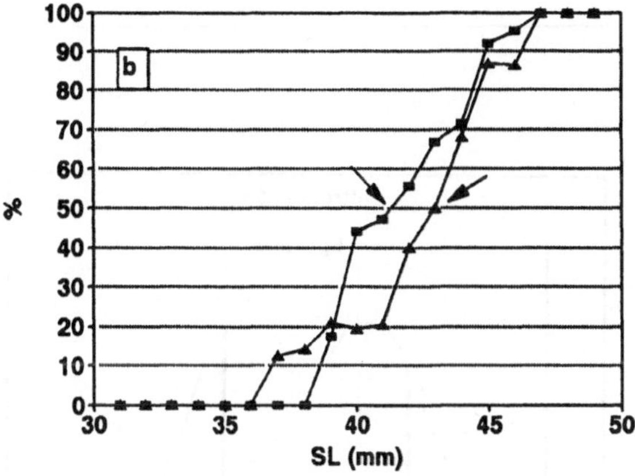

Fıg 13 Percentage of ındıvıduals that are sexually mature a – male *Pseudobarbus afer* (n = 568 squares) and female *P afer* (n = 1002, trıangles) and b – male *P asper* (n = 1385, squares) and female *P asper* (n = 1230, trıangles)

tions of each specıes correspondıng to their different environmental factors (Fig. 14).

There ıs a cyclical demand for both energy and materials to operate the 'reproductive machinery' of a fish (Miller 1984). Within the overall reproductive style there ıs a temporal pattern in amplitude, frequency and dırectıon of the fuelling of reproduction. The 'reproductıve machınery' of *P. asper* requires more fuelling (hıgher GSI) channelled into many small eggs for a longer reproductive season. *P. afer* and *P. asper* therefore were found to have distinct differences ın the temporal pattern of allocation of resources for reproductıon and also in the way these resources were partitioned.

P. afer was found to be more precocial whereas *P. asper* was the more altrıcial species (Cambray 1992). Shifts ın parental ınvestment per offspring

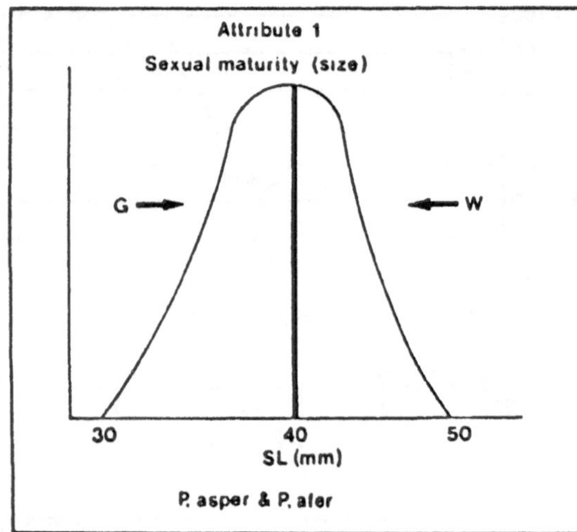

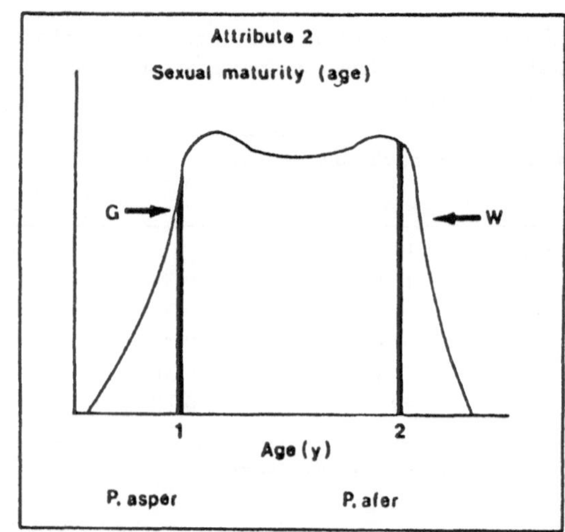

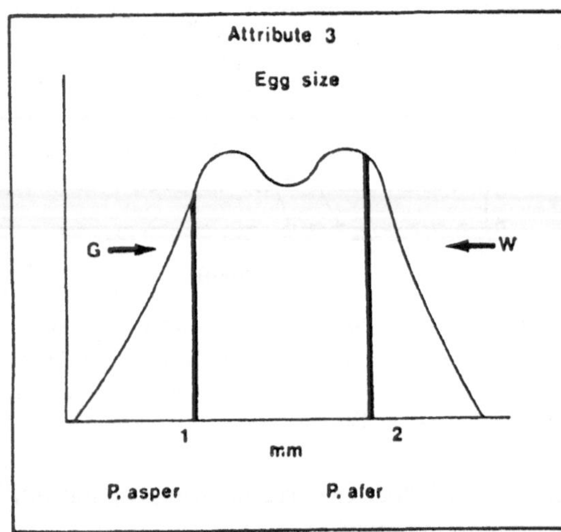

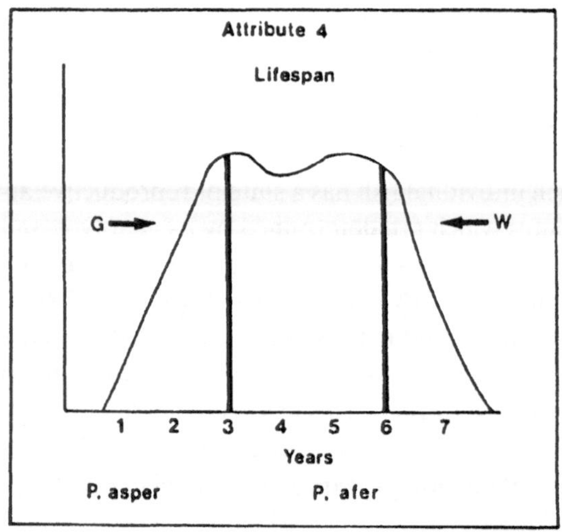

Fig. 14. Variability of several of the life-history attributes of *Pseudobarbus afer* and *P. asper.* Diagrammatic reproductive style of *P. afer* and *P. asper* showing the range of phenotypic expression (illustrated by the curves) of each attribute for the sister species pair. The phenotypic option expressed by *P. asper* and *P. afer* is shown by the thick vertical lines and depends on the environmental conditions (W = Wit River and G = Groot River environments) experienced by each of the sister species (modified from Wootton 1990).

may have resulted in alternative life-history trajectories for *P. afer* and *P. asper.* It was hypothesized that the investment per offspring was important and as noted by Reid (1985, p. 251) '. . . the earlier in development a change occurred, the greater the evolutionary saltation might be'. Prior to having any evolutionary impact these epigenetically introduced changes lead to alternative life histories which can occur even in a clutch from the same parent (Balon 1980, 1981b, 1985, 1990). The more pronounced events (e.g. phylogenetic divergence) are stronger 'reverberations' of the same epigenetic process (Balon 1990). The amount of parental investment per offspring is the earliest event which 'fuels' development and ultimately influences the life-history trajectory of that individual.

It is hypothesized here that a population of '*P. afer*' occurred in the Groot River and over time coevolved as the river changed from perennial, clear, oligotrophic with a low conductivity to an intermit-

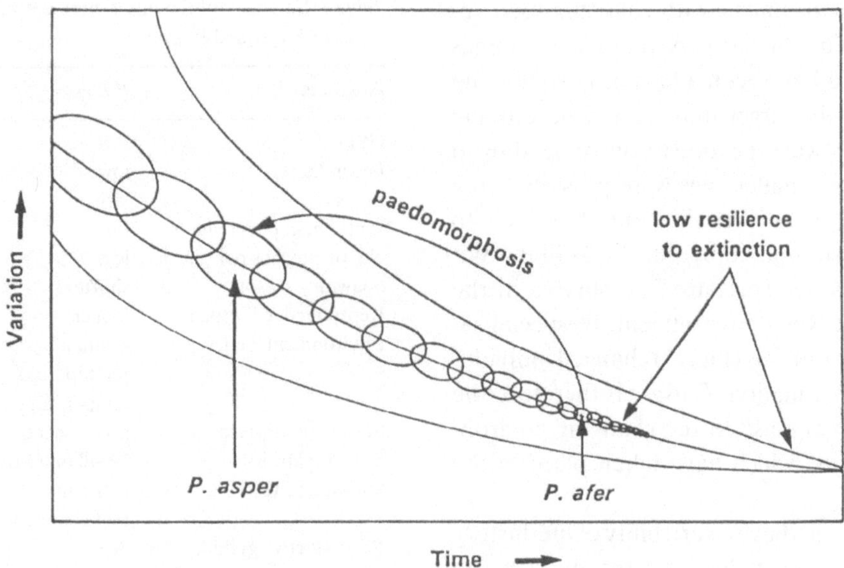

Fig 15 Stabilized trajectory for genetic and epigenetic variation in a succession of reproductive lineages towards a precocial *Pseudobarbus afer P asper* is hypothesized as having undergone an epigenetic retreat via paedomorphosis to a species that is more altricial and having more variability (adapted from Balon 1990)

tent, turbidity and eutrophic environment with a high conductivity. The precocial *P. afer* individual was vulnerable to these changes and the eventual extinction of the redfin minnows in this environment was prevented by the survival of the juvenilized ('despecialized') progeny. As the nature of the Groot River changed, individuals with altricial attributes were selected for and survived. The fish had longer guts (Skelton 1988) and could utilize smaller and less nutritious food items. The formation of smaller scales (Skelton 1988) may have been due to epigenetic influences, such as salinity. There may be epigenetic induction of squamation development in fishes by mechanical constraints (Sire & Arnulf 1990). The differences and ranges of scale sizes (meristic count variability) which Skelton (1980) found in the *P. afer/P. asper* complex may be due to salinity and/or other epigenetic influences in the different environments. With the increase in turbidity and enrichment of the Groot River, the investment per progeny of the '*P. afer*' population became less and there were more eggs which resulted in higher relative fecundities. This may have led to more variability in the number of eggs per spawning and an increased number of spawnings per season per fish. In individuals which were adapted in this way the embryos hatched in a shorter time and therefore more survived as the river flowed irregularly and for shorter periods. Over time a species with smaller scales, longer guts (food availability – quality and quantity?) and a different pigmentation pattern (turbidity?) evolved which had a tolerance to high salinity levels and poor visibility, i.e. *P. asper.* It was not necessary for a set of altricial and a set of precocial eggs to be spawned each generation as suggested by Balon (1990). There is size variability in batches of eggs that are spawned but it has not been found here to be in two distinct groups, one altricial set and one precocial set. The dichotomy we see today in the redfin minnows of the Gamtoos River system, *P. afer* and *P. asper,* may not be the result of the production of two forms within each generation but with the production of variability within each generation, which would suggest a range of forms, here called the altricial to precocial range of forms. This is not as simple as the altricial/precocial pair idea but may be more relevant in this case as a means to ensure the continuation of variability. Variability in egg size in the two environments of the Gamtoos River system probably plays an important evolutionary role and is seen as adaptive in a fluctuating environment. The changing

Groot River environment with changing food resources favoured a higher survival of smaller eggs and there would have been a higher mortality due to stranding of the larger more slowly developing eggs. So not only were the food resources leading to smaller eggs, the smaller eggs were possibly being selected for simply because they were less likely to be stranded or smothered by silt. *P. afer* in the Wit River form was overspecialized to survive in the changing Groot River environment. Despecialization by paedomorphosis (Fig. 15) changed individuals within this population. *P. asper* is therefore the paedomorphic response to the changing environmental conditions which have taken place in the Groot River.

Many fish species have a variability of life-history attributes such as fecundity, growth rate, size and age at sexual maturity. Eventually individuals within a population may adapt to a long-term change through alterations in gene and genotype frequencies. The environmental change (Wit River – Groot River or Groot River – past to Groot River – present) may have acted as an evolutionary factor causing changes in the gene pool of the population through the process of natural selection and/or by epigenetic mechanisms. The *Pseudobarbus* population in the Karoo stream had to make some changes to phenotypic options in their reproductive style to be successful in the changing environment of the Karoo stream (Table 7). The adaptive and ecological significance of such changes in phenotypic options of life-history attributes are directly related to the two distinct environments. *P. asper* have shown the potential flexibility of the *Pseudobarbus* species in their response to environmental change in life-history attributes such as age at maturity and increased reproductive effort per season. There may be two different selection pressures for multiple spawning. In the *P. asper* population there is a need to spread the risk of losing an entire year class whereas in *P. afer* there is a need to spread the number of fish larvae over time with regard to scarce food resources and possibly limited spawning sites (Cambray 1992).

The time courses of the two mechanisms of adaptation, phenotypic variability and genetic selection, are different (Wootton 1990). Quick changes can be made by phenotypic adaptation within a generation. Egg size changes in the *Pseudobarbus* may have been rapid and such phenotypic adaptations to the immediate environment may have occurred within a generation. This would be the phenotypic option at the individual level in response to the encounter with an enriched environment. A phenotype is an information gathering and transmitting device and epigenesis 'creates' new phenotypes (Balon 1990). The genetic adaptations, such as salinity tolerance in the Groot River could occur between generations. It is a 'strategic' response observable at the level of the population (gene pool), which is driven by the adaptive differences between individuals (Wootton 1990).

It is therefore suggested that the food resources utilized by the female parent influenced the size and number of offspring and that, within a batch of eggs, there was a range of sizes. In contrast to what was happening in the Groot River, the many scattered and isolated present day populations of *P. afer* (Albany Museum records) inhabited relatively stable, Cape Fold Mountain stream environments and did not speciate. The ancestor of *P. asper* in the Groot River speciated as the environment changed resulting in the present *P. asper* population. *P. asper* was probably formed as the result of paedomorphosis and was, as Balon (1990) termed it, an epigenetic retreat to a less specialized early ontogeny, with a

Table 7. Summary of the differences between the reproductive styles of *P. afer* and *P. asper.*

Attribute	*P. afer*	*P. asper*
GSI	low	high
Fecundity	low	high
Egg size	large	small
Investment per clutch	low	high
No. of clutches per season	less	more
Spawning season	shorter	longer
Reproductive lifespan	longer	shorter
Environmental cues	rainfall photoperiod temperature	water flow photoperiod temperature
Sexual dimorphism	pronounced	reduced
Size at maturity	39–40 mm SL	41–43 mm SL
Spawning site	riffles in mid-channel	riffles in mid-channel
Reproductive guild	A.1.3	A.1.3

shorter embryonic period, as shown by the early life-history study (Cambray 1992). The trajectory depicted in Figure 15 would have led to the extinction of a precocial form of the ancestor of *P. asper* in the present day Groot River.

The physiological tolerances of the two species are viewed as a key character which separates *P. afer* and *P. asper.* Can the salinity tolerance of *P. asper* be explained by epigenetic mechanisms? It is unlikely that in every generation some fish are produced which are a little more tolerant to high salinity levels compared to a second batch. It is more likely that the ancestral *Pseudobarbus* species had progeny which were adapted to and survived increased salinity levels. However, the very fact that paedomorphosis occurred, may indicate that reverting to the more altricial (variable) form carried with it the advantage of a higher tolerance to salinity levels. *P. asper* therefore did not have to adapt by genetic adaptation to higher salinity levels but instead it increased its potential for tolerance by paedomorphosis (i.e. the potential tolerance was already in the parental gene pool). This tolerance level was possibly present in the history of the lineage. It is of interest to note here that a member of what Skelton (1980) proposed as the outgroup of the *Pseudobarbus* group, *Barbus anoplus,* co-occurs with *P. asper* in the Groot River. *B. anoplus* is tolerant of salinity fluctuations, and it does not occur in the Wit River with the sister species, *P. afer* (Cambray 1992).

All the *Pseudobarbus* species studied to date are non-guarders of their non-adhesive eggs and young, open substrate benthic spawners on coarse bottoms (rocks) and have photophobic free embryos (Cambray & Meyer 1988, Cambray 1992). Riverine spawning sites indicate a conservative tendency and may represent a phylogenetic constraint within the lineage as compared to the more variable attributes such as egg size which are under environmental control within the limits expressed by the genotype.

No one reproductive pattern is universally adaptive and instead there should be selection for a combination of life-history characteristics in a life-history style which confers the highest fitness (Ware 1984). The ability of the *Pseudobarbus* lineage to respond adaptively to environmental changes is most clearly seen in *P. asper. P. asper* is meristically and morphologically the more derived of the sister species pair (Skelton 1980). In its reproductive style it is also derived and atypical compared to the other *Pseudobarbus* species studied to date (Cambray & Stuart 1985, Cambray & Meyer 1988, Cambray 1992).

In the Gamtoos River system the *Pseudobarbus* were faced with a marked geomorphological bifurcation, the Cape Fold Mountain streams and the Karoo stream. *P. asper* was able to survive in the variable Karoo stream due to its ability to become an altricial form distinct from the more precocial form of *P. afer* which had life-history attributes which were adapted to the relatively stable, clear mountain streams.

Acknowledgements

The Director of the Albany Museum approved this study. The Department of Cape Nature Conservation and the Foundation for Research Development provided funding. Pat Black helped in the laboratory and prepared some of the figures. Eve Cambray is thanked for field assistance, laboratory help and for helpful comments on the drafts of this paper. Mike Bruton offered valuable comments. Garth and Conrad Cambray helped with the field work. This paper is number three in the series Aquatic studies of the Baviaanskloof.

References cited

Bagenal, T.B. 1978. Aspects of fish fecundity. pp. 75–101. *In:* S.D. Gerking (ed.) Ecology of Freshwater Fish Production, Blackwell, Oxford.

Balon, E.K. 1975. Reproductive guilds of fishes: a proposal and definition. J. Fish. Res. Board Can. 32: 821–864.

Balon, E.K. 1980. Early ontogeny of lake charr, *Salvelinus (Cristivomer) namaycush.* pp. 485–562. *In:* E.K.Balon (ed.) Charrs: Salmonid Fishes of the genus *Salvelinus,* Perspectives in Vertebrate Science 1, Dr W. Junk Publishers, The Hague.

Balon, E.K. 1981a. Additions and amendments to the classification of reproductive styles in fishes. Env. Biol. Fish. 6: 377–389.

Balon, E.K. 1981b. Saltatory processes and altricial to precocial forms in the ontogeny of fishes. Amer. Zool. 21: 567–590.

Balon, E.K. 1985. Early life histories of fishes: new developmental, ecological and evolutionary perspectives Developments in Env Biol Fish 5, Dr W Junk Publishers, Dordrecht 280 pp

Balon, E K 1988a Tao of life universality of dichotomy in biology 1 The mystic awareness Rivista di Biologia/Biology Forum 81 185–231

Balon, E K 1988b Tao of life universality of dichotomy in biology 2 The epigenetic mechanisms Rivista di Biologia/Biology Forum 81 339–380

Balon, E K 1990 Epigenesis of an epigeneticist the development of some alternative concepts on the early ontogeny and evolution of fishes Guelph Ichthyol Rev 1 1–48

Breder, C M & D E Rosen 1966 Modes of reproduction in fishes Natural History Press, New York 941 pp

Bruton, M N 1989 The ecological significance of alternative life-history styles pp 503–553 *In* Bruton (ed) Alternative Life-History Styles of Animals, Perspectives in Vertebrate Science 6, Kluwer Academic Publishers, Dordrecht

Burt, A , D L Kramer, K Nakatsuru & C Spry 1988 The tempo of reproduction in *Hyphessobrycon pulchripinnis* (Characidae), with a discussion on the biology of 'multiple spawning' in fishes Env Biol Fish 22 15–27

Cambray, J A 1982 The life-history strategy of a minnow, *Barbus anoplus,* in a man-made lake in South Africa M Sc Thesis, Rhodes University, Grahamstown 248 pp

Cambray, J A 1991 The effects on fish spawning and management implications of impoundment water releases in an intermittent South African river Regul Rivers 6 39–52

Cambray, J A 1992 A comparative study of the life histories of the sister species, *Pseudobarbus afer* and *Pseudobarbus asper,* in the Gamtoos River system, South Africa Ph D Thesis, Rhodes University, Grahamstown 276 pp

Cambray, J A & M N Bruton 1984 The reproductive strategy of a barb, *Barbus anoplus* (Pisces Cyprinidae), colonizing a man-made lake in South Africa J Zool (Lond) 204 143–168

Cambray, J A & K Meyer 1988 Early ontogeny of an endangered, relict, cold-water cyprinid from Lesotho, *Oreodaimon quathlambae* (Barnard, 1938) Rev Hydrobiol trop 21 309–333

Cambray, J A & C T Stuart 1985 Aspects of the biology of a rare redfin minnow, *Barbus burchelli* (Pisces, Cyprinidae), from South Africa S Afr J Zool 20 155–165

Carlander, K D 1950 Handbook of freshwater fishery biology C Brown, Dubuque 752 pp

Heins, D C , G E Gunning & J D Williams 1980 Reproduction and population structure of an undescribed species of *Notropis* (Pisces Cyprinidae) from Mobile Bay Drainage, Alabama Copeia 1980 822–830

Hubbs, C 1985 Darter reproductive seasons Copeia 1985 56–68

LeCren, E D 1951 The length-weight relationship and seasonal cycle in gonad weight and condition in the perch (*Perca fluviatilis*) J Anim Ecol 20 201–219

Mann, R H K , C A Mills & D T Crisp 1984 Geographical variation in the life-history tactics of some species of freshwater fish pp 171–186 *In* G W Potts & R J Wootton (ed) Fish Reproduction Strategies and Tactics, Academic Press, London

Miller, P J 1984 The tokology of gobioid fishes pp 119–153 *In* G W Potts & R J Wootton (ed) Fish Reproduction Strategies and Tactics, Academic Press, London

Miller, R G 1980 Simultaneous statistical inference 2nd edn Springer, New York 299 pp

Nikolsky, G V 1963 The ecology of fishes Academic Press, London 352 pp

Phillips, G L 1969 Accuracy of fecundity estimates for the minnow *Chrosomus erthrogaster* (Cyprinidae) Trans Amer Fish Soc 3 524–526

Reid, R G B 1985 Evolutionary theory the unfinished synthesis Croom Helm, London 405 pp

Ricklefs, R E 1979 Adaptation, constraint, and compromise in avian postnatal development Biol Rev 54 269–290

Sire, J Y & I Arnulf 1990 The development of squamation in four teleostean fishes with a survey of the literature Jap J Ichthyol 37 133–143

Skelton, P H 1980 Systematics and biogeography of the redfin *Barbus* species (Pisces Cyprinidae) from southern Africa Ph D Thesis, Rhodes University, Grahamstown 416 pp

Skelton, P H 1987 South African Red Data Book – Fishes S Afr Nat Sci Prog Rep 137 199 pp

Skelton, P H 1988 A taxonomic revision of the redfin minnows (Pisces, Cyprinidae) from southern Africa Ann Cape Prov Mus (nat Hist) 16(10) 201–307

Sokal, R R & F J Rohlf 1973 Introduction to biostatistics W H Freeman, San Francisco 368 pp

Stearns, S C & R E Crandall 1984 Plasticity for age and size at sexual maturity a life-history response to unavoidable stress pp 13–33 *In* G W Potts & R J Wootton (ed) Fish Reproduction Strategies and Tactics, Academic Press, London

Ware, D M 1984 Fitness of different reproductive strategies in teleost fishes pp 349–366 *In* G W Potts & R J Wootton (ed) Fish Reproduction Strategies and Tactics, Academic Press, London

Wootton, R J 1984 Introduction strategies and tactics in fish reproduction pp 1–12 *In* G W Potts & R J Wootton (ed) Fish Reproduction Strategies and Tactics, Academic Press, London

Wootton, R J 1990 Ecology of teleost fishes Chapman & Hall, London 404 pp

Zar, J H 1984 Biostatistical analysis, second edition Prentice-Hall, London 718 pp

Environmental Biology of Fishes **41**: 269–286, 1994.

Reproduction in an aggregating grouper, the red hind, *Epinephelus guttatus*

Yvonne Sadovy[1], Aida Rosario & Ana Román
Fisheries Research Laboratory, Department of Natural Resources, PO Box 3665, Mayagüez, Puerto Rico 00680, U.S.A.
[1] *Current address and correspondence: Department of Zoology, University of Hong Kong, Pokfulam Road, Hong Kong*

Received 5.4.1993 Accepted 12.10.1993

Key words: Spawning, Serranidae, Pisces, Puerto Rico, Protogyny

Synopsis

We examined the reproductive pattern of an aggregating grouper, the red hind, *Epinephelus guttatus*, in Puerto Rico. Macroscopic and histological examination of gonads confirmed that, although mature, ripe ovaries are found over a three-month period, spawning activity is limited to about 2 weeks each year. Females are determinate spawners and individuals are able to spawn more than once during the course of the annual spawning season. The red hind is protogynous and 50% maturity is attained at 215 mm fork length. In western Puerto Rico, spawning occurs in aggregations at several sites within loosely defined areas located towards the edge of the insular platform. Sex ratios of individuals taken by hook and line at one of the aggregation areas over a consecutive six-year period, suggest considerable intra- and inter-annual variation most likely attributable to a combination of differential ingress and egress by males and females during the course of an aggregation and to fluctuations in recruitment, differential mortality by sex and sex change among years. A comparative assessment of the reproductive patterns of seven western Atlantic *Epinephelus* spp. suggests that aggregation-spawning is associated with medium- to large-sized groupers, while smaller groupers do not aggregate. Mating systems vary among congeners in association with trends in male-female size dimorphism, sexual pattern and sperm competition. The short-term, localized, nature of spawning aggregations renders species with this reproductive mode particularly vulnerable to fishing pressure at spawning sites. Aggregating species, therefore, may require special management consideration.

Introduction

The groupers (Epinephelini: Serranidae) of the western Atlantic are typically thought of as species which aggregate to spawn and exhibit protogyny whereby all males derive from adult females via sex change. However, as knowledge of these serranids accumulates it is becoming clear that they display a diverse array of mating systems and reproductive patterns. They range from pair-spawning, non-aggregating protogynous species which reproduce for extended periods annually (graysby, *E. cruentatus*), to those which spawn in multi-male clusters ('group'-spawning) within short-term aggregations each year and are, at least partially, gonochoristic (e.g. Nassau grouper, *Epinephelus striatus*) (Nagelkerken 1979, Colin 1992, Sadovy & Colin unpublished data, Sadovy et al. 1994). Because of such diversity, groupers promise a wealth of comparative data with which to organize information and to generate and test hypotheses concerning the functional significance of interspecific differences in reproductive patterns and aggregating behaviour (Clutton-Brock & Harvey 1984). Further details on reproductive dynamics such as location and duration of spawning activity, aggregation sex ratios, mating system, spawning frequency, etc., would provide a framework for better understanding the range of reproductive patterns in this group of fishes.

The red hind, *Epinephelus guttatus* (Linnaeus, 1758), is a grouper of particular interest because its reproductive biology appears to lie between extremes exhibited by its congeners. For example, it is protogynous, forms short-term spawning aggregations and pair-spawns in single-male/multi-female clusters (Burnett-Herkes 1975, Colin et al. 1987, Sadovy et al. 1992, Shapiro et al. 1993a). Individuals may live for 18 years or more and, during the non-reproductive season, live in overlapping home ranges (García-Moliner 1986). Detailed information on reproductive activity during the course of the reproductive season, adult sex ratios and mating system of the red hind, however, is scarce.

The red hind is also of interest because of its considerable commercial significance throughout the islands of the Caribbean and warm-temperate Atlantic. The red hind is believed to be commercially overexploited in many areas, including Puerto Rico and the US Virgin Islands (Beets & Friedlander 1992, Sadovy & Figuerola 1992, Sadovy 1994a). In particular there is a concern over the possible impact of heavy fishing pressure on spawning aggregations. Evaluation of the potential for such impact requires a better understanding of the reproductive patterns of this species especially regarding mating styles, changes in operational (reproductive) sex ratio over time and the potential for sperm limitation (Emlen & Oring 1977, Bannerot et al. 1987).

The aim of this study was to characterize reproduction in the red hind in western Puerto Rico to address some of these questions through: (1) identification of spawning sites; (2) determination of the timing and duration of annual spawning activity; (3) determination of female spawning frequency and size of sexual maturation; (4) evaluation of the potential for sperm competition; (5) confirmation of the sexual pattern; and (6) evaluation of size compositions and adult sex ratios in annual spawning aggregations.

Methods

Several aggregation areas in western Puerto Rico were identified and one area was sampled over a six-year period. Reproductive activity was determined through gonad sampling using two approaches, given in detail below. First, gonads were removed from individuals collected from spawning aggregations. They were examined macroscopically to establish operational sex ratios (OSRs) and to study spawning frequency and gonad weight over the course of an annual aggregation. Second, individuals were sampled monthly over a multi-year period and gonads were prepared histologically, thereby to determine the annual cycle of reproduction, size of sexual maturation, sexual pattern and to validate the macroscopic approach.

Aggregation sampling

Four traditionally exploited red hind spawning areas in western Puerto Rico were identified by informal interviews and contacts with local fishers and biologists and through sampling programs carried out by the Fisheries Research Laboratory (FRL) of the Puerto Rico Department of Natural Resources. One area off southwestern Puerto Rico was identified by Colin et al. (1987). Three additional aggregation areas were identified off the west coast. One of these, Bajo el Cico (approximately 3 × 6 km), located 26 km west of Mayagüez (Fig. 1), was sampled six years consecutively (1987–1992) at the time of the aggregation by FRL personnel using hook and line and fishing throughout the area according to fish availability; hook and line samples accurately represent the sizes and sexes of aggregating fishes (Shapiro et al. 1993a). Sampling each year was initiated in January when fishers first reported catches of ripe females and continued over a minimum of four consecutive weeks until spawning had terminated, as judged by declining catches and a predominance of females with post-spawning ovaries.

Fish were measured to the nearest mm (fork length, FL; although fork length and total length are essentially equivalent in the red hind, length mea-

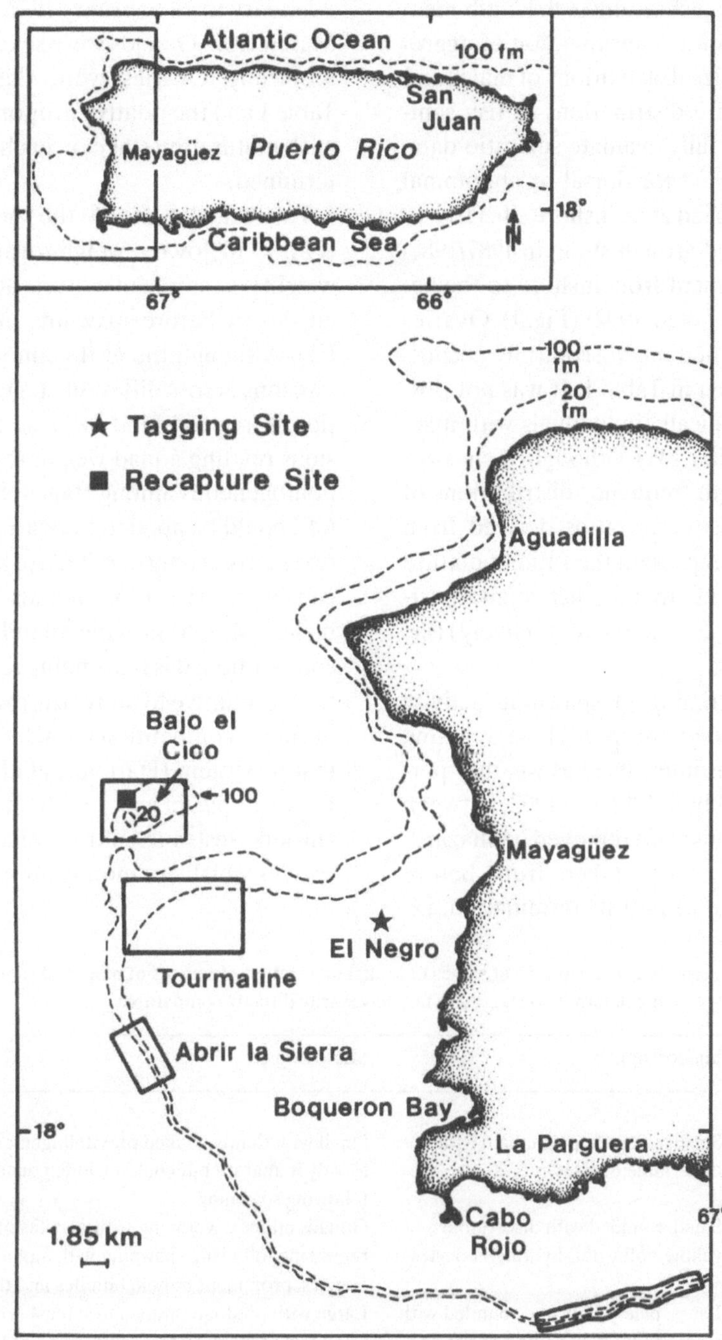

Fig 1 Map of Puerto Rico and detail of west and southwest coasts showing locations of known red hind spawning aggregation areas indicated by the four boxes The tagging (star) and recapture (square) sites of a single animal are marked The narrow area south of La Parguera is described in Colin et al (1987) and in Shapiro et al (1993a) The 20 and 100 fathom depth contours are indicated (1 fathom = 1 83 m)

surements taken were technically fork length measurements) and sexed for determination of aggregation sex ratios and size distributions of males and females. Details on size distributions of fish sampled are necessary to fully evaluate sex ratio data. Fish were also examined for dorsal or abdominal tags that had been applied at an inshore site (El Negro) during an age and growth study in 1987/1988, for evidence of movement from inshore to the aggregation area (Sadovy et al. 1992) (Fig. 1). Ovaries and testes were classified macroscopically according to the criteria given in Table 1. It was not possible to sex macroscopically individuals with inactive gonads (i.e. F1 and M1) although their sizes were included in length frequency distributions of aggregation-caught fish. Sex ratios derived from these data, therefore, represent the ratio of mature males (M2) to maturing, mature, active and post-spawning females (F2, F3 and F4, respectively) (i.e. OSR) in aggregations.

To determine the timing of spawning activity within an annual aggregation period, aggregating fishes were sampled as intensively as weather permitted, 1–4 times weekly, in 1991 and 1992, between 8:00 and 14:00 h. Gonads were removed haphazardly from subsamples of fish taken from before spawning had initiated to past its termination, i.e. 9 January to 28 February 1991, and 22 January to 11 March 1992. Gonads were sexed and staged macroscopically. Ovaries were classified according to Table 1 and the relative proportion of individuals in each maturation stage on each sample date was determined.

For calculation of the gonadosomatic index (GSI = 100[ovary weight/ovary weight + somatic weight]) a sample of mature, active gonads was taken shortly before spawning initiated (gonad stage F3). Although the GSI cannot be applied to comparisons across different stages of ovarian maturation in the red hind because the slopes of regressions relating gonad weight to body weight are not homogeneous among stages (Shapiro et al. 1993b), GSI could be applied to ovaries in the present context because only one stage of maturation was being considered (i.e. F3). Gonads were removed from males (M2), to provide an index of possible sperm competition; it is reasonable to predict that ripe testis size, relative to body size, can be used an an index of sperm competition to make inferences about the mating system (Harcourt et al. 1981, Warner 1984). For example, in species which spawn in multi-male clusters, testes should be relatively larger than in species which are monogamous or spawn in single-

Table 1. Macroscopic and microscopic descriptions of stages of sexual maturation of gonads of male and female red hind, *Epinephelus guttatus*. Macro- and microscopically determined stages are not necessarily directly comparable.

Stage of maturation	Macroscopic	Microscopic
Ovaries		
F1 (inactive)	Small, compact, pale cream with no discernible oocytes	Small with tightly packed previtellogenic stage 1 and 2 oocytes. Mostly immature but could include gonads which have regressed following spawning
F2 (maturing)	Small, rounded with discernible yellow, yolky non-hydrated oocytes	Gonads either developing with oocytes of stages 1, 2 and 3 or regressing following spawning with alpha-atretic vitellogenic oocytes, prominent muscle bundles and thick gonad wall
F3 (mature active)	Large, pale yellow and rounded with many large vitellogenic oocytes and few to many clear (hydrated) oocytes	Large with predominantly late stage 4 (vitellogenic) oocytes and often few to many hydrated (stage 5) oocytes, gonad wall thin
F4 (early post-spawn)	Large and flaccid with scattered vitellogenic oocytes	Large and disorganized with post-ovulatory follicles, alpha-atretic vitellogenic and scattered vitellogenic oocytes
Testes:		
M1 (inactive)	Indistinguishable macroscopically from F1 females	Early stages of spermatogenesis, gonad small and compact with gonia and seminiferous tubules
M2 (active)	Large, white with exit of sperm on pressure, flaccid if post-spawning	All stages of spermatogenesis are equal, or later stages dominate. Post-spawning testes are disorganized with empty lumina

male groups. Gonads were blotted dry and weighed to 0.01 g.

To establish the minimum number of times individual females are capable of spawning, 8 gonads, taken at intervals throughout the 1991 sampling period and frozen, were examined for oocyte diameter frequencies; the progression of distinct modes of oocyte diameter in a series of females over the course of the aggregating period would suggest that individual females may spawn more than once each year, assuming that spawning frequency is not size- or age-dependent (Hunter et al. 1992). A subsample of each gonad was removed from the centre of one gonadal lobe and the oocytes separated manually from the gonadal tunica. Using a dissecting microscope, the maximum diameters of oocytes equal to and greater than 0.1 mm were measured until approximately 600 oocytes had been processed. Maximum diameter had to be estimated for some of the larger hydrated oocytes because of their highly irregular form following defrosting. The eight gonads were also prepared histologically.

Annual reproductive cycle and sexual pattern: gonadal histology

Gonads were processed histologically to establish the annual sequence of oocyte maturation and ovarian development, to validate macroscopic determination of stages of gonadal maturation, to identify individuals undergoing sexual transition, and to determine the minimum size of sexual maturation of females. Gonad samples were taken monthly by FRL personnel with hook and line between September 1987 and January 1989. Additional samples (22% of total) were taken between February 1989 and February 1991 from fish caught commercially, both by hook and line and fish trap (3.18 cm mesh), on the insular platform of western Puerto Rico. Fish were measured in mm FL and weighed to the nearest gram. Gonads were removed, blotted dry, fixed in Davidson's solution, embedded in paraffin, sectioned at 7–9 μm and stained in hematoxylin and eosin. Small gonads were sectioned in their entirety. The mid-sections of large gonads were removed and sectioned transversely; the stage of sexual maturation is uniform throughout the organ (Shapiro et al. 1993b). Gonads (with the exception of 11 in poor condition) were classified microscopically into one of four stages of sexual maturation for females (F1–F4), two stages of maturation for males (M1 and M2) (Table 1) or as transitionals. Classifications were made on the basis of constituent stages of oocyte or spermatocyte development (Moe 1969, Wallace & Selman 1981, Shapiro et al. 1993b): previtellogenic oocytes are stages 1 and 2; cortical alveolus (yolk vesicle) oocytes are stage 3; vitellogenic oocytes are stage 4, and hydrated oocytes are stage 5. Early stage degenerating vitellogenic oocytes undergoing alpha-atresia were classified according to Bretschneider & Duyvene deWit (1947) (i.e. disintegration and vacuolation of nucleus, yolk globules and dissolution of zona radiata). Gonads of individuals believed to be undergoing sexual transition from female to male (i.e. transitionals) consisted of degenerating vitellogenic oocytes and/or muscle/connective tissue bundles (features indicative of prior spawning as a female according to Shapiro et al. 1993b) together with scattered areas of spermatogenic tissue. To determine the smallest size class in which 50% of females were sexually mature, a maturity curve of the percent of maturing and mature females (F2, F3 and F4), as determined histologically from monthly samples and pooled for all years, was developed. Statistical tests were taken from Sokal & Rohlf (1969).

Results

Aggregation areas, sex ratios and sizes

Descriptions of aggregation areas indicated that these were quite extensive (i.e. up to several square kilometers), each encompassing several sites of red hind concentration which varied in exact location from year to year. The principal aggregation areas off the west coast of Puerto Rico were identified as Tourmaline and Bajo el Cico (Fig. 1). Abrir la Sierra was considered by fishers to be of lesser significance as an aggregation area but has been increasingly ex-

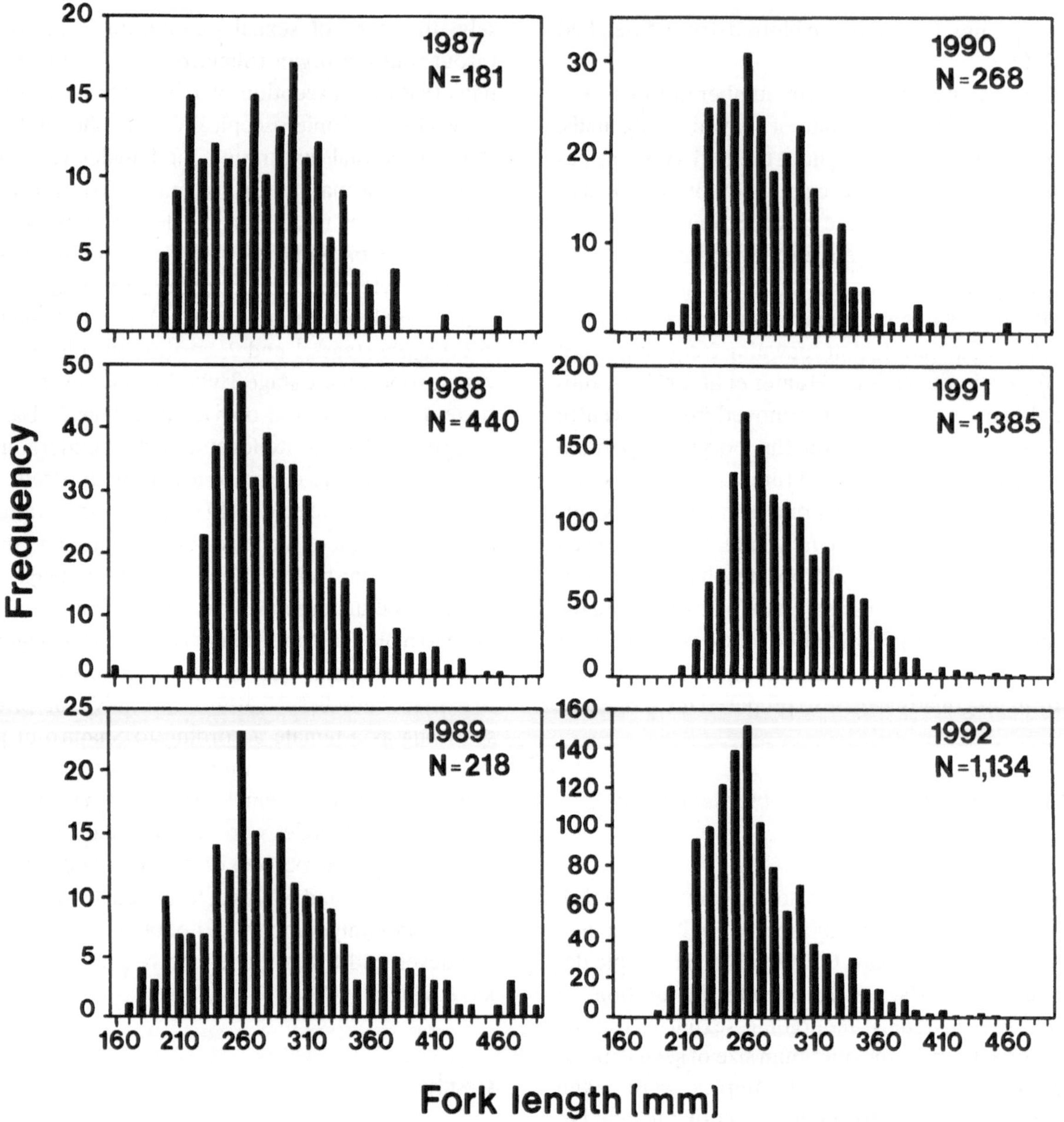

Fig. 2. Annual size-frequency distributions of red hind collected by hook and line at the spawning aggregation area of Bajo el Cico between 1987 and 1992. Sample size (N) of all fish taken is given for each year.

ploited in recent years, while landings from the Tourmaline site have reportedly declined.

Size-frequency distributions of all individuals (active and inactive) taken at the Bajo el Cico site, from depths ranging from 20–100 m, during the aggregation period between 1987–1992 are shown in Figure 2. Details of sample dates and sample sizes for each year are given in Table 2. Variability in the relative numbers of small and large individuals were notable among years (Fig. 2). For example, differences in size-frequency distributions between 1988, 1989 and 1990 were due, at least in part, to a relatively large number of individuals smaller than 230 mm in 1989, and larger individuals were increasingly absent from 1989 onwards. Also of note were the high percentages of sexually inactive (F1 and M1) individuals in aggregations, ranging from 13–36% of the total fish sampled (Table 2). Most of

Table 2 Sampling period, number of sampling days, dates of January and February full moons, sample sizes (N) for all aggregating fish sampled (active and inactive), number of inactive (F1 + M1) fish, and operational sex ratios (male female) and sample sizes (N) of maturing and active (M2, F2, F3, F4) fish from spawning aggregations of the red hind, *Epinephelus guttatus*, at Bajo el Cico off western Puerto Rico for 1987 to 1992

Year	Sampling dates (days)	Full moon	N	N (F1 + M1)	Sex	N ratio
1987	23 Jan – 20 Feb (9)	15 Jan/13 Feb	181	65	1 115	116
1988	14 Jan – 25 Feb (15)	5 Jan/ 2 Feb	440	70	1 4	370
1989	18 Jan – 20 Feb (15)	21 Jan/20 Feb	218	69	1 20	149
1990	9 Jan – 13 Feb (9)	11 Jan/ 9 Feb	268	71	1 98	197
1991	9 Jan – 28 Feb (20)	30 Jan/28 Feb	1385	181	1 10	1204
1992	22 Jan – 11 Mar (23)	21 Jan/18 Feb	1134	170	1 11	964

these fish were smaller than the smallest class in which 100 percent of females were mature (i.e. 285 mm FL see below).

Aggregation sex ratios (OSRs) varied widely and significantly ($p < 0.005$, chi-square) among all years, except between 1991 and 1992 which did not differ. Sex ratios in all years differed from unity ($p < 0.005$, chi-square) (Table 2) and also varied significantly during the course of a single aggregation period (Fig. 3). For example, in 1991 sex ratios from 16 daily samples taken between 22 January, when spawning was imminent, and 28 February, ranged in extreme from 1:4.4 (M:F) to all-female with the least female-biased sex ratios occurring shortly before the full moons of January and February.

During aggregation sampling in 1991, an individual tagged at an inshore reef (El Negro) in October 1989, for an age and growth study, was recaptured by FRL personnel at the Bajo el Cico site (Fig. 1) (Sadovy et al. 1992). The female had travelled a minimum distance of 18 km, and must have crossed water of at least 194 m depth to reach the aggregation site. This individual evidently did not move to the nearest aggregation area (i.e. Tourmaline) relative to its tagging site. Out of 139 tagged fish, eight were recaptured, seven at the tagging site up to 11

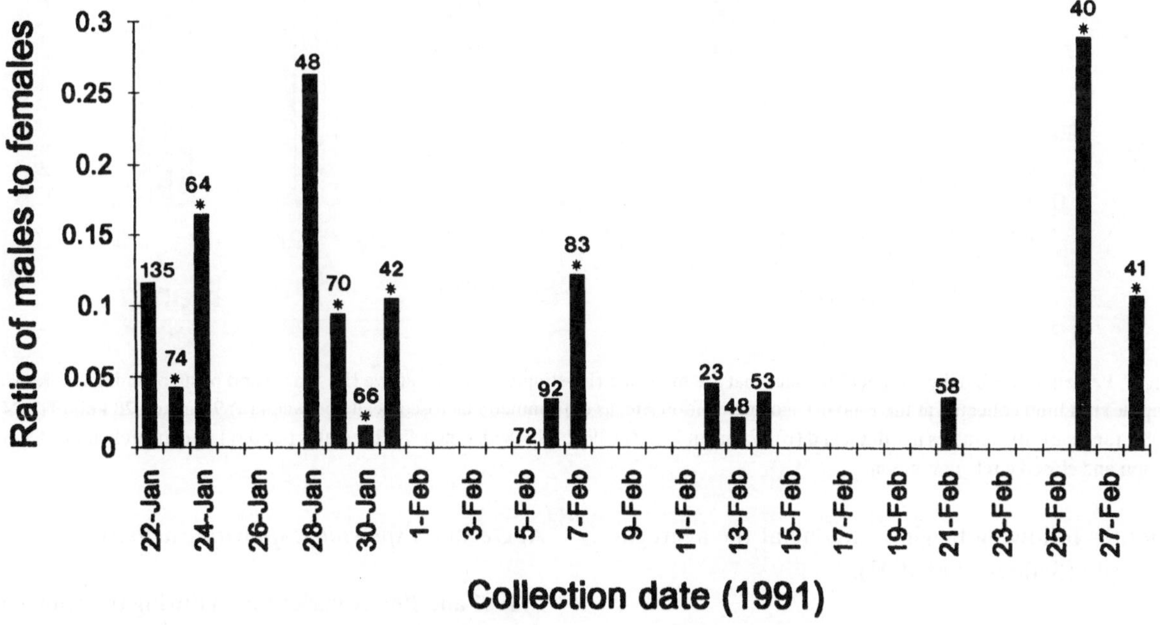

Fig 3 Ratio of active (M2) males to maturing and active (F2, F3 and F4) females for 16 sampling days in 1991 Number of fish caught each day appears above each column and asterisks denote that the sex ratio differs significantly ($p < 0\ 05$, chi-square) from the preceding sample date Full moon occurred on 30 January and 28 February

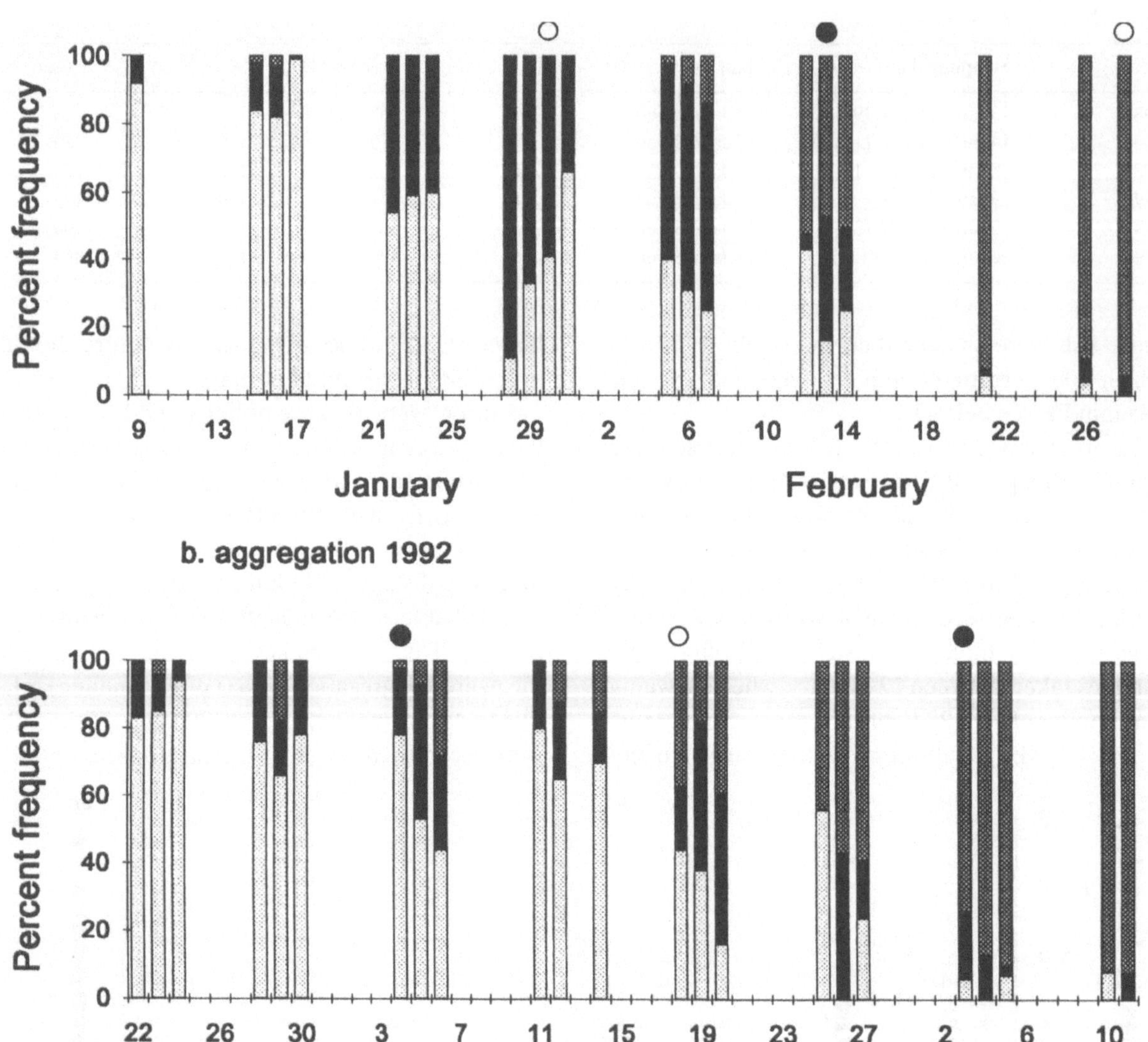

Fig. 4. Percent of each daily sample of females that are maturing (F2-stipple), mature, active (F3-black) and post-spawning (F4-reverse stipple) red hind collected at the Bajo el Cico aggregation site, as determined macroscopically between (a) 9 Jan and 28 Feb 1991 (N = 1087, range: 21–104 females per day), and (b) 22 Jan and 11 Mar 1992 (N = 890, range: 7–71 females per day). Open circle indicates full moon and closed circle new moon.

months following tagging and one at the aggregation site (Sadovy et al. 1992).

Aggregation spawning: spawning activity

In 1991 and 1992, ovaries taken during the course of the spawning seasons and staged macroscopically as F2, F3 or F4 were plotted in terms of daily percentage of each gonad stage (Fig. 4a). Data from

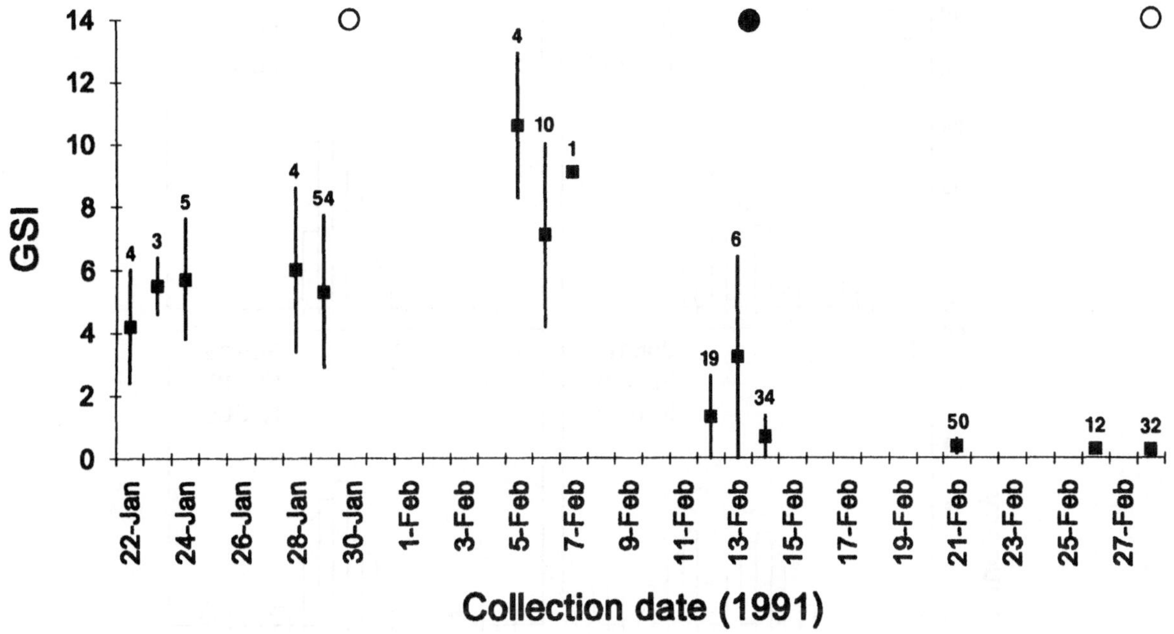

Fig 5 Mean gonadosomatic index (GSI) and standard deviation for ovaries taken from females captured between 22 Jan and 28 Feb 1991 Sample sizes are given for each day Open circle indicates full moon and closed circle new moon

1991 indicate that, while limited spawning occurred in mid-January as evidenced by a few post-spawning ovaries, principal spawning activity occurred in the week prior to the new moon and possibly for up to a week thereafter. In 1992, limited spawning activity occurred during late January and early February but the principal spawning period occurred in mid- to late February, from shortly before, to a week or two after the full moon (Fig. 4b). Peak spawning activity in 1992, as for 1991, occurred over a limited period of a few weeks.

Analysis of GSI and oocyte diameters from ovaries collected during the 1991 spawning season support the pattern of spawning indicated by macroscopically staged ovaries. The GSI increased steadily from 22 January until about 7 February and dropped rapidly over the following week indicating that the majority of spawning took place during the week preceding the new moon (Fig. 5); no GSI data are available for sampling days 30 and 31 January.

Oocyte diameter analyses indicated that stage 1 and 2 (previtellogenic) oocytes were < 0.1 mm (these oocytes were numerous but were not counted), stage 3 oocytes ranged in diameter from > 0.1 to < 0.6 mm, stage 4 oocytes ranged from approximately > 0.5 to < 1.0 mm, and hydrated oocytes were generally > 0.8 mm in diameter. Prior to spawning, ovaries were comprised of oocytes at various stages of vitellogenesis, sometimes exhibiting distinct modes (Fig. 6a, b, c, d). Substantial numbers of hydrated oocytes appeared in early February when two distinct modes of advanced stage oocytes (vitellogenic and hydrated) became evident (Fig. 6d). Spawning took place over the following 1–2 weeks, through the new moon phase, with ovulation of hydrated oocytes (Fig. 6e) and successive hydration and ovulation of vitellogenic oocytes (Fig. 6f, g). This progression of modes suggested that individual females may spawn more than once during the course of an aggregation. By late February, most females had spawned and some ovaries retained a few unspawned vitellogenic oocytes which subsequently underwent resorption (Fig. 6h). There was no evidence for recruitment of non-vitellogenic oocytes during the course of the spawning season. This observation, combined with the absence of vitellogenic oocytes outside of the spawning season (see below), suggest that annual fecundity in the red hind is determinate i.e. only those oocytes entering vitellogenesis by the beginning of the spawning period may potentially be spawned. This means that potential annual fecundity in this spe-

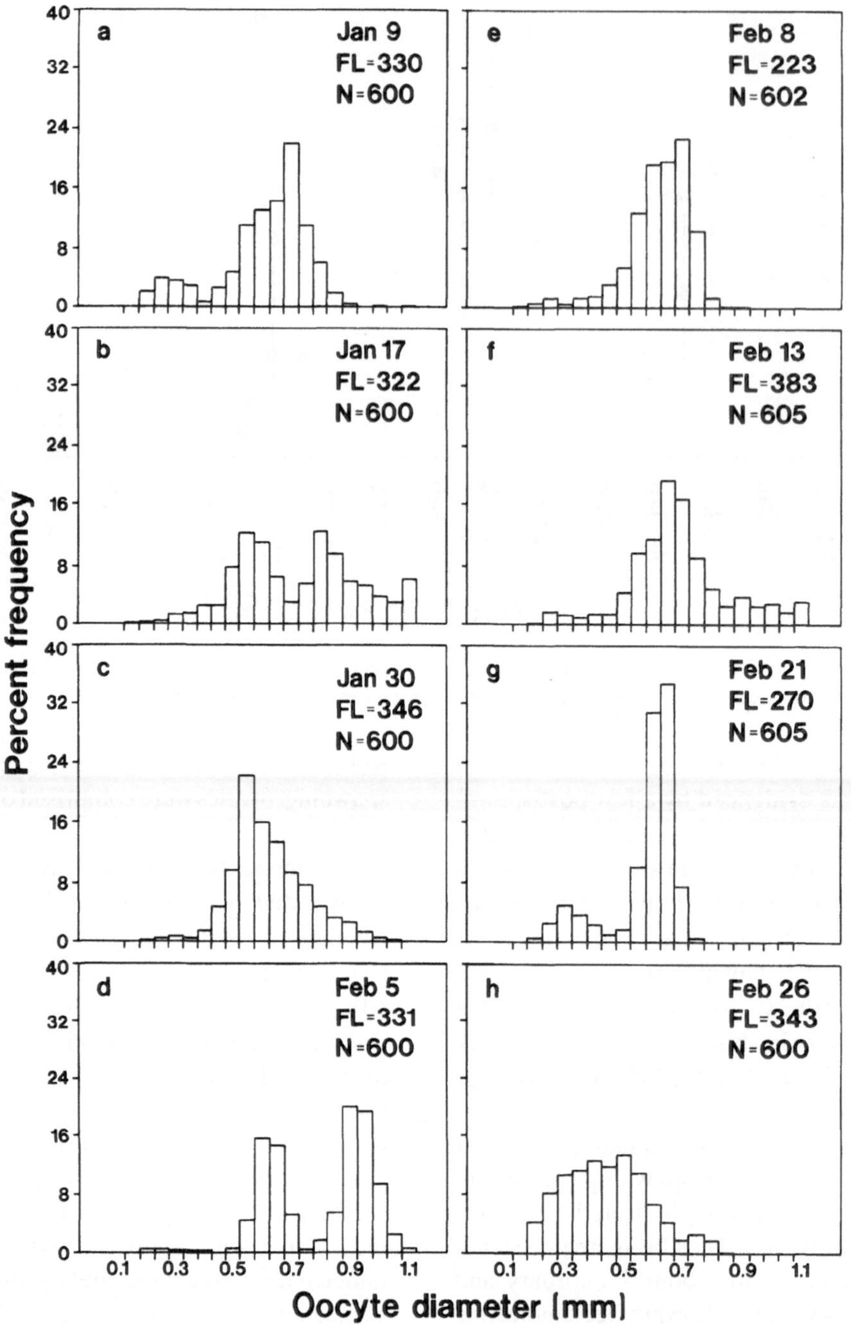

Fig. 6. Frequency distributions of the percent of oocytes in different classes of oocyte diameter (mm) in ovaries from 8 red hind taken at the Bajo el Cico spawning aggregation site between 9 Jan and 26 Feb 1991. Fork length, sample date and number of oocytes counted are given for each fish.

cies may be determined by counting vitellogenic oocytes shortly prior to the initiation of spawning.

Gonad weights of males and females captured shortly prior to spawning were examined to provide a measure of the potential for sperm competition inferred from testis size; ovarian weights provide an indication of the size that gonads may attain in fish of different lengths. The relationships between gonad weight (GW) and body weight (BW) for both males and females were significant (females: GW =

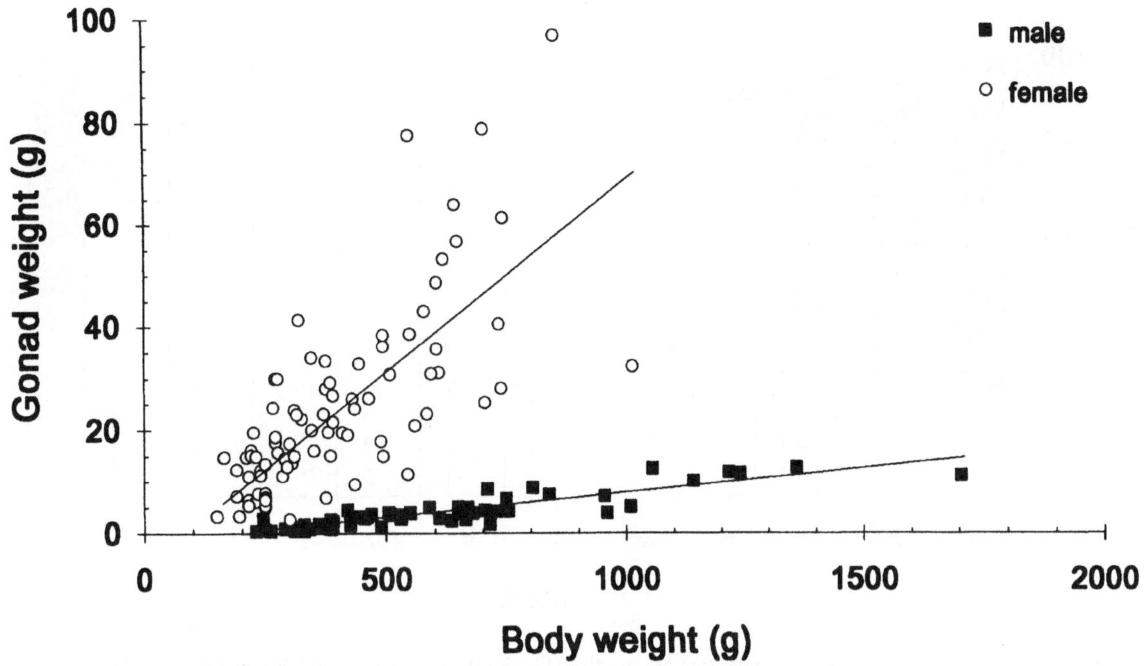

Fig. 7. Relationships between gonad weight (g) and body weight (g) for male (N = 58) and female (N = 85) red hinds taken at the Bajo el Cico spawning aggregation area between 15 Jan and 6 Feb 1991.

– 4.646 + 0.072 BW, $r^2 = 0.52$, N = 85; males: GW = – 1.619 + 0.009 BW, $r^2 = 0.79$, N = 58, Fig. 7). Shortly prior to spawning, ovaries weighed approximately 10% of body weight, while testes represented < 1%.

Annual reproductive cycle: histological analyses

A total of 380 ovaries, taken between 1987 and 1991, were classified for stage of sexual maturation. Mature, active (F2) females were collected predominantly in December, January and February; post-spawning ovaries predominated in March (Fig. 8). Typically, post-spawning ovaries contained few atretic vitellogenic oocytes, indicating that most vitellogenic oocytes are likely to be spawned during the reproductive season, although occasionally ovaries contained substantial numbers of degenerating yolky oocytes.

Between spawning periods, ovaries regressed to a resting state comprised of previtellogenic oocytes. As post-spawning time elapsed, progressively fewer remnants of degenerated and resorbing oocytes persisted; these had largely disappeared by May. Progressively smaller bundles of muscle and connective tissue, indicators of prior spawning activity, were observed until October. Oocytes in early vitellogenesis (stage 3) were first observed in November and later stage vitellogenic oocytes (stage 4) were not observed until December. On the basis of the occurrence of vitellogenic oocytes, annual reproduction occurred sometime between December and March, although the more detailed analyses clearly demonstrated that spawning activity lasts for only a few weeks. The co-occurrence of indications of prior spawning activity (post-ovulatory follicles) with vitellogenic and hydrated oocytes in the same gonad on 6 February 1991 (FL = 251 mm) also indicated that females were capable of spawning more than once. Since by mid-February of 1991, hydrated oocytes were relatively scarce and post-ovulatory follicles and oocytes undergoing alpha-atresia were relatively common, the histological data support the GSI and oocyte diameter data in indicating a short spawning season in this species.

The smallest mature, active, female was 195 mm FL, with 50% of individuals mature at 215 mm FL and all mature by 285 mm FL (Fig. 9). Although it was not possible to distinguish with certainty between F2 females which had already spawned and

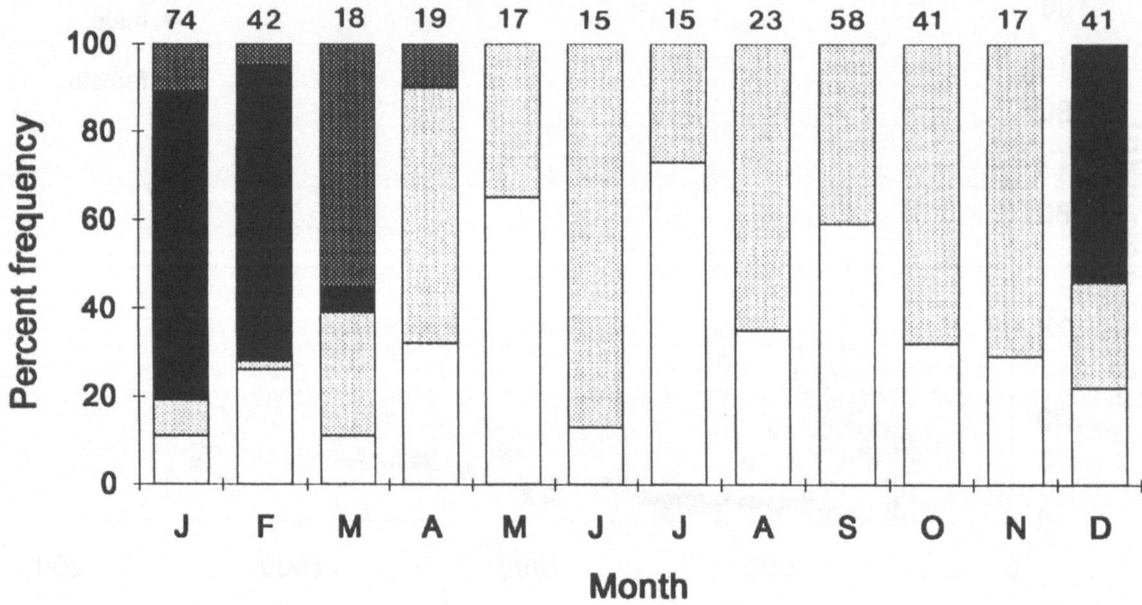

Fig. 8. Percent of each monthly sample of females that are inactive (F1-white), maturing (F2-stipple), mature, active (F3-black) and post-spawning (F4-reverse stipple) female red hind taken between 1987 and 1991 as determined histologically. The number of females sampled each month appears above each column (N = 380).

those which were maturing for the first time, the overlap in size ranges of F2 females with F3 and F4 (Fig. 10) females strongly indicates that F2 ovaries are regressed or entering yet another maturation

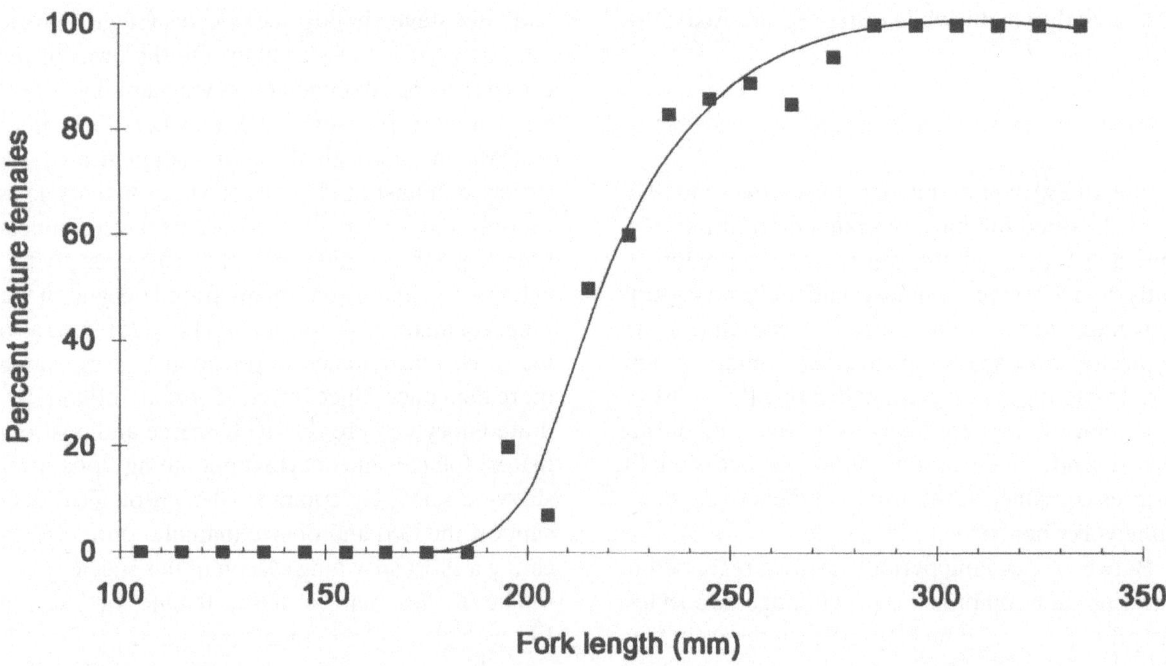

Fig. 9. Maturity curve showing the percent of female red hind of maturity stages F2, F3 and F4 in 10 mm FL size classes as determined histologically (N = 260).

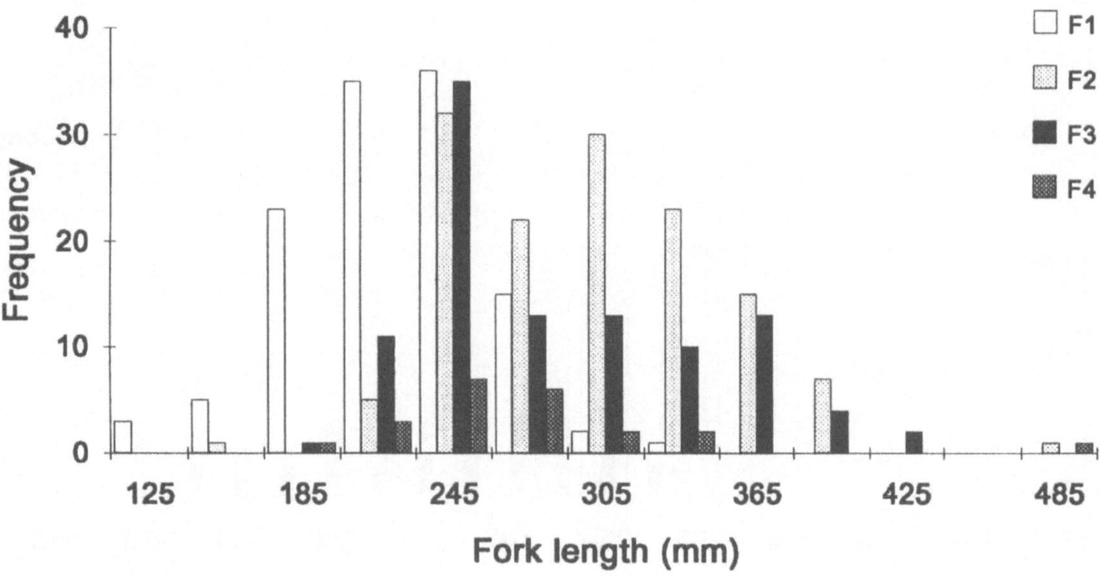

Fig. 10. Size-frequency distributions of inactive (F1-white, N = 120), maturing (F2-stipple, N = 136), mature, active (F3-black, N = 102) and post-spawning (F4-reverse stipple, N = 22) female red hind collected between 1987 and 1991 in western Puerto Rico as determined histologically (N = 380).

cycle rather than maturing for the first time. The size range of F1 individuals, on the other hand, differs significantly from those of the other maturity classes (F2, F3, F4) (Kolmogorov-Smirnov: D = 0.481, $p < 0.01$) suggesting that this microscopically determined maturity stage predominantly represents sexually immature gonads, rather than regressed mature ovaries.

Based on monthly collections taken throughout the 12 calendar months between 1987 and 1991, 510 individuals were sexed and an additional seven were classified as sexually transitional (Fig. 11). Females (N = 390) ranged in size from 110 to 480 mm FL, males (N = 120) from 245 to 510 mm FL, and transitionals (N = 7) ranged from 273–345 mm FL. Males were significantly larger than females (Kolmogorov-Smirnov: D = 0.270, $p < 0.01$). Transitionals were taken in June (N = 1), August (N = 1), September (N = 4) and October (N = 1). The overall sex ratio (inactive and active) of these samples was 1:3.25 (M:F). Inactive males (M1) were found between April and October (N = 57) and active males (M2) from October to April (N = 62). Inactive males were not significantly smaller than active males (Kolmogorov-Smirnov: D = 0.009, NS).

Based on the incidence of seven sexually transitional individuals and the presence of other characteristics typical of female to male sex change (Sadovy & Shapiro 1987), protogyny in the red hind in western Puerto Rico is confirmed (Sadovy et al. 1992, Shapiro et al. 1993b). Since none of the juvenile females exhibited spermatogenic tissue and males were larger than the smallest mature females, the direct development of males from the juvenile phase, if it does occur (Sadovy et al. 1992), is likely to be rare.

Discussion

The red hind is a protogynous hermaphrodite which spawns in well-defined aggregation areas where fish concentrate in patches over the substrate. Specific sites of fish concentrations within identified areas may vary from year to year and even during the course of an annual aggregation (Shapiro et al. 1993a). Aggregation areas are generally located in the vicinity of the edge of the insular platform but may also extend some kilometers shoreward. Individuals do not necessarily move to the nearest aggregation area from a given resident reef. Since collections during aggregation months

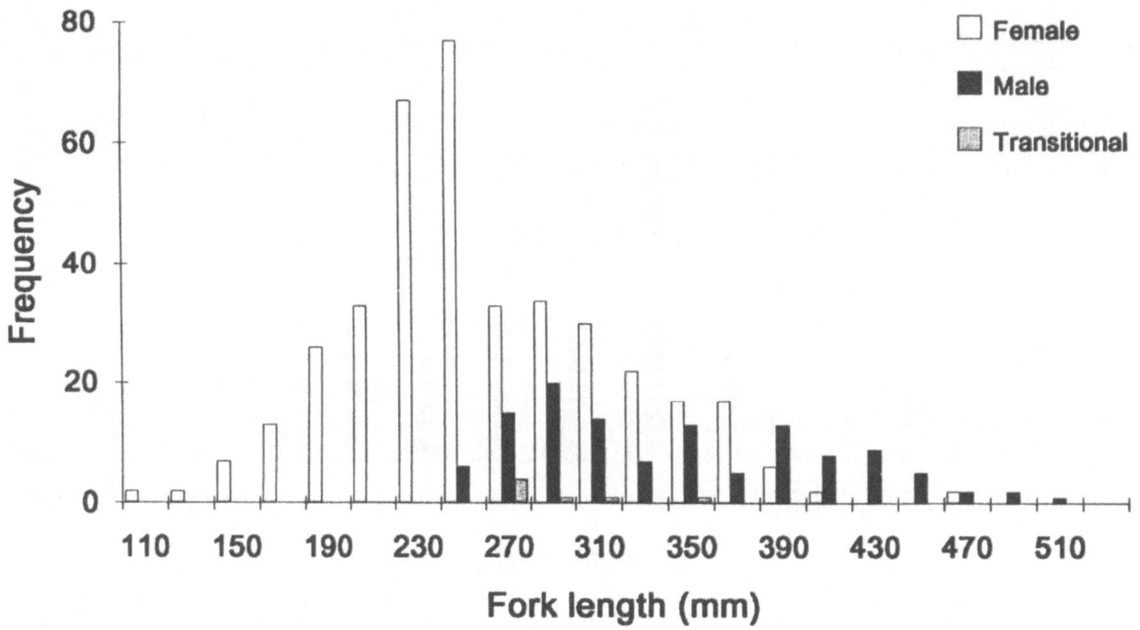

Fig. 11. Size-frequency distributions of female (white, N = 390), male (black, N = 120) and transitional (stipple, N = 7) red hind collected between 1987 and 1991 in western Puerto Rico, as determined histologically.

were generally made within the vicinity of spawning areas, it is not known whether spawning also occurs in non-aggregation areas. Although ripe ovaries have been reported over the course of several months of the year (e.g. Burnett-Herkes 1975, Shapiro et al. 1993b), spawning activity is restricted to a considerably shorter period, approximately two weeks, through either the full or new moon phases between January and March in Puerto Rico. Previous work has indicated that spawning is most typically associated with the full moon in January and/or February in Puerto Rico and the U.S. Virgin Islands, and during the summer months with no clear lunar trend in Bermuda (e.g. Burnett-Herkes 1975, Colin et al. 1987, Shapiro et al. 1993b, Beets & Friedlander 1992).

Females exhibit determinate fecundity and individuals may spawn more than once during the annual spawning period. Following the reproductive season, ovaries occasionally retain low numbers of vitellogenic oocytes but soon regress to a uniformly previtellogenic state. Individuals in early stages of sexual transition were noted in low numbers between June and October and ranged in size from 273–345 mm FL. The size ranges of transitionals, females and males suggest that the majority, if not all, individuals reproduce as females before sex change.

If testis size may be used as an index of sperm competition, the small size of testes, relative to male body weight (i.e. < 1%) suggests a mating system with little or no sperm competition; the potential for a considerable increase in testis size in the red hind is indicated by the large size of ripe ovaries 10%) relative to body weight. This conclusion is consistent with reports of pair-spawning in single-male/multi-female clusters (Colin et al. 1987, Shapiro et al. 1993a), a mating system in which sperm competition is unlikely (Clutton-Brock & Harvey 1984). The relatively small size of ripe testes in the red hind contrasts sharply with the large testis size (10%) reported for the Nassau grouper, *Epinephelus striatus.* The Nassau grouper spawns in multi-male clusters in which sperm competition is likely to occur (Colin 1992, Sadovy & Colin unpublished data).

Most individuals collected at aggregations in this study, had maturing or mature gonads, between 13 and 36% were sexually inactive. These individuals were characteristically among the smallest individuals taken from annual aggregations and almost all were smaller than 285 mm FL, the smallest size at which all females are sexually mature. Since relatively more inactive individuals occurred in years when the operational sex ratio was more female-biased (i.e. 1987, 1989 and 1990) than in other years, it is possible that recent recruitment of young females into the fishery (as in 1989) resulted in a pulses of young, sexually inactive, females in these years.

Operational sex ratios within aggregations were female-biased, with significant intra- and inter-annual variation. To evaluate such variation, temporal and spatial factors that could influence sex ratios must be considered. For example, in 1991, significant changes in sex ratio were noted over the course of the spawning period. While the small sample sizes and temporal clustering of samples limit the conclusions to be drawn, the data suggest an influx of males, or egress of females, at the time of the full moon. Variations in sex ratios of the red hind were also noted by Olsen & LaPlace (1979), with an influx of males at the January full moon of the 1975 aggregation in the U.S. Virgin Islands. Burnett-Herkes (1975) noted very different aggregation sex ratios in two samples of red hind (1:35 and 1:3) (M:F) taken four days apart by fishers in Bermuda, although sample sizes were small (N = 72 and N = 55, respectively). In the Indo-Pacific grouper, *Plectropomus areolatus,* males were reported to precede females to the aggregation site (Johannes 1988). It appears, therefore, that in the red hind and also in other groupers, there may be differential movement of males and females into, and out of, aggregations. Such temporal variation should be factored into studies that examine reproductive dynamics, sex ratios and fish sizes within grouper aggregations.

Intra-annual variability in sex ratios is unlikely, on the other hand, to be attributable soley to sampling errors such as sex-related differences in spatial distribution or in the taking of bait. Data from a study of a red hind aggregation area in southwestern Puerto Rico in 1984 indicated that sex ratios did not differ over a one week study period between: (a) fish caught by research vessels and by commercial fishers; (b) fish speared in identified clusters underwater and surface-caught fish at the same site; (c) fish caught at one study site and taken from other sites in the area and (d) fish speared in clusters or as solitary individuals (Shapiro et al. 1993a).

Given temporal intra-annual variability in sex ratios, inter-annual changes should be evaluated with caution, especially where sample sizes are small or sampling days are few. Between year variations in sex ratios not attributable to sampling error may be caused by any of a number of factors: annual variation in recruitment of females into the fishery; differential mortality of females and males through fishing activity outside of the aggregation; sexual transitions of females to males, which essentially translate to a higher 'natural' mortality of females; and differential migration of the sexes to a specific aggregation area between years. For example, large recruitment events into the fishery likely produce increased female bias in subsequent years. On the other hand, sex changes from female to male, all else being equal, would result in a relative increase in males. Determination of the causes of inter-annual variations in sex ratios requires evaluation of aspects of population demography such as what factors induce sex change, reproductive versus non-reproductive sex ratios, etc.

One approach to understanding the reproductive patterns of the red hind relative to those of closely related congeners, involves interspecific comparisons of reproductive traits which provide a framework for posing adaptive questions about function and associations (Clutton-Brock & Harvey 1984). Comparisons among phylogenetically similar species, especially if these overlap in distribution, can be particularly valuable because phylogenetic and geographic influences are minimized. They may be used to establish general trends and to indicate relationships between behavioural and ecological traits and provide a valuable means of testing hypotheses.

The reproductive patterns among seven western Atlantic species of *Epinephelus* are shown in Table 3. Smaller species such as *Epinephelus fulvus* and *E. cruentatus* do not aggregate, are protogynous and spawn in single-male/multi-female social units. Males are on average larger than females, spawning activity may last a couple of months with no apparent lunar component and sperm competition is not indicated (Smith 1959, Thompson & Munro 1978, Nagelkerken 1979, Sadovy et al. 1994, Colin person-

Table 3 Associations of maximum species fork length (FL), occurrence of aggregating behaviour during spawning, mating system (single-male mating groups or clusters versus multi-male mating clusters), sexual pattern (Protogyny with all males derived from females, or Protogyny/gonochorism with some or many males derived from juveniles rather than from adult females), presence of sperm competition as inferred from size of ripe testes for seven species of western Atlantic grouper

Epinephelus spp	Max FL (mm)	Aggregates	Mating system	Sexual pattern	Sperm competition
E fulvus	300	no	single-male	p	
E cruentatus	300	no	single-male	p	no
E guttatus	550	yes	single-male	p	no
E adscensionis	550	yes			
E morio	800	yes		p	no
E striatus	800	yes	multi-male	p/g	yes
E itajara	2000	yes		p/g?	

al observation). Intermediate and larger-sized species all aggregate to spawn. In *E. guttatus* and *E. morio,* and possibly also in *E. adscensionis,* aggregations are patchily distributed over relatively broad physical areas. Both *E. guttatus* and *E. morio* are protogynous, sexually dimorphic, and the small size of ripe testes indicates the absence of sperm competition in both species (Moe 1969, Bannerot 1984, Colin et al. 1987, Koenig personal communication). The red hind spawns for about two weeks annually over the full or new moon phases, apparently in single-male/multi-female clusters. In the larger Nassau grouper, on the other hand, spawning events consist of brief, female-led, multi-male clusters arising from within discrete, highly localized, aggregations (Colin 1992). The species is not sexually dimorphic for size, sperm competition is suggested by the large testes sizes and the spawning season lasts for a few days over one or two full moon periods each year (Colin 1992, Sadovy & Colin unpublished data). The sexual pattern of this species, and possibly also of the jewfish, *E. itajara,* includes a component of gonochorism whereby at least some of the males develop directly from juveniles without passing through an adult female phase and undergoing sex change (Colin et al. 1987, Colin 1992, Bullock et al. 1992, Colin 1994, Sadovy & Colin unpublished data).

These trends in reproductive patterns suggest several associations: (a) between aggregation spawning, body size and possibly also, length of annual reproductive activity, and (b) among sexual pattern, mating system, sexual dimorphism, and, if reflected in relative testis size, sperm competition. The apparent differences in mating system and sexual pattern between the closely related aggregating species, the red and Nassau groupers, however, indicate that aggregating behaviour and type of mating system do not necessarily covary.

Aggregating behaviour occurs in larger *Epinephelus* spp. and is not known for the smaller species. Larger species are presumably more capable of migrating the distances required to aggregate and may also be less vulnerable to the predation that such movements likely incur (Thresher 1984). Establishing cause and effect, however, is difficult. Aggregation spawning may simply be necessary to enable males and females to meet (Shapiro et al. 1993a) in sexually dimorphic species which exhibit size and depth-related relationships with factors such as food preferences or shelter requirements. For example, depth-related size variations have been noted for the red hind but not for the coney, *E. fulvus*, in western Puerto Rico (ANOVA for two depth strata, 37–92 m and > 92 m; coney – $F_{1,46} = 0.11$, $p = 0.75$, NS; red hind – $F_{1,50} = 7.07$, $p = 0.01$) (FRL unpublished data). Alternatively, aggregating behaviour may be a specialisation of larger groupers which confers direct reproductive advantage by coordinating spawning among adults, or a survival advantage to propagules (see Thresher 1984 and Shapiro et al. 1993a for discussions).

The length of the spawning season is inevitably linked in some way to aggregating behaviour since aggregations are characteristically short-lived. However, whether the length of spawning season within a specific geographic area is latitude- or aggregation-related, or primarily dependent on other factors, needs to be explored.

Associations among sexual pattern, mating system and sexual dimorphism are evident and not unexpected. Studies on a broad range of species have shown relationships between sexual dimorphism and degree of polygyny because of a competitive advantage to larger males, and between reproductive advantage to larger males and protogyny (Ghiselin 1969, Clutton-Brock & Harvey 1984, Warner 1984). We would expect little or no sperm competition in single- male mating groups (with one or more females) and sperm competition in species which typically form multi-male spawning groups; systematic studies in primates and in cervids have demonstrated this to be the case and the same pattern applies to fishes (Clutton-Brock & Harvey 1984, Warner 1984).

Additional details on the reproductive patterns of western Atlantic groupers are needed to fully apply the comparative approach. Although predictions may be made based on the present analysis, such as single-male mating clusters in *E. morio* and absence of sperm competition in *E. fulvus,* more details on a greater number of western Atlantic species are needed to address questions of ultimate and proximate causes of spawning aggregations and reproductive patterns. Information, for example, is needed on the duration of spawning activity each year (i.e. not just the number of months that vitellogenic oocytes are observed) and what determines that duration, mating systems and sexual patterns.

The concentrated nature and short duration of the spawning aggregations of groupers render these species especially susceptible to heavy fishing pressure. A number of aggregations have disappeared and declines in many others have been reported (e.g. Olsen & LaPlace 1979, Carter et al. 1994, Sadovy 1994b). Such declines appear to be especially severe in species that aggregate in highly concentrated, discrete locations, such as the jewfish, *E. itajara,* and the Nassau grouper, *E. striatus* (Olsen & LaPlace 1979, Sadovy 1994b). Although red hind aggregations are more patchily distributed than those of its larger congeners, aggregations have nonetheless declined in Bermuda, Puerto Rico and the US Virgin Islands (Beets & Friedlander 1992, Sadovy 1994a). Given the apparent importance of aggregation spawning to annual reproduction in this species (Colin 1992, Shapiro 1987) and the likely disruption of aggregation mating units by fishing activity, it is clearly important that excessive exploitation be avoided and that aggregations be carefully monitored.

Acknowledgements

We dedicate our work to Rosemary Lowe-McConnell and Ethelwynn Trewavas and especially to Eugenie Clark for their inspiration to those who follow in their wake. We are most grateful to the Exploration Team of the Fisheries Research Laboratory and to local fishers Wilfredo Velez and Santiago Velez for invaluable assistance in sample collection. Valuable comments on the manuscript were provided by Richard Appeldoorn, Jim Beets, Doug Shapiro and two anonymous reviewers. The Caribbean Fishery Management Council partially funded the collection and processing of samples and the preparation of the manuscript. Miguel Figuerola and George Mitcheson have assisted in numerous ways throughout this study. Bonny Bower-Dennis drafted several figures.

References cited

Bannerot, S P 1984 The dynamics of exploited groupers (Serranidae) an investigation of the protogynous hermaphroditic reproductive strategy Ph D Dissertation, University of Miami, Coral Gables 393 pp

Bannerot, S P, W W Fox & J E Powers 1987 Reproductive strategies and the management of snappers and groupers in the Gulf of Mexico and Caribbean pp 561–603 *In* J J Polovina & S Ralston (ed) Tropical Snappers and Groupers Biology and Fisheries Management, Westview Press, Boulder

Beets J & A Friedlander 1992 Stock analysis and management strategies for red hind, *Epinephelus guttatus,* in the U S Virgin Islands Proc Gulf Carib Fish Inst 42 66–80

Bretschneider, L J & J J Duyvene deWit 1947 Sexual endocrinology of non-mammalian vertebrates Monographs on the progress of research in Holland during the war, Vol 2, Elsevier, New York 147 pp

Bullock, L H , M D Murphy, M F Godcharles & M E Mitchell 1992 Age, growth, and reproduction of jewfish *Epinephelus itajara* in the eastern Gulf of Mexico U S Fish Bull 90 243–249

Burnett-Herkes, J 1975 Contribution to the biology of the red hind, *Epinephelus guttatus*, a commercially important serranid fish from the tropical western Atlantic Ph D Dissertation, University of Miami, Coral Gables 154 pp

Carter, J , G J Marrow & V Pryor 1994 Aspects of the ecology and reproduction of Nassau grouper, *Epinephelus striatus,* off the coast of Belize, Central America Proc Gulf Carib Inst 43 64–110

Clutton-Brock, T H & P H Harvey 1984 Comparative approaches to investigating adaptation pp 7–29 *In* J R Krebs & N B Davies (ed) Behavioural Ecology, An Evolutionary Approach, 2nd ed , Sinauer Associates, Sunderland

Colin, P L 1992 Reproduction in the Nassau grouper, *Epinephelus striatus* (Pisces Serranidae) and its relationship to environmental conditions Env Biol Fish 34 357–377

Colin, P L 1994 Preliminary investigations of reproductive activity of the jewfish, *Epinephelus itajara* (Pisces Serranidae) Proc Gulf Carib Fish Inst 43 137–146

Colin, P L , D Y Shapiro & D Weiler 1987 Aspects of the reproduction of two species of groupers, *Epinephelus guttatus* and *E striatus* in the West Indies Bull Mar Sci 40 220–230

Emlen, S T & L W Oring 1977 Ecology, sexual selection and the evolution of mating systems Science 197 215–223

Garcia-Moliner, G E 1986 Aspects of the social spacing, reproduction and sex reversal in the red hind, *Epinephelus guttatus* M Sc Thesis, University of Puerto Rico, Mayaguez 104 pp

Ghiselin, M T 1969 The evolution of hermaphroditism among animals Quart Rev Biol 44 189–208

Harcourt, A H , P H Harvey, S G Larson & R V Short 1981 Testis weight, body weight and breeding system in primates Nature 293 55–57

Hunter, J R , B J Macewicz, N C Lo & C A Kimbrell 1992 Fecundity spawning and maturity of female Dover sole *Microstomus pacificus,* with an evaluation of assumptions and precision U S Fish Bull 90 101–128

Johannes, R E 1988 Spawning aggregation of the grouper, *Plectropomus areolatus* (Ruppel) in the Solomon Islands Proc 6th Int Coral Reef Symp 2 751–755

Moe, M A 1969 Biology of the red grouper, *Epinephelus morio* (Valenciennes), from the eastern Gulf of Mexico Florida Dept Natural Resources Papers Series 10 95 pp

Nagelkerken, W P 1979 Biology of the graysby, *Epinephelus cruentatus,* of the coral reef of Curaçao Studies on the fauna of Curaçao and other Caribbean Islands 60 1–118

Olsen, D A & J A LaPlace 1979 A study of a Virgin Islands grouper fishery based on a breeding aggregation Proc Gulf Carib Fish Inst 31 130–144

Sadovy, Y 1994a Grouper stocks of the western central Atlantic the need for management and management needs Proc Gulf Carib Fish Inst 43 43–63

Sadovy, Y 1994b The case of the disappearing grouper *Epinephelus striatus,* the Nassau grouper, in the Caribbean and western Atlantic Proc Gulf Carib Fish Inst 45 (in press)

Sadovy, Y & M Figuerola 1992 The status of the red hind fishery in Puerto Rico and St Thomas, as determined by yield-per-recruit analysis Proc Gulf Carib Fish Inst 42 23–38

Sadovy, Y & D Y Shapiro 1987 Criteria for the diagnosis of hermaphroditism in fishes Copeia 1987 136–156

Sadovy, Y , M Figuerola & A Roman 1992 Age and growth of red hind, *Epinephelus guttatus,* in Puerto Rico and St Thomas U S Fish Bull 90 516–528

Sadovy, Y, P L Colin & M Domeier 1994 Aggregation and spawning in the tiger grouper, *Mycteroperca tigris* (Pisces Serranidae) Copeia (in press)

Shapiro, D Y 1987 Reproduction in groupers pp 295–327 *In* J J Polovina & S Ralston (ed) Tropical Snappers and Groupers Biology and Fisheries Management, Westview Press, Boulder

Shapiro, D Y & Y Sadovy & M A McGehee 1993a Size, composition, and spatial structure of the annual spawning aggregation of the red hind, *Epinephelus guttatus* (Pisces Serranidae) Copeia 1993 367–374

Shapiro, D Y , Y Sadovy & M A McGehee 1993b Periodicity of sex change and reproduction in the red hind, *Epinephelus guttatus* a protogynous grouper Bull Mar Sci 53 399–406

Smith, C L 1959 Hermaphroditism in some serranid fishes from Bermuda Papers of the Michigan Academy of Science Arts and Letters 111–119

Sokal, R R & F J Rohlf 1969 Biometry W H Freeman San Francisco 776 pp

Thompson, R & J L Munro 1978 Aspects of the biology and ecology of Caribbean reef fishes Serranidae (hinds and groupers) J Fish Biol 12 115–146

Thresher, R E 1984 Reproduction in reef fishes T F H Publications, Neptune City 399 pp

Wallace, R A & K Selman 1981 Cellular and dynamic aspects of oocyte growth in teleosts Amer Zool 72 326–343

Warner, R R 1984 Mating behavior and hermaphroditism in coral reef fishes Amer Sci 72 128–136

Environmental Biology of Fishes **41**: 287–299, 1994.

Reproductive biology and systematics of phallostethid fishes as revealed by gonad structure

Harry J. Grier[1] & Lynne R. Parenti[2]
[1] *Florida Marine Research Institute, 100 Eighth Avenue S. E., St. Petersburg, FL 33701-5095, U.S.A.*
[2] *Division of Fishes, National Museum of Natural History, Smithsonian Institution, Washington, D.C. 20560, U.S.A.*

Received 9.11.1992 Accepted 9.8.1993

Key words: Phallostethidae, Atherinomorpha, Testis, Spermatozeugmata, Internal fertilization, Ovary, Histology of museum specimens

Synopsis

Testis and ovary structure was examined histologically in seven of the 19 species in the three tribes of the teleost fish family Phallostethidae, series Atherinomorpha. These diminutive species have testes in which spermatogonia are restricted to the distal ends of lobules, a diagnostic character of atherinomorphs. Sperm in the ovarian lumen and chorionic attachment filaments on eggs confirms observations that phallostethids are internally fertilizing and lay fertilized eggs. The immense number of sperm in ovarian cavities means that all, or nearly all, ovulated oocytes will be fertilized. As revealed in histological sections, testicular ducts in most phallostethids examined contain 'granular' secretions that have not been reported in any other atherinomorphs. Species in the tribes Neostethini and Gulaphallini form unique spermatozeugmata that differ from those of other internally fertilizing atherinomorphs examined in that they have sperm nuclei that are oriented towards one side of the sperm bundle. Spermatozeugmata are not formed in species in the tribe Phallostethini. A unique spermatozeugmatum is interpreted as being a diagnostic character of phallostethids that has been lost or modified in phallostethins. Gonads of phallostethids and hypothesized close relatives are posterior and posteroventral to the gut rather than dorsal to the gut, as they are in most other fishes. Museum specimens preserved over sixty-five years ago are as useful for demonstrating gonad histology as are those preserved in the past few years.

Introduction

Phallostethids are a group of southeast Asian brackish and freshwater fishes that can be distinguished from all other teleosts by a complex, bilaterally asymmetric, subcephalic copulatory organ called the priapium, which is used to transfer sperm from the male to the female reproductive tract (Regan 1913, 1916). Phallostethids are small to minute; mature adults range from about 14 to 37 mm SL (Roberts 1971a, b, Parenti 1989). Diagnostic specializations of phallostethids, in addition to reproductive styles, include modifications associated with small size, such as a reduced hyobranchial apparatus and reduced or absent first dorsal fin. In the most recent taxonomic revision of the Phallostethidae, Parenti (1989) recognized 19 species in four genera, in three tribes and two subfamilies (Table 1).

Phallostethid fishes were first reported by Duncker (1904) and described by Regan (1913). Yet, probably because of their small size and the dif-

Table 1. Classification of atherinomorph fishes (after Rosen & Parenti 1981, Parenti 1989, 1993).

Series Atherinomorpha
 Division I
  family Bedotiidae
  family Melanotaeniidae
  family Pseudomugilidae
  family Telmatherinidae
  family Atherinidae
  family Isonidae
  superfamily Phallostethoidea
   family Phallostethidae
    subfamily Phallostethinae
     tribe Phallostethini
      genus *Phallostethus*
      genus *Phenacostethus*
     tribe Neostethini
      genus *Neostethus*
    subfamily Gulaphallinae
     tribe Gulaphallini
      genus *Gulaphallus*
   family Dentatherinidae
 Division II
  order Cyprinodontiformes
  order Beloniformes

ficulty of obtaining live specimens for laboratory study, phallostethid reproductive biology and the histological structure of reproductive organs is known incompletely, and only from a few species (e.g. Regan 1916, Smith 1927, Villadolid & Manacop 1934, Manacop 1936, Munro & Mok 1990). Given limitations of the techniques of histology and light microscopy available to him, Regan (1916) made remarkably accurate observations, noting that one phallostethid species, *Neostethus lankesteri,* forms sperm bundles, whereas another, *Phallostethus dunckeri*, does not. Regan (1916) did not precisely describe the sperm bundles, however, and called them spermatophores rather than spermatozeugmata. These two types of sperm bundles differ in that spermatophores are encapsulated whereas spermatozeugmata are naked. Regan (1916) also concluded that phallostethid sperm bundles were similar to those found in the cyprinodontiform family Poeciliidae; with more information available to us, we are now able to distinguish between them (see below).

Whether phallostethids are oviparous or viviparous was open to speculation until Smith (1927) reported that female phallostethids lay fertilized eggs. Observations by Villadolid & Manacop (1934) on breeding *Gulaphallus mirabilis,* and by Manacop (1936) on *Gulaphallus falcifer,* confirmed that during copulation an external bone of the priapium is used by the male to clasp the female. In neostethins and gulaphallins, the elongate clasping bone is the ctenactinium, a homologue of a pelvic-fin ray. In phallostethins, the hook-like clasping bone is the toxactinium, of uncertain homology (see Parenti 1989). A second component of the priapium, including a fleshy seminal papilla, is used to transfer sperm to the female reproductive tract.

In internally fertilizing atherinomorphs such as the cyprinodontiform families Poeciliidae and Goodeidae (Grier 1981), and the beloniform family Hemiramphidae, genera *Zenarchopterus* (Grier & Collette 1987) and *Dermogenys* (Grier unpublished), naked sperm bundles (spermatozeugmata) are transferred from male to female. Spermatozeugmatum morphology in each of these families is unique, suggesting that spermatozeugmata are not homologous and that internal fertilization evolved independently in different atherinomorph groups, a conclusion supported by osteology and other data (see Parenti 1981, for cyprinodontiforms). In some atherinomorphs, such as the cyprinodontiform family Anablepidae (including Jenynsiidae, following Parenti 1981), tubular structures of the male anal fin are used to transfer sperm, and either individual sperm or partial spermatozeugmata are transferred to the female during copulation (Grier et al. 1981). In one atherinomorph species, the beloniform *Horaichthys setnai,* the male transfers sperm to the female in encapsulated sperm bundles (spermatophores). Structure of the spermatophore (Kulkarni 1940; fig. 12, 13) and testis (Grier 1984; fig. 2) of *H. setnai* has been described and illustrated.

Grier et al. (1980) described two different testis types in teleost fishes: testes with a restricted or unrestricted distribution of spermatogonia. The restricted testis type was reported in all atherinomorphs whose testis structure was examined, including a single phallostethid, *Gulaphallus mirabilis* (Grier et al. 1980: Table 1). Munro & Mok (1990) confirmed this testis type in *Phenacostethus smithi.*

Phylogeny of chordate testis structure was reinterpreted recently, and a revised terminology was developed (Grier 1992, 1993). Three types of testis structure are now recognized in teleost fishes: anastomosing tubular with an unrestricted spermatogonial distribution, lobular with spermatogonia distributed along the lobule lengths (unrestricted), or lobular with spermatogonia restricted to distal ends of lobules that terminate at the testis periphery (restricted). The last character, spermatogonia restricted to distal ends of lobules, is a more precise description of testis type unique to atherinomorph fishes; it is one of several synapomorphies that support atherinomorph monophyly (Rosen & Parenti 1981, Parenti 1993).

Our investigation of gonad morphology in the Phallostethidae was undertaken to answer the following questions: (1) what type of testis do phallostethids have; (2) do phallostethids form spermatozeugmata in the testis, and, if so, how might these spermatozeugmata differ from those of other atherinomorphs; and (3) how are sperm stored in the phallostethid female reproductive tract?

Materials and methods

Gonad structure of seven phallostethid species, representing three genera from all three tribes (Table 1), was examined histologically. Nomenclature used to describe phallostethid testis structure is that of Grier (1993). Gonads examined in this study are from two adult male and two adult female specimens of each species which were deposited in the California Academy of Sciences (CAS) and the National Museum of Natural History (USNM), Smithsonian Institution. Specimens bearing catalog numbers with the prefix CAS-SU are from the Stanford University fish collection now housed at CAS. Table 2 contains names of phallostethid species examined histologically, their catalog numbers, collector, year collected, and localities. To the best of our knowledge, all were initially fixed in formalin of unrecorded percentages, perhaps sometimes with buffers, and were transferred to 70–75% ethanol for museum storage. All phallostethids that were examined histologically were embedded in glycolmethacrylate (Polysciences). They were transferred from the storage ethanol to 95% ethanol, then to a 1:1 mixture of ethanol:glycolmethacrylate, and finally through two changes of 100% glycolmethacrylate before being embedded. In most cases, only gonads were embedded. Owing to their small adult size of less than 20 mm SL, whole specimens of both *Phenacostethus smithi* and *P. posthon* were examined, however. Glass knives were used to cut 3.5 μm thick sections on a retracting LKB Historange® microtome. Sections were floated on water and subsequently mounted onto acid-cleaned microslides. Mounted sections were stained with periodic acid Schiff (PAS), hematoxylin and metanil yellow (PAS/MY) (Quintero-Hunter et al. 1991), hematoxylin and eosin, or thionin (P. Nagle personal communication). Photographs were taken

Table 2 Phallostethid species examined histologically

Tribe/Species	Catalog No	Collector (Year)	Locality
Phallostethini			
Phenacostethus smithi	USNM 88667	H Smith (1927)	Bangkok, Thailand
	CAS-SU 35957	H Smith (1931)	Bangkok, Thailand
Phenacostethus posthon	USNM 229302	T Roberts (1973)	Muar R , W Malaysia
Neostethini			
Neostethus borneensis	USNM 321316	L Parenti (1991)	Dolhakim R , Brunei
Neostethus lankesteri	CAS-SU 67162	E Alfred (1966)	Seletar R Singapore
Neostethus bicornis	CAS-SU 35783	A Herre (1937)	Kranji R , Singapore
Gulaphallini			
Gulaphallus mirabilis	CAS 50721	T Roberts (1976)	Olo Creek, Luzon, Philippines
Gulaphallus bikolanus	CAS 50722	T Roberts (1976)	Guinobatan R , Luzon, Philippines

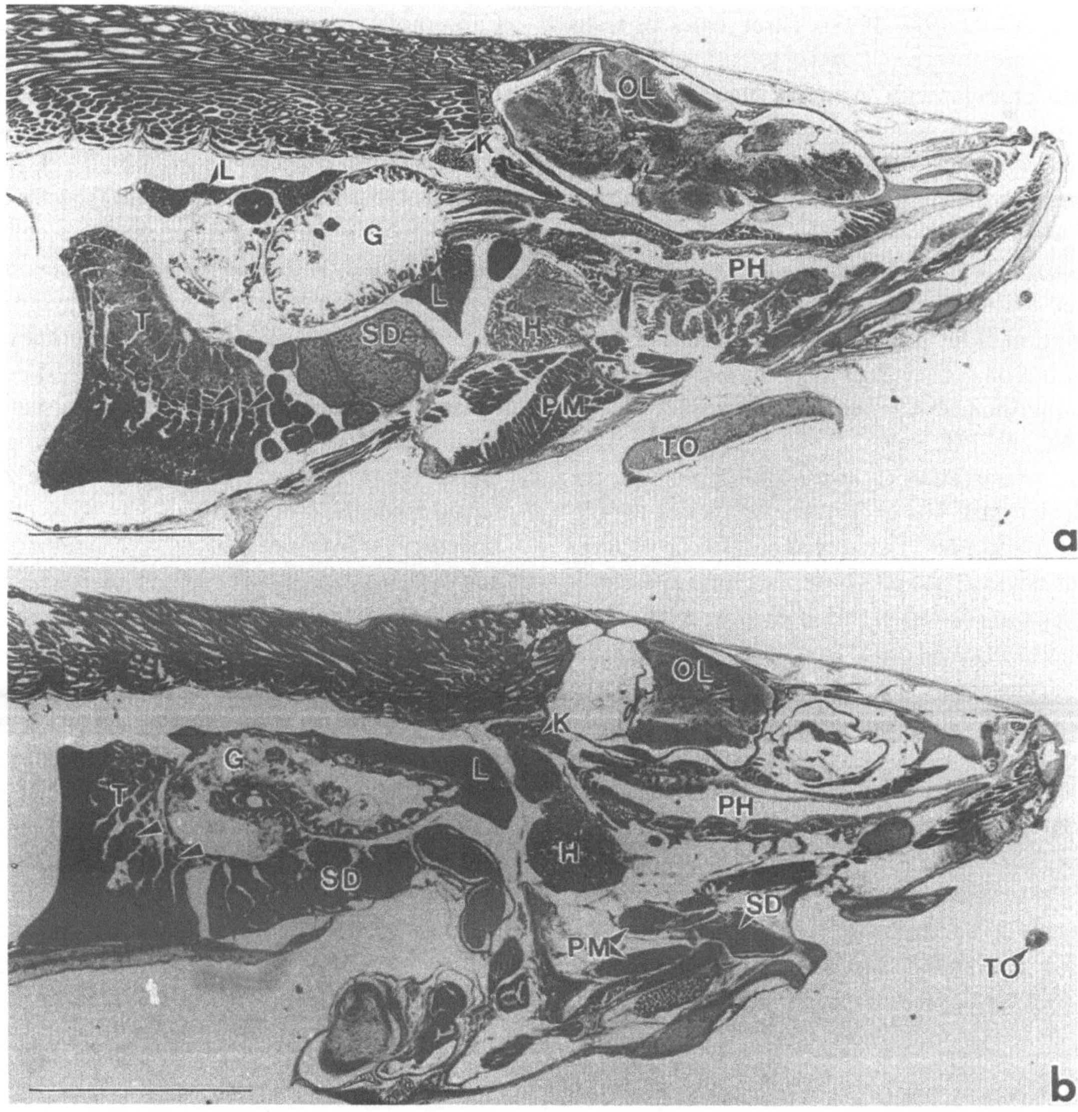

Fig. 1. Longitudinal sections through *Phenacostethus smithi* (a) and *P. posthon* (b), respectively. In each species, the gut (G) is antero-dorsal to the lobular testis (T) and the sperm duct (SD), which leads to the priapium. The section of the sperm duct in *P. posthon* is just ventral to the pharyngeal cavity (PH). Part of the priapium, the toxactinium (TO), a hook-like bone used to clasp females, is also sectioned. Arrows indicate testicular lobes in which circular spermatocysts are located. PM = priapial musculature; L = liver; H = heart; K = kidney; OL = optic lobe of brain. Bars = 0.5 mm.

through a green filter on Kodak T Max® 100 film using an Olympus Vanox® microscope. Thionin-stained sections showed the clearest cytological detail and were used, unless otherwise noted, for illustrations.

An extensive amount of preserved material, including phallostethid and outgroup fishes that had been cleared and stained to show skeletal structure, was examined by LRP and reported on in recent studies on phallostethid comparative anatomy and systematics (Parenti 1984, 1986a, b, c, 1989). Catalog numbers of comparative material examined in this study are given in the text and in Table 2.

Results

Generalizations apply to testis structure of phallostethids. Here, these generalizations are documented by specific species examples, but they apply to all species examined, and any observed structural variations between species are noted. In all examined phallostethids, the testis is a single, unpaired organ composed of nonbranching lobules and a well-developed efferent duct system. Lobule distal ends terminate at the testis periphery as illustrated for *Phenacostethus smithi, P. smithi, Gulaphallus bikolanus, Gulaphallus mirabilis,* and *Neostethus bicornis* (Fig. 1a–b, 2d, 3a, c, 4a). In the lobules of all phallostethids, spermatogenesis and spermiogenesis occur within spermatocysts composed of maturing germ cells surrounded by Sertoli cell processes (Fig. 2a, c, 3b, 4b). Only mature spermatocysts, those that

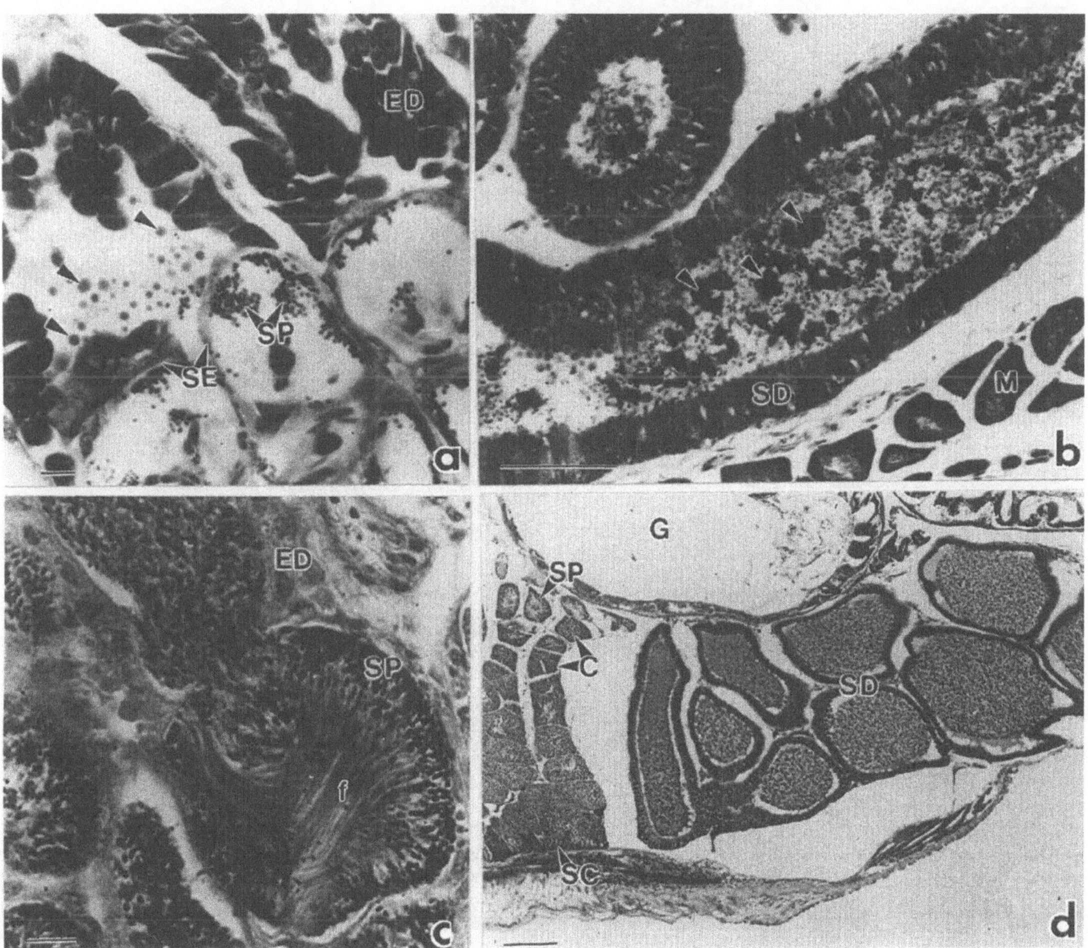

Fig 2 a- *Phenacostethus smithi* testis illustrating sperm (SP) in spermatocysts bordered by Sertoli cell (SE) cytoplasm Note that sperm nuclei are oriented towards one side of the spermatocyst At spermiation, sperm are released into the efferent ducts (ED), which also contain granules (arrows) Bar = 100 µm b- The sperm duct (SD) of *Phenacostethus smithi* with sperm, the nuclei of which are located between aspherical granules (arrows) Height of sperm duct epithelium ranges between 20 and 30 µm M = body wall musculature Bar = 50 µm c- Spermiation in *Phenacostethus posthon* Sperm (SP) are being voided from a spermatocyst into the efferent duct (ED) Most sperm nuclei are oriented towards one side of the spermatocyst, and sperm flagella (f) fill the center of the spermatocyst Bar = 10 µm d- Testis (T) and sperm duct (SD) of *Phenacostethus posthon* illustrating spermatocysts (C) within lobules Note that spermatocysts (SC) containing early stages of sperm development are at the periphery of the testis and that those containing mature sperm (SP) are dorsal – in the region of the efferent ducts Sperm nuclei fill the sperm duct (SD) G = Gut Bar = 100 µm

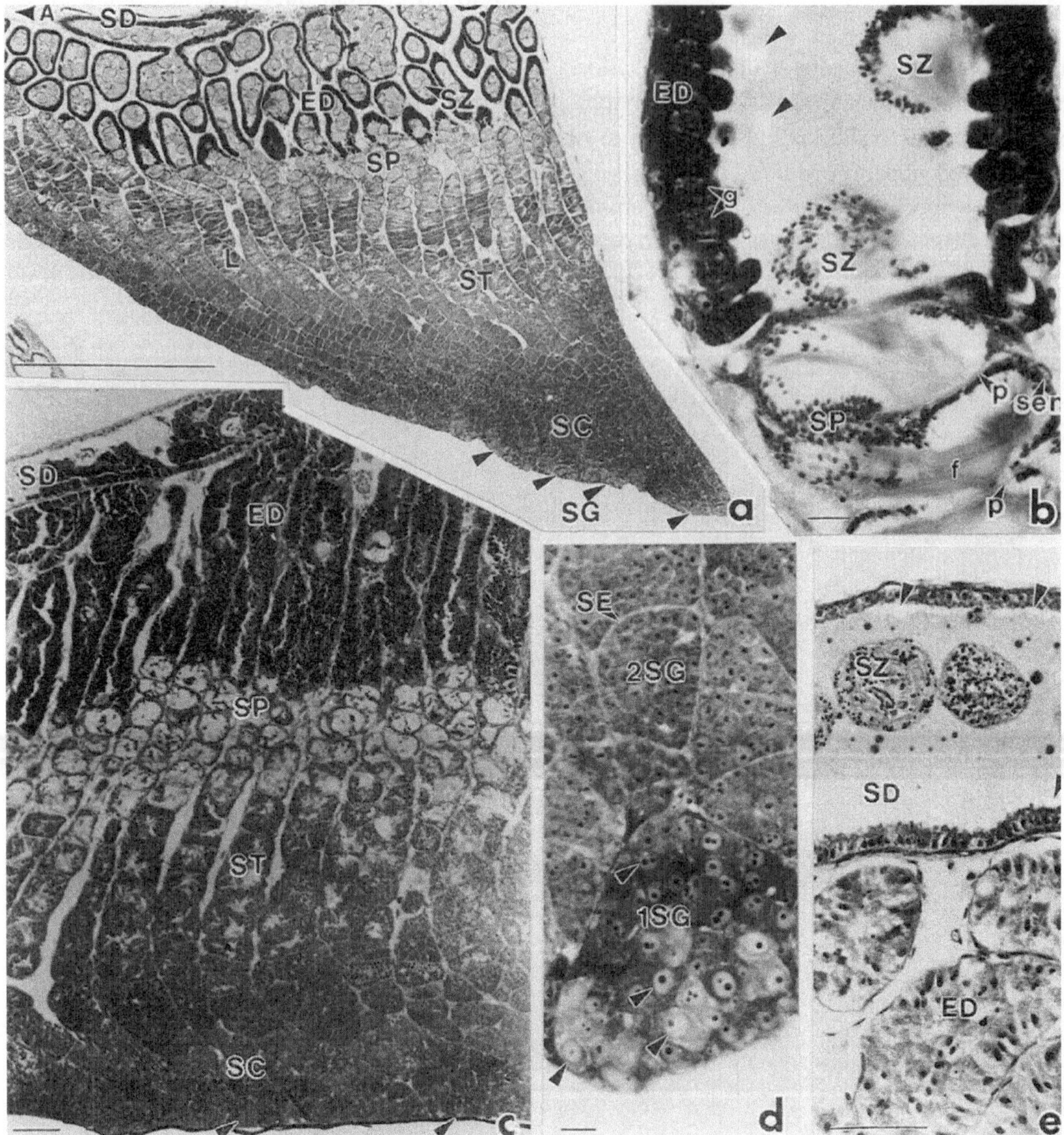

Fig. 3. a- The testis of *Gulaphallus bikolanus* showing spermatogonia (SG) in the distal ends of lobules (L) at the posteroventral testis periphery. Spermatocysts with more advanced stages of sperm maturation (spermatocytes [SC], spermatids [ST] and sperm [SP]) are located progressively closer to efferent ducts (ED). Spermatozeugmata (SZ) are found within both the efferent ducts and the dorsal sperm duct (SD). Anterior is indicated by (< A). Bar = 1 mm. b- Spermatocysts about to undergo spermiation in testis of *Gulaphallus bikolanus.* Thin Sertoli cell processes (p) form the borders of the spermatocysts. Sperm nuclei are located at, but not uniformly distributed around, the periphery of the spermatocyst, a distribution reflected in the morphology of spermatozeugmata (SZ). Efferent duct cells (ED) are columnar and have a reticulated area interpreted as Golgi apparatus (g). The ducts are filled with a lightly staining secretory product, the resolution of which is enhanced where fixation effects cause it to become separated from the duct cells (arrows). sen = nucleus of Sertoli cell, SP = sperm. Bar = 10 µm. c- The lobular testis of *Gulaphallus mirabilis* showing regionalization of spermatocysts with those containing progressively later stages of germ cell maturation (spermatocytes [SC]; spermatids [ST]; and sperm [SP] located progressively closer to efferent ducts (ED). Spermatogonia (arrows) are located along the ventral edge of the testis. SD = sperm duct. Bar = 0.1 mm. d- Periphery of the testis of *Gulaphallus mirabilis* showing clusters of primary spermatogonia (1SG) and illustrating four generations (arrows) differentiated from each other by nuclear diameter and intensity of cytoplasmic staining with thionin. Secondary spermatogonia (2SG) are found within spermatocysts, the borders of which are formed by Sertoli cells (SE). Bar = 10 µm. e- Dorsal part of the testis of *Gulaphallus mirabilis* illustrating the efferent ducts (ED) and spermatozeugmata (SZ) in the sperm duct (SD). Secretory product in the sperm duct has separated from duct cells (arrows). The quality of fixation is judged to be marginal because the efferent duct cells are poorly defined. Note the spherical granules in the sperm duct. Bar = 50 µm.

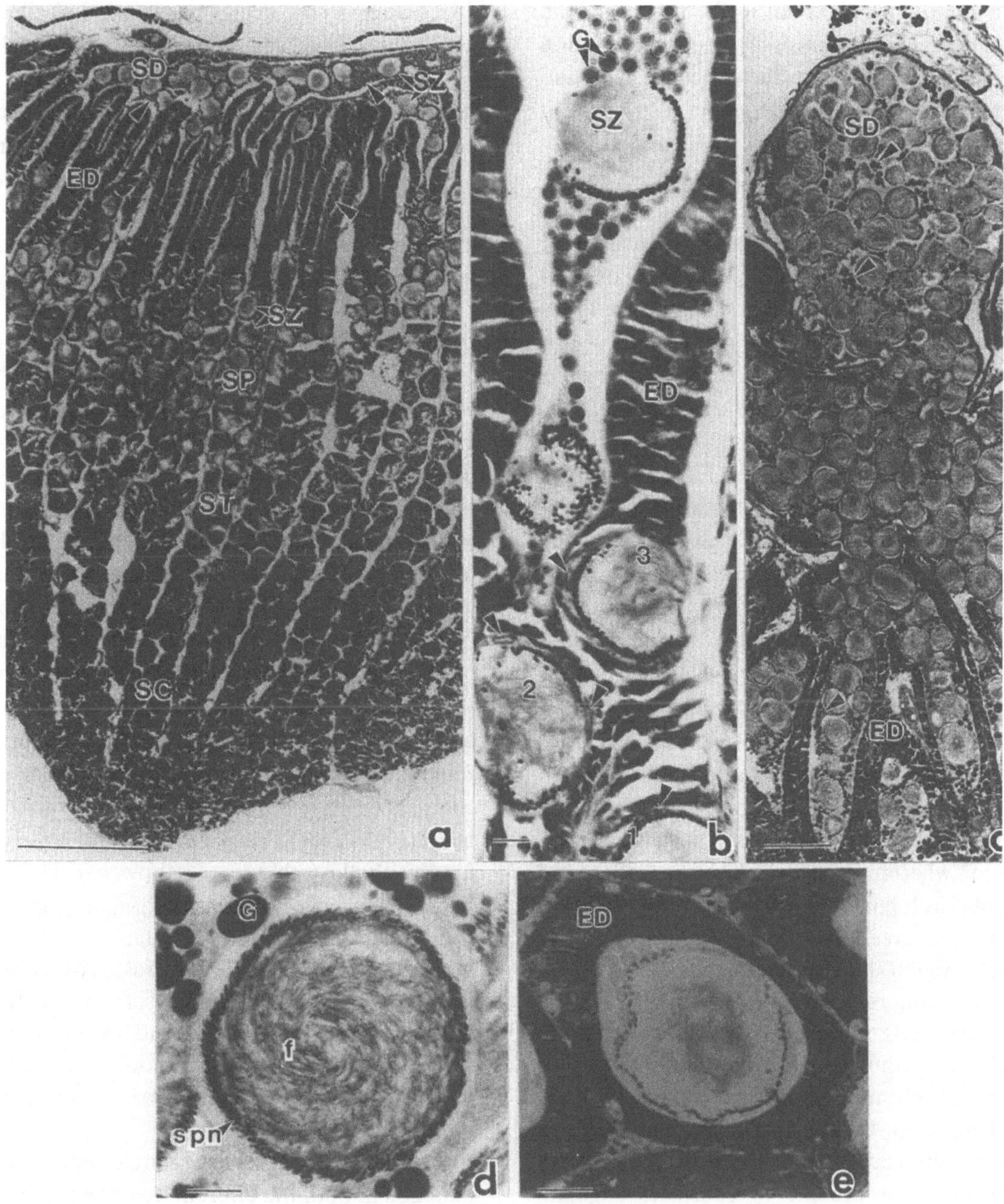

Fig. 4. a- The testis of *Neostethus bicornis* showing spermatocysts within numerous lobules. Spermatocysts with spermatids (ST) and sperm (SP) are located closer to the efferent ducts (ED) than those with spermatocytes (SC). Within both the efferent ducts and sperm ducts (SD), spermatozeugmata (SZ) and numerous, spherical granules (arrows) are present. Bar = 1 mm. b- Spermiation in *Neostethus bicornis.* Some spermatocysts (1, 2, 3) are shown transgressing the columnar cell wall of the efferent duct (ED). Stain characteristics of Sertoli cell nuclei (arrows) and cytoplasm bordering the spermatocysts are distinctly different from those of the efferent duct cells. Numerous spherical granules (G) and spermatozeugmata (SZ) are located within the efferent ducts. Note that the distribution of sperm nuclei in the spermatocysts is reflected in their distribution in the spermatozeugmata, i.e. along the periphery and to one side. Bar = 10 µm. c- Spherical spermatozeugmata fill the efferent ducts (ED) and sperm duct (SD) in *Neostethus lankesteri* Spherical granules (arrows) are present between the spermatozeugmata. Bar = 100 µm. d- A spermatozeugmatum from *Neostethus lankesteri* sectioned in such a way that sperm nuclei (spn) are oriented towards the periphery and sperm flagella are oriented (f) towards the center. Granules (G) within a granular secretory product surround the spermatozeugmatum. Bar = 10 µm. e- A single spermatozeugmatum from the testis of *Neostethus borneensis* illustrating peripheral sperm nuclei surrounded by light-staining granules. ED = efferent duct. Bar = 50 µm.

contain sperm, are observed in the regions of the efferent ducts (Fig. 2a, 3a–c, 4a) where spermiation (release of sperm into the ducts) occurs. All phallostethid testes have spermatogonia restricted entirely to the lobule distal ends as illustrated in *G. bikolanus* (Fig. 3a) and *G. mirabilis* (Fig. 3c–d). Primary spermatogonia always occur in clusters rather than in spermatocysts (Fig. 3d). Different nuclear diameters and differential thionin staining of cytoplasm (not observed in other staining methods used) suggest that at least four generations of primary spermatogonia are present in *G. mirabilis* (Fig. 3d). Primary spermatogonia become secondary spermatogonia when a group of them is surrounded by Sertoli cell processes (Fig. 3d) to form a spermatocyst in which spermatogenesis and spermiogenesis take place. Sertoli cells compose the borders of spermatocysts (Fig. 2a, 3b, d, 4b).

In *Phenacostethus smithi, P. posthon,* and *Gulaphallus bikolanus*, testicular lobules tend to terminate posteroventrally (Fig. 1a–b, 3a). In contrast, lobules terminate along the ventral edge of the testis in *Gulaphallus mirabilis* (Fig. 3c), but in all species spermatogonia are at the distal termini of the lobules. Within the testicular lobules of all phallostethids, germ cell maturation exhibits a typical distribution wherein successively more advanced stages of development (spermatocytes, spermatids, sperm) are located progressively closer to the efferent duct system (Fig. 3a, c, 4a).

Documenting relationships that might exist between germ cells, particularly between sperm and Sertoli cells, requires ultrastructural methods not used in our light microscope study. We note, however, that sperm nuclei become distributed to one side of mature spermatocysts (Fig. 2a, c, 3b, 4b), a distribution also present in spermatozeugmata (Fig. 3b, 4b). The difference between these two structures is that a spermatocyst is defined by an encompassing Sertoli cell layer (Fig. 2a, 3a) whereas the latter are not (Fig. 3b, e, 4b–e).

Spermatozeugmata are not formed in testes of the two smallest phallostethids examined, *Phenacostethus smithi* and *P. posthon* (adults less than 20 mm SL) (Fig. 1a–b, 2a–d). In both species, lobules containing developing sperm extend as efferent ducts, all of which are connected to the sperm duct leading to the exterior as in other phallostethids. Within lobules, mature spermatocysts are located near the efferent ducts (Fig. 2a, c) and have sperm nuclei oriented around only part of the cyst periphery; sperm flagella extend into the cyst interior. This character is found in the other phallostethids examined. However, at spermiation, or release of sperm from the spermatocyst into efferent ducts, free spermatozoa are liberated (Fig. 2c).

Instead of free spermatozoa, as observed in *P. smithi* and *P. posthon*, efferent ducts and sperm ducts in both *Gulaphallus* (Fig. 3a–b, e) and *Neostethus* (Fig. 4a–c) contain spermatozeugmata. These unencapsulated or naked sperm bundles are formed at spermiation (Fig. 3b, 4b), a process that involves either release of sperm associated as spermatozeugmata into the efferent ducts (Fig. 3b), or movement of a spermatocyst through the columnar layer of efferent duct cells prior to spermatozeugmatum release (Fig. 4b). Sperm within a spermatocyst form a single spermatozeugmatum that is stored in the efferent and sperm ducts (Fig. 3a, 4a, c). Depending on plane of sectioning of spermatozeugmata, sperm nuclei frequently appear to encompass entirely the peripheries of spermatozeugmata (Fig. 4d–e). This is not a natural observation for any phallostethid species, however. The typical structure of phallostethid spermatozeugmata is shown in Figures 3b and 4b for *Gulaphallus bikolanus* and *Neostethus bicornis*, respectively. Sperm nuclei cluster towards one side of a spermatozeugmatum, and their flagella extend towards the other side. Individual flagella could not be resolved in the material studied; they are grayish masses in micrographs (Fig. 3b).

We observed aspherical granules between spermatozoa in both the efferent ducts and sperm duct in *P. smithi* (Fig. 2a–b). These granules are lacking in the testes and sperm duct of *P. posthon* (Fig. 2c–d). Spherical granules occur in testicular ducts of *G. mirabilis* (Fig. 3e) and are extremely numerous in *N. bicornis* (Fig. 4a–b), *N. lankesteri* (Fig. 4c–d), and *N. borneensis* (Fig. 4e). In all species, testicular ducts contained fluid secretions. Presence of putative Golgi apparati in the columnar efferent duct cells (Fig. 3b) is consistent with the probable secretory nature of these cells. We assume that duct cells

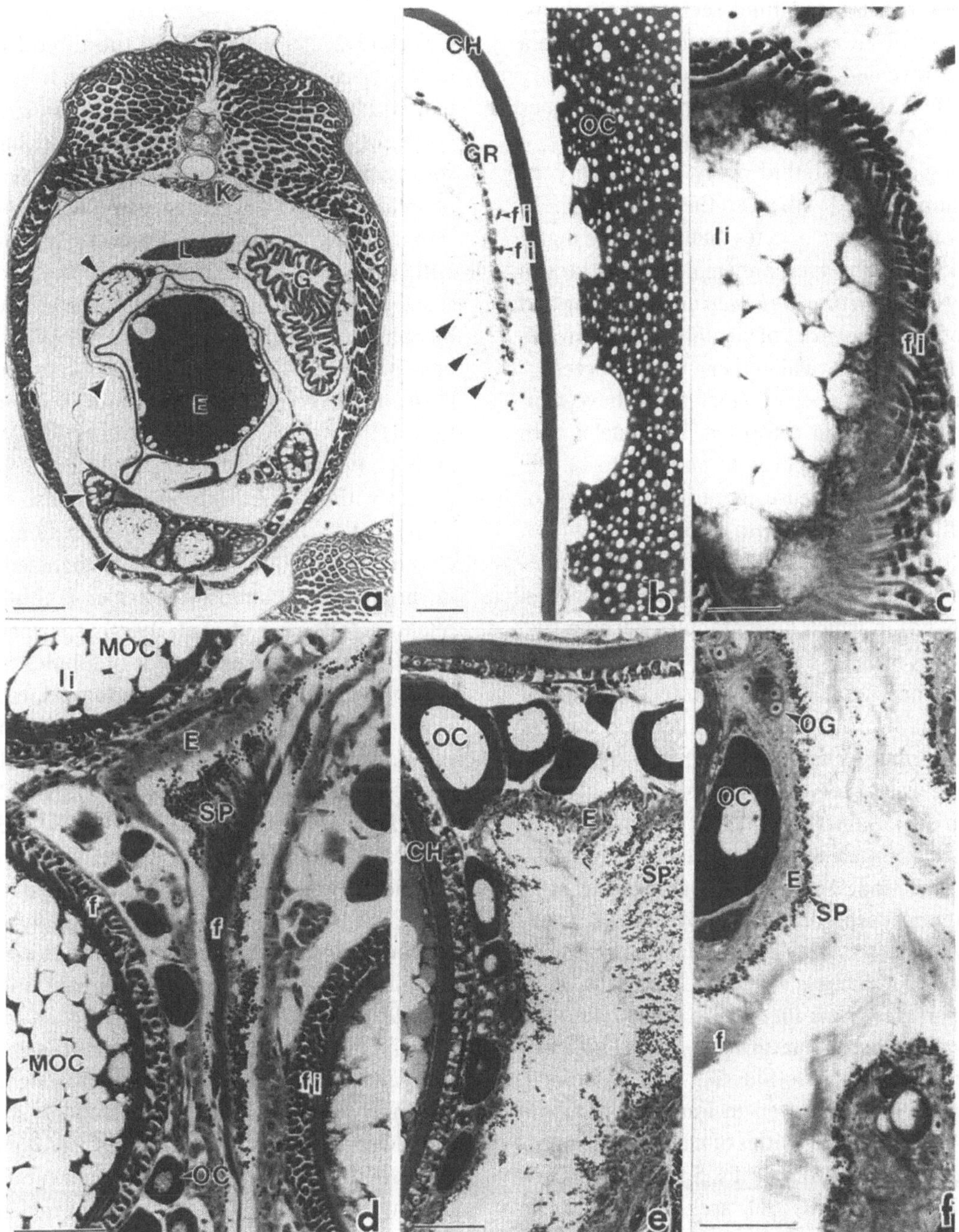

Fig 5 a Cross section through the body of a female *Phenacostethus smithi* illustrating the ventral position of the single ovary relative to the viscera gut (G) kidney (K) and liver (L) An ovulated egg (E) filled with dark staining yolk is located within the ovarian cavity Small arrows indicate developing oocytes filled with liposomes larger arrow points to area enlarged in b Bar – 1 mm b Surface of an ovulated oocyte (OC) in *Phenacostethus smithi* Protein yolk is densely staining and contains numerous lipid droplets (clear areas) Exterior to the oocyte envelope (CH) is the desquamating granulosa cell layer (GR) Oocyte filaments (fi) and sperm nuclei (arrows) are present Bar = 10 µm c Tangential view of the surface of a developing oocyte in *Phenacostethus smithi* similar to the oocytes indicated by small arrows in a Large lipid yolk droplets (li) are present in the oocyte cytoplasm and filaments (fi) course over its surface Bar = 10 µm d Stored sperm (SP) in ovary of *Neostethus bicornis* Sperm nuclei are oriented towards the ovarian cavity epithelium (E) and flagella (f) trail into the lumen Lipid yolk (li) is observed in maturing oocytes (MOC) the surfaces of which are covered with numerous filaments (fi) nonvitellogenic oocytes (OC) are also present Bar – 50 µm e Ovarian cavity of a female *Gulaphallus bikolanus* filled with stored sperm (SP) the nuclei of which are oriented towards the cavity epithelium (E) Immature nonvitellogenic oocytes (OC) are present CH envelope of one oocyte in final maturation Bar = 50 µm f Sperm (SP) in the ovarian cavity of *Gulaphallus mirabilis* have nuclei oriented towards the cavity epithelium (E) Flagella (f) extend into lumen Oogonia (OG) and a nonvitellogenic oocyte with dark staining cytoplasm (OC) are present within the ovarian stroma Bar 50 µm

are responsible for secretion of the thionin-positive granules. Granular and fluid secretory products, however, were always PAS-negative. The thionin technique revealed presence of the fluid secretory product (Fig. 3b, e, 4d) better than did techniques using PAS/MY and hematoxylin and eosin.

The single phallostethid ovary has a cavity, or ovarian lumen, lined with an epithelium (Fig. 5e–f), and a stroma in which oocytes mature. In the material studied, we observed only maturing oocytes in the ovary of *Neostethus lankesteri* and maturing and mature oocytes in ovaries of all other species. In one species, *P. smithi*, an ovulated egg was observed.

Nonvitellogenic oocytes (Fig. 5c–f) have uniformly dense-staining cytoplasm and nuclei with numerous nucleoli. Lipid yolk appears to be the first yolk formed because maturing oocyte cytoplasm is filled completely with clear liposomes (Fig. 5a, c–d). Oocyte maturation secondarily involves formation of a dense-staining, probably protein yolk (Fig. 5a) in which lipid granules become suspended (Fig. 5b). This may not be true of all phallostethids; ovarian material examined contained few fully mature oocytes. Most maturing oocytes contained primarily lipid yolk (Fig. 5a, c–d). An ovulated egg was observed in the ovary of only one specimen of *P. smithi* (Fig. 5a). The immense size of this oocyte, compared to the entire body cross section and to sectioned internal organs, is remarkable. The periphery of this oocyte (Fig. 5b) had a desquamating granulosa cell layer, external to which were sperm nuclei. Filaments were attached to the outer chorion surface. These filaments extended into the granulosa cell layer of maturing oocytes (Fig. 5c–d) in ovaries of all phallostethids that were examined. Sperm were present, often in immense numbers (Fig. 5d–f), in ovarian cavities of all female phallostethids examined histologically. Sperm nuclei oriented primarily towards, but rarely appeared to be within recesses of, the epithelial cell layer of the ovarian cavity (Fig. 5d–f).

Phallostethid testes and ovaries are single, not paired (see also Munro & Mok 1990). The phallostethid priapium, urogenital ducts, and anus are located anteriorly, posteroventral to the pharyngeal cavity (Fig. 1a–b). Phallostethid gonads are posterior and posteroventral to the gut (Fig. 1a–b, 5a).

Discussion

Grier (1992, 1993) recognized the three following testes types in teleosts: anastomosing tubular, lobular with an unrestricted distribution of spermatogonia, or lobular with a restricted distribution of spermatogonia. The restricted lobular testis has been found so far only in fishes of the teleost series Atherinomorpha. Rosen & Parenti (1981) and Parenti (1993) used the character of spermatogonia restricted to the distal end of the testis as a synapomorphy of atherinomorph fishes. Phallostethids have been classified as atherinomorph fishes on the basis of osteological and other morphological features (Rosen & Parenti 1981, Parenti 1984, White et al. 1984, Ivantsoff et al. 1987). Inclusion of phallostethids within the Atherinomorpha is also supported by testis structure. In all species examined, including those in three of the four recognized genera in three tribes (Phallostethini, Neostethini, Gulaphallini, following Parenti 1989), spermatogonia are restricted to the distal ends of lobules.

In most atherinomorphs, gonads are dorsal to the gut, but in phallostethids they are posteroventral to the gut. By comparison, in the so-called Old World silversides *Dentatherina merceri* (USNM 230374) and *Hypoatherina temmincki* (USNM 323208, USNM 323211), there is a single testis wrapped around the gut. This rotation of the gonads relative to the gut is correlated with the relatively anterior anus and urogenital openings and the elongate and anteriorly expanded first anal pterygiophore (Parenti 1984, 1989).

Numerous patterns of sperm transfer have evolved in atherinomorphs. Spermatozeugmata are common to species that have atubular anal fins modified for sperm transfer, as in the Poeciliidae and Goodeidae (Grier 1981). When a tubular anal fin is used to transfer sperm between male and female, partial spermatozeugmata, as in *Anableps dowi* (Grier et al. 1981), or free spermatozoa, as in *A. anableps* and *Jenynsia lineata* (Grier et al. 1981), may be transferred. In the atherinid *Labidesthes sicculus,* the genital duct extends as a short tube through which free sperm are transferred to the female reproductive tract; spermatozeugmata are not formed (Grier et al. 1990).

Phallostethids are unique among fishes in that they have an intromittent organ, the priapium, that is formed from the pelvic fin and parts of the pectoral fin, rather than from the anal fin (Parenti 1989) Phallostethids transfer either free spermatozoa or spermatozeugmata from the male to the female reproductive tract Species in the tribe Phallostethini have free sperm, whereas spermatozeugmata are formed in Neostethini and Gulaphallini Spermatozeugmata in *Neostethus* and *Gulaphallus* are morphologically unique and differ from those in the cyprinodontiform families Poeciliidae, Anablepidae, and Goodeidae (Grier 1981) Structure of phallostethid spermatozeugmata, as in the Poeciliidae and Goodeidae (Grier et al 1978, Grier 1981), is reflected by the grouping of sperm in mature spermatocysts prior to spermiation In the Poeciliidae and Goodeidae, ultrastructural techniques have revealed that specific associations between sperm and Sertoli cells form during spermiogenesis (Grier et al 1978, Grier 1981) Confirmation of similar associations in phallostethid testes awaits application of these techniques

Parenti (1989) considered two possible phylogenetic relationships among phallostethid tribes Phallostethini and Neostethini are sister tribes and Gulaphallini is their plesiomorphic sister group (Fig 6a), or, Neostethini and Gulaphallini are sister tribes and Phallostethini is their plesiomorphic sister group (Fig 6b) The first hypothesis of relationships was supported by more characters in her phylogenetic analysis and, therefore, was used as the basis of a reclassification (Parenti 1989, Table 1) Presence of unique spermatozeugmata in which sperm nuclei are oriented towards one side of the sperm bundle may be a synapomorphy of neostethins and gulaphallins, hence providing additional support for the second hypothesis (Fig 6b) Alternatively, spermatozeugmata may represent a diagnostic character of all phallostethids that has been lost or modified in phallostethins That spermatozeugmata have been lost in phallostethins, all of which have a well-developed seminal papilla, is supported by speculation on the evolution of sperm packaging Males in the atherinomorph families Anablepidae (including Jenynsiidae) and at least one species of Atherinidae, discussed above, have

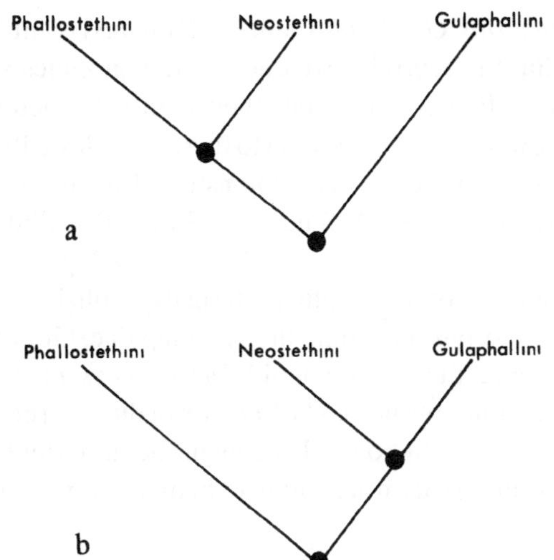

Fig 6 Two alternative cladograms of relationships among the three phallostethid tribes following Parenti (1989 Fig 7)

tubular structures for sperm transfer Anablepids may form spermatozeugmata, whereas atherinids do not Direct transfer of sperm from male to female through a tubular structure need not involve a mechanism for sperm packaging The papillary portion of the phallostethid priapium is prominent in phallostethins (e g Roberts 1971b) in which free sperm, not spermatozeugmata, are transferred The papillary portion of the priapium may be functionally equivalent to a tubular gonopodium Hence, hypothesized loss of spermatozeugmata in phallostethins is not in conflict with the first scheme of phallostethid relationships (Fig 6a)

Phallostethid gonads are posterior and posteroventral to the gut Hypothesized close relatives of phallostethids, a subgroup of so-called Old World silversides including the family Dentatherinidae and some members of the family Atherinidae and Telmatherinidae (Table 1, see also Parenti 1984, 1989, Ivantsoff et al 1987), have the gut and gonads similarly rotated A complete survey of relative position of the gut and gonads in silverside fishes is beyond the scope of this project, but such a survey is expected to reveal additional characters that may help to resolve phylogenetic relationships among Old World silversides

Efferent and sperm ducts in phallostethid fish are secretory Of interest is the formation of aspherical (*P smithi*) or spherical (*N borneensis, N lankesteri,*

N. bicornis, G. mirabilis, and *G. bikolanus*) granules within the reproductive ducts. These granules are always PAS-negative, but they stain with metanil yellow or eosin. In addition to these granules, a fluid secretory product was demonstrated in the testis ducts that is also PAS-negative. In the Poeciliidae, efferent duct secretory products are PAS-positive, but this is not true of other internally fertilizing atherinomorphs, including the Anablepidae, Goodeidae, and the hemiramphid *Dermogenys pusillus* (Grier unpublished). The function of duct secretory products is unknown. They might be important in preventing sperm activation or in final sperm maturation.

Neither spermatozeugmata nor granules were seen in reproductive tracts of phallostethid females. These products must break down during or after sperm transfer. Granular duct secretions are unique to phallostethids; histological evidence indicates that they are secretory products of columnar efferent cells and sperm duct cells. Observations made by Smith (1927) on *Neostethus* and by Villadolid & Manacop (1934) and Manacop (1936) on *Gulaphallus* that phallostethids lay fertilized eggs are confirmed. Sperm may be present in immense numbers (Fig. 5d–f) in ovarian cavities of female phallostethids, which means it is unlikely that an ovulated oocyte would not be fertilized. Oocyte filaments, extending from the chorion of maturing oocytes in all species examined, function to keep eggs attached to plants or substrate after they are laid (Villadolid & Manacop 1934: pl. 2, fig. 4; Manacop 1936: pl. 2, fig. 1; Kulkarni 1940: fig. 15). Atherinomorphs with internal fertilization followed by internal gestation do not form oocyte filaments. We conclude that the mode of reproduction in phallostethids is internal fertilization followed by the laying of fertilized eggs, a mode also common to the atherinomorphs *Horaichthys setnai* (Kulkarni 1940), *Tomeurus gracilis* (Rosen 1964), and *Labidesthes sicculus* (Grier et al. 1990), all of which possess oocyte filaments.

We are encouraged by the discovery that museum specimens preserved over sixty-five years ago, in 1927, are as useful for demonstrating gonad histology as those preserved in the past few years (Table 2). Quality of histological preparation seems to be dependent upon quality of initial preservation (Fig. 3e). Furthermore, embedding in plastic (glycol methacrylate) yields far better results than traditional embedding in paraffin for at least two reasons (Grier, Reese & Quintero-Hunter unpublished). First, it is possible to cut thinner sections of plastic-embedded tissue, 3.5 µm rather than 5 to 8 µm thick, which gives better resolution of tissue detail. Second, shrinkage or swelling artifacts, seen as spaces between tissue components, are minimized. We conclude that formalin-fixed, alcohol-stored teleost fish specimens can be kept for an indefinite period and still be expected to reveal natural histological detail.

Acknowledgements

We have been influenced and inspired by Ethelwynn Trewavas, Rosemary Lowe-McConnell, and Eugenie Clark. Ethelwynn, in particular, has encouraged our research on phallostethid fishes, and we dedicate this publication to her with our heartfelt thanks. A Smithsonian Institution (SI) Scholarly Studies grant to LRP supported fieldwork in Brunei in 1991. Fishes were collected through the cooperation and collaboration of Haji Matussin bin Omar and Marina Wong, Muzium Brunei. An SI short-term visitor grant allowed HJG to visit the USNM in May 1992. Jeffrey C. Howe (USNM) provided technical assistance. We thank W.N. Eschmeyer, T. Iwamoto, D. Catania, and P. Sonoda (CAS) for providing specimens on loan, information, and logistical support to LRP at CAS. Bruce B. Collette kindly read and commented on an earlier version of the manuscript. We thank Ruth Reese, David Camp, Judy Leiby, and Llyn French for editing the final manuscript and Iliana Quintero-Hunter, Pam Nagle, and Catalina Erika Brown for histological assistance.

References cited

Duncker, G. 1904. Die Fische der malayischen Halbinsel. Mitt. Nat. Mus. Hamburg 21: 133–207.

Grier, H.J. 1981. Cellular organization of the testis and spermatogenesis in fishes. Amer. Zool. 21: 345–357.

Grier, H J 1984 Testis structure and formation of spermatophores in the atherinomorph teleost *Horaichthys setnai* Copeia 1984 833–839

Grier, H J 1992 Chordate testis the extracellular matrix hypothesis J Exp Zool 261 151–160

Grier, H J 1993 Comparative organization of Sertoli cells including the Sertoli cell barrier pp 704–739 *In* L D Russell & M D Griswold (ed) The Sertoli Cell, Cache River Press, Clearwater

Grier, H J , J R Burns & J A Flores 1981 Testis structure in three species of teleosts with tubular gonopodia Copeia 1981 797–801

Grier, H J & B B Collette 1987 Unique spermatozeugmata in testes of halfbeaks of the genus *Zenarchopterus* (Teleostei Hemiramphidae) Copeia 1987 300–311

Grier, H J , J M Fitzsimons & J R Linton 1978 Structure and ultrastructure of the testis and sperm formation in goodeid teleosts J Morphol 156 419–438

Grier, H J , J R Linton, J F Leatherland & V L deVlaming 1980 Structural evidence for two different testicular types in teleost fishes Amer J Anat 159 331–345

Grier, H J , D P Moody & B C Cowell 1990 Internal fertilization and sperm morphology in the brook silverside, *Labidesthes sicculus* (Cope) Copeia 1990 221–226

Ivantsoff, W , B Said & A Williams 1987 Systematic position of the family Dentatherinidae in relationship to Phallostethidae and Atherinidae Copeia 1987 649–658

Kulkarni, C V 1940 On the systematic position, structural modifications, bionomics, and development of a remarkable new family of cyprinodont fishes from the province of Bombay Rec Indian Mus (Calcutta) 42 379–423

Manacop, P R 1936 A new phallostethid fish with notes on its development Philipp J Sci 59 375–381

Munro, A D & E Y M Mok 1990 Occurence [sic] of the phallostethid fish *Phenacostethus smithi* Myers in southern Johor, Peninsular Malaysia, with some observations on its anatomy and ecology Raffles Bull Zool 38 219–239

Parenti, L R 1981 A phylogenetic and biogeographic analysis of cyprinodontiform fishes (Teleostei, Atherinomorpha) Bull Amer Mus Nat Hist 168 335–557

Parenti, L R 1984 On the relationships of phallostethid fishes (Atherinomorpha), with notes on the anatomy of *Phallostethus dunckeri* Regan, 1913 Amer Mus Novit 2779 1–12

Parenti, L R 1986a Bilateral asymmetry in phallostethid fishes (Atherinomorpha), with description of a new species from Sarawak Proc Calif Acad Sci 44(12) 225–236

Parenti, L R 1986b Homology of pelvic fin structures in female phallostethid fishes (Atherinomorpha, Phallostethidae) Copeia 1986 305–310

Parenti, L R 1986c The phylogenetic significance of bone types in euteleost fishes Zool J Linn Soc 87 37–51

Parenti, L R 1989 A phylogenetic revision of the phallostethid fishes (Atherinomorpha, Phallostethidae) Proc Calif Acad Sci 46 243–277

Parenti, L R 1993 Relationships of atherinomorph fishes (Teleostei) Bull Mar Sci 52 170–196

Quintero-Hunter, I , H J Grier & M Muscato 1991 Enhancement of histological detail using metanil yellow as counterstain in periodic acid Schiff's hematoxylin staining of glycol methacrylate tissue sections Biotech Histochem 66 169–172

Regan, C T 1913 *Phallostethus dunckeri,* a remarkable new cyprinodont fish from Johore Ann Mag Nat Hist 12 548–555

Regan, C T 1916 The morphology of the cyprinodont fishes of the subfamily Phallostethinae with descriptions of a new genus and two new species Proc Lond Zool Soc 1916 1–26

Roberts, T R 1971a Osteology of the Malaysian phallostethid fish *Ceratostethus bicornis* with a discussion of the evolution of remarkable structural novelties in its jaws and external genitalia Bull Mus Comp Zool 142 393–418

Roberts, T R 1971b The fishes of the Malaysian family Phallostethidae (Atheriniformes) Breviora 374 1–27

Rosen, D E 1964 The relationships and taxonomic position of the halfbeaks, killifishes, silversides, and their relatives Bull Amer Mus Nat Hist 127 217–268

Rosen, D E & L R Parenti 1981 Relationships of *Oryzias,* and the groups of atherinomorph fishes Amer Mus Novit 2719 1–25

Smith, H M 1927 The fish *Neostethus* in Siam Science 65 353–355

Villadolid, D V & P R Manacop 1934 (issued 1935) The Philippine Phallostethidae, a description of a new species, and a report on the biology of *Gulaphallus mirabilis* Herre Philippine J Sci 55 193–220

White, B N , R J Lavenberg & G E McGowen 1984 Atheriniformes development and relationships pp 355–362 *In* H G Moser, W J Richards, D M Cohen, M P Fahay, A W Kendall, Jr & S L Richardson (ed) Ontogeny and Systematics of Fishes, Amer Soc Ichthyol Herp Spec Publ No 1, Lawrence

Lynne Parenti at the 1992 ASIH meeting in Champaign-Urbana Photograph by E K Balon

Environmental Biology of Fishes **41**: 301–309, 1994.

Does gonad structure reflect sexual pattern in all gobiid fishes?

Kathleen S. Cole[1], D. Ross Robertson[2] & Alcibiades A. Cedeno[2]
[1] *Department of Biology, Bishop's University, Lennoxville, Quebec J1M 1Z7, Canada*
[2] *Smithsonian Tropical Research Institution (Panama), Unit 0948, APO AA 34002-0948, U.S.A.*

Received 9.2.1993 Accepted 1.3.1994

Key words: Ontogeny, Protogyny, *Gobiosoma*, Gobiidae

Synopsis

In immature and adult females of protogynous gobies, small distinctive masses of cells associated with the ovarian wall develop into testis-associated glandular structures during sex change. These precursive accessory gonadal structures, or pAGS, have been found in females of known protogynous goby species, but not among gonochoric goby species, suggesting that their presence can be used as a species-specific indicator of protogyny within the family. However, a detailed examination of a developmental series of ovaries in two gonochoric species, *Gobiosoma illecebrosum* and *G. saucrum,* revealed the presence of a gonadal feature previously thought to be restricted to protogynous gobies. Among immature females of both species, pAGS-like structures having a similar appearance and placement as functional pAGS of protogynous gobies were found. In female *G. illecebrosum*, the size of these structures among immatures progressively decreased with maturation and were absent in all but the smallest adult females. A similar pattern was evident in a small sample of *G. saucrum.* Population demography based on field collections showed that *G. illecebrosum* exhibits sex ratios and male and female size-frequency distributions typical of gonochores and laboratory experiments indicated that final sexual identity was unaffected by social environment during the juvenile period. Thus, the presence of pAGS in juvenile female *G. illecebrosum* is not related to an ability to change sex at that ontogenic interval. Whether the transient pAGS observed here are vestiges of an ancestral protogynous condition is unknown. Based on their presence among immatures in two gonochore gobies, however, only the presence of pAGS in adult females should be used to predict protogyny among gobies.

Introduction

As far as is known, males of all fishes in the suborder Gobioidei share unique glandular structures typically associated with the sperm duct of the testis. These have variously been referred to as seminal vesicles (Egami 1960, Arai 1964), sperm-duct glands (Miller 1984) and, in some protogynous gobies, where they derive from cell masses located in the ovarian wall, accessory gonadal structures (AGS – Cole & Robertson 1988). The precursive cell masses (pAGS; illustrated in Fig. 1a) which develop into AGS during sex change have been found in all females examined (both immature and mature) among 11 experimentally confirmed protogynous goby species within five genera (Cole 1983, 1988, 1990, Cole & Robertson 1988, Cole & Shapiro 1990, Cole unpublished data) but not among a small number of similarly examined females of four gonochore goby species (Cole 1988). Consequently, Cole (1988) postulated that among gobies the presence of pAGS associated with the ovary should be considered a reliable indicator of protogyny.

Hermaphroditism has now been reported for 13

gobiid genera, including *Gobiodon* and *Paragobiodon* (Lassig 1977), *Coryphopterus* and *Gobiosoma* (Robertson & Justines 1982), *Lythrypnus* (Cole 1988), *Pleurosicya, Bryaninops, Luposicya* (Fishelson 1989) and *Lophogobius, Fusigobius, Eviota, Trimma* and *Priolepis* (Cole 1990). In seven genera in which more than one species has been examined (*Gobiodon, Paragobiodon, Lophogobius, Lythrypnus, Fusigobius, Trimma, Priolepis* and *Coryphopterus*), hermaphroditism has been found in all species (n = 33) examined to date (Cole unpublished data, Cole & Hoese unpublished data). This includes *Coryphopterus,* in which hermaphroditism has been found in all 10 of the 11 described species that have been examined (one rare species remains unexamined). Universal protogyny among examined congeners within the above genera suggests that hermaphroditism in gobiids may be a shared trait among closely related species and, possibly, an ancestral condition (Cole & Shapiro 1990).

One gobiid genus, however, does not follow this pattern. In *Gobiosoma,* protogyny has been reported for one species, *G. multifasciatum* (Robertson & Justines 1982) but is absent in three others, *G. saucrum, G. illecebrosum* (Robertson & Justines 1982) and *G. evelynae* (Cole 1988). If hermaphroditism is ancestral in *Gobiosoma*, species such as *G. saucrum, illecebrosum* and *evelynae* which demonstrate no functional hermaphroditism may nevertheless reveal some features of gonad structure typically restricted to protogynous species. Given that the only consistent gonadal feature of protogyny in gobiids is the presence of pAGS associated with the ovary of females prior to sex change (Cole & Robertson 1988, Cole & Shapiro 1990), it seems likely that any structural anomalies in gonochore species that are closely related to hermaphroditic species will be found associated with the ovary.

Previous work (Robertson & Justines 1982) indicated that *G. illecebrosum* is a gonochore with a 1:1 sex ratio and no sex-change potential among adult females. However, with the occurrence of protogyny in a closely-related species and the proposition that hermaphroditism may be an ancestral condition in *Gobiosoma*, the gonochore designation for *G. illecebrosum* deserves closer examination. Previous experiments that tested for sex change ability in gonochoric species of *Gobiosoma* were performed only with adult females. Since in some other protogynous fishes (i.e. Scaridae – Robertson & Warner 1978) females appear capable of transforming the ovary to a testis prior to maturity, it is possible that immature females of some 'gonochore' gobies could have similar precocious sexual lability. If so, one might expect such lability in apparently gonochoric species that have protogynous congeners.

This paper describes an investigation of several aspects of ovarian morphology and sexual development in *G. illecebrosum* and, to a lesser extent, *G. saucrum* and *G. multifasciatum.* The first part examines the histostructure of ovaries of immature and mature females of these three species for the possible presence in *G. illecebrosum* and *G. saucrum* of structural features typically associated with ovaries of protogynous goby species, including *G. multifasciatum*. The second part presents information on population sexual demography based on field collections and results of a series of laboratory experiments with *G. illecebrosum* that were designed to test for: (a) sex change potential in adult and subadult females and adult males, and (b) lability in the development of sexual identity by undifferentiated immature fish. We predicted that if *G. illecebrosum* is a strict gonochore, both immatures and adults from field collections and laboratory experiments would exhibit a 1:1 primary sex ratio, irrespective of the natural or experimental social environment.

Methods and materials

Field collections

A large field collection of *G. illecebrosum* was carried out in August 1993 to obtain baseline data on sex-ratios and size-frequency distributions. Individual *G. illecebrosum* were collected from small patch reefs in the vicinity of the Smithsonian Tropical Research Institute's (S.T.R.I.) field station in the San Blas Islands, Panama (Lat. 9 34′ N, Long. 78 58″W) using a fish anesthetic quinaldine sulfate and a dip net. *Gobiosoma illecebrosum* live in small groups, or aggregations, on isolated coral heads. As the social structure and mating systems of these groups and whether, in fact, they comprise cohesive social groups, is unknown, all individuals collected from a single coral head henceforth will simply be referred to as an aggregation. Collected individuals were killed by over-anesthetization immediately after collection, preserved in Dietrich's fixative, then examined with a dissecting microscope to establish sex based on genital papilla structure. In this species, the male genital papilla is elongate with a pointed terminus, has a small genital pore at the apex, and often has melanocytes scattered along the length; the female genital papilla is shorter, square to rectangular in outline with a blunt terminus, has a broad genital pore at the apex and exhibits few, or no, melanocytes (Cole & Robertson unpublished data). Individuals were classified as male or female accordingly, then measured (mm, standard length). The resulting data were then combined with information taken from similarly collected individuals previously collected from the same locale in 1980 (Robertson & Justines 1982) to provide information on sex ratio and size-frequency distributions for males and females.

Histological examination

In separate collections from those described above, individual *G. illecebrosum* were collected from small patch reefs in the vicinity of the S.T.R.I. San Blas field station with quinaldine sulfate. In the first collection, made in October 1988, all fish (n = 93) were killed immediately after collection, preserved in Dietrich's fixative, decalcified in Fisher's Cal-Ex solution and embedded in toto in paraplast. The posterior portion of the body was then serially sectioned at 10 µm and all sections were mounted, stained with Harris' haematoxylin and eosin and viewed with a light microscope. This series was used to examine gonad structure in sexually undifferentiated fish, immatures, adult females and adult males. Some additional immature fish (n = 28) obtained from subsequent collections were treated similarly and added to the histological sample for a total of 121 histologically examined specimens.

Gobiosoma multifasciatum and *G. saucrum* of a range of sizes including both immatures and adults (see Results) were collected at the same site and treated in the same manner as the *G. illecebrosum* prepared for histological examination.

Rearing experiments

In separate collections from the above, additional *G. illecebrosum,* including adults, sexually distinct immatures and smaller, sexually indistinct immatures, were collected between October 1988 and June 1989, transported in insulated containers to S.T.R.I.'s Naos Marine Laboratory in Panama City and within 24 h of initial capture were placed in aquaria for the different experimental treatments.

In each experimental replicate one or more fish, depending on the treatment (see below) were maintained in a visually isolated 50 l aquarium provided with rocks and coral skeletons for shelter, flow-through sea water, aeration and a natural photoperiod. A superabundance of freshly hatched brine shrimp nauplii was added to each aquarium daily as food for the test fish. Each experiment ran between 1–2 months until all immatures had surpassed the minimum size at which maturity occurs in the wild (21 mm SL, see Results).

To establish experimental groups, individuals were first divided into juveniles and adults according to size (based on size of first gamete production according to histological information described above and presented in Results). The sex of individuals having sexually differentiated papillae were further identified as male or female, as described above. Most individuals up to 12 mm standard length (SL) had sexually undifferentiated papillae and could not be sexed externally.

Five series of experiments were run:

1. *Groups of adult or subadult females:* To substantiate previous reports of an absence of protogyny in this species, we established three experimental groups each comprising three adult-sized females and six groups each comprising three immature-sized fish that had female-shaped genital papillae, for 3 weeks.

2. *Group of immature males:* Up to now, protogyny is the only known pattern of sequential sex

change among gobies but, as a preliminary test of sexual lability among adult fish with male papillae, we kept one group of five immature males together for two months.

3. *Small immatures in varying social environments:* In these experiments, test fish were all less than 10 mm standard length (i.e. individuals of a size that typically do not have sexually distinct germ cells within the gonad, as described in Results) and had undifferentiated genital papillae.

(a) *Solitary immatures:* 27 randomly chosen fish were reared singly.

(b) *Pairs of immatures:* 44 randomly constructed pairs were reared to maturity.

(c) *Single immature and single adult:* single, randomly chosen immatures were reared either with an adult male (n = 16) or an adult female (n = 18).

Each of these experiments ran for 5–6 weeks.

Results

Population demography of G. illecebrosum

Thirteen aggregations collected in 1980 by Robertson & Justines (1982) and an additional 21 aggregations collected in August 1993 were examined for adult (i.e. ≥ 21 mm SL) sex ratios. The sex ratios for 1980 (18 female, 20 male) did not differ significantly from that of 1993 (41 female, 34 male) (Yates adjusted $X^2 = 0.29$, $p = 0.59$, Sokal & Rohlf 1981). In the combined sample of 113 adults, the adult sex ratio (59 female, 54 male) did not differ significantly from 1:1 ($G = 0.105$, $p = 0.75$, G-test for goodness of fit, Sokal & Rohlf 1981), and the two sexes did not differ significantly from one another in terms of size-frequency distributions (Kolmogorov-Smirnov two sample test, $p = 0.77$).

Similarly, female:male sex ratios among juveniles measuring 17–20 mm SL collected in 1993 (35 female, 25 male) did not differ significantly from those collected in the 1980 sample (17 female, 12 male; $X^2 = 0.15$, $p = 0.70$) and combined, did not differ significantly from a 1:1 sex ratio ($G = 1.67$, $p = 0.20$).

In terms of sexual composition of aggregations, seven aggregations collected in August 1993 and seven more collected in 1980 (Robertson & Justines 1982) included four or more adults (i.e. individuals ≥ 21 mm SL). In the former, 22 were male and 22 were female while in the latter 16 were male and 22 were female.

Gonad structure of field-collected immatures and adults of both sexes of G. illecebrosum

Of 93 fish collected in October 1988, 28 exhibited a short, blunt papilla typical of female gobies. Thirty-three other fish had an elongate, pointed papilla characteristic of males in other goby species (Miller 1984). In all cases, histological examination of the gonad verified these sexual designations. This sex ratio was not significantly different from 1:1 (Yates corrected $X^2 = 0.072$, $p = 0.79$). The remaining 32 fish had a small, immature papilla having no sexually distinct features.

Of these 32 immature fish, plus 28 other individuals with undifferentiated papillae obtained from subsequent collections which were all examined histologically, eight ranging in size from 9–11 mm SL had clearly differentiated gonadal lobes but germ cell identity was indistinct and sex could not be assigned. Ten individuals ranging from 13–18 mm SL were immature males. Their gonads consisted of two small testicular lobes having seminiferous tubules lined with crypts of spermatogonia and spermatocytes, but no spermatozoa. Associated with each testicular lobe was a smaller body having anastomosing, cell-lined lumina but little other structural differentiation. These bodies joined with their respective testicular lobes posteriorly at the point of union of the two testicular lobes and were identical in appearance to typical early-stage accessory gonadal structures (i.e. sperm duct glands sensu Miller 1984) found in immature males of other gonochoric goby species (Cole unpublished data). No oocytes or other ovarian features were evident in any of the immature testes observed in this sample.

Twenty-one individuals ranging from 10–22 mm SL were immature females and their bilobed ovaries contained previtellogenic oocytes in various stages of development, as well as localized protruberances of the ventral ovarian wall in the region of

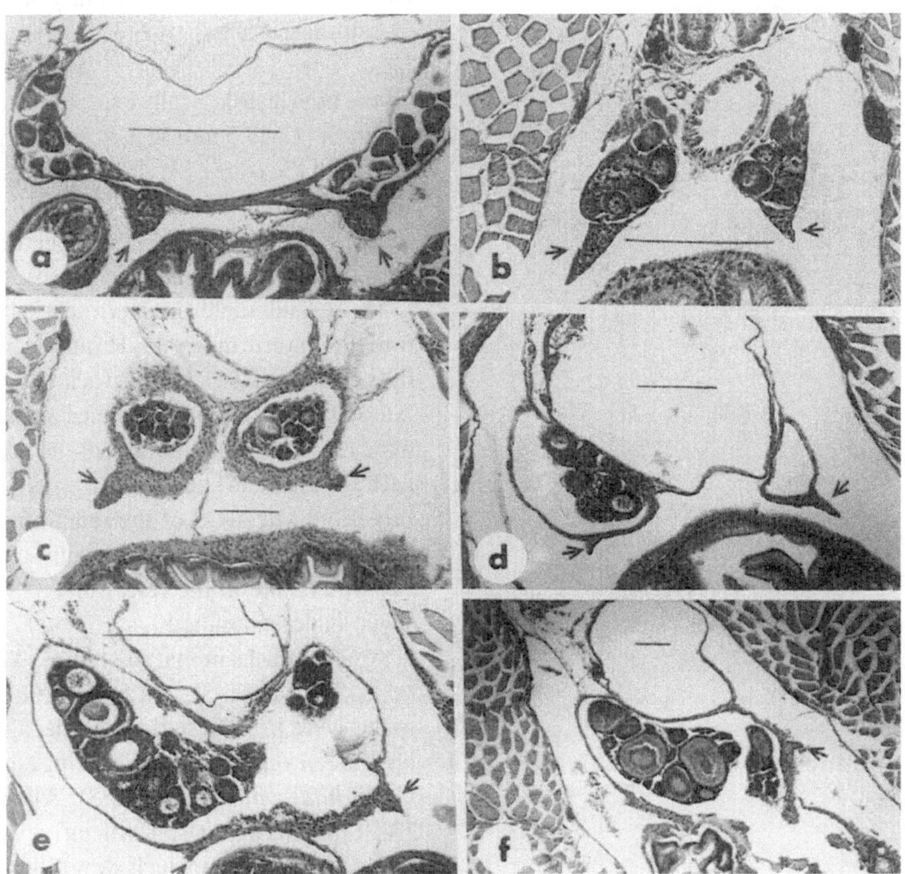

Fig 1 Ovarian structure in a protogynous goby and in immature and early adult *Gobiosoma illecebrosum* a) – cross sectional view of the posterior portion of the ovary including associated pAGS (indicated by arrows) in the protogynous goby *Coryphopterus hyalinus* b–f *G illecebrosum* b) ovary of immature female (11 8 mm SL) showing (arrow) large pAGS like structures associated with the ventral wall in the posterior region close to the point of union of the two lobes c) – pAGS like structures of immature female 19 1 mm SL d) – reduced pAGS like structures of slightly larger immature female 20 4 mm SL e) – remnant pAGS like structures of small adult female 23 6 mm SL f) remnant pAGS like structures of slightly larger adult female 24 2 mm SL Bar is 100 μm

the union of the two ovarian lobes (Fig 1b–d) These structures, which were quite large in the smallest immature females (Fig 1b–c), were similar in both appearance and location to precursive cell masses found in females of protogynous goby species (Fig 1a) and described elsewhere (Cole 1988, 1990) which, upon sex change, develop into accessory gonadal structures

The remaining fish (17 females and 4 males) were all adult While localized enlargements of the ventral ovarian wall were large in small immature females (Fig 1b), these cell masses became both rela tively and absolutely smaller (Fig 1c,d) as females approached the size of maturity Based on the pres ence of vitellogenic oocytes in the ovary of the smallest adult female present in this sample, we estimated the size for first maturity for females to be 21 mm SL Remnants of these cell masses were still visible in three adult females (21, 23 and 24 mm SL, Fig 1e, f) but were absent in all of the remaining 14 adult females whose size ranged from 21–32 mm SL (Fig 2) Neither immature nor adult females exhibited any testicular tissue, either in the form of seminiferous tubules or unorganized crypts of spermatocytes, within the body of the ovary, or accessory gonadal structures

No oocytes were evident within the testes proper

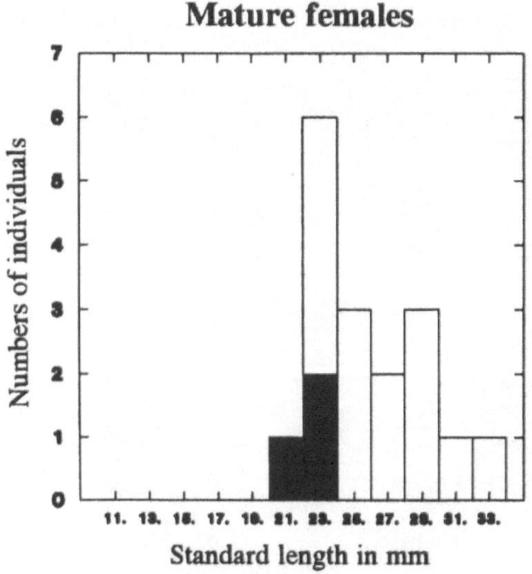

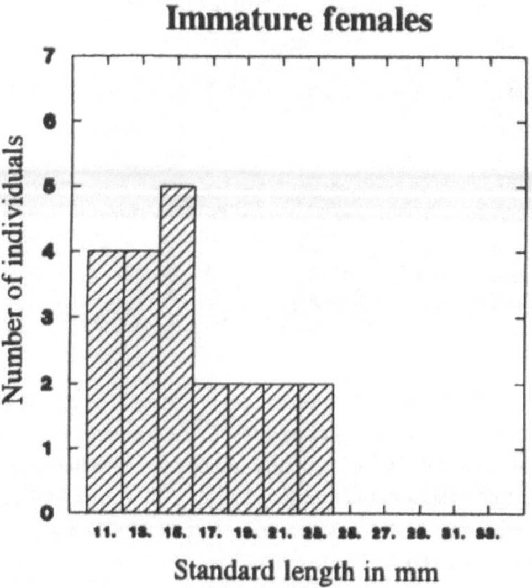

Fig. 2. Bar histogram illustrating the occurrence of pAGS-like structures among (i) immature (n = 21); and (ii) adult (n = 17), female *G. illecebrosum*. Horizontal scale represents increasing size (standard length) in 2 mm increments. Open bar, females with no pAGS-like structures; black bar, females with vestigal pAGS-like structures; hatched bar, females with pAGS-like structures.

or the associated AGS of the four adult males (18–23 mm SL) examined in this sample.

Gonad structure of G. saucrum *and* G. multifasciatum

Of the nine histologically examined adult female *G. saucrum* (13–17.5 mm SL), none had pAGS associated with the ovary. Six of 10 immature females (9–10 mm SL) had distinct pAGS and the remaining four had smaller pAGS-like structures associated with the caudo-ventral ovarian wall.

Three adult male *G. saucrum* (15–19 mm SL) and four immature males (9–11 mm SL) all exhibited fully differentiated AGS associated with the testis. Among all of the mature males and one immature male (11 mm SL) the lumina of the AGS were filled with an acellular, acidophillic (i.e. eosin-staining) secretion. The AGS of the remaining three immatures were fully formed but inactive. No ovarian tissue was visible in the testes or AGS of any of the seven males examined.

Among three immature female *G. multifasciatum* of 12–14 mm SL, all had pAGS. No adult females were present in this sample, although pAGS have been reported elsewhere in adult females of *G. multifasciatum* (Cole 1988). All three males of 19–21 mm SL were adult with large, well-developed AGS, the lumina of which were filled with spermatozoa. No ovarian tissue was visible.

Experimental rearings of G. illecebrosum

Robertson & Justines (1982) previously ran experiments in which groups of 4–6 adult females were maintained together for 3–5 weeks. Since the number of replicates of those experiments was small (n = 4) and they found one male among the females at the end of the experiments, we set up some additional all-female groups to verify gonochorism. Upon histological examination of the six groups each consisting of three immature females, and the three groups each made up of three adult females, no individuals showed any indication of any partial or complete sex change, ovarian degeneration or early-stage testicular tissue. In the single group of five immature males, no individual showed any signs of testicular degeneration or the appearance of early-stage ovarian tissue in their gonads.

Of 27 immatures reared in isolation, 12 were female and 15 male. This ratio is not significantly different from 1:1 (Yates corrected $X^2 = 0.143$, $p = 0.71$). Among the 44 pairs of immatures reared to maturity, 20% (n = 9) were made up of two females, 18% (n = 8) of two males and 62% (n = 27) of a female and male, for a total of 45 females and 43 males. This distribution was not significantly different from that expected by chance (1F:2MF:1M) from random assemblages of pairs from a population with a 1:1 sex ratio ($G = 1.19$, $p = 0.60$). Of 18 immatures reared with an adult female, 33% were female and 67% were male; of 16 immatures reared with an adult male, 31% were female and 69% were male. Neither of these sex ratios was significantly different from 1:1 (with adult female, $G = 2.039$, $p = 0.15$; with adult male, $G = 2.306$, $p = 0.13$) and did not differ from each other (G test for independence, $G = 0.017$, $p = 1.0$, Sokal & Rohlf 1981).

Discussion

In various species of sequentially hermaphroditic fishes, including gobies, sex-change can be induced in some adult individuals by a change in the social environment (Fishelson 1970, 1975, Robertson 1972, Fricke & Fricke 1977, Warner 1978, 1988, Shapiro 1979, 1981, 1987, Shapiro & Lubbock 1980, Robertson & Justines 1982, Fricke 1983, Ross 1983, Cole & Robertson 1988). By the same token, one might expect that variation in social conditions during development from an undifferentiated juvenile state to adulthood should expose lability in gonadal development that may be present in a species, particularly one with close relatives that do show adult sexual lability. In our experiments we varied the social conditions under which differentiated adult *G. illecebrosum* were maintained and under which both differentiated and undifferentiated juveniles developed to adulthood. We predicted that if *G. illecebrosum* is a strict gonochore with a 1:1 primary sex ratio that lacks lability in sexual potential, then no sex-change should occur among either groups of mature females, immature females or mature males; and that the adult sex-ratio of fish that developed from an undifferentiated state to maturity should not differ from 1:1 regardless of the social situation in which they were reared.

The results of the present study support and extend those of Robertson & Justines (1982) which indicated a lack of sexual lability in adult *G. illecebrosum*. Comparisons of adult sex-ratios and size-frequency distributions from field collections presented here demonstrate that *G. illecebrosum* exhibits male and female size-frequency distributions and a 1:1 sex-ratio typical of gonochore fish species. In addition, the failure of sexually undifferentiated juveniles that were reared to maturity under similar, and varying, social conditions, and of experimental groups of adults, to show signs of any developmental plasticity indicates that *G. illecebrosum* is an illabile gonochore with a 1:1 sex ratio.

Since sample sizes in each of the experiments we ran were not large, the power of each statistical test to detect a deviation from a 1:1 sex ratio is not high. In particular, it should be noted that the sex ratio of single juveniles reared with a female was (non-significantly) male biased. Considering the experimental social situation, this bias is in the direction one would expect if sexual differentiation was plastic. However, we do not give much weight to this result since the experiment in which juveniles were reared with a male also produced a (non-significant) male bias. Further, it is possible to combine the results of a series of statistical tests that aim to test the same hypothesis. We used this 'combination of probabilities' (Sokal & Rohlf 1981) technique to test whether the four experiments in aggregate produced sex ratios different from 1:1 and obtained a probability value of 0.75, indicating that the non-significant sex ratio biases in each individual experiment do not indicate an overall sex ratio different from 1:1.

Immature and mature females of all protogynous gobies so far studied bear distinctive cell masses associated with the ovary, termed precursive accessory gonadal structures (pAGS) which, during sex-change, develop into accessory glandular structures similar to sperm duct glands (sensu Miller 1984) described for gonochoric gobies (Cole & Robertson 1988, Cole 1988, 1990, Cole & Shapiro 1990, Cole unpublished data). As we expected, pAGS were present in the ovaries of immature, protogynous *G.*

multifasciatum and no such structures were present among adult females of the two examined gonochore species, *G. illecebrosum* and *G. saucrum*. However, among immature females of these latter two species, ovary-associated structures that seem to be similar, both in appearance and location on the ovary, to pAGS, were found. This indicates that both *G. illecebrosum* and *G. saucrum*, while functional gonochores, possess some hermaphroditic features in the immature ovary.

The presence of pAGS-like structures among immature females of *G. illecebrosum* evidently is not related to an ability to change sex at that stage, since our experiments do not indicate that there is any lability in the sexual development of *G. illecebrosum*. This suggests that the pAGS-like structures found in immature *G. illecebrosum* and *G. saucrum* are either unresponsive, or not competent to respond, to changes that activate pAGS in sex-changing gobies. The progressive diminution of the pAGS-like structures as females of *G. illecebrosum* approach the size of maturity, and their absence in all but the smallest mature females, is striking. Were these structures merely extensions of the ovarian wall, maturation should have had a neutral or positive effect on their development. Their gradual reduction in absolute size with approaching maturity and ultimate disappearance among adults suggests that endogenous changes associated with maturation in females are antagonistic to their retention. This would be consistent with the fate of testis-associated tissue which presumably requires a certain level of circulating androgens for continued maintenance or further development.

According to Cole (1988), the presence of pAGS-like structures associated with the ovary in gobiid species is a reliable indicator for protogyny. However, based on our findings here, it is evident that this postulation requires modification. Clearly, gonad structure, particularly among immatures, does not necessarily reflect functional sexual pattern, at least in some *Gobiosoma* and possibly in other gobies. Consequently, only the presence of pAGS in *adult* females may reliably predict protogyny among gobies.

The development and subsequent disappearance of pAGS-like structures in two gonochore *Gobiosoma* species is intriguing. It is tempting to speculate that these species, like present-day *G. multifasciatum*, were once protogynous and that this sexual pattern has been secondarily lost. If so, the transient pAGS observed here are simply ontogenetic remnants of a former labile sexual pattern and, as such, their ephemeral presence during development, if observed in other gonochore goby species, may be useful indicators for ancestral protogyny. However, inferring ancestral states and directionality for evolutionary transitions on the basis of a single morphological feature is risky at best, particularly in this case since little is known about gonad development in fishes in general, and gobies in particular.

In order to make educated guesses regarding ancestral sexual patterns in *G. illecebrosum* and *G. saucrum*, we first need to establish generalities for patterns of gonad development in *Gobiosoma*, including examinations of gonad development at the cellular level for both gonochoric and protogynous species. Comparative work on other gonochoric and protogynous gobies, and other teleosts (see also Fishelson 1992), is also essential. As a family, gobies exhibit a variety of sexual patterns including gonochorism, simultaneous hermaphroditism (St. Mary 1994) and protogyny (i.e. Cole 1990). In order to understand the evolution of this diversity, the existence of patterns in the distribution of gonochorism and protogyny among gobiids, and the role of phylogeny in that distribution, need to be determined. Unfortunately, phylogenetic relationships within this most speciose family of marine fishes (Nelson 1984) are still poorly understood (Hoese 1984, Birdsong et al. 1988). Finally, details of the life histories and mating systems of gonochoric and protogynous gobies will need to be elucidated in order to understand the role of selective forces that have favored the development of protogyny in other fishes (e.g. Charnov 1982, Warner 1978, 1988), in producing this diversity of sexual patterns among the gobiids.

Acknowledgements

We thank the Kuna General Congress and the Government of the Republic of Panama for permission to carry out this study in the San Blas Islands, Marvalee Wake for helpful conversations regarding the difficulties in determining the evolutionary status of morphological features in vertebrates and two anonymous reviewers. Photographic assistance was generously provided by C. Coleby, Royal Ontario Museum and the Photographic Department of the Smithsonian Tropical Research Institute. J. Porter and R. van Hulst provided assistance with manuscript preparation, and statistical analyses and graphics, respectively. This research was supported by a Smithsonian Short-term Fellowship, a Bishop's University Senate Research Grant awarded to KSC and general STRI research funds. Portions of this manuscript were written while the senior author was supported by a National Science and Engineering Research Council of Canada Operating Grant.

References cited

Arai, R 1964 Sex characters of Japanese gobioid fishes (I) Bull Nat Sci Mus Tokyo 7 295–306

Birdsong, R S, E O Murdy & F Pezold 1988 A study of vertebral column and median fin osteology in gobioid fishes with comments on gobioid relationships Bull Mar Sci 42 174–214

Charnov, E 1982 Alternative life-histories in protogynous fishes a general evolutionary theory Mar Ecol Prog Ser 9 305–309

Cole, K S 1983 Protogynous hermaphroditism in a temperate territorial marine goby, *Coryphopterus nicholsi* Copeia 1983 809–812

Cole, K S 1988 Predicting the potential for sex-change on the basis of ovarian structure, in gobiid fishes Copeia 1988 1082–1086

Cole, K S 1990 Patterns of gonad structure in hermaphroditic gobies (Teleostei Gobiidae) Env Biol Fish 28 125–142

Cole, K S & D R Robertson 1988 Protogyny in a Caribbean reef goby, *Coryphopterus personatus* gonad ontogeny and social influences on sex change Bull Mar Sci 42 317–333

Cole, K S & D Y Shapiro 1990 Gonad structure and hermaphroditism in the gobiid genus *Coryphopterus* (Teleostei Gobiidae) Copeia 1990 966–973

Egami, N 1960 Comparative morphology of the sex characters in several species of Japanese gobies, with reference to the effects of sex steroids on the characters J Fac Sci Univ Tokyo, Ser IV 9 67–100

Fishelson, L 1970 Protogynous sex reversal in the fish *Anthias squamipinnis* (Teleostei, Anthiidae) regulated by the presence or absence of a male fish Nature 227 90–91

Fishelson, L 1975 Ecology and physiology of sex reversal in *Anthias squamipinnis* (Peters), (Teleostei, Anthiidae) pp 284–294 *In* R Reinboth (ed) Intersexuality in the Animal Kingdom, Springer Verlag, New York

Fishelson, L 1989 Bisexuality and pedogenesis in gobies (Gobiidae Teleostei) and other fish, or why so many little fish in tropical seas? Senckenbergia marit 20 147–169

Fishelson, L 1992 Comparative gonad morphology and sexuality of the Muraenidae (Pisces, Teleostei) Copeia 1992 197–209

Fricke, H W 1983 Social control of sex field experiments with the anemonefish *Amphiprion bicinctus* Z Tierpsychol 61 71–77

Fricke, H W & S Fricke 1977 Monogamy and sex change by aggressive dominance in a coral reef fish Nature 266 830–832

Hoese, D F 1984 Gobioidei relationships pp 588–591 *In* H G Moser et al (ed) Ontogeny and Systematics of Fishes, Spec Pub No 1, Am Soc of Ichthy and Herpet Allen Press, Lawrence

Lassig, B R 1977 Socioecological strategies adopted by obligate coral-dwelling fishes pp 565–570 *In* D L Taylor (ed) Proc Third Intern Coral Reef Symp I, Biology, Miami

Miller, P J 1984 The tokology of gobioid fishes pp 119–153 *In* G W Potts & R J Wootton (ed) Fish Reproduction Strategies and Tactics, Academic Press, New York

Nelson, J S 1984 Fishes of the world, 2nd ed John Wiley and Sons, New York 416 pp

Robertson, D R 1972 Social control of sex reversal in a coral reef fish Science 177 1007–1009

Robertson, D R & G Justines 1982 Protogynous hermaphroditism and gonochorism in four Caribbean reef gobies Env Biol Fish 7 137–142

Robertson, D R & R R Warner 1978 Sexual patterns in the labroid fishes of the western Caribbean, II The parrot fishes (Scaridae) Smithsonian Contributions to Zoology 255 1–26

Ross, R M, G S Losey & M Diamond 1983 Sex change in a coral reef fish dependence of stimulation and inhibition on relative size Science 221 574–575

St Mary, C M 1994 Novel sexual patterns in two simultaneously hermaphroditic gobies, *Lythrypnus dalli* and *Lythrypnus zebra* Copeia (in press)

Shapiro, D Y 1979 Social behavior, group structure and the control of sex reversal in hermaphroditic fish Advances in the Study of Behavior 10 43–102

Shapiro, D Y 1981 Size, maturation and the social control of sex reversal in the coral reef fish *Anthias squamipinnis* (Peters) J Zool (Lond) 193 105–128

Shapiro, D Y 1987 Differentiation and evolution of sex change in fishes BioScience 37 490–497

Shapiro, D Y & R Lubbock 1980 Group sex ratio and sex reversal J Theor Biol 82 411–426

Sokal, R R & F J Rohlf 1981 Biometry The principles and practice of statistics in biological research, 2nd ed W H Freeman and Company, San Francisco 859 pp

Warner, R R 1978 The evolution of hermaphroditism and unisexuality in aquatic and terrestrial animals pp 78–101 *In* E S Reese & F J Lighter (ed) Contrasts in Behavior John Wiley & Sons, New York

Warner, R R 1988 Sex change in fishes hypotheses, evidence and objectives Env Biol Fish 22 81–90

Kassı Cole and her typıcal smıle Photograph by M Keenleysıde

Environmental Biology of Fishes **41**: 311–329, 1994.

The early ontogeny of the southern mouthbrooder, *Pseudocrenilabrus philander* (Pisces, Cichlidae)

Kathleen K. Holden & Michael N. Bruton
J.L.B. Smith Institute of Ichthyology, Private Bag 1015, Grahamstown 6140, South Africa

Received 28.10.1992 Accepted 16.8.1993

Key words: African fish, Dwarf bream, Eggs, Embryos, Respiration, Functional anatomy, Haplochromine, Tilapiine, Direct development, Life-history model

Synopsis

The early development of the southern mouthbrooder, *Pseudocrenilabrus philander,* is documented from activation until the early stages of the juvenile period. The duration of the embryonic period is about 14 days at 25° C. Development is direct and there is accelerated exogenous feeding into the embryonic period. The pattern of development and the timing of ontogenetic events and structure formation are a reflection of both internal and external environmental conditions. During mouthbrooding, oxygen uptake is facilitated by embryonic respiratory plexuses and flapping of the pectoral fins. At the time of first release from the buccal cavity, the embryos are in an advanced state of development. The switch-over from the temporary embryonic respiratory system to the adult branchial system has occurred. The yolksac serves as a supplemental source of nutrition as the embryos develop their external food-gathering abilities. The skeletal and sensory systems are sufficiently developed to allow the young to return to the safety of the female's buccal cavity. Pigmentation may provide disruptive colouration. The rate and pattern of development of another mouthbrooding cichlid, *Oreochromis mossambicus,* is similar to that of *P. philander* despite their phylogenetic differences, and may be a consequence of similar life-history styles.

Introduction

The southern mouthbrooder, *Pseudocrenilabrus philander* (Weber, 1897), occurs in the Zambezi, Limpopo, upper Zaire and Orange River basins, in the river systems of Natal, southern Mozambique and the Okavango Swamps, as well as in the sink-holes of Namibia (Bell-Cross 1966, Trewavas 1973, Skelton et al. 1985). It is found in backwaters, shallow lagoons, swamps, small streams, rivers and springs. Although it is present in natural lakes and artificial impoundments (Balon 1974a, b), it usually confines itself to marginal, shallow regions or swampy areas. In large rivers it prefers slow-flowing regions and it also occurs in estuaries (Bruton 1980, Whitfield 1980, Loiselle 1982, see also Greenwood 1989). It is a eurytopic, generalized, riverine haplochromine which has attributes which may be similar to those of the ancestral haplochromines from which the lacustrine species, as well as some of the derived haplochromines, may have arisen (Poll 1967, Trewavas 1973, Van Couvering 1982). Of all the haplochromines, the *P. philander* superspecies is considered the most ecologically versatile (Loiselle 1982). Its generalized characteristics include its widespread distribution and its tolerance for wide fluctuations in pH, salinity, temperature and hardness (Loiselle 1982, Ribbink in Loiselle 1982).

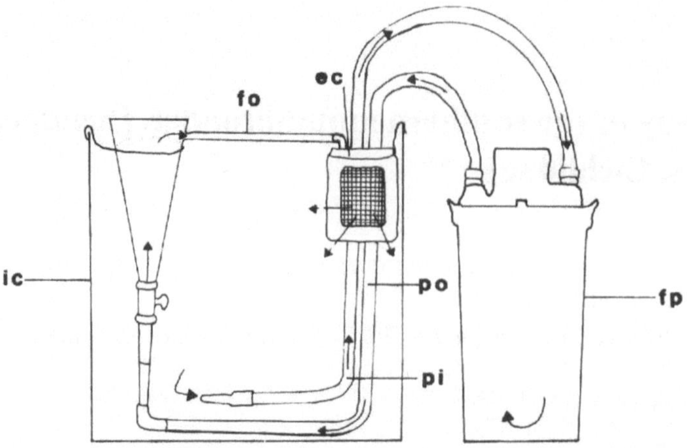

Fig. 1. Side view of the incubating system. Arrows indicate the direction of the water flow. The outflow of the water coming from the pump churned the eggs/embryos at the bottom of the funnel. The embryos swam through the funnel outflow into the embryo chamber once they had developed sufficient swimming abilities (ec = embryo chamber; fo = funnel outflow; fp = filter/pump; ic = incubating tank; pi = pump inflow; po = pump outflow). Drawing modified from Holden & Bruton (1992).

P. philander lays relatively numerous, small eggs and has a shorter incubation period than lacustrine, haplochromine species (Fryer & Iles 1972). Although *P. philander* is not threatened (Skelton 1987) some of the geographically isolated populations may be at risk (Ribbink & Twentyman-Jones 1989, Skelton 1990, de Villiers et al. 1992). Current research suggests that *P. philander* may be in some of these isolated populations an incipient species (de Villiers et al. 1992).

Several authors consider *P. philander* to have plesiomorphic features which suggest that it might be an ancestral haplochromine (Poll 1967, Trewavas 1973, Van Couvering 1982). Greenwood (1989), however, questions this conclusion, suggesting instead that at least some of the ancestral features may be neotenic. Regardless of its phylogenetic relationships, *P. philander* is regarded by all authors to be a eurytopic, riverine, generalized haplochromine species. Balon (1977) described the early development of *Labeotropheus* sp., a specialized, stenotopic haplochromine. The present paper provides a baseline for comparisons of the early ontogeny of a generalized, and possibly ancestral, haplochromine. This is especially important as adaptive radiation is not confined to adult features but also includes early life-history characteristics (Fryer & Iles 1972).

Materials and methods

The parent fish were collected from a wild population in a stream near Durban, South Africa (29°56′ S; 30°59′ E). The descriptions of the early ontogeny were derived from two spawnings of different females with the same male. The female: male ratios in the breeding tanks were 11:1 and 9:1. The methods of Balon & Flegler-Balon (1985) were followed and were adjusted for mouthbrooding species as outlined below and described in Holden & Bruton (1992).

Samples were taken from the time of activation until the juvenile period at varying time intervals depending on the rate of development (sampling intervals were more frequent during the early stages). Activation was considered to be the midway point between the time of first egg deposition and the last sperm uptake by the female, and was used for ageing purposes. The specimens were placed in a large depression slide and positioned under a Nikon SMZ-10 stereo-zoom microscope with a Microflex HFX-II photomicrographic attachment. Drawings were made using a drawing tube attachment. A fibre optic light source (Fi L151) was used for reflected light in order to avoid overheating the specimens during the microscopic observations. Specimens were placed in vials for preservation and stored in a buffered formalin solution (1.8 g sodium phosphate monobasic and 1.8 g anhydrous sodium phosphate dibasic in 1 l of 5% formalin).

The breeding and incubation tanks were housed in a temperature-controlled room with a photoperiod of 14 h light: 10 h dark and the water temperature was maintained at 25 ± 0.5° C. The eggs were removed from the female's buccal cavity immediately after spawning. The eggs and embryos were incubated in a 60 l glass aquarium with heaters which were connected to an electronic relay box. Temperature control was maintained via a contact thermometer which was connected, in series, to the heaters through the relay box. A simulated mouthbrooding action was obtained by connecting steep-sided separating funnels to the outflow of a Fluval 102 outside filter/pump (Fig. 1). The outlet of the funnels flowed through transparent plastic tubing into small plastic jars covered with fine netting, which were suspended in the incubator tank. The incubation tank was contained in a darkened enclosure in order to simulate the lower light levels expected in a female's buccal cavity. Eighteen days and two hours after activation, embryos were removed from the darkened enclosure and placed in breeding baskets in a tank with a photoperiod of 14 h light: 10 h dark.

Older, active specimens were anaesthetized with a 1 ppt solution of MS222 (tricaine methane sulfonate). A few drops were added to the depression slide after recording behavioural and morphological information such as eye, jaw and fin movements and heart rate. Heart rate was determined by timing 10 or 20 beats with a stopwatch. Two to four counts per individual were taken and, in some cases, heart rate for more than one specimen of a given age was recorded.

Food was added to the incubating funnels (at age 8 d) in advance of the expected time of first exogenous feeding. Commercially produced fish food (Tetra Min Baby Fish Food 'E' and Liquifry No. 1) and live brine shrimp nauplii were frequently added to the funnels and specimens were removed periodically to check visually for food in the gut.

Specimens were cleared with trypsin and stained with alcian blue for mucopolysaccharides (cartilage) and with alizarin red-S for calcium phosphate (bone). The procedure used was that of Potthoff (1984) with some slight modifications. Because alcian blue solutions have the potential to decalcify bone, some specimens were stained with alizarin red-S only and compared with the double-stained specimens.

Composites of the developmental illustrations were derived from slides, drawings, and cleared and stained specimens. Embryonic lengths were measured from drawings of live specimens. As the size and shape of the yolk and yolksac changed over time it was not possible to determine yolk volume. Additionally, the ovoid shape of the eggs made a mathematical formulation to determine volume inadequate. Therefore, yolk area was measured from lateral drawings of live specimens to demonstrate yolk absorption. These data were used as an indication of relative changes in rates, sizes and/or patterns of development.

We have chosen to follow the terminology of ontogenetic events and intervals of Balon (1975, 1990), and its applicability to another mouthbrooder with a similar developmental style is discussed in Holden & Bruton (1992). The embryo period is subdivided into a cleavage, an embryo and a free-embryo phase. The cleavage phase begins with activation and ends once organogenesis begins. During the embryo phase, intense organogenesis occurs and hatching begins. Hatching of almost all the clutch members frees them from the confines of the egg envelope and marks the beginning of the free embryo phase. During the embryonic period the major

Table 1. Mean maximum and minimum diameters of *Pseudocrenilabrus philander* eggs from two clutches (STD = standard deviation).

Clutch	n	Mean max. length (mm)	STD	Mean min. length (mm)	STD
1	22	2.20	0.273	1.59	0.124
2	23	2.50	0.073	1.70	0.072
Total	45	2.33	0.238	1.65	0.113

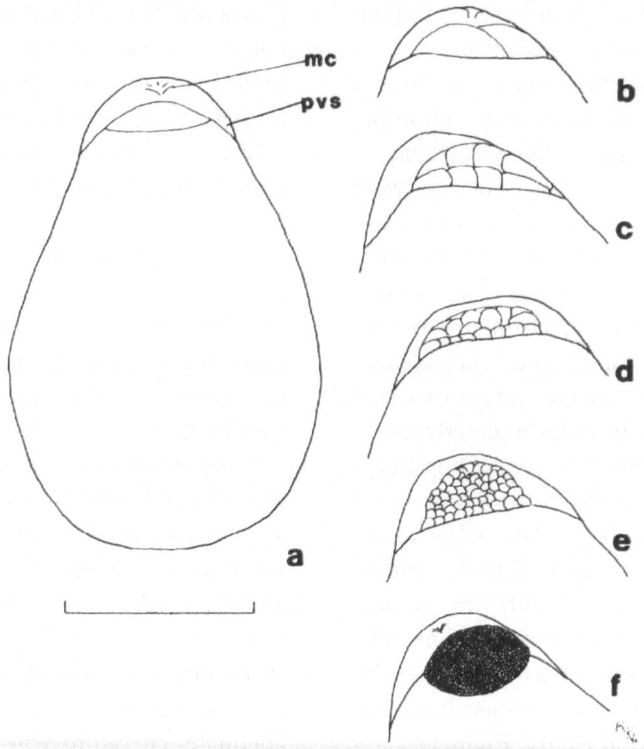

Fig. 2. Cleavage phase: a – age 00:01; b – age 00:02; c – age 00:04; d – age 00:05; e – age 00:07; f – age 00:11 (mc = micropyle; pvs = perivitelline space). Scale = 1 mm.

or exclusive source of nutrition is endogenous via the yolk. Once the yolk has been absorbed and the nutritional requirements are met by exogenous sources, the young fish is considered to be either a larva or a juvenile. In the case of a mouthbrooder, which has direct development, the young fish retain a large yolksac once first exogenous feeding begins and an interval of mixed feeding exists. The level of development at this time is not that of a juvenile and there is no larval period. We, therefore, have chosen to call this the free-embryo phase (for an expanded argument on the terminology see Holden & Bruton 1992).

The age of the specimens was denoted as days: hours after activation, and indicates the beginning of the sampling time when the specimen was removed from the incubator or the breeding basket. The age is also expressed in temperature units (TU = degree-days or temperature units) and was calculated by multiplying age (hours/24) by temperature (25° C). The main sources of information used for naming blood vessels and skeletal structures were Cunningham & Balon (1985) and Holden & Bruton (1992).

Results

Description of development

The eggs of *P. philander* are ovoid with the longitudinal axis longer than the transverse axis and the animal pole narrower than the vegetal pole. The yellow, opaque yolk is of uniform consistency without oil globules. Table 1 lists the mean maximum and minimum diameters of eggs from two clutches.

Embryo period

Cleavage phase 00:00-01:12. During the first hour, the cytoplasm forms a one-celled blastodisc and the perivitelline space becomes visible at the animal

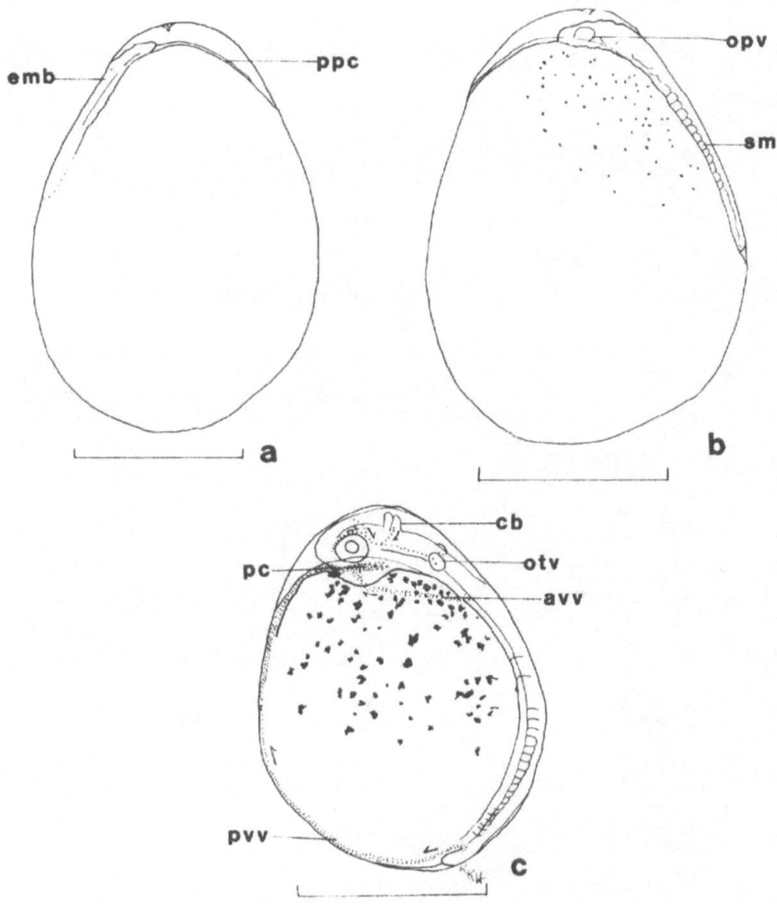

Fig. 3. Embryo phase: a – age 01:02; b – age 02:12; c – age 03:00 (avv = anterior vitelline vein; cb = cerebellum; emb = embryonic shield; opv = optic vesicle; otv = otic vesicle; pc = pericardial cavity; ppc = presumptive pericardial cavity; pvv = posterior vitelline vein; sm = somites). Scales = 1 mm.

pole (Fig. 2a). By age 00:02 (TU = 2) first division takes place and the resultant two cells extend well above the yolk (Fig. 2b). At age 00:03 (TU = 3) there are four cells, and at age 00:04 (TU = 4) the eight cells (Fig. 2c) divide into 16 cells. The first horizontal division occurs at age 00:05 (TU = 5) (Fig. 2d) and within the next hour there is a 64-celled blastodisc (see Fig. 2e). In some cases the blastodisc lies skewed to one side of the yolk. By age 00:11 (TU = 11), there are innumerable cells of undetermined size (Fig. 2f). Between the ages of 00:12 and 00:13 (TU = 12, 13) the periblast forms below the lip of the blastodisc.

At age 00:14 (TU = 14) the periblast becomes broader and more obvious and epiboly begins. By age 01:02 (TU = 27) the embryonic shield lies to one side of the yolk and a layer of less dense, translucent cells is at the opposing side. Over the next several hours, the translucent area becomes a thin layer over the apex of the yolk and the blastodisc extends to the equator.

Embryo phase 01:12-04:23. At age 01:12 (TU = 38) swelling in the anterior region of the embryonic shield and a faint line along its axis indicate that organogenesis has begun. The presumptive pericardium forms a narrow, transparent chamber along the dorsal surface of the yolk (Fig. 3a). After eight hours (TU = 46), rudimentary optic vesicles form elliptical bulges and at least eight somites are present. Melanophores occur along the yolk adjacent to the embryonic body. Body pigments occur on the

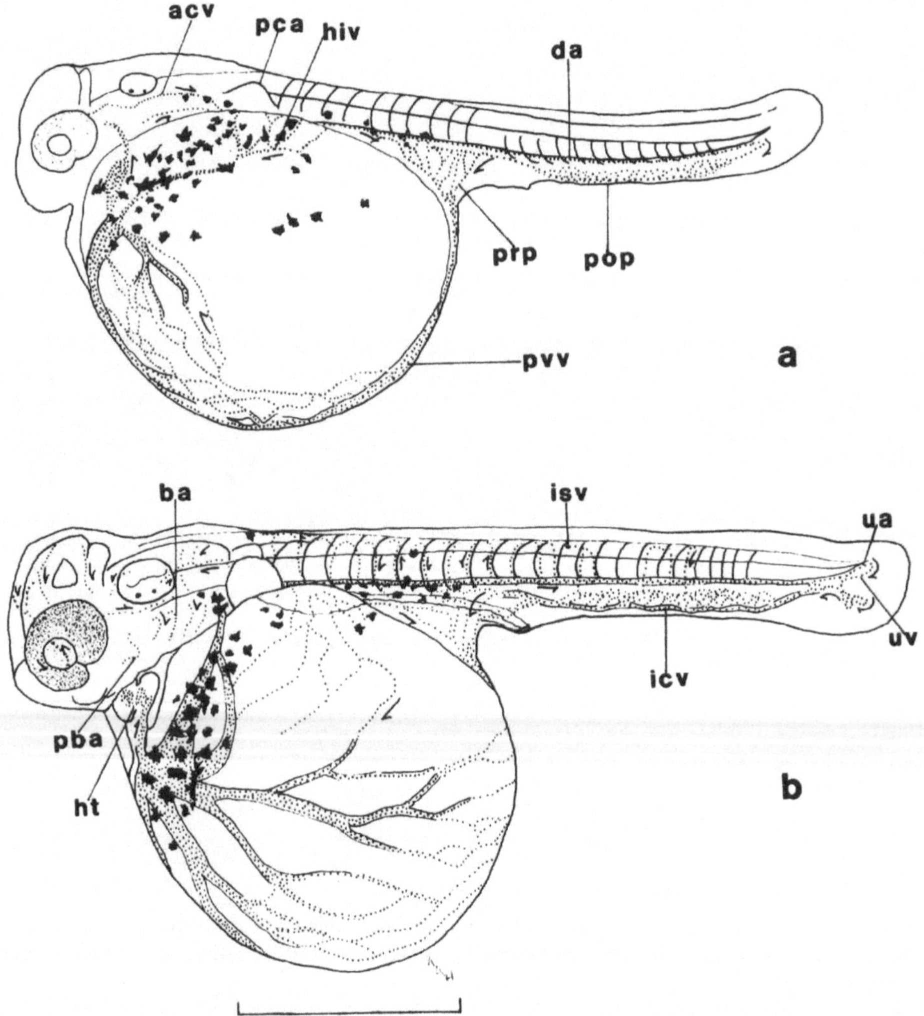

Fig 4 a – Embryo phase age 04 03, b – Free-embryo phase age 05 11 (acv = anterior cardinal vein, ba = branchial arteries, da = dorsal aorta, hiv = hepatic/intestinal vitelline system, ht = heart-tube, icv = inferior caudal vein, isv = intersegmental blood vessels, pba = pseudobranchial artery, pca = pectoral fin anlagen, pop = postanal respiratory plexus, prp = pre-anal respiratory plexus, pvv = posterior vitelline vein, ua = urostylar artery, uv = urostylar vein) Scale = 1 mm

first few somites at age 02:00 (TU = 50) (Fig. 3b). Six hours later (TU = 56), otic vesicles and a heart-tube are visible and the number of somites increases to 17. By age 02:09 (TU = 59) one otolith per side is present, eye lenses form, body melanophores extend to two thirds the body length and undercutting of the tail mound begins. Constrictions in the anterior region make the three major components of the brain distinguishable (the prosencephalon, mesencephalon and rhombencephalon) and cardiac contractions of 115 beats min^{-1} were recorded for one specimen.

By age 03:00 (TU = 75), two otoliths per side are present, pigment cells extend half-way down over the yolksac and the cerebellum is distinguishable. Blood flow follows a circular route around the eyes and empties into the anterior cardinal veins which flow into the anterior vitelline veins (Fig. 3c). Blood from the pre-anal finfold enters the posterior vitelline vein which flows along the ventral surface of the yolksac where it branches into smaller vessels

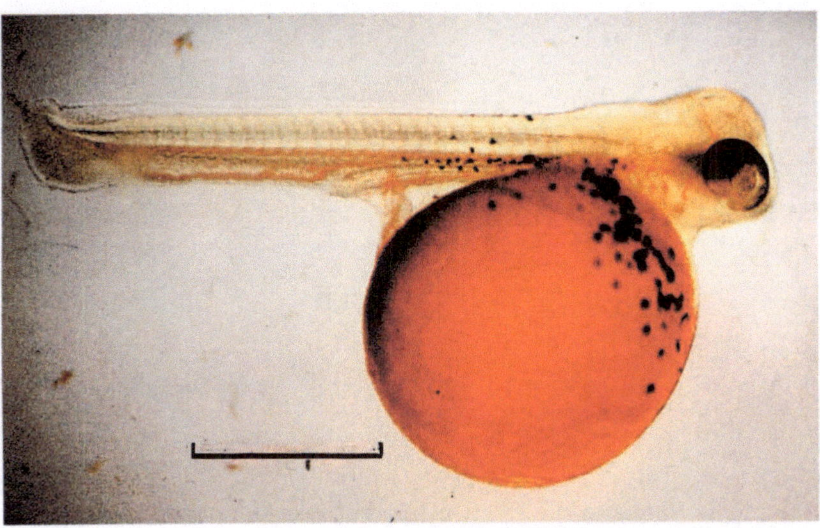

Fig 5 Free embryo at age 05 05 from right side Scale 1 mm

and continues dorsally to the heart-tube. The tail is undercut from the yolksac and body muscle contractions occur.

Three hours later (TU = 78), the posterior vitelline vein branches shortly after it enters the yolksac, the heart-tube begins twisting and the dorsal aorta is visible. At age 03:11 (TU = 86) a transparent chamber is visible in the mesencephalon, the yolksac melanophores are larger and there are 26–27 somites. The dorsal aorta forms a loop in the tail region where it flows anteriorly into the inferior caudal vein. Twelve hours later (TU = 99), hatching begins and the extent of vascularization is more obvious. Additional vessels around the eyes and in the brain region form loops before emptying into the anterior cardinal veins. The inferior caudal vein runs along the distal margin of the postanal finfold. A rudimentary respiratory network is present in the tissue of both the pre- and postanal finfolds. The pre-anal finfold receives blood from the vessels of the postanal finfold and from an artery running posteriorly above the presumptive visceral cavity. The dorsal aorta has a second caudal loop and a third one is forming.

At age 04:03 (TU = 103) pigment extends along the body/yolksac interface and spreads ventrally on the yolksac. The eyes are shaded gray and pectoral fin anlagen are present. Veins enter the yolksac from the right and left sides of the embryo in the vicinity of the presumptive visceral cavity and join the anterior vitelline veins (Fig. 4a). This indicates the beginning of the hepatic and intestinal vitelline system. The left vessels are larger than those entering on the right side (see Fig. 5).

Free-embryo phase (04:23-14:00). By age 04:23 (TU = 124) almost all the embryos have hatched. The head is free from the yolksac and the longitudinal axis is straighter. The entire surface of the eye is lightly and evenly pigmented, the pectoral fin anlagen begin to rotate perpendicularly to the body line, and labyrinth formation in the dorsal regions of the otic capsules begins. Four branchial arteries are visible, the urostylar artery and vein are slightly anterior to the tip of the notochord, and intersegmental vessels in the trunk area form single loops. The density of vessels on the median finfold increases and their pattern is more complex. Blood from the posterior, anterior and hepatic/intestinal vitelline systems forms a sheet-like movement prior to entering the heart-tube. The hepatic/intestinal vitelline veins on the left side form a more complex pattern than on the right. Six hours later (TU = 130), mesenchyme aggregations form along the dorsal and caudal finfolds, the pseudobranchial artery lies anterior to the branchial arteries, vascularization on the lateral portions of the yolksac occurs and internal folding in the foregut is evident. The intersegmental

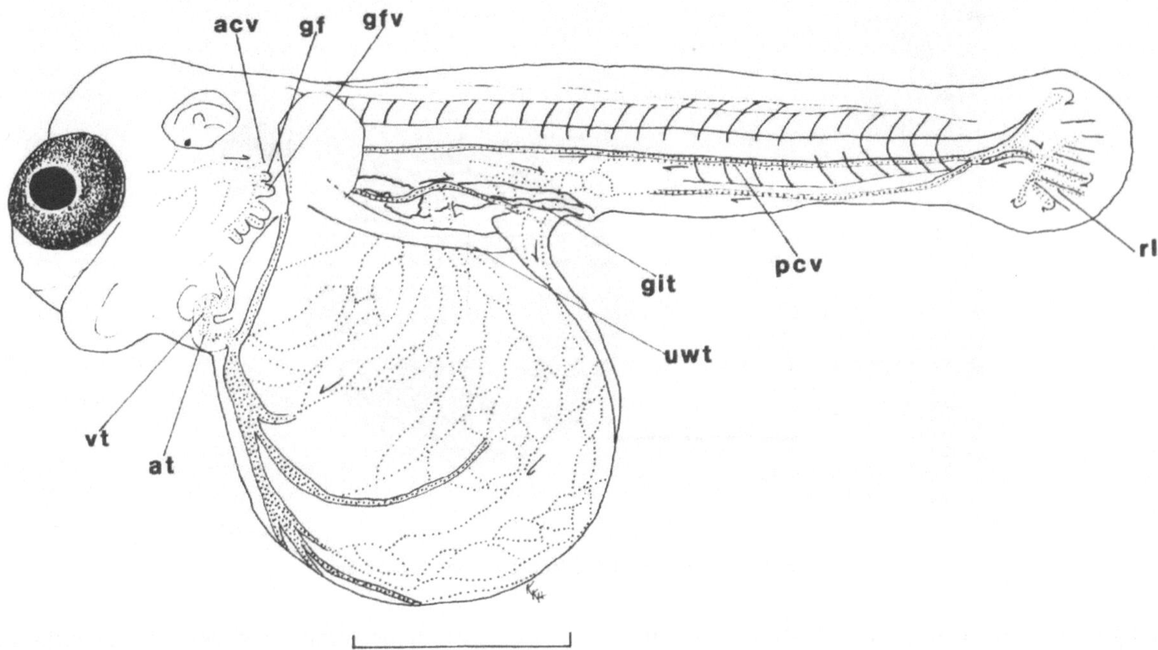

Fig 6 Free-embryo phase age 06 23 (acv = anterior cardinal vein, at = atrium, gf = gill filaments, gfv = gill filament blood vessels, git = gastro-intestinal tract, pcv = profundal caudal vein, rl = radial loop, uwt = unidentified white tissue, vt = ventricle) Scale = 1 mm

vessels extend to the end of the caudal peduncle and the intersegmental veins empty into the postanal plexus (Fig. 4b, 5).

By age 05:23 (TU = 149) a thickened area of white tissue lies along the dorsal rim of the yolksac and constrictions form along the gastro-intestinal tract. Constrictions of the heart-tube make the sinus venosus, the atrium, the ventricle and the bulbous arteriosus distinctive as separate rudimentary compartments. The nares are well developed and the mouth is open but does not move. The first three gill arches are pouch-like with some filaments present. There is a reduction in the central regions of the postanal respiratory plexus. The yolksac respiratory plexus reaches maximum development and the pattern of the veins is symmetrical on both sides. The trabeculae appear as two pale blue lines which fork around the anterior tip of the notochord. Meckel's cartilage, the angulo-articular, the gill arches, the pectoral fin anlagen and the process of the opercle retain some alcian blue stain. The tissue of the caudal finfold is striated; a precursor to ray formation. The cleithrum appears as a thin, transparent line.

At age 06:05 (TU = 155) the eyes, including the lenses, are darkly pigmented and contain iridocytes. Pigment cells appear above the mid-brain, and dorsal body melanophores extend from posterior of the rhombencephalon to mid-body. The shape of the yolksac is oval and mesenchymal aggregations form in the caudal finfold and the pectoral fin buds. Circulation begins in the gill filaments. Five radial loops are present in the developing caudal circulatory system. The inferior caudal vein lies along the ventral body line and the respiratory plexus occurs only in the posterior regions of the postanal finfold. Muscular development and coiling begins in the gastro-intestinal tract which contains a lumen but is not open at the vent. The gall bladder is visible on the right side. Jaw movements begin. Chondrification begins in the ceratohyal, the interhyal, the hyo-symplectic, the palatoquadrate and along the ventral, distal rim of the otic capsule.

At age 06:23 (TU = 174) the pigment cells above the brain are more extensive and stellate-like Melanophores extend along the ventral body line into the caudal finfold. The pectoral fin bud lies at its final position perpendicular to the body axis and

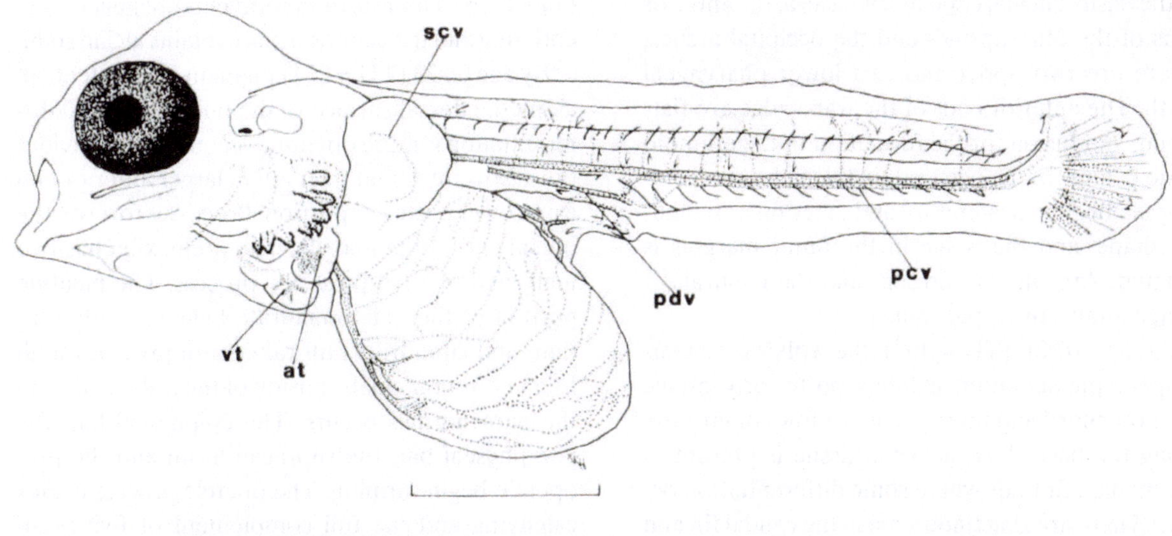

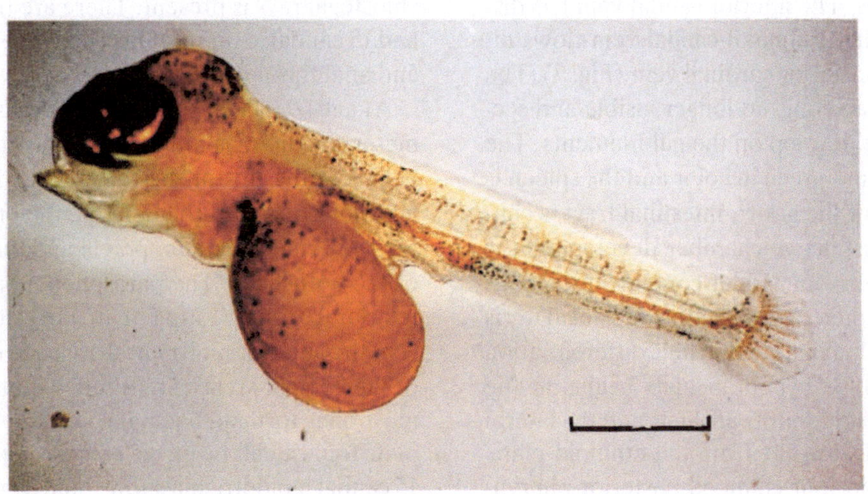

Fig 7 Free-embryo phase age 07 21 (at = atrium, pcv = profundal caudal vein, pdv = posterior cardinal vein, scv = subclavian vein, vt = ventricle) Scales = 1 mm

constrictions along the caudal peduncle indicate the beginning of finfold differentiation, especially between the caudal fin and the median finfolds. Four rudimentary actinotrichia exist in the caudal fin and six radial loops are present. The profundal caudal vein is present in the anterior regions of the trunk and all that remains of the postanal plexus is the inferior caudal vein (Fig. 6). The anterior vitelline vein flows along the anterior margin of the yolksac.

By age 07:06 (TU = 181) the profundal caudal vein extends almost to the end of the caudal peduncle and there is a large reduction in the number of vessels in the pre-anal finfold. The atrium lies dorsal to the ventricle, where muscular development is visible. The subclavian veins form a single loop in the pectoral fin buds. Two branchiostegal rays are present, the posterior region of the angulo-articular begins to elongate, a jagged area forms dorsal to Meckel's cartilage and the maxilla appears as a thin line. All the above structures retained some alcian stain. Other skeletal structures beginning to chondrify are the four ceratobranchials, the anterior tip of the basibranchial copulae, the sclera, the anterior rims of the otic capsules and the occipital arches. There are two upper and two lower pharyngeal teeth. The anterior ends of the trabeculae are flattened, and fusion for the formation of the ethmoid plate begins. In the pectoral fin buds, the coracoid-scapula begins to chondrify and takes on a triangular shape, and the tissue in the outer margins is striated. Arcualia are present and the hypural region contains three pale blue plates.

At age 07:21 (TU = 197) the yolksac is tear-shaped, pigmentation extends posteriorly above the notochord and there is pale yellow colouration along the back. Mesenchymal tissue is present in the median finfolds where some differentiation occurs. There are 11 actinotrichia in the caudal fin and nine radial loops. The inferior caudal vein has disappeared and the profundal caudal vein flows directly into the posterior cardinal vein (Fig. 7). The anterior vitelline vein is no longer visible and secondary lamellae develop on the gill filaments. The gall bladder is dark green in color and the spleen is visible. Coiling of the gastro-intestinal tract is considerable but distinctive chamber development is not evident. The swimbladder contains thick walls with a narrow lumen. A fourth branchiostegal ray is forming, the premaxilla is a thin line, differentiation of the presumptive hypobranchials begins at the ventral ends of the ceratobranchials and the lateral ethmoid extends dorsally from the ethmoid plate. One anterior and four clumped posterior pharyngeal teeth per side take up alizarin red-S stain in their tips. There are 3–4 lower pharyngeal teeth per side. The outer walls, some internal structures and the posterior rim of the otic capsules begin to chondrify. The opercle takes on a fan-shaped configuration and both it and the dorsal rim of the subopercle are blue. The 27 presumptive neural arches vary from pale blue to transparent in colour from the anterior to the posterior. Some presumptive haemal arches are barely visible.

At age 08:05 (TU = 205) two dentary teeth are present. There is an increase in the number of pharyngeal teeth, all of which have begun to calcify. Double filaments form on, and chondrification begins in, the gill filaments of the first three ceratobranchials. The fourth epibranchial begins differentiating and the actinosal plate retains alcian stain.

By age 08:23 (TU = 224) pigment cells extend all along the dorsal surface to the tail, and finfold differentiation is more obvious. A constriction divides the gastro-intestinal tract into a larger anterior and a narrower posterior portion. Processes form on the dorsal end of the maxilla. The premaxilla has two teeth and a dorso-posterior process. The palatine portion of the palatoquadrate begins to differentiate and chondrify. Gill rakers are present on all four ceratobranchials. Fusion of the trabeculae and the parachordals occurs. The epiphyseal bar, the paraphyseal bar, the supracleithrum and the preopercle begin forming. The opercle process begins calcifying and the full complement of five branchiostegal rays is present. There are three pectoral and 15 caudal actinotrichia. The tissue of the dorsal and anal finfolds is striated.

At age 09:08 (TU = 233), 10 dorsal and four anal mesenchymal rays are present. Slight calcification begins in the tips of the dentary teeth, the dorso-anterior portion of the dentary, the branchiostegal rays, the rim of the subopercle and the anterior tip of the notochord. The parasphenoid and the post-temporal begin to form but are not calcified. All except the anterior nine neural arches are fused. Most of the haemal arches are fused, with the posterior-most ones forming haemal spines. In the actinosal plate four radials begin differentiating. Seven of the 15 caudal lepidotrichia show some degree of calcification. The pre-anal respiratory network has disappeared and blood flows directly from the profundal caudal vein into vessels around the gastro-intestinal tract. The anterior end of the gastro-intestinal tract contains internal, finger-like projections.

At age 10:00 (TU = 250) pigmentation occurs along the lateral flanks, especially in the anterior, and extends to the tail and along most of the caudal lepidotrichia (Fig. 8a). Blood flows into the second lamellae of the gill filaments and along the pectoral actinotrichia. The blood vessels of the yolksac are relatively equal in width and form a fine network. There is a noticeable increase in the degree and occurrence of calcification. Some of the new structures which begin calcifying are the premaxillary teeth, the articulation points of the angulo-articular and

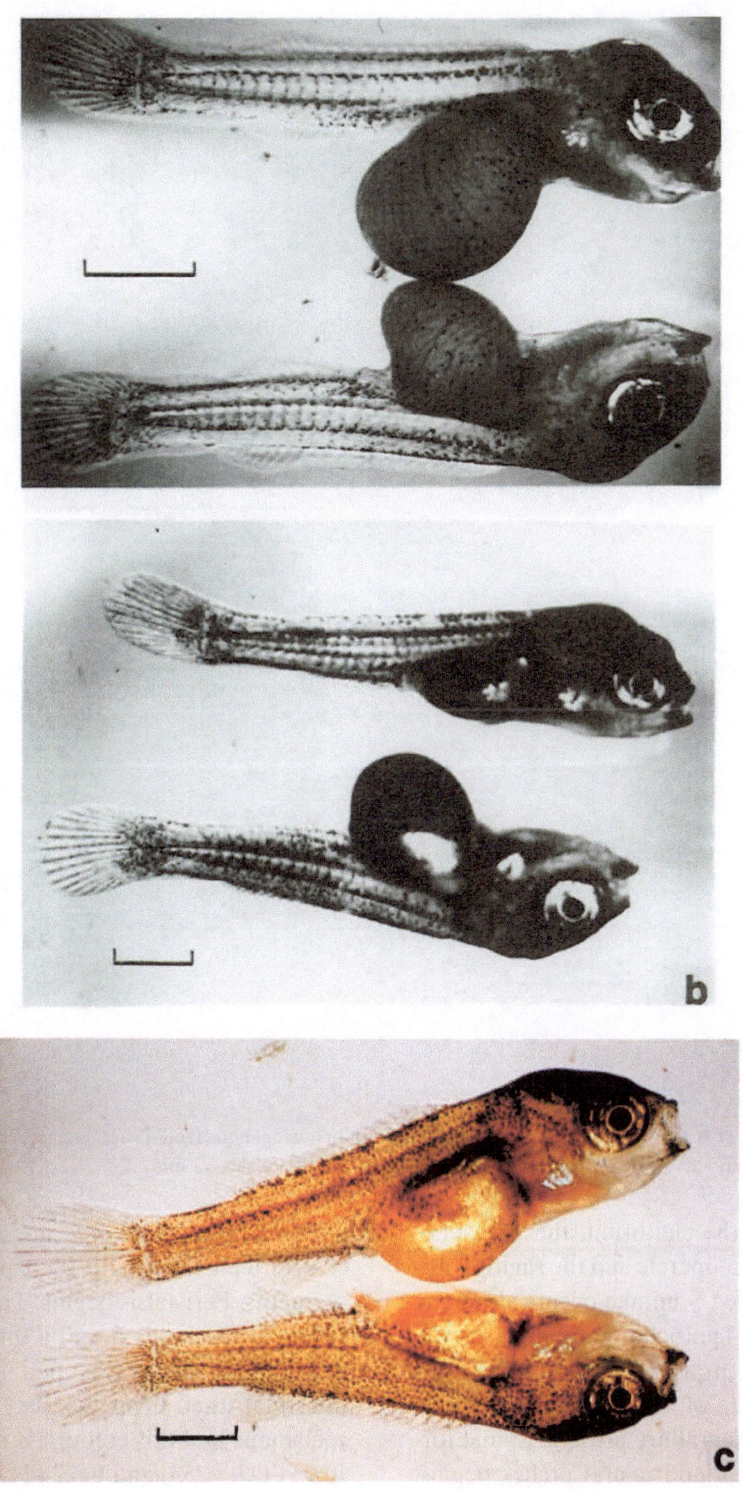

Fig. 8 a – age 10.00. b – age 12.00. c – juvenile at age 14.00. Note the variation in the size of the yolksac and the rapid decline in yolksac size over time. Scales: 1 mm.

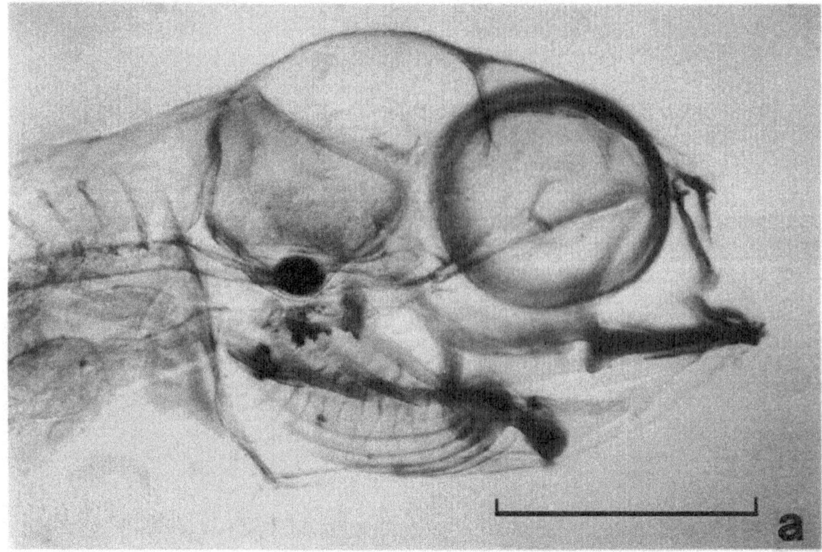

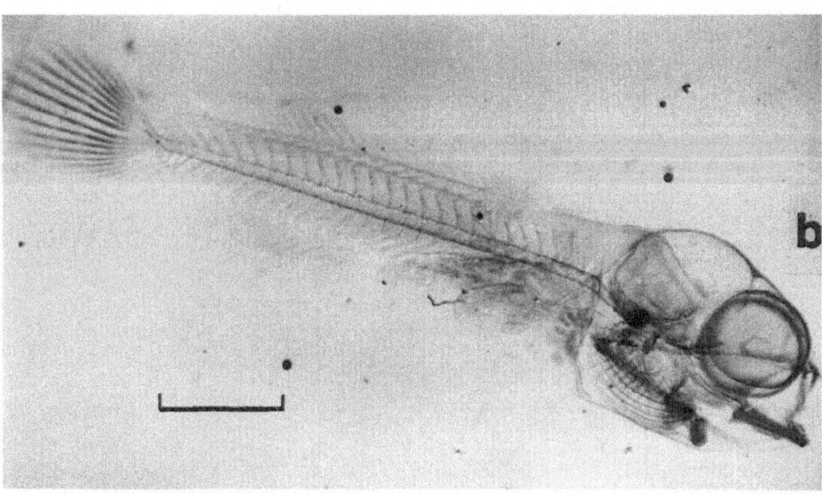

Fig. 9. Skeletal development at age 11:00. This age is coincidental with first exogenous feeding and release, and illustrates the level of development of the structures associated with feeding (a) and locomotion (b). Scales = 1 mm.

the palatoquadrate, the cleithrum, the parasphenoid, the rim of the preopercle and the sagittal otoliths. Slight alizarin red-S uptake occurs along the margins of the central notochord and the urostyle. Vertical lines in the central regions of the notochord indicate the beginning of vertebral development. One epural and uroneural are present. Spinal formation on the haemal and neural arches begins. The pectoral fins are fan-shaped with scalloped margins. There are eight, six and 19 anal, pectoral and dorsal actinotrichia, respectively. Median fin proximal pterygiophores are barely visible. The central lepidotrichia in the caudal fin have three segments. Peristalsis begins. There is a yellow substance in the hindgut and a thin, transparent exudate extends out of the open lumen of the gastrointestinal tract. Over the next 24 h several of the specimens had this colourless exudate and at ages 10:03 (TU = 253) and 10:22 (TU = 273) dark particulate material was found in the exudate of two embryos. Thus, first exogenous feeding begins between ages 10:00 and 11:00.

At age 11:00 (TU = 275) iridocytes are present on the gill covers and the descending body wall (Fig. 8b), and the embryos take a pale blue/green colouration. All that remains of the median finfolds is a narrow fold between the caudal fin and the presumptive dorsal and anal fins. Pelvic fin anlagen form along the flanks of the descending body wall. From this age onwards individual variation becomes more apparent and there are marked differences in the degree of yolksac enclosure (Fig. 8b–c).

By 12:00 (TU = 300) there are two radial loops of vessels along some of the caudal lepidotrichia. Marked increases in calcification occur at this time. The premaxilla and the maxilla are joined at their dorsal ends by a calcifying plate-like structure. Dechondrification occurs in the central areas of the hyo-symplectic and the palatoquadrate. Calcification begins along the margins of these two structures, in the central regions of the ceratohyals and in the urohyals, the hypohyals, the pharyngeal bones and the pharyngobranchials. In the neurocranial region, alizarin uptake occurs in the bases of the occipital arches, in the parasphenoid up to its junction with the ethmoid plate, the tip of the notochord along the neurocranial floor and in all three otoliths. The vomer is rough and ridged on its ventral surface. Along the notochord the first and the seventeen posterior-most vertebrae, the first two neural arches, the posterior neural/haemal arches and the uroneural are all calcified to some degree. Staining occurs along the dorsal rim of the preopercle where three pores begin forming. All neural and haemal spines are present. There are three hypurals that retain some alizarin and a parhypural is present. The pelvic basipterygium appears as an undifferentiated blue line. The pectoral lepidotrichia begin segmenting. Figure 9 illustrates overall skeletal development of a specimen at age 11:00. All but one embryo congregated at the bottom of the separating funnel with their heads oriented towards the inflow of the water. Some embryos have large, bulbous, highly vascularized yolksacs. In others a remnant of the yolksac with a few blood vessels is visible ventrally through a narrow gap between the descending body walls (Fig. 8b).

By age 13:01 (TU = 362), finfold differentiation is almost complete and the embryos begin to resemble juveniles. In most specimens the yolksac does not extend below the ventral contour and what little remains of the yolksac can be seen through a narrow gap on the ventral surface of the embryo. Blood elements in fine vessels flow on the yolksac remnant. Melanophores occur along the pectoral and dorsal lepidotrichia. Many of the embryos are free-swimming in the water column of the funnel and some make their way into the cups at the end of the outflow tubes.

Juvenile period 14:00 +

By 14:00 (TU = 350) melanophores form along the anal lepidotrichia and pink chromatophores appear. All the vertebrae are distinguishable as individual units and fin differentiation is complete (Fig. 10). Calcification begins in the orbital commissures, the vomer, the postcleithra and in the dorsal, anal and pectoral lepidotrichia. Anterior distal pterygiophores begin forming in both median fins. The pelvic basipterygium begins differentiating and there are four pelvic actinotrichia. Chondrification begins on the neurocranial roof. Some dorsal and anal lepidotrichia have two and three segments, respectively, and the caudal lepidotrichia have up to five segments. The full complement of median lepidotrichia and spines is reached (D XIV-10, A III-8). Midlateral neuromasts extend anteriorly along the caudal peduncle to the posterior of the yolksac.

By 15:21 (TU = 397) neuromasts extend to the top of the head and scales begin to form on the caudal peduncle. The outer walls of the otic capsules, the roof of the neurocranium, the pterygiophores and the pelvic lepidotrichia begin calcifying. By age 17:00 (TU = 425) circulation begins in the median fins and, within 24 h, at least two and up to four rows of scales form along the caudal peduncle. By this time, even though the yolksac does not protrude ventrally, there is still a gap between the descending body walls of some specimens. A female that had been left to incubate her eggs was observed releasing and recalling the young to her buccal cavity 17 to 18 days after spawning.

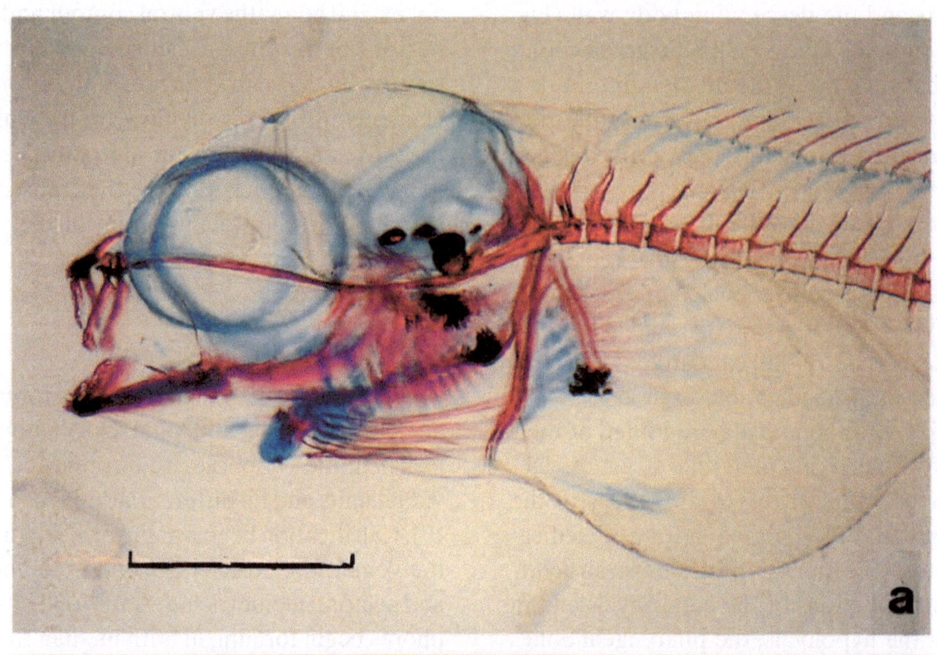

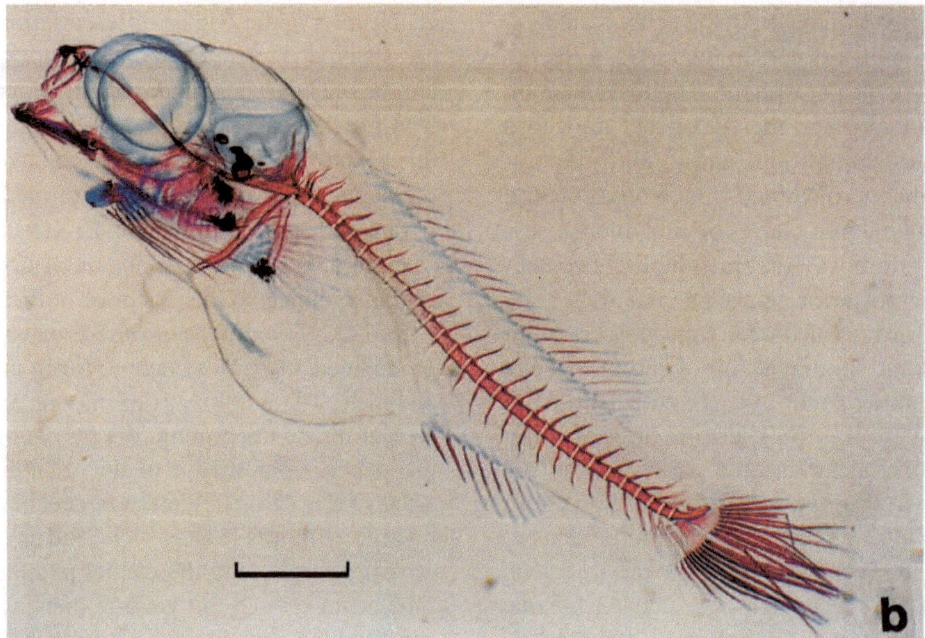

Fig 10 Juvenile at age 14 00 depicting the advanced state of development of the skeletal system Note the completion of finfold differentiation and yolksac enclosure, and the vertebrae as distinctive units Scales = 1 mm

Heart rate, growth and yolk area

Individual variation in yolk area is great, which may obscure trends over time. Yolk area changes little until halfway through the free-embryo phase (Fig. 11). Embryo length increases gradually over the same period. Yolk area decreases markedly once the embryos begin feeding exogenously but em-

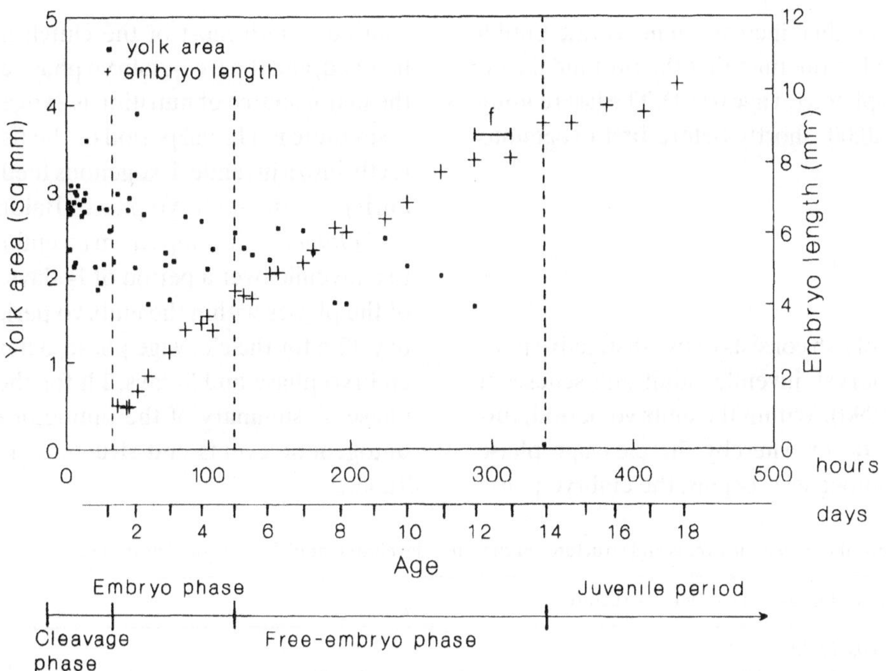

Fig 11 Changes in embryonic length (mm) and yolk area (mm²) of *Pseudocrenilabrus philander* over time Vertical lines represent phase and period boundaries (f = age at first exogenous feeding)

bryonic growth continues more gradually into the juvenile period.

During the embryo period, no pattern in heart rate over age is apparent. Heart rate increases gradually and peaks at age 05:05 (Fig. 12), which coincides with hatching and a notable increase in circulatory and respiratory structures. A subsequent marked drop coincides with a reduction in the post-anal finfold respiratory plexus and an increase in the vitelline respiratory plexus (at age 05:23). This is

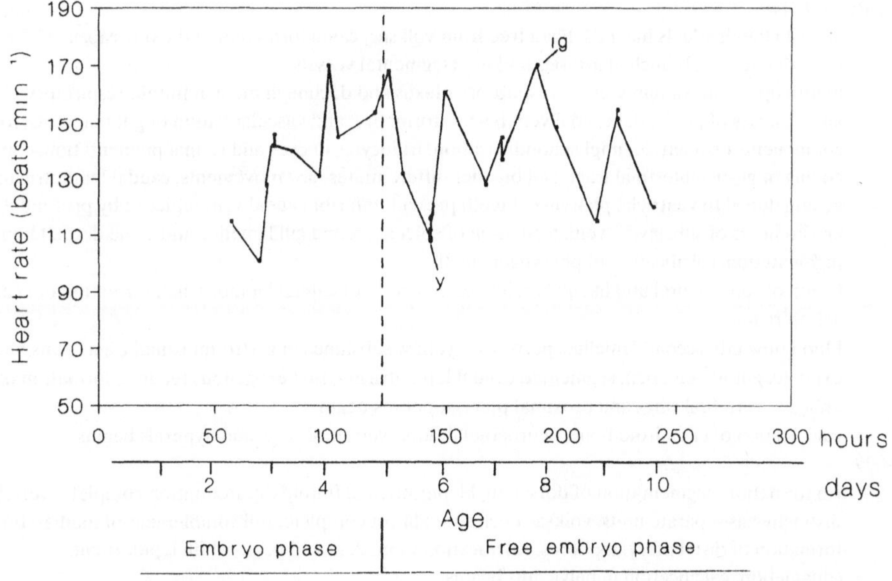

Fig 12 Heart rate of *Pseudocrenilabrus philander* in the embryo and free-embryo phases The vertical line represents the boundary between the embryo and free-embryo phase (ig = inferior caudal vein gone, y = maximum yolksac respiratory plexus)

followed by a gradual increase in heart rate until a peak is reached at the time that the profundal caudal vein is completed (at age 07:21). The last reading is low at age 10:00, shortly before first exogenous feeding.

Discussion

The ontogeny of fish consists of five distinctive periods; embryo, larval, juvenile, adult and senescent (Balon 1975, 1990). Within the embryo period, further divisions occur whereby the cleavage phase lasts until organogenesis begins, the embryo phase continues until most of the clutch members have hatched, and the free-embryo phase continues until the major source of nutrition is exogenous. In some cases there is a larval period or the fish develops directly into a juvenile. Exogenous feeding may begin during the free-embryo phase (Balon 1990).

P. philander develops from an embryo directly into a juvenile over a period of 14 days. The duration of the phases within the embryo period are about 1 day, 12 h for the cleavage phase, 3 days, 11 h for the embryo phase and 9 days, 1 h for the free-embryo phase. A summary of the timing of the important ontogenetic events and structures is presented in Table 2.

Table 2 The timing of ontogenetic events and structures in the early development of *Pseudocrenilabrus philander*

Age (days hours)	Ontogenetic event or structure
Cleavage phase 00 00–01 12	
00 01	bipolar differentiation, perivitelline space formation, hardening of the egg envelope
00 01–00 12	cleavage, periblast formation
00 14	epiboly begins
Embryo phase 01 12–04 23	
01 12	neurulation, organogenesis, presumptive pericardial cavity forms
01 20–02 06	optic and otic vesicles, and heart-tube formation, yolksac and body pigmentation, 17 somite pairs
02 09–03 00	cardiac and muscular contractions, two otoliths per side, formation of eye lenses, three brain components, head blood vessels, anterior and posterior vitelline veins, elongation and separation of tail, cerebellum distinguishable
03 23–04 03	hatching begins, formation of second caudal loop, hepatic/intestinal vitelline respiratory system and pectoral fin anlagen, ventral finfold circulation increases to form a network, eye pigmentation begins
Free-embryo phase 04 23–14 00	
04 23	almost all individuals hatched, head free from yolksac, canal formation in the otic capsules, formation of urostylar vessels, branchial arteries and intersegmental vessels
05 23–06 23	mouth opens, maximum yolksac respiratory plexus and decline in median finfold respiratory plexii, anastomoses of profundal caudal vein begins, formation and vascularization of gill filaments, four heart components distinctive, simple chondrification, iridocytes in eyes and retinal pigmentation, constriction and coiling of gastro-intestinal tract, gall bladder differentiates, jaw movements, caudal lepidotrichia form
07 06–07 21	atrium dorsal to ventricle, pharyngeal teeth present, inferior caudal vein replaced by profundal caudal vein, anastomoses of subclavian vein, formation of spleen, second gill lamellae and some dermal bones, yellow pigmentation, calcification of pharyngeal teeth
09 08	fusion of some neural and haemal arches, calcification of caudal lepidotrichia, formation of caudal and anal actinotrichia
10 00–11 00	blood flow into second lamellae, peristalsis, yellow substance in gastro-intestinal tract, transparent thread exudate, gut lumen open, segmented caudal lepidotrichia, first exogenous feeding, formation of pelvic fin anlagen, vertebral rings and proximal pterygiophores begin
12 00$13 00	calcification of chondroid bone, neurocranial floor, vertebral rings and hypurals begins
Juvenile period 14 00	
14 00	rib formation, segmentation of dorsal/anal lepidotrichia, finfold differentiation complete, vertebral rings distinctive as separate units, yolksac enclosure almost complete, full complement of median fin lepidotrichia, formation of distal pterygiophores, calcification of median and pectoral fin lepidotrichia
15 21	squamation, calcification in pelvic fins begins

During the cleavage phase, epiboly begins at age 00:14 and the embryonic shield forms. Neurulation and organogenesis mark the beginning of the embryo phase at age 01:12 until most of the embryos have hatched at age 04:23. During this phase, the basic body form from which rudimentary organ systems develop is established. The nervous system and the brain are relatively well developed while sensory organs (except for the eyes) are simple. The development of the temporary embryonic respiratory system is a major morphological event during this phase coupled with the inundation of blood vessels throughout the body, and the beginning of cardiac contractions. During the first portion of the free-embryo phase, permanent adult respiratory organs begin to develop and increase in complexity as the embryonic finfold plexuses decline. In the circulatory system, there is a marked increase in the vascularization of the entire embryo and differentiation of the heart-tube. The organs of the gastrointestinal tract begin to differentiate and coil. Basic skeletal development occurs, particularly in the suspensorium, the hyoid arches and the pectoral girdle. By the end of this phase, temporary embryonic respiratory structures are replaced by the permanent adult structures. Although the ventral gap between the opposing sides of the descending body walls is not completely closed, a residual yolksac lies within the visceral cavity. Except for the gonads, all the adult structures are present. There is an interval of mixed feeding from ages 10:00 to 11:00 until age 14:00. During this time a rapid change from predominantly endogenous to almost exclusively exogenous feeding occurs. The free-swimming abilities of the embryos indicate that skeletal components required for locomotion and feeding are functional. Further differentiation and calcification of the skeletal system continues into the juvenile period.

Evans (in Ribbink 1975) found that allopatric populations of *P. philander* differed in egg size and size of newly released young. The mean maximum and minimum lengths of eggs from the Durban population in her study were 2.5 and 1.8 mm, respectively. The adult fish in this study were from a Durban population and the mean egg lengths from two clutches were 2.2 and 2.5 mm by 1.6 and 1.7 mm. Embryos in the latest stages of development, when yolksac absorption neared completion, measured 8.95 mm [as in the Durban population of Evans (Ribbink 1975)]. Lengths of embryos at a similar level of development were 7.7–9.1 mm in this study.

The pattern of development and the timing of important ontogenetic events clearly reflect the environmental conditions experienced by the young southern mouthbrooder. The development of the extensive, temporary, respiratory plexuses on both the ventral finfold and the yolksac augments gaseous exchange within the buccal cavity of the female where oxygen levels are presumably low (Fig. 4) (Balon 1977). Flapping of the pectoral fins, which develop early and rapidly, creates a water current over the vitelline plexus, thus facilitating respiration (Fishelson 1966). Prior to the time of first release, between 11 and 14 days at 25° C (Ribbink 1971), the temporary respiratory structures on the ventral finfold are no longer present and the switchover to the permanent adult branchial system has occurred (Fig. 7). Higher oxygen concentrations outside the buccal cavity probably reduce the necessity for extensive additional respiratory structures. The vitelline plexus would supplement oxygen uptake should there be a deficiency. However, as there is a rapid decrease in the size of the yolksac between ages 11:00 and 14:00, and variation in yolk sac size is large, its role in gaseous exchange is probably minimal (Fig. 8). In addition, the time of first release is variable.

During the interval of mixed feeding the yolk is primarily a supplemental source of nutrition. Peristaltic movements and the expulsion of fluid through the vent just prior to release indicate that the digestive system is functional and capable of processing an exogenous food source. The remaining yolk may assure some nutritional intake (Balon 1977) while the young fish improve their food-gathering abilities.

The problem of predation pressure on the young after release, although not a consideration during the incubation period, is resolved by several characteristics of early ontogeny. The level of development of the skeletal and sensory systems is sufficient for avoiding predators by hiding or returning to the buccal cavity of the female. Pigmentation over the entire body may provide disruptive colouration in the vegetated areas where the young are released (Fig. 8, 9). By the time the female ceases guarding, the young have reached a juvenile state and the skeletal system is well developed (Fig. 8c, 10).

Life-history traits also reflect the co-evolution of this species with its habitat. The carotenoid content of the yolk, and the churning of the eggs and embryos within the buccal cavity, increase oxygen utilization potential (see Balon 1991 for information about the oxidative role of carotenoids). Mouthbrooding, recall behaviour and schooling of the young fish are traits which reduce predation risk. The mobility of the brooding female allows movement into areas of optimal or preferred habitat conditions.

The rate and pattern of development of the Mozambique tilapia *Oreochromis mossambicus* (Holden & Bruton 1992) and *P. philander* are similar. The differences in the duration of the cleavage and embryo phases are 1 h and 6 h, respectively. In the free-embryo phase there is a difference in duration of 1 day, 7 hours. Hatching begins 8 h earlier in *O. mossambicus* but, with both species, continues for a period of about 24 h. Completion of the development of the digestive system followed by first exogenous feeding occurs between the ages of 10:00 and 11:00 for both species. The interval of mixed feeding and the beginning of the juvenile period differ by 1 day (age 14:00 for *P. philander* and 15:00 for *O. mossambicus*). The differences mentioned may be attributed to differing sampling times, small sample sizes (in most cases only one individual), subjectivity about the timing of events, and individual variation. Closer examination of the early development of these two species may prove otherwise, and is the subject of another study.

Several of the life-history characteristics of the two species are similar. Males build concave nests in arenas by excavating depressions in the river or lake bottom to which females are attracted for spawning (Ribbink 1971, 1975, Bruton & Boltt 1975, Bruton 1979, Trewavas 1983). Immediately after spawning, the female leaves the nest and the male attempts to spawn with other females. Females school together while brooding and the young are released in vegetated, warm shallow water (or in the case of *P. philander*, the young may also be released into deeper water; Ribbink 1975). In Lake Sibaya, young *O. mossambicus* may be released over totally barren sandy terraces (Bruton & Boltt 1975, Bruton 1979). The females continue to guard and recall the young to their buccal cavity for several days.

Several life-history traits are not shared by the two species. *P. philander* adults are smaller, the males build smaller nests, and the females produce larger eggs relative to body size, mature at smaller sizes and have smaller clutches (Ribbink 1971, 1975, Loiselle 1982, Trewavas 1983). Overlapping life-history characteristics are incubation time, interbrood time, age at maturation and size of newly released young. Since many of these traits are dependent on environmental conditions (e.g. temperature and food availability), and both these species are phenotypically plastic, it is difficult to determine the significance or the magnitude of these differences and their influence on early development.

Both species exhibit direct development with accelerated exogenous feeding into the embryonic period (Balon 1990, Holden & Bruton 1992). The southern mouthbrooder, like the Mozambique tilapia, belongs to the reproductive guild of mouthbrooders without buccal feeding (Balon 1990). It is therefore not surprising that their early development is similar, despite the phylogenetic differences between the tilapiine and haplochromine cichlids. The similar environmental conditions which the eggs and embryos are subjected to may have resulted in both species adopting the same eco-ethological and eco-morphological developmental styles (see Holden & Bruton 1992).

Acknowledgements

We thank the staff of the J.L.B. Smith Institute of Ichthyology for their support throughout this project. Thanks go to R.E. Stobbs for technical assistance, and to A.J. Ribbink and G.S. Merron for their useful comments on the manuscript. Eugene Balon and Christine Flegler-Balon shared their time and knowledge generously. This project was funded by a grant from the Foundation for Research Development to M.N. Bruton.

References cited

Balon, E K 1974a Fishes of Lake Kariba, Africa length-weight relationship, a pictorial guide T F H Publications, Neptune City 144 pp

Balon, E K 1974b Fish production of a tropical ecosystem pp 249–748 *In* E K Balon & A G Coche (ed) Lake Kariba A Man-Made Tropical Ecosystem in Central Africa, Monographiae Biologicae 24, Dr W Junk Publishers, The Hague

Balon, E K 1975 Terminology of intervals in fish development J Fish Res Board Can 32 1663–1670

Balon, E K 1977 Early ontogeny of *Labeotropheus* Ahl, 1927 (Mbuna, Cichlidae, Lake Malawi), with a discussion on advanced protective styles in fish reproduction and development Env Biol Fish 2 147–176

Balon, E K 1990 Epigenesis of an epigeneticist the development of some alternative concepts on the early ontogeny and evolution of fishes Guelph Ichthyol Rev 1 1–48

Balon, E K 1991 Probable evolution of the coelacanth's reproductive style lecithotrophy and oral feeding embryos in cichlid fishes and in *Latimeria chalumnae* pp 249–256 *In* J A Musick, M N Bruton & E K Balon (ed) The Biology of *Latimeria chalumnae* and Evolution of Coelacanths, Kluwer Academic Publishers, Dordrecht

Balon, E K & C Flegler-Balon 1985 Microscopic techniques for studies of early ontogeny in fishes problems and methods of composite descriptions pp 33–55 *In* E K Balon (ed) Early Life Histories of Fishes New Developmental, Ecological and Evolutionary Perspectives, Dev in Env Biol Fish 5, Dr W Junk Publishers, Dordrecht

Bell-Cross, G 1966 The distribution of fishes in Central Africa Fish Res Bull Zambia 4 3–20

Bruton, M N 1979 The fishes of Lake Sibaya pp 162–245 *In* B R Allanson (ed) Lake Sibaya, Dr W Junk Publishers, The Hague

Bruton, M N 1980 An outline of the ecology of the Mgobozeleni lake-system at Sodwana, with emphasis on the mangrove community pp 408–426 *In* M N Bruton & K H Cooper (ed) Studies on the Ecology of Maputaland, Rhodes University, Grahamstown

Bruton, M N & R E Boltt 1975 Aspects of the biology of *Tilapia mossambica* Peters (Pisces Cichlidae) in a natural freshwater lake (Lake Sibaya, South Africa) J Fish Biol 7 423–445

Cunningham, J E R & E K Balon 1985 Early ontogeny of *Adinia xenica* (Pisces, Cyprinodontiformes) 1 The development of embryos in hiding Env Biol Fish 14 115–166

de Villiers, D L, E H Harley & A J Ribbink 1992 Mitochondrial DNA restriction enzyme variation in allopatric populations of *Pseudocrenilabrus philander* (Pisces Cichlidae) S Afr J Sci 88 96–99

Fishelson, L 1966 Untersuchungen zur vergleichenden Entwicklungsgeschichte der Gattung *Tilapia* (Cichlidae, Teleostei) Zool Jahrb Anat 83 571–656

Fryer, G & T D Iles 1972 The cichlid fishes of the Great Lakes of Africa their biology and evolution T F H Publications, Neptune City 641 pp

Greenwood, P H 1989 The taxonomic status and phylogenetic relationships of *Pseudocrenilabrus* Fowler (Teleostei, Cichlidae) Ichthyological Bulletin of the J L B Smith Institute of Ichthyology 54 1–16

Holden, K K & M N Bruton 1992 A life-history approach to the early ontogeny of the Mozambique tilapia *Oreochromis mossambicus* (Pisces, Cichlidae) S Afr J Zool 27 173–191

Loiselle, P V 1982 *Pseudocrenilabrus* the dwarf African mouthbrooder Part two The *Pseudocrenilabrus ventralis* and the *Pseudocrenilabrus philander* complex Freshwat mar Aquat 5 (2) 30–34, 66–71, 75

Poll, M 1967 Contribution a la faune ichthyologique de l'Angola Comp Diam Angola Publ cult 75 1–379

Potthoff, T 1984 Clearing and staining techniques pp 35–37 *In* H G Moses, W J Richards D M Cohen, M P Fahay, A W Dendal & S L Richardson (ed) Ontogeny and Systematics of Fishes, Ahlstrom Symposium 1983, La Jolla, Special Publications 1, Amer Soc Ichthyol & Herpetol Lawrence

Ribbink, A J 1971 The behaviour of *Hemihaplochromis philander*, a South African cichlid fish Zool afr 6 263–288

Ribbink, A J 1975 A contribution to the understanding of the ethology of the cichlids of southern Africa Ph D Thesis, Rhodes University Grahamstown 211 pp

Ribbink, A J & V Twentyman-Jones 1989 Captive propagation as a conservation tool pp 145–150 *In* M -D Crapon de Caprona & B Fritzsch (ed) Ann Musce Royal de L'Afrique Centrale 57, Tervuren

Skelton, P H , M N Bruton, G S Merron & B C W van der Waal 1985 The fishes of the Okavango drainage system in Angola, South West Africa and Botswana taxonomy and distribution Ichthyological Bulletin of the J L B Smith Institute of Ichthyology 50 1–21

Skelton, P H 1987 South African red data book – fishes South African National Scientific Programmes Report 137 1–199

Skelton, P H 1990 The status of fishes from sinkholes and caves in Namibia Namibia Scientific Society 42 75–83

Trewavas, E 1973 II A new species of cichlid fishes of rivers Quanza and Bengo, Angola with a list of the known Cichlidae of these rivers and a note on *Pseudocrenilabrus natalensis* Fowler Bull Br Mus nat Hist 25 27–37

Trewavas, E 1983 Tilapiine fishes of the genera *Sarotherodon, Oreochromis* and *Danakilia* British Museum (Natural History), London 583 pp

Van Coevering, J A H 1982 Fossil cichlid fishes of Africa Spec pap palaeont Ass 29 1–103

Whitfield, A K 1980 A checklist of the fish species recorded from Maputaland estuarine systems pp 204–229 *In* M N Bruton & K H Cooper (ed) Studies on the Ecology of Maputaland Rhodes University Grahamstown

Environmental Biology of Fishes **41**: 330, 1994.

Fish imagery in art 70: Polynesian Maori carving of the Maui myth

Robert M. McDowall
National Institute of Water and Atmospheric Research, P.O. Box 8602, Christchurch, New Zealand

New Zealand's Polynesian Maori had no written history, though many legends were recorded in stylised wood carvings in meeting houses, fortifications and other structures. A myth widespread among Polynesian peoples, including New Zealand Maori, relates to a man named Maui, to whom legend attributes many miraculous feats. One of these myths tells how Maui was in a canoe, fishing with his brothers. They would not let him use their gear, but Maui happened to have his grandmother's jawbone with him, which he used as a hook. He hooked a tremendous fish, and when he hauled it to the surface it became the North Island of New Zealand, still known to Maori as Te Ika a Maui (Maui's fish). Some traditions say that the canoe was stranded on a mountain, but later slid into the sea, to become New Zealand's South Island (Buck 1949).

This myth is depicted here by Maori carver Riki Manuel. The wood carving, of kauri (*Agathis australis*), comprises traditional images in base relief, showing Maui holding a line connected to a bone-carving hook in the fish's mouth. Maui's eyes are small paua (abalone, *Haliotis australis*). Weaving in behind the figure and including traditional Maori motifs, is Maui's fish (Te Ika a Maui) painted in modern paints, and in blue and green, pigments that were not available to traditional, Maori. The fish is mythical and cannot be identified with known species. So this modern (1990s) carving uses both traditional Maori images and materials and contemporary paint pigments to portray one of the classical myths of New Zealand Maori mythology. The carving, now owned by the author, is 45 by 22 by 5 cm.

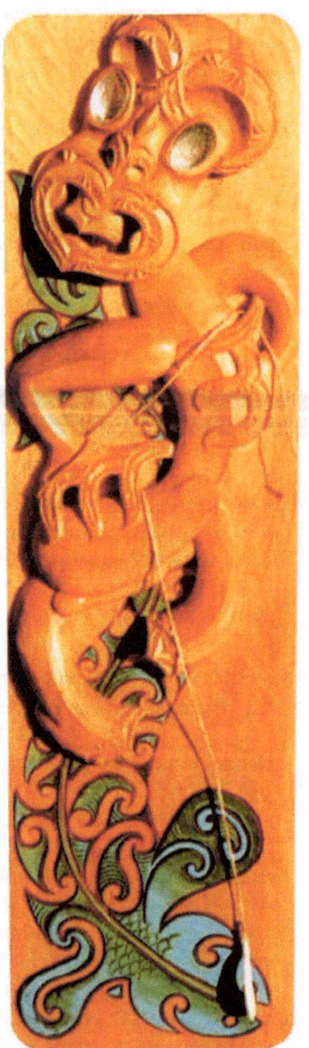

Buck, P.H. 1949. The coming of the Maori. Whitcombe and Tombs, Christchurch. 551 pp.

Environmental Biology of Fishes **41**: 331–367, 1994.

Alternative life histories of the genus *Lucania:* 2. Early ontogeny of *L. goodei,* the bluefin killifish

Stephen S. Crawford & Eugene K. Balon
Institute of Ichthyology and Department of Zoology, University of Guelph, Guelph, Ontario N1G 2W1, Canada

Received 24.11.1993 Accepted 30.3.1994

Key words: Development, Saltatory ontogeny, Ecomorphology, Cleavage, Embryo, Larva, Juvenile, Cyprinodontidae, *L. parva*

Synopsis

This is the second of three papers devoted to the interpretation of morphological development in altricial and precocial species within the genus *Lucania.* The focus of this paper was the early life history of the bluefin killifish, *Lucania goodei.* Reproductively mature specimens were collected in the run below Newport Spring, north of St. Marks National Wildlife Refuge in northwest Florida. These specimens were transported to the laboratory, where they served as brood stock for specimens described in this study. Offspring were reared under controlled conditions and were described according to the theory of saltatory ontogeny, which gives a sampling design based on morphological, rather than chronological, progression. The morphological development of these offspring is described on the basis of detailed illustrations, photomicrographs, and measurements of mensural and meristic characters. This account of early ontogeny, in combination with a corresponding study on the early ontogeny of the rainwater killifish, *L. parva* (Crawford & Balon 1994a), establishes the empirical basis for an altricial-precocial life history model (Crawford & Balon 1994b).

Introduction

Two species are currently recognized in the North American genus *Lucania:* the rainwater killifish, *L. parva* (Baird, 1855) and the bluefin killifish, *L. goodei* Jordan, 1880 (Hubbs & Miller 1965, Lee et al. 1980). Some other details of the taxonomic relationships within this genus are presented in the study of early ontogeny in *L. parva* by Crawford & Balon (1994a).

Lucania goodei is largely confined to the inland waters of Florida (Hubbs et al. 1943, Lee et al. 1980, Loftus & Kushlan 1987), in a manner similar to many other fish species that are endemic to Florida (Briggs 1958). *Lucania goodei* is almost always found in freshwater creeks, rivers and lakes; often close to hypoxic springs of the Floridan Aquifer (Tagatz 1968, Dineen 1974). Occasionally, they have been collected in mildly brackish waters, usually during a flush of fresh water (Hubbs et al. 1943, Kilby 1955, Tabb & Manning 1961, Loftus & Kushlan 1987).

Very little information is available on the early ontogeny of *L. goodei.* Kuntz (1916) published brief notes and a few illustrations of *L. goodei* embryos and larvae, with comments on the general similarity between *L. goodei* and *L. parva.* Foster (1967) also presented general notes and illustrations of a few developmental states of *L. goodei,* in the context of a family-wide comparison of ecology and behavior. Veenstra (1987) compared the early development of *L. goodei* and *L. parva;* however, he focused primarily on mensural and meristic measures of devel-

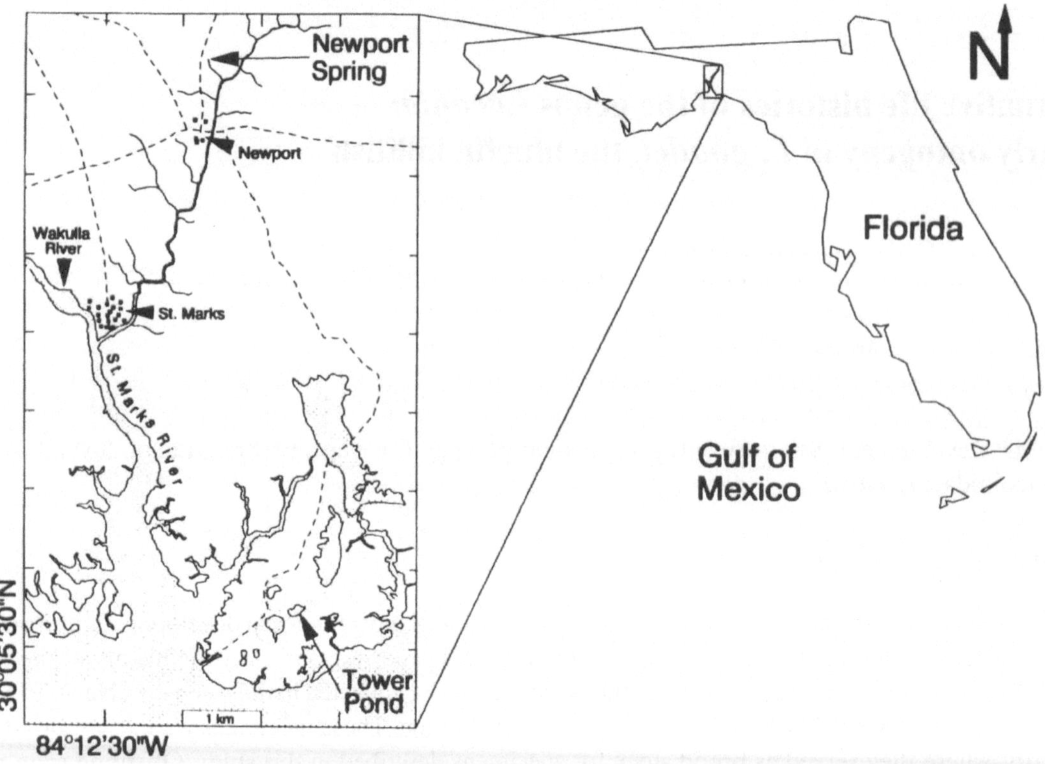

Fig. 1. Sampling locations for wild, adult *Lucania goodei* at Newport Spring, Florida. Wild *L. parva* were sampled at Tower Pond, St. Marks National Wildlife Refuge (see Crawford & Balon 1994a, b).

opment, rather than providing an interpretive account supported by extensive descriptions.

The specific objective of this study was to give a detailed description of the early ontogeny of *L. goodei*. This description is the second instalment of a three-part, comparative series on the early ontogeny of fishes in the genus *Lucania*. The first instalment of the series was an investigation of the early ontogeny in *L. parva* (Crawford & Balon 1994a) collected from a location near the source of *L. goodei* for this study. The third and final instalment of the series (Crawford & Balon 1994b) focusses on selected information provided in the first two papers and brings them together for the purpose of comparison.

Material and methods

The techniques involved in this study were identical to those described in the paper on *L. parva* ontogeny (Crawford & Balon 1994a). Thus, for the sake of brevity, we will repeat only the most general informations here, and direct interested readers to the first instalment of the series for additional details.

Adult specimens of *L. goodei* were collected with a Deka 3000 electrofisher (Deka-Gerätebau, Rudolf Mühlenbein, Germany) from the creek below Newport Spring, Florida (30° 12′ 30″ N, 84° 10′ 30″ W, Fig. 1, 2). Fifty adult specimens (25 males, 25 females, Fig. 3) were collected between 2–7 March 1990 and transported live to the laboratory to produce offspring for the study of morphological development. An additional fifty specimens (25 males, 25 females) were collected in order to describe the morphology of wild adults.

Individual females were transferred to 31 l holding aquaria (50 × 25 × 25 cm deep). Individual males were transferred to 21 l breeding aquaria (35 × 20 × 30 cm deep) that were held in a large water bath (118 × 43 × 45 cm deep) with circulating water heated to 25 ± 1° C. All aquaria in this study

Fig 2 Photographs of the habitat at Newport Spring in March 1990 a – collections of *L goodei* were obtained by electrofishing the run below the small waterfall at the dam (approximately 0 5 m height), b – most of the fish were collected at the edge of submerged vegetation, over a sandy substrate

Fig 3 Photograph of adult *L goodei* collected from Newport Spring, Florida female (above) and male (below)

were maintained at 25 ± 1° C under a 16 h : 8 h (light : dark) photoperiod with 2.5 ppt brackish water created by mixing filtered well-water (Hodson & Sprague 1975) and Instant Ocean salts. Twice daily, adults were fed to satiation with Murex frozen brine shrimp, *Artemia salina.*

Individual females were arbitrarily selected for exposure to the isolated males. Each of the pairs was allowed physical contact for 30 min between 13:00 and 15:00 h. During this period, all mating clasps were timed to the nearest minute. All matings took place between 1 April and 31 August 1990. Activated eggs were transferred to individual 1 ml meshed baskets suspended in an aerated 9.8 l incubation aquarium (42 × 26 × 9 cm deep) heated by a Braun Thermomix 1420 heater to 25 ± 0.5° C. The baskets were checked three times daily for decomposing eggs which were discarded, and for newly hatched embryos which were transferred to larger 100 ml mesh baskets suspended in a 48 l incubation aquarium (30 × 28 × 58 cm deep) that was heated to 25 ± 1° C. Twice daily, offspring in these larger incubator baskets were presented with an excess of freshly-hatched, live brine shrimp nauplii.

Randomly-selected offspring were removed from the incubator aquaria at sampling ages according to the schedule given in Table 1. This sampling design was based on the appearance of morphological structures and/or behaviours that characterize the various steps and periods of fish ontogeny (Balon 1985, 1986). Using this life-history model, we were able to distinguish between 'morphological state' (i.e. progress of structural complexity) and 'chronological state' (i.e. age in units of time). In all cases, the developmental age of specimens is given in the format days : hours : minutes.

Drawings were made from a dorsal perspective using a Zeiss Stemi SV8 stereomicroscope with a mounted drawing tube, and from a variety of lateral perspectives with a C&D Scientific Instruments projection microscope. Photomicrographs were taken of specimens (B&W Ilford FP4 125 ASA, Color Ektachrome ET160 slides) using Minolta SRT 200 and XE 5 35 mm cameras on the Zeiss and/or C&D microscopes. Verbal and illustrative descriptions of each step were made on the basis of composites drawn from drawings and photographs of live specimens, as well as cleared and stained specimens. All preserved specimens, in buffered 10% formalin solution from the beginning and middle of each developmental step, were cleared and stained following the methods described by Balon & Flegler-Balon (1985).

Table 1 Classification of developmental periods, phases and steps of *Lucania goodei* based on the appearance of morphological and behavioural characters (Cunningham & Balon 1985) Sampling intervals were maintained until characters of the next step were recognized

Period	Phase	Step	Sampling interval	Morphological characters
Embryo	Cleavage egg	C1	15 min	- unactivated egg - perivitelline space - blastodisc
		C2	30 min	- cleavage
		C3	1 h	- epiboly - embryonic shield
	Embryo	E1	1 h	- optic vesicles - germ ring closure
		E2	2 h	- somites - blood elements - otic vesicles - heart
		E3	2 h	- blood circulation - vitelline network - pectoral fins
		E4	4 h	- segmental vessels - hepatic vein - olfactory vesicles
		E5	4 h	- caudal finfold circulation - dorsal/anal finfolds - caudal rays
		E6	8 h	- subclavian vein - gut - profundal vein
	Free embryo	F1	8 h	- hatching - swimbladder
Larva		L1	12 h	- first feeding
		L2	12 h	- dorsal/anal rays - ribs
		L3	24 h	- pelvic fins
		L4	24 h	- scales
Juvenile	Juvenile	J		- finfold resorption

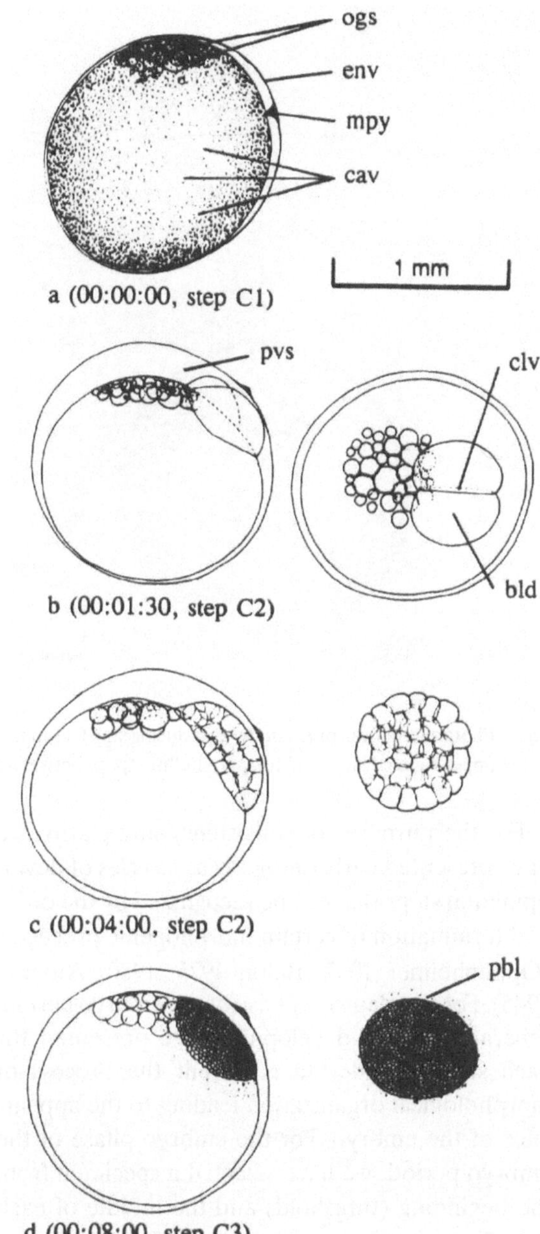

Fig 4 Illustrations of cleavage phase *L goodei* specimens, lateral (left) and dorsal (right) perspectives a – C1 unactivated egg, b – C2 2-cell cleavage egg, c – C2 16-cell cleavage egg, d – C3 cleavage egg at onset of epiboly (bld = blastodisc, cav = cortical alveoli, clv = cleavage furrow env = egg envelopes mpy = micropyle, ogs = oil globules pbl = periblast, pvs = perivitelline space)

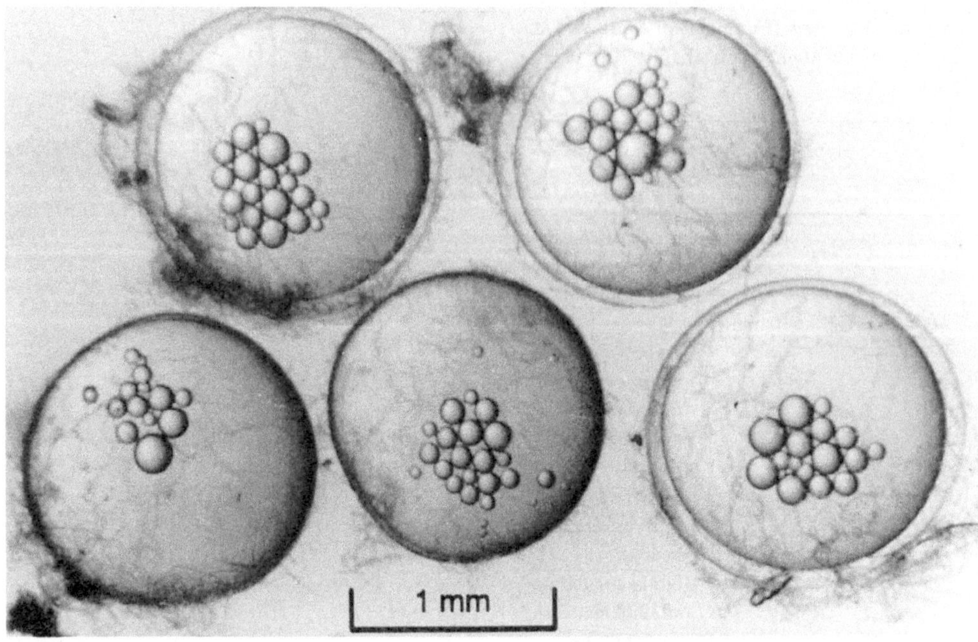

Fig 5 Photomicrograph of *L goodei* eggs during step C1 (approximately 00 00 05 after release) from a dorsal perspective The three eggs in the upper right corner exhibit a perivitelline space between the inner cytoplasm and the outer egg envelopes, indicating activation

For the purposes of consistency and clarity, we have presented early ontogeny as a series of developmental steps that can be recognized by the onset and termination of certain morphogenic processes (Oppenheimer 1937, Balon 1975, 1980, Alberch 1985). For the cleavage phase of the embryo period, several stages of development are presented for each step, in order to represent the process of morphological organization leading to the appearance of the embryo. For the embryo phase of the embryo period, we have selected a specimen from the beginning (threshold) and the middle of each step. In most cases, important morphological characters of these specimens have been illustrated from both a lateral and dorsal perspective. For the larva period, specimens from the beginning and middle of each step are also presented, but these illustrations were all made from the lateral perspective. Except where indicated, quantitative measures are expressed in the text as means (minimum-maximum), and these values were taken from 25 specimens sampled for each of the steps in the series.

It should be noted that our description of early ontogeny in *L. goodei* has been organized to facilitate comparison with our description of early ontogeny in *L. parva* (Crawford & Balon 1994a). For this reason, we decided to present all figures and tables in the same order as the previous paper – despite the awkwardness of some figure citations in this text.

Results

Courtship and spawning

Ovarian eggs collected from wild adult females were in a variety of oogenic states. Wild females contained an average of 54.1 vitellogenic ova. Regression analysis indicated a weak, but statistically significant ($p < 0.05$), linear relationship between the number of ovarian eggs and standard length of *L. goodei* females (# eggs = 2.5 (SL mm) – 29.8, r^2 = 0.285).

Spawning behaviour of *L. goodei* under laboratory conditions was very similar to that described for

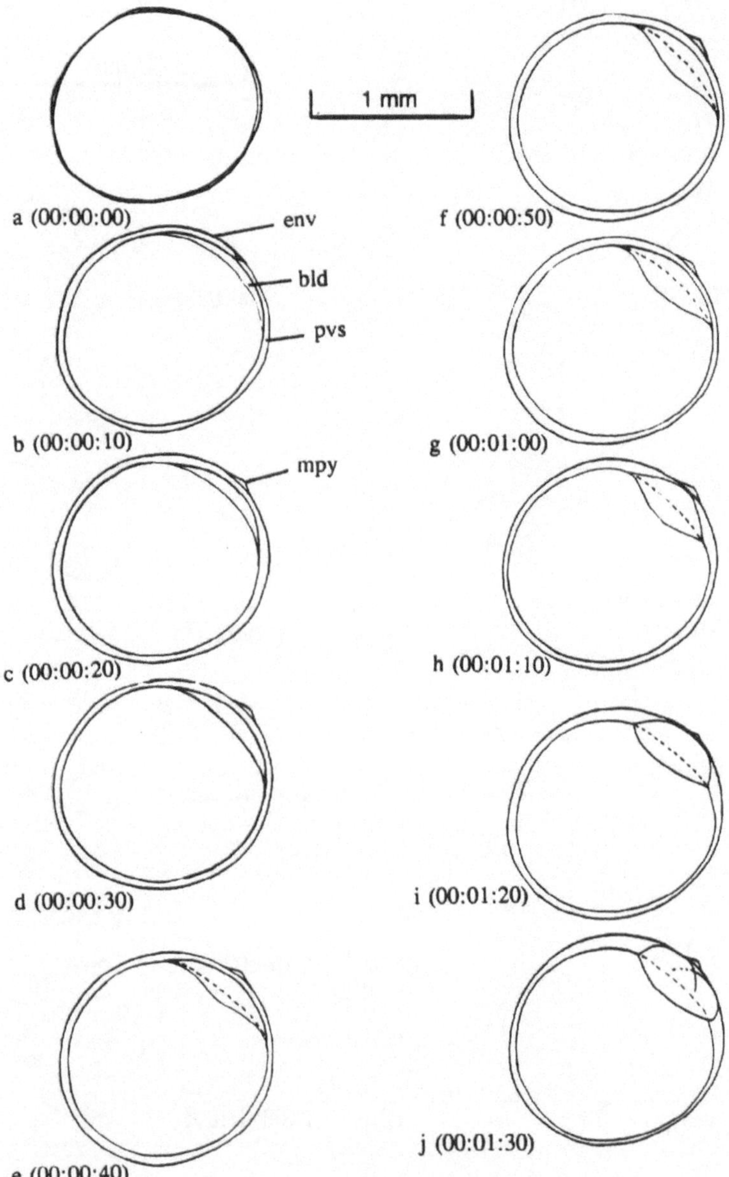

Fig 6 Illustrations of *L goodei* blastodisc formation during step C1, from time-lapse microphotographs of a single specimen (lateral perspective) Photographs were made every 10 or 20 minutes until the onset of blastodisc cleavage (step C2) (bld = blastodisc, env = egg envelopes, mpy = micropyle, pvs = perivitelline space)

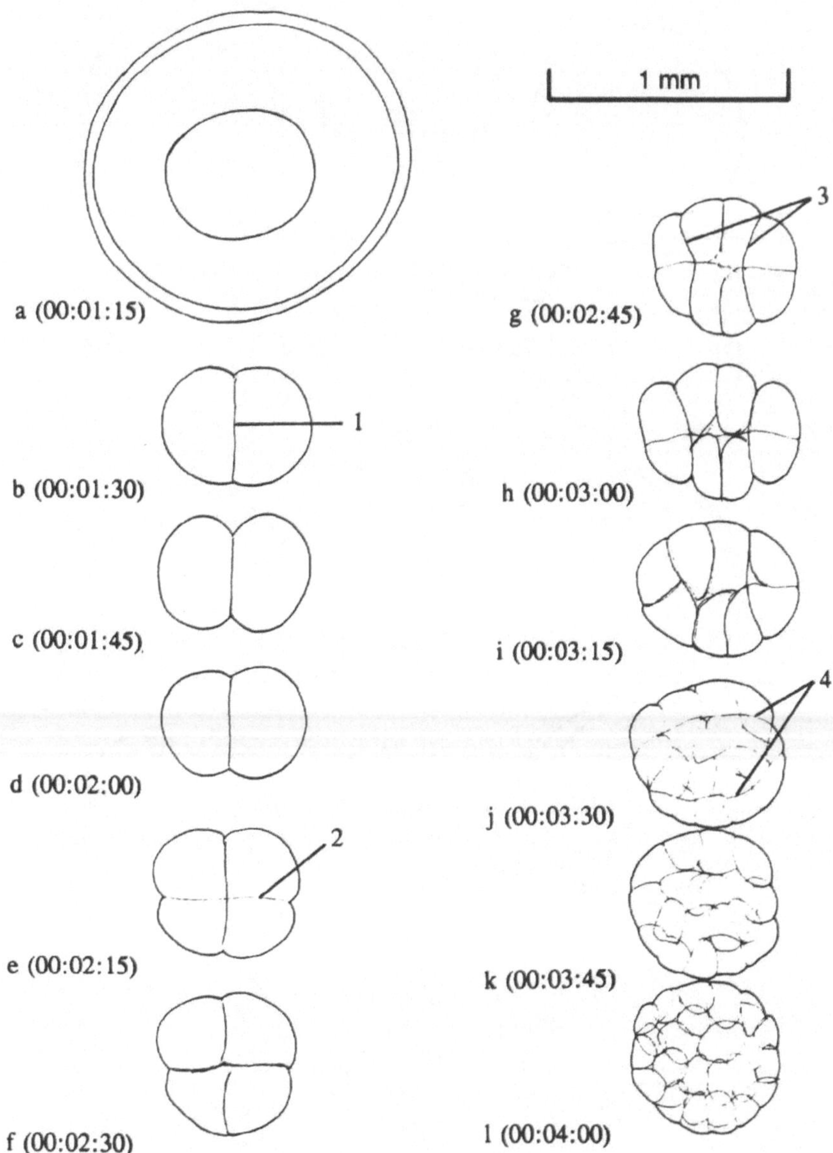

Fig. 7. The first four cleavages of a *L. goodei* blastodisc during step C2, taken from time-lapse microphotographs of a single specimen (dorsal perspective). Egg envelopes and yolk are shown only in the egg at 00:01:15 (1, 2, 3, 4 = first four cleavage furrows).

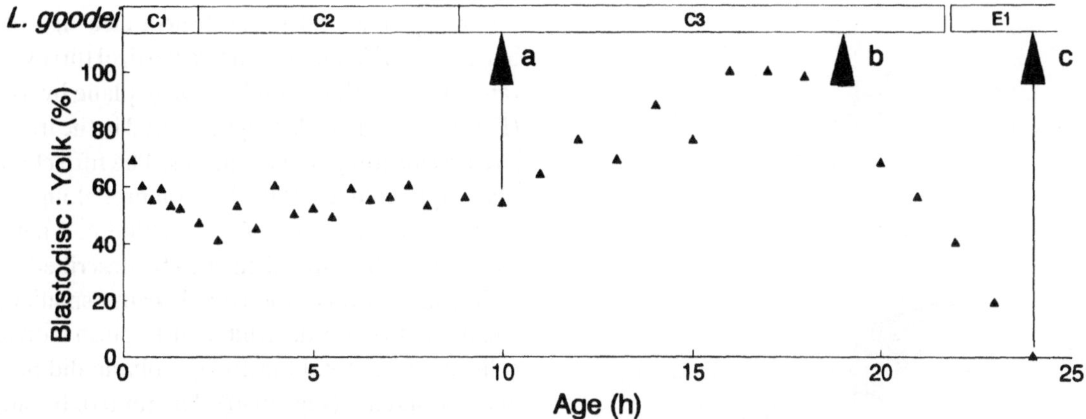

Fig. 8. Changes in *L. goodei* blastodisc width:yolk width ratio during the steps of cleavage phase (top bar) measured as a function of age (a = onset of gradual blastodisc expansion during the first half of epiboly, b = onset of rapid blastodisc expansion during second half of epiboly, c = completion of epiboly).

L. parva by Crawford & Balon (1994a) and for many members of the family Cyprinodontidae (Foster 1967). One clear distinction between *L. goodei* and *L. parva* was the relatively non-aggressive behaviour of *L. goodei* males during interactions with conspecifics.

Female *L. goodei* did not release all of their eggs at one time, but rather released clutches of eggs during a series of mating events. Female *L. goodei* spawning under laboratory conditions released an average of 6.8 eggs per clutch. There was no significant ($p > 0.05$) linear relationship between clutch size and maternal standard length.

Description of ontogeny

Period: Embryo
Phase: Cleavage egg
Step: C1 (00:00:00–00:01:30)

Unactivated eggs sampled from female *L. goodei* averaged 1.316 (1.22–1.44) mm in diameter, with an average dry mass of 0.300 (0.25–0.35) mg. The egg envelopes fit quite closely to the oocyte membrane (Fig. 4a), with an outer envelope that was sparsely covered with very adhesive filaments (Fig. 5). These filaments have not been included in the drawings in order to enhance the clarity of embryonic structures. A single, inverted micropyle was located in the dorsal hemisphere of the egg envelope, with a micropylar papilla that appeared to contact the surface of the cytoplasm. Numerous cortical alveoli covered the surface of the translucent yolk-cytoplasm matrix which contained an average of 22.8 (12–39) oil globules with a mean diameter of 0.104 (0.03–0.24) mm (Fig. 4a). Undisturbed eggs would usually roll until the oil globules aggregated within the viscous matrix at a polar antigravity location, with the micropyle slightly off center.

Activation of *L. goodei* eggs was recognized by the formation of a perivitelline space, caused by separation of the egg envelopes from the oocyte membrane (Fig. 4b, 5). An average of 61.2 (12–100)% of all eggs in the *L. goodei* clutches formed a perivitelline space. By 00:00:20, activated eggs had acquired a transparent appearance as the cortical alveoli erupted during formation of the perivitelline space (Fig. 5, 6).While the perivitelline space was forming, the cytoplasm of the *L. goodei* egg accumulated beneath the micropyle to form a blastodisc (Fig. 6b–i). As the height of the blastodisc increased and its width decreased, the cytoplasm took on the shape of a biconvex cap (Fig. 6j). Using the equations described by Crawford et al. (1994), the average proportion of blastodisc volume to yolk volume was estimated to be 3.58 (1.6–9.1)% at the end of step C1.

Step: C2 (00:01:30–00:08:00)

Cleavage in *L. goodei* eggs began at 00:01:30 with meroblastic division of the blastodisc (Fig. 7). The

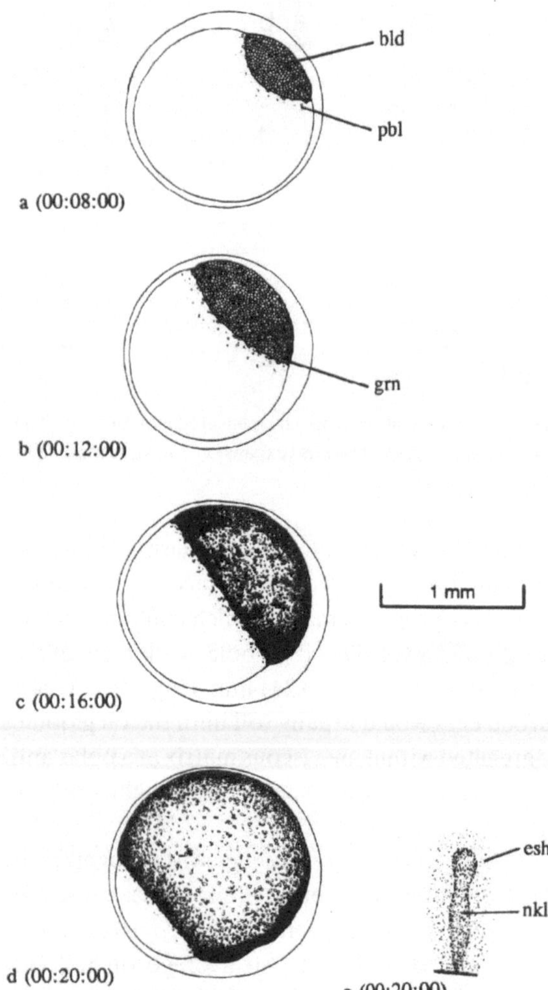

Fig 9 Epiboly and formation of the embryonic shield in *L. goodei* a–d – lateral perspective, e – ventral perspective (bld = blastodisc, esh = embryonic shield, grn = germ ring, nkl = neural keel, pbl = periblast)

first cleavage furrow divided the cytoplasmic mass into two cells of approximately the same size (Fig. 4b, 7b), although these cell volumes were not measured. This cleavage occurred in the vertical (meridional) plane relative to the base of the blastodisc. The second cleavage furrow formed at 00:02:15, in a vertical plane that was perpendicular to the first cleavage furrow (Fig. 7e). The third cleavage of the blastodisc took place at 0:02:45, with two cleavage furrows forming parallel to the first plane of cleavage. The 8-celled blastodisc was elongated along the plane of second cleavage, while retaining bilateral symmetry (Fig. 7f–h). The fourth cleavage of the blastodisc took the form of two vertical furrows that ran parallel to the second cleavage plane by 0:04:00 (Fig. 4c, 7i–l). This cleavage resulted in the irregular size and location of blastomeres. The fifth cleavage plane appeared to cut the blastomeres along a horizontal plane, however, the resulting cell patterns were not sufficiently distinct to be described.

During cleavage, the cells became smaller and smaller until it was difficult to distinguish individual cells, however total blastodisc volume did not appear to increase (Fig. 4b–d). The ratio of blastodisc width to yolk width remained at approximately 60% throughout step C2 (Fig. 8). Approximately 6 hours after the onset of activation, the blastodisc began to flatten over the surface of the yolk. By 00:08:00 the blastodisc had obtained the biconvex shape of an inverted saucer (Fig. 4d).

Step: C3 (00:08:00–00:22:00)

By the age of 00:08:00, most *L. goodei* embryos exhibited the first indications of epiboly, with the appearance of the periblast; a band of cytoplasm and nuclei that spanned the circumference of the blastodisc edge (Fig. 9a).The rim of the blastodisc also thickened to form a germ ring that moved out over the surface of the yolk (Fig. 9b). Over the next 4 hours, an expanding band of nuclei in the periblast continued to move ahead of the advancing germ ring. By approximately 00:16:00 the germ ring had reached the yolk equator, and the periblast had been reduced in width (Fig. 9c). The final four hours of epiboly brought about rapid changes in egg morphology. First, the migration of cells over the yolk caused a thinning in the central region of the blastoderm. Many of these cells began to concentrate along a meridional axis to form a thickened plate of blastoderm; the embryonic shield. By 00:20:00 a dense neural keel of cells had settled down along the midline of the embryonic shield (Fig. 9d, e). During the latter stages of germ ring closure over the yolk, the first indications of the embryo were recognized. Cytoplasm was bulging to form a yolk plug and the embryonic shield appeared to have two regions of swelling; one smaller node at the trailing edge of the shield and one longer region that ended at the margin of the germ ring. By this

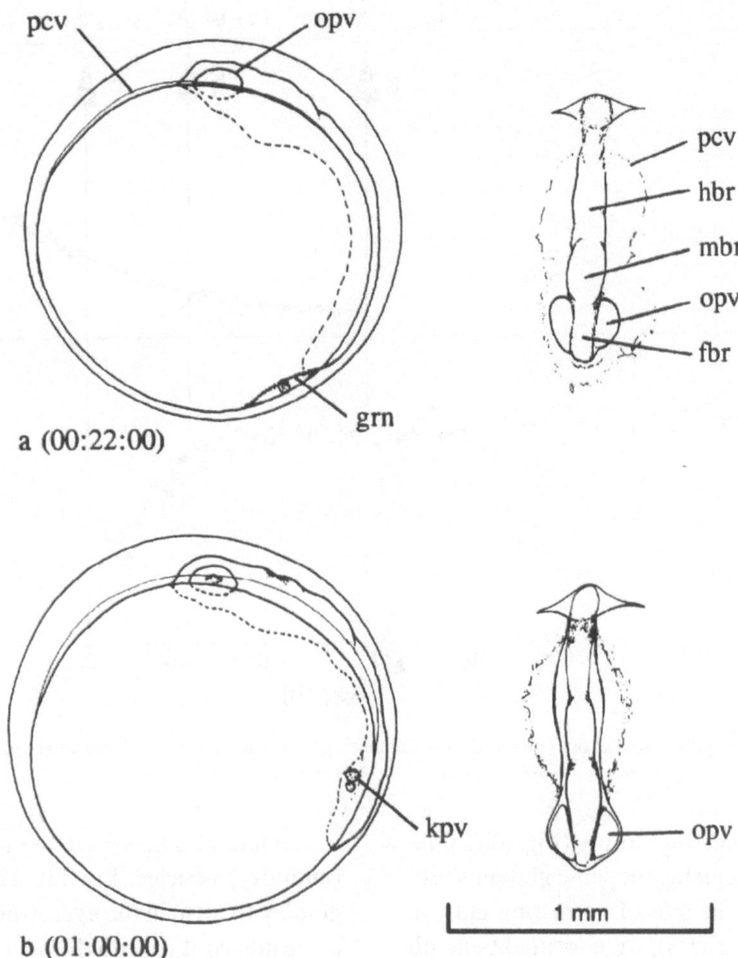

Fig 10 Embryos of *L. goodei* during step E1 from lateral and dorsal perspectives (fbr = forebrain, grn = germ ring, hbr = hindbrain, kpv = Kupffer's vesicle, mbr = midbrain, opv = optic vesicle, pcv = pericardial cavity)

time, the oil globules of most specimens were fixed in dispersed locations throughout the yolk (not shown in drawings). The process of epiboly had taken 12 hours to reach this stage, however the speed of cellular migration during the second half was much greater than during the first half (Fig. 8a–b, b–c).

Phase: Embryo
Step: E1 (00:22:00–01:02:00)

The embryo of *L. goodei* was distinguishable 22 hours after activation. This observation was based on the formation of primary neural rudiments at the anterior end of the body (Fig. 10a). At this age, the embryo still extended to the germ ring, which had covered most of the yolk mass. Slight constrictions in the head region identified the differentiation of the embryonic brain, and paired optic vesicles were observed as lateral outgrowths of the forebrain walls. From a lateral perspective, the body lay in a depression within the yolk, and the head was twice the depth of the tail. A thin layer of ectoderm (the presumptive pericardium) extended from anterior of the head, around the embryo laterally, to the blastoderm on the surface of the yolk. Numerous, small pigment cells had formed along the edge of this layer and were scattered over the yolk surface. Total length of the embryo was approximately one quarter the circumference of the yolk. Epiboly had been completed in most specimens by the age of 01:00:00, with the posterior tip of the embryo situa-

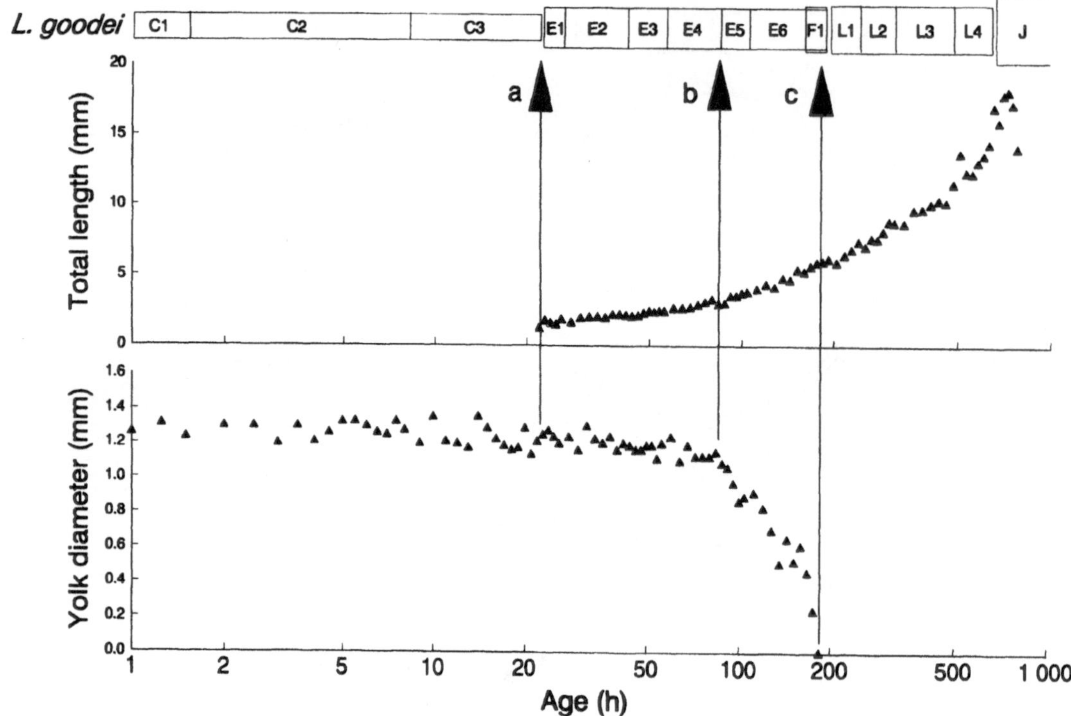

Fig. 11. Total body length and yolk diameter of *L. goodei* during the cleavage egg, embryo, free embryo and larva steps (top bar) measured as a function of logarithmic age (a = first distinction of embryo, b = increase in yolk utilization, c = exhaustion of yolk supply).

ted over the site of germ ring closure (Fig. 10b). One or two globules, presumably the same globules previously observed at the site of germ ring closure (perhaps Kupffer's vesicles), were located beneath the tail. From a lateral perspective, the pericardial wall had lifted and expanded anteriorly. By this age, the embryo had grown in length to one third the circumference of the yolk. There was little utilization of the yolk during this step (Fig. 11a), a trend that would continue until the vitelline transport pathways began to form.

Step: E2 (01:02:00–01:18:00)

Somite formation in *L. goodei* was first detected at the age of 01:02:00 (Fig. 13a, 14a). The somites formed at the midpoint along the middle embryonic axis, and were added to the posterior end of the series. A well defined notochord extended from the hindbrain to the tail, which was beginning to grow at a more rapid rate than in the previous step.

Within two to six hours of the first somites, the embryos exhibited lateral thickenings of tissue between the hindbrain and the most anterior somites. These lateral plates were the rudiments of the otic (auditory) vesicles (Fig. 13b, 17a). Lenses were beginning to form in the eyes, where the optic vesicles came into contact with the overlying tissue. A newly formed brain region lay between the optic lobes and extended into the forebrain where it was flanked by a pair of olfactory vesicles.

The distribution of melanophores had expanded over the embryo and yolk by 01:10:00 (Fig. 16). These cells were distributed sparsely over the surface of the yolk and body of the embryo. The heart was recognized as a small cone between 01:10:00 and 01:14:00, during a rapid increase in somite number. Soon after it appeared, the heart began to beat sporadically. Body growth decelerated somewhat during the next 4 hours, as blood elements began to accumulate anterior of the heart and in the tail where the posterior vitelline vein would exit the body. By the end of step E2, the *L. goodei* embryos

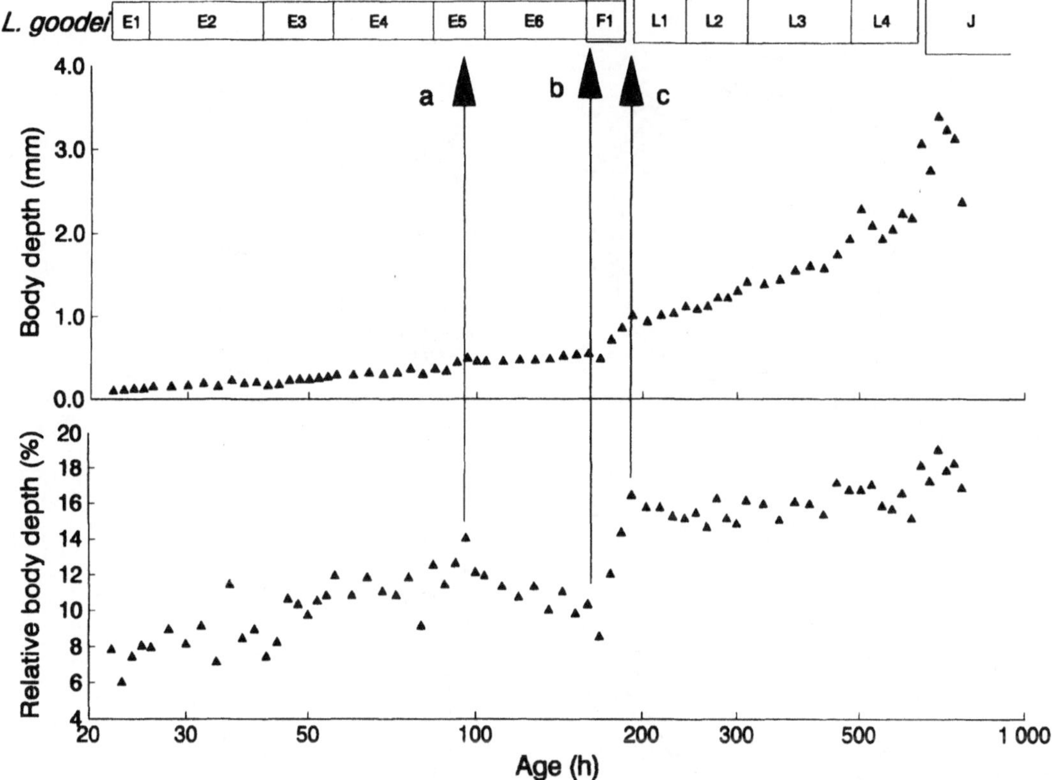

Fig. 12. Absolute and relative (% TL) body depth of *L. goodei* during the embryo, free embryo and larva steps (top bar) measured as a function of logarithmic age (a = cessation of growth in body depth leading to a decline in relative body depth, b = onset of rapid increase in absolute and relative body depth, c = achievement of definitive relative body depth).

exhibited approximately half the full complement of somites (Fig. 14).

Step: E3 (01:18:00–02:08:00)

At 01:18:00 in the embryonic development of *L. goodei,* clumps of blood cells began to pulse back and forth along the dorsal aorta and out to the yolk surface at the site of the most posterior somites (Fig. 18a). The heart exhibited a bend from dorsal left to right, with a constriction in the middle of the bend. This constriction distinguished the anterior atrium and posterior ventricle, which beat in biphasic contractions without apparent flow of blood. The embryonic brain had continued to differentiate, especially the conspicuous fourth ventricle with a sharply defined rhombic lip at its anterior margin. Melanophores appeared to be concentrated posterolateral to the head at this age.

Within four hours, blood elements were observed to move from the head through the paired dorsal arteries and along the common dorsal artery to a location near the posterior somites (Fig. 18b). However, since several additional somites had formed, the blood flow had to loop from the posterior somites via the caudal vein, back to the original site of blood departure. Blood flowed out onto the yolk surface through the posterior vitelline vein, branching out to form a relatively simple network of capillaries that was centered on the anterior surface of the yolk. The capillaries of the vitelline network rejoined to form the sinus venosus, a collecting vessel immediately anterior to the heart. As the blood vessels formed over the yolk surface, new branches seemed to extend between clusters of pigment cells.

By 02:02:00, the heart began to beat regularly (approx. 100 beats min^{-1}), pumping more blood as the body grew and additional somites formed (Fig. 18c). The caudal vein extended posteriorly over 12 somite pairs to feed the vitelline network which had expanded in flow, number of branches, and cover-

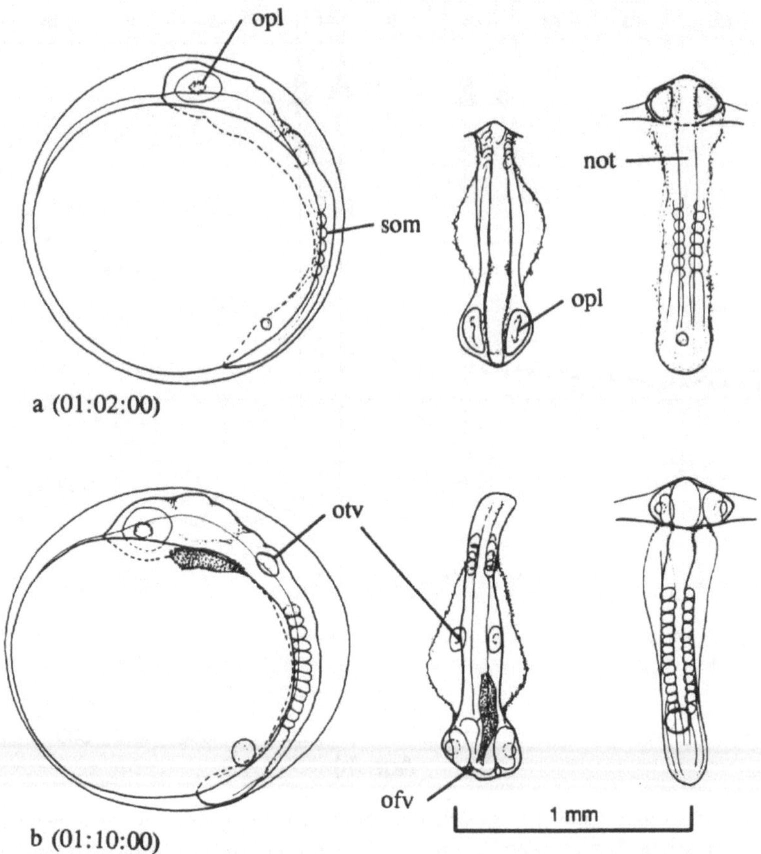

Fig 13 Embryos of *L goodei* during step E2 from lateral (left), dorsal (center) and posterior (right) perspectives (not = notochord, ofv = olfactory vesicle, opl = optic lens, otv = otic vesicle, som = somite)

age of the yolk surface. Left and right anterior vitelline veins also carried blood from the cardinal veins, departing the body immediately posterior of the otic vesicles and the newly formed pectoral fin rudiments. These anterior vitelline veins were observed to either feed the sinus venosus directly, or to join the existing branches of the vitelline network. Larger melanophores covered the yolk surface, while smaller pigment cells were scattered throughout the head and tail (Fig. 19, 20). At the end of step E3, *L. goodei* embryos averaged 2.371 (2.18–2.50) mm in total length and were approaching the full complement of 28–29 body segments (Fig. 14).

Step: E4 (02:08:00–03:12:00)

Embryos of *L. goodei* at age 02:08:00 exhibited an increase in the flow of blood through the vitelline capillary network (Fig. 21a). The intestinal loop of the caudal vein remained at the 10th somite position, the same region of the tail where the first segmental vessels were observed carrying blood from the dorsal aorta, around the somites and back to the caudal vein. There was no rigid pattern to the direction of segmental flow, as individual vessels carried blood in anterior, posterior and/or lateral directions.

The extent of the vitelline network reached a maximum within 8 hours, as a greater volume of blood was produced by the embryo and shunted from the body out over the yolk. The flow of blood through the left anterior vitelline vein was augmented by the appearance of a hepatic vein at age 02:16:00 (Fig. 21b). This vein drained the future site of gut and liver formation and suggested the elaboration of tissue within these organs. Pigmentation clusters were deposited along the larger vitelline

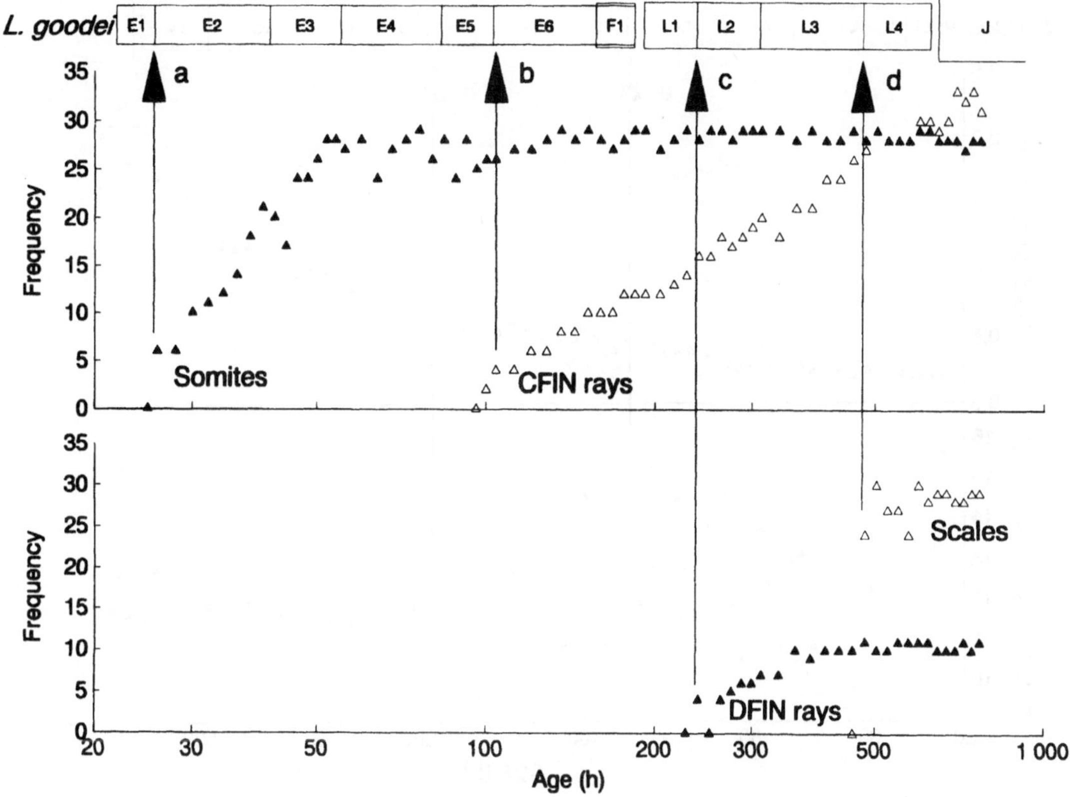

Fig. 14. Frequency of somite pairs, caudal fin (CFIN) rays, dorsal fin (DFIN) rays and vertical scale rows of *L. goodei* during the embryo, free embryo and larva steps (top bar) measured as a function of logarithmic age (first distinction of a = somites, b = caudal fin rays, c = dorsal fin rays, d = scales).

circulatory vessels, especially around the sinus venosus and the origin of the left and right vitelline veins, both of which were darkly pigmented (Fig. 24).

The notochord consisted of vacuolated, disc-shaped cells arranged in a regular, linear manner. By 03:08:00, the spleen rudiment was recognized as a red sphere observed within the embryonic body, immediately adjacent to the hepatic vein. Spasmodic body contractions were more frequent and more intense than during the last step, involving progressively more somites toward the posterior region of the tail. By the end of step E4, the embryos averaged 2.95 (2.6–3.3) mm in length, approximately 60% of yolk circumference. The latter stages of step E4 in *L. goodei* marked the beginning of dramatic increases in yolk utilization, as measured by yolk diameter (Fig. 11b).

Step: E5 (03:12:00–04:08:00)

Lucania goodei embryos at age 03:12:00 exhibited blood circulation in the caudal finfold, a feature that is characteristic of step E5 (Fig. 25a). The loop in the caudal finfold was initially a simple snare which was twisted over itself before returning to feed the caudal vein. Growth of the head had drawn the rounded pectoral fins to a location anterior of the anterior vitelline veins, and had given a maximum relative head depth for both embryo and larva periods (Fig. 15a). The first sign of the liver was observed as a green nodule within the embryonic body, anterior and ventral to the spleen. Otoliths appeared as one or two dark granules in the otic vesicles. An increase in the number and size of melanophores had occurred on the eyes, posterior to the head and to a lesser extent, throughout the tail.

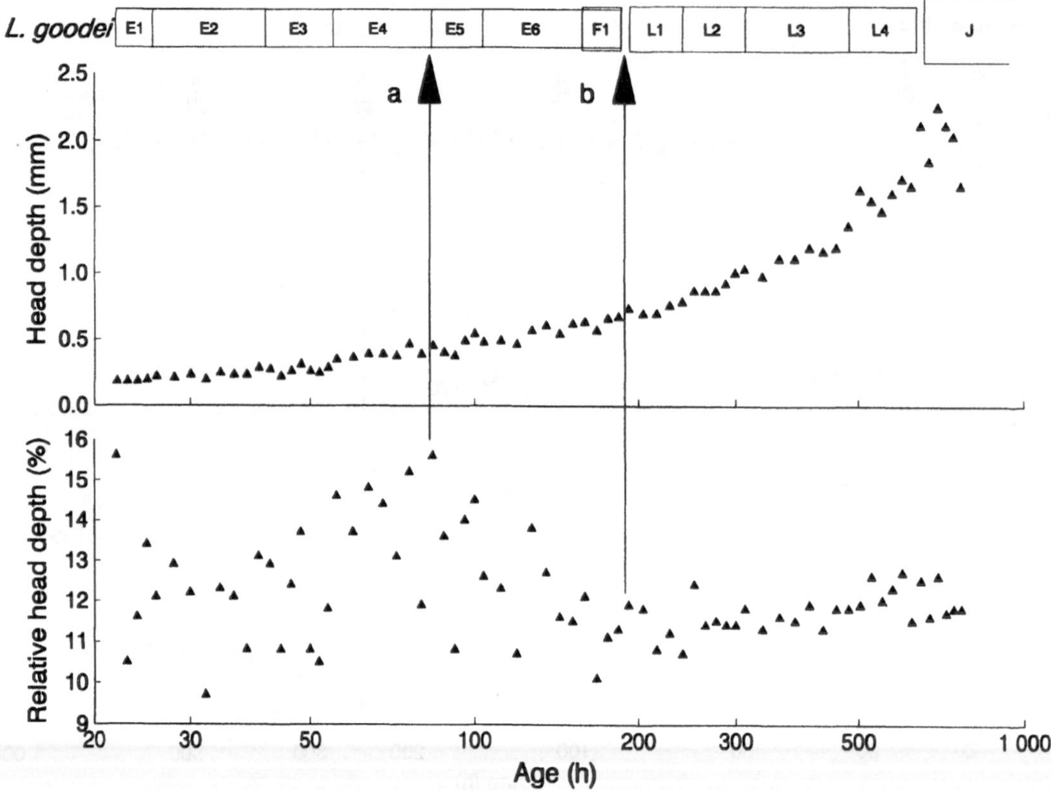

Fig. 15. Absolute and relative (% TL) head depth of *L. goodei* during the embryo, free embryo and larva steps (top bar) measured as a function of logarithmic age (a = maximum relative head depth, b = achievement of definite relative head depth).

There were a few small melanophores scattered along the pectoral fins.

By 04:00:00 the caudal snare had bifurcated into two loops that curled dorso-ventrally from the tip of caudal artery before rejoining as the caudal vein (Fig. 25b). In addition, a new blood vessel, the coeliac-mesenteric artery had appeared just posterior to the exit of the perivitelline veins, draining blood directly from the caudal artery. The tail itself had continued to grow, forming both dorsal and anal finfolds. The dorsal finfold originated directly above the origin of the anal finfold at the posterior junction of yolk and body. At this age, approximately 20 of the anterior vertebral segments exhibited rudiments of neural arches, with paired haemal arches after the first 3 segments. There was no evidence of chondrification or calcification anywhere in the embryonic body.

Step: E6 (04:08:00–06:16:00)

At the age of 04:08:00, *L. goodei* embryos exhibited the first indications of fin rays in the caudal fin (Fig. 26a, 14b). One or two pairs of capillaries in this fin originated from the arterial arch at the posterior tip of the notochord, in a bilaterally symmetrical manner. The efferent length of capillaries always turned toward midline to form the afferent vessel. Each rudimentary fin ray was bounded on the distal (from midline) edge by the efferent vessel, and on the proximal edge by the afferent vessel. The afferent vessels collected into a common vessel, which in turn carried blood to the caudal vein. Just posterior of the caudal vein, where it looped ventrally to feed the posterior vitelline vein, was the first indication of the urinary bladder (Fig. 26a). The first indications of a simple gut were also observed in specimens at this age. All vertebrae exhibited nonstaining extensions of paired neural and haemal arches. The only parts of the embryonic body to retain alcian blue stain at this age were the pair of concretions in each otolith. The head began to grow at this age, giving a relative otic length that would be car-

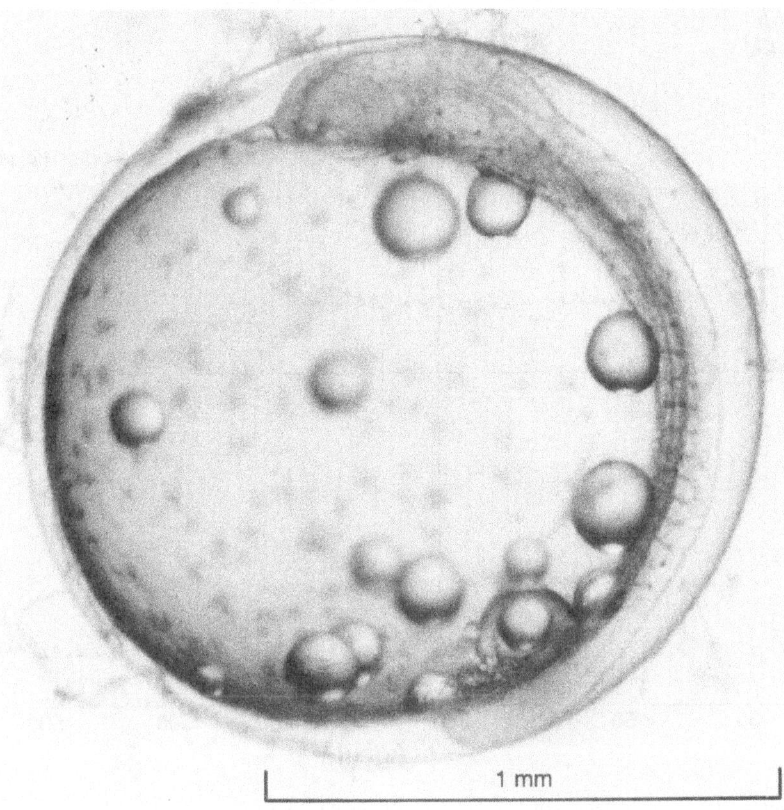

Fig 16 Photomicrograph of a *L goodei* embryo during step E2 (01 10 00) from a lateral perspective Approximately 12 somites can be seen in this specimen Large branched melanophores are scattered over the yolk surface (most out of focus) while smaller melanophores can be seen on the body especially in the head region

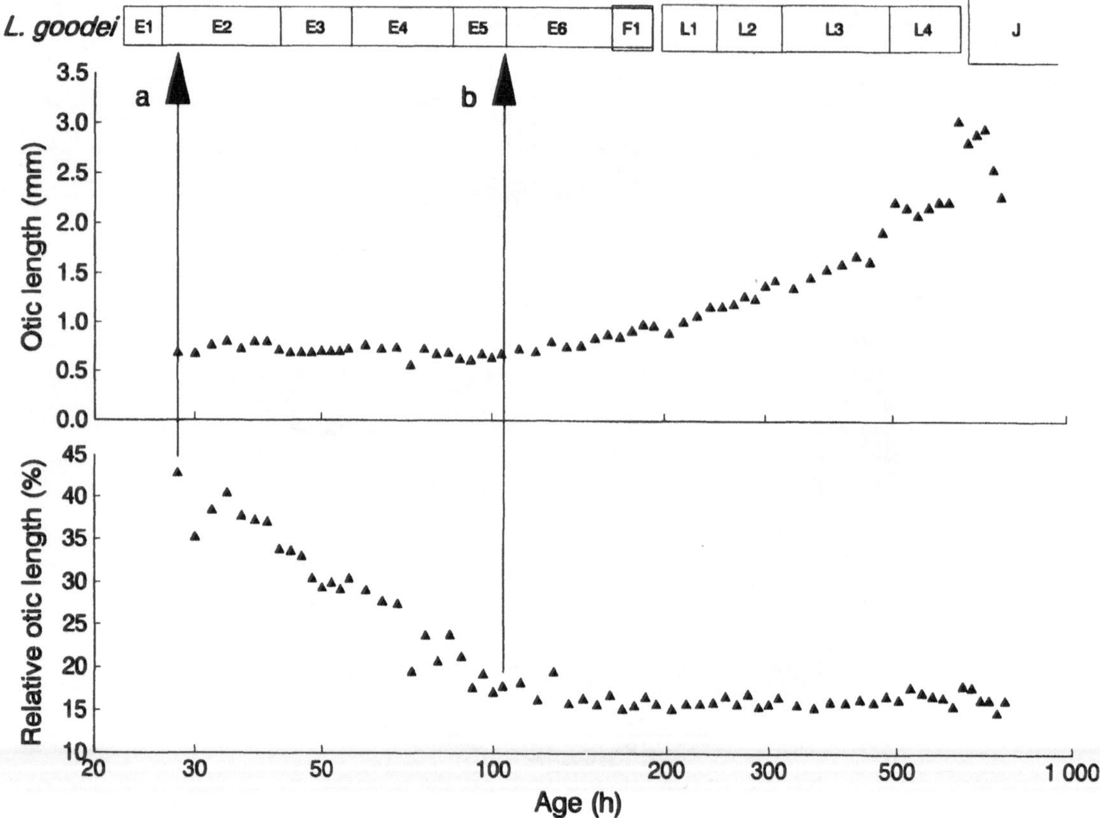

Fig. 17. Absolute and relative (% TL) otic length of *L. goodei* during the embryo, free embryo and larva steps (top bar) measured as a function of logarithmic age (a = first distinction of otic vesicles, b = achievement of definite relative otic length).

ried to the juvenile phenotype (Fig. 17b). The eyes were noticeably darker by this step, and the mid-dorsal and midventral lines showed concentrations of small and large branched melanophores (Fig. 27).

In addition to blood circulation in the caudal fin, a new blood vessel had also appeared in the fan-shaped pectoral fins where the subclavian vein entered from the ventral base of the fin and looped parallel to the border (Fig. 26a). Embryos with a subclavian vein in the pectoral fins also exhibited sporadic fluttering of these fins. Occasionally, the tail and pectoral fins engaged in rhythmic movements, but these were infrequent and brief. Eye movements were also observed for the first time in embryos of this developmental state. The pericardial wall had receded to the anterior region of the lower jaw, while the eyes were almost completely free.

At 05:00:00 there were six to eight rays in the caudal fin, each with blood circulation described earlier (Fig. 26b). The first indications of the profundal caudal vein formed at the posterior end of the body, although most of the blood still reached the posterior vitelline veins via the inferior caudal vein. By this age, the dorsal and anal finfolds had almost doubled in depth. These embryos also exhibited the first signs of gill formation, as filaments formed in rows directly below the otic region. The paired neural and haemal spines of the vertebrae had reached far into the dorsal somites and beyond the ventral caudal aorta, respectively. There was a continued lateral and ventral retreat of the pericardial wall that completely exposed the well-defined lower jaw. The mouth opened occasionally, but not regularly and not in co-ordination with the fins.

Primary structures within the *L. goodei* neurocranium exhibited the first indication of chondrification during step E6 (Fig. 28). The base of the neurocranium was formed by the trabeculae, while the

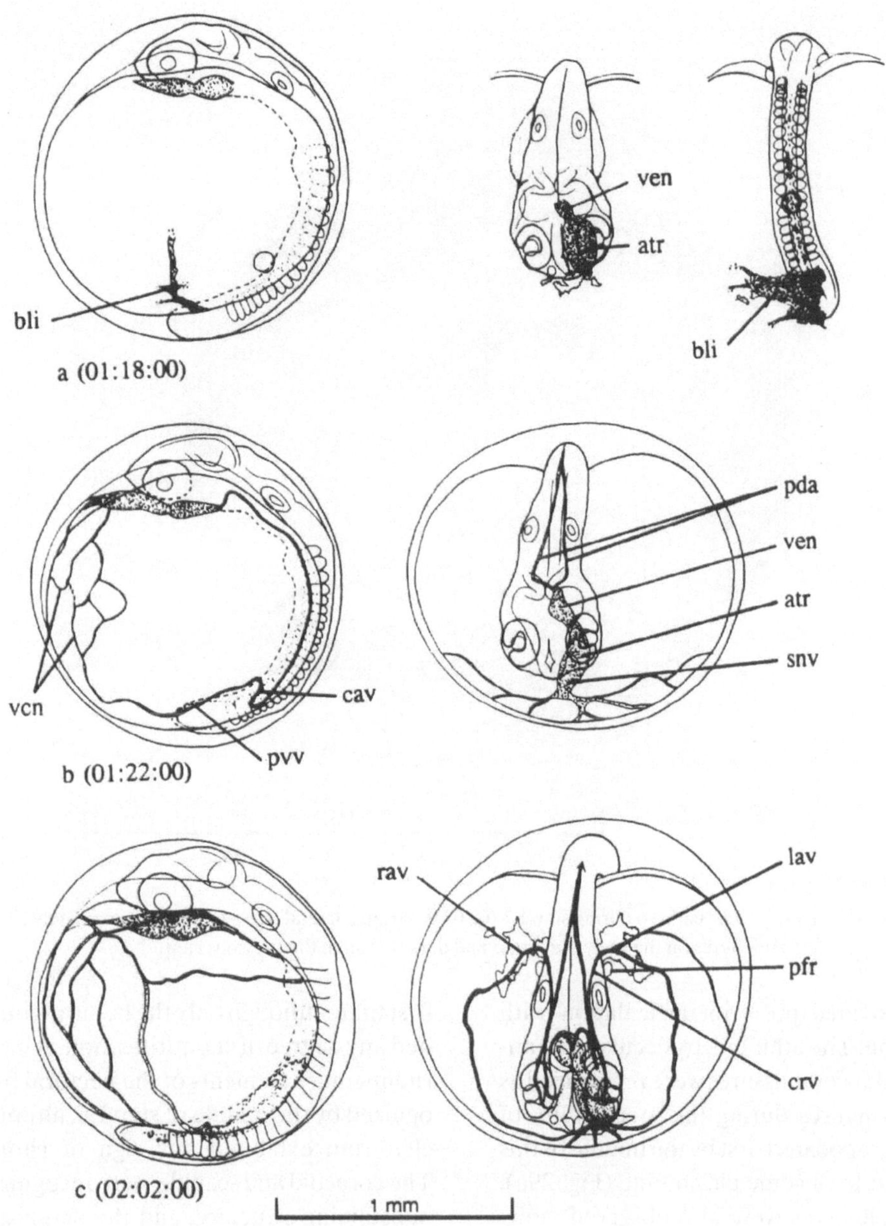

Fig. 18. Embryos of *L. goodei* during step E3 from lateral, dorsal and posterior perspectives (atr = atrium, bli = blood island, cav = caudal vein, crv = cardinal vein, lav = left anterior vitelline vein, pda = paired dorsal arteries, pfr = pectoral fin rudiment, pvv = posterior vitelline vein, rav = right anterior vitelline vein, snv = sinus venosus, vcn = vitelline capillary network, ven = ventricle).

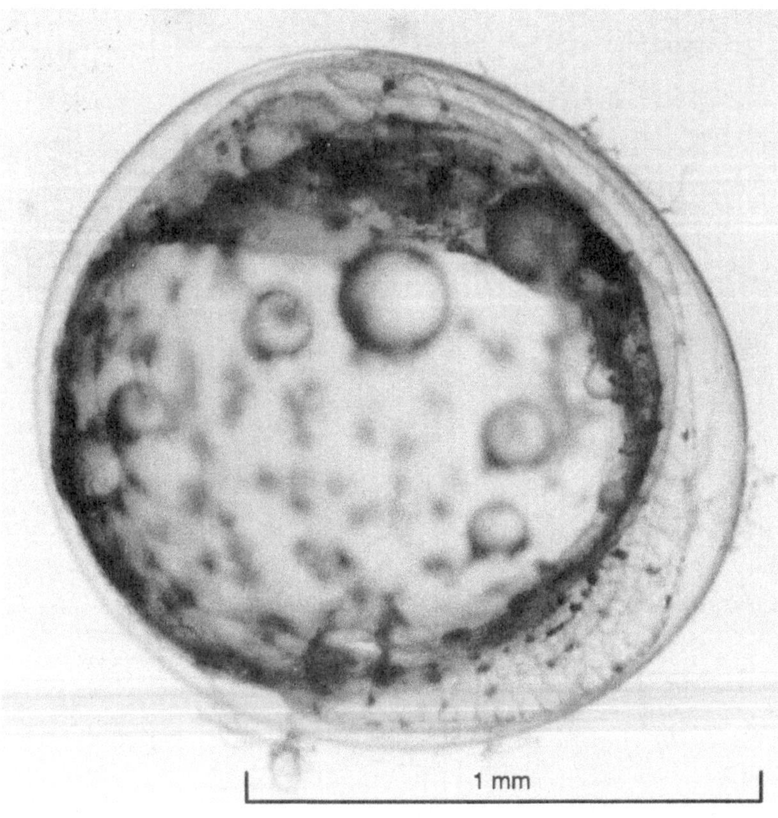

Fig 19 Photomicrograph of a *L goodei* embryo during step E3 (02 02 00) from a lateral perspective Approximately 23 somites can be seen in this specimen The melanophores on the body are larger and darker than in the previous step

parachordal provided posterior articulation with the axial skeleton. The ethmoid, trabeculae and anterior basicapsular commissures were recognized as non-staining structures during this step. Many of the jaw elements appeared first by the middle of this step, retaining at least some alcian blue (Fig. 29a). The opercular elements were also observed, however these did not retain any stain. The first pharyngeal structures to exhibit signs of chondrification were the basihyal, ceratohyal and series of ceratobranchials by the middle of step E6. The pharyngeal teeth and branchiostegal rays were observed as the first pharyngeal components, although both of these were non-staining.

Small neural and haemal processes formed on the notochord, and exhibited light blue stain by the middle of the step (Fig. 30a). The hypural complex and caudal fin lepidotrichia were observed for the first time, although only the hypural complex exhibited any degree of chondrification. Almost all of the rudimentary elements of the pectoral fins were recognized by the middle of step E6, although only the cleithrum exhibited any sign of chondrification. The coracoid and scapula were recognized as a joint nonstaining structure, and the actinosal plate had not yet formed separate subunits.

The yolk of *L. goodei* embryos continued to be consumed at a rapid rate during step E6, but the reserves would not be exhausted until the next step (Fig. 11). Embryo otic length, measured as the length between snout and otic vesicle, increased for the first time since otoliths could be distinguished at the beginning of step E1 (Fig. 17). As a result, the ratio of otic length to body length remained at approximately 17%. Total body depth held constant during step E6, despite continued increase in body

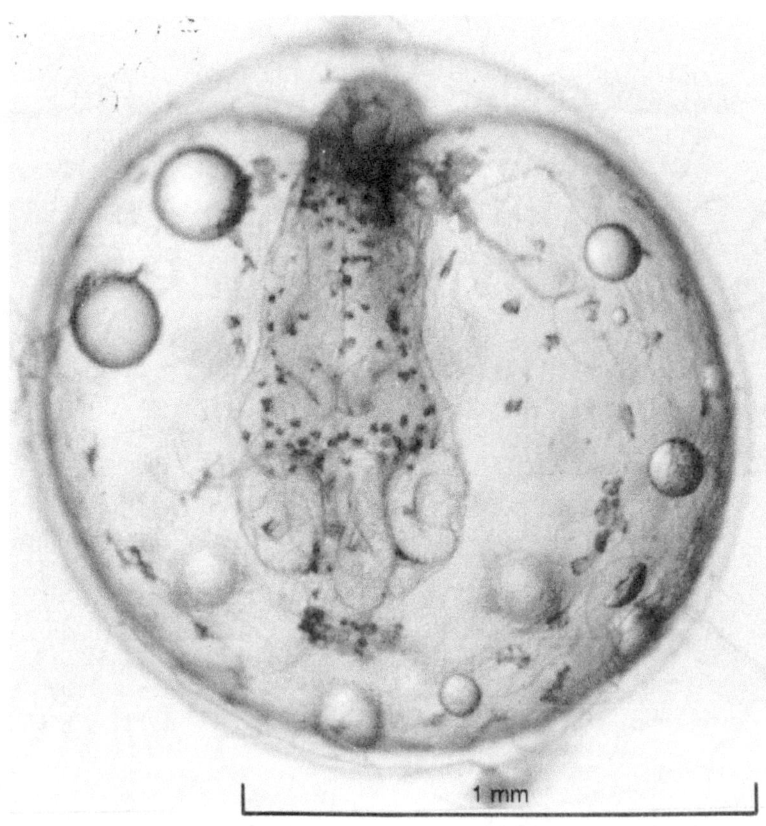

Fig 20 Photomicrograph of a *L goodei* embryo during step E3 (02 00 00) from a dorsal perspective Clusters of melanin can be seen near the embryonic body at the departure of the anterior vitelline veins from the embryonic body near the sinus venosus, while melanophores can be seen on the head and yolksac

length (Fig. 12). This phenomenon was probably due to offsetting decreases in yolk depth and increases in depth of the embryonic body. As a result, the embryos exhibited a decrease in relative body depth that continued until the drastic body changes observed immediately before first exogenous feeding (Fig. 12b).

Phase: Free embryo
Step: F1 (06:16:00–07:16:00)

The median age at hatching for *L. goodei* was 06:16:00, and ranged from 06:00:00 to 07:08:00. The embryo thrashed its body and tail about, until the mechanical force of the movements ruptured a crack or hole in the egg envelopes, through which the embryo became free. A newly hatched embryo exhibited 8 to 12 fin rays with 3 segments in the serrated caudal fin, as well as a completely developed profundal caudal vein, running parallel to the inferior caudal vein (Fig. 31a). A small volume of yolk was still observed in embryos at this state (Fig. 11), however a withdrawal of the yolk wall 'pulled' the heart caudally, such that the sinus venosus was posterior to the atrium, which was posterior to the ventricle. In addition, the retreat of the yolk wall began to draw the vitelline vessels together at the yolk ventrum, as indicated by a dark streak of pigmentation. There had also been an increase in the concentration of melanophores along the dorsal surface of the head, along the pectoral fins and to a lesser extent, over the lateral surface of the tail. A few branched melanophores were observed in both dorsal and anal finfolds (Fig. 32).

The free embryos of *L goodei* would lie on the

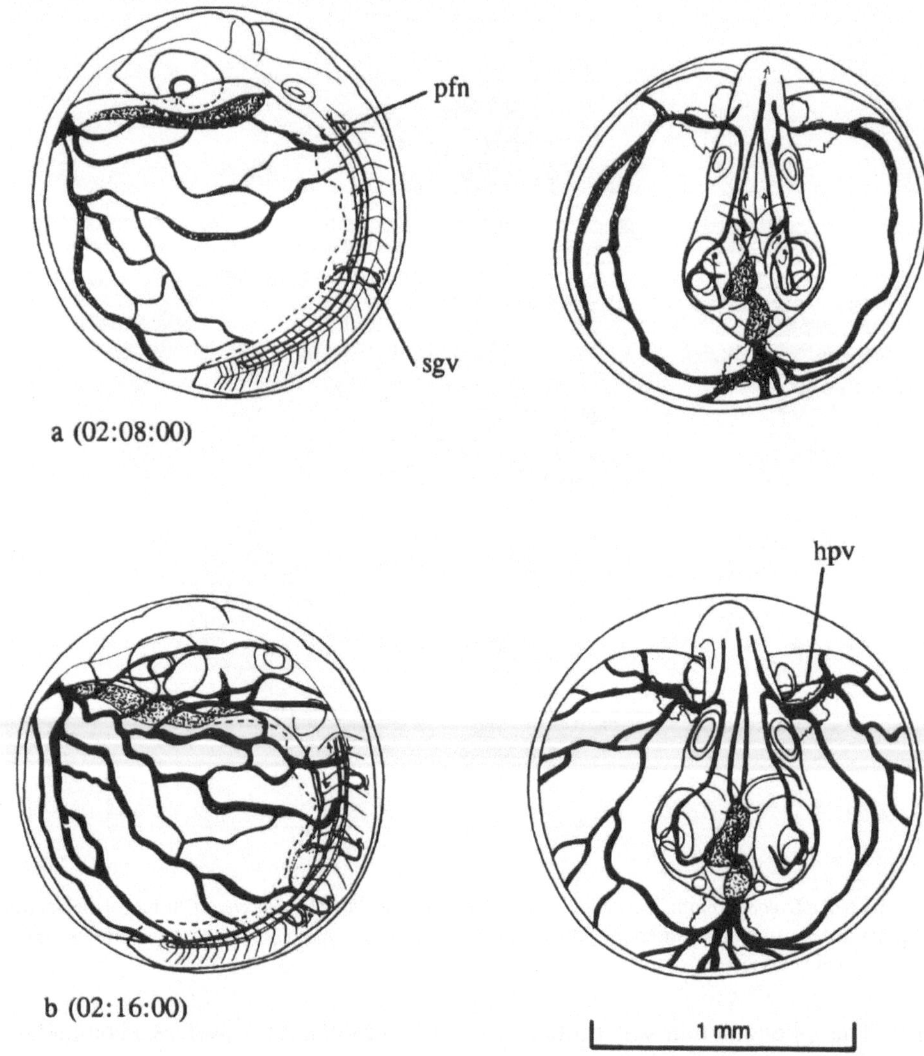

Fig. 21. Embryos of *L. goodei* during step E4 from lateral (left) and dorsal (right) perspectives (hpv = hepatic vein, pfn = pectoral fin, sgv = segmental vessel).

bottom of the incubation cups, rising occasionally to the water surface. We did not observe any embryos swallowing gas bubbles, however only free embryos with a gas-filled swimbladder were observed to swim with co-ordinated movements of the mouth, gill covers, pectoral fins and tail. The sinus vensosus, atrium and ventricle were oriented in a line that led directly to the gills, which were already well-formed.

Many elements within the neurocranium of F1 *L. goodei* free embryos were added since the previous step, and most of these structures retained some alcian blue (Fig. 28). The neurocranial structures which appeared for the first time were the orbital commissures, epiphyseal bar, and the parasphenoid at the base. Several structural elements of the jaw were present and at least partially chondrified at hatching in *L. goodei* (Fig. 29a). The exceptions to this observation were the non-staining jaw teeth, premaxillary, and maxillary. The simple pharyngeal complex continued to become chondrified during this step, with the addition of the non-staining epibranchials. The axial skeleton and pectoral fins also

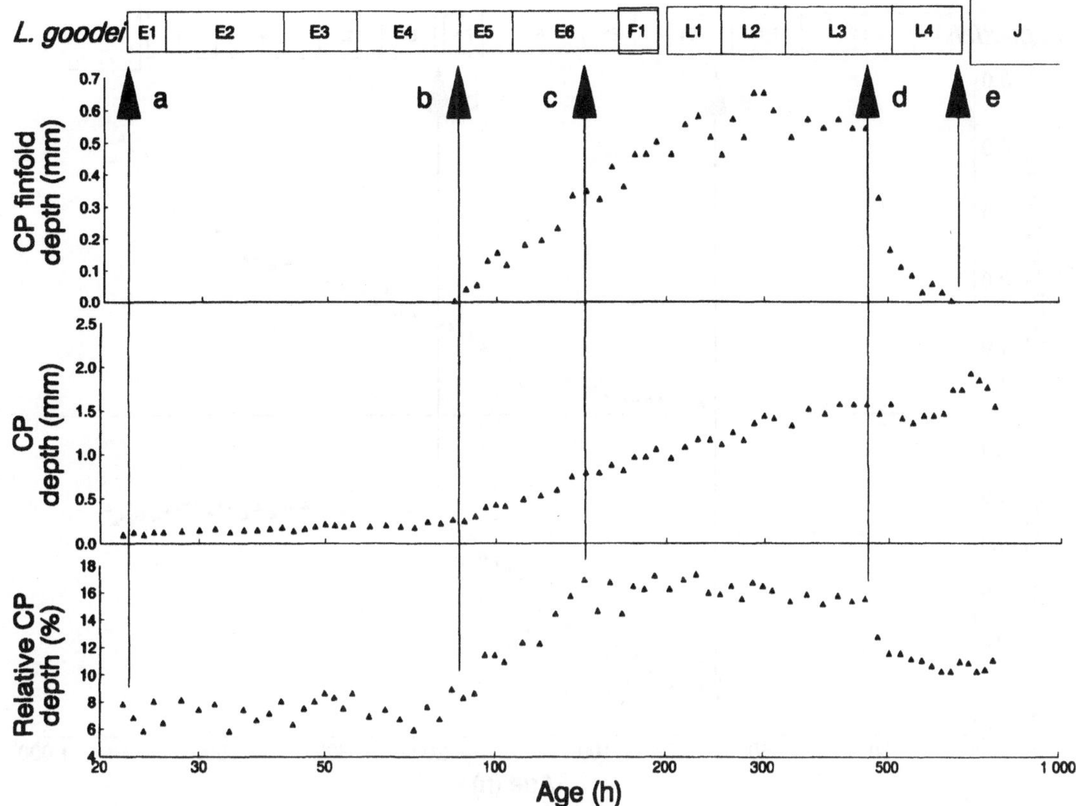

Fig. 22. Absolute depth of caudal peduncle (CP) finfold (dorsal = anal) at middle of caudal peduncle, and absolute and relative (% TL) depths of the caudal peduncle of *L. goodei* during embryo, free embryo and larva steps (top bar) measured as a function of logarithmic age (a = first distinction of caudal peduncle, b = first distinction of finfolds, c = achievement of maximum relative CP depth, d = onset of rapid finfold resorption, e = completion of finfold at midpoint).

remained structurally similar to previous stages (Fig. 30a, b).

Yolk diameter declined rapidly during this step as the endogenous supplies of the embryo dwindled (Fig. 11c). Both absolute and relative measures of body depth increased dramatically (Fig. 12b–c), probably as a function of swimbladder inflation. Although the peduncle finfold continued to increase in depth, the relative depth of the caudal peduncle reached a maximum of approximately 16% SL (Fig. 22c). This relative depth of the caudal peduncle was held until absorption of the caudal finfold.

Period: Larva

Step: L1 (07:16:00–10:00:00)

Larvae of *L. goodei* were first observed with brine shrimp in their guts at a median age of 07:16:00 (07:00:00–08:16:00), and the median time lag between hatching and first feeding was 16 hours. A simple gut appeared by this age, as a tube that led straight to the anus, anterior to the origin of the anal finfold (Fig. 31b). The relative head depth and relative length of the caudal fin of the larvae seemed to be set near the juvenile phenotype after the occurrence of first exogenous feeding (Fig. 15b, 23b). At 09:00:00 a convoluted intestine stuffed with brine shrimp nauplii filled the body cavity. By this age, the yolk supply had been completely exhausted (Fig. 31c, 11c), with no trace of vitelline vessels to be found. In addition, a concentration of small melanophores had appeared along the midlateral surface of the tail. Larger, branched melanophores were observed over the jaws and snout. Guanine deposits were recognized on the surface of the swimbladder, eyes and gill covers (Fig. 33).

The neurocranium and jaws of first feeding *L. goodei* were not drastically different from those of the free embryo (Fig. 28, 29a). However, the lateral

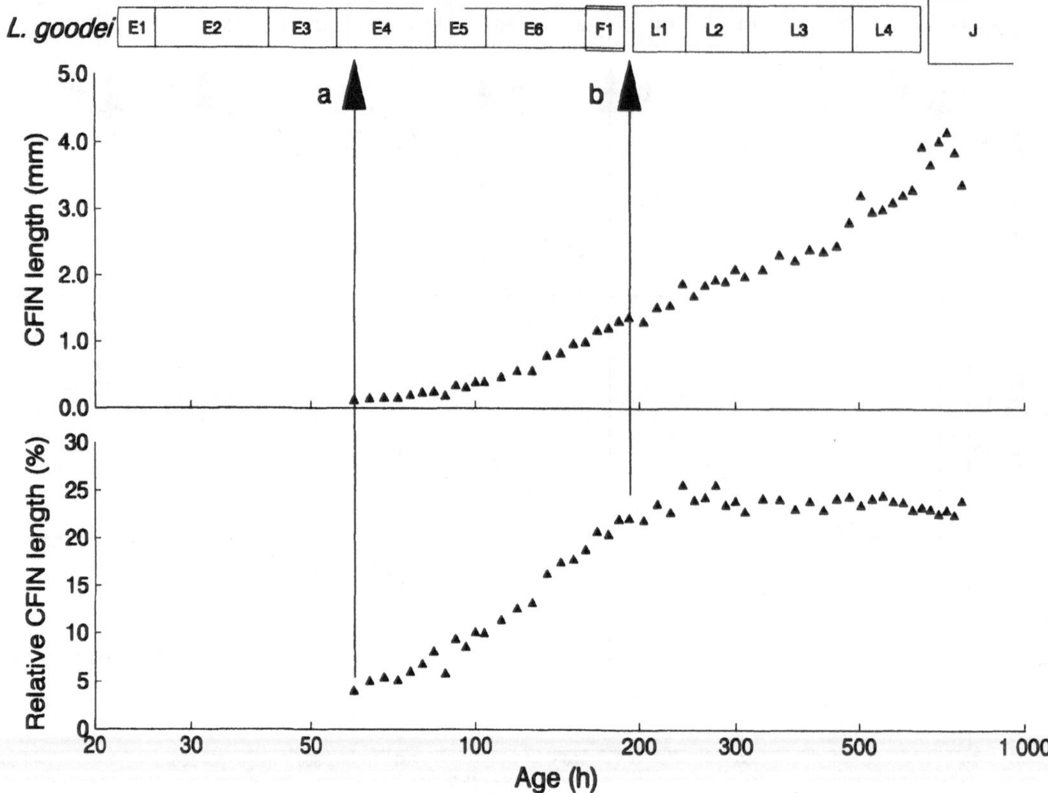

Fig. 23. Absolute and relative (% TL) caudal fin (CFIN) length of *L. goodei* during the embryo, free embryo and larva steps (top bar) measured as a function of logarithmic age (a = first distinction of caudal fin, b = achievement of definite relative caudal fin length).

ethmoid processes and eye rings did appear for the first time by the beginning of step L1. By the middle of the step, the pharyngobranchial, pharyngeal teeth and branchiostegal rays all exhibited the first retention of alizarin red, indicating the onset of calcification. In the axial skeleton, the neural and haemal processes of the vertebrae, and the caudal lepidotrichia all retained some alizarin red by the middle of the step (Fig. 30a). All of the rudimentary structures of the pectoral fins exhibited signs of chondrification, except for the cleithrum which already retained alizarin red.

Step: L2 (10:00:00–13:00:00)

The larvae of *L. goodei* exhibited the first indications of both dorsal and anal fin rays by age 10:00:00 (Fig. 34a). It should be noted that rays of both fins developed in the absence of blood vessels along their margins. Another structure that appeared for the first time was the series of ribs extending from the four most anterior vertebrae. These processes extended laterally over the elongated swimbladder before curving ventrally. Thick zones of notochordal sheath appeared between all of the vertebrae, as rings anterior to the base of the neural and haemal spines. Over the next 48 hours, the rays of both dorsal and anal fins continued to increase in number and length, causing the finfolds to form clear outlines of the definitive fins (Fig. 34b). Large, branched melanophores appeared among the dorsal and anal finrays. Concentrated lines of small melanophores had appeared along the middorsal, midlateral and midventral lines.

The elements that formed the rudimentary neurocranium of *L. goodei* (i.e. parasphenoid, parachordal and basicapsular commissures) were the first neurocranial structures to exhibit signs of calcification (Fig. 28, 35). In addition, the lateral walls of the neurocranium stained blue for the first time. Step L2 marked the calcification of jaw bones, namely the premaxillary, maxillary and dentary (Fig. 29a). There were no pharyngeal structures ex-

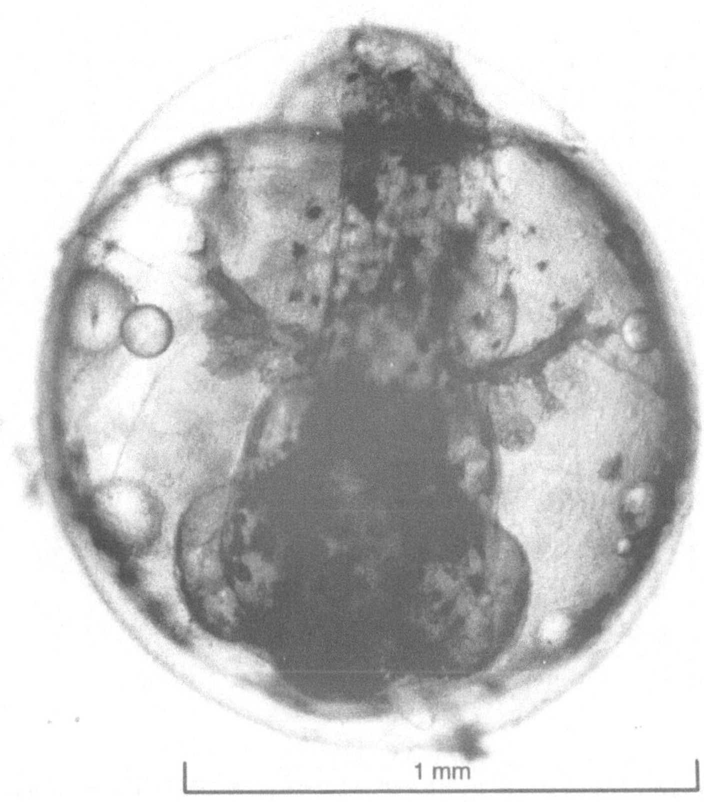

Fig 24 Photomicrograph of a *L goodei* embryo during step E4 (03 04 00) from a dorsal perspective Melanophores are distributed throughout the body, and are especially dense on the head, and covering the departure branches of the anterior yolksac vessels

hibiting signs of calcification, relative to the previous step, however chondrified gill filament supports were observed by the middle of the step. Virtually all of the remaining axial elements were recognized by the middle of step L2, either as structures which were non-staining or retaining only alcian blue (Fig. 30a). The new structures observed during this step included ribs, and the pterygiophores and lepidotrichia of the dorsal and anal fins. With the exception of non-staining lepidotrichia, the pectoral fins remained mostly unchanged from the previous step.

Step: L3 (13:00:00–20:00:00)

At the age of 13:00:00, *L. goodei* larvae were marked by the appearance of paired pelvic fin rudiments, just anterior to the anus (Fig. 36a). The pelvic fins were quite small, however the pectoral fin had enlarged considerably since the last part of step L2. The dorsal and anal fin rays extended to the edge of their respective finfolds, giving a serrated appearance to the edge of the fins. In addition, pterygiophores had formed below the bases of the first four or five lepidotrichia in both the dorsal and anal fins. These pterygiophores slanted at an acute angle to contact the base of their respective fin rays.

By 15:00:00, the dorsal and anal fins exhibited a full complement of 11 or 12 rays (Fig. 14). The ribs had grown in size and number to extend around the swimbladder, which had continued to enlarge and elongate. At age 17:00:00 the larvae exhibited the first indications of finfold intrusion, immediately posterior to the dorsal and anal fins (Fig. 36b, 22). Guanine deposits continued to accumulate on the surface of the eyes, gill covers and gut (Fig. 37).

During step L3 in *L goodei* development, approximately half of the existing neurocranial ele-

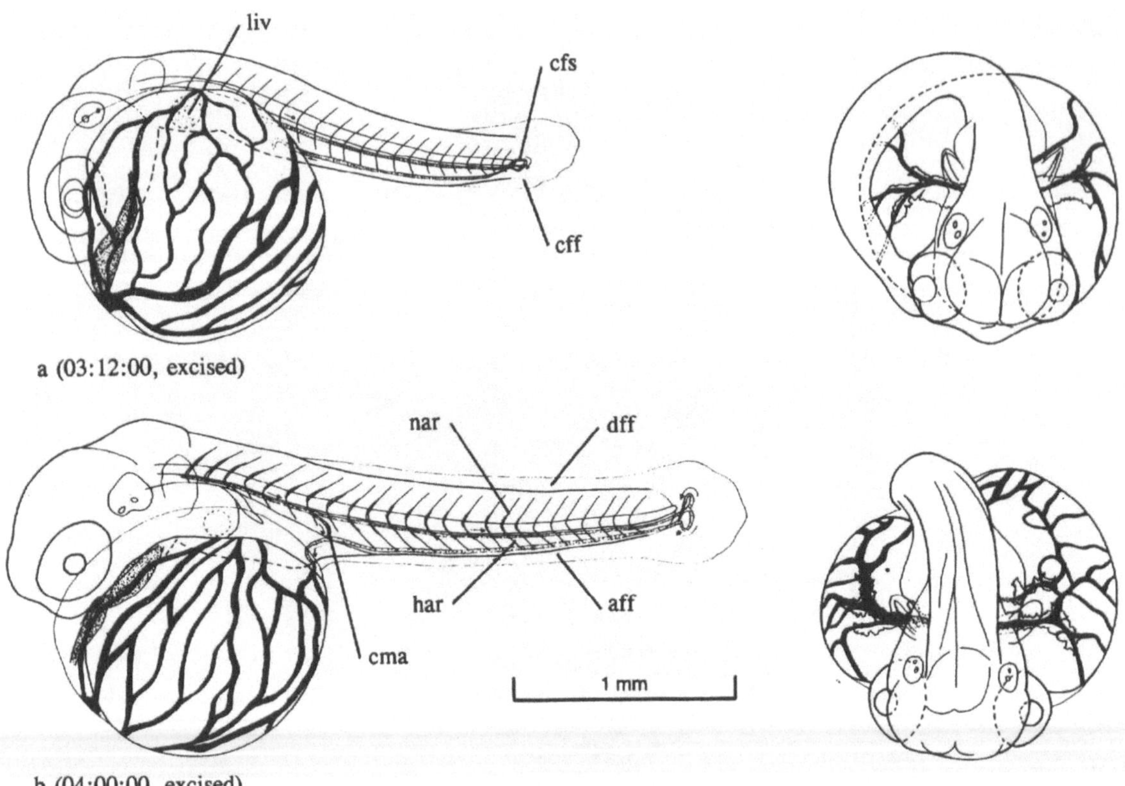

Fig. 25. Excised embryos of *L. goodei* during step E5 from lateral and dorsal perspectives (aff = anal finfold, cff = caudal fin fold, cfs = caudal fin snare, cma = coeliac-mesenteric artery, dff = dorsal finfold, har = haemal arch, nar = neural arch).

ments exhibited some degree of calcification (Fig. 28, 38). Virtually all of the jaw, opercular and pharyngeal bones exhibited advanced calcification with little or no cartilage remaining (Fig. 29a, b). The axial skeleton remained similar to the previous step, except for the hypural complex which exhibited the first signs of calcification (Fig. 30a). Finally, the first rudiments of the pelvic lepidotrichia were apparent, however these structures were non-staining.

Step: L4 (20:00:00–27:00:00)

Scalation in *L. goodei* larvae began at the age of 20:00:00, with one horizontal row of scales, centered below the posterior base of the dorsal and anal fins (Fig. 39a, 14d). At this time, the dorsal and anal finfolds began a dramatic resorption into the caudal peduncle, proceeding from anterior to posterior (Fig. 39a–b; 22d–e). By 23:00:00 the scales had continued to form both anteriorly and posteriorly to the initial scale positions. The lateral scale count quickly approached the full complement of 29 (27–31) scales. Only remnants of the dorsal and anal finfolds remained at the posterior region of the caudal peduncle (Fig. 39b). A broad patch of melanophores had appeared on the dorsal surface of the head, and the midlateral pigmentation stripe was becoming darker (see Fig. 41). By the middle of step L4, all of the neurocranial elements were present, most of which were calcified with little or no cartilage remaining (Fig. 40). The same observation held for the jaws, gill covers and pharyngeal structures (Fig. 29). The cartilaginous pterygiophores of the pectoral fins, and the pectoral girdle were finally observed by the onset of step L4 (Fig. 30).

Period: Juvenile
Phase: Juvenile
Step: J (27:00:00+)

The juvenile phenotype of *L. goodei,* as defined by the complete resorption of the median finfold,

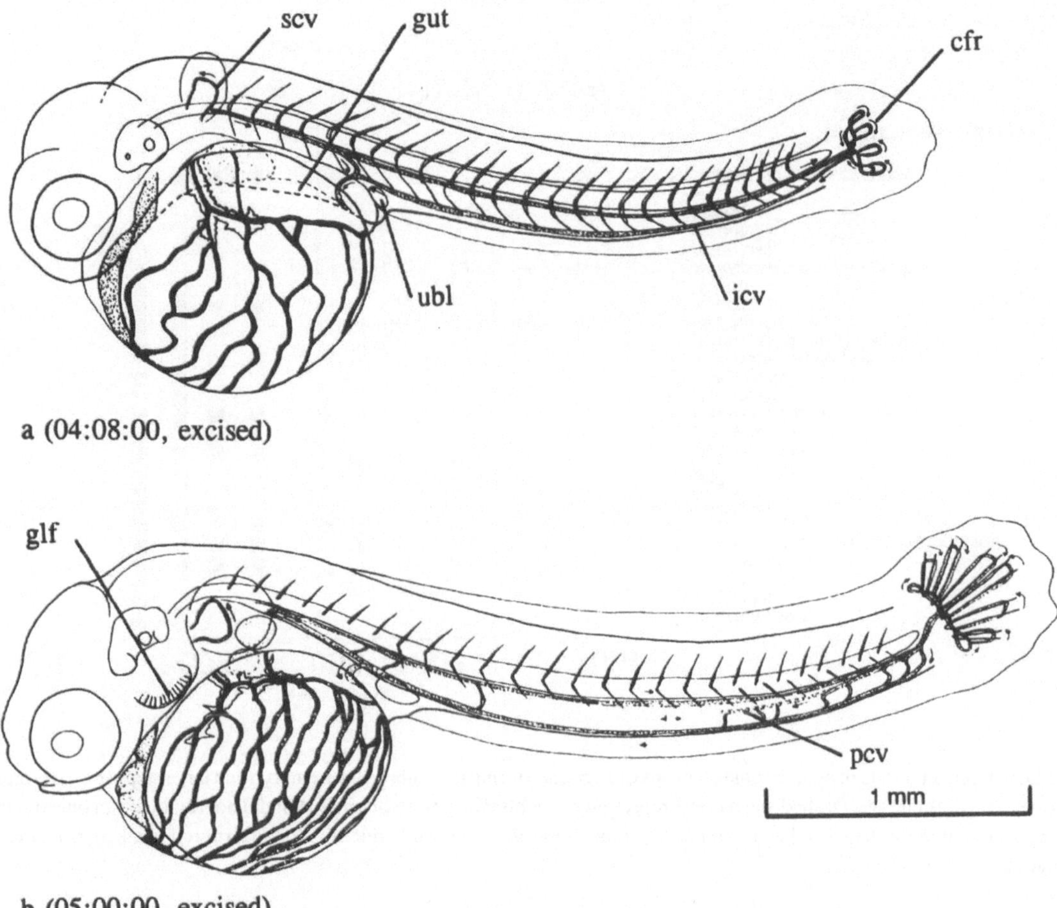

Fig. 26. Excised embryos of *L. goodei* during step E6 from a lateral perspective (cfr = caudal fin ray, glf = gill filament, gut = simple gut, icv = inferior caudal vein, pcv = profundal caudal vein, scv = subclavian vein, ubl = urinary bladder).

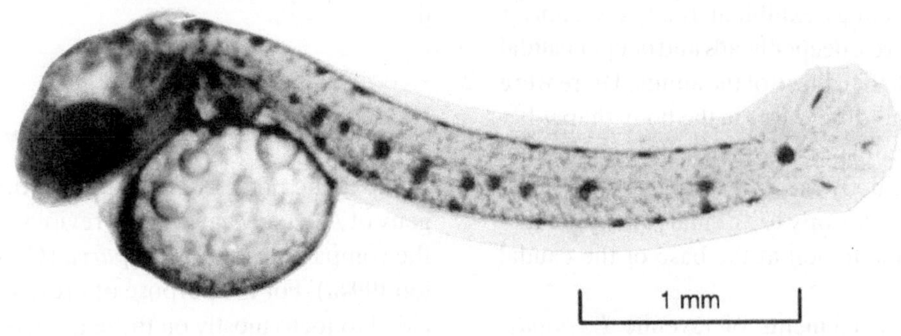

Fig. 27. Photomicrograph of a *L. goodei* embryo excised from the egg envelopes during step E6 (05:00:00) from a lateral perspective.

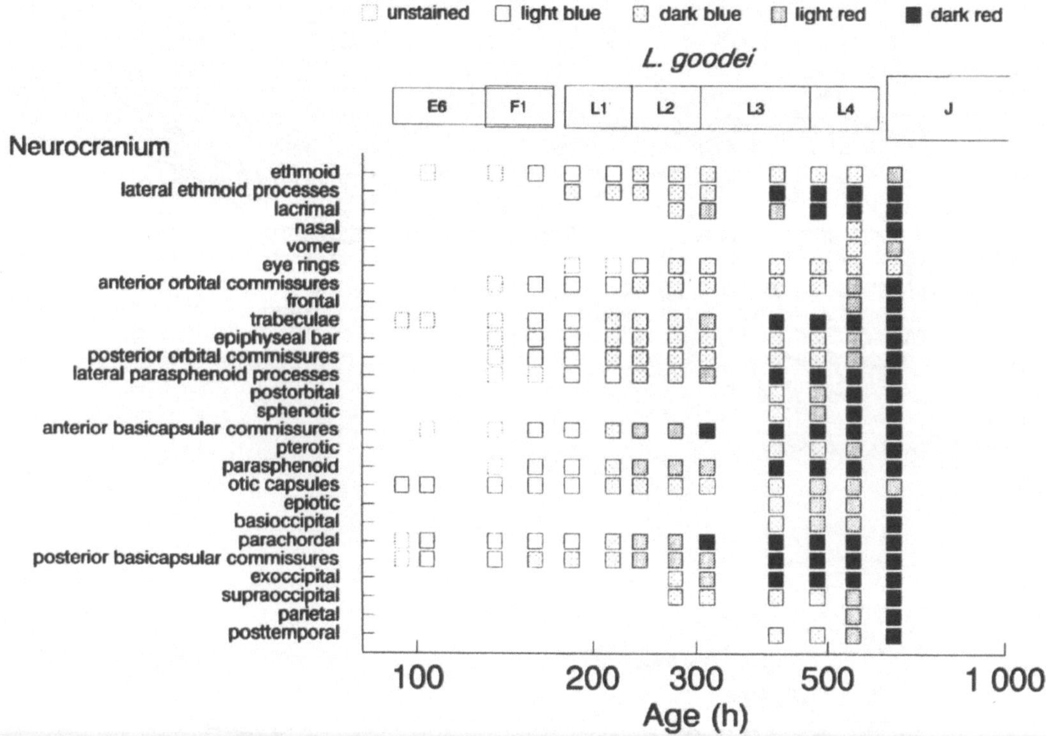

Fig. 28. Developmental state of neurocranial elements of *L. goodei* during the embryo, free embryo and larva steps (top bar) measured as a function of logarithmic age. Dashed, open boxes represent elements which were observed but did not retain alcian blue or alizarin red stain. Retention of alcian blue is taken as an indication of chondrification, while retention of alizarin red is taken as an indication of calcification.

was typically achieved at the age of 27:00:00 (Fig. 39c). By this time, the juveniles were approximately 36% of the standard length of the wild adults (Table 2). The other characters that were noticeably different between laboratory-reared juveniles and wild adults, included changes in relative head shape and tail depth. Juveniles exhibited relatively shorter snouts, larger eyes, deeper heads and deeper caudal peduncle depth than those of the adults. There were no other obvious differences in the body shapes between the juveniles and adults. The midlateral stripe along the body was very distinct by this stage, often leading posteriorly to an enlarged spot of melanophore pigmentation at the base of the caudal fin (Fig. 41).

Some skeletal elements of juvenile *L. goodei* were almost completely calcified, as seen by the deep retention of alizarin red stain (Fig. 28–30). However, many other structures were observed to be predominantly cartilaginous, including the eye rings, Meckel's cartilage, gill filament supports, coracoid, actinosts, as well as the pterygiophores and lepidotrichia of the dorsal, anal and pectoral fins. Thus, by the age at which *L. goodei* achieved the 'definitive' phenotype, there were still many structures that would undergo the process of calcification.

Discussion

Many of the processes described in the early ontogeny of *L. goodei* have been previously discussed in the companion study of *L. parva* (Crawford & Balon 1994a). For the purpose of brevity, we have decided to focus mostly on those aspects of *L. goodei* development that are different from those of *L. parva.* A direct comparison of early ontogenies in the genus *Lucania* is reserved for the third and final instalment of this series (Crawford & Balon 1994b).

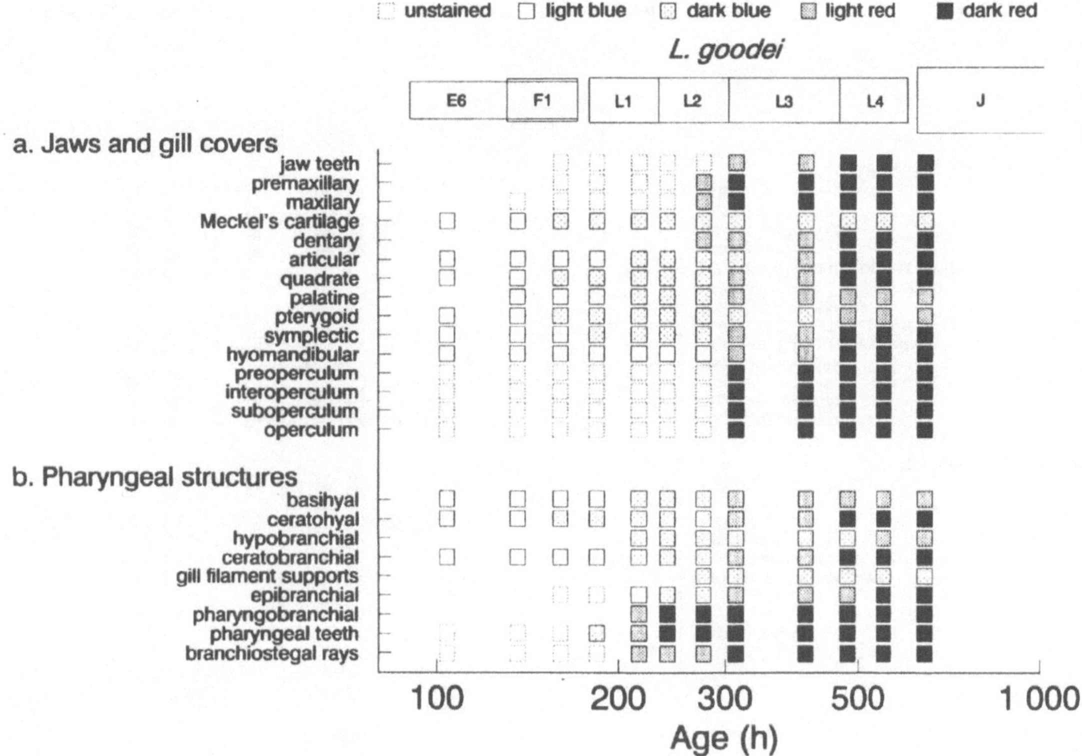

Fig 29 Developmental state of a – jaws and gill covers, and b – pharyngeal structures of *L goodei* during the embryo, free embryo and larva steps (top bar) measured as a function of logarithmic age Dashed, open boxes represent elements which were observed but did not retain alcian blue or alizarin red stain

The courtship behavior of *L. goodei* is similar to that described by Foster (1967) for most North American cyprinodontids, with the exception that *L. goodei* seem to be much less aggressive than other species during courtship (Sommer 1978). If killifish aggression is associated with the defense of a spatially limiting resource (e.g. territoriality), then a lack of aggression on *L. goodei* may suggest an abundance of such resources in the wild.

Some populations of *L. goodei* have been reported to exhibit spawning from January to September, or perhaps even year round (Foster 1967, Arndt 1971). Given the highly consistent environmental conditions of Newport Spring over time (Rosenau et al. 1977, Crawford & Balon 1994b), it is quite possible that this particular population could exhibit year-round reproduction. It should be noted that the capacity of *L. goodei* to reproduce is not unlimited, since some females held under laboratory conditions died after six months of intensive spawning (Foster 1967).

The size of *L. goodei* eggs observed in this study corresponds to the data presented by Foster (1967) for the same species. Foster (1967) also commented on the appearance of 'adhesive threads' at one end of the eggs; evidence that supports our observation that these filaments are sparsely distributed over the surface of the egg envelopes. These adhesive filaments may serve in some capacity to prevent the eggs from being swept downstream by the current generated at Newport Spring. Finally, Foster (1967) reported that *L. goodei* eggs exhibited 10–12 oil globules, while we found an average of 22.8 oil globules per egg. This difference may be an artefact of counting techniques, or it may reflect some population difference in the biochemical basis of oil deposition or yolk viscosity. During activation, the perivitelline space of *L. goodei* eggs seemed to be relatively smaller than those described for other cyprinodon-

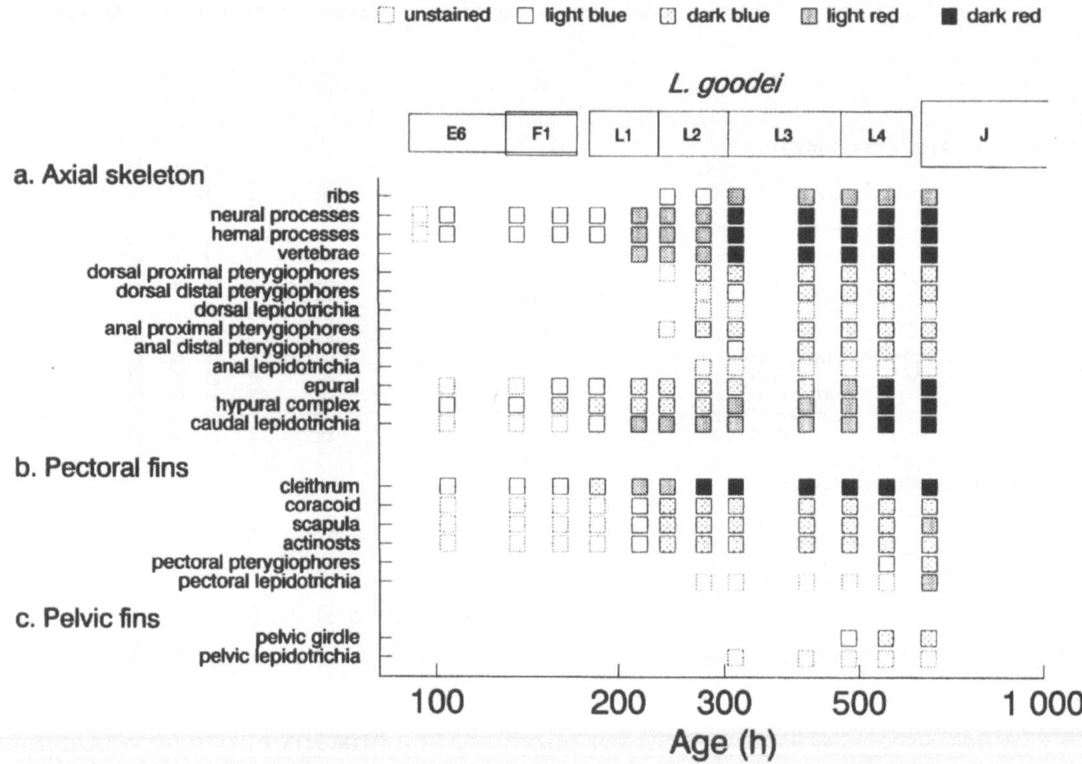

Fig 30 Developmental state of a – axial skeleton, b – pectoral fins, and c – pelvic fins of *L goodei* during the embryo, free embryo and larva steps (top bar) measured as a function of logarithmic age Dashed, open boxes represent elements which were observed but did not retain alcian blue or alizarin red stain

tid species (Foster 1967). This difference may reflect species differences in egg envelope size, quantity of yolk investment, density of cortical alveoli distribution, or some combination thereof (Laale 1980).

The distinction of the embryonic *L. goodei* body, as defined by appearance of optic vesicles, did not take place until germ ring closure at the end of epiboly. This is similar to the descriptions of other species such as *Fundulus heteroclitus* (Oppenheimer 1937), but is late compared to other cyprinodontids, especially those with smaller yolk investments (Foster 1967). The idea that surface area of the yolk may have an important effect on the rate of epiboly is supported by our observation that the cell layer 'thinned out' during cell migration. If the relationship between yolk surface area and epibolic migration is valid, it would support the claim that relative timing of morphogenic processes (Table 3) is not rigidly fixed, but rather may depend on local factors such as yolk investment, which may in turn depend on the nutritional state of the female during vitellogenesis (Balon 1989, 1990).

The formation of an extensive vitelline circulatory network in *L. goodei* embryos may reflect the hypoxic environment in which these fish survive and reproduce. Foster (1967) noted that many cyprinodontids inhabit waters that are seasonally or perpetually poor in oxygen, and that a complex vitelline circulatory network may function to maximize embryonic oxygen uptake in these environments. We explore this particular feature of early ontogeny in the genus *Lucania* in the final paper of this series (Crawford & Balon 1994b).

Chondrification of some skeletal elements in *L. goodei* began before hatching of the embryo, but calcification of most structures was not observed until late in larva development (i.e. step L3). This observation suggests that while free embryos of this species are equipped with functional elements for feeding and locomotion, these elements are not

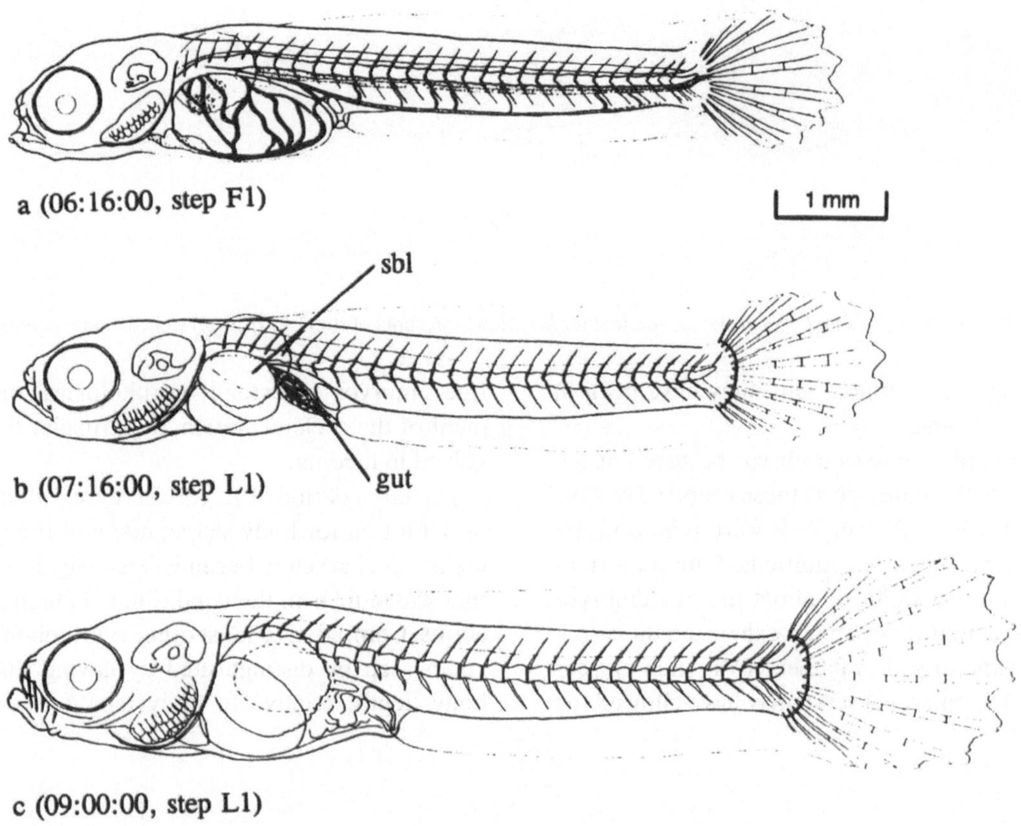

Fig 31 Free embryo and larvae of *L goodei* during steps F1 and L1 from a lateral perspective (gut = convoluted gut, sbl = swimbladder)

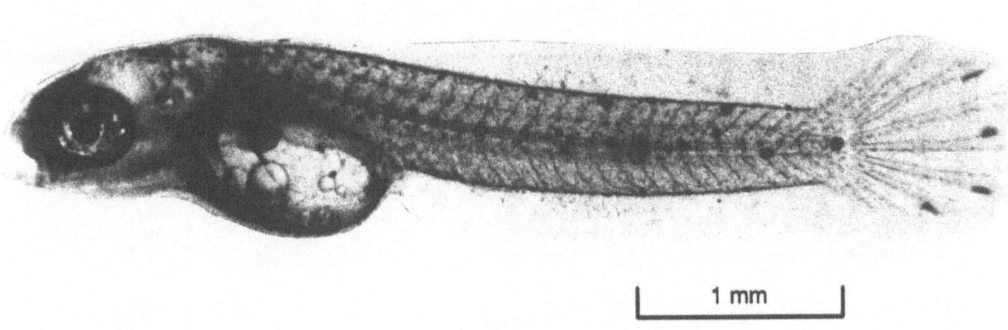

Fig 32 Photomicrograph of a *L goodei* free embryo after hatching at the threshold of step F1 (06 16 00) from a lateral perspective

Fig 33 Photomicrograph of a *L goodei* larva after first feeding at the threshold of step L1 (07 16 00) from a lateral perspective

morphologically 'defined' until they have been in use for some time.

Hatching of *L. goodei* embryos occurred at approximately the same age as those reported by Foster (1967), where the embryos were reared under similar environmental conditions. One important observation that we made about the free embryos was the fact that they had not exhausted their yolk supply by the time they hatched from their egg envelopes. This energy reserve may have allowed the free embryos to extend morphological development of the skeletal system, in particular those involved in feeding.

The larva period of *L. goodei* represented a time of definition for body shape; many of the relative mensural characters became established in a state that was similar to the condition exhibited by wild adults. Compared to most other cyprinodontids, *L. goodei* can be distinguished by having shallower body depth, relative to body length (see Foster

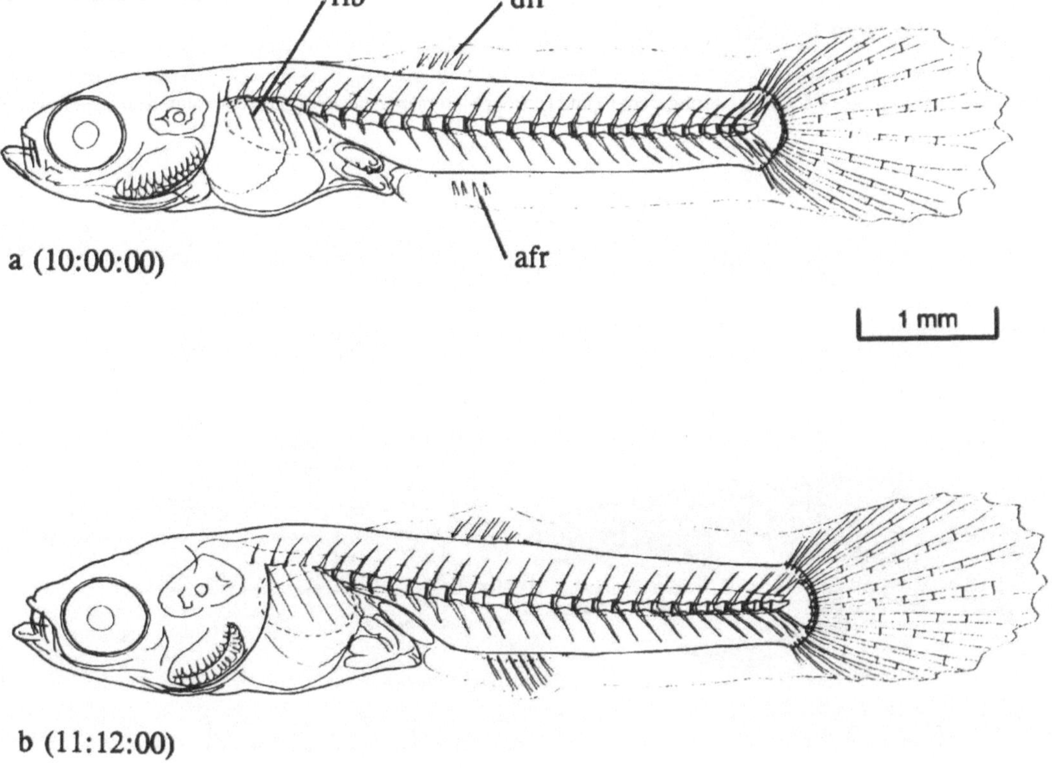

Fig 34 Larvae of *L goodei* during step L2 from a lateral perspective (afr = anal fin ray, dfr = dorsal fin ray, rib = rib)

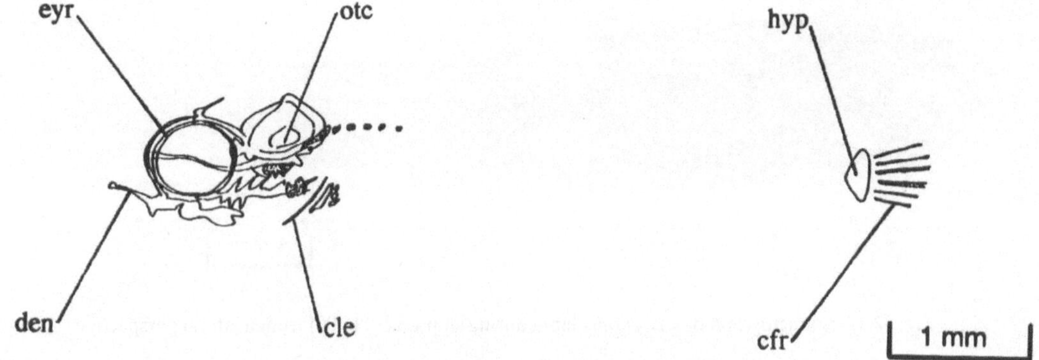

Fig. 35. Cleared and stained larva of *L. goodei* during step L2 (10:00:00) from a lateral perspective. Unstippled structures were stained with alcian blue (mucopolysaccharide in cartilage), stippled structures were stained with alizarin red (calcium phosphate in bone) (cle = cleithrum, cfr = caudal fin ray, den = dentary, eyr = eye ring, hyp = hypural complex, otc = otic capsule).

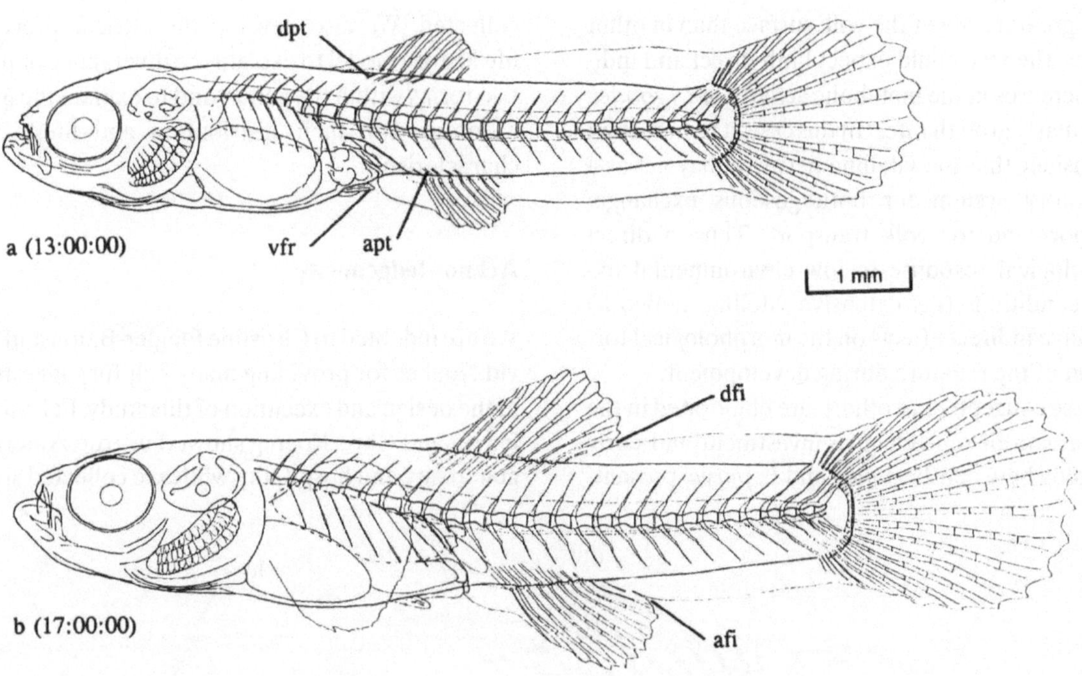

Fig. 36. Larvae of *L. goodei* during step L3 from a lateral perspective (afi = anal finfold identation, apt = anal pterygiophore, dfi = dorsal finfold identation, dpt = dorsal pterygiophore, vfr = pelvic fin ray).

Fig 37 Photomicrograph of a *L goodei* larva during step L3 (18 16 00) from a lateral perspective

1967). We suspect that such a phenotypic difference can be explained in terms of allometric growth rates, which in turn can be affected by the quantity and quality of yolk investment. If the vitelline networks of *L. goodei* embryos extend over an absolutely greater area of the yolk surface than in other species, then we could expect both direct and indirect increases in the metabolic activity of *L. goodei,* particularly growth rates. In this case, it is important to consider that the vitelline network may act as a circulatory system for both gaseous exchange/transport and for yolk transport. Thus, a direct physiological response to low environmental oxygen conditions (i.e. extensive vitelline network) may have indirect effects on the morphological formation of the creature during development.

These concepts, and others, are elaborated in the comparison of reproductive investment and early ontogeny between *L. goodei* and *L. parva,* presented in the third and final instalment of this series (Crawford & Balon 1994b). We examine the developmental differences exhibited between these two species and interpret them with respect to differences in the environments from which they were collected. We also develop the altricial-precocial life history model to explain the divergence of these two forms within the genus, and to explain the gross differences in their distribution and life-history characteristics.

Acknowledgements

We are indebted to Christine Flegler-Balon and David Noakes for providing many helpful suggestions in the design and execution of this study. Felicia Coleman and Chris Koenig showed us true American generosity and hospitality while we collected speci-

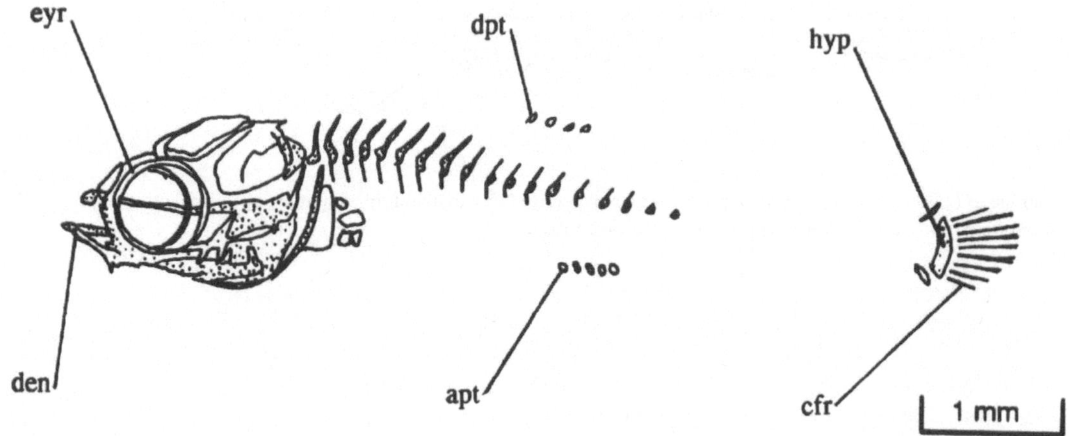

Fig 38 Cleared and stained larva of *L goodei* during step L3 (13 00 00) from a lateral perspective Unstippled structures were stained with alcian blue, stippled structures were stained with alizarin red (apt = anal pterygiophore, cfr = caudal fin ray, den = dentary, dpt = dorsal pterygiophore, eyr = eye ring, hyp = hypural complex)

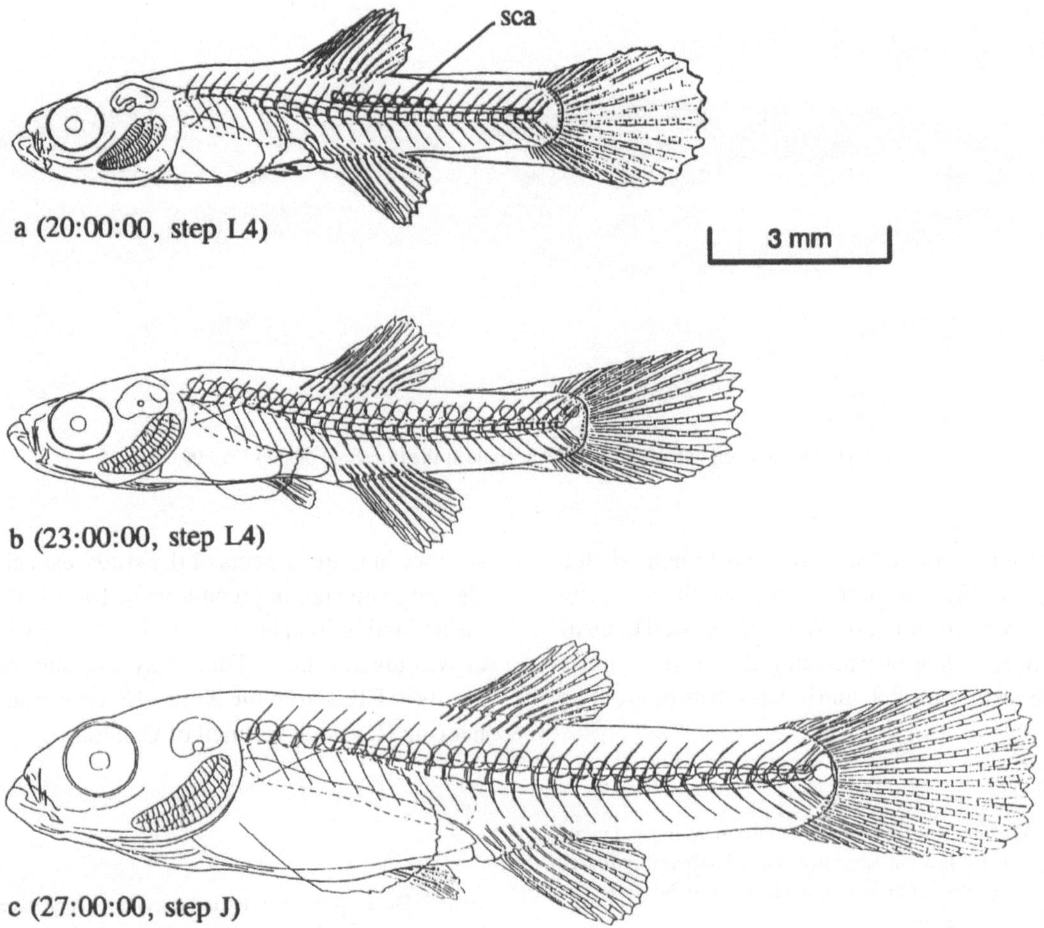

Fig. 39. Larvae and a juvenile of *L. goodei* during steps L4 and J from a lateral perspective (sca = scale). Scale formation began posterior of mid-body and proceeded anteriorly and posteriorly. Only single horizontal rows of scales are indicated on these specimens, although successive dorsal and ventral rows formed concurrently, but at slightly slower rates than at midline.

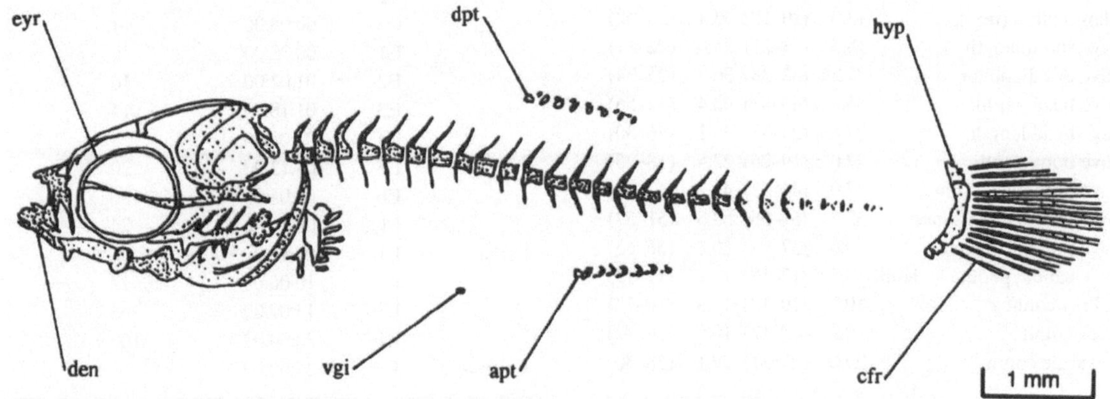

Fig. 40. Cleared and stained larva of *L goodei* during step L4 (20:00:00) from a lateral perspective. Unstippled structures were stained with alcian blue, stippled structures were stained with alizarin red (apt = anal pterygiophore, cfr = caudal fin ray, den = dentary, dpt = dorsal pterygiophore, eyr = eye ring, hyp = hypural complex, vgi = pelvic girdle).

Fig. 41. Photomicrograph of a *L. goodei* juvenile at the threshold of step J (27:00:00).

mens from the field. Mark Wiercinski helped electrofish on the collecting trips. Skuli Skúlason, Dwight Watson and Terry Wheeler showed keen interest in the ideas surrounding the altricial-precocial life history model. marie Rush was of great assistance in many aspects of the study, especially the clearing and staining techniques. Todd Buttenham performed his magic in the darkroom. Patrice Baker was always there. This study was supported by grants to EKB from the Natural Sciences and Engineering Research Council of Canada.

Table 2. A comparison of mean (min–max) mensural and meristic characters for juvenile specimens (step J, laboratory reared, N = 25) and pooled adult male and female (wild, N = 50) specimens of *Lucania goodei*.

Character	Juvenile		Adult	
	mean	(min–max)	mean	(min–max)
Standard length (mm)	12.0	(10–15)	33.4	(26–42)
Relative snout length	18.3	(13–21)	25.6	(22–31)
Relative eye diameter	35.2	(32–38)	30.3	(25–34)
Relative head depth	55.4	(46–63)	48.4	(42–56)
Relative head length	27.6	(24–32)	28.1	(26–30)
Relative body depth	22.0	(19–26)	22.5	(18–25)
Relative prepelvic distance	47.0	(44–52)	46.2	(42–50)
Relative predorsal distance	52.7	(51–58)	53.5	(51–57)
Relative preanal distance	59.0	(57–65)	60.8	(56–65)
Relative caudal peduncle depth	13.5	(12–17)	11.9	(11–13)
Dorsal ray count	10.5	(10–12)	11.3	(10–13)
Anal ray count	10.2	(10–12)	10.6	(10–12)
Lateral scale count	29.0	(27–31)	29.1	(28–30)

Table 3. Developmental rates for *Lucania goodei*, based on the thresholds of developmental steps defined in Table 1.

Period	Step	Age (day:hour:min)	Duration (hours)
Embryo	C1	00:00:00	1.5
	C2	00:01:30	6.5
	C3	00:08:00	14
	E1	00:22:00	4
	E2	01:02:00	16
	E3	01:18:00	14
	E4	02:08:00	28
	E5	03:12:00	20
	E6	04:08:00	56
	F1	06:16:00	24
Larva	L1	07:16:00	56
	L2	10:00:00	72
	L3	13:00:00	168
	L4	20:00:00	168
Juvenile	J	27:00:00	

References cited

Alberch, P 1985 Problems with the interpretation of developmental sequences Syst Zool 34 46–58

Arndt, R G E 1971 Ecology and behavior of the cyprinodont fishes *Adinia xenica, Lucania goodei, Leptolucania ommata* Ph D Thesis, Cornell University, Ithaca 344 pp

Balon, E K 1975 Terminology of intervals in fish development J Fish Res Board Can 32 1663–1670

Balon, E K 1980 Early ontogeny of the lake charr, *Salvelinus (Cristivomer) namaycush,* pp 485–562 *In* E K Balon (ed) Charrs Salmonid Fishes of the Genus *Salvelinus,* Perspectives in Vertebrate Science 1, Dr W Junk Publishers, The Hague

Balon, E K 1985 The theory of saltatory ontogeny and life history models revisited pp 13–30 *In* E K Balon (ed) Early Life Histories of Fishes New Developmental, Ecological and Evolutionary Perspectives, Dev in Env Biol Fish 5, Dr W Junk Publishers, Dordrecht

Balon, E K 1986 Saltatory ontogeny and evolution Rivista di Biologia 79 151–190

Balon, E K 1989 The epigenetic mechanisms of bifurcation and alternative life-history styles pp 467–501 *In* M N Bruton (ed) Alternative Life-History Styles of Animals, Kluwer Academic Publishers, Dordrecht

Balon, E K 1990 Epigenesis of an epigeneticist the development of some alternative concepts on the early ontogeny and evolution of fishes Guelph Ichthyol Rev 1 1–42

Balon, E K & C Flegler-Balon 1985 Microscopic techniques for studies of early ontogeny in fishes problems and methods of composite descriptions pp 33–55 *In* E K Balon (ed) Early Life Histories of Fishes New Developmental, Ecological and Evolutionary Perspectives, Dev in Env Biol Fish 5, Dr W Junk Publishers, Dordrecht

Briggs, J C 1958 A list of Florida fishes and their distribution Bull Flor State Mus Biol Sci 2 223–318

Crawford, S S 1993 Ecomorphological comparison of early ontogeny in species of the genus *Lucania* (Pisces Cyprinodontidae) Ph D Thesis, University of Guelph, Guelph 264 pp

Crawford, S S & E K Balon 1994a Alternative life histories of the genus *Lucania* 1 Early ontogeny of *L parva,* the rainwater killifish Env Biol Fish 40 349–389

Crawford, S S & E K Balon 1994b Alternative life histories of the genus *Lucania* 3 An ecomorphological explanation of altricial (*L parva*) and precocial (*L goodei*) forms Env Biol Fish 41 369–402

Crawford, S S, E K Balon & K S McCann 1994 A mathematical technique for estimating blastodisc yolk volume ratios in spherical eggs, with an example from fishes of the genus *Lucania* Can J Zool (submitted)

Cunningham, J E R & E K Balon 1985 Early ontogeny of *Adinia xenica* (Pisces, Cyprinodontiformes) 1 The development of embryos in hiding Env Biol Fish 14 115–166

Dineen, J W 1974 Fishes of the Everglades pp 375–385 *In* P J Gleason (ed) Environments of South Florida Present and Past, Miami Geol Soc Memoir 2, Miami

Foster, N R 1967 Comparative studies on the biology of killifishes (Pisces, Cyprinodontidae) Ph D Thesis Cornell University, Ithaca 369 pp

Hodson, P V & J B Sprague 1975 Temperature-induced changes in acute toxicity of zinc to Atlantic salmon (*Salmo salar*) J Fish Res Board Can 32 1–10

Hubbs, C L & R R Miller 1965 Studies of cyprinodont fishes 22 Variation in *Lucania parva,* its establishment in western United States, and description of a new species from an interior basin in Coahuila, Mexico Misc Publ, Museum of Zoology, University of Michigan 127 1–104

Hubbs, C L, B W Walker & R E Johnson 1943 Hybridization in nature between species of American cyprinodont fishes Contrib Lab Vert Biol Univ Mich 23 1–21

Kilby, J D 1955 The fishes of two gulf coastal marsh areas of Florida Tulane Stud Zool 2 175–247

Kuntz, A 1916 Notes on the embryology and larval development of five species of teleostean fishes Bull U S Bureau Fish 34 407–429

Laale, H W 1980 The perivitelline space and egg envelopes of bony fishes a review Copeia 1980 210–226

Lee, D S, C R Gilbert, C H Hocutt, R E Jenkins, D E McAllister & J R Stauffer Jr 1980 Atlas of North American freshwater fishes North Carolina State Museum of Natural History, Raleigh 867 pp

Loftus, W F & J A Kushlan 1987 Freshwater fishes of Southern Florida Bull Flor State Mus Biol Sci 31 1–344

Oppenheimer, J M 1937 The normal stages of *Fundulus heteroclitus* Anat Rec 68 1–15

Rosenau, J C, G L Faulkner C W Hendry Jr & R W Hull 1977 Springs of Florida State of Florida Department of Natural Resources, Bulletin Number 31 461 pp

Sommer, W 1980 The blue-fin, *Lucania goodei,* a coldwater top minnow Trop Fish Hobb 28 46–56

Tabb, D C & R B Manning 1961 A checklist of the flora and fauna of northern Florida Bay and adjacent brackishwaters of the Florida mainland collected during the period July 1957, through September 1960 Bull Mar Sci Gulf and Caribbean 11 552–649

Tagatz, M E 1968 Fishes of the St Johns River, Florida Quart J Flor Acad Sci 30 25–50

Veenstra, R S 1987 Ecomorphological comparisons of early development in an altricial (*Lucania parva*) and in a precocial (*Lucania goodei*) cyprinodont M Sc Thesis, University of Guelph, Guelph 95 pp

Environmental Biology of Fishes **41**: 369–402, 1994.

Alternative life histories of the genus *Lucania:* 3. An ecomorphological explanation of altricial (*L. parva*) and precocial (*L. goodei*) species

Stephen S. Crawford & Eugene K. Balon
Institute of Ichthyology and Department of Zoology, University of Guelph, Guelph, Ontario N1G 2W1, Canada

Received 24.11.1993 Accepted 14.3.1994

Key words: Development, Saltatory ontogeny, Ecomorphology, Speciation, Evolution, Cleavage, Embryo, Larva, Juvenile, Cyprinodontidae, Alprehost, Altricial-precocial model

Synopsis

Important differences were observed in the early ontogenies of *Lucania parva* and *Lucania goodei.* These differences can be explained in terms of the altricial-precocial model of speciation. *Lucania parva* can be recognized as an altricial form that produces many eggs with relatively little yolk investment, compared to the more precocial *L. goodei.* Many of the differences observed in embryo, larva and juvenile specimens appear to be related to these differences in gamete investment. Accelerated developmental rates in the precocial form suggest that paedomorphosis is an important proximate mechanism in the bifurcation of alternative life-history styles in this genus. Some morphological characteristics, such as vitelline circulation and body shape, may be transformations associated with the particular environmental conditions in which the animals must develop and survive. Our observations suggest that these two species in the genus *Lucania* have followed different ontogenic trajectories in response to prevailing environmental conditions.

Introduction

This paper is the third and final instalment of a comparative series on early ontogeny in the genus *Lucania.* The first instalment in the series (Crawford & Balon 1994a) was a detailed description and interpretation of early ontogeny of the rainwater killifish, *L. parva.* The second instalment in the series (Crawford & Balon 1994b) was a companion study on the early ontogeny of the bluefin killifish, *L. goodei.* The reader is strongly urged to review these papers before proceeding with the interspecific comparison given below.

Saltatory ontogeny

In its latest form, the theory of saltatory ontogeny (Balon 1986a) describes a progression of stabilized (homeorhetic) states during the continuous course of development. These states are thought to represent the natural tendency of developing systems to resist fluctuations in energy state. Changes in structure and function of an organism accumulate until developmental constraints are encountered. At this point, a more rapid but less stable transition in structure or function (i.e. threshold) takes a developing organism to a new stabilized state. In this way, the smaller, quantitative changes that accumulate during the stabilized steps may be transformed into an important, qualitative difference in the organism

(Waddington 1975, Balon 1989). The endogenous developmental rhythm described by Gorodilov (1992) may be an additional manifestation of saltatory ontogeny. Together, the alternation of these steps and thresholds during ontogeny can be recognized as a saltatory phenomenon (e.g. Lampl et al. 1992).

For the purpose of this study, we accept the theory of saltatory ontogeny (Balon 1986a) as a working model of morphological development. This theory allows us to view developing systems from the general perspective of energetic stability and self-organization, while providing a practical method for classifying life-history intervals.

Life-history model of ontogeny

Balon (1975, 1981, 1990) has developed a life-history model of ontogeny in fishes as a hierarchical representation of development, based on the theory of saltatory ontogeny. The application of this theory yields a description of ontogeny that is more realistic than using the 'normal stages' of traditional embryology. We employ a form of the life-history model that has been used to describe development in a variety of fish species (e.g. Balon 1977, McElman & Balon 1979, Paine & Balon 1984, Cunningham & Balon 1985, Goto 1990, Holden & Bruton 1992). In this model, the term 'stage' is used to mean an instantaneous state of ontogeny, while the terms 'step', 'phase' and 'period' refer to ontogenic intervals of increasing duration (Balon 1985a). A sequence of steps comprise a phase which, with other phases, comprises an ontogenic period. In this manner, the particular developmental state of an individual can be identified within a biologically meaningful, hierarchical system of classification.

The life-history model also provides a consistent basis for comparing ontogenic patterns, especially between groups that exhibit differences in developmental rates. Thresholds may occur at different developmental ages or states, and this variation can lead to very confusing interpretations of ontogenic processes. However, by describing structural development with a common scale, it is possible to identify the thresholds at which ontogenic trajectories depart (Bruton 1989, Wake 1990). In this manner, the theory of saltatory ontogeny leads us to view the process of ontogeny as an integrated component within the process of phylogeny (Balon 1989).

Altricial-precocial life-history styles

Given the life-history model of ontogeny, it is possible to classify the various modes of ontogeny in different species (Balon 1981). A detailed knowledge of intra- and inter-specific differences in ontogeny may help us to make inferences about the mechanisms responsible for phylogenetic separation of new species within a lineage (Oster & Alberch 1982). In addition, knowledge of ontogenic mechanisms underlying speciation in the past may enable us to predict developmental possibilities that could be expected in the future.

Ontogenic comparisons of various species have led to the recognition of alternative life-history styles that are adapted to local environmental conditions (e.g. Bruton 1989). These comparisons, placed in the context of the theory of saltatory ontogeny, have led to the concept of altricial-precocial homeorhetic states, or 'alprehost' as described by Balon (1989, 1990). In this model, shifts in various life-history parameters during less stable thresholds can result in the bifurcation of alternate evolutionary pathways during ontogeny. The altricial form is usually recognized as being smaller and less morphologically advanced at particular thresholds, compared to the precocial form (Bruton 1989, especially his Table 1, Duellman 1989, Balon 1990). These alternative forms have been reported in several species of fish (e.g. Balon 1980e, 1984, Noakes & Balon 1982, Liem & Kaufman 1984, Noakes et al. 1989, James & Bruton 1992) and other vertebrates (e.g. Duellman 1989, Perrin 1989). It has been suggested that small differences between altricial and precocial forms in each generation can become exaggerated over time, perhaps resulting in the bifurcation of species that are successful in different environments (Løvtrup 1989, Balon 1990).

Environmental conditions and life-history variables

The consequences of developmental bifurcations can be viewed as evolutionary alternatives to local environmental conditions. In this study, we define environmental stability as the degree of constancy and predictability of changes in local environmental conditions (Bruton 1989, Holm 1989). Other investigations have shown that altricial forms are typically associated with unstable or colonial environments, where competition for resources is usually less intense than in more stable environments (May 1974, Balon 1981, 1986b, Noakes & Balon 1982, Stearns 1982, Bruton 1989). These studies have concentrated on the relationships between environmental stability and various attributes of altricial and precocial forms. We have selected two of these relationships and the causal mechanisms that have been hypothesized to account for them:

1. Reproductive investment
If environmental instability reduces the probability of offspring survival, then organisms in less stable environments should offset higher mortality rates by increasing the number of gametes produced. If the energy available for reproduction is also limited, then organisms in less stable environments should also exhibit a decline in energy investment per offspring, concurrent with increases in gamete number. Therefore, altricial females inhabiting less stable environments should produce a greater number of smaller eggs, relative to those in more stable environments (Balon 1979, 1990, Stearns 1982, Bruton 1989).

In addition, if parental energy investment in the gametes limits the extent to which offspring may develop, then offspring with greater energy supplies should be able to progress further through ontogeny before the beginning of exogenous feeding (Balon 1977, 1986b, Bruton 1989). This difference in developmental progress may be recognized in the morphological state of offspring at the time when parental energy investment has been exhausted. Therefore, altricial offspring developing in less stable environments, with smaller eggs, should exhaust their endogenous energy supply at an 'earlier' developmental state, relative to those in more stable environments (Flegler-Balon 1989, Ward 1989).

2. Rate of morphological development
If the risk of mortality decreases with morphological development, then offspring that develop under conditions associated with higher mortality should exhibit accelerated rates of development (Stearns 1980). In this manner, the organisms could decrease the overall risk of mortality during their lifetime by reducing the window of highest risk. Therefore, altricial offspring developing in less stable environments should develop at a more rapid rate than those in more stable environments (Bruton 1989).

An evolutionary model

By looking at developmental similarities and differences among closely related species, we can recognize the evolutionary processes that have drawn them apart (Alberch 1985). Using this knowledge, we can also extend our reach into the future and begin to make predictions about potential evolutionary pathways. For these reasons, we have compared the morphological development of two closely-related North American killifishes, *Lucania parva* and *L. goodei.* These species inhabit very different environments, and represent an exceptional 'natural experiment' with which to explore the life-history model described above (Freeman 1982, cf. Constanz 1979).

The order Cyprinodontiformes is an extremely large and diverse group of teleost fishes, native to all tropical and sub-tropical waters, except those of Australia (Foster 1967). In total, the order comprises several hundred species, belonging to five major families. The Cyprinodontidae is the largest family among the Cyprinodontiformes, with more than 80 genera and 600 species, approximately 50 of which are native to North America (Foster 1967, McClane 1978). Members of this family are small, elongate fishes with terminal or superior mouths, protractile premaxillaries, cycloid scales and rounded caudal fins. In contrast to other families in the order, cyprinodontid fishes exhibit an oviparous rather than viviparous reproductive style. Within the family Cyprinodontidae, 14 genera belong to the subfamily Fundulinae, including the monophyletic groups (Parenti 1981): *Fundulus, Adinia, Leptolucania* and *Lucania.*

Fishes of the genus *Lucania* Girard, 1859 are distinguished from all other cyprinodontids by the presence of an independent cartilage block between the interarcual cartilage and the articulation point of the second pharyngobranchial element (Parenti 1981). There are currently two species in the genus: the rainwater killifish, *Lucania parva* (Baird, 1855) and the bluefin killifish, *Lucania goodei* Jordan, 1880 (Hubbs & Miller 1965, Lee et al. 1980).

There are very clear differences in the geographic distributions and habitats of these two species. *Lucania parva* is found in saltmarshes and estuaries along the Atlantic and Gulf Coasts of the United States, in a manner that is similar to other temperate cyprinodontiforms such as *Fundulus, Menidia* and *Floridichthys* (Duggins 1980). While they may survive full seawater (Breder 1948, Chris Koenig personal communication), they prefer brackish waters near a regular supply of fresh water (Jordan & Evermann 1896, Kilby 1955, Hubbs & Miller 1965, Springer & Woodburn 1960). *Lucania parva* is rarely found in completely fresh water (Hildebrand & Schroeder 1928, Foster 1967), however this species does exhibit some inland penetration of headwater creeks and coastal rivers in the Florida peninsula (Hubbs & Miller 1965, Burgess et al. 1977, Loftus & Kushlan 1987).

In contrast, the distribution of *L. goodei* is largely confined to the inland waters of Florida (Hubbs et al. 1943, Lee et al. 1980, Loftus & Kushlan 1987), in a manner that is similar to many other endemic Florida fish species (Briggs 1958). *Lucania goodei* is almost always found in freshwater creeks, rivers and lakes usually close to hypoxic springs of the Floridan Aquifer (Tagatz 1968, Ager 1971, Dineen 1974). Occasionally, *L. goodei* has also been collected in mildly brackish waters, usually during a flush of fresh water (Hubbs et al. 1943, Kilby 1955, Tabb & Manning 1961, Loftus & Kushlan 1987).

There are only a few locations where investigators have collected *L. parva* and *L. goodei* together (e.g. Foster 1967, Gilbert 1978, Loftus & Kushlan 1987). In these cases, *L. parva* had moved inland to the typical freshwater habitat of *L. goodei*, in water particularly high in calcium salts. The two species have also been collected together at the limits of the *L. goodei* distribution, near brackish coastal waters (Hubbs et al. 1943, Foster 1967, Loftus & Kushlan 1987). Natural hybridization between the two species is extremely uncommon, occurring only when the abundance of *L. parva* is much greater than that of *L. goodei* (e.g. Hubbs et al. 1943). Despite the similarity of reproductive behaviour in the two species (Foster 1967), the pugnacity of *L. parva* males may be partly responsible for this reproductive isolation. The rarity of natural hybrids and the difficulty of laboratory crosses (personal observation) combine to support the claim of reproductive isolation between these species.

Given the ranges and habitats of these two species, it should be possible to locate populations that are in close geographic proximity, yet which live under very different conditions of environmental stability. A comparison of morphological development in *L. parva* and *L. goodei* holds potential for exploring the life-history theories described above. To date, only a handful of studies have attempted to compare directly the biology of *L. parva* and *L. goodei.*

Hubbs et al. (1943) remarked on the allopatric tendencies of *L. parva* and *L. goodei* in a special publication that focussed on the rarity of natural hybridization between the two species. Loftus & Kushlan (1987) also described differences between geographic distributions of these two species, as part of a larger survey of fishes in Florida. However, the environmental conditions that might be associated with observed differences in distribution were not explored in these studies.

In an extensive description of morphological variation in *L. parva,* Hubbs & Miller (1965) made a few direct references to *L. goodei.* Aside from clear differences in adult pigmentation, they recognized very few mensural or meristic differences between the two species. Even the morphology of jaw teeth, a character that had traditionally been used to distinguish *L. parva* and *L. goodei,* appeared to be questionable in this respect. Unfortunately, this comparison of the two species was overshadowed by the great 'ecophenotypic plasticity' exhibited by *L. parva* over its wide geographic distribution (Hubbs & Miller 1965). The issue of environmental stability was not directly addressed by the authors.

Both Foster (1967) and Arndt (1971) presented excellent, general descriptions of the ecology and ethology of *L. parva* and *L. goodei*, especially with reference to other members of the subfamily Fundulinae. Neither of these authors compared directly the life histories of these two species, nor did they consider environmental stability as an important factor in the evolution of the genus.

Dunson & Travis (1991) attempted to describe the effect of environmental conditions on competition between *L. parva* and *L. goodei* in an experimental situation. Their data seem to support the idea that *L. goodei* exhibited a much lower tolerance of salinity, compared to *L. parva.* However, poor experimental design and uncontrolled sources of mortality prevent any general conclusions from this study. Duggins et al. (1983) and Quanwei (1986) both found little evidence of genetic differentiation between these two species. Once again, these authors did not attempt to relate their findings to environmental or morphological variation.

With respect to morphological development, our knowledge of *L. parva* and *L. goodei* is extremely limited. Aside from anecdotal descriptions of field collections, there have been only three studies that provided information on the early ontogeny of these species. Kuntz (1916) described some states of morphological development in *L. parva,* especially as they compared to those of *Cyprinodon variegatus.* Foster (1967) provided notes on the development of both *L. parva* and *L. goodei* in his species descriptions, however morphological states were not directly compared. Veenstra (1987) described changes in body form of *L. parva* and *L. goodei* developing under laboratory conditions, however he did not attempt to relate the development of morphological variation to conditions of environmental stability.

Objective

The general objective of this study is to determine whether the life-history patterns of *L. parva* and *L. goodei* can be explained in terms of the altricial-precocial model of speciation. To accomplish this goal, we provide detailed morphological descriptions of adult specimens from the wild. After examining wild adults, we turn our attention to the descriptions of early ontogeny for each species, based on offspring reared under laboratory conditions (Crawford & Balon 1994a, b). Specifically, we describe the differences in morphological development of the two species, with the purpose of identifying patterns and proximate mechanisms of phenotypic differentiation. Interspecific differences in morphological development will be discussed in relation to environmental conditions in the wild. We also determine if the interspecific differences between *L. parva* and *L. goodei* are consistent with the predictions of life-history theories described above. Finally, we attempt to develop hypotheses regarding the evolutionary history of speciation in the genus *Lucania.*

Materials and methods

The following description is intended to give the reader an overview of techniques employed in the collection, rearing and description of *L. goodei* and *L. parva* for this study. For additional details, the reader should refer to the first two instalments of this comparative series (Crawford & Balon 1994a, b).

Adult specimens of *Lucania parva* were collected with a 15 m seine net (approximately 1 cm^2 mesh) from the drainage sluice between Tower Pond and Picnic Pond in St. Marks Wildlife Refuge, Florida (30° 05′ 15″ N, 84° 09′ 30″ W, Fig. 1). Adult specimens of *Lucania goodei* were collected with a Deka 3000 electrofisher (Deka-Gerätebau, Rudolf Mühlenbein, Germany) from the creek below Newport Spring, Florida (30° 12′ 30″ N, 84° 10′ 30″ W, Fig. 1). The temperature, dissolved oxygen and salinity of water at both sites were measured with a YSI con-

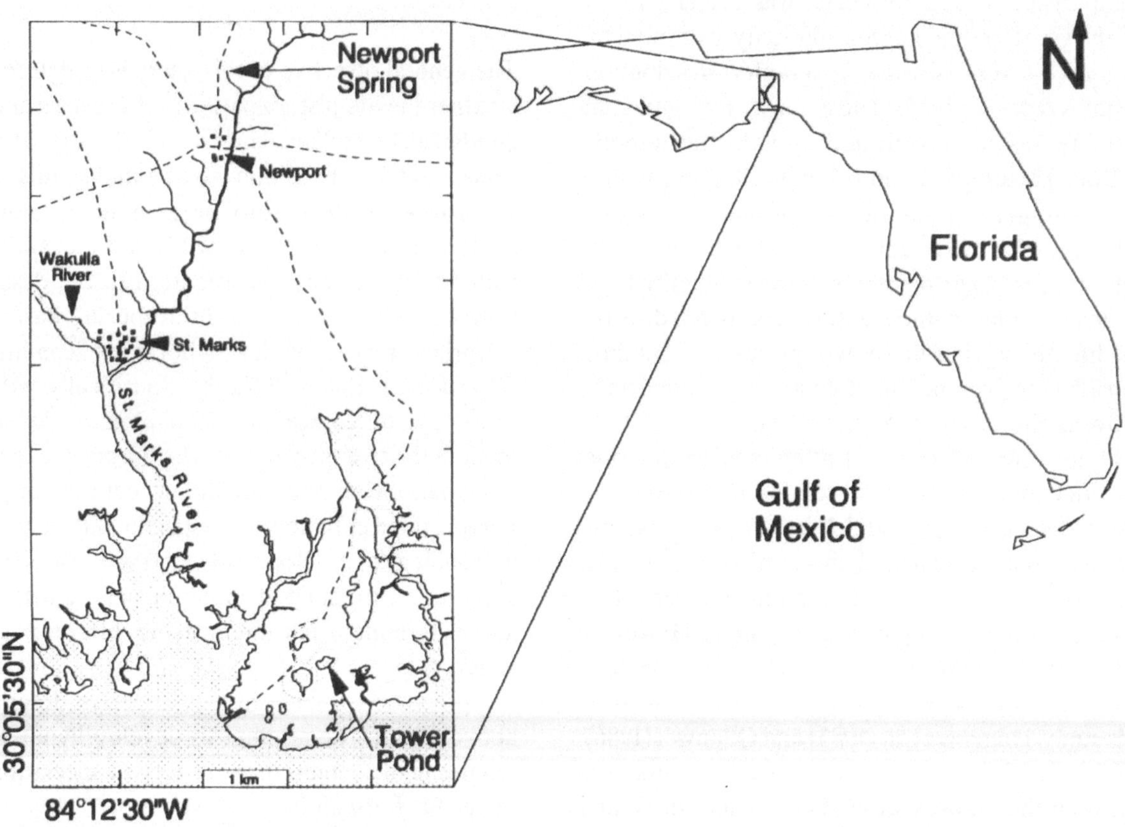

Fig. 1. Sampling locations for wild, adult *Lucania parva* at Tower Pond, St. Marks National Wildlife Refuge and *L. goodei* at Newport Spring, Florida.

ductivity bridge, monthly from March 1989 to March 1990.

Fifty adult specimens (25 males, 25 females) of both species were collected between 2–7 March 1990 and transported live to the laboratory to produce offspring for the study of morphological development. All aquaria in this study were maintained at 25 ±1° C under a 16 h:8 h (light:dark) photoperiod schedule with 2.5 ppt brackish water created by mixing filtered well-water (chemically described by Hodson & Sprague 1975) and Instant Ocean salts. Twice daily, adults were fed to satiation with Murex frozen brine shrimp, *Artemia salina*. An additional fifty specimens (25 males, 25 females) of both species were collected in March 1991 for a morphological comparison of adults from wild populations.

Activated eggs produced during mating events were transferred to individual 1 ml meshed baskets suspended in an aerated 9.8 l incubation aquarium (42 × 26 × 9 cm deep) heated by a Braun Thermomix 1420 heater to 25 ± 0.5° C. The baskets were checked three times daily for decomposing eggs which were discarded, and for newly hatched embryos which were transferred to 100 ml mesh baskets suspended in a 48 l incubation aquarium (30 × 28 × 58 cm deep) that was heated to 25 ±1° C. Twice daily, offspring in these larger incubator baskets were presented with an excess of freshly-hatched, live brine shrimp nauplii.

Randomly selected offspring were removed from the incubator aquaria at sampling ages that conformed to the schedule given in Table 1. This sampling design was based on the appearance of morphological structures and/or behaviours that characterize the various steps and periods in the life-history model of ontogeny (Balon 1986a). Using this model, we were able to distinguish between 'morphological state' (i.e. progress of structural complexity) and 'chronological state' (i.e. age in units of time). In all cases, the developmental age of specimens is given in the format days:hours:minutes.

Mensural and meristic measurements of eggs, embryos, larvae, and juveniles were made from drawings or photomicrographs of live specimens taken at various magnifications, following the methods of Hubbs & Lagler (1958), Hubbs & Miller (1965), Mansueti & Hardy (1967) and Duggins (1980). We used simple descriptive statistics for the comparison of morphological characters between *L. parva* and *L. goodei*. Graphical techniques were used to examine the normality of data sets, where necessary. For the morphological comparison of wild adults, we used a principal components analysis to identify characters that accounted for variation within, and between, species. We also used statistical tests to detect differences in the means (Student's t-test) and variances (F-test) of selected morphological characters. Initially, two-tailed tests were performed to determine whether significant ($p < 0.05$) differences existed between the two samples, followed by one-tailed tests to support the claim that the measurements from one sample were significantly ($p < 0.05$) greater than the other. For the sake of brevity and clarity in the text, we have used the term 'significant' exclusively to indicate differences that were statistically significant (p <

Table 1. Classification of developmental periods, phases and steps of *L. goodei* and *L. parva* based on the appearance of morphological and behavioural characters (Cunningham & Balon 1985). Sampling intervals were maintained until characters of the next step were recognized.

Period	Phase	Step	Sampling interval	Morphological characters
Embryo	Cleavage egg	C1	15 min	– unactivated egg – perivitelline space – blastodisc
		C2	30 min	– cleavage
		C3	1 h	– epiboly – embryonic shield
	Embryo	E1	1 h	– optic vesicles – germ ring closure
		E2	2 h	– somites – blood elements – otic vesicles – heart
		E3	2 h	– blood circulation – vitelline network – pectoral fins
		E4	4 h	– segmental vessels – hepatic vein – olfactory vesicles
		E5	4 h	– caudal finfold circulation – dorsal/anal finfolds – caudal rays
		E6	8 h	– subclavian vein – gut – profundal vein
	Free embryo	F1	8 h	– hatching – swimbladder
Larva		L1	12 h	– first feeding
		L2	12 h	– dorsal/anal rays – ribs
		L3	24 h	– pelvic fins
		L4	24 h	– scales
Juvenile	Juvenile	J		– finfold resorption

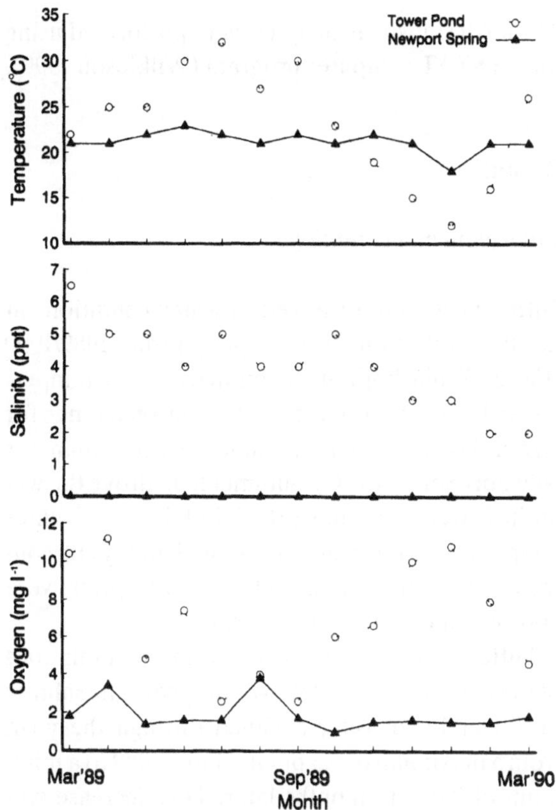

Fig. 2. Physical conditions of the aquatic environments at Tower Pond (*L. parva*) and Newport Spring (*L. goodei*), Florida from March 1989 to March 1990.

0.05). All statistical analyses were performed using the SYSTAT computer program (Wilkinson 1990).

Results

Environmental conditions

Differences were observed in water conditions at the two collection sites sampled during 1989/1990 (Fig. 2). While both sites exhibited average temperatures between 20 and 25° C, the seasonal range for Tower Pond (20° C) was much greater than for Newport Spring (5° C). Summer heat drove the water in Tower Pond above 30° C in July, while winter temperatures fell below 15° C in January. In contrast, water temperatures at Newport Spring were approximately 21° C all year round.

Differences were also observed in water salinities of the two collection sites over the year. The salinity of Tower Pond water declined through the year, from a maximum of 6.5 ppt in March 1989 to a minimum of 2 ppt 12 months later. This decrease was probably due to the combined effect of above-average rainfall in St. Marks Refuge in 1989, and the lack of compensatory saltwater pumping into the system (Red Gidden personal communication). These changes at Tower Pond may have been small in comparison to other habitats where *L. parva* is found, but the seasonal range of salinity at Tower Pond was still great in comparison to the consistently fresh water that flowed year-round at Newport Spring.

Dissolved oxygen measurements also revealed differences between Tower Pond and Newport Spring. First, oxygen levels fluctuated over a relatively wide range in Tower Pond, with a winter maximum of 11.2 mg l^{-1} and a summer minimum of 2.6 mg l^{-1}. In contrast, the water from Newport Spring was consistently low in dissolved oxygen (approximately 2.0 mg l^{-1}) during the entire year. We recognize the possibility that the flow of water immediately below the Newport Spring pool (including a small waterfall) may have increased the dissolved oxygen concentration, although we did not examine this phenomenon.

Table 2. Factor loadings from a principal components analysis of variation among 12 mensural characters for wild adult specimens of *L. parva* and *L. goodei*. All characters were measured as absolute values. The three factor loadings that dominated the second principal component (PC2) are indicated with an asterisk.

Mensural character	PC1	PC2	PC3
Standard length	0.966	0.138	0.108
Snout length	0.745	0.226	- 0.593
Eye diameter	0.825	- 0.283	- 0.078
Head length	0.919	0.153	- 0.230
Head depth	0.905	- 0.293*	- 0.151
Head width	0.901	0.229	- 0.016
Body depth	0.895	- 0.381*	0.121
Body width	0.905	0.088	0.314
Predorsal distance	0.956	0.207	0.107
Preventral distance	0.957	0.198	0.126
Preanal distance	0.945	0.196	0.192
Caudal peduncle depth	0.774	- 0.595*	- 0.043
Percent total variation	79.9	7.8	5.2

Adult morphology

A principal components analysis was performed on the absolute values of 12 standard morphological characters (Table 2) for 25 adult males and 25 adult females of *L. parva* and *L. goodei* collected in the wild. The first two components of variation demonstrated clearly the sexual and species dimorphism within, and between, these species.

The first principal component (PC1) accounted for almost 80% of total variation in the data, with high factor loadings observed for each character (Table 2). The factor scores for PC1 exhibited a very strong linear correlation with standard length ($PC1 = 0.243(SL)–8.322$, $r^2 = 0.932$, $p < 0.001$, Fig. 3). In both *L. parva* and *L. goodei,* females were significantly longer (standard length) than males. An examination of length-frequency distributions of wild specimens suggested that both samples were comprised of a single year class.

After the general effect of body size had been removed by PC1, the second principal component (PC2) accounted for 7.8% of the total variation, and approximately 39% of the remaining variation (Table 2). This axis was dominated by relatively strong, negative factor loadings for characters that described general depth of the body (i.e. head depth,

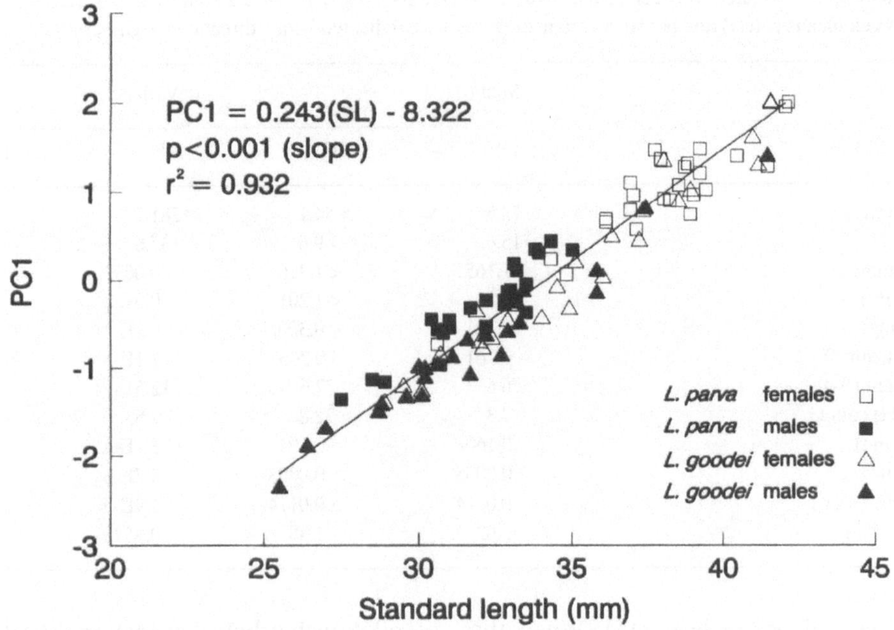

Fig 3 A linear regression of factors loadings resulting from the first component of variation (PC1) and standard length The principal components analysis was performed on 12 morphological characters (Table 2) for 25 females (open symbols) and 25 males (closed symbols) of both *L parva* (squares) and *L goodei* (triangles) The solid line represents the linear regression given on the diagram

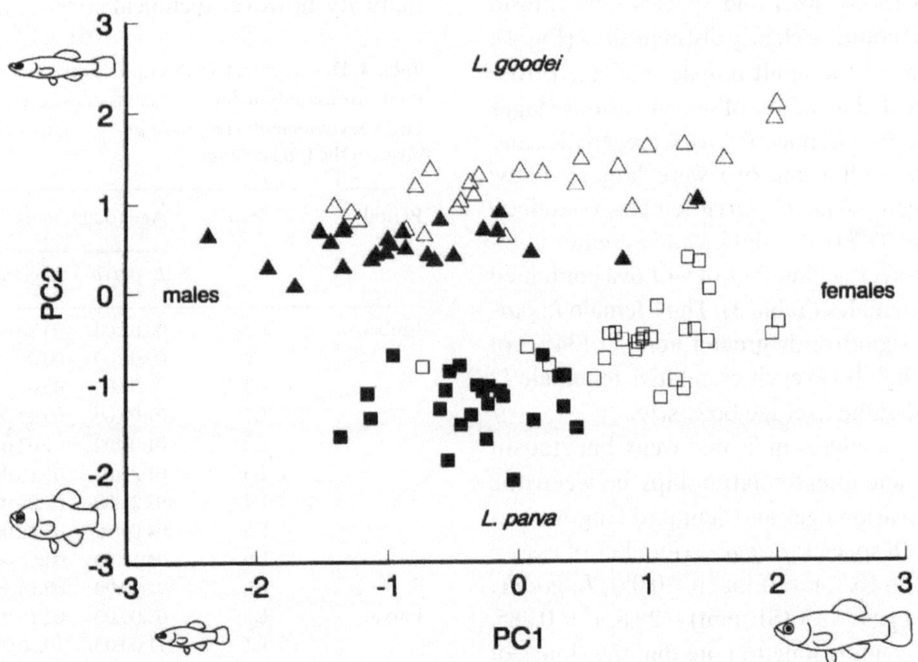

Fig 4 The relationship of the first two principal components of variation based on 12 morphological characters (Table 2) for 25 females (open symbols) and 25 males (closed symbols) of both *L parva* (squares) and *L goodei* (triangles) The first principal component (PC1) describes general body size, while the second principal component describes body depth

Table 3. A comparison of characters between 25 unactivated eggs (step C1) of *L. parva* (LP) and *L. goodei* (LG). Significant ($p < 0.05$) differences between means (t-test) and between variances (F-test) are indicated with a directional sign (< or >).

Character	Mean		Variance	
	LP	LG	LP	LG
Ovarian egg count	72.5	> 54.1	281.7	224.5
Clutch size	15.6	> 9.8	137.6	84.1
Egg diameter (mm)	1.165	< 1.316	0.0037	0.0041
Egg volume (mm^3)	0.834	< 1.201	0.0162	0.0308
Egg dry mass (mg)	0.231	< 0.300	6.8E-4	6.5E-4
Egg density (mg mm^{-3})	0.261	0.246	1.1E-3	7.1E-4
Egg water content (%)	76.6	77.6	12.51	> 3.87
Oil globule (OG) count	7.3	< 22.8	7.88	< 46.17
OG diameter (mm)	0.160	> 0.104	8.1E-4	> 1.7E-4
OG volume (mm^3)	0.0035	> 0.0008	2.7E-6	> 7.3E-8
Total OG volume (mm^3)	0.0214	> 0.0174	1.9E-5	< 5.9E-6
OG/egg volume (%)	2.62	> 1.48	0.6595	> 0.0882

body depth, caudal peduncle depth). When the specimens were grouped by species, both sexes of *L. parva* were found to be significantly deeper in body than those of *L. goodei.* When the first and second components of variation were plotted against each other, the combined effects of sexual dimorphism (body size) and species dimorphism (body depth) could be clearly distinguished (Fig. 4).

Dissections of the adult females collected from the wild revealed a variety of ova at various stages of oogenesis. Previtellogenic ova were smaller and white, while vitellogenic ova were larger, yellow and translucent. Wild *L. parva* females contained an average of 72.5 ova, a significantly greater number than the average number of 54.1 ova contained in *L. goodei* females (Table 3). Thus, female *L. parva* carried a significantly greater number (34%) of eggs within their body cavities, relative to female *L. goodei* of the same average body size.

Regression analyses indicated weak but statistically significant linear relationships between the number of ovarian eggs and standard length of females for both species; *L. parva* (number of eggs = 2.5 (SL mm) – 16.3, $r^2 = 0.152$, $p < 0.05$), *L. goodei* (number of eggs = 2.5 (SL mm) – 29.8, $r^2 = 0.285$, $p < 0.05$). It is interesting to note that the slopes of these relationships were the same for both species, while the negative y-intercept for *L. parva* was approximately one half that for *L. goodei.* If this linear relationship between maternal body size and egg capacity extended back to first egg production, then *L. parva* would begin producing eggs at a smaller body size, and perhaps at a younger age and state of development, relative to *L. goodei.* It is possible to test this prediction by rearing offspring to sexual maturity, however technical constraints prohibited

Table 4. Developmental rates for *L. parva* and *L. goodei,* based on the thresholds of developmental steps defined in Table 1. The lag in developmental rate of *L. parva,* compared to *L. goodei,* is given in the final column.

Period	Step	Age (dd:hh:mm)		Lag (h)
		L. parva	*L. goodei*	
Embryo	C1	00:00:00	00:00:00	
	C2	00:02:00	00:01:30	0.5
	C3	00:09:00	00:08:00	1
	E1	01:00:00	00:22:00	2
	E2	01:04:00	01:02:00	2
	E3	01:22:00	01:18:00	4
	E4	02:20:00	02:08:00	8
	E5	04:00:00	03:12:00	12
	E6	04:16:00	04:08:00	8
	F1	07:00:00	06:16:00	8
Larva	L1	08:00:00	07:16:00	8
	L2	11:00:00	10:00:00	24
	L3	15:00:00	13:00:00	48
	L4	19:00:00	20:00:00	– 24
Juvenile	J	32:00:00	27:00:00	120

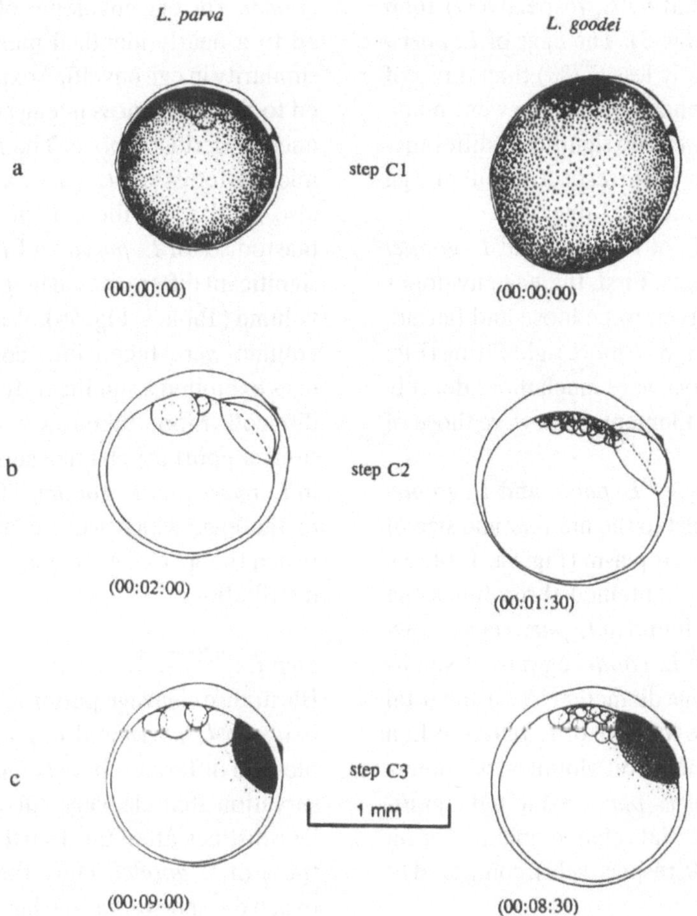

Fig. 5. Illustrations of *L. parva* and *L. goodei* during the cleavage phase of the embryo period from a lateral perspective (a – step C1 unactivated eggs, b – step C2 first cleavage, c – step C3 onset of epiboly).

us from doing so. There was no significant linear relationship between maternal body size and egg size for either species.

Comparative morphological development

Developmental rates

One of the most striking differences in the ontogenies of *L. parva* and *L. goodei* is in the rate of development (Table 4). From the first step onward, *L. parva* took longer than *L. goodei* to reach virtually all of the selected developmental thresholds. At the beginning of the embryo phase (step E1), *L. parva* offspring were approximately 2 hours older than *L. goodei.* This difference increased to 8 hours by step E4, and remained nearly the same until the beginning of the larva period (step L1). Differences between the two species increased during the larva period, as *L. goodei* offspring development continued to accelerate, relative to those of *L. parva.* A single exception to this trend was observed at the beginning of step L4 when *L. parva* exhibited the first scales slightly before *L. goodei.* In the end, offspring of *L. parva* were observed to achieve the juvenile phenotype (step J) a full 120 hours (5 days) after those of *L. goodei.*

Step C1

There were clear differences between the species with respect to egg size. The average diameter and volume of unactivated *L. parva* eggs were significantly smaller (89% and 69%, respectively) than those of *L. goodei* (Table 3). The eggs of *L. parva* also weighed significantly less (77%) than those of *L. goodei*. However, when corrections were made for egg volume, there were no significant differences between the species with respect to either egg density or egg water content (Table 3).

The unactivated eggs of *L. parva* and *L. goodei* differed in many respects. First, the egg envelopes of *L. parva* were observed to be loose and flaccid, while those of *L. goodei* were more tight fitting (Fig. 5a). The eggs of *L. parva* were much more densely covered with adhesive filaments than were those of *L. goodei*.

The unactivated eggs of *L. parva* and *L. goodei* also differed with respect to the number and size of oil globules within the cytoplasm (Fig. 5a, Table 3). The eggs of *L. goodei* contained three times the number of oil globules found in *L. parva* eggs, however the oil globules in *L. goodei* eggs were significantly smaller in average diameter (65%) and total volume (23%), relative to those of *L. parva*. When we summed the volume of oil globules, we determined that the eggs of *L. parva* exhibited significantly greater absolute and relative volumes of oil globules (19% and 44%, respectively), compared to those of *L. goodei*.

The eggs of *L. parva* exhibited a significantly higher (30%) activation rate (as indicated by perivitelline space formation) than the eggs of *L. goodei*. However, in most respects the actual process of activation was very similar between *L. parva* and *L. goodei*. The egg envelopes of both species expanded to a nearly identical maximum diameter. This similarity in egg envelope expansion might be related to the differences in egg size and envelope looseness described above. The formation of cytoplasmic blastodiscs in *L. parva* and *L. goodei* eggs was also similar. By the end of step C1, pre-cleavage blastodiscs of *L. parva* and *L. goodei* exhibited no significant differences in height, width or calculated volume (Table 5, Fig. 5b). When differences in yolk volume were taken into consideration, *L. parva* eggs exhibited a significantly greater (37%) blastodisc:yolk ratio. Taken as a whole, these observations support the claim of similar oogenic processes in *L. parva* and *L. goodei*. The only major features of the eggs which were consistently different between the species were yolk volume and oil globule distribution.

Step C2

Blastodisc cleavage patterns were generally similar between *L. parva* and *L. goodei*. The only morphological difference between the species was the observation that cleavage furrows in *L. parva* were less distinct after the fourth cleavage, relative to those of *L. goodei*. Thus, the major difference between the species was the lag in developmental rate, where *L. parva* eggs formed cleavage furrows at later ages than those of *L. goodei*. This was the first clear indication of a difference in developmental rates; a difference that would increase as ontogeny progressed. There was, however, one observation

Table 5. A comparison of characters between 25 cleavage eggs (step C2) of *L. parva* (LP) and *L. goodei* (LG). Significant ($p < 0.05$) differences between means (t-test) and between variances (F-test) are indicated with a directional sign (< or >).

Character	Mean		Variance	
	LP	LG	LP	LG
Clutch size	12.0	> 6.8	99.0	> 33.9
Activation rate (%)	79.4	> 61.1	602.9	1027.2
Yolk diameter (mm)	1.060	< 1.201	2.6E-3	2.7E-3
Yolk volume (mm^3)	0.628	< 0.911	7.7E-3	1.4E-2
Yolk shrinkage (%)	6.96	6.95	13.460	10.445
Blastodisc height (mm)	0.153	0.146	1.0E-3	1.0E-3
Blastodisc width (mm)	0.614	0.645	6.0E-3	8.0E-3
Blastodisc volume (mm^3)	0.029	0.031	2.2E-3	2.0E-4
Blastodisc:yolk ratio (%)	4.8	> 3.5	6.6E-4	3.5E-4

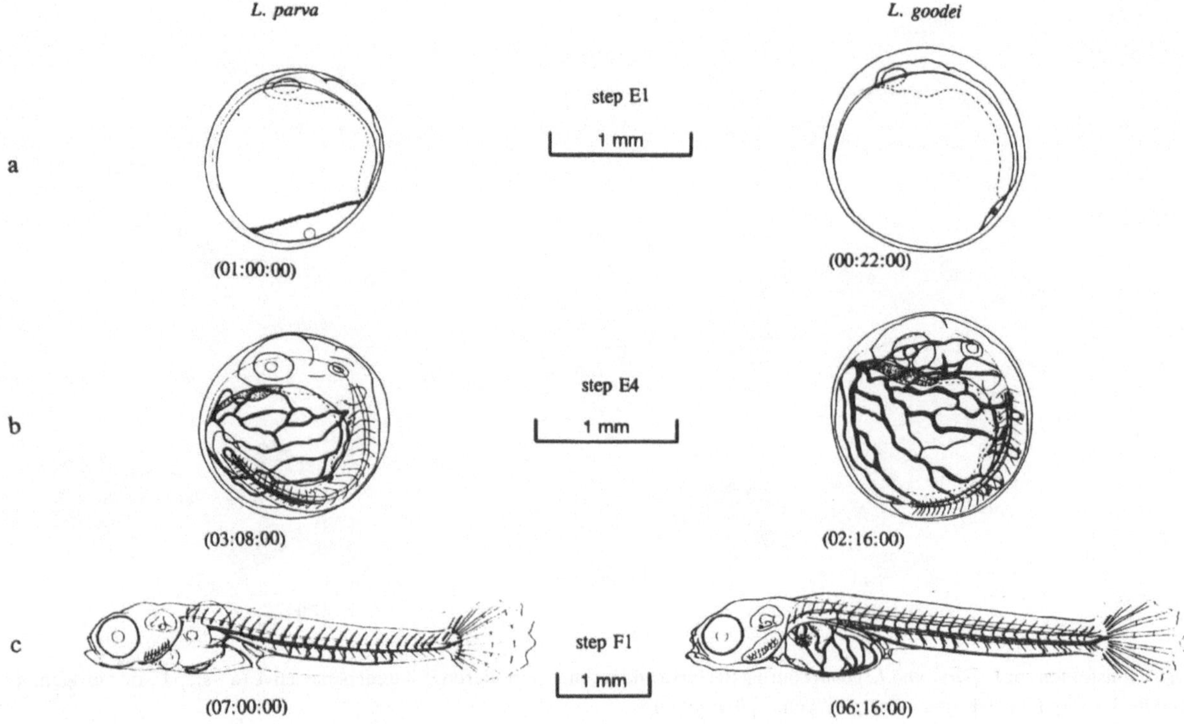

Fig. 6. Illustrations of *L. parva* and *L. goodei* during the embryo period from a lateral perspective (a – step El embryonic body, b – step E4 segmental blood vessels, c – step Fl free embryo).

during the cleavage of *L. parva* and *L. goodei* blastodiscs that was unexpected. As the cells of *L. parva* eggs began to flatten out over the yolk, just prior to epiboly, we often noted the appearance of 'globules' at a location that was polar opposite to the blastodisc (i.e. future site of germ ring closure). Later observations suggested that these structures may have been involved in the formation of Kupffer's vesicle(s).

Steps C3–El

At the onset of epiboly, blastodiscs of *L. parva* and *L. goodei* appeared similar in both size and shape (Fig. 5c). However, as the blastoderm migrated over the surface of the yolk, several differences between the species emerged. First, the progress of periblast advance was more rapid in *L. goodei,* despite a greater yolk volume to cover. Although we did not measure the depth of tissue over the yolk surface, it appeared to be stretched more thinly over the yolk of the *L. goodei* embryos. By the end of the cleavage phase, *L. goodei* embryos exhibited optic vesicles earlier, both chronologically and morphologically (i.e. in terms of epiboly), relative to those of *L. parva* (Fig. 6a). It is interesting to note that the size of these embryos, measured as total length, was not significantly different between the two species at this state.

Steps E2–Fl

The two hour lag continued through to the beginning of step E2, as indicated by the first appearance of somites. The only major difference between the two species during this step was the early appearance (both chronologically and morphologically) of the heart in *L. goodei* (01:10:00), relative to its appearance in *L. parva* (01:18:00).

Perhaps the most important difference between *L. parva* and *L. goodei* during step E3 was in the formation of an embryonic circulatory system. At the onset of vitelline blood flow, a four hour lag existed between these species. Within the next four hours, *L. goodei* formed anterior vitelline veins, structures that did not form in *L. parva* embryos un-

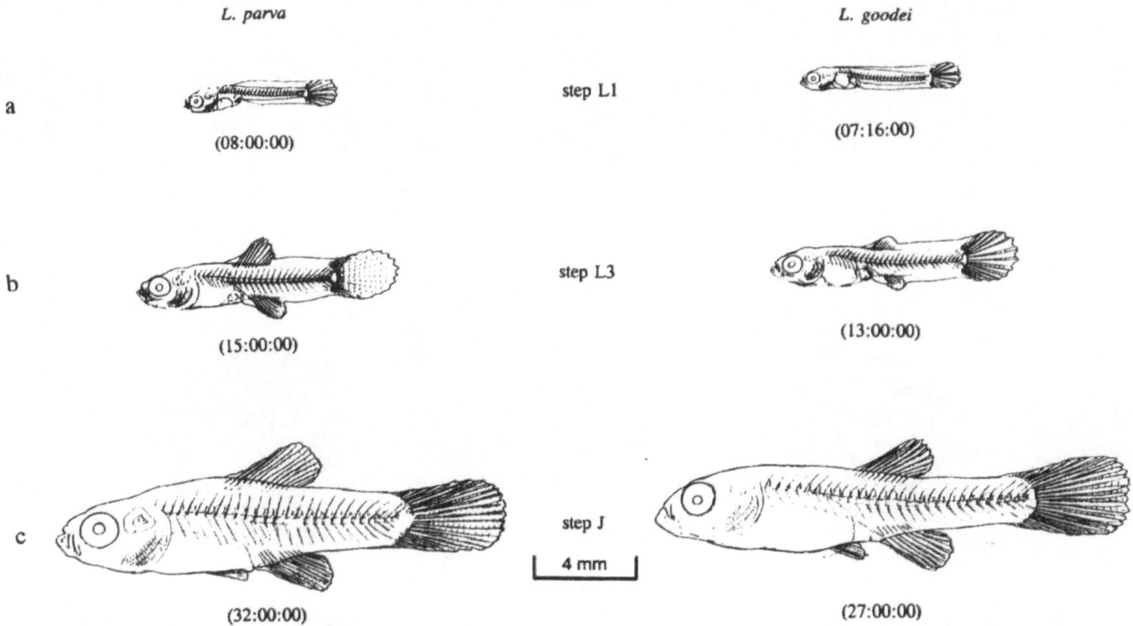

Fig. 7. Illustrations of *L. parva* and *L. goodei* during the larva and juvenile period from a lateral perspective (a – step L1 first exogenous feeding, b – step L3 pelvic fins, c – step J juvenile phenotype).

til 10 hours after the beginning of the step. In addition to forming earlier, the vitelline circulatory network of *L. goodei* embryos was also much larger; extending over a greater yolk surface area than in *L. parva* embryos (Fig. 6b). As the blood vessels formed over the yolk surface, new branches seemed to extend from one cluster of pigment cells to the next. Although we did not quantify the density and distribution of these cells, it appeared that both the number and size of vitelline capillaries was greater in *L. goodei* than in *L. parva.*

During the interval between onset of blood flow (step E3) and hatching (step F1), there was an exponential decline of yolk diameter in both *L. parva* and *L. goodei.* While the rates of yolk utilization seemed similar between the two species, *L. parva* embryos exhausted their yolk supply at an earlier chronological age and significantly smaller body size (i.e. total length), relative to *L. goodei.* It is important to note that despite being smaller at hatching (Fig. 6c), the free embryos of *L. parva* displayed evidence of considerably more chondrification in all existing skeletal systems, relative to *L. goodei* (Crawford & Balon 1994a, b).

Steps L1–J

Given the chronological and morphological differences of the free embryos, it was no surprise to find that the first-feeding larvae of *L. goodei* were younger and larger than those of *L. parva* (Fig. 7a). The species difference in developmental rates continued throughout the larva period, with the exception of scale formation (step L4) where *L. parva* ex-

Table 6 Factor loadings from a principal components analysis of variation among 7 mensural characters for laboratory-reared larvae (step L1, N = 25), laboratory-reared juveniles (step J, N = 25) and wild adult specimens (N = 25 males, 25 females) of *L. parva* and *L. goodei.* All characters were measured as absolute values. The three factor loadings that dominated the second principal component (PC2) are indicated with an asterisk.

Morphological character	PC1	PC2	PC3
Standard length	0.995	0.049	– 0.044
Snout length	0.986	0.137*	0.090
Eye diameter	0.993	0.029	– 0.091
Head length	0.995	0.083	– 0.012
Head depth	0.996	– 0.040	0.012
Body depth	0 990	– 0.126*	0.010
Caudal peduncle depth	0.988	– 0.133*	0.036
Percent total variation	98.4	0.92	0.29

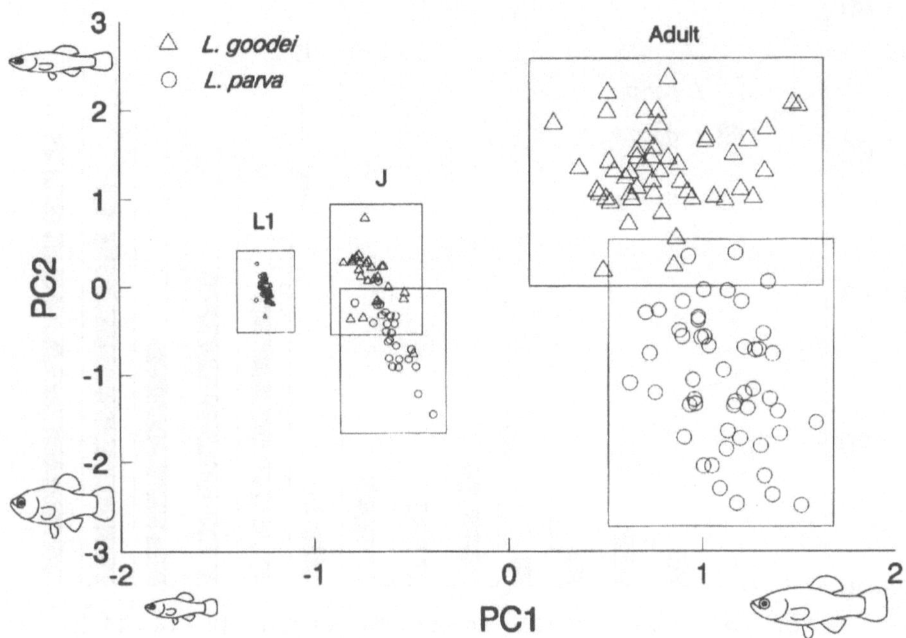

Fig. 8. The relationship of the first two principal components of variation based on 7 morphological characters (Table 6) for *L. parva* (circles) and *L. goodei* (triangles) during development. A total of 25 specimens were measured for both species at step L1 (first exogenous feeding larva) and step J (juvenile phenotype) that were reared under the same laboratory conditions. A total of 50 wild adult specimens were measured for both species.

hibited the first scales approximately one day earlier than *L. goodei* (Table 4). In the end, *L. goodei* achieved the juvenile phenotype (step J) by 27:00:00, five days earlier than *L. parva* (Fig. 7c).

Perhaps the most important difference in the larval development of *L. parva* and *L. goodei* was in the establishment of relative body shape. In order to control for growth in body size during this period, we performed a principal components analysis on absolute morphological characters for larvae (step L1) and juveniles (step J) from the developmental series, and adults collected from the wild (Table 6). For more information regarding the application of this statistical technique, see the analysis of adult morphology given above.

The first axis accounted for 98.4% of total variation in the data set, and was strongly associated with body size (Table 6). The second axis accounted for 0.92% of the total variation, or approximately 50% of the remaining variation, by describing differences in general body depth (head depth, body depth, caudal peduncle depth) and snout length. Thus, once variation due to body size had been removed from the data set, specimens with deeper bodies and shorter snouts were distinguished from specimens with more shallow bodies and longer snouts.

In a similar manner to the analysis of adult morphology, we plotted the factor scores for the first two principal components of variation (Fig. 8). The larvae of both species were observed to overlap with each other at an intermediate range of relative body depth and snout length (Fig. 7a). By the juvenile phenotype, *L. parva* and *L. goodei* had begun to exhibit a species dimorphism along the second component of variation, with approximately one third overlap between the species. In this case, juvenile *L. parva* were deeper in body, with shorter snouts, relative to those of *L. goodei* (Fig. 7c). Finally, the adults of the two species were almost completely separated on the basis of relative body depth and snout length.

It should be noted that the distinction of adult *L. parva* and *L. goodei* in this analysis was similar to that described for the investigation of adult morphology independent of larval or juvenile morphology (cf. Fig. 4). In order to examine the influence of

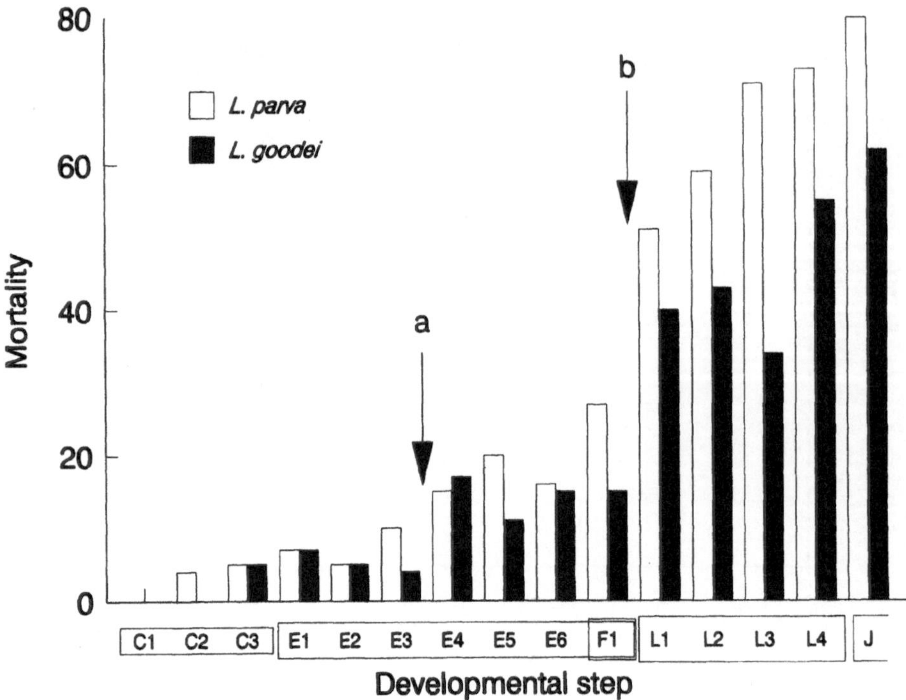

Fig. 9. Mortality of laboratory-reared offspring of *L. parva* (open bars) and *L. goodei* (closed bars) over 15 developmental steps (see Table 1). Mortality is defined as the number of randomly selected offspring that did not survive to meet the sample requirement of N = 25 living specimens (a – increase in mortality at the onset of blood circulation, b – increase in mortality at first exogenous feeding).

adults on the combined analysis, we performed another principal components analysis on the larvae and juveniles, without the adults. The resulting factor scores and factor plot were similar to those generated by the analysis with the adults included.

Thus, while completing the formation of most structures, the late larvae of *L. parva* and *L. goodei* began to exhibit differences in body depth and head shape. The changes in body shape marked the beginning of allometric growth that would lead to the morphological distinction of adult *L. parva* and *L. goodei* collected in the wild.

Mortality

One by-product of the sampling design in the developmental series was a crude estimate of offspring mortality under laboratory conditions. For the purpose of this study, mortality was defined as: the number of specimens designated for sampling at each step, that were not alive by the time of sampling; sampling at each step continued until 25 living specimens had been described. The levels of mortality for both species, over the entire sampling period, are given in Figure 9.

For both *L. parva* and *L. goodei,* mortality rates increased over the developmental steps. However, there were two intervals in the sampling design associated with relatively large increases in mortality. From the cleavage phase (steps C1–C3) through to the early embryonic phase, mortality was between 0 and 10 specimens sample^{-1}. However, after the onset of blood flow (step E3), the mortality increased to approximately 15–20 specimens sample^{-1} (Fig. 9a). This level of mortality did not change until first feeding (step L1), after which mortality levels were once again doubled (Fig. 9b). The mortality rates of both species appeared to increase during the larva period.

There were also interspecific differences in levels of mortality during early ontogeny. After the first increase in mortality at the onset of blood flow, the mortality of *L. parva* increased while that of *L. goodei* did not. By step F1, the mortality of *L. parva* offspring was almost twice that of *L. goodei* offspring. The absolute difference in mortality between the species was maintained during the larva period. By the definitive phenotype (juvenile, step J), the mortality estimate for *L. parva* (82 sample^{-1}) was 32% greater than that for *L. goodei* (62 sample^{-1}).

Discussion

Environmental conditions

Although we collected adult *L. parva* and *L. goodei* within 10 km of each other, the environments that they inhabited were different. In particular, there were differences between the stability of environmental conditions (i.e. variability and predictability) at Tower Pond and Newport Spring. While we did not attempt to sample different populations or habitats for the two species, there was no indication from the literature that our sampling locations were drastically atypical.

The conditions at Tower Pond were highly variable, as one might expect from a habitat that is near an exposed interface of fresh- and sea-water conditions. Lack of forest canopy for the standing water may result in diurnal and seasonal temperature fluxes. Warm summer water in Tower Pond, in combination with algal blooms and decomposition, regularly results in catastrophic fish mortality (Red Gidden personal communication). Local rainfall and overland water flow into Florida lagoons can be extremely variable, both seasonally and annually. More than 80% of the yearly precipitation can fall in two months during the wet season, while entire drainage systems can vanish during the dry season (Kushlan 1974, Loftus & Kushlan 1987). During winter, fish in habitats like Tower Pond are also subjected to predation by several species of migratory waterfowl (Foster 1967, Gidden[1]).

In addition to these seasonal fluctuations in Tower Pond, the inhabitants must also cope with a variety of unpredictable changes in environmental conditions. The marshes of St. Marks Wildlife Refuge are exposed to the full force of hurricanes and tornadoes that irregularly affect the Gulf coast. In 1985, tidal floods caused by Hurricane Kate overtopped the levees at Tower Pond and dumped tons of seawater and plant detritus, leading to a massive fish kill (Gidden[1]). The next hurricane to affect Tower Pond did not occur until 1988, and it only slightly changed the salinity of water in the system (Gidden[1]). In addition to natural catastrophes, Tower Pond is also subjected to the effects of management practices employed by the U.S. Fish and Wildlife Service. One of the principle objectives for the St. Marks Wildlife Refuge is to provide appropriate feeding grounds for migratory waterfowl. In order to control aquatic vegetation, the manager may change the water levels or salinity of impounded water to have an effect on particular plant species. In 1987, the refuge staff attempted to control a particular plant species by keeping Tower Pond completely drained from June until September, then reflooded it with 25% seawater in late October (Gidden[1]). The application of these management techniques is not consistent, either within or between years.

In contrast, the environmental conditions at Newport Spring did not exhibit the range or the unpredictability of changes observed at Tower Pond. The quantity and quality of water flowing from this artesian spring are remarkably constant over both daily and seasonal cycles. The same phenomenon has been observed in virtually all of the other springs that draw from the Floridan Aquifer (Rosenau et al. 1977). The low levels of dissolved oxygen observed at Newport Spring are probably due to the fact that this water had not been previously exposed to atmospheric oxygen. In addition, the surrounding forest would also serve to buffer the stream from the sources of environmental variability described for Tower Pond. Finally, the spring is not managed, and with the exception of the occasional human swimmer, the site is rarely visited (Red Gidden personal communication).

The laboratory conditions under which we maintained the developing offspring were not intended to approximate the conditions that would be found at either sampling location. Instead, we selected a

[1] Gidden, C. 1985–1988. St. Marks National Wildlife Refuge year-end reports, U.S. Department of the Interior, Fish and Wildlife Service. St. Marks, Florida.

combination of temperature and salinity that was intermediate between those described at the two sampling locations, and which would allow for the successful development of offspring from both species. We may be criticized for selecting an arbitrary, intermediate suite of environmental conditions for spawning and rearing offspring, however we believe that the first step toward an effective ontogenic comparison is to eliminate first-order sources of variation from the environment. Obviously, those interested in further extending descriptions of development from the laboratory to the wild, should take the appropriate precaution of documenting the effect of environmental variation on developmental processes. In this study, we have simply attempted to relate differences in morphological development under constant laboratory conditions, to recorded differences between natural environmental conditions at the sampling locations. Given the biological and technical constraints of such a study, we believe that this is a reasonable basis for interpreting our results.

Adult morphology

The principal components analysis of morphological variation among wild specimens revealed species dimorphism between *L. parva* and *L. goodei*, as well as sexual dimorphism within each species. These dimorphisms may be directly related to the environmental conditions in which the adults had been collected.

The species dimorphism that existed between *L. parva* and *L. goodei* was recognized as a difference in body depth. In this study, as in the report of Hubbs et al. (1943), both sexes of *L. parva* were observed to be much deeper than those of *L. goodei.* Fish that inhabit lagoon or pond environments, such as *L. parva,* often exhibit a greater degree of lateral compression which aids in stability and locomotion (Gatz 1979, Mahon 1984, Balon et al. 1986). In contrast, fish that inhabit flowing riverine environments, such as *L. goodei,* often exhibit more slender body shapes that present less resistance to the current. There is some independent support for this ecomorphological interpretation of body depth dimorphism in the family Cyprinodontidae. Hubbs & Miller (1965) identified a consistent decrease in the body depth of *L. parva* collected in Florida along a gradient from static brackish water to flowing freshwater. Schoenherr (1977) compared populations of the viviparous poecilid *Poeciliopsis occidentalis* in different environments, and suggested that differences in body morphology might be attributed to the physical demands of swimming in flowing water. Finally, Foster (1967) suggested that the lateral pigmentation stripe exhibited by *L. goodei* may reflect the tendency of this species to school in riverine currents (see also Hildebrand & Schroeder 1928). Thus, the dimorphism in body shape between *L. parva* and *L. goodei* may directly reflect the different demands imposed by the aquatic environments in which they live.

The sexual dimorphism that existed between males and females of *L. parva* and *L. goodei* was recognized as a difference in body size. In both species, the females were observed to be larger than the males. The same observation has been made previously for both *L. parva* (Hildebrand & Schroeder 1928, Jordan & Evermann 1896, Hubbs & Miller 1965, Parenti 1981) and for *L. goodei* (Foster 1967, Parenti 1981). This dimorphism in body size may reflect differences between the energy allocation of males and females. Males of both species may utilize more energy in non-growth behaviour such as territoriality and courtship (Foster 1967). In contrast, females may utilize more energy in body growth, perhaps to maximize egg production as in *Fundulus heteroclitus* (Armstrong & Child 1965).

Reproductive investment

The differences that we found in the 'instantaneous indicators' of reproductive investments of *L. parva* and *L. goodei* were consistent with predictions of the altricial-precocial life-history theory. Female *L. parva* from the less stable environment produced a greater number of smaller eggs than female *L. goodei* from the more stable environment. It should be noted that we did not attempt to estimate the seasonal or life-long reproductive investment of these females. For the purposes of the following discussion we will assume that the 'instantaneous indicators' of reproductive investment examined in this study apply throughout the reproductive season of each species.

The observed differences in reproductive investment between the two species may reflect a higher risk of offspring mortality in the less stable environment (Duellmann 1989). Krumholz (1963 in Schoenherr 1977) found that female *Gambusia manni* that were exposed to higher mortality through predation, produced greater numbers of eggs. A greater number of offspring produced by *L. parva* may ensure that some offspring survive in Tower Pond, especially in the event of catastrophic mortality. However, if the total energy available for reproduction is limited in both of these species, then there should be an inverse relationship between the number of eggs produced and the amount of energy invested in each egg (Stearns 1982, Blaxter 1988, Flegler-Balon 1989, Balon 1990). This relationship was clearly demonstrated by the smaller eggs in *L. parva,* relative to those in *L. goodei.* In this comparison, we were able to use egg size as an estimate of energy investment, since there was no indication of species differences in the density of the yolk (see Balon 1985b). Foster (1967) also reported smaller egg diameters in *L. parva* (1.23 mm) than in *L. goodei* (1.35), but he did not discuss the potential importance of this difference in reproductive investment.

Together, the observations of greater number of eggs and smaller egg size in *L. parva* are consistent with the hypothesis that organisms in less stable environments should produce a greater number of offspring, each with a smaller investment of energy. This kind of response to environmental conditions has been observed in other species of fish. For example, Lowe-McConnell (1982) found that reproductive investment in two subspecies of *Oreochromis shiranus* depended on local environmental conditions; females produced smaller, more numerous eggs in harsher abiotic conditions. In addition, Lowe-McConnell (1982) also detected shifts in this pattern of reproductive investment that were responsive to short-term changes in environmental stability.

In order to interpret differences in reproductive investment from the perspective of the altricial-precocial life-history theory, there must be a relationship between reproductive investment and 'quality' of the offspring (Scrimshaw 1944, Blaxter 1988, Flegler-Balon 1989). We found that offspring of *L. parva* exhausted their yolk supplies at smaller body sizes and at less advanced morphological states, compared to the offspring of *L. goodei.* This observation supports the existence of a causal relationship between energy investment and quality of the offspring, measured in terms of size and morphological state. Several authors have suggested that larger, advanced offspring are better foragers, more resistant to starvation, and less susceptible to predation (e.g. Balon 1975, 1981, Blaxter 1988, Heming & Buddington 1988, Flegler-Balon 1989). We did not attempt to relate indicators of reproductive investment to the availability of energy resources (e.g. primary and secondary productivity) at the sampling locations, although this would likely be a fruitful exercise.

It is possible that lower mortality, associated with a more stable environment, may have released the ancestors of *L. goodei* from the necessity of generating maximum numbers of offspring. As a result, there could have been an increase in the amount of energy available to each offspring, giving rise to larger eggs with more yolk. These larger eggs could have supported development for a longer period of time and carried the offspring to a more advanced state before exhaustion of endogenous nutrients. In this manner, the reproductive investment patterns of *L. parva* and *L. goodei* may be directly related to conditions of environmental stability, through the risk of offspring mortality.

Morphological development

Obvious differences were recognized in the morphological development of *L. parva* and *L. goodei.* It is important to stress that these differences occurred under controlled conditions in the laboratory. Every attempt was made to ensure that the parents and offspring of both species were exposed to identical and constant environmental stimuli, in order to focus on the inherited components of morphological development. In this way, we should be able to identify morphological characters that are transmitted directly from one generation to the next. We have dealt with the development of intra- and interspecific morphological variability under these controlled laboratory conditions in a separate paper (Crawford & Balon 1994c).

Developmental rate

Offspring of *L. goodei* were observed to reach most morphological states at chronologically younger ages, relative to the offspring of *L. parva.* This observation contradicts the hypothesis that organisms living in an unstable environment will minimize the risk of higher mortality by accelerating developmental rates, relative to those in a stable environment (Freeman 1982, Loftus & Kushlan 1987). This contradiction warrants a re-examination of the factors that are thought to affect developmental rate.

First, there may be a causal relationship between egg size and developmental rate. Some interspecific studies have reported an inverse relationship between egg size and developmental rate (e.g. Blaxter 1988, Pepin & Myers 1991). Heming & Buddington (1988) suggested that this relationship is a reflection of increased rates of yolk absorption in larger eggs with greater absolute surface areas. According to this explanation, the offspring of *L. goodei,* by virtue of producing larger eggs than *L. parva,* could be expected to absorb their yolk reserves at a greater rate. This accelerated absorption of yolk could be translated into accelerated rates of cell division and organogenesis (Sinervo & McEdward 1988). In this way, the paradox of accelerated developmental rates in more stable environments may be a simple by-product of larger eggs.

Egg size is not the only determinant of developmental rate among the cyprinodontids. According to Koster (1948), offspring of *Cyprinodon variegatus* (egg diameter 1.2–1.4 mm; Kuntz 1916, Breder 1948) have an accelerated rate of development, compared to those of *Fundulus heteroclitus* (egg diameter 1.9–2.1 mm; Solberg 1938, Armstrong & Child 1965). In another comparison, offspring of *Adinia xenica* (egg diameter 1.8–2.2 mm; Koenig & Livingston 1976, Cunningham & Balon 1985) have the same general egg size as offspring of *F. heteroclitus,* yet they exhibit a faster rate of morphological development. These exceptions to the rule 'larger eggs, faster development' suggest that other factors may also affect the developmental rate of *L. parva* and *L. goodei.*

Local environmental conditions are very important in the determination of developmental rates among cyprinodontid species. Water temperature is known to have an important and direct effect on developmental rates in most morphological systems (Solberg 1938, Kinne 1960). A strong relationship also exists between dissolved oxygen and developmental rates of killifish embryos and larvae (Lewis 1970, Koenig & Livingston 1976, Cunningham & Balon 1985). Kinne & Kinne (1962) showed that retardation of developmental rates in *Cyprinodon macularius* can be caused by reductions in the oxygen saturation and absorption coefficients of the ambient water. In contrast, experiments have demonstrated an inverse relationship between water salinity and general rates of development in cyprinodontids (Kinne 1960, Kinne & Kinne 1962, Lewis 1970).

While *L. goodei* exhibited a faster rate of development than *L. parva,* there were notable exceptions. For instance, the onset of scale formation (step L4) was recognized approximately 24 h earlier in *L. parva* than in *L. goodei.* This suggests a decoupling of morphological development in the systems that we used to identify the steps of ontogeny (e.g. pelvic fins, scales, median finfolds). Another example of decoupling was observed in the development of skeletal elements. At any given corresponding morphological state (external), the skeletal elements of *L. goodei* were less advanced in chondrification or calcification compared to those of *L. parva.* Although *L. parva* reached the various ontogenic thresholds at later ages than *L. goodei,* the process of skeletal development in *L. parva* occurred at a more rapid rate, in both relative and absolute terms. Possible explanations for acceleration of skeletal development in *L. parva* are offered later in the discussion of hatching and feeding.

Decoupling of developmental rates (heterochrony) has been proposed as an important mechanism in the origin and speciation of the family Cyprinodontidae. Several authors (Rosen 1964, Foster 1967, Parenti 1981, Freeman 1982) have suggested that the family was derived through heterochrony from common ancestry with the flying fishes (Exocoetidae) and silversides (Atherinidae). Slight changes in the developmental rates of different morphological systems, triggered by minor genetic or environmental stimuli, may have the potential to yield major changes at the phenotypic level. In this way, heterochronic shifts in ontogeny could represent a proximate mechanism to account for the phenotypic differentiation of the two species (Gould 1977, Balon 1980e, Noakes et al. 1989).

Heterochrony at the level of the whole organism (Bruton 1989) also offers the possibility of morphological rejuvenilization (paedomorphosis) to species that otherwise might have become locked into a trajectory of narrowing phenotypic variation (Alberch et al. 1979, Balon 1980e, 1986a, Hall 1984, Meyer 1987). This ontogenic phenomenon can potentially serve as an important source of morphological variation for evolutionary lineages in which genetic variation has been lost (Maderson et al. 1982).

Foster (1967) recognized several morphological characters of the cyprinodontids that suggest the occurrence of paedomorphosis from an ancestral exocoetid lineage, including: functional pronephros, unfused lower pharyngeal bones and poorly developed lateral line scales. Within the genus *Lucania,* the species differences observed in early morphological development can largely be attributed to the relative acceleration of developmental rates in *L. goodei,* compared to *L. parva.* It is possible that paedomorphosis is a proximate mechanism for generating the generalized form by prolonging the less-specialized juvenile period (Balon 1985b, Bruton 1989).

Hubbs & Miller (1965) considered shallow body depth and large eye size to be 'juvenile' characteristics of *L. parva* that were retained by individuals in the populations that inhabited the inland, flowing waters of Florida. It is possible that both *L. parva* and *L. goodei* can achieve a body form that is suited to riverine conditions, through relatively simple changes in the developmental rates of different morphological systems.

Flexibility of developmental rates may also have been a key factor in the exploitation of harsh environments by killifish. Many species of oviparous cyprinodontids have the peculiar ability to survive through periods of desiccation by entering one or more intervals of extended developmental arrest at different steps in embryonic development (Peters 1965, Wourms 1964, 1972a, b, c, Parenti 1981).

Unactivated eggs

The differences in egg size between *L. parva* and *L. goodei* were discussed in the previous section on reproductive investment. However, there were other morphological differences in the unactivated eggs of these two species that may be related to environmental conditions in the wild.

There was a clear difference between species with respect to the number and size of oil globules that existed in the cytoplasm. *Lucania goodei* eggs exhibited a greater number (mean = 22.8) of smaller oil globules than those of *L. parva* (mean = 7.3). However, the total oil volume of *L. goodei* eggs was calculated to be less than that for *L. parva.* These observations contradict the remark of Foster (1967) that the eggs of both species exhibited 10–12 'medium-sized' oil globules. Perhaps oil globule number and size are features of the egg that are also influenced by local environmental conditions.

Species in the subfamily Fundulinae generally exhibit a greater number of smaller oil globules, compared to those in the families Rivulinae or Cyprinodontinae (Foster 1967). This suggests that the process of oil globule investment may be related to the greater yolk invested in funduline eggs. Oil globules are located among the yolk platelets and contain triglycerides, proteins, and occasionally carotenoid pigments and wax esters (Heming & Buddington 1988). Due to their chemical composition, oil globules may serve as a specialized form of nourishment, in addition to the yolk (Blaxter 1988). If oil globules do represent an energetic investment from parent to offspring, then the inverse relationship between yolk investment and oil investment may reflect some physiological constraint on the process of egg production. However, it is also possible that the oil globules do not represent an available source of energy for development in either species. Trinkaus (1967) experimentally removed the oil globules from *F. heteroclitus* eggs and found no general effect on the morphological development of embryos.

It has been suggested that oil globules may be important in controlling the buoyancy of embryos (Balon 1975, Blaxter 1988, Mommsen & Walsh 1988). However, despite higher absolute and relative volumes of oil in the *L. parva* eggs, the eggs of both species were seen to be negatively buoyant in brackish water from 1 to 15 ppt (personal observation). Perhaps the positively buoyant force of brackish water, combined with a higher parental investment of oil, serves to reduce the possibility that *L. parva* eggs will sink into poorly oxygenated waters at Tower Pond. The issue of buoyancy may not be as important for *L. goodei* embryos in Newport Spring, especially if these embryos possess the ability to withstand consistently hypoxic conditions.

There were also morphological differences in the egg envelopes of *L. parva* and *L. goodei*. First, the envelopes of *L. parva* eggs appeared to be more flaccid and wrinkled, compared to those of *L. goodei*. This difference in appearance may actually reflect a fundamental similarity in the egg morphology of both species. Specifically, if female *L. parva* and *L. goodei* surround different yolk investments with the same size of egg envelope, then we could expect to observe more flaccid, wrinkled egg envelopes around the smaller eggs (Laale 1980). This explanation is supported by our time-lapse observations of activation, in which the egg envelope diameters of the two species swelled to the same maximum diameter. These observations support the claim that aside from yolk volume, the reproductive investments of *L. parva* and *L. goodei* are remarkably similar.

Once again, we are confronted with the possibility that a small change that precedes the process of ontogeny (e.g. oogenesis), can be responsible for much greater effects at the phenotypic level (Balon 1981, 1986a, Noakes et al. 1989). This is similar to the phenomenon that chaos theorists have called 'the butterfly effect' (sensitive dependence on initial conditions), where small changes at the beginning of a developmental process may cascade and amplify through the system, ultimately resulting in much greater changes (Gleick 1987). Other examples of early morphological differences between *L. parva* and *L. goodei* that seem to have amplified effects later in development, will be discussed below.

One final difference in the morphology of unactivated eggs in the two species requires some consideration. We observed that the outer egg envelope of *L. parva* eggs was usually covered with a much denser mat of adhesive fibers, compared to that of *L. goodei* eggs. This observation was also made by Foster (1967). These fibers provide a means of attachment to virtually any kind of substrate. It is possible that the higher concentration of these fibers on *L. parva* eggs, perhaps in combination with less negative buoyancy, serves to prevent the eggs from sinking to the bottom of Tower Pond, where hypoxic conditions may prevail (Foster 1967, Anderson 1974, Balon 1975, Laale 1980). As noted above, there may be little chance of avoiding low dissolved oxygen concentrations in Newport Spring, regardless of location in the water column. In addition to providing a means of physical attachment to the substrate in Tower Pond, the adhesive fibers on *L. parva* envelopes may also serve to retain moisture if the eggs are temporarily exposed to the atmosphere during low water conditions (Brummett & Dumont 1981). Once again, this property would probably not be required for *L. goodei* embryos, developing in the consistent flow of Newport Spring.

Activation and cleavage

Given the differences in yolk investment between the two species, it might seem peculiar to find that the smaller eggs of *L. parva* exhibited a 30% higher activation rate (as indicated by perivitelline space formation) than the eggs of *L. goodei*. This suggests that larger eggs alone do not ensure greater survival during early ontogeny. In addition, local environmental conditions have been shown to affect activation rates of some species in the laboratory. Trinkaus (1967) demonstrated that activation of *F. heteroclitus* eggs occurs in full sea water but not in 30% diluted sea water. Armstrong & Child (1965) induced activation of *F. heteroclitus* eggs simply by exposing them to sea water. In this case, 'normal' development proceeded all the way to blastodisc formation, before retardation and eventual termination.

The formation of a perivitelline space in *L. parva*

and *L. goodei* eggs occurred while cortical alveoli on the cytoplasm surface had begun to erupt (see Balon 1980a, 1985a). In other species of cyprinodontids, the alveoli erupt in a progressive wave that originates from the micropylar region (Wessels & Swartz 1953, Huver 1956, Armstrong & Child 1965, Koenig & Livingston 1976). The release of cortical material was probably responsible for some increase of the egg envelope diameter during the first 30 minutes of activation (Huver 1960). Laale (1980) has noted that at the time of activation, many teleost eggs are also permeable to water and may allow small molecules to be drawn into the perivitelline space by osmosis.

Accumulation of the blastodisc cytoplasm occurred beneath the micropyle in both species, usually in a position that was slightly off the gravitational plane. We did not observe any indication of 'cytoplasmic streaming' as reported for *L. parva* by Kuntz (1916), and for *F. heteroclitus* by Oppenheimer (1937), Solberg (1938) and Huver (1960). No visible recognition of 'cytoplasmic streaming' was reported in *A. xenica* (Koenig & Livingston 1976) nor in another study of *F. heteroclitus* (Armstrong & Child 1965). The reasons for these disparate observations on the formation of cyprinodontid blastodiscs remain unclear (but see Cunningham & Balon 1985). Regardless of the mechanism, the similarity in blastodisc volumes reinforces the assertion that yolk volume is the only major difference in reproductive investment between *L. parva* and *L. goodei*. As a result, the eggs of *L. parva* exhibited a higher ratio of cytoplasm:yolk volume than those of *L. goodei*, a condition that is generally associated with a more altricial evolutionary trajectory (Balon 1979, 1981, Flegler-Balon 1989).

Finally, we observed that the cleavage furrows in *L. parva* appeared less distinct than those in *L. goodei* (cf. Oppenheimer 1937). It is possible that this difference in the furrows could have been caused by a different physico-chemical composition of the cytoplasm. Kuntz (1916) reported that the cleavage of *L. parva* blastodiscs was less symmetrical than in *C. variegatus*, possibly in reference to the indistinct furrows in the former species. Løvtrup (1982) and Balon (1989) have suggested that asymmetrical cleavage patterns may be the earliest determinants of cell position and tissue fate.

Epiboly and embryo formation

Despite having a larger surface area of yolk, epiboly proceeded at a faster rate in *L. goodei* than in *L. parva*. Given the similar volumes of cytoplasm in the two species, it is not surprising that the cell mass of *L. goodei* appeared to be sparsely distributed as it migrated over the yolk surface. It is possible that this difference in blastoderm thickness between the species could also be related to differences in the development of the vitelline circulatory network. If such a relationship existed, it would reinforce the idea that minor differences between the reproductive investment of the two species can be amplified during ontogeny.

The first signs of an embryonic body in *L. parva* were observed at a relatively advanced state of epiboly, compared to *L. goodei*. Furthermore, the similarity in body size between the two species suggests that the apparent differences in blastoderm distribution did not constrain cell migration toward the embryonic axis. Despite having a larger yolk surface over which to move the same volume of cytoplasm, the eggs of *L. goodei* completed epiboly faster, while forming a similar-sized embryo in less time. Together, these observations support the idea that the initial formation of the embryo is controlled more by factors associated with the cytoplasm than the yolk mass (cf. Scrimshaw 1944).

Formation of the optic vesicles and somites was similar in embryos of the two species, and appeared to be related to the accumulation of cells along the embryonic axis prior to germ ring closure. However, in other species of cyprinodontids these structures do not usually appear until just before or after germ ring closure (e.g. *F. heteroclitus* in Solberg 1938; *A. xenica* in Koenig & Livingston 1976, Cunningham & Balon 1985). Such variation in the timing of early morphogenesis may have an important effect on the development of differences in relative body shape between species. Oppenheimer (1937) also reported the early formation of optic vesicles and somites in *F. heteroclitus* under unspecified 'peculiar environmental conditions'. This suggests that environmental differences in the wild may also have an important effect on early morphogenesis in *L. parva* and *L. goodei.*

We were intrigued by the formation of droplets at the future site of germ ring closure in both species. Although they seemed similar in consistency to the oil globules, there was no indication of their existence until the onset of epiboly. In addition, their ventral location was in stark contrast to the oil globules that had previously 'floated' to the dorsal surface of the yolk. Armstrong & Child (1965) and Koenig & Livingston (1976) also reported the formation of these droplets in *F. heteroclitus* and *A. xenica,* respectively. Given that the tail of cyprinodontids usually forms at the site of germ ring closure (Oppenheimer 1937, Solberg 1938), it is possible that these droplets became associated with the embryonic body as Kupffer's vesicle(s), or cellular debris at the 'yolk plug' (e.g. Balon 1980a).

Blood circulation and yolk absorption

In teleosts, the primary rudiments of vitelline blood vessels are laid down as clusters of mesenchymal cells distributed over the yolk surface (Balinsky 1981). In *L. goodei* embryos, these clusters were observed to form more closely together than in embryos of *L. parva.* As a result, when branches of newly-formed blood vessels extended from cluster to cluster (Kuntz 1916, Armstrong & Child 1965, Koenig & Livingston 1976, Cunningham & Balon 1985), the number and density of vitelline circulatory vessels appeared to be higher in *L. goodei* than in *L. parva.* It is possible that this difference in the formation of vitelline vessels could have been affected by differences in blastoderm migration during epiboly. Specifically, there may have been a relationship between thickness of the blastoderm (which is likely related to yolk volume) and the distribution of cell clusters on the yolk surface.

It is not surprising to find that other investigators have reported a dramatic rise in the oxygen consumption of embryos at the onset of circulatory vessel formation (e.g. Scott & Kellicott 1916, Boyd 1928, Rombough 1988). This phenomenon might be explained simply by the presence of circulatory vessels that actively take up oxygen. The demand for oxygen would be exaggerated under hypoxic conditions; conditions which have the general effect of retarding developmental rates (Rombough 1988). While the embryos of both *L. parva* and *L. goodei* must face the challenge of obtaining sufficient oxygen for development in the wild, *L. goodei* must also contend with consistently hypoxic conditions such as those at Newport Spring.

There are several differences in the circulatory development of these species that may be related to differences in the oxygen regime of their natural habitats. First, due to an accelerated rate of circulatory development, the vitelline network appeared at an earlier morphological state in *L. goodei,* compared to *L. parva.* The vitelline network has been shown to be an important site of gas transfer for respiration in many species of fish (e.g. Kryzhanovsky 1933, 1934a, b, Balon 1959). Some Russian investigators have previously recognized the relationship between vitelline network development and the oxygen conditions under which embryos survive (Nikolsky 1963, Soin 1968, Smirnov 1975). Unfortunately, there is little information about the details of respiration systems in embryonic fish, especially at the biochemical level (e.g. hemoglobin). Second, *L. goodei* embryos appeared to exhibit a greater flow of blood through the vitelline network, relative to those of *L. parva.* Such an increase in flow rate could have enhanced the rates of oxygen uptake on the yolk surface, as well as providing physical stimuli required for further bifurcations of the vitelline network (Balon 1980a, Balinsky 1981). Finally, the larger yolks of *L. goodei* embryos provided a greater absolute surface area for this gas transfer, compared to the smaller yolks of *L. parva* (Balon 1959, 1975, 1985b). This direct effect of a larger yolk investment may have an important influence on the ability of *L. goodei* embryos to survive and develop under hypoxic conditions in the wild.

In addition to providing an arena for gas transfer, the vitelline network also serves as a mechanism for delivering nutrients to the sites of differentiation and growth. Prior to the formation of a vitelline network, yolk depletion rates for both *L. parva* and *L. goodei* were relatively slow. However, as the vitelline network formed, the yolk supply began to decline rapidly. In this case, it is likely that the vitelline circulatory network serves to meet the increased protein requirements for expanded metabolic activity during the later steps of embryogenesis. Heming & Buddington (1988) suggested that the absorption rate of yolk is directly related to the size of the egg surface area, presumably due to the potential for a larger vitelline network. Although individual absorption rates were not examined in this study, *L. goodei* embryos seemed to utilize their yolk supply at a relatively greater rate than *L. parva* embryos. However, due to the smaller absolute volume of yolk in *L. parva,* they exhausted their supply at an earlier chronological age and morphological state, compared to those of *L. goodei.*

There is evidence of progressive elaboration of vitelline blood circulation during the evolution of species in the order Atheriniformes. Species of the families Oryziatidae, Atherinidae and Exocoetidae usually have embryos with a single pair of veins that traverse the vitelline surface without any secondary branches (Foster 1967). In contrast, embryos of the family Cyprinodontidae exhibit great diversity in vitelline network development, ranging from a few simple branches to an intricate plexus. It is possible that this elaboration of vitelline circulation is simply a by-product of a trend towards expanded yolk investment. However, it is also possible that a trend toward increased yolk volume is itself a by-product of a trend towards expanded vitelline blood circulation (cf. Matsuda 1987). In the case of *L. parva* and *L. goodei,* it seems that the interrelationships between environmental demands, reproductive investments and morphological development have settled into two complex, but stable trajectories.

Hatching and first feeding

Under laboratory conditions, *L. goodei* embryos were observed to hatch at chronologically earlier ages, at larger body sizes, and with more yolk than *L. parva* embryos. These differences may be partly explained by the physico-chemical mechanisms that are associated with hatching in cyprinodontid fishes. First, specialized cells in the mouth and gill cavities have been associated with the secretion of hatching enzymes (Armstrong 1936, Kaighn 1964, Yamamoto 1967). These enzymes have the effect of digesting the inner layer of the egg envelope, leaving it thin and fragile. In addition, cyprinodontid embryos on the verge of hatching typically begin to thrash their bodies within the egg envelopes (Armstrong 1936, Foster 1967). Together, these two events allow the embryo to rupture the egg envelopes and become a free embryo.

It has been suggested that a low concentration of dissolved oxygen within the perivitelline space induces the opercular beating (causing enzyme secretion) and tail thrashing that lead to hatching (Armstrong 1936, Foster 1967, Rombough 1988). This claim is supported by the inhibitory effect of high dissolved oxygen concentrations on the time to hatching in *F. heteroclitus* (Milkman 1954). In the present study, it is possible that the levels of dissolved oxygen in the perivitelline space surrounding *L. goodei* embryos declined below some critical level, thus triggering the hatching process. It is also possible that *L. parva* embryos did not reach this critical oxygen level until later, due to a generally slower rate of development. Wild *L. goodei* embryos may hatch even earlier than those in the laboratory, especially in consistently hypoxic conditions such as those at Newport Spring.

There may be factors other than ambient oxygen concentrations that affect the age at which cyprinodontid embryos hatch. Several authors have suggested that the exhaustion of available yolk may induce hatching by causing the same physiological responses as low oxygen concentrations (Armstrong 1936, Solberg 1938, Oppenheimer 1937, Armstrong & Child 1965). In the present study, *L. parva* embryos with a slower developmental rate than those of *L. goodei,* may have exhausted their yolk supply long before experiencing a critical concentration of dissolved oxygen.

Under laboratory conditions, *L. goodei* larvae were observed to begin exogenous feeding at chronologically earlier ages and larger body sizes than *L. parva* larvae. It is important to realize that the time between hatching and feeding was the same (8 hours) for both species. It is possible that the time required for free embryos to begin feeding is determined by some factor that is not affected by differences between the developmental rates of the two species. For example, embryos probably have to achieve neutral buoyancy by inflating their swimbladders in order to effectively forage for prey (Foster 1967).

The skeletal elements of *L. goodei* at first feeding were morphologically less advanced than those of *L. parva* at the same state. This trend stands in direct contrast to the generally accelerated morphological development of *L. goodei*, compared to *L. parva*. The decoupling and retardation of skeletal development has been recognized as an indication of paedomorphic speciation of cyprinodontids (Foster 1967, Parenti 1981). It is possible that the relative retardation of skeletal development in *L. goodei* is an extension of this evolutionary process. However, it is also possible that the rate of skeletal development is directly associated with functional needs of the free embryos.

Under laboratory conditions, *L. parva* embryos exhausted their yolk supply shortly after hatching, while *L. goodei* embryos did not exhaust their yolk supply until just before first exogenous feeding. It is possible that the total energy invested in *L. parva* offspring is near the absolute minimum required to produce an organism that can feed itself (see Flegler-Balon 1989). This idea is supported by the higher mortality rates observed among *L. parva*, compared to *L. goodei*, during the transition to exogenous feeding. Perhaps the relative acceleration of skeletal development in *L. parva*, particularly the neurocranium and jaws, increases the probability of successfully handling food items (Balon 1959, Blaxter 1988). In this way, the decoupling of developmental rates may be a response to changes in form and function, which in turn may be caused by relatively simple changes in reproductive investment.

Meristic and mensural characters

The larva period of early ontogeny in cyprinodontid species is generally considered to be an interval during which the offspring acquires the necessary energy to complete development of the definitive phenotype (Balon 1981, 1986b, Freeman 1982, Flegler-Balon 1989). Many vital systems of the definitive phenotype are already functional by the onset of first exogenous feeding, however most meristic structures and mensural dimensions continue to develop. These characters are particularly vulnerable to environmental effects, given that the larvae move about in their environment and feed on exogenous sources of nourishment.

There were no consistent differences in the development of final meristic counts between laboratory-reared *L. parva* and *L. goodei*. The juvenile dorsal and anal fin ray counts were slightly greater in *L. parva* than *L. goodei*. In contrast, the scale counts of *L. goodei* were usually greater than those in *L. parva*. These observations contradict the claim that meristic counts are simply determined by general rates of development (Hubbs 1926, Gabriel 1944, Fahy 1978, Lindsey 1978) or egg size (Lindsey 1988). The factors that determine meristic characters in *L. parva* and *L. goodei* remain unclear.

Aside from the differentiation of several meristic characters, the larva period is also characterized by the determination of body shape. Several authors have stressed the importance of allometric growth during larval development (e.g. Kuntz 1916, Fuiman 1984, Blaxter 1988). Despite the observed differences between developmental rates in *L. parva* and *L. goodei*, their growth rates were quite similar under laboratory conditions. The growth rates for these species could be very different under environmental conditions in the wild. Kinne (1960) found that the growth rates in different populations of *Cyprinodon macularius* and *C. variegatus* were predominantly affected by local environmental conditions, rather than being genetically determined. Stearns (1980) found that mosquitofish, *Gambusia affinis*, that were exposed to freshwater early in development, produced offspring with slower growth rates, relative to those reared entirely in brackish water. These observations suggest that environmental conditions could have an effect on the development of body shape in the genus *Lucania*.

The principal components analysis of mensural characters for larvae and juveniles of both *L. parva* and *L. goodei* showed that the species dimorphism recognized in wild adults is heritable. Hubbs & Miller (1965) remarked that the relatively shallow body depth and relatively large eyes found in *L. parva* from inland waters, were 'juvenile' characters of the species. Hubbs et al. (1943) observed that these same characters varied within and between species from one location to the next. Our developmental comparison of *L. parva* and *L. goodei* suggests that this dimorphism is a direct result of a difference in developmental rates between the species. The mechanism underlying this shift in developmental rates has not been clearly identified, however it is possibly related to the difference in reproductive investment between the two species. If this relationship exists, then the environmental differences between unstable brackish ponds and stable springs in Florida may have had a profound effect on the evolution of species in the genus *Lucania.*

Altricial-precocial life-history styles

This study has focused mainly on the morphological development of *L. parva* and *L. goodei* under controlled conditions, with references to environmental conditions in the wild. Many embryological studies end at this point, without hypothesizing about the ultimate factors that may have affected, or been affected by, changes in ontogeny. Similarly, conventional life-history theories focus on general patterns in population characteristics, often without explaining the causal factors that could account for these patterns. These explanations will elude evolutionary biologists until they have some basis in theory for translating reproductive investments into the organisms that make up populations. The differences observed in the early ontogeny of the species in this study can be interpreted by the altricial-precocial model of speciation (Balon 1980a, b, 1989). The purpose of this model is to provide clues to the proximate mechanisms that link various life-history patterns.

Taken together, the life-history characteristics exhibited by *L. goodei* can be considered to be more precocial than those of *L. parva.* Adult female *L. goodei* produced significantly fewer eggs, with significantly more yolk. The offspring of *L. goodei* developed at a more rapid rate than those of *L. parva,* reaching the definitive (juvenile) phenotype at an earlier age, with lower mortality and with a different body shape. These differences in life-history styles may have important effects on the ability of these species to meet the particular demands of their different environments (Bruton 1989).

Many authors have hypothesized that the freshwater springs of Florida provided fishes with a refuge from extreme environmental stress during the Pleistocene period (e.g. Gunter 1961, Foster 1967, Loftus & Kushlan 1987). Burgess & Franz (1978) suggested that the ancestors of many coastal species avoided the interglacial floods of the peninsula by taking refuge near the springs. Duggins (1980) postulated that a common ancestor of *Lucania* was displaced from the Florida coast by high temperatures associated with interglacial climatic warming. Regardless of the decisive environmental factor, the refuge hypothesis describes a situation where the ancestors of *L. goodei* took environmental shelter in the hypoxic, relatively cool, freshwaters of the springs (Kushlan & Lodge 1974, Loftus & Kushlan 1987).

Although we know virtually nothing about the ancestral lineage of *L. goodei,* it may have been very similar to the relatively altricial form of *L. parva.* It is possible that the stable conditions of the freshwater refuges may have demanded the evolutionary changes in reproductive investment and morphological development observed in modern *L. goodei.* Some species of salmonids and cichlids have been shown to adopt a more altricial life-history style when environmental changes allow the colonization of seasonal habitats (Balon 1980a–e, 1984, Noakes & Balon 1982, Winemiller 1989). Increased levels of phenotypic variation, such as those described for wild *L. parva,* have also been associated with colonizing and early successional species (Bradshaw 1965, Hickman 1975, Mahon & Balon 1977, Bruton 1989).

Notwithstanding the environmental conditions that might have separated the two species many years ago (cf. Winemiller 1992), we are still confronted with questions regarding the current lack of sympatry between *L. parva* and *L. goodei* (Hubbs et al. 1943, Hubbs & Miller 1965, Foster 1967, Loftus & Kushlan 1987). The modern differences in distribution between *L. parva* and *L. goodei* may be related to differences in competitive abilities between altricial and precocial species (Balon 1980a, e, 1989). Echelle et al. (1972) and Echelle (1973) have suggested that some species of cyprinodontids (e.g. *Cyprinodon macularius*) are poor competitors and are most successful in harsh environments with few other species. It is possible that the precocial *L. goodei* is a poor competitor in coastal waters, compared to the altricial *L. parva,* and it may have avoided direct competition by seeking isolation in the springs (Balon 1980d, 1984, Løvtrup 1989). However, Loftus & Kushlan (1987) have reported that *L. goodei* is one of the most abundant fishes in hypoxic Everglades swamps, along with *Gambusia affinis, Heterandria formosa* and *Poecilia latipinna.* Given this observation, it seems unlikely that *L. goodei* has great difficulty in competing with ecologically-similar species in a freshwater environment.

Several authors have also suggested that the lack of sympatry between *L. parva* and *L. goodei* is a direct reflection of differences in the osmoregulatory abilities of these species. Many species in the family Cyprinodontidae are known for their ability to osmoregulate under extreme conditions (Gunter 1956, 1961, Hillyard 1981). Loftus & Kushlan (1987) considered *L. parva* as a 'secondary freshwater' species by virtue of the fact that it has been found in a wide variety of salinities. However, the inland penetration of marine species is generally thought to be constrained by physiological processes for living in a hypotonic, freshwater environment (Loftus & Kushlan 1987). In a lagoon study by Winemiller & Leslie (1992), 62% of the marine species were not found at the mouth of an estuarine lagoon, and 86% of the species did not penetrate further inland. While *L. parva* seems to prefer brackish waters, it seems to avoid waters lacking salt . . . except in Florida.

The ability of *L. parva* to successfully invade certain freshwater rivers in Florida may be related to the chemical composition of the water in those rivers. Foster (1967) suggested that the collection of *L. parva* in varying salinities over seasons may be explained by an anadromous reproductive pattern in which *L. parva* migrates to less saline water to breed, then returns to brackish waters afterwards (Hildebrand & Schroeder 1928, Beck & Massmann 1951). Physiologists have shown that the presence of calcium salts in 'fresh' water can increase the ability of euryhaline fish to osmoregulate (e.g. Breder 1934, Black 1957). Florida springwater is very high in sodium, calcium and chloride ions, due to sediments that were laid down around the Pamlico terrace during the Pleistocene period (Rosenau et al. 1977, Burgess & Franz 1978). Some authors have argued that the inland distribution of marine species in Florida is strongly associated with areas that fringed the Pamlico terrace (e.g. Odum 1953, Loftus & Kushlan 1987). Species such as *L. parva* have been able to move great distances up certain 'freshwater' rivers of the Florida peninsula (Odum 1953, Foster 1967, Tagatz 1968, Gilbert 1978).

What prevents *L. goodei* from colonizing the coastal waters of Florida? Despite having an altricial ancestor that was probably euryhaline (Parenti 1981), the precocial *L. goodei* may have secondarily lost the ability to osmoregulate over a wide range of salinities (see Griffith 1974). The transition from 'secondary freshwater' to 'primary freshwater' existence has been observed in many of the more temperate killifish species, especially in the genus *Fundulus* (Griffith 1974). Briggs (1958) suggested that the low topography of the Floridan terrace, associated with low rates of stream capture, has made emigration very difficult for primary freshwater fishes in this region. This difficulty in emigration may also explain the endemism of many freshwater Florida fishes, such as *L. goodei.* It is also possible that suitable habitats for *L. goodei* exist outside their current distribution, while the colonization paths leading to these habitats do not exist (Kilby 1955, Gilbert 1978). In this manner, a lack of osmoregulatory ability may be a key factor in preventing *L. goodei* from moving further toward the coast and colonizing the brackish waters where *L. parva* is found (see Bruton 1989).

If the alleged differences in osmoregulatory ability between *L. parva* and *L. goodei* have been effectively neutralized by the high mineral content of Floridan rivers, what keeps *L. parva* from colonizing the springs? The solution to this evolutionary riddle might be found in the comparison of early ontogeny. Differences in the reproductive investment and morphological development of *L. parva* and *L. goodei* are consistent with the generation of altricial and precocial alternatives within the genus. In this case, enhanced vitellogenesis in *L. goodei* could be associated with the precocial development of morphological characters. However, the differences in reproductive investment and morphological development may have actually been driven by factors other than those associated with environmental stability (Holm 1989). Lee et al. (1980) suggested that *L. goodei* occur predominantly in waters with extremely low dissolved oxygen concentrations. While the adults of both *L. parva* and *L. goodei* are morphologically and physiologically suited to hypoxic environments (Lewis 1970), it is possible that developing offspring of *L. parva* cannot survive and develop under the consistently hypoxic conditions of the Florida springs. Colonization of these springs may be impossible if the population is not supported by local recruitment. In contrast, the offspring of *L. goodei* may be able to survive and develop in these environments by virtue of an expanded vitelline circulatory network, that forms over the larger surface area provided by larger eggs. In this case, the absolute dimensions of the reproductive investment may be more important than the energy investment of the egg. Thus, the hypoxic conditions in the Florida springs may be a key factor in preventing *L. parva* from moving further inland and colonizing the freshwaters where *L. goodei* is found.

The hypothesis described above is intended to generate testable predictions regarding the abilities of *L. parva* and *L. goodei* to develop and survive under different environmental conditions. If, through specialization, *L. goodei* has seondarily lost the ability to osmoregulate in brackish waters, we would expect the enhanced performance of *L. parva* in tolerance tests at higher salinities. Likewise, if *L. parva* has not acquired the ability to osmoregulate under truly freshwater conditions (i.e. controlled for various ion concentrations), then we would expect the enhanced performance of *L. goodei* in tolerance tests at lower salinities. More importantly, the postulated relationship between the vitelline circulatory network and the ability to withstand hypoxic conditions at the springs inhabited by *L. goodei*, is also open to critical experimentation. For example, if the vitelline network of a *L. parva* offspring is insufficient to allow for development and survival under hypoxic conditions, then we would expect smaller eggs of *L. goodei*, with smaller vitelline networks, to experience greater developmental stress and mortality. Indeed, by selecting eggs from both species on the basis of egg diameter, we could establish a balanced 2×2 experimental contingency to test this prediction. This design could also allow for a closer examination of the relationship between egg size and the complexity and density of the vitelline circulatory network. Finally, this type of experiment would allow us to more closely examine the hypothesized relationship between yolk investment and developmental rates. Although heterochrony is a complex component of ontogeny in the genus *Lucania,* a rigorous experimental design based on the descriptions of morphological development can lead to the discovery of proximate mechanisms that control this evolutionary phenomenon.

Acknowledgements

We are indebted to Christine Flegler-Balon and David Noakes for providing many helpful suggestions in the design and execution of this study. Felicia Coleman helped us to collect specimens and data on environmental conditions from the field. Mark Wiercinski took the dangerous end of the seine on the collecting trips. Skuli Skúlason, Dwight Watson and Terry Wheeler showed keen interest in the ideas surrounding the altricial-precocial life history model. Patrice Baker was always there. This study was supported by grants to EKB from the Natural Sciences and Engineering Research Council of Canada.

References cited

Ager, L A 1971 The fishes of Lake Okeechobee, Florida Quart J Florida Acad Sci 34 53–62

Alberch, P 1985 Problems with the interpretation of developmental sequences Syst Zool 34 46–58

Alberch, P, S J Gould, G F Oster & D B Wake 1979 Size and shape in ontogeny and phylogeny Paleobiol 5 296–317

Anderson, E 1974 Comparative aspects of the ultrastructure of the female gamete Int Rev Cytol Suppl 4 1–70

Armstrong, P B 1936 Mechanism of hatching in *Fundulus heteroclitus* Biol Bull 71 407

Armstrong, P B & J S Child 1965 Stages in the normal development of *Fundulus heteroclitus* Biol Bull 128 143–168

Arndt, R G E 1971 Ecology and behavior of the cyprinodont fishes *Adinia xenica, Lucania goodei, Leptolucania ommata* Ph D Thesis, Cornell University, Ithaca 344 pp

Balinsky, B I 1981 An introduction to embryology, 5th edition Saunders College Publishing, Toronto 768 pp

Balon, E K 1959 Die embryonale und larvale Entwicklung der Donauzope (*Abramis ballerus* subsp) Biologicke prace 5 1–87

Balon, E K 1975 Terminology of intervals in fish development J Fish Res Board Can 32 1663–1670

Balon, E K 1977 Early ontogeny of *Labeotrophus* Ahl, 1927 (Mbuna, Cichlidae, Lake Malawi), with a discussion on advanced protective styles in fish reproduction and development Env Biol Fish 2 147–176

Balon, E K 1979 The theory of saltation and its application in the ontogeny of fishes steps and thresholds Env Biol Fish 4 97–101

Balon, E K 1980a Early ontogeny of the lake charr, *Salvelinus (Cristivomer) namaycush* pp 485–562 *In* E K Balon (ed) Charrs Salmonid Fishes of the Genus *Salvelinus,* Perspectives in Vertebrate Science 1, Dr W Junk Publishers, The Hague

Balon, E K 1980b Early ontogeny of the North American landlocked arctic charr – *Salvelinus (Salvelinus) alpinus oquassa* pp 568–606 *In* E K Balon (ed) Charrs Salmonid Fishes of the Genus *Salvelinus,* Perspectives in Vertebrate Science 1, Dr W Junk Publishers, The Hague

Balon, E K 1980c Early ontogeny of the European landlocked arctic charr – *Salvelinus (Salvelinus) alpinus alpinus* pp 607–630 *In* E K Balon (ed) Charrs Salmonid Fishes of the Genus *Salvelinus,* Perspectives in Vertebrate Science 1, Dr W Junk Publishers, The Hague

Balon, E K 1980d Early ontogeny of the brook charr – *Salvelinus (Salvelinus) fontinalis* pp 631–666 *In* E K Balon (ed) Charrs Salmonid Fishes of the Genus *Salvelinus,* Perspectives in Vertebrate Science 1, Dr W Junk Publishers, The Hague

Balon, E K 1980e Comparative ontogeny of charrs pp 703–720 *In* E K Balon (ed) Charrs Salmonid Fishes of the Genus *Salvelinus,* Perspectives in Vertebrate Science 1, Dr W Junk Publishers, The Hague

Balon, E K 1981 Saltatory processes and altricial to precocial forms in the ontogeny of fishes Amer Zool 21 573–596

Balon, E K 1984 Life histories of Arctic charrs an epigenetic explanation of their invading ability and evolution pp 109–141 *In* Johnson & B L Burns (ed) Biology of the Arctic Charr, University of Manitoba Press, Winnipeg

Balon, E K 1985a The theory of saltatory ontogeny and life history models revisited pp 13–30 *In* E K Balon (ed) Early Life Histories of Fishes New Developmental, Ecological and Evolutionary Perspectives, Dev in Env Biol Fish 5, Dr W Junk Publishers, Dordrecht

Balon, E K 1985b Reflections on epigenetic mechanisms hypotheses and case histories pp 239–270 *In* E K Balon (ed) Early Life Histories of Fishes New Developmental, Ecological and Evolutionary Perspectives, Dev in Env Biol Fish 5, Dr W Junk Publishers, Dordrecht

Balon, E K 1986a Saltatory ontogeny and evolution Rivista di Biologia 79 151–190

Balon, E K 1986b Types of feeding in the ontogeny of fishes and the life-history model Env Biol Fish 16 11–24

Balon, E K 1989 The epigenetic mechanisms of bifurcation and alternative life-history styles pp 467–501 *In* M N Bruton (ed) Alternative Life-History Styles of Animals, Perspectives in Vertebrate Science 6, Kluwer Academic Publishers, Dordrecht

Balon, E K 1990 Epigenesis of an epigeneticist the development of some alternative concepts on the early ontogeny and evolution of fishes Guelph Ichthyol Rev 1 1–42

Balon, E K , S S Crawford & A Lelek 1986 Fish communities of the upper Danube River (Germany, Austria) prior to the new Rhein-Main-Donau connection Env Biol Fish 15 243–271

Beck, W R & W H Massmann 1951 Migratory behavior of the rainwater fish, *Lucania parva,* in the York River Virginia Copeia 1951 176

Black, V S 1957 Excretion and osmoregulation pp 163–205 *In* M E Brown (ed) The Physiology of Fishes Volume 1, Metabolism, Academic Press, New York

Blaxter, J H S 1988 Pattern and variety in development pp 1–58 *In* W S Hoar & D J Randall (ed) Fish Physiology, Vol 11a, The Physiology of Developing Fish Eggs and Larvae Academic Press, Toronto

Boyd, M 1928 A comparison of the oxygen consumption of unfertilized and fertilized eggs of *Fundulus heteroclitus* Biol Bull 55 92–98

Bradshaw, A D 1965 Evolutionary significance of phenotypic plasticity in plants Adv Genetics 13 115–155

Breder, C M Jr 1934 Ecology of an oceanic freshwater lake, Andros Island, Bahamas, with special reference to its fishes Zoologica 8 57–88

Breder, C M Jr 1948 Field book of marine fishes of the Atlantic coast G P Putnam and Sons, New York 332 pp

Briggs, J C 1958 A list of Florida fishes and their distribution Bull Flor State Mus Biol Sci 2 223–318

Brummett A R & J N Dumont 1981 A comparison of chorions from eggs of northern and southern populations of *Fundulus heteroclitus* Copeia 1981 607–614

Bruton, M N 1989 The ecological significance of alternative life-history styles pp 503–553 *In* M N Bruton (ed) Alternative Life-history Styles of Animals, Perspectives in Vertebrate Science 6, Kluwer Academic Publishers, Dordrecht

Burgess, G H & R Franz 1978 Zoogeography of the aquatic fauna of the St Johns River system with comments on adjacent peninsular faunas Amer Midl Nat 100 160–170

Burgess, G H , C R Gilbert, V Guillory & D C Taphorn 1977 Distributional notes on some north Florida freshwater fishes Florida Sci 40 33–41

Constanz, G D 1979 Life history patterns of a livebearing fish in contrasting environments Oecologia 40 189–201

Crawford, S S 1993 Ecomorphological comparison of early ontogeny in species of the genus *Lucania* (Pisces Cyprinodontidae) Ph D Thesis, University of Guelph, Guelph 264 pp

Crawford, S S & E K Balon 1994a Alternative life histories of the genus *Lucania* 1 Early ontogeny of *L parva,* the rainwater killifish Env Biol Fish 40 349–389

Crawford, S S & E K Balon 1994b Alternative life histories of the genus *Lucania* 2 Early ontogeny of *L goodei*, the bluefin killifish Env Biol Fish 41 331–367

Crawford, S S & E K Balon 1994c Development of morphological variation in fishes of the genus *Lucania* Copeia (in press)

Cunningham, J E R & E K Balon 1985 Early ontogeny of *Adinia xenica* (Pisces, Cyprinodontiformes) 1 The development of embryos in hiding Env Biol Fish 14 115–166

Dineen, J W 1974 Fishes of the Everglades pp 375–385 *In* P J Gleason (ed) Environments of South Florida Present and Past, Miami Geol Soc Memoir 2, Miami

Duellman, W E 1989 Alternative life-history styles in anuran amphibians evolutionary and ecological implications pp 101–126 *In* M N Bruton (ed) Alternative Life-History Styles of Animals, Perspectives in Vertebrate Science 6, Kluwer Academic Publishers, Dordrecht

Duggins, C F 1980 Systematics and zoogeography of *Lucania parva, Floridichthys,* and *Menidia* (Osteichthyes Atheriniformes) in Florida, the Gulf of Mexico Ph D Thesis, The Florida State University, Tallahassee 168 pp

Duggins, C F Jr , A A Karlin & K G Reylea 1983 Electrophoretic variation in the killifish genus *Lucania* Copeia 1983 564–570

Dunson, W A & J Travis 1991 The role of abiotic factors in community organization Amer Nat 138 1067–1091

Echelle, A A 1973 Behavior of the pupfish *Cyprinodon rubrofluviatilis* Copeia 1973 68–76

Echelle, A A , A F Echelle & L G Hill 1972 Interspecific interactions and limiting factors of abundance and distribution in the Red River pupfish (*Cyprinodon rubrofluviatilis*) Amer Midl Nat 88 109–130

Fahy, W E 1978 The influence of crowding upon the total number of vertebrae developing in *Fundulus majalis* (Walbaum) J Cons int Explor Mer 38 252–256

Flegler-Balon, C 1989 Direct and indirect development in fishes – examples of alternative life-history styles pp 71–100 *In* M N Bruton (ed) Alternative Life-History Styles of Animals, Perspectives in Vertebrate Science 6, Kluwer Academic Publishers, Dordrecht

Foster, N R 1967 Comparative studies on the biology of killifishes (Pisces, Cyprinodontidae) Ph D Thesis, Cornell University, Ithaca 369 pp

Freeman, G 1982 What does the comparative study of development tell us about evolution? pp 155–167 *In* J T Bonner (ed) Evolution and Development, Springer-Verlag, New York

Fuiman, L A 1984 Ostariophysi development and relationships pp 126–137 *In* H G Moser, W J Richards, D M Cohen, M P Fahay, A W Kendall Jr & S L Richardson (ed) Ontogeny and Systematics of Fishes, American Society of Ichthyologists and Herpetologists, Special Publication Number 1, Allen Press Inc , Lawrence

Gabriel, M L 1944 Factors affecting the number and form of vertebrae in *Fundulus heteroclitus* J Exp Zool 95 105–147

Gatz, A J 1979 Community organization in fishes as indicated by morphological features Ecology 60 711–718

Gilbert, C R (ed) 1978 Rare and endangered biota of Florida, Volume 4 Fishes University Presses of Florida, Gainesville 58 pp

Gleick, J 1987 Chaos, making a new science Penguin Books, New York 352 pp

Gorodilov, Y N 1992 Rhythmic processes in lower vertebrate embryogenesis and their role for developmental control Zool Sci 9 1101–1111

Goto, A 1990 Alternative life-history styles of Japanese freshwater sculpins revisited Env Biol Fish 28 101–112

Gould, S J 1977 Ontogeny and phylogeny Harvard University Press, Cambridge 501 pp

Griffith, R W 1974 Environment and salinity tolerance in the genus *Fundulus* Copeia 1974 319–331

Gunter, G 1956 A revised list of euryhaline fishes of North and Middle America Amer Midl Nat 56 345–354

Gunter, G 1961 Some relations of estuarine organisms to salinity Limnol Oceanogr 6 182–190

Hall, B K 1984 Developmental processes underlying heterochrony as an evolutionary mechanism Can J Zool 62 1–7

Heming, T A & R K Buddington 1988 Yolk absorption in embryonic and larval fishes pp 407–446 *In* W S Hoar & D J Randall (ed) Fish Physiology Volume 11A, The Physiology of Developing Fish, Eggs and Larvae, Academic Press, Toronto

Hickman, J C 1975 Environmental unpredictability and plastic energy allocation strategies in the annual *Polygonum cascadense* (Polygonaceae) J Ecol 63 689–701

Hildebrand, S F & W C Schroeder 1928 Fishes of Chesapeake Bay Bull U S Bur Fish 43 1–366 (1972 Reprint by T F H Publications, Neptune 388 pp)

Hillyard, S D 1981 Energy metabolism and osmoregulation in desert fishes pp 385–140 *In* R J Naiman & D L Soltz (ed) Fishes in North American Deserts John Wiley and Sons, Toronto

Hodson, P V & J B Sprague 1975 Temperature-induced changes in acute toxicity of zinc to Atlantic salmon (*Salmo salar*) J Fish Res Board Can 32 1–10

Holden, K K & M N Bruton 1992 A life-history approach to the early ontogeny of the Mozambique tilapia *Oreochromis mossambicus* (Pisces, Cichlidae) S Afr J Zool 27 173–191

Holm, E 1989 Environmental restraints and life strategies a habitat templet matrix pp 197–208 *In* M N Bruton (ed) Alternative Life-History Styles of Animals Perspectives in Vertebrate Science 6 Kluwer Academic Publishers, Dordrecht

Hubbs, C L 1926 The structural consequences of modifications of the developmental rate in fishes, considered in reference to certain problems of evolution Amer Nat 60 57–81

Hubbs, C L & K F Lagler 1958 Fishes of the Great Lakes region The University of Michigan Press, Ann Arbor 213 pp

Hubbs, C L & R R Miller 1965 Studies of cyprinodont fishes XXII Variation in *Lucania parva* its establishment in western United States, and description of a new species from an interior basin in Coahuila, Mexico Misc Publ Museum of Zoology, University of Michigan 127 1–104

Hubbs, C L B W Walker & R E Johnson 1943 Hybridization

in nature between species of American cyprinodont fishes Contrib Lab Vert Biol Univ Mich 23 1–21

Huver, C W 1956 The relation of the cortex to the formation of the perivitelline space in the eggs of *Fundulus heteroclitus* Biol Bull 111 304

Huver, C W 1960 The stage at fertilization of the egg of *Fundulus heteroclitus* Biol Bull 119 320

James, N P E & M N Bruton 1992 Alternative life-history traits associated with reproduction in *Oreochromis mossambicus* (Pisces Cichlidae) in small water bodies of the eastern Cape, South Africa Env Biol Fish 34 379–392

Jordan, D S & B W Evermann 1896–1900 The fishes of North and Middle America Bull U S Nat Mus 47 1–3313

Kaighn, M E 1964 A biochemical study of the hatching process in *Fundulus heteroclitus* Devel Biol 9 56–80

Kilby, J D 1955 The fishes of two gulf coastal marsh areas of Florida Tulane Stud Zool 2 175–247

Kinne, O 1960 Growth, food intake, and food conversion in a euryplastic fish exposed to different temperatures and salinities Physiol Zool 33 288–317

Kinne, O & E M Kinne 1962 Effects of salinity and oxygen on developmental rates in a cyprinodont fish Nature 193 1097–1098

Koenig, C C & R J Livingston 1976 The embryological development of the diamond killifish (*Adinia xenica*) Copeia 1976 435–445

Koster, W J 1948 Notes on the spawning activities and the young stages of *Plancterus kansae* (Garman) Copeia 1948 25–33

Kryzhanovsky, S G 1933 The respiratory organs of fish larvae (Teleostomie) and the pseudobranchs Tr Lab Evol Morphol 1 5–104 (In Russian)

Kryzhanovsky, S G 1934a Die Atmungsorpane der Fischlarven (Teleostomi) Zool Jahrb, Abt Anat Ontog der Tiere 58 21–60

Kryzhanovsky, S G 1934b Die Pseudobranchie (Morphologie und biologische Bedeutung) Zool Jahrb, Abt Anat Ontog der Tiere 58 171–238

Kuntz, A 1916 Notes on the embryology and larval development of five species of teleostean fishes Bull U S Bureau Fish 34 407–429

Kushlan, J A 1974 Effects of a natural fish kill on the water quality, plankton, and fish population of a pond in the Big Cypress Swamp, Florida Trans Amer Fish Soc 103 235–243

Kushlan, J A & T E Lodge 1974 Ecological and distributional notes on the freshwater fish of southern Florida Florida Scientist 2 110–128

Laale, H W 1980 The perivitelline space and egg envelopes of bony fishes a review Copeia 1980 210–226

Lampl, M, J D Veldhuis & M L Johnson 1992 Saltation and stasis a model of human growth Science 258 801–803

Lee, D S, C R Gilbert, C H Hocutt, R E Jenkins, D E McAllister & J R Stauffer Jr 1980 Atlas of North American freshwater fishes North Carolina State Museum of Natural History, Raleigh 867 pp

Lewis, W M Jr 1970 Morphological adaptations of cyprinodontoids for inhabiting oxygen deficient waters Copeia 1970 319–326

Liem, K F & L S Kaufman 1984 Intraspecific macroevolution functional biology of the polymorphic cichlid species *Cichlasoma minckleyi* pp 203–215 *In* A A Echelle & I Kornfield (ed) Evolution of Fish Species Flocks, University of Maine Press, Orono

Lindsey, C C 1978 Form, function, and locomotory habits in fish pp 1–100 *In* W S Hoar & D J Randall (ed) Fish Physiology, Volume 7, Academic Press, New York

Loftus, W F & J A Kushlan 1987 Freshwater fishes of southern Florida Bull Flor State Mus Biol Sci 31 344 pp

Løvtrup, S 1982 The four theories of evolution Rivista di Biologia 75 53–66, 231–272, 385–409

Løvtrup, S 1989 On divergent and progressive evolution pp 55–69 *In* M N Bruton (ed) Alternative Life-history Styles of Animals, Perspectives in Vertebrate Science 6, Kluwer Academic Publishers, Dordrecht

Lowe-McConnell, R H 1982 Tilapias in fish communities pp 83–113 *In* R S V Pullin & R H Lowe-McConnell (ed) The Biology and Culture of Tilapias, ICLARM Conference Proceedings 7, Manila

Maderson, P F A, P Alberch, B C Goodwin, S J Gould, A Hoffman, J D Murray, D M Raup, A de Ricqles, A Seilacher, G P Wagner & D B Wake 1982 The role of development in macroevolutionary change group report pp 279–312 *In* J T Bonner (ed) Evolution and Development, Springer-Verlag, Heidelberg

Mahon, R 1984 Divergent structure in fish taxocenes of north temperate streams Can J Fish Aquat Sci 41 330–350

Mahon, R & E K Balon 1977 Fish community structure in lakeshore lagoons on Long Point, Lake Erie, Canada Env Biol Fish 2 71–82

Mansueti, A J & J D Hardy Jr 1967 Development of fishes of the Chesapeake Bay region an atlas of egg, larval, and juvenile stages Part I Natural Resources Institute, University of Maryland, Baltimore 202 pp

Matsuda, R 1987 Animal evolution in changing environments with special reference to abnormal metamorphosis John Wiley and Sons, New York 355 pp

May, R M 1974 Ecosystem patterns in randomly fluctuating environments pp 1–50 *In* R Rosen & F M Snell (ed) Progress in Theoretical Biology, Volume 3, Academic Press, New York

McClane, A J 1978 McClane's field guide to saltwater fishes of North America Henry Holt and Company, New York 283 pp

McElman, J F & E K Balon 1979 Early ontogeny of walleye, *Stizostedion vitreum,* with steps of saltatory development Env Biol Fish 4 309–348

Meyer, A 1987 Phenotypic plasticity and heterochrony in *Cichlasoma managuense* (Pisces, Cichlidae) and their implications for speciation in cichlid fishes Evolution 41 1357–1369

Milkman, R 1954 Controlled observation of hatching in *Fundulus* Biol Bull 107 300

Mommsen, T P & P J Walsh 1988 Vitellogenesis and oocyte assembly pp 347–406 *In* W S Hoar & D J Randall (ed) Fish

Physiology, Volume 11A, The Physiology of Developing Fish, Eggs and Larvae, Academic Press, Toronto

Nikolsky, G V 1963 The ecology of fishes Academic Press, New York 352 pp

Noakes, D L G & E K Balon 1982 Life histories of tilapias an evolutionary perspective pp 61–82 *In* R S V Pullin & R H Lowe-McConnell (ed) The Biology and Culture of Tilapias ICLARM Conference Proceedings 7, Manila

Noakes, D L G , S Skulason & S S Snorrason 1989 Alternative life-history styles in salmonine fishes with emphasis on arctic charr, *Salvelinus alpinus* pp 329–346 *In* M N Bruton (ed) Alternative Life-History Styles of Animals, Perspectives in Vertebrate Science 6, Kluwer Academic Publishers, Dordrecht

Odum, H T 1953 Factors controlling marine invasion into Florida freshwaters Bull Mar Sci Gulf Caribb 3 134–156

Oppenheimer, J M 1937 The normal stages of *Fundulus heteroclitus* Anat Rec 68 1–15

Oster, G & P Alberch 1982 Evolution and bifurcation of developmental programs Evolution 36 444–459

Paine, M D & E K Balon 1984 Early development of the northern logperch, *Percina caprodes semifasciata,* according to the theory of saltatory development Env Biol Fish 11 173–190

Parenti, L R 1981 A phylogenetic and biogeographic analysis of cyprinodontiform fishes (Teleostei, Atherinomorpha) Bull Amer Mus Nat Hist 168 341–557

Pepin, P & R A Myers 1991 Significance of egg and larval size to recruitment variability of temperate marine fish Can J Fish Aquat Sci 48 1820–1828

Perrin, M R 1989 Alternative life-history styles of small mammals pp 209–242 *In* M N Bruton (ed) Alternative Life-History Styles of Animals, Perspectives in Vertebrate Science 6, Kluwer Academic Publishers, Dordrecht

Peters, N 1965 Diapause und embryonale Missbildung bei eierlegenden Zahnkarpfen Roux' Archiv fur Entwicklungsmechanik 156 75–87

Quanwei, X 1986 A study on isozymes in *Lucania* Scientia Sinica 29 618–622

Rombough, PJ 1988 Respiratory gas exchange, aerobic metabolism, and effects of hypoxia during early life pp 59–161 *In* W S Hoar & D J Randall (ed) Fish Physiology, Volume 11A, The Physiology of Developing Fish, Eggs and Larvae, Academic Press, Toronto

Rosen, D E 1964 The relationships and taxonomic position of the halfbeaks, killifishes silversides, and their relatives Bull Amer Mus Nat Hist 127 217–268

Rosenau, J C , G L Faulkner, C W Hendry Jr & R W Hull 1977 Springs of Florida State of Florida Department of Natural Resources, Bulletin Number 31 461 pp

Schoenherr A A 1977 Density dependent and density independent regulation of reproduction in the Gila topminnow, *Poeciliopsis occidentalis* (Baird and Girard) Ecology 58 438–444

Scott C G & W E Kellicott 1916 The consumption of oxygen during the development of *Fundulus heteroclitus* Anat Rec 11 531

Scrimshaw, N S 1944 Embryonic growth in the viviparous poeciliid, *Heterandria formosae* Biol Bull 87 37–51

Sinervo, B & L B McEdward 1988 Developmental consequences of an evolutionary change in egg size an experimental test Evolution 42 885–899

Smirnov, A I 1975 Biology, reproduction and development of Pacific salmons Izd Moskovskogo Univ , Moskva 335 pp (In Russian)

Soin, S G 1968 Adaptational features in fish ontogeny Israel Program for Scientific Traslations, Jerusalem, 1971 72 pp

Solberg, A N 1938 The development of a bony fish Progr Fish Cult 40 1–19

Springer, V G & K D Woodburn 1960 An ecological study of the fishes of the Tampa Bay area Profess Pap Series 1, Florida State Board of Conservation Marine Lab 104 pp

Stearns, S C 1980 A new view of life-history evolution Oikos 35 266–281

Stearns, S C 1982 The role of development in the evolution of life histories pp 237–258 *In* J T Bonner (ed) Evolution and Development, Springer-Verlag, Heidelberg

Tabb, D C & R B Manning 1961 A checklist of the flora and fauna of northern Florida Bay and adjacent brackishwaters of the Florida mainland collected during the period July, 1957, through September 1960 Bull Mar Sci Gulf and Caribbean 11 552–649

Tagatz, M E 1968 Fishes of the St Johns River, Florida Quart J Flor Acad Sci 30 25–50

Trinkaus, J P 1967 Procurement, maintenance and use of *Fundulus* eggs pp 113–122 *In* F H Wilt & N K Wessells (ed) Methods in Developmental Biology, T Y Crowell Co , New York

Veenstra, R S 1987 Ecomorphological comparisons of early development in an altricial (*Lucania parva*) and in a precocial (*Lucania goodei*) cyprinodont M Sc Thesis, University of Guelph, Guelph 95 pp

Waddington, C H 1975 The evolution of an evolutionist Edinburgh University Press, Edinburgh 328 pp

Wake, M H 1990 The evolution of integration of biological systems an evolutionary perspective through studies on cells, tis sues and organs Amer Zool 30 897–906

Ward, D 1989 Allometry and the breeding biology of some plovers pp 371–384 *In* M N Bruton (ed) Alternative Life-History Styles of Animals, Perspectives in Vertebrate Science 6, Kluwer Academic Publishers, Dordrecht

Wessels, N K & F J Swartz 1953 Relation of the micropyle to cortical changes at fertilization in the egg of *Fundulus hetero clitus* Anat Rec 117 557–558

Wilkinson, L 1990 SYSTAT The system for statistics SYSTAT Inc , Evanston 677 pp

Winemiller, K O 1989 Patterns of variation in life history among South American fishes in seasonal environments Oecologia 81 225–241

Winemiller, K O 1992 Ecomorphology of freshwater fishes Nat Geograph Research and Exploration 8 308–327

Winemiller, K O & M A Leslie 1992 Fish assemblages across a

complex, tropical freshwater/marine ecotone. Env. Biol. Fish. 34: 29–50.

Wourms, J.P. 1964. Comparative observations on the early embryology of *Nothobranchius taeniopygus* (Hilgendorf) and *Aplocheilichthys pumilis* (Boulenger) with special reference to the problem of naturally occuring embryonic diapause in teleost fishes. East African Freshwater Fish. Res. Org. Ann. Rep. 68–73.

Wourms, J.P. 1972a. Developmental biology of annual fishes. I. Stages in the normal development of *Austofundulus myersi* Dahl. J. Exp. Zool. 182: 143–168.

Wourms, J.P. 1972b. Developmental biology of annual fishes. II. Naturally occurring dispersion and reaggregation of blastomeres during the development of annual fish eggs. J. Exp. Zool. 182: 169–200.

Wourms, J.P. 1972c. Developmental biology of annual fishes. III. Pre-embryonic and embryonic diapause of variable duration in the eggs of annual fishes. J. Exp. Zool. 182: 389–414.

Yamamoto, T. 1967. Medaka. pp. 101–111. *In:* F.W. Wilt & N.K. Wessels (ed.) Methods in Developmental Biology, T.Y. Crowell Company, New York.

Environmental Biology of Fishes **41**: 403–414, 1994.

Intraspecific variation in sand-diving and predator avoidance behavior of green razorfish, *Xyrichtys splendens* (Pisces, Labridae): effect on courtship and mating success

Simon C. Nemtzov
Department of Ecology and Evolution, State University of New York, Stony Brook, NY 11794-5245, U.S.A.
Present address: Department of Zoology, Tel Aviv University, Ramat Aviv, Israel

Received 14.10.1992 Accepted 2.4.1993

Key words: Caribbean Sea, Fish behavior, Phenotypic plasticity, Predation pressure, Predation risk, Reproductive success, Wrasse

Synopsis

Green razorfish are Caribbean wrasses that live in harems on shallow sand or seagrass beds, which offer little cover for predator avoidance (PA). Field observations showed that non-conspecific fishes that intruded were either attacked, ignored, or actively avoided. Food competitors and small piscivores were attacked by male razorfish. Razorfish PA behaviors varied among three habitats with different substratum compositions, suggesting that these fish possess phenotypic plasticity for PA behavior. In a rocky-rubble habitat, razorfish dove into the coarse sand for PA, but most sand dives observed there were to soften a small site for future PA. In a sandbed habitat, they hid among coral branches and dove into the sand when attacked; few maintenance dives were observed as soft sand was widespread. In a seagrass habitat, they hid among blades of grass for PA, and dove into the sand less frequently than at the other sites. Some female razorfish that were transferred among habitats adopted PA behaviors similar to those of females in the new site, while others did not, suggesting that behavioral plasticity is not universal in this species. Razorfish spawned lower in the water column in the presence of natural predators and a predator model, than when these were absent. When the predator model was introduced into a male's territory during spawning periods, there was a reduction in his courtship rate, but not in the number of spawns he achieved. Predation pressure may reduce males' long-term fitness by causing decreased courtship rates which can facilitate sex change in harem females.

Introduction

Animals must avoid predators or they will generally lose future fitness (ignoring any possible inclusive fitness). With such a high price to pay for being eaten, there is necessarily a very strong selective pressure on animals to avoid being killed (Endler 1986). Animals have evolved a wide variety of behavioral and morphological adaptations to avoid potential predators (Vermeij 1987). For prey species, phenotypic plasticity itself may be the most important weapon in the predator-prey arms race (Dawkins & Krebs 1983). Although evolutionary flexibility does not always equal behavioral adaptability (Gause 1942), many animals do adjust their behavior in ecological time in response to predation risk (Lima & Dill 1990).

The effect of predators on the evolution of morphology, physiology, and behavior has been documented in many prey fish species (e.g. Feder &

Lauder 1986, Simenstad & Caillet 1986, Kerfoot & Sih 1987, Hixon 1991, Lott 1991, Magnhagen 1991). Unlike coral-reef fishes that use the structural heterogeneity of the reef for shelter (Ehrlich 1975), fishes on sand flats must seek shelter from predators in other ways (McGovern & Clark 1991). Some of the adaptations for avoiding predators in this habitat, which have been studied by Clark and her co-workers, include burrow construction (sand tilefish *Malacanthus plumieri,* Clark et al. 1988), armor (sea moth *Eurypegasus draconis,* Clark 1983a), toxicity (Moses sole *Pardachirus marmoratus,* Clark & George 1979), mimicry (black razorfish *Xyrichtys niger,* Clark 1983a, b), cryptic coloration (spotted sandperch *Parapercis hexophthalma,* Clark et al. 1991), group living (garden eel *Gorgasia sillneri,* Clark et al. 1990), and sand-diving (five-fingered razorfish *X. pentadactylus,* Clark 1983b).

For some of these fishes, modification of the substratum to create a predator avoidance (PA) site represents a substantial investment. This investment increases the value of the territory, so that an individual will be more likely to defend it (Davies & Houston 1984). Individuals of species that inhabit diverse habitats can show intraspecific differences in social behaviors due to the variation in ecological conditions among the sites (Lott 1991). Thus, the necessity for the construction and maintenance of PA sites can lead to intraspecific differences in both the spatial distribution of animals and their social behavior.

Razorfish of the labrid genus *Xyrichtys* live on open sand flats away from coral reefs (Clark 1983b, Nemtzov 1982, 1985, 1992, Victor 1987) and are therefore subject to great risk of predation from piscivorous fish (hereafter referred to simply as risk). The absence of cover in this habitat has apparently led to the evolution of their unusual head shape, which enables sand-diving (Clark 1983b), and their colonial behavior, which facilitates group vigilance for predators (Victor 1987). Similarly, living in this high risk habitat has necessitated adjustment to local conditions in ecological time, as in foraging mode, spawning behavior and territoriality (Nemtzov 1992). Thus, the specific ecological conditions, in the face of high levels of risk, represent an important influence on many aspects of razorfish biology. In this report I present observations of intraspecific variation in PA behavior by green razorfishes in different habitats.

Because they live in exposed habitats, razorfishes are also ideal subjects for investigating how predation risk affects courtship behavior and mating success. Animals must often expose themselves to some risk of predation in the course of many important activities. During courtship the tradeoffs between survival and PA become particularly obvious (Magnhagen 1990, 1991, Williams 1989). In many animals conspicuous courtship behaviors, coloration, chemicals, or songs are correlated with male mating success, but they can expose the male to great risk of predation (e.g. Endler 1983, Lloyd 1984, Ryan et al. 1981).

By maintaining a harem, a male may experience lower risk by alleviating the need to search for mates, while still achieving relatively high mating success (Robinson 1986). The reproductive success of a male with a harem is dependent largely upon the number of females in his harem (Clutton-Brock 1988), which is generally considered to be a function of the male's ability to control either females or resources important to them (Emlen & Oring 1977), which in turn can be a function of the amount of risk he is willing to take (e.g. Gosling 1986). Thus, risk of predation can affect reproductive success (Magnhagen 1990) even for males with harems.

Adaptations for accomplishing mating in the presence of predators have been well documented for only a few species, e.g. water strider *Gerris remigis* (Sih et al. 1990), fireflies *Photinus* spp. (Lloyd 1984), guppy *Poecilia reticulata* (Farr 1975, Endler 1982, 1987, Luyten & Liley 1985), and Túngara frog *Physalaemus pustulosus* (Ryan 1985). This report seeks to fulfill a need (Lima & Dill 1990, Magnhagen 1991) for field studies that show how predation risk affects individual reproductive behavior in ecological time.

Methods

Study species and study sites

Green razorfish, *X. splendens,* are small (mean

standard length 7–9 cm), sexually dichromatic wrasses (family Labridae) endemic to the western Atlantic (Randall 1965) that generally live over shallow, flat, grassy and sandy areas (Nemtzov 1992). Like all tropical labrids, green razorfish engage in broadcast (sometimes called pelagic) mating in which pairs swim rapidly toward the surface and together release gametes into the water at the apex of this spawning rush, producing planktonic larvae which receive no parental care. Spawning occurs daily, year-round (E. Clark personal communication) during a late afternoon spawning period. Green razorfish undergo protogynous sex change (Roede 1972, Nemtzov 1992).

Razorfish were observed in the field from May to August, 1989 to 1991, and in January 1991, at three habitats around the southern Caribbean island of Bonaire, Netherlands Antilles (12° N, 68° W). One site (Lac Bay) was a shallow lagoon (depth = 4–6 m) with a *Thalassia* grassbed, the second site (Playa Bengé) had a soft sandbed (depth = 8–10 m), and the third (Kralendijk harbor) had a bottom covered with coarse sand, small rocks, and coral rubble (depth = 6–8 m).

Field observations of behaviors

I recorded behaviors of at least 200 unmarked razorfish, and of 157 individual razorfish which were identified by natural markings or by plastic anchor tags (Floy Tag & Manufacturing, Seattle) injected into the dorsal musculature. SCUBA divers using handnets caught the fish, measured and tagged them (without anesthetic) while underwater, and kept them for 15–60 min in a holding net before release to ensure recovery from the tagging procedure. There was no detectable difference in behavior between tagged and untagged fish. Data on unidentified fish were pooled, whereas data on recognizable individuals are presented as individual means.

Stationary SCUBA divers observed recognized razorfish as focal individuals during replicate 15-min observations on at least three separate days. The observer used an underwater slate to record the frequency with which potential predators passed by, and the frequency of predator avoidance (PA) behaviors by the razorfish. PA involved either taking refuge while exhibiting barred coloration, or diving in the sand to avoid attack. Distributions of observation periods with various numbers of PA behaviors were analyzed with G-tests (Sokal & Rohlf 1981) performed on untransformed counts. In all 2×2 G-tests of independence, I applied Yates' correction for continuity to yield a more conservative test (Sokal & Rohlf 1981).

In order to determine the fish's height above the substratum I watched focal individuals during replicate 15-min observations. I divided vertical space into 0.5-m intervals with the substratum at zero, and estimated the vertical position of the fish to be within one of these intervals every 30 sec. I used the upper limit of the interval as a score to represent that interval, e.g. a fish swimming between 0.5 and 1.0 m above the substratum received a score of 1.0. Fish that had dived below the sand received a height score of zero. The 30 data points from each 15-min observation period were analyzed for temporal autocorrelation with Markov matrices and G-tests; none was found.

To record mating success, an observer watched male razorfish as focal individuals during the entire daily afternoon spawning period for at least three days using SCUBA. We recorded the frequency of spawns and courtship behaviors during 2.0–2.5 h of continuous observation divided into successive 15-min intervals. We recorded the frequency of two courtship displays, *flips* and *rises* (Nemtzov 1992). A *flip* is when a male swims quickly past a female while rolling its body (flipping) onto its side and swimming with an undulating motion for 1–2 sec. A *rise* is when the male swims slowly toward the surface for 1–2 m while vigorously vibrating the caudal half of his body. The observer also estimated the height of the apex of each spawning rush, noted what time of day the spawning period began (onset of courtship behavior) and the time when each spawning rush occurred. A mean spawning time was calculated as the sum of the number of minutes after the beginning of the spawning period for each spawning rush observed, divided by the total number of spawns.

Fig 1 Underwater photograph of the model of a predatory fish used to test for the influence of predation risk on male green razorfish courtship and mating success at the rocky-rubble site The model had the following measurements total length 71 cm, fork length 65 cm, maximum depth 25 cm, tail height 34 cm Photo by S Kajiura

Transfer of females among sites

Five female razorfish were transferred from the rocky-rubble habitat to the sandbed site, and 14 females were transferred from the sandbed to the rocky-rubble site. Fish were transported in an insulated cooler where water temperature rose < 2° C during the transfers, which lasted 60–70 min. Transferred fish were kept in a holding net at their new site for 30 min before release and were each observed by one diver during the first 30 min after release when possible. Behaviors of transferred fish were then recorded almost daily for three weeks after the transfers.

Physical parameters

I calculated the coarseness of the sand at each site from five replicate 50-ml samples, collected with a portable submersible corer, dried at 60° C to zero weight loss, and then passed through successive stainless steel sieves with 2, 1, and 0.5 mm meshes. By the Wentworth scale of particle sizes, sand samples with larger proportions (by weight) of smaller grains were considered to be of finer texture (Levinton 1987).

I estimated horizontal visibility in the water to the nearest 1 m, as the distance that a black-and-white Secchi disk (diameter 20 cm) held 1 m above the bottom was visible across the study site. I considered the horizontal visibility for each day's spawning period as the mean of two readings taken at the beginning and end of each spawning period.

Predator model experiment

In order to investigate the influence of an increase in risk on courtship displays and reproductive success, a model predator was towed through a male's territory at the rocky-rubble site during replicate spawning periods. Due to logistic problems, this experiment was performed on only one male. The model of a large barjack, *Caranx ruber,* was constructed out of aluminum foil on a wire mesh frame (Fig. 1). A snorkeler on the surface controlled the model with monofilament lines attached to its snout and tail. A preliminary test had shown that the presence of a snorkeler over the male's territory without the predator model elicited no PA responses ($n = 3$, 15 min observational periods).

Before the introduction of the model, I observed the male for the entire spawning period on three successive days as a control. For the experiment, the snorkeler brought the model through the male's territory repeatedly (with a mean interval of 28 sec) during the entire spawning period while I recorded the time of spawns, and the frequency of courtship and predator avoidance behaviors. I repeated the experiment with the same male on three successive days.

Table 1 Responses of green razorfish to other fishes common in their vicinity

Actively avoided by green razorfish
- Tarpon (*Megalops atlanticus,* Elopidae)
- Jacks (Carangidae)
- Tunas (Scombridae)
- Snappers (Lutjanidae)
- Groupers (Serranidae)
- Great barracuda (*Sphyraena barracuda,* Sphyraenidae)

Usually attacked by green razorfish
- Gold-spotted snake eel (*Myrichthys oculatus,* Ophichthidae)
- Lizardfishes (Synodontidae)
- Peacock flounder (*Bothus lunatus,* Bothidae)
- Slippery dick (*Halichoeres bivittatus,* Labridae)
- Rosy razorfish (*Xyrichtys martinicensis,* Labridae)

Usually ignored by green razorfish
- Bonefish (*Albula vulpes,* Albulidae)
- Herrings (Clupeidae)
- Anchovies (Engraulidae)
- Garden eel (*Heteroconger halis,* Heterocongridae)
- Goatfishes (Mullidae)
- Mullets (Mugilidae)
- Palometa (*Trachinotus goodei,* Carangidae)
- Sand tilefish (*Malacanthus plumieri,* Malacanthidae)
- Yellowfin mojorra (*Gerres cinereus,* Gerreidae)
- Bandtail puffer (*Sphoeroides spengleri,* Tetraodontidae)
- Filefishes and triggerfishes (Balistidae)
- Parrotfishes (Scaridae)
- Most wrasses (Labridae)
  - Pearly razorfish, *Xyrichtys novacula*
  - Yellowhead wrasse, *Halichoeres garnoti*
  - Clown wrasse, *H maculipinna*
  - Rainbow wrasse, *H pictus*
  - Bluehead wrasse, *Thalasomma bifasciatum*

Results and discussion

Responses to encroaching non-conspecific fishes

Green razorfish exhibited three types of behavior in response to fishes that came into their vicinity. Most species of fish were ignored, some were actively attacked, and a few were avoided (Table 1). Fishes that were ignored were either species smaller than green razorfish, such as bluehead wrasse, or species whose diet did not generally overlap with the razorfish's, such as parrotfishes (Randall 1967). Females attacked intruders only very rarely; virtually all observed attacks were by males. Females avoided predators more frequently than males did because the females were apparently at greater risk than males due to their smaller size. Similarly, sexual asymmetry in predation risk was reported in the guppy, *Poecilia reticulata* (Magurran & Nowak 1991), and in some sticklebacks (Whoriskey & FitzGerald 1985).

Fishes that were attacked can be classified either as food competitors or as potential predators. The male chased away slippery dicks and rosy razorfish whose diets were similar to the razorfish's (Randall 1967). Interspecific defense of food resources is apparently important for razorfish because they are site-restricted. Other species that were attacked (ophichthid eels and synodontid lizardfishes) are

Table 2 Distribution of the percentage of 15-min observation periods with different numbers of predator avoidance (PA) responses by male and female green razorfish from three different sites

	Observation periods	PA responses per observation period					
		0	1	2	3	4	>4
Males (pooled sites)	691	70.9	19.8	6.8	1.7	0.3	0.4
Grassbed	227	79.3	14.5	4.8	0.9	0.4	0
Sandbed	145	60.7	29.0	8.3	2.1	0	0
Rocky rubble	319	69.6	19.4	7.5	2.2	0.3	0.9
Females (pooled sites)	520	56.7	27.9	10.0	3.1	1.3	1.0
Grassbed	34	85.3	8.8	5.9	0	0	0
Sandbed	254	52.4	31.1	11.8	2.8	1.2	0.8
Rocky-rubble	232	57.3	27.2	8.6	3.9	1.7	1.3

piscivores (Randall 1967). Even small members of these groups were larger than most of the razorfish and were potential predators. By taking on the role of defense, males may have reduced some costs for the females in the harems, which would facilitate polygyny by lowering the polygyny threshold (Orians 1969). The male's role as 'policeman' was also evident when he interposed himself between females within the harem who were engaged in aggressive interactions (Nemtzov 1992). This 'policeman' role for males has been observed in other haremic species (e.g. Baird 1988, de Waal 1986).

The species of fish from which razorfish hid were all piscivores (Randall 1967), larger than green razorfish. When sites were pooled, females exhibited PA behaviors significantly more often than males did (Table 2, $G = 28.7$, $p < 0.001$). Males and females had no significant differences in PA frequency in response to potential predators at the grassbed and sandbed sites, (G-tests, $p > 0.05$), however, females had significantly more frequent PA responses than males did at the rocky-rubble site ($G = 10.1$, $p < 0.025$). This difference between the sexes in only one habitat may be due to the exceptionally large size of the males (standard length > 12 cm) at this site (Nemtzov 1992).

Variation in behaviors among sites

The feeding behavior and responses of individual *X. splendens* to potential predators differed among the three sites. At the rocky-rubble habitat, females fed on plankton high in the water column (Nemtzov 1992). When a potential predator first approached, both sexes moved downward next to an algae-covered rock or piece of coral rubble where they held the body in a tight curve while hovering over a distinct depression in the sand while exhibiting 6–10 distinct dark bars on the body. If actually attacked, they dove into the depression in the sand ('dive site') directly below them next to the rock.

At the sandbed site, where they were also planktivorous, razorfish moved down out of the water column at the approach of potential predators and hid among the swaying arms of gorgonian corals (e.g. *Muricea muricata, Plexaura homomalla, Eunicea* spp.), or under the branches of a staghorn coral, *Acropora cervicornis,* while exhibiting the barred coloration described above. There were no clear depressions in the sand below where the fish hid. If a predator attacked, the razorfish dove into the sandy bottom wherever they were.

At the grassbed site, the green razorfish fed on benthos and emerged from among the grass blades only for spawning (Nemtzov 1992). When potential predators approached, fish exhibited the barred coloration described above. If predators approached slowly, razorfish usually moved deeper into the grassbed, and hovered among the grass blades with their bodies held in a tight curve. Due to the presence of the grass, I could not determine if there was a distinct depression in the sand directly below them. If a potential predator attacked, green razorfish dove head-first into the sand right below wherever they were hovering.

When predators were in the vicinity, green razorfish dove into the sand only when attacked. Sand-diving was apparently quite successful as we never saw a razorfish caught by a fast moving piscivore during > 1000 man-hours of field observation.

Both males and females showed significant differences in frequency of PA responses among the three sites (Table 2, males, $G = 18.2$, $p < 0.01$; females, $G = 18.8$, $p < 0.005$). Razorfish dove into the sand to avoid predators most often at the sandbed

Table 3. Sand dives and predator avoidance (PA) responses by green razorfish in three habitats (mean ± SE).

	Sandbed	Grassbed	Rocky-rubble
Observation periods (15 min)	329	104	498
PA responses (sand dive or hide) per observation period	0.57 ± 0.06	0.17 ± 0.04	0.45 ± 0.05
PA sand dives per observation period	0.23 ± 0.02	0.15 ± 0.03	0.05 ± 0.01
Percentage of all PA responses that were sand dives	10.1	22.2	2.7

site and at the rocky-rubble site, and less often at the grassbed site (Table 3). Sand-diving was the most commonly used PA response at the grassbed site, used less at the sandbed site, and used least at the rocky-rubble site. After predator attacks, razorfish stayed under the sand for 252 ± 24 sec (mean ± SE, $n = 15$ timed sand dives), excluding one unusual sand dive in which a male stayed under the sand for 61 min.

The first act of PA for fish at all the sites, was to become less conspicuous by moving out of the water column, or by changing color. By hovering over the sand with the body held tightly curved, the fish was in the C-start or S-start positions which allow for fast response (Eaton & Hackett 1984). Sand-diving was only used as a last resort, and constituted a relatively small percentage of PA responses at all sites. While hiding among grass blades or branches of a coral, a razorfish could see the potential predator, but once under the sand the individual had little information about the predator's activities. It is likely that sand dives for PA were not without risk and therefore, they were used rarely.

At the rocky-rubble site, sand dives constituted a much lower proportion of PA responses than at the other two sites, although the frequency of predator responses was similar at the sandbed site (Table 3). The reluctance of fish at the rocky-rubble site to dive was likely due to the difference in composition of the sand at that site. Sand samples taken outside of the dive sites showed that the sand was softest at the sandbed habitat and least soft at the rocky-rubble habitat. The mean percentage (by weight) of sand particles < 0.5 mm in diameter was 89.2, 77.2 and 63.9 for the sandbed, grassbed and rocky-rubble habitats, respectively. These three values were significantly different from one another (multiple pairwise comparisons, $p < 0.05$).

The coarser sand at the rocky-rubble site left fish that dove close to the surface, whereas at the sandbed site razorfish moved easily through the sand itself. Similarly, fish that attempted to dive into the rocky-rubble substratum in a place where a dive site had not been prepared had great difficulty in penetrating the sand (personal observation). Fish at the rocky-rubble site may have used sand-diving as a form of PA less often than fish at the other sites because of the higher risk in being under sand through which they could not move.

Although sand-diving at the rocky-rubble site was not used often for PA, there were still more sand dives at this site per hour of observation than at any other site. Almost 90% of these dives were not for PA, a larger proportion than at any other site. Because of the coarseness of the sand in this habitat, the fish apparently had to maintain a dive site by repeatedly diving into the sand making the dive site a soft, easily-penetrated spot in the substratum. The softer texture of the sand, which I could easily feel when I pressed my fingers into the dive site, may be due to more water being mixed with the sand in this spot or, as Clark (1983b) suggested, the softness may be caused by mucus from the fish's body (which could also serve as an olfactory cue in locating the dive site).

At the sandbed site there were no clear depressions in the sand, presumably because razorfish did not maintain dive sites there.

Females that were transferred from the grassbed to the rocky-rubble habitat adopted sand diving PA behaviors similar to those of the resident females (Nemtzov 1992). When potential predators appeared, these transferred fish hid next to rocks and did not attempt to hide within nearby tufts of algae. In contrast, females transferred from the sandbed to the rocky-rubble site behaved similarly to residents of their new habitat (Table 4); they stayed close to a depression in the sand (dive site) and did not wander freely around the habitat. The transfer experiments suggest the presence of phenotypic plasticity in these razorfish in terms of this PA behavior.

Table 4. Sand dives in the absence of predators by green razorfish in different habitats. The 'Transferred' group are fish observed at the sandbed site after having been transferred from the rocky-rubble site.

Group	Sand dives observed	Dives in the absence of predators, n (%)
Sandbed	27	7 (25.9%)
Rocky-rubble	52	46 (88.5%)
Transferred	10	9 (90.0%)

Sand dives in the absence of predators

Razorfish also dove in the sand when potential predators were not in the vicinity and there was no apparent threat. These dives, which were apparently for maintenance of the dive site, differed from PA dives in that fish rose slowly from the sand, and the dives were often repeated in quick succession. During one field season I recorded the proportion of sand dives that occurred in the absence of predators at two sites during 153.5 hours of observation (Table 4). Fish at the rocky-rubble site had a significantly higher proportion of sand dives in the absence of predators (maintenance dives) than did fish at the sandbed site ($G = 23.7$, $p < 0.001$). Also, the proportion of these dives exhibited by females that had been transferred to the sandbed from the rocky-rubble site (Table 4), was not significantly different from that of fish in their source habitat ($G = 1.3$, $p = 0.5$), but the proportion was significantly different from that of fish in the new habitat ($G = 8.8$, $p < 0.025$).

Dive site maintenance was apparently unnecessary in the sandbed habitat because the sand was soft everywhere. However, fish that were transferred there from the rocky-rubble site to the sandbed site continued to dive in the sand (in the absence of predators) at a rate similar to that of fish from their source habitat (Table 4). In contrast, fish transferred from the grassbed to the rocky-rubble site did not maintain their old PA behaviors (Nemtzov 1992). This lack of behavioral plasticity by some transferred fish could be caused by genetic differences among stocks or, more likely, by canalization of behaviors at an early age. It seems apparent, therefore, that flexibility in some PA behaviors is not universal in this species, even though it is likely to be generally advantageous (Sih 1987). Similarly, Victor (1987) observed female *X. martinicensis* engage in repeated, slow, vibrating rises out of dive sites in a habitat where the sand was soft, and dive site maintenance was presumably not necessary.

Predation pressure and spawning

Visits by naturally occurring potential predators significantly affected razorfish spawning behavior. Two male razorfish mated at a lower height when natural predators were prevalent during spawning periods (male A, 3.4 ± 0.2 m [mean height ± SE] without predators and 1.9 ± 0.1 m with, $n = 12$ and 3 matings, respectively; male B, 2.8 ± 0.2 m without and 1.3 ± 0.2 m with, $n = 12$ and 3 matings). Although the two males mated at different mean heights overall, they both mated significantly lower than usual when the predators were present, and this difference was due to the predators, independent of the males (ANOVA, $p < 0.025$).

Introduction of an artificial predator (Fig. 1) into the territory of one male caused a significant difference in its behavior. Comparison of two consecutive 15-min observational periods outside of spawning times (first without and then with the model) showed that the male swam lower in the water column when the model was present, than during the control observation period; height score 0.68 ± 0.24 (mean ± SD) with, and 0.92 ± 0.37 without the predator model ($F_{[1,58]} = 16.4$, $p < 0.001$).

When the predator model was presented during spawning periods, the male swam lower in the water column, height score 1.57 ± 0.52, and 1.07 ± 0.57, recorded consecutively without, and then with the predator model, respectively ($F_{]1,58]} = 24.5$, $p < 0.001$). There was no significant difference in horizontal visibility among the spawning periods with and without the predator model, 22.8 ± 2.3 m, and 21.8 ± 2.0 m, respectively ($F_{[1,4]} = 0.42$, $p > 0.7$).

When the predator model was present during spawning periods, the male courted significantly less often than during control spawning periods (Fig. 2). There was a significant difference between the experimental and control observations for *flips* ($T = 0$, $p < 0.025$), but not for *rises* ($T = 5$, $p > 0.05$, Wilcoxon matched-pair signed-rank test [Sokal & Rohlf 1981]). I could not detect a significant difference in the frequency of *rises*, which may be due to their very low frequencies. During the same spawning periods, however, there was no significant difference ($F_{[1,4]} = 0.49$, $p > 0.5$) between the number of matings obtained by the focal male in the presence or absence of the predator model, 21.7 ± 1.7 matings ($n = 3$ days), and 19.7 ± 5.4 matings ($n = 3$), respectively. The height of the apex of the spawning rush

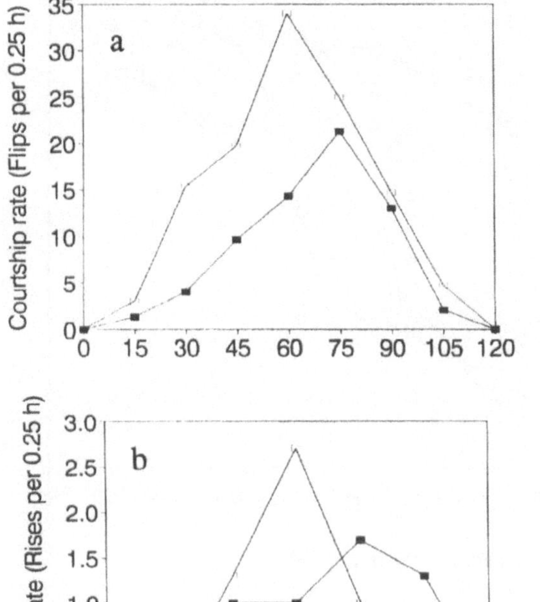

Fig. 2. Courtship rate by a male green razorfish under a simulated increase in predation risk: a – *flips*, b – *rises*. The means from three replicate spawning periods are plotted from experimental (■) and control (□) observations. Note the greatly different scales of the ordinates of a and b.

was significantly closer to the substratum ($F_{[1,117]}$ = 47.8, p < 0.001) when the model was present than when there was no predator model; mean height of the spawning rush was 2.44 ± 0.48 m (n = 65 matings), and 2.86 ± 0.47 m (n = 54), with and without the predator model, respectively. The differences in *flips*, *rises*, matings, and mating heights, when the model was present or absent, was not attributable to variation among replicate spawning periods as there was no significant variation among replicates for any of these observations (ANOVA's, p > 0.05).

Reduction in the level of courtship displays will generally exact a severe cost in mating success for males of many species, e.g. Túngara frogs (Ryan 1985). Yet reduction in courtship by this male green razorfish seemed to carry no immediate fitness cost as he was still able to achieve the same number of matings each day. Although mating success seemed unaffected, increased predation risk could still exact fitness costs because of the lower height at which gametes were released. A substantial cost could result from mating too low if at various heights above the substratum there were differences in the current's ability to carry eggs away from the reef, or in egg predators' ability to find them. Although little is known about what happens to fertilized fish eggs soon after they are released into the plankton (Leis 1991), a higher spawning rush apparently improves dispersal (Keenleyside 1979, Thresher 1984).

Although there was no significant difference between the number of matings achieved by the male with or without the predator model, the peak of courtship activity occurred later in the spawning period when the model was present (Fig. 2). Similarly, matings occurred significantly later in the spawning period when the predator model was present (Fig. 3, D = 0.42, p < 0.001, Kolmogorov-Smirnov two-sample test). The mean spawning time was significantly later ($F_{[1,130]}$ = 72.9, p < 0.001) on days with the model (76.8 ± 9.9 min, n = 65 matings), than on days without the model (59.4 ± 21.1 min, n = 67). Delayed courtship in the presence of predators was also observed in guppies (Endler 1987). The reason for delay may be that light intensity diminishes rapidly close to sunset in the tropics, so a delay in courtship can significantly improve safety with only minimal chance of lowering mating success (Endler 1987). Presumably, as evening approached, conspicuous displays directed at site-restricted females within the male's harem would become less noticeable to distant predatory fish, but probably not to the nearby females.

Conclusions

Green razorfish altered their reproductive behavior in ways that apparently reduced their risk. Razorfish at the rocky-rubble site and the sandbed site swam and mated lower in the water column when potential predators were present. In the clear Caribbean water, spawning rushes very far above the substratum and high numbers of conspicuous courtship displays presumably subjected them to high

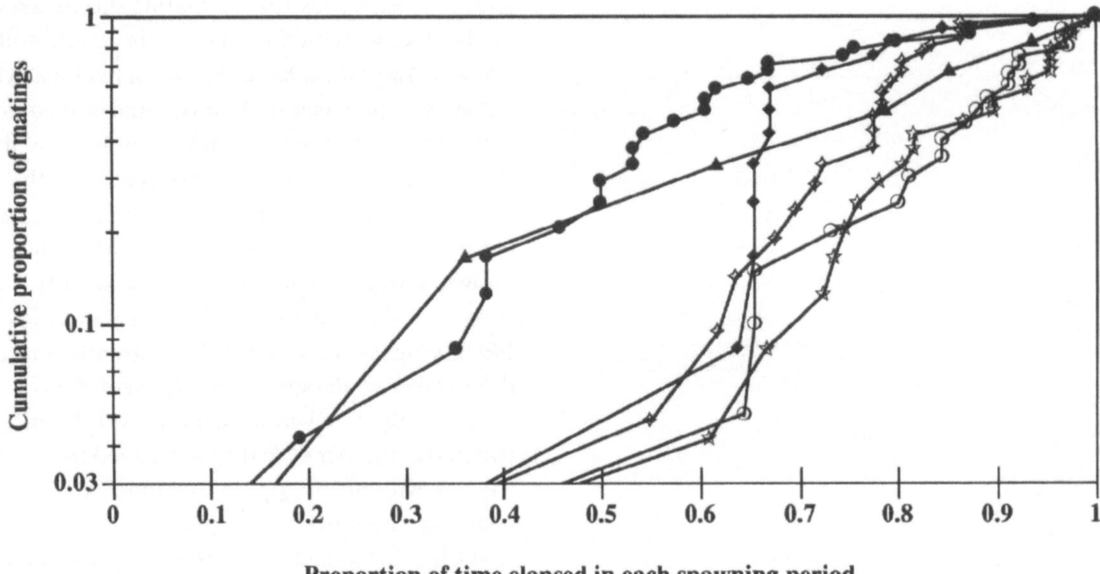

Fig. 3. Log-linear plot showing when matings occurred in spawning periods (n = 3) with the predator model (open symbols), and in spawning periods (n = 3) without the model (filled symbols).

risk. Visibility was high at both sites and the responses were similar, suggesting that the observed changes in behavior were in response to the visits by potential predators. Similarly, when predation risk was increased artificially, the subject razorfish swam and mated lower in the water column, and reduced the frequency of the major courtship display (*flips*).

Although mating success was unaffected by high levels of predation pressure, long-term male reproductive success could be reduced by lowered courtship rates. Vigorous courtship may seem to be unnecessary in these haremic fish where the females are ready to mate every afternoon. Yet the males expend much energy, and increase their exposure to predation risk, with elaborate courtship displays. The primary benefits of their courtship may only be realized in the long term. Courting during the spawning period may be useful for immediate arousal and synchrony, but it may be even more important for its long-term effects on the social system of these territorial males and their site-restricted females. Females that do not receive stimuli from males at a prescribed rate are more likely to change sex (Fricke & Fricke 1977) which can reduce the male's long-term fitness by reducing the number of females in the harem and by increasing the number of potential male competitors. In all four cases where I have seen female green razorfish change sex in the wild without manipulation, they first positioned themselves as far from the center of the harem as possible thereby reducing contact with the male (Nemtzov 1992). If females leave the harem or change sex early when courtship is reduced, then the male's long-term reproductive success will suffer even if his immediate mating success appears unaffected. This points to the importance of taking both short- and long-term effects into account when studying the impact of predation on reproductive behavior.

My observations have shown the importance of predator risk in shaping many aspects of green razorfish biology. Some of their adaptations are obviously genetically fixed (e.g. head shape). However, some other adaptations (e.g. dive site maintenance behavior) are apparently more variable, within and among individuals, and may represent a phenotypically plastic response to local environmental conditions.

This study shows that razorfish are sensitive to predation pressure, and adjust their behavior to lower this risk. Although this result might be expected to be common in many species, it has been poorly documented, especially when cost of survival also exacts a severe cost of reproduction (Bell & Koufpanou 1986). By reducing courtship rates and lowering the height of the spawning rush, male green razorfish appear to act in ways that maximize their fitness by optimizing their risk of predation against the success of their mating effort.

Acknowledgements

This paper is dedicated to Eugenie Clark with my thanks for her introducing me to the study of the behavioral ecology of the fishes of the 'undersea desert'. For generous financial support of my field research I acknowledge the following: the Smithsonian Tropical Research Institute, the Lerner-Gray Fund for Marine Research, the Explorer's Club, the Raney Fund, and Sigma Xi. I thank Todd Boettcher, Linnea Bredenberg, Kristianne Grief, Stephen Kajiura, Chris Lompart, and Heather Roffey for expert assistance in the field. I thank the government of Bonaire for permission to conduct this study in their marine park, and to Roberto Henson, Dee Scarr, Jules van Rooij, and Heinrich Bruggermann for assistance in Bonaire. I thank Michael Bell, Robert Cowen, Susan Foster, Charles Janson, and George Williams for helpful comments on earlier versions of this paper. This paper represents contribution no. 861 from the Graduate Program in Ecology and Evolution at the State University of New York at Stony Brook. The research presented here formed part of a dissertation presented to that program in partial fulfillment of the Ph.D. degree.

References cited

Baird, T A 1988 Female and male territoriality and mating system of the sand tilefish, *Malacanthus plumieri* Env Biol Fish 22· 110–116

Bell, G & V Koufpanou 1986 The cost of reproduction pp 83–131 *In* R Dawkins & M Ridley (ed) Oxford Surveys in Evolutionary Biology, vol 3, Oxford University Press, New York

Clark, E 1983a Hidden life of an undersea desert National Geographic 164 128–144

Clark, E 1983b Sand-diving behavior and territoriality of the Red Sea razorfish, *Xyrichtys pentadactylus* Bull Inst Oceanogr Fish , Cairo 9 225–242

Clark, E & A George 1979 Toxic soles, *Pardachirus marmoratus* from the Red Sea and *P pavoninus* from Japan, with notes on other species Env Biol Fish 4 104–125

Clark, E , J S Rabin & S Holderman 1988 Reproductive behavior and social organization in the sand tilefish, *Malacanthus plumieri* Env Biol Fish 22 273–286

Clark, E , M Pohle & J [S] Rabin 1991 Stability and flexibility through community dynamics of the spotted sandperch Nation Geogr Res 7 138–155

Clark, E , J F Pohle & D C Shen 1990 Ecology and population dynamics of garden eels at Râs Mohammed, Red Sea Nation Geogr Res Expl 6 306–318

Clutton-Brock, T H (ed) 1988 Reproductive success studies of individual variation in contrasting breeding systems University of Chicago Press, Chicago 538 pp

Davies, N B & A I Houston 1984 Territory economics pp 148–169 *In* J R Krebs & N B Davies (ed) Behavioural Ecology, 2nd ed , Sinauer Associates, Sunderland

Dawkins, R & J R Krebs 1983 Arms races between and within species Proc Royal Soc London B205 489–511

Eaton, R C & J T Hackett 1984 The role of the Mauthner cell in fast-starts involving escape in teleost fishes pp 213–266 *In* R C Eaton (ed) Neural Mechanisms of Startle Behavior, Plenum Press, New York

Ehrlich, P R 1975 The population biology of coral reef fishes Ann Rev Ecol Syst 6 211–247

Emlen, S T & L W Oring 1977 Ecology, sexual selection, and the evolution of mating systems Science 197 215–223

Endler, J A 1982 Convergent and divergent effects of natural selection on color patterns in two fish faunas Evolution 36 178–188

Endler, J A 1983 Natural and sexual selection on color patterns in poeciliid fishes Env Biol Fish 9 173–190

Endler, J A 1986 Defense against predators pp 109–134 *In* M E Feder & G V Lauder (ed) Predator-Prey Relationships Perspectives and Approaches from the Study of Lower Vertebrates, University of Chicago Press, Chicago

Endler, J A 1987 Predation, light intensity, and courtship behaviour in *Poecilia reticulata* (Pisces Poeciliidae) Anim Behav 35 1376–1385

Farr, J A 1975 The role of predation in the evolution of social behavior in the guppy, *Poecilia reticulata* (Pisces, Poeciliidae) Evolution 29 151–158

Feder, M E & G V Lauder (ed) 1986 Predator-prey relationships perspectives and approaches from the study of lower vertebrates University of Chicago Press, Chicago 198 pp

Fricke, H W & S Fricke 1977 Monogamy and sex change by aggressive dominance in coral reef fish Nature 266 830–832

Gause, G F 1942 The relation of adaptability to adaptation Q Rev Biol 17 99–114

Gosling, L M 1986 The evolution of mating strategies in male antelopes pp 244–281 *In* D I Rubenstein & R W Wrangham (ed) Ecological Aspects of Social Evolution Birds and Mammals, Princeton University Press, Princeton

Hixon, M A 1991 Predation as a process structuring coral reef fish communities pp 475–508 *In* PF Sale (ed) The Ecology of Fishes on Coral Reefs, Academic Press, San Diego

Keenleyside, M H A 1979 Diversity and adaptation in fish behaviour Springer-Verlag, Berlin 208 pp

Kerfoot, W C & A Sih (ed) 1987 Predation direct and indirect impacts on aquatic communities University Press of New England, Hanover 386 pp

Leis, J M 1991 The pelagic stage of reef fishes the larval biology of coral reef fishes pp 183–230 *In* PF Sale (ed) The Ecology of Fishes on Coral Reefs, Academic Press, San Diego

Levinton, J S 1987 Marine ecology Prentice-Hall, Englewood Cliffs 526 pp

Lima, S L & L M Dill 1990 Behavioral decisions made under the risk of predation a review and prospectus Can J Zool 68 618–640

Lloyd, J E 1984 On deception, a way of all flesh, and firefly signaling and systematics pp 48–84 *In* R Dawkins & M Ridley (ed) Oxford Surveys in Evolutionary Biology, vol 1, Oxford University Press, New York

Lott, D F 1991 Intraspecific variation in the social systems of wild vertebrates Cambridge University Press, Cambridge 238 pp

Luyten, PH & N R Liley 1985 Geographic variation in the sexual behavior of the guppy, *Poecilia reticulata* (Peters) Behaviour 95 164–179

Magnhagen, C 1990 Reproduction under predation risk in the sand goby, *Pomatoschistus minutus,* and the black goby, *Gobius niger* the effect of age and longevity Behav Ecol Sociobiol 26 331–335

Magnhagen, C 1991 Predation risk as a cost of reproduction Trends Ecol Evol 6 183–186

Magurran, A E & M A Nowak 1991 Another battle of the sexes the consequences of sexual asymmetry in mating costs and predation risk in the guppy, *Poecilia reticulata* Proc Royal Soc Lond B246 31–38

McGovern, A & E Clark 1991 The desert beneath the sea Scholastic Press, New York 48 pp

Nemtzov, S C 1982 Monandric protogynous hermaphroditism in the Red Sea razorfish *Xyrichtys pentadactylus* (Teleostei, Labridae) M Sc Thesis, University of Maryland, College Park 53 pp

Nemtzov, S C 1985 Social control of sex change in the Red Sea razorfish *Xyrichtys pentadactylus* (Teleostei, Labridae) Env Biol Fish 14 199–211

Nemtzov, S C 1992 Intraspecific variation in the social and mating behavior of Caribbean razorfishes (Teleostei, Labridae) Ph D Dissertation, State University of New York, Stony Brook 200 pp

Orians, G H 1969 On the evolution of mating systems in birds and mammals Amer Nat 103 589–603

Randall, J E 1965 A review of the razorfish genus *Hemipteronotus* (Labridae) of the Atlantic Ocean Copeia 1965 487–501

Randall, J E 1967 Food habits of reef fishes of the West Indies Stud Trop Oceanogr 5 665–847

Robinson, S K 1986 The evolution of social behavior and mating systems in the blackbirds (Icterinae) pp 175–200 *In* D I Rubenstein & R W Wrangham (ed) Ecological Aspects of Social Evolution Birds and Mammals, Princeton University Press, Princeton

Roede, M J 1972 Color as related to size, sex, and behavior in seven Caribbean labrid species (genera *Thalassoma, Halichoeres* and *Hemipteronotus*) Stud Fauna Curaçao Caribb Isl 42 1–166

Ryan, M J 1985 The Tungara frog a study in sexual selection and communication University of Chicago Press, Chicago 230 pp

Ryan, M J , M D Tuttle & L F Taft 1981 The costs and benefits of frog chorusing behavior Behav Ecol Sociobiol 8 273–278

Sih, A 1987 Predators and prey lifestyles an evolutionary and ecological overview pp 203–224 *In* W C Kerfoot & A Sih (ed) Predation Direct and Indirect Impacts on Aquatic Communities, University Press of New England, Hanover

Sih, A , J Krupa & S Travers 1990 An experimental study on the effects of predation risk and feeding regime on the mating behavior of the water strider Amer Nat 135 284–290

Simenstad, C A & G M Cailliet (ed) 1986 Contemporary studies on fish feeding the proceedings of GUTSHOP '84 Dev Env Biol Fish 7, Dr W Junk Publishers, Dordrecht 334 pp

Sokal, R R & F J Rohlf 1981 Biometry 2nd ed, W H Freeman San Francisco 859 pp

Thresher, R E 1984 Reproduction in reef fishes T F H Publications, Neptune City 400 pp

Vermeij, G L 1987 Evolution and escalation an ecological history of life Princeton University Press, Princeton 527 pp

Victor, B C 1987 The mating system of the Caribbean rosy razorfish, *Xyrichtys martinicensis* Bull Mar Sci 40 152–160

Waal, F B M de 1986 Dynamics of social relationships pp 421–429 *In* B B Smuts, D L Cheney, R M Seyfarth, R W Wrangham & T T Struhsaker (ed) Primate Societies, University of Chicago Press, Chicago

Whoriskey, F G & G J FitzGerald 1985 The effects of bird predation on an estuarine stickleback (Pisces Gasterosteidae) community Can J Zool 63 301–307

Williams, G C 1989 A sociobiological expansion of *Evolution and Ethics* pp 179–214 *In* J Paradis & G C Williams (ed) Evolution and Ethics T H Huxley's *Evolution and Ethics* with New Essays on its Victorian and Sociobiological Context, Princeton University Press, Princeton

Environmental Biology of Fishes **41**: 415–422, 1994.

Social system of an inshore stock of the red hind grouper, *Epinephelus guttatus* (Pisces : Serranidae)

Douglas Y. Shapiro[1,3], Graciela Garcia-Moliner[1,4] & Yvonne Sadovy[2]
[1] *Department of Marine Sciences, University of Puerto Rico, Box 5000, Mayagüez, PR 00681, U.S.A.*
[2] *Fisheries Research Laboratory, Box 3665, Mayagüez, PR 00680, U.S.A.*
[3] *Present address: Department of Biology, Eastern Michigan University, Ypsilanti, MI 48197, U.S.A. To whom correspondence should be addressed.*
[4] *Present address: Graduate School of Oceanography, University of Rhode Island, Narragansett, RI 02882, U.S.A.*

Received 13.8.1992 Accepted 24.4.1993

Key words: Home range, Sex ratio, Protogyny, Sex change, Spawning

Synopsis

According to sex allocation theory, the decision by a female in protogynous fish species to change sex or not should be influenced by, among other things, the mating sex ratio during spawning periods and/or by factors that vary directly with the spawning sex ratio, such as relative rates of behavioral interaction with males and females outside of spawning periods. In groupers that only spawn during a few weeks of the year in large aggregations, individuals must assess the relative value of changing sex or not entirely within the aggregation unless the social system during the remainder of the year provides a behavioral equivalent of the mating sex ratio. Fifty-five individuals of the red hind, *Epinephelus guttatus,* were tagged and repeatedly located during a 152-day period within a 100 × 100 m grid on a shallow forereef off southwestern Puerto Rico. The home ranges of 22 tagged individuals sighted 10 or more times were 112–5636 m^2 in area. Individual home ranges overlapped with the home ranges of 1–18 other individuals. Home ranges of small fish were not clustered within the borders of the home range of larger fish, i.e. fish did not form spatially defined social units. At the end of the study, 31 tagged individuals remained on the grid together with five newly sighted fish. All 36 individuals proved on histological examination to be females similar in size to females in the spawning aggregation of the following year. The sex ratio of this all-female inshore stock differed significantly from the sex ratio of that spawning aggregation. Hence, information predicting the reproductive value of a sex change is not available to females in the inshore stock during nonspawning months.

Introduction

In protogynous hermaphrodites, adult individuals change sex from female to male (Atz 1964). In theory, sex change is advantageous to a female when its expected reproductive success as a male exceeds its remaining expected reproductive success as a female (Charnov 1982). Although lifetime reproductive success results both from the longevity and instantaneous spawning rate of the individual, in protogynous fishes the most studied factor influencing reproductive success is the mating system (e.g. Warner et al. 1975, Shapiro 1988a, Ross 1990). If a female could change sex and reproduce as a male with suf-

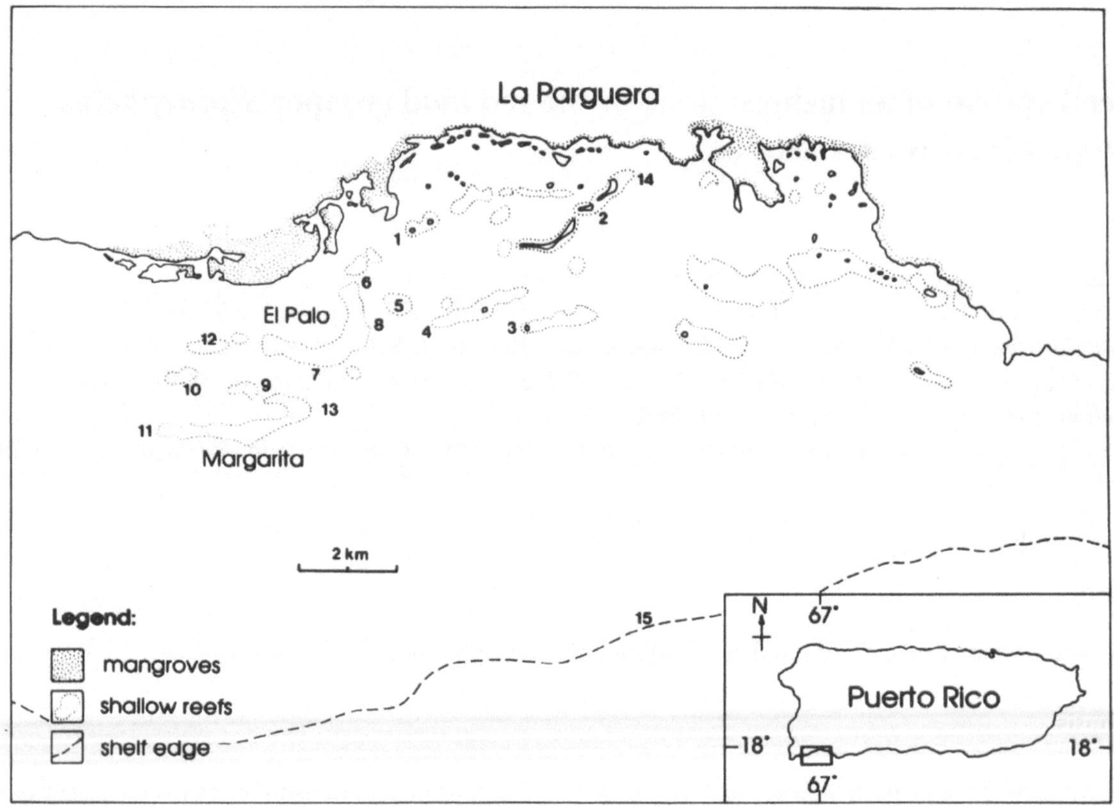

Fig. 1. Large patch reefs near La Parguera, Puerto Rico. Reefs 1–14 were surveyed for *E. guttatus*; home ranges were measured on reef 7 (El Palo); fish were removed and sexed from reefs 7, 9, and 11 (El Palo, Corona, and a patch reef near Margarita). Location 15 is one of two nearby spawning aggregation sites (the other is off the map to the west).

ficient females to more than offset the costs of gender change, the individual should alter sex (Shapiro 1989). Consequently, we expect a relationship between mating systems with female-biased sex ratios and protogynous sexuality (Charnov & Bull 1989).

The mating sex ratio may influence the physiological 'decision' to change sex directly or indirectly. In species that spawn frequently, e.g. daily, weekly, or even monthly throughout much of the year, the mating sex ratio is directly and repeatedly available to influence sex change. In species that spawn infrequently or during limited times of the year, the mating sex ratio may influence sex change indirectly whenever the relative frequency of interacting with males and females accurately reflects the mating sex ratio. Social systems reflecting the mating sex ratio include stable social groups and other spatial arrangements in which the probability of spawning with the opposite sex is similar to the probability of interacting with it outside of spawning periods. In these cases, sex change could presumably be influenced by the mating sex ratio during the mating season and/or by the behavioral or demographic equivalent of that ratio during non-mating periods.

Some protogynous groupers spawn in large aggregations during highly restricted periods of the year, sometimes as short as 1 week (Smith 1972, Colin et al. 1987, Shapiro 1987, Carter et al. 1993, Colin 1992). If, during the remainder of the year, the social system does not provide information about the mating sex ratio within the spawning aggregation, and if the decision to change sex is influenced by the spawning sex ratio, the only opportunity available to females to assess the relative reproductive values of continuing as a female or switching to a male lies within the aggregation once yearly. An important issue, then, is the degree of match between the social system outside of the aggregation and the mating system within it.

The red hind, *Epinephelus guttatus,* a protogynous grouper (Smith 1959, Burnett-Herkes 1975, Sadovy et al. 1992), spawns at the edge of the insular shelf off the southwest coast of Puerto Rico. Spawning aggregations form during approximately 1 week around the full moon each January, and infrequently during approximately 1 week around the full moon in February (Colin et al. 1987, Shapiro et al. 1993a). Within the aggregation, several females or 1 male and several females form spatial clusters. The sex ratio of intact clusters of individuals speared in a 1984 aggregation averaged 5.6 females per male (n = 34 individuals from 12 clusters), while the overall aggregation sex ratio for specimens collected with this and other techniques (n = 190) was 6.6 females per male (Shapiro et al. 1993a). Spawning has only been observed rarely, but the courtship behavior of males suggests that males spawn with females in the clusters with which they are associated (Colin et al. 1987, Shapiro et al. 1993a).

The purpose of this study is to describe the social system of a shallow, inshore stock of *E. guttatus* and to compare it with the known spatial structure and sexual composition of the local spawning aggregation the following January (Shapiro et al. 1993a). The approach was to examine the location and size of home ranges of marked individuals and then to capture and sex them. We expected that a male and several females would co-occupy a home range, resulting in a polygynous social system and a moderately female-biased sex ratio. This expectation stemmed from the social system of groupers that are not aggregate spawners (Donaldson 1991, Shpigel & Fishelson 1991), from the social structure of spawning aggregations of this species (Shapiro et al. 1993a), and from an assumption that the simplest social system would be one that is uniform throughout the year, both within and outside the spawning aggregation.

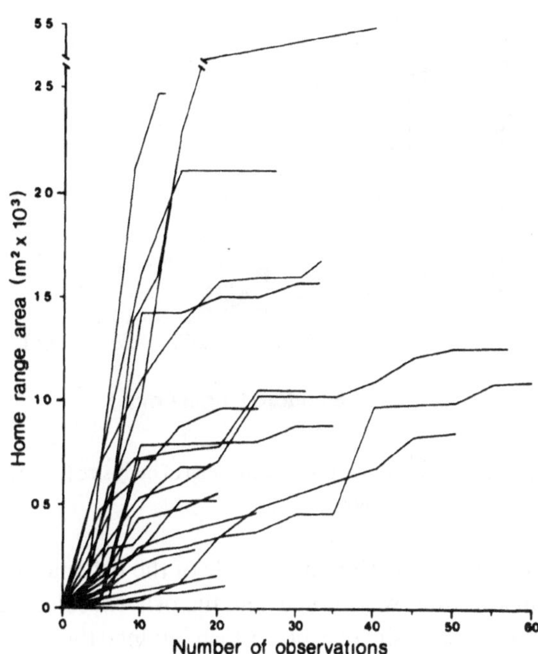

Fig 2 Home range area (m^2) as a function of number of observations for 22 female *E guttatus* in a 100 × 100 m grid on the El Palo forereef

Methods

Fourteen large patch reefs near La Parguera, Puerto Rico, were surveyed initially and three with dense stocks that were infrequently fished were selected for study (Garcia-Moliner 1986): El Palo, Corona, and a patch reef near Margarita (Fig. 1). The three study reefs were approximately 8.5 km from each of 2 known spawning aggregation sites on the edge of the insular shelf, one to the southeast and one to the southwest of the study reefs. Midway along the El Palo forereef, a 100 × 100 m grid of polyethylene rope was laid on the bottom in depths of 6–10 m (average 8 m). The grid was subdivided into nine 33 × 33 m squares. Bottom topography consisted of patches of *Acropora cervicornis,* sand, soft corals, gorgonians, and variously sized coral aggregates.

Individual *E. guttatus* were caught within the grid by divers dangling small hooks baited with squid in front of the fish. The standard length (SL) of captured individuals was immediately measured to the nearest mm, a colored Floy 'spaghetti' tag was inserted at one of three positions into the dorsal mus-

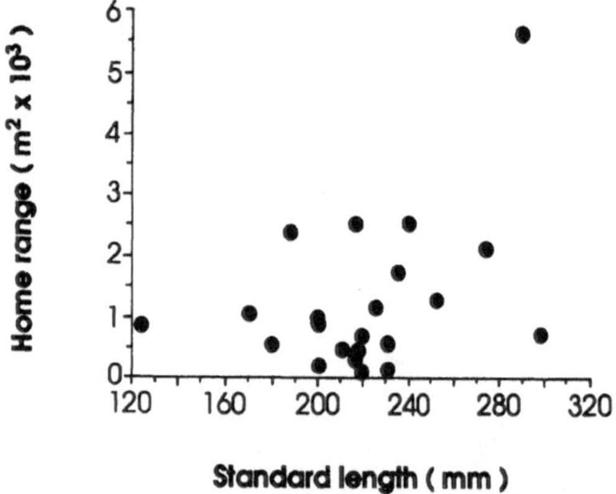

Fig. 3. Home range area (m^2) as a function of body size (mm SL) for 22 female *E. guttatus.*

culature on the right or left side of the fish, natural tail markings were noted, and the fish was released at the precise site of capture. Fish retained their tags 1–152 days, with a mean of 37 days. When a fish lost its tag or an untagged fish appeared for the first time on the grid, it was captured and retagged. Fish with lost tags were identified by location of the tag scar, body size, and tail markings. All red hinds (N = 55) observed in the grid during the study were tagged.

From April to August 1983, inclusive, the location of each fish within the grid was recorded by two divers during 2.5 h morning dives 4 or 5 days a week, totalling approximately 600 observation hours over 152 days. The starting point on the grid and the direction of swimming by divers (parallel to grid lines) were selected haphazardly in each dive. Divers then swam slowly back and forth across the grid 1.5 m off the bottom. When a fish was seen it was identified and its location recorded as the X and Y distances from the nearest grid lines. Locations were plotted later on a map of the grid in the laboratory.

Home ranges were drawn for each fish using the convex polygon method (Brown 1975). When home range area was plotted against number of sightings, the curves for most individuals began to flatten by the tenth sighting (Fig. 2). Consequently, for final analysis, data were used only from individuals (n = 22) observed on at least ten dives (Garcia-Moliner 1986).

In the third week of September, all red hinds remaining in the grid were speared or captured on hook-and-line by divers (N = 36). Gonads were removed, fixed for 24–48 h in paraformaldehyde, washed overnight, dehydrated in ethanol, cleared, embedded in paraplast, sectioned at 6–7 μm, mounted on slides, and stained with Harris' hematoxylin and eosin. Individuals were sexed by microscopic examination of the slides. All individuals from the two additional patch reefs (Fig. 1) were speared, in November 1983 on the Margarita patch reef and in June 1984 on Corona, and sexed using the same histological technqiues.

Results

During any single dive, 0–16 tagged fish (mean = five fish) were observed. For the 22 fish observed at least ten times, individuals were seen on 13–57 days, with the median fish observed on 23 days. The residence time, i.e. number of days between day of tagging and last sighting, was 54–143 days. Sightings were not equally distributed among the nine grid squares ($\chi^2 = 109.8$, df = 8, $p < 0.0001$), with fewer sightings in squares dominated by open sand bot-

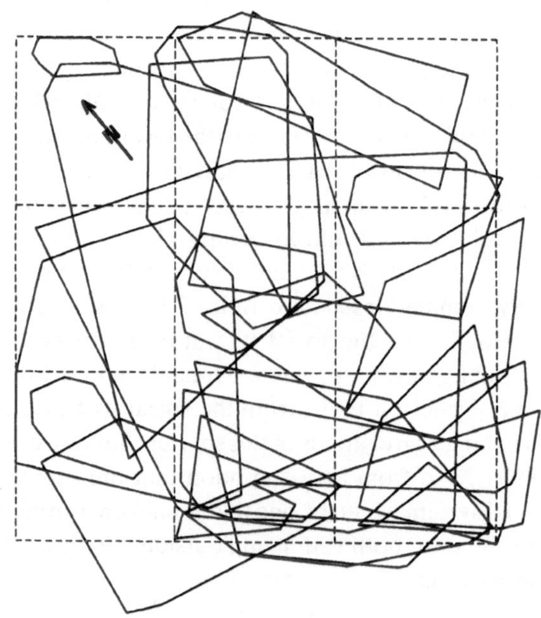

Fig. 4. Home ranges of 22 female *E. guttatus* in a 100 × 100 m grid on the El Palo forereef. Each polygon represents the outer limits of an individual's home range.

tom and more sightings in squares with much coral cover.

Home ranges for fish observed at least ten times (n = 22) were 112–5636 m^2 (median = 862 m^2). These fish measured 124–298 mm SL (median = 217 mm). Fish size correlated neither with the number of times individuals were seen on the grid (Spearman $r_s = 0.148$, n = 22, NS) nor with home range area (Spearman $r_s = 0.262$, n = 22, NS). When the data were examined with linear regression (Fig. 3), an almost significant trend appeared between home range area and body size ($r^2 = 0.164$, p = 0.06). However, the trend was primarily due to a single outlier with a very large home range. When the outlier value was removed, the trend disappeared ($r^2 = 0.024$, p = 0.51). Overall, then, home range area was not a function of body size.

Virtually the entire area of the grid was incorporated into the cumulative home ranges of the 22 individuals (Fig. 4). Home ranges overlapped heavily. The number of home ranges overlapped by the home range of each fish ranged from 1–18, with most fish overlapping the home range of 6–10 other fish (Fig. 5). The number of home ranges overlapped by an individual correlated significantly with the individual's body size (Spearman $r_s = 0.43$, df = 20, $p < 0.05$). Otherwise, we found no pattern to the overlaps, e.g. the home ranges of several smaller fish did not fall within the home range of one large fish, as one might find in a polygynous social system.

At the end of home range observations, 36 fish were collected from the grid, 31 of which had previously been tagged. All 36 were gonadally female. In addition, 23 fish were collected from a small patch reef near Margarita and 23 from Corona. These fish represented all individuals seen on both reefs and ranged in size from 116–268 mm SL, with median 186 mm on Corona and from 146–262 mm, with median 180 mm, on the Margarita patch reef. All 46 were gonadally female.

Discussion

The social system of this inshore stock, at least that portion of it closely associated with patch reefs, was weakly structured. Individuals moved over areas

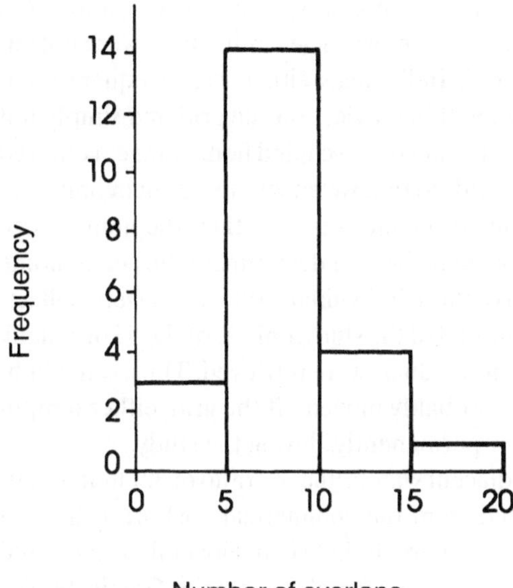

Fig. 5. Frequency distribution for the number of home ranges overlapped by the home range of each of 22 female *E. guttatus* in a 100 × 100 m grid on the El Palo forereef.

occupied by as many as 18 other individuals. This degree of home range overlap might suggest that *E. guttatus* is highly social. However, behavioral interactions among individuals on the grid were rare, and most sightings were of single fish. Individuals tended to remain alone in one position for many minutes at a time, often partially hidden in crevices, behind coral aggregates, or at cleaning stations, and we observed nothing to indicate that individuals were influenced by the movements of other individuals on the grid.

Surprisingly, all individuals on the grid and on two other patch reefs were female. While these individuals were statistically somewhat smaller overall than females from the spawning aggregation of January 1984 (for individuals placed in three size classes, $\chi^2 = 35.4$, df = 2, $p < 0.001$), all inshore females fell within the same size range as females in the aggregation (Shapiro et al. 1993a). The all-female nature of the inshore, patch-reef stock contrasted sharply with the sex ratio (n = 190) of 6.6 females per male in the January 1984 aggregation ($\chi^2 = 10.4$, df = 1, $p < 0.01$).

Do females move regularly between shallow patch reefs and slightly deeper areas where they encounter males in numbers proportional to the mating sex ratios within the spawning aggregation? On any single dive most tagged individuals were not encountered. Individuals either were frequently ensconced within crevices on the grid and simply not visible or some fish occupied home ranges centered off the grid, so that we encountered them only sporadically. If the latter is true, then the home range sizes provided here underestimate the actual home ranges of these individuals. When fish were collected at the end of the study, only 31 of the 55 originally tagged individuals were retrieved. Thus, some individuals probably moved off the grid, either temporarily or permanently, during the study.

In adjacent waters, the sex ratio of 321 individuals obtained from the commercial, inshore fishery at biweekly intervals between November 1982 and June 1984 was 11.3 females per male (Garcia-Moliner 1986). While this sample contained significantly more males than the all-female stock on the three patch reefs ($\chi^2 = 5.82$, df = 1, $p < 0.02$), the inshore sex ratio tended to be higher than the spawning sex ratio (n = 190) of 6.6 females per male in the 1984 aggregation ($\chi^2 = 2.86$, df = 1, $p = 0.09$). When data from a longer, 29 month sample (N = 609, including the 321 specimens from the above data) were considered (Shapiro et al. 1993b), the sex ratio in non-aggregation months from the commercial, inshore fishery (10.8 females per male, n = 463) was significantly higher than the combined sex ratio from three successive, annual aggregations (4.9 females per male; $\chi^2 = 7.22$, df = 1, $p < 0.01$). Consequently, even if females moved off the grid and into adjacent waters the sex ratio encountered there would not reflect the spawning sex ratio in aggregations. During the 11 nonspawning months of the year, males were probably located over the edge of the insular shelf in deeper water.

Thus, sex ratio and social system differed substantially between the inshore stock and the spawning aggregation. Information predicting expected mating success for a female that changed sex and became a male is simply not available to females in the inshore stock outside of the aggregation itself.

The persistence of tagged females on all-female reefs in inshore areas raises the question of the proximate cause of sex change. Why did females not change sex in the absence of a male? In all other protogynous hermaphrodites for which there are data, sex change is induced by alterations in the behavioral interactions between the sexes, generally by separating females from males (Ross 1990). One possible explanation for persistence of females is that sex change is not controlled simply by the absence of a male but by disappearance of a male that was previously present (Shapiro 1988a). If juveniles settled inshore initially and remained there until they were ready to spawn for the first time, and if the inshore females of our study had not yet experienced a spawning aggregation, then these females would have had little prior exposure to males. All-female social units are known in one other protogynous serranid, which spawns in stable social groups. In that species, sex change occurs immediately after male disappearance from bisexual groups but only after lengthy delays in all-female groups (Suzuki et al. 1978, Shapiro 1984, 1988b).

This explanation needs additional evaluation, however. The size range of inshore females from the three patch reefs fell within the size range of aggregating females, so some or all of the inshore fish may have participated in the previous spawning aggregation. In another study of this species, at least one small, tagged female migrated from a shallow, inshore site to a spawning aggregation (Sadovy et al. 1992). None of the gonads of our inshore, patch reef females contained muscle and connective tissue bundles indicative of prior spawning (Shapiro et al. 1993b). However, such bundles generally remain within the gonads for fewer than 160 days after spawning and most of our inshore specimens were collected more than 160 days after spawning. Thus, we cannot be certain whether these females had prior exposure to males or not.

A second possibility is that sex change in this species is not controlled by separation of females from a male. In an experiment designed to simulate conditions in which females of other protogynous species were induced to change sex, male and female red hinds were captured from within a spawning aggregation and held in cages in the sea for up to 18 weeks. Each cage contained one male and six adult females. After one week, the male was removed from experimental cages but not from control cages. Eighteen weeks later, the incidence of sex change did not differ between experimental and control cages. Thus, separation of females from males under these conditions was not successful at inducing sex change, at least within the time frame of the experiment (Figuerola 1987).

Our results point to the need for additional studies of both the proximate and ultimate causes of sex change in this species. According to existing theory and our results, the information needed for inshore individuals to evaluate when to change sex is available only within the spawning aggregation. However, one prediction that follows from this result proved false: transitional individuals, as determined by histological examination of the gonads, were not clustered in time soon after the aggregation, as predicted, but were found throughout the year (Shapiro et al. 1993b). The dilemma seems resolvable but only with data on the social system of deeper stocks than have thus far been studied, i.e. in regions containing males, and with further study of the earliest stages of sexual transition. Such data might reveal a social system in deep water that does reflect the mating sex ratio or might indicate that early onset of sexual transition (prior to the appearance of histologically recognizable testicular tissue) did, in fact, occur predominantly soon after the spawning aggregation.

Acknowledgements

The study was conducted at the Department of Marine Sciences, University of Puerto Rico, Mayaguez. Work was supported by NOAA National Sea Grant Program Office, Department of Commerce, under Grant No. UPR-SG 04-F-158-44030 (Project No. R/LR-06-87-DSP2), the NOAA Undersea Research Program, and NIH grant SO6RR-08103. The U.S. Government is authorized to produce and distribute reprints for governmental purposes, notwithstanding any copyright notation that may appear hereon. Special thanks to all the divers involved in this study and to the fishermen of La Parguera.

References cited

Atz, J W 1964 Intersexuality in fishes pp 142–232 *In* C N Armstrong & A J Marshall (ed) Intersexuality in Vertebrates, Including Man, Academic Press, London

Brown, J L 1975 The evolution of behavior W W Norton and Co , New York 761 pp

Burnett-Herkes, J 1975 Contribution to the biology of the red hind, *Epinephelus guttatus*, a commercially important serranid fish from the tropical western Atlantic Ph D Thesis, Univesity of Miami, Coral Gables 154 pp

Carter, J , G J Marrow & V Pryor 1993 Aspects of the ecology and reproduction of Nassau grouper, *Epinephelus striatus*, off the coast of Belize, Central America Proc Gulf Carib Fish Inst 43 (in press)

Charnov, E L 1982 The theory of sex allocation Princeton University Press, Princeton 355 pp

Charnov, E L & J J Bull 1989 Non-fisherian sex ratios with sex change and environmental sex determination Nature 338 148–150

Colin, P L 1992 Reproduction of the Nassau grouper, *Epinephelus striatus* (Pisces Serranidae) with relationship to environmental conditions Env Biol Fish 34 357–377

Colin, P L , D Y Shapiro & D Weiler 1987 Aspects of the reproduction of two groupers, *Epinephelus guttatus* and *E striatus* in the West Indies Bull Mar Sci 40 220–230

Donaldson, T J 1991 Courtship and spawning behavior of the grouper *Cephalopholis spiloparaea* pp 13 (abstract) *In* 22nd International Ethology Conference Satellite Meeting on Reproductive Behavior and Ecology of Marine Fishes and Other Animals, Okinawa

Figuerola, M 1987 Plasma levels of sex steroids and hormonal correlates of sex change in five species of protogynous hermaphroditic fishes M Sc Thesis, University of Puerto Rico, Mayaguez 87 pp

Garcia-Moliner, G E 1986 Aspects of the social spacing, reproduction and sex reversal in the red hind *Epinephelus guttatus* M Sc Thesis, University of Puerto Rico Mayaguez 104 pp

Ross, R M 1990 The evolution of sex-change mechanisms in fishes Env Biol Fish 29 81–93

Sadovy, Y, M Figuerola & A Roman 1992 Age and growth of red hind, *Epinephelus guttatus*, in Puerto Rico and St Thomas U S Fish Bull 90 516–528

Shapiro, D Y 1984 Sex reversal and sociodemographic processes in coral reef fishes pp 103–118 *In* G W Potts & R J Wooton (ed) Fish Reproduction Strategies and Tactics, Academic Press, London

Shapiro, D Y 1987 Reproduction in groupers pp 295–327 *In* J J Polovina & S Ralston (ed) Tropical Snappers and Groupers Biology and Fisheries Management, Westview Press, Boulder

Shapiro, D Y 1988a Behavioural influences on gene structure and other new ideas concerning sex change in fishes Env Biol Fish 23 283–297

Shapiro, D Y 1988b Variation of group composition and spatial structure with group size in a sex-changing fish. Anim. Behav. 36: 140–149.

Shapiro, D.Y. 1989. Sex change as an alternative life-history style. pp. 177–195. *In:* M.N. Bruton (ed.) Alternative Life-History Styles of Animals, Kluwer Academic Publishers, Dordrecht.

Shapiro, D.Y., Y. Sadovy & M.A. McGehee. 1993a. Size, composition, and spatial structure of the annual spawning aggregation of the red hind, *Epinephelus guttatus* (Pisces : Serranidae). Copeia 1993: 367–374.

Shapiro, D.Y., Y. Sadovy & M.A. McGehee. 1993b. Periodicity of sex change and reproduction in the red hind, *Epinephelus guttatus,* a protogynous grouper. Bull. Mar. Sci. 53 (in press).

Shpigel, M. & L. Fishelson. 1991. Territoriality and associated behaviour of three species of the genus *Cephalopholis* (Pisces, Serranidae) in the Gulf of Aqaba (Red Sea). J. Fish Biol. 38: 887–896.

Smith, C.L. 1959. Hermaphroditism in some serranid fishes from Bermuda. Pap. Michigan Acad. Sci. 44: 111–119.

Smith, C.L. 1972. A spawning aggregation of Nassau grouper *Epinephelus striatus* (Bloch). Trans. Amer. Fish. Soc. 101: 257–261.

Suzuki, K., K. Kobayashi, S. Hioki & T. Sakamoto. 1978. Ecological studies of the anthiine fish, *Franzia squamipinnis,* in Suruga Bay, Japan. Japan. J. Ichthyol. 25: 124–140.

Warner, R.R., D.R. Robertson & E.G. Leigh. 1975. Sex change and sexual selection. Science 190: 633–638.

The junior author Yvonne Sadovy at the 1992 ASIH meeting at Champaign-Urbana. Photograph by E.K. Balon.

Environmental Biology of Fishes **41**: 423–437, 1994.

Quantitative analysis of published data on the growth, metabolism, food consumption, and related features of the red-bellied piranha, *Serrasalmus nattereri* (Characidae)

Daniel Pauly
International Center for Living Aquatic Resources Management (ICLARM), MC P.O. Box 2631 Makati, Metro Manila 0718, Philippines

Received 29.3.1993 Accepted 11.8.1993

Key words: Length-frequency analysis, Length-weight relationships, Von Bertalanffy model, Diet composition, Anthropophagy, Brazil, Guyana, Venezuela, Amazon, Rupununi, Orinoco, Characiformes

Synopsis

A tentative set of growth parameters of the von Bertalanffy growth equation were estimated for the red-bellied piranha, *Serrasalmus nattereri*, a common characid of the Amazonas and adjacent floodplains, based on length-frequency data collected by R.H. Lowe-McConnell in Guyana. These parameters and related statistics are then used, along with published data from metabolic, field and feeding experiment data to estimate the relative food consumption of a population of *S. nattereri.* This is complemented with biological data assembled from the scattered literature on *S. nattereri* to provide a 'snapshot' of this species.

Introduction

This contribution is to consolidate and interpret some published data on the red-bellied piranha, *Serrasalmus nattereri* (Kner, 1860), an abundant species of South American floodplains (Fig. 1, Table 1).

The author has no personal experience with this fish – except for having seen it in various public and private aquaria. However, this is a species for which a large, albeit very scattered literature exists. Thus, this contribution may illustrate how small bits of information distributed throughout the literature can be consolidated into a synoptic 'snapshot' to provide a basis for more comprehensive studies (Rosa 1965) or for entry into FishBase, the computerized encyclopedia on fish.[1]

Nomenclature

Serrasalmus nattereri (family Characidae) was originally described by Kner (1860) as *Pygocentrus nattereri,* and is named after Johann Natterer (1787–1843), an Austrian naturalist who sampled Brazilian animals for nearly 18 years.[2] Synonyms include *P. altus, P. stigmaterythraeus, Rooseveltiella natteri* and *Serrasalmo piranha* (Riehl & Baensch 1991).

[1] Froese, R. 1990. FISHBASE: an information system to support fisheries and aquaculturc research. Fishbyte 8: 21–24, and Pauly, D. & R. Froese. 1991. FishBase: assembling information on fish. Naga, ICLARM Q. 14: 10–11.

[2] Anon. 1845. Johann Natterer. Neuer Nekrolog der Deutschen 21: 1843.

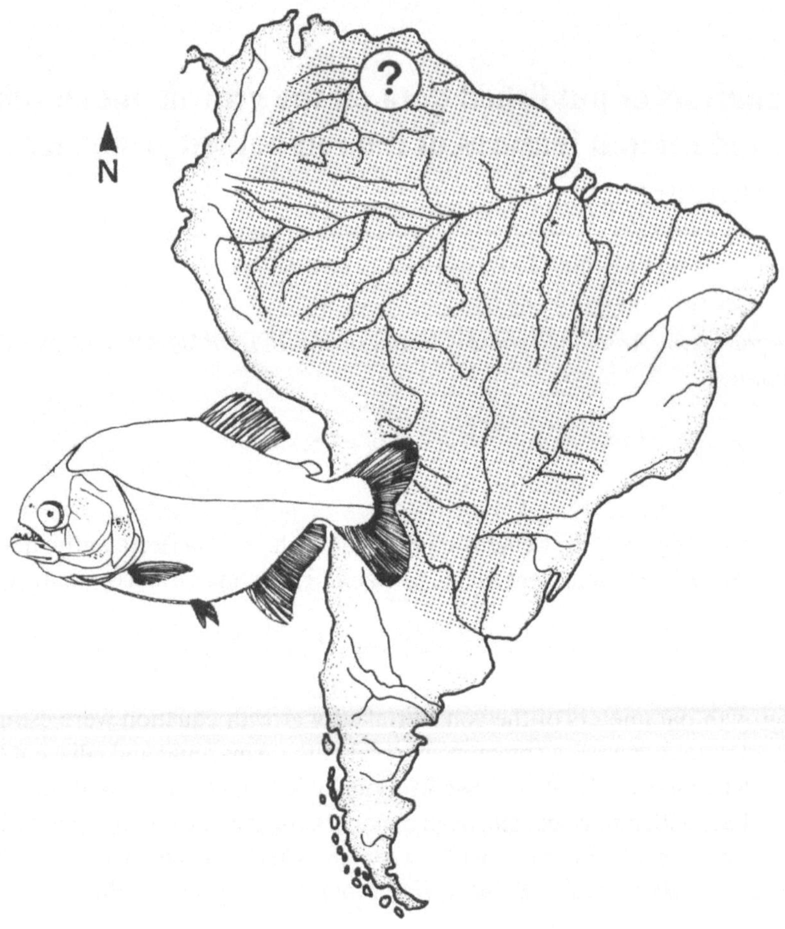

Fig 1 Distribution of the red-bellied piranha *Serrasalmus nattereri* in South America (modified from Braga 1975 and Schulte 1988) Note question mark for the Orinoco basin, where *S nattereri* is replaced by *S notatus,* a close relative or synonym

The taxonomic status and hence the distribution of this and related piranhas are not well-established and some closely related species, such as *S. notatus,* the 'caribe colorao' of the Orinoco River Basin, may be synonyms (Géry 1977, Schulte 1988).

Alternatively, what is now considered a single widespread species ('*S. nattereri*') may end up being split into several species with narrower ranges (Schulte 1988, Riehl & Baensch 1991), and again bearing Kner's original generic name (Machado 1985).

Given its broad natural range, and its use as aquarium fish, *S. nattereri* has a number of common names, notably 'palometa' (Argentina, Bolivia), 'paña' (Peru), 'palometa de rio' (Uruguay), 'caribe boca de locha' (Venezuela); 'red pirai' (Guyana); 'Natterers Sàgesalmler' (German). In the Cuiabá rivers and the Pantanal, *S. nattereri* is called 'piranha-queixcuda', i.e. 'big jawed' (I. Sazima personal communication), but in most of Brazil, the common name is 'piranha caju', or cashew-fruit piranha, because of the color similarity between cashew fruit and the fish when both are ripe (Goulding 1981).

Growth and natural mortality

Formal studies on the growth of *S. nattereri,* or of other piranhas for that matter, do not appear to have been conducted. However, Lowe-McConnell (1964) presented various information which when re-expressed as in Table 2 allows estimation of the parameters of the von Bertalanffy growth function (VBGF, von Bertalanffy 1938), which has the form

$$L_t = L_\infty (1-e^{K(t\ t_o)}), \quad (1)$$

where L_t is the length at age t, L_∞ the mean size the fish would reach if they were to grow indefinitely, K a growth coefficient of dimension time^{1} and t_o is (theoretical) age at which L = 0. Estimation of these parameters was done with the program of Gaschutz et al.,[3] with very high weighting factors for the extreme lengths (1 and 26 cm in Table 2); this resulted in the following estimates: SL_∞ = 26 (in cm), K = 0.893 year^{1} and t_o = – 0.05 year.

The corresponding growth curve and data points

[3] Gaschutz, G , D Pauly & N David 1980 A versatile BASIC program for fitting weight and seasonally oscillating length-frequency data Int Counc Explor Mer, Coun Meet 1980/D 6, Stat Cttee 14 pp

Table 1 Selected occurrence records of and biological information on red-bellied piranha *Serrasalmus nattereri* and its close relative *S notatus* (Venezuela only)

Occurrence	Maximum reported length (cm)[a]	Remarks	Source
Brazil			
Cuiaba River, Matto Grosso	–	Origin of type specimens	Kner (1860)
Pantanal, Matto Grosso	24 (SL)	Common in creeks and interconnected ponds, where it influences distribution and feeding of other fish	Sazima & Machado (1990)
Tocantin River	25 (SL)	Common in lakes adjacent to and slow-flowing segments of this river and similar rivers, where it functions as a 'grand predateur'	Dos Santos et al (1984), De Merona et al (1987)
Rio Machado & Rio Negro	–	Rare, occurring predominantly in areas of high primary production, replaced by *S rhombeus* in nutrient-poor areas	Goulding (1980)
Rio Madeira	–	Most abundant predator, caught from February to May with pole and line, tears gillnets	Goulding (1981)
Peru			
Amazon Basin	?	Details in Gery (1964b, not seen)	Ortega & Vari (1986)
Bolivia			
Rio Mamore	26 (SL)	Largest specimen caught (= 1 05 kg)	Lauzanne & Loubens (1985)
Venezuela			
Orinoco Delta	31 5 (TL)	Largest specimen caught (= 1 0 kg)	Novoa et al (1982), Novoa & Ramos (1978)
Orinoco, Apure State	30 (SL)	Most abundant piranha in the Llano	same as above
	–	'where common [*S notatus*] may have a pervasive effect on the spatial structuring of fish communities'	Winnemiller (1989)
Orinoco, Middle course	48 (TL)	Such large specimens do not appear to occur in *S nattereri* (except in neighboring 'Guyane'?)	same as above
Guiana/Guyane			
Rupununi River System	30 (TL)	Peak migration at dawn and early morning, locked in savanna ponds during dry season	Lowe-McConnell (1964, 1975)
'Guyane'	43 (?)	A large, aggressive 'form'	Gery (1964a)
Various countries			
Aquaria	30 (TL?)	optimum 25–27° C, water hardness up to 10° H, pH 6 8	Paysan (1975), Franke (1978), Axelrod et al (1987)

[a] SL = Standard Length, i e , from the tip of the snout to the end of the caudal peduncle, TL = Total length, i e , from the tip of the snout to the end of the (longest) caudal fin lobe

Table 2. Summary of information on the growth of *S. nattereri* in Guyana (extracted from Lowe-McConnell 1964).

Age (month)	Standard length (cm)	Remarks
0	1	Approximate length at hatching, at the onset of the rains, in mid-May
4	7	Modal length of fish caught at Karenambo, in mid-September 1957 (range 4–9 cm)
7	12	Mean length of fish caught in January 1960, and resulting from *late* rains (June 1959), hence assumed one month younger than 'May' fishes
12	16	Length at first maturity in May (i.e., at 1 year of age)
∞	26	Maximum size of *S. nattereri*, converted from TL ≈ 30 cm (Table 1)

are shown in Figure 2 together with rainfall data (adapted from Fig. 2 in Lowe-McConnell 1964) showing the relationship between spawning/hatching, growth and the seasonal cycle of rains in the Rupununi savanna district, Guyana.

Nico & Taphorn (1986) wrote that '. . . the Orinoco red-bellied piranha reaches 5 to 8 inches SL by their second rainy season'. This statement (which refers to *S. notatus*) implies a (mid-range) length of 16.5 cm SL at an age of about 1 year, very close to the values in Table 2 and Figure 2. The growth curve in Figure 2 is also confirmed by Schulte (1988) who reports that in aquaria, *S. nattereri* reaches 4.5 cm after 2 months and that 'when aged eight months, the largest fish were 120 mm (5 inches) long'.

Table 3 presents data for establishing a length-weight relationship in red-bellied piranha; given its broad range, no attempt was made to derive a precise allometric relationship. Rather, isometry shall be assumed (i.e., an exponent = 3), leading to

$$W = 0.028\ (TL)^3 = 0.043\ (SL)^3, \qquad (2)$$

where W is in g live weight and length in cm, and where TL ≈ 0.87 * SL.

Thus, the growth in weight of *S. nattereri* (and of *S. notatus* in locations such as the Orinoco Delta, where a length of 30 cm TL is not usually exceeded) can be described from

$$W_t = 756\ (1-e^{-0.893(t\ +\ 0.05)})^3. \qquad (3)$$

The estimates of asymptotic sizes ($SL_\infty = 26$, TL = 30 cm, $W_\infty = 756$ g) do not preclude that larger and heavier red-bellied piranha do occur. Rather, these values are in line with the definition of L_∞ (and W_∞) as *means* (see above and Pauly 1984).

The estimate of asymptotic length (TL) and of K, and a mean annual temperature of 28° C, entered into the empirical equation of Pauly (1980), leads to an estimate of $M = 1.66\ year^{-1}$, implying that about 81% of a stock of juvenile and/or adult *S. nattereri* will die annually of natural causes (Fig. 2). Here again, no account is taken of seasonal changes, and hence this natural mortality estimate must be seen as referring mainly to between-, and less to within-year changes.

Metabolic rate

Experiments on the oxygen consumption of piranhas appear to have been conducted only by Braga (1975), and a summary of his results may be found in Table 4; note that various experimental details are lacking, notably on the activity of the fish. Nevertheless, these data were analyzed using a multiple (log) linear regression which yielded, for prediction of the metabolic rate (C, in $mgO_2 \cdot h^{-1}$) in small *S. nattereri*, the model

$$C = 0.387 * W^{0.539} * O_2^{1.13}, \qquad (4)$$

where W is the live weight of the fish in g, and O_2 is the oxygen content of the water, in $mg\ l^{-1}$. The overall fit is good (R = 0.950 and see Fig. 3a); the standard errors of the exponents are 0.163 and 0.205, respectively, for 4 degrees of freedom. Given the small range of weights considered here, the relatively large standard errors about the estimates, and the low number of degrees of freedom, it would not be appropriate to assume that the slope linking O_2 consumption and body weight is, in *S. natteri*, significantly different from that proposed by Winberg

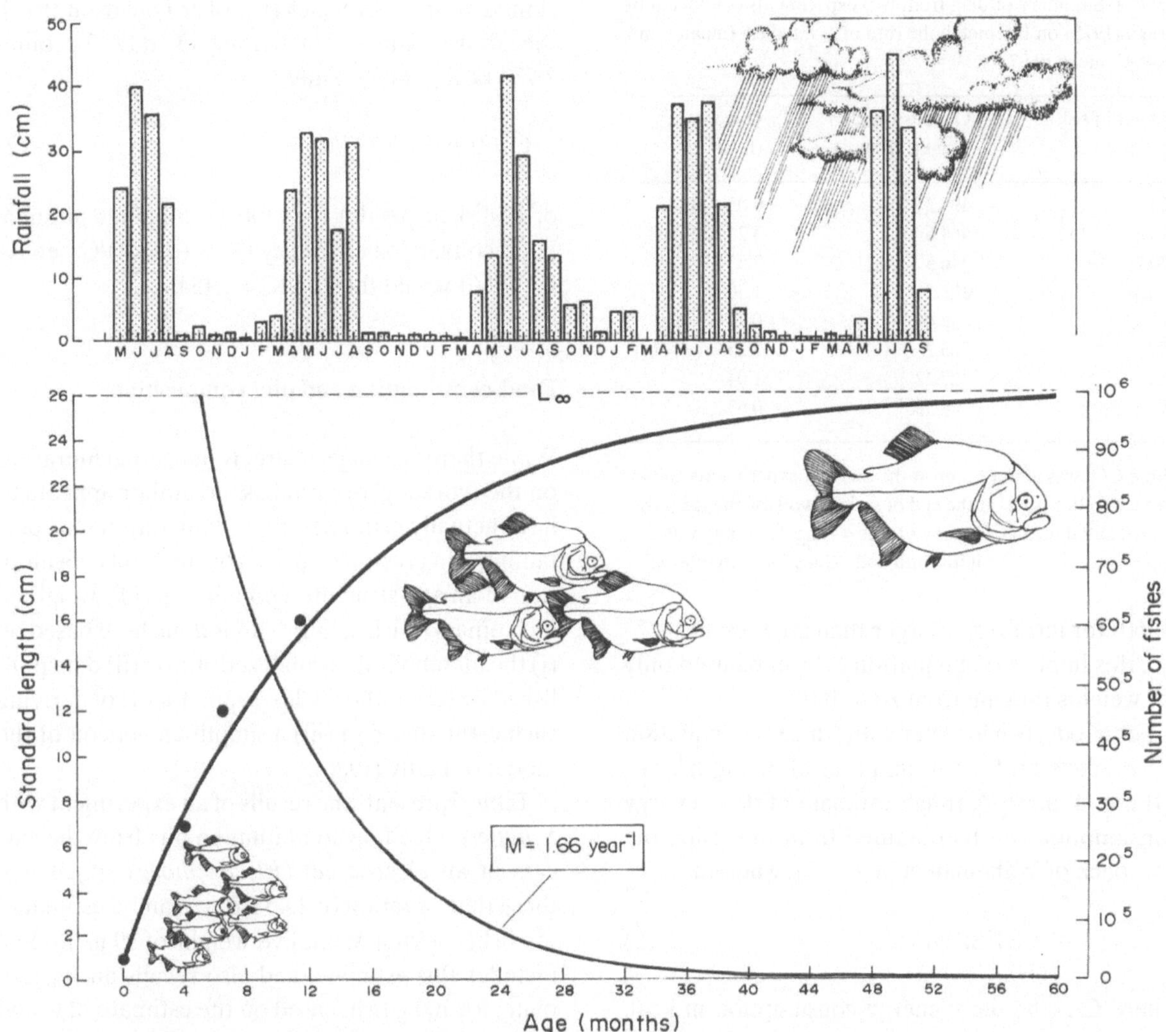

Fig. 2. Growth and (natural) mortality of red-bellied piranha *S. nattereri,* as inferred from data in Table 2. Note that both growth and natural mortality are probably subjected to seasonal oscillations, linked with rainfall, a process not investigated here.

Table 3. Data for establishing a length-weight relationship in red-bellied piranha (*S. nattereri* and *S. notatus;* data referring to the latter are marked with an *).

SL	Length (cm) TL	Weight (live, g)	Condition factor[a] (SL/W)	(TL/W)	Source of L, W data[b]
–	31.5*	1000*	–	3.12*	Novoa et al. (1982)
–	25.0	400	–	2.56	Dos Santos et al. (1984)
–	–	–	4.44[c]	–	De Merona et al. (1987)
–	9.0	19.0	–	2.61	Bellamy (1968)
6.8*	–	13.3*	4.23*	2.8*	Nico (1990)
Mean condition factors:			4.3	2.8	This study

[a] c.f. = a·100, where a is the multiplicative factor in a length-weight relationship of the form $W = a \cdot L^3$

[b] To ensure that a broadly-based L/W relationship emerges, only one L/W data pair were taken from each source.

[c] Average of 4 values for different months and sites.

Table 4. Summary of data from two experiments conducted by Braga (1975) on the metabolic rate of *S. nattereri* (mean temp. 28.4° C, mean pH 7.5)[a].

Weight of fish (g)	O_2 consumption ($mg\ kg^{-1} \cdot hour^{-1}$)	O_2 content[b] ($mg\ l^{-1}$)
20.8	496.2	6.05
20.8	464.6	4.24
20.8	346.5	2.53
20.8	181.3	1.30
20.8	56.06	0.71
158	203.2	3.23
158	27.22	1.04
158	4.386	0.63

[a] Free CO_2 was about 1 ppm at the onset of experiments and increased to 16.5 mg l^{-1} at the end of series involving the 158 g fish; the weighted mean CO_2 content was ≈ 7 mg l^{-1}; salinity was 0‰.
[b] Geometric means of initial and end values for each interval.

(1960) for most fishes larger than guppies, i.e., 0.7–0.8; this implies that equation (4) can be used only for weights ranging from 20 to 160 g.

For a 100 g fish in water with 6 mg O_2 l^{-1}, equation (4) predicts an O_2 consumption of 35 mg h^{-1}, i.e., 841 mg O_2 day^{-1}. A rough estimate of daily energy consumption can be obtained from this using the approach of Wakeman et al. (1979), wherein

$$C = (\Delta W + RESP)/0.75, \quad (5)$$

where C is the daily energy consumption in kcal, ΔW the energy content of the (daily) growth increment, and RESP is the oxygen consumption.

The first derivative (i.e., growth rate) of the von Bertalanffy equation in terms of wet weight is

$$dw/dt = 3KW((W_\infty/W)^{1/b}-1). \quad (6)$$

This, solved for $W_\infty = 756$ g, $K = 0.893/365 = 0.00245$ day^{-1}, and $b = 3$, gives for a 100 g fish a daily growth increment of 0.706 g, corresponding to 0.706 kcal if the calorific value of fish wet weight is set equal to unity (Brett & Blackburn 1978). The available information on body composition of 'piranha caju' flesh (Junk 1976, in Smith 1979) is 8.2% fat, 15.0% protein, and 4.4% ash, not very different from values reported from other fishes (Bykov 1983). If an oxycalorific equivalent of 0.00325 kcal mg^{-1} O_2 is assumed, as in other fishes (Elliot & Davidson 1975), the above estimate of 841 mg O_2 day^{-1} becomes 2.733 kcal per day. Thus

$$C = 0.706 + 2.733/0.75 \quad (7)$$

or 4.585 kcal per day for a 100 g red-bellied piranha. Food conversion efficiency ($K_1 = (dw/dt)/C$, see Ivlev 1966) would then be $K_1 = 0.154$.

Food consumption and diet composition

While there is a huge, if largely anecdotal literature on the 'voracity' of piranhas, no author appears to have actually estimated their daily ration (R_d, pertaining to a given size group) or the food consumption of an age-structured population (Q), weighted by biomass (B), i.e., Q/B. This is done here based on (i) the metabolic data analyzed above, (ii) data published by two authors who stopped short of deriving such estimates, and (iii) a simplified version of the model of Pauly (1986).

Table 5 presents the results of an experiment with *S. nattereri* feeding ad libitum on bits from the carcass of an electric eel (*Electrophorus* sp.). From these data, a ration of 2.46 g day^{-1} can be estimated for fishes with a mean live weight of 19 g. Table 6 includes this estimate, and also recalls an R_d estimate, for 100 g fish, based on the estimate of C and K_1 in the above section on metabolism.

Figure 4 presents a diurnal feeding cycle in young *S. notatus*, based on stomach content data in Nico (1990), fitted with Model I of Sainsbury (1986), as modified in Jarre et al. (1991) and implemented by Jarre et al.[4] The fit is excellent, and leads to the following estimates: beginning of feeding period ≈ 02:00 h; end of feeding period ≈ 14:00 h; feeding rate ≈ 1% of body weight per hour; instantaneous stomach evacuation rate = 9.65% of stomach content per hour.

[4] Jarre, A., M.L. Palomares, F.C. Gayanilo, Jr. & D. Pauly. 1992. A user's manual for MAXIMS, a computer program for estimating the food consumption of fishes from diel stomach contents data and population parameters. Vers. 1.0. ICLARM Software 4. 27 pp.

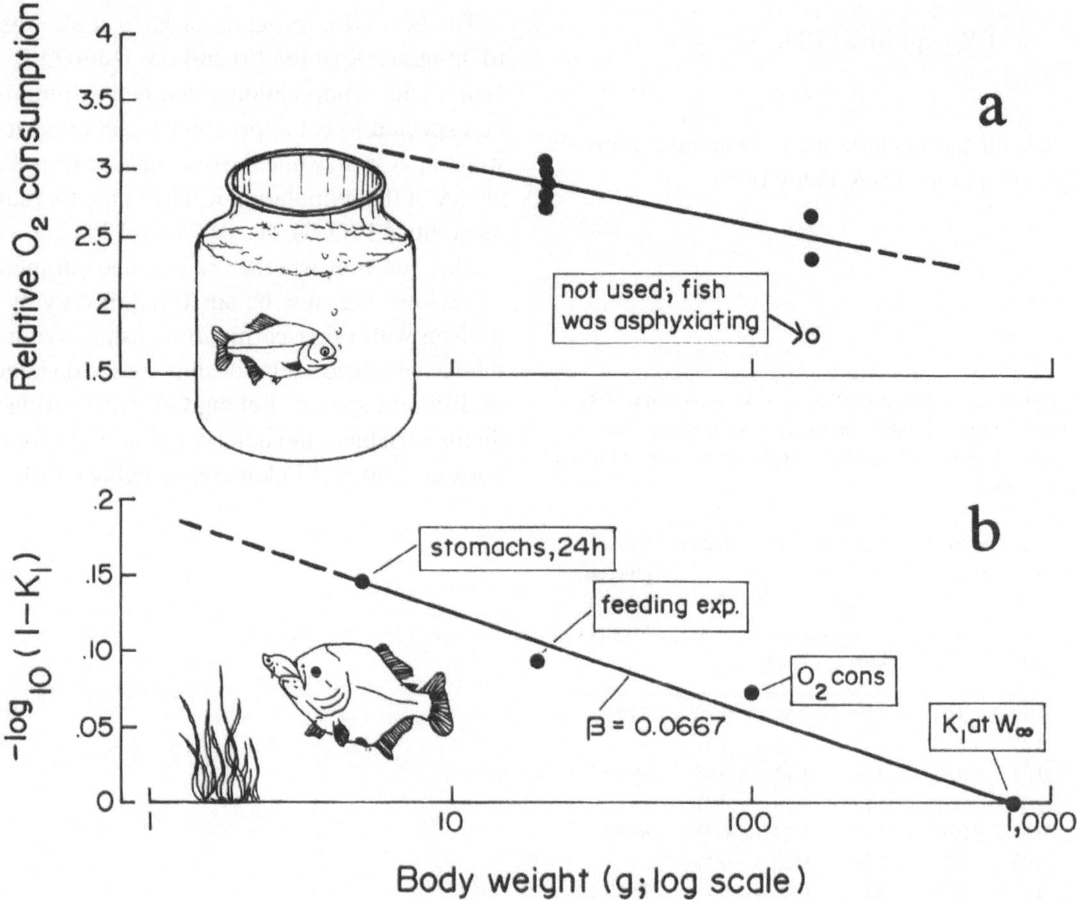

Fig. 3. Weight dependence of two physiological processes in red-bellied piranha: a – Relationship between oxygen consumption and body weight (after account has been taken of different ambient O_2 levels, see equation 7. b – Relationship between a log-transformation of gross food conversion efficiency (K_1) and body weight (based on data in Table 6).

The feeding rate, multiplied by the feeding period leads to a ration estimate of 12.14% BWD for *S. notatus* with a mean length of 4.9 cm SL, i.e., a mean weight of 5 g (Table 6).

Table 6 presents the K_1 and W data used here to estimate the parameter β of a general relationship linking fishes, food conversion efficiency (K_1) and body weight (W) in the form

$$K_1 = 1-(W/W_\infty)^\beta, \tag{8}$$

where K_1 = growth increment/food consumed, for any body weight between hatching and asymptotic size (Pauly 1986, Silvert & Pauly 1987). Estimation of β was done here as shown in Figure 3b, i.e., via a plot of $-\log_{10}(1-K_1)$ vs. $\log_{10}$ W, with the X intercept of the abscissa forced through $\log_{10}(W_\infty)$, and whose slope, with sign changed, provides an estimate of $\beta = 0.067$.

As might be seen, this plot shows that the three available estimates of R_d and the 4 estimates of K_1 are mutually compatible, despite the widely different type of data and models used for their estimation. The relative food consumption (Q/B) of an age-structured population of red-bellied piranhas can thus be estimated using

$$Q/B = \int_{t=0}^{t=\infty} [(dw/dt)exp(-M(t-t_o)))^{b\beta}[dt/ \tag{9}$$

$$\int_{t=0}^{t=\infty} [W_t \cdot \exp(-M(t-t_r))]dt.$$

for which all parameters are as estimated above (Pauly 1986, Palomares & Pauly 1989).

Table 5. Data for the estimation of ration in *S. nattereri* of 18–20 g, feeding ad libitum with bits from the carcass of an electric eel, with additional data on blood glucose contents (as read off Fig. 1 in Bellamy 1968).

Time (hours)	Day one food intake (g per 100 g)		Day two food intake (g per 100 g)		Mean	Blood glucose (mg%) Means
	Low	High	Low	High		
0800	3.49	3.94	2.19	3.58	3.30	56.75[a]
0900	0.93	1.34	1.33	1.75	1.34	
1000	0.22	0.47	0.88	1.12	0.67	86.50
1100	0.07	0.19	0.19	0.38	0.21	
1200	0.38	0.42	0.22	0.29	0.33	89.00
1300	0.08	0.14	0.05	0.18	0.11	
1400	0.11	0.28	0.12	0.22	0.18	76.50
1500	0.08	0.19	0.02	0.25	0.14	
1600	1.32	1.67	1.49	2.00	1.62	81.00
1700	0.78	0.94	0.92	1.03	0.92	
1800	1.43	1.79	1.06	1.34	1.41	90.50
1900	1.02	1.16	0.28	0.57	0.76	
2000	0.78	0.82	0.31	0.42	0.58	87.50
2100	0.08	0.27	0.09	0.26	0.18	
2200	0.17	0.28	0.11	0.17	0.18	77.00
2300	0.08	0.13	0.17	0.17	0.14	
2400	0.17	0.22	0.11	0.11	0.15	69.00[b]
0100	0.19	0.31	–	–	0.25	
0200	0.12	0.22	–	–	61.00	
0300	0.04	0.11	–	–	0.08	
0400	0.04	0.04	–	–	0.04	66.00
0500	0.02	0.02	–	–	0.02	
0600	0.05	0.05	–	–	0.05	56.00
0700	0.58	1.27	–	–	0.93	
Average food intake (per hour) –		–	0.57		–	–

[a] Based on means (58.5 & 55.0) for 08:00 h, at beginning and end of 24-hour cycle.

[b] Value interpolated linearly.

The MAXIMS program of Jarre et al.[4] was used to integrate equation (9) and this led to Q/B = 17.8 year^{-1}. Thus, a population of red-bellied piranha can be expected to eat approximately 18 times its own weight per year, while the overall food transfer efficiency of that population will be equal to: (natural) mortality * 1/(Q/B), i.e., 0.093, or 9.3%.

The question whether red-bellied piranhas are 'voracious' can now be answered, if only by comparison with other carnivorous fishes. We use for this four empirical relationships, based on hundreds of different species, linking Q/B with various predictor variables, and adjusted to account for carnivory, i.e., those of Palomares & Pauly (1989):

Table 6. Data for estimating the growth conversion efficiency (K_1) and related statistics of red-bellied piranha in nature.

Weight[a]	dw/dt[b]	Ration (g day^{-1}) observed/predicted	K_1[c]	Source
5	0.159	0.558/0.588	0.285	Fig. 1
19	0.473	2.46/1.54	0.192	Table 4
100	0.706	4.58/5.59	0.154	Equation (6)
756	0[d]	–/127.7[e]	0[d]	See text and[e]

[a] Live weight in g; the largest value is W_∞.

[b] First derivative of equation (2).

[c] Conventional definition; K_1 = growth increment/food ingested (Ivlev 1966); here K_1 = (dw/dt)/Rd.

[d] By definition, since fish at W_∞ do not grow.

[e] Estimated by solving Equation 5 for a value of W slightly smaller than W_∞, and dividing by the corresponding value of dw/dt.

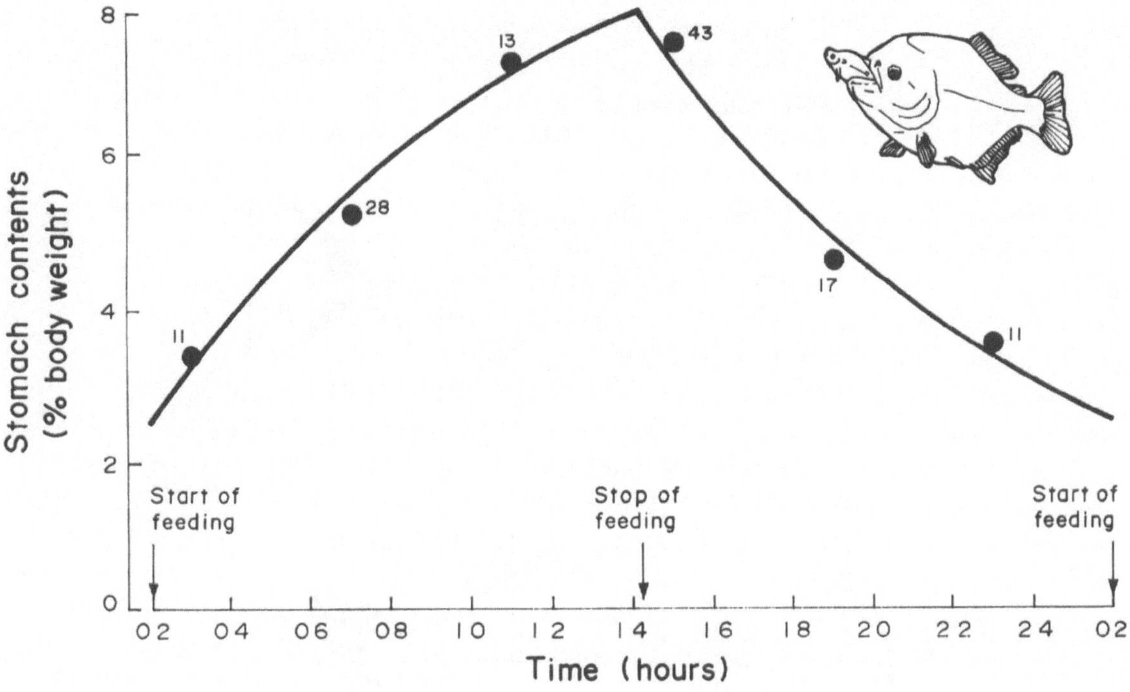

Fig 4 Diurnal feeding cycle of 5 g red-bellied piranha (*S notatus*), with data points from Figure 1 in Nico (1990) [fitted with model and software of Jarre et al (1991,[4]), respectively] The representation of *S notatus* grasping its prey after a chase (a characin, *Astyanax bimaculatus*) is adapted from Sazima & Machado (1990)

$$Q/B = 3.06 * T^{0.612} * W_{\infty}^{-0.202} * A^{0.516}, \quad (10)$$

Pauly (1989):

$$Q/B = 0.790 * T^{0.444} * W_{\infty}^{-0.115} * A^{0.427} * D^{0.577} * P^{-0.464}, \quad (11)$$

Palomares (1991):

$$Q/B = 1.82 * T^{0.759} * W_{\infty}^{-0.165} * A^{0.405}, \quad (12)$$

and Christensen & Pauly[5]:

$$Q/B = 10^{6.4} * 0.0313^{Tk} * W_{\infty}^{-0.168}, \quad (13)$$

where T = mean annual water temperature (here 28° C); T_k = temperature transformed, i.e. Tk = 1,000/(T/T+273.1); W (or W_{∞}) = (asymptotic) live weight, in g (here 756); A or A′ = aspect ratio of the caudal fin (here A = 3.7 and A′ = 2.8, see Fig. 5); D = standard length over maximum body depth (here see Fig. 5); P = depth of caudal peduncle over maximum body depth (here 0.19, see Fig. 5).

Palomares (1991) demonstrated, based on a large number of cases from both types of environments, that the Q/B values of marine and freshwater fishes are not significantly different when account is taken of food type, temperature and of morphological variables, and hence all four equations presented above can be applied to red-bellied piranhas.

The four equations above predict values of Q/B = 12.1, 7.6, 11.6 and 8.3 year^{-1}, with a mean of 9.9 year^{-1}. With an estimated Q/B of 17.8 year^{-1}, red-bellied piranhas consume about twice as much as would be expected based on their size and shape and the temperature of their habitat. Thus *S. natteri* may indeed be described as 'voracious'.

All studies so far conducted on *S. notatus* (e.g., Nico & Taphorn 1988, Winemiller 1989) and on *S. nattereri* (e.g., Bonetto et al. 1967, Braga 1975) confirm the strong tendency of red-bellied piranhas to

[5] Christensen, V. & D Pauly 1992 A guide to the ECOPATH II software system (version 2 1) ICLARM Software 6 72 pp

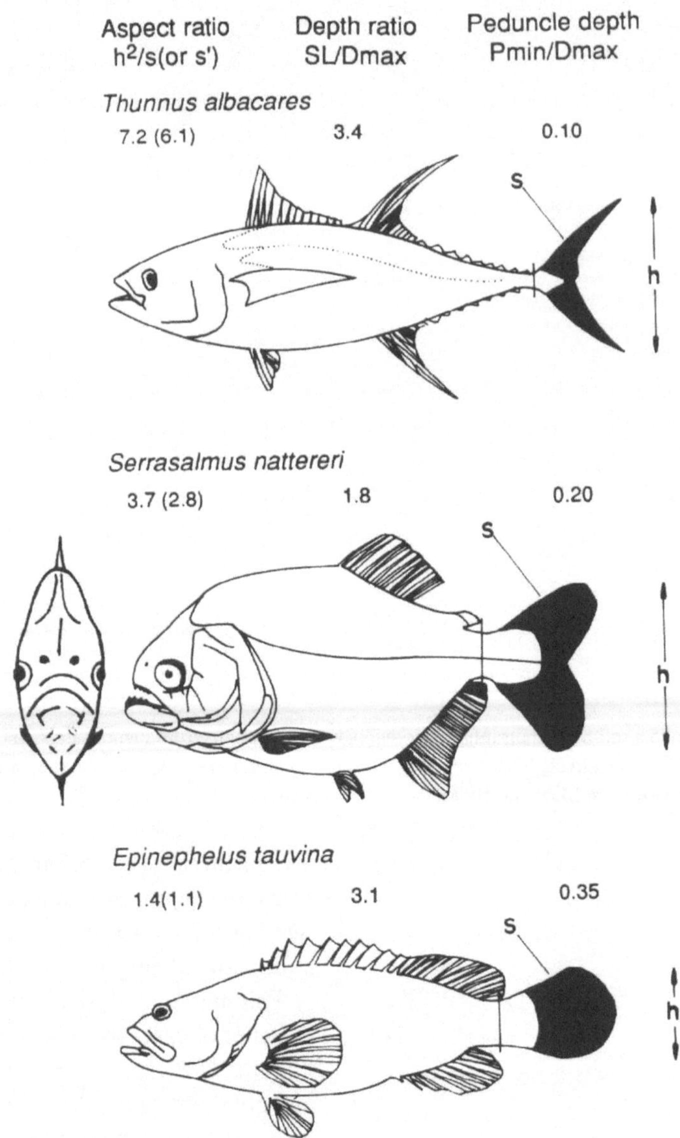

Fig. 5. Comparison of shape between three types of piscivorous fishes, with emphasis on three indices that can be used to identify these types: (i) Aspect ratio, defined by h^2/s (or h^2/s', in brackets, with s' including the surface area up to the thinnest section of the caudal peduncle or P_{min}; (ii) standard length over maximum depth (SL/D_{max}); and (iii) P_{min}/D_{max}. Note intermediate position of *S. nattereri* in two of these indices and high value of SL/D_{max}. The frontal view of *S. nattereri* is from Figure 1 in Sazima & Machado (1990), the lateral view is adapted from Kner (1860).

feed on whole fishes and/or pieces thereof (Table 7). Other foods are taken, however, and these include arthropods (insects, crustaceans), molluscs and small vertebrates or parts thereof, as well as small amounts of plant materials (and see the Discussion for mammals as food items of piranhas).

Reproduction

Schulte (1988) discusses the reproduction of red-bellied piranhas in some details; the following account of reproduction in *S. nattereri* in aquaria was adapted, however, because of its conciseness from Riehl & Baensch (1991):

Table 7 Some reported morphological and behavioral adaptations for carnivory by red-bellied piranha *S nattereri* [author's comments in square brackets]

Item	Source
Relatively short intestine, Length of intestine/SL ≈ 1 1	Luengo (1965), Jegu & Dos Santos (1988)
Highly evolved auditory capacity	Stabentheiner (1988), Nico (1990)
Adults feed mainly at dusk and dawn [as is common among piscivores]	Sazima & Machado (1990), Bellamy (1968), Hobson (1972)
(Daytime) 'lurking', then 'dashing' [as also implied by relatively low caudal fin aspect ratio and high caudal peduncle]	Sazima & Machado (1990) [see Fig 5 and text]
Teeth replacement on alternating sides of jaw, allowing continuous feeding	Sazima & Machado (1990)
Hierarchies within 'packs' (i e , small schools) [as also occurs, e g , in wolves]	Zbinden (1973)

'Reproduction generally occurs after "new" water with a neutral pH and a hardness of 6° H has been added; the males, dig plate-sized pits in gravel, into which the eggs are deposited. Spawning occurs from 4–5 AM. The male defends the spawn; for 24 h, he is supported in this by the female, after which he drives her away (if the eggs are removed, the male spawns with another female of the same school 2–3 days later). The 500–1000 eggs are transparent-golden, and stick to the gravel. The larvae hatch after 8 days, and start feeding after 4–5 days, i.e. once their yolk sac is consumed. The juveniles have black spots and their only red coloration is on a spot near and on the lower part of the operculum, and on the anal fin.'

Unfortunately, this description is too concise to unequivocally assign *S. nattereri* to one of the ethological-ecological groups proposed by Balon (1990), i.e., future research will have to determine whether they are indeed 'phytophil clutch tenders', the most likely category (see also Table 8).

Other information on the reproduction of red-bellied piranha is compiled in Table 8; note that the available size at first maturity (in Table 2), divided by the asymptotic size (Table 1 and equation 1) leads to $L_m/L_\infty = 0.67$, i.e., to an estimate of 'reproductive load' that is compatible with values reported from other fishes (Cushing 1981, Pauly 1984).

Thermal tolerance

Braga (1975) conducted a set of experiments on the cold and heat tolerance of *S. nattereri*, of unknown weight, summarized in Figure 6.

As might be seen, the temperature range be-

Table 8 Selected information on the reproduction of red-bellied piranha (*Serrasalmus nattereri* and *S notatus*)

Item	Source
No visible differences between ♀ and ♂ in the specimens sent by J Natterer from Brazil	Kner (1860)
No reliable external sex difference in *S nattereri*	Paysan (1975)
In '*Serrasalmus* sp aff *nattereri*', reported to occur in the Orinoco Basin (Venezuela, Guyana), the males have heads that are more 'bull-like', but are more slender than the females	Riehl & Baensch (1991)
Mature specimens are found from March to June (esp in April) in the Orinoco River (*S notatus*)	Novoa et al (1982)
In *S nattereri*, spawning generally occurs in May, at the onset of the rains, eggs are laid on tree roots trailing in the waters, and are guarded, reproductive success (i e recruitment) may strongly vary from year to year depending on how the savanna was flooded	Lowe-McConnell (1964, 1975)
'A single spawning may produce 4 000–5 000 large eggs which adhere to the plants and are not attacked by the parent fishes They hatch in 9 to 10 days'	Mills & Vevers (1989)
Reproductive load, i e , $L_m/L_\infty = 0\ 65$	See Table 2 and text
Detailed account of reproduction in the aquarium	Schulte (1988)

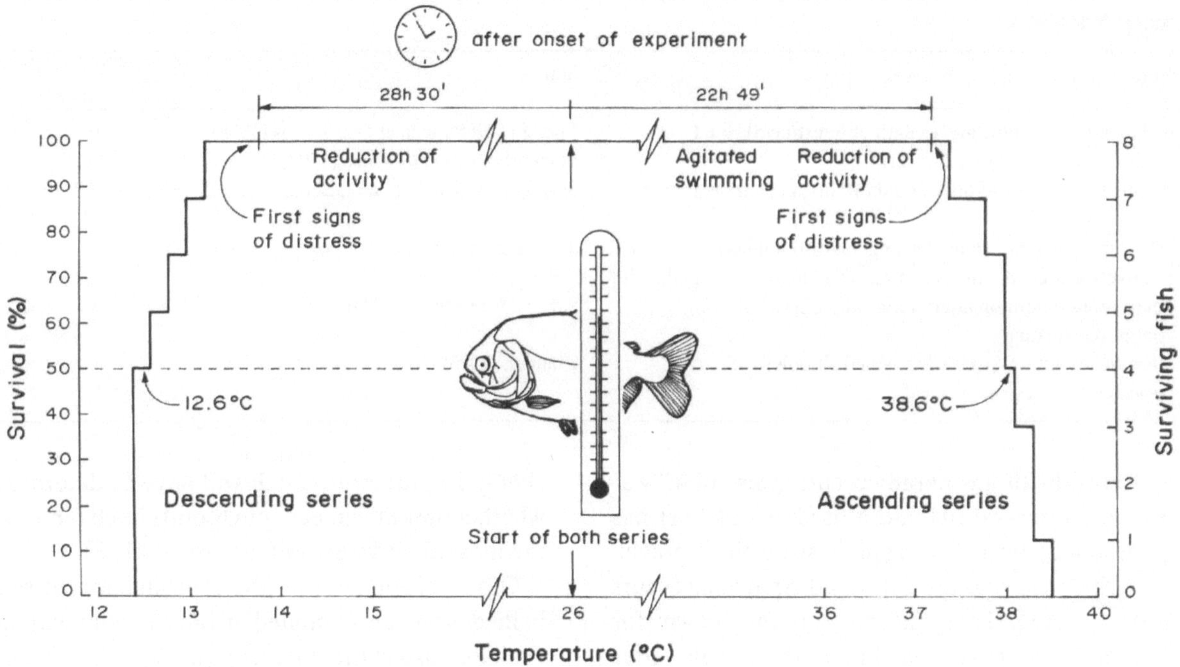

Fig. 6. Graphical summary of experiments by Braga (1975) on the thermal tolerance of *S. nattereri.*

tween the two lethal limits was 24° C. The upper limit (38.6° C) does not have obvious ecological implications, as extremely high temperatures probably occur only under conditions where other factors (notably O_2 and food) are likely to be limiting as well. On the other hand, the lower limit of 12.6° C is ecologically interesting because it clearly impacts on the southward expansion of *S. nattereri.* Indeed, Bonetto et al. (1967) show that winter mortalities are the main limiting factor for *S. nattereri* (and some other piranhas) in the middle and particularly in the lower Parana River, where winter temperatures drop well below 10° C.

Parasites

Thatcher (1991) reported the following parasites from *S. nattereri*– Monogenoidea: *Amphithecium brachycirrum; A. calycinum; A. camelum; A. catalaoensis; A. falcatum; A. junki; Anacanthorus anacanthorus; A. brazilensis; A. maltai; A. neotropicalis; A. reginae; A. rondonensis; A. thatcheri; Cleidodiscus amazonensis; C. piranhus; C. serrasalmus; Notothecium aegidatum; N. mizellei; Notozothecium penetrarum; N. minor; Urocleidus crescentis; U. orthus.* Nematoda: *Spirocamallanus inopinatus.* Copepoda: *Rhinergasilus piranhus.* Branchiura: *Argulus multicolor; Argulus* sp.; *Dolops bidentata; D. carvalhoi;* and *D. longicauda.*

Discussion

Figure 7 is an attempt to illustrate, for the anatomy of *S. nattereri*, what the text has attempted with regard to its biology and ecology: a 'reconstruction' of the red-bellied piranha from various elements scattered in the literature. Such reconstruction, synthetic as it might be, appears to require a second-order synthesis, presenting the species as a whole, i.e., an attempt to answer the question: 'What is a red-bellied piranha?'

The Fisheries Administrative Order 'prohibiting the importation and/or possession of any live pira-

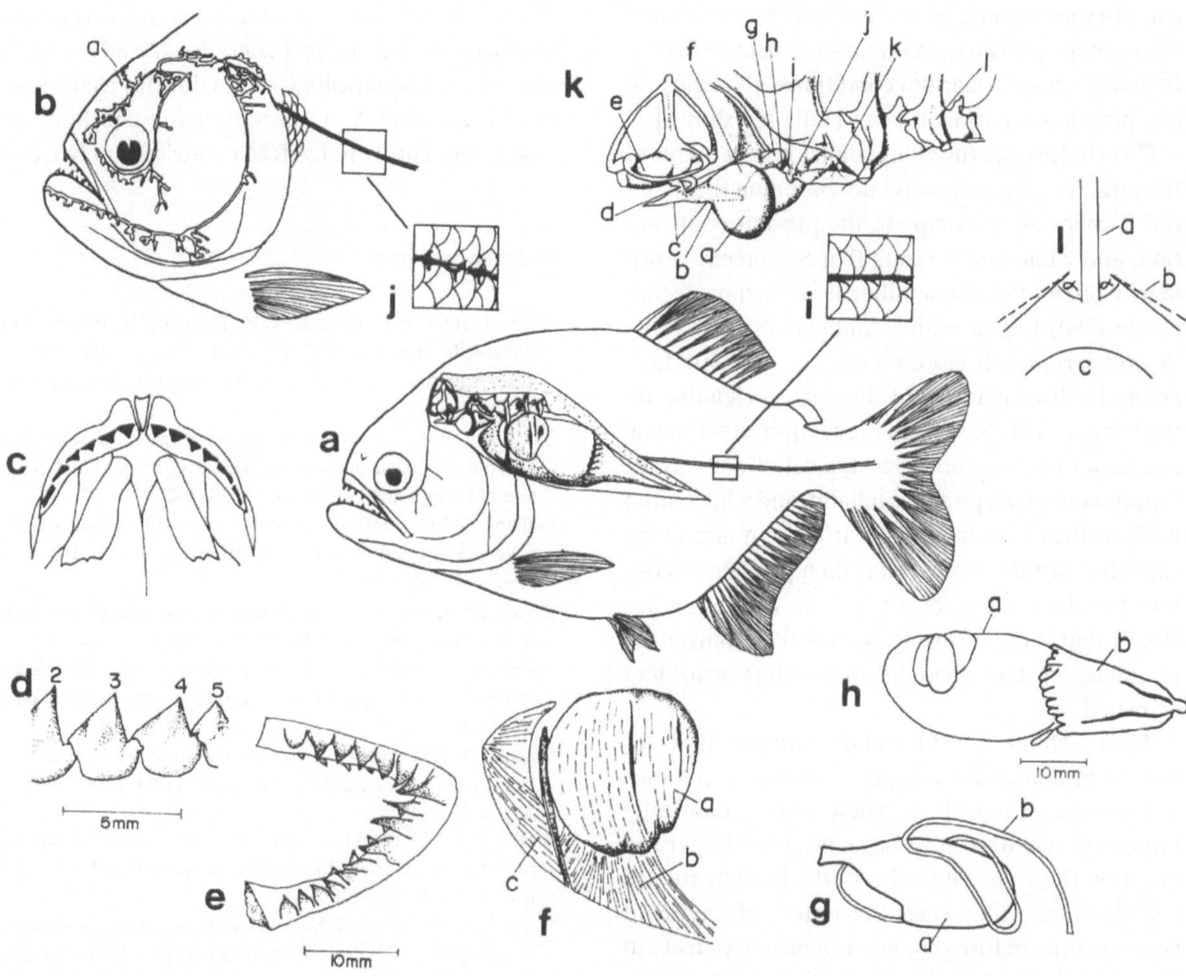

Fig. 7. Aspects of the functional anatomy of red-bellied piranha *Serrasalmus nattereri*: a – adult specimen, with ribs, vertebral processes and head bones partly removed, to show auditory and drumming apparatus (adapted from Kner 1860 and Stabentheiner 1988); b – head, showing lateral line canals; a = junction canal between the left and right sides of the head (adapted from Stabentheiner 1988); c – upper premaxillary, showing position of teeth (adapted from Géry 1964a); d – teeth on side of lower jaw, numbered from median tooth (from Jégu & Dos Santos 1988); e – external lateral view of gill rakers on first branchial arch (source: as in d-); f – drumming apparatus; a = gas bladder; b = tendon of drumming muscle; c = *tunica externa*, to which *tripus* is attached (see K) (source: as in b-); g – alimentary tract; a = stomach, b = intestine (source: as in d-); h – left lateral view of gall bladder; a = cranial sac; b = caudal sac (sources: as in b- and d-); i and j – magnified details of the trunk canal of the lateral line (source: as in b-); k – details of the auditory apparatus, a = *legana;* b = *foramen socculo-lagenaris;* c = *sacculus;* d = *canalis transversus*; e = *utriculus*; f = *sinus imper;* g = *scaphium*, h = *claustrum* ; i = *ligamenti*; j = *intercalarium;* k = *tripus*; l = *vertebra* (source: as in b-); l – dorsal view (schematic); a = *saccular macula;* b = *lagenar macula;* c = gas bladder. The angle (α) ranges from 50 to 55° (source: as in b-).

nha' in the Philippines[6] defines piranha as 'fishes with lacerating teeth and strong set of well-developed mandibles with which to take bites out of the flesh of its victims, usually found in northern South America. They are found in northern South America. They are strictly freshwater species, sturdy and could adapt easily to new environment, even in confinement under aquarium conditions.' [Needless to say, piranhas are available in Manila pet shops.]

Clearly, this and the many similar restatements of the piranha's ferocity won't do. Based on the material I reviewed to complete the present contribution, and while emphasizing that *S. nattereri* is primarily piscivore (Winnemiller 1989, Sazima & Machado 1990) I agree with Schulte (1988) who views '*S. nattereri* and its close relatives as the only fast-acting health squad [with] the task of cleaning up the waters'. This 'task' is mainly required following the rather large-scale floods, typical of the various floodplains which piranhas inhabit, and which often leads to the cadavers of terrestrial mammals being carried down the rivers. This, then provides a context for the contention of Sazima & Guimaraes (1987) that most cases of humans skeletonized by piranhas, in fact, pertain to persons who had drowned.

Thus Schulte (1988) rightly stresses that 'of course, piranhas attack living creatures, sometimes sick or injured ones [. . .]. Their other ecologically important function, however, tends to be largely ignored in these sensational reports. In their role as carrion-eaters, they clear the waters of any dead creature long before they get a chance to putrefy in the warm water. It is this extremely important role played by piranhas in the ecosystems of the South American rivers and streams that has so far hardly been brought to the general public's attention with the necessary emphasis'.

And to scientists as well.

[6] Anon 1982 Updated Index to Presidential Decrees on Fisheries Bureau of Fisheries and Aquatic Resources, Quezon City 124 pp

Acknowledgements

I would like to thank R. Lowe-McConnell, R. Welcomme and P. Bayley for sending photocopies of hard-to-access publications, A. Contemprate for his excellent translations of my ideas into the graphics presented herein, R. Froese for the advice of an experienced aquariologist, I. Sazima for a constructive review and V. Christensen for checking my equations. This is ICLARM Contribution No. 819.

References cited

Axelrod, H R, W E Burgess, C W Emmens, N Pronek, J G Walls & R Hunziker 1987 Dr Axelrod's mini-atlas of freshwater aquarium fishes Tropical Fish Hobbyist, Neptune City 992 pp

Balon, E K 1990 Epigenesis of an epigeneticist the development of some alternative concepts on the early ontogeny and evolution of fishes Guelph Ichthyol Rev 1 1–48

Bellamy, D 1968 Metabolism of the red piranha (*Rooseveltiella nattereri*) in relation to feeding behaviour Comp Biochem Physiol 25 343–347

Bertalanffy, L von 1938 A quantitative theory of organic growth (inquiries on growth laws II) Human Biology 10 181–213

Bonetto, A, C Pignalberi & E Cordiviola 1967 Las "Palometas" o "Pirañas" de las aguas del Parana medio Acta Zool Lilloana 23 45–66

Braga, R A 1975 Ecologia e etologia de piranhas no nordeste do Brasil (Pisces – *Serrasalmus* Lacépède, 1803) Biol Caerense Agron 15/16 1–268

Brett, J R & J M Blackburn 1978 Metabolic rate and energy expenditure of the spiny dogfish, *Squalus acanthias* J Fish Res Board Can 35 816–821

Bykov, V P 1983 Marine fishes chemical composition and processing properties Amerind Publishing Co, New Delhi 322 pp

Cushing, D H 1981 Fisheries biology a study in population dynamics 2nd ed University of Wisconsin Press, Madison 295 pp

De Merona, B, J Lopes de Carvalho & M M Bittencourt 1987 Les effets immediats de la fermeture du barrage de Tucurui (Brésil) sur l'ichtyofaune en aval Rev Hydrobiol Trop 20 73–84

Dos Santos, G M, M Jégu & B de Merona 1984 Catálogo de peixes comerciais do Baixo Rio Tocantins Eletronorte/INPA, Manaus 83 pp

Elliott, J M & W Davidson 1975 Energy equivalents of oxygen consumption in animal energetics Oecologia 19 195–201

Franke, H J 1978 Gattungen der characoidei – *Serrasalmus* pp 525–526 *In* G Sterba (ed) Enzyklopadie der Aquaristik und speziellen Ichthyologie, Verlag J Neumann-Neudamm, Melsungen

Gery, J 1964a Contributions à l'etude des poissons Characoides – 27 systématique et évolution de quelques piranhas (*Serrasalmus*) Vie et Milieu 14 597–617

Géry, J. 1964b. Poissons characoides de l'Amazone péruvienne (résultats scientifiques de l'expédition Amazone-Ucayali du Dr. L.H. Lüling, 1959–1960). Beiträge zur Neotropischen Fauna 4: 1–44.

Géry, J. 1977. Characoids of the world. T.F.H. Publications, Neptune City. 672 pp.

Goulding, M. 1980. The fishes and the forest: explorations in Amazonian natural history. University of California Press, Los Angeles. 280 pp.

Goulding, M. 1981. Man and fisheries on an Amazon frontier. Dr W. Junk Publishers, The Hague. 137 pp.

Hobson, E.S. 1972. Activity of Hawaiian reef fishes during the evening and morning transitions between daylight and darkness. U.S. Fish. Bull. 70: 715–740.

Ivlev, V.S. 1966. The biological productivity of waters. [translated by W.E. Ricker]. J. Fish. Res. Board Can. 23: 1717–1759.

Jarre, A., M.L. Palomares, M.L. Soriano, V.C. Sambilay, Jr. & D. Pauly. 1991. Some new analytical and comparative methods for estimating the food consumption of fishes. ICES Mar. Sci. Symp. 193: 99–108.

Jégu, M. & G. dos Santos. 1988. Le genre *Serrasalmus* (Pisces, Serrasalmidae) dans le bas Tocantin (Brésil, Pará) avec la description d'une espèce nouvelle, *S. geryi*, du bassin Araguaia-Tocantins. Rev. Hydrobiol. trop. 21: 239–274.

Junk, W.J. 1976. Biologia de água doce e pesca interior. p. 105. *In:* Relatorio Anual de INPA, Instituto Nacional de Pesquisas da Amazonia, Manaus.

Kner, R. 1860. Zur Familie der Characiden. III. Folge der ichthyologischen Beiträge. Denkschr. Kaiserlichen Akademie Wiss. (Math-Naturwiss. Classe) 18: 9–62.

Lauzanne, L. & G. Loubens. 1985. Peces del Rio Mamoré. ORSTOM, Trav. Doc. 192, Paris. 116 pp.

Lowe-McConnell, R.H. 1964. The fishes of the Rupununi Savana district of British Guyana, South America. J. Linn. Soc. (Zool.) 45: 103–144.

Lowe-McConnell, R.H. 1975. Fish communities in tropical freshwaters: their distribution, ecology and evolution. Longman, London. 338 pp.

Luengo, J.A. 1965. La longitud del tubo digestivo de *Prochilodus reticulatus* y *Serrasalmus nattereri* en relacion con sus habitos alimentarios. Physis (Buenos Aires) 25: 371–373.

Machado, A. 1985. Estudios sobre la subfamilia Serrasalminae. Parte III: sobre el estatus genérico y relaciones filogenéticas de los géneros *Pygpritis, Pygocentrus, Pristobrycon* y *Serrasalmus* (Teleostei-Characidae-Serrasalminae). Acta Biol. Venez. 12: 19–42.

Marlier, G. 1967. Ecological studies on some lakes in the Amazon Valley. Amazoniana 1: 91–115.

Marlier, G. 1968. Les poissons du lac Redondo et leur régime alimentaires; les chaines trophiques du lac Redondo; les poissons du Rio Prêto de Eva. Cadernos da Amazonia (INPA), Manaus 11: 21–57.

Mills, D. & G. Vevers. 1989. The Tetra encyclopedia of freshwater aquarium fishes. Tetra Press, Morris Plains. 208 pp.

Nico, L.G. 1990. Feeding chronology of juvenile piranhas, *Pygocentrus notatus,* in the Venezuelan Llanos. Env. Biol. Fish. 29: 51–57.

Nico, L.G. & D.C. Taphorn. 1986. Those bitin' fish from South America. Tropical Fish Hobbyist 34(Feb.): 24–27, 30–34, 36, 40–41, and 56–57.

Nico, L.G. & D.C. Taphorn. 1988. Food habits of piranhas in the low Llanos of Venezuela. Biotropica 20: 311–321.

Novoa, D.F., F. Cervigón & F. Ramos. 1982. Catalogo de los recursos pesqueros del Delta del Orinoco. pp. 261–360. *In:* D.F. Novoa (ed.) Los Recursos Pesqueros del Rio Orinoco y su Explotacion, Corporacion Venezolana de Guyana, Editorial Arte, Caracas.

Novoa, D.F. & F. Ramos. 1978. Las pesquerias comerciales del Río Orinico. Corporacion Venezolana de Guyana. Editorial Arte, Caracas. 161 pp.

Ortega, H. & R.P. Vari. 1986. Annotated checklist of the freshwater fishes of Peru. Smithsonian Contributions to Zoology 437. 25 pp.

Palomares, M.L. 1991. La consommation de nourriture chez les poissons: étude comparatrice, mise au point d'un modèle prédictif et application à l'étude des réseaux trophiques. Ph.D. Dissertation, Institut National Polytechnique de Toulouse, Toulouse. 211 pp.

Palomares, M.L. & D. Pauly. 1989. A multiple regression model for predicting the food consumption of marine fish populations. Aust. J. Mar. Freshw. Res. 40: 259–284.

Pauly, D. 1980. On the interrelationships between natural mortality, growth parameters and mean environmental temperature in 175 fish stocks. J. Cons. CIEM 39: 175–192.

Pauly, D. 1984. A mechanism for the juvenile-to-adult transition in fishes. J. Cons. CIEM 41: 280–284.

Pauly, D. 1986. A simple method for estimating the food consumption of fish populations from growth data and food conversion experiments. U.S. Fish. Bull. 84: 827–840.

Pauly, D. 1989. Food consumption by tropical and temperate marine fishes: some generalizations. J. Fish Biol. 35 (Suppl. A): 11–20.

Paysan, K. 1975. The Hamlyn guide of aquarium fishes. Hamlyn, London. 239 pp.

Riehl, R. & H.A. Baensch. 1991. Mergus Aquarien-Atlas: das aktuelle Nachschlagewerk der Aquaristik. Verlag für Natur. und Heimtierkunde Hans A. Baensch, Melle. Vol. 1, 992 pp. and Vol. 2, 1216 pp.

Rosa, H. Jr. 1965. Preparation of synopses on the biology of species of living aquatic organisms. FAO Fisheries Synopsis, No. 1, Rev. 1. 84 pp.

Sainsbury, K.J. 1986. Estimation of food consumption from field observations of fish feeding cycles. J. Fish Biol. 29: 23–36.

Sazima, I. & S. de Andrades Guimaraes. 1987. Scavenging on human corpses as a source for stories about man-eating piranhas. Env. Biol. Fish. 20: 75–77.

Sazima, I. & F.A. Machado. 1990. Underwater observations of piranhas in western Brazil. Env. Biol. Fish. 28: 17–31.

Schulte, W. 1988. Piranhas in the aquarium. T.F.H. Publications, Neptune City. 128 pp.

Silvert, W. & D. Pauly. 1987. On the compatibility of a new expression for gross conversion efficiency and the von Bertalanffy growth equation. U.S. Fish. Bull. 85: 139–140.

Smith, N. 1979. A pesca no rio Amazonas. Instituto Nacional de Pesquisa da Amazonia, Manaus. (number of pages not available).

Stabentheiner, A. 1988. Correlation between hearing and sound production in piranhas. J. Comp. Physiol. A 162: 67–76.

Thatcher, V.E. 1991. Amazon fish parasites. Amazoniana 11 (3/4): 263–571.

Wakeman, J.M., C.R. Arnold, D.E. Wohlschlag & S.C. Rabalais. 1979. Oxygen consumption, energy expenditure and growth of the red snapper (*Lutjanus campecheanus*). Trans. Amer. Fish. Soc. 108: 288–292.

Winberg, G.G. 1960. Rate of metabolism and food requirements of fishes. Fish. Res. Board Can. Transl. Ser. (194). 239 pp.

Winemiller, K.O. 1989. Ontogenic diet shift and resource partitioning among piscivorous fishes in the Venezuelan Llanos. Env. Biol. Fish. 26: 177–199.

Zbinden, K. 1973. Verhaltensstudien an *Serrasalmus nattereri*. Rev. Suisse Zool. 80: 521–542.

Environmental Biology of Fishes **41**: 438, 1994.

Fish imagery in art 71: Holland-Scholer's *The Story Joan Told*

Marilyn A. Moyle
612 Eisenhower St., Davis, CA 95616, U.S.A.

The Story Joan Told was inspired by Karla Holland-Scholer's teacher, Joan Brown, a professor of art at the University of California at Berkeley. In this small ceramic sculpture a women in a wet suit sits in a chair, holding a large fish on her lap. Such fish were a personal symbol for Brown, who was a competitive swimmer as well as an artist. She swam from Alcatraz Island to San Francisco and made several paintings based on this event. This sculpture honors Brown's achievements and includes many of the symbols Brown liked to use in her own paintings – the fish, cat, and Alcatraz Island. Brown died in India in 1990, but her legacy to her students is still very much alive. She encouraged her students to make art based on stories, dreams, myths, and folktales. Holland-Scholer is an artist and teacher who lives in Davis, California.

The Story Joan Told (1993, oil and wax on clay, 41 × 41 × 10 cm) is used courtesy of the John Natsoulas Gallery, Davis, California.

Environmental Biology of Fishes **41**: 439–442, 1994.

An overview of water pollution and fishes

John B. Sprague
Sprague Associates Ltd., 474 Old Scott Road, Saltspring Island, British Columbia V8K 2L7, Canada

Lloyd, R. 1992. Pollution and freshwater fish. Fishing News Books (Blackwell Scientific Publications Ltd), Oxford. 176 pp. £ 25.00 (paper)

This book is relatively short and easy to assimilate. It is an explanation of water pollution problems, how they affect fish and other aquatic organisms, approaches that lead to improvements and those that are not so successful. The chapters indicate the comprehensive coverage, and include: defining pollution; how to test for lethal and sublethal toxicity; toxic action of common pollutants; effects of mixtures; water quality standards; regulatory methods; and economic value of fisheries. The book grew from lectures which Richard Lloyd delivered as Buckman Professor under the auspices of the Buckman Foundation which devotes itself to economic fish culture.

The book's purpose is 'to introduce the technical non-specialist to the complexities of the subject through a balanced overall view.' It is '... a personal view, based on experience gained and research carried out, during the past 40 years.' The value of the book is in this broad outlook. Lloyd was one of the earliest of researchers in fish toxicology, when that started to emerge as a discrete field of endeavour in the 1950s. By the end of his career with the U.K. civil service he was advising at the highest levels of government, and he must be the only person awarded the Order of the British Empire for his services as a fish toxicologist and water pollution biologist. Obviously, he has the credentials to give advice.

Lloyd *says* that this book is for people working in related fields, i.e. non-specialists in water pollution, and the style and content of the book satisfy that purpose. However, that statement of a goal is deceptively modest, because this apparently simple book covers many sophisticated principles that are often neglected. There are many places and situations in the world where the water pollution picture would brighten, if we specialists embraced all of the principles. An example is Lloyd's outline of the benefits of carrying out tests of acute toxicity by documenting the relation between concentration and exposure-time. That allows estimation of a time-independent threshold for lethal action, not just the lethal concentration for an arbitrary exposure-time. The threshold is a relatively robust estimate of toxicity since it is a concentration of some significance, viz. the concentration that a fish can *just* deal with, by excreting or detoxifying a chemical as fast as it enters the body. Estimation of thresholds might be regarded as an old-fashioned idea because it has been largely ignored during two to three recent decades of accumulating toxicity data, particularly in publications from North America which constitute a large part of the database. I pretty well gave up campaigning for flexible exposure times as a lost cause, after attempting to promulgate the idea in a 1969 review article. Lloyd continues to foster the concept in 1992, and he is exactly in step with a 'new' trend in the literature which emphasizes the importance of more complete time/concentration analyses (Newman & Aplin 1992).

Another example is Lloyd's emphasis on the limitations of the current approach in toxicity research which places near-total emphasis on determining effects in relation to concentration of toxicant in the water. Certainly that is effective for accumulating data of a practical nature, but there is ample indication that the information has very limited value for formulating general and predictive relationships. Pollutants cover a wide range of molecules of diverse size and nature, and diverse behaviour in water and on biological membranes. Lloyd points out the areas where greater advances can be made by relating toxic effects to the degree of accumulation within the body of the fish. This too is a current frontier of research in aquatic toxicology, explored by only a few investigators (e.g., McCarty et al. 1992) who have demonstrated the great integrative power of the approach. Lloyd also lists some potential difficulties with this technique.

In keeping with the book's purpose of giving general orientation and advice, there are very few citations of the literature. This is the opposite of a detailed, reference-studded review of individual research findings. There are, however, key references which provide a three-page bibliography on major topics; those are mostly recent and useful, and some of them were not previously known to me.

In a few places, Lloyd fills in details and explains the mechanisms and reasons behind particularly complex or important phenomena of aquatic toxicology, and these explanations are particularly interesting in the areas where he himself has done some of the research. For example, there are 9.5 interesting pages describing why the common pollutant ammonia behaves in such a complicated fashion towards fishes. This includes the probable role of respiratory carbon dioxide in governing pH of water at the gill surface, and hence the ionic form of ammonia.

The latter parts of the book deal with pollution control and regulatory matters. This is primarily directed towards European readers, and the examples draw on Lloyd's experience with development of water quality standards in the U.K., EEC and by the European Inland Fisheries Advisory Commission. There was no coverage of the extensive parallel strategies which developed in the U.S.A. Here again, however, the book covers general principles that are relevant everywhere. In some fields such as biological monitoring and basing regulations on biotic classification, most of us in North America would do well to borrow from Europe including the U.K. Some parallel details of legislation were striking; it was clear that the words of the key anti-pollution section of the Canadian Fisheries Act were lifted from Britain's Freshwater Fisheries Protection Act which dates from more than 100 years ago.

The discussion of the legal framework for protecting water quality (Chapter 8) contains a prime example of basic lessons that should be understood everywhere. Lloyd points to some principles developed in a report of a British Royal Commission in 1912 that have clearly retained their validity. Unfortunately the principles still may not always be comprehended by officials responsible for regulations against water pollution, apparently the case with recent Canadian federal regulations for pulp and paper mill effluents (Sprague 1990, 1991).

The Royal Commission of 1912 outlined two quite separate, albeit closely related, approaches for applying regulatory limits. From Lloyd's quotation of the 1912 document: 'A chemical standard may be applied . . . to the contaminating discharge by itself, or to the stream which has received the discharge. ... it would seem logical that standards should be applied not to sewage liquors or effluents alone but to such discharges under ordinary conditions, i.e. when mixed with river water.'

So there are two control tactics, the first of which is the *best-technology* tactic, described by Lloyd with the following words: 'Effluent standards can be set on the basis of the performance of the best treatment plant available, taking into account the economic feasibility of such processes.'

That first tactic is one of good housekeeping, since it applies equally to any size of industrial or municipal operation discharging to any size of waterbody. It maintains reasonable levels of effluent quality, prevents 'pollution havens' on large bodies of water, but does not necessarily protect small waterbodies. It is the second or *water quality* tactic that endeavours to directly protect fish and other organisms: 'Effluent standards can be set to ensure that the water quality standards in the receiving water are not exceeded.'

In this second tactic, although the allowable discharge is judged by the concentration suitable for fish after dilution in the receiving water (the water quality criterion, objective, or standard), the control efforts are most conveniently applied to the effluent itself (at the end of the pipe). This water quality tactic could also use toxicity testing of the effluent, with some requirement such as absence of sublethal effects following the estimated dilution in the receiving water. That would still be a predictive control measure, even though it utilized a biological measurement.

Both of those tactics should be used in controlling pollution, because sometimes one will be more restrictive and sometimes the other. Attempting to regulate by only one of the tactics leads at best to needless discharge of contaminants, and at worst to gross or calamitous pollution at a particular location. The nature of the two tactics, and the need for both of them, appear transparently simple, and yet there appear to be many administrators and even scientists, who apparently have not stopped to consider the tactics, or somehow fail to comprehend them. For those cases of poorly designed anti-pollution legislation or regulations which I have seen, the usual deficiency has been failure to discern the need to have both tactics in place. For example, recent Canadian pulp and paper effluent regulations did not list any requirements under the second (water quality) tactic, although there exists an excellent set of Canadian water quality guidelines (CCREM 1987). Lloyd shows that the two tactics for pollution control are not mutually exclusive, and points out the disadvantages of emphasizing either one over the other, although he does not give a strong statement that both must be used.

I would emphasize that both control tactics are merely predictive, because they are based on chemical measurements of concentrations, which are then compared with established standards or technical criteria for 'safe' concentrations in the environment. The predictions could be wrong, because of such things as complex behaviour of chemicals in water, combined effects of several pollutants, or simple lack of toxicological knowledge. It is imperative that any effective control program should include a third tactic of *biological monitoring*.

The biological monitoring tactic cannot provide any control at the time of discharge, but can only check afterwards, whether a discharge has had an effect. Biological monitoring must check the quality of living components of the receiving waterbody, at least at the whole-organism level, and preferably at the community level, to see whether the two control tactics have been successful, i.e. whether the predicted 'safe' conditions are in fact being achieved. That is the primary role of biological monitoring. Other monitoring may include conventional chemical measurements in water, or measurements of dangerous contaminants in sediments or tissues of resident fish and other organisms; these evaluations involve predictive associations rather than definitive biological assessments of community health.

Lloyd covers the advantages of biological monitoring although he does not stress the unique role as a third tactic in an overall pollution control strategy. Europeans are well-advanced in using biological surveys as a tool for classifying pollutional status, and rapid developments are taking place, even since Lloyd wrote his chapters. Britain is not only considering the adoption of a scheme for classifying sections of river by their biotic communities, but is also contemplating use of the system for setting objectives. In effect, that would establish biological standards of water quality (NRA 1991, 1992). It makes excellent sense, and I wonder why we biologists have tolerated, in so many jurisdictions, the dominance of chemical water quality standards as the only formal method of judging the cleanliness of bodies of water.

References

CCREM 1987 Canadian water quality guidelines Canadian Council of Resource and Environment Ministers, Task Force on Water Quality Guidelines Environment Canada, Ottawa, Ontario

McCarty, L S, D Mackay, A D Smith, G W Ozburn & D G Dixon 1992 Residue-based interpretation of toxicity and bioconcentration QSARs from aquatic bioassays neutral narcotic organics Environ Toxicol Chem 11 917–930

Newman, M C & M S Aplin 1992 Enhancing toxicity data interpretation and prediction of ecological risk with survival time modelling an illustration using sodium chloride toxicity to mosquitofish *Gambusia holbrooki* Aquatic Toxicol 23 85–96

NRA 1991 Proposals for statutory water quality objectives Report of the National Rivers Authority National Rivers Authority, Bristol NRA Water Quality Series No 5 99 pp

NRA 1992 Recommendations for a scheme of water quality

classification for setting statutory water quality objectives. National Rivers Authority, Bristol. 19 pp.

Sprague, J.B. 1990. Environmentally desirable approaches for regulating effluents from pulp mills. Water Sc. Technol. 24: 361–371.

Sprague, J.B. 1991. Suzuki meets Godzilla – a tale of Canadian effluent regulations. pp. 6–28. *In:* P. Chapman, F. Bishay, E. Power, K. Hall, L. Harding, D. McLeay, M. Nassichuk & W. Knapp (ed.) Proc. Seventeenth Annual Aquatic Toxicity Workshop. Can. Tech. Rept. Fish. Aquat. Sci. No. 1774.

Species index*

* Prepared by M.N. Bruton and J. Pote

Subject index*

* Prepared by M.N. Bruton and J. Pote